MW00710242

Electrical Construction Databook

Robert B. Hickey, P.E.

McGraw-Hill
New York Chicago San Francisco Lisbon London Madrid
Mexico City Milan New Delhi San Juan Seoul
Singapore Sydney Toronto

Cataloging-in-Publication Data is on file with the Library of Congress

McGraw-Hill

*A Division of The **McGraw-Hill** Companies*

1 2 3 4 5 6 7 8 9 0 PBT/PBT 0 9 8 7 6 5 4 3 2 1

ISBN 0-07-137349-7

The sponsoring editor for this book was Larry Hager, the editing supervisor was Steven Melvin, and the production supervisor was Sherri Souffrance. It was set in New Century Schoolbook per the MHT design by Wayne A. Palmer of McGraw-Hill Professional's Hightstown, N.J., composition unit.

Phoenix Color/Book Technology was printer and binder.

This book is printed on recycled, acid-free paper containing a minmum of 50% recycled, de-inked fiber.

To Pat, my wife, whose love and support made this book possible

Contents

Introduction xxvii
Acknowledgments xxix

Section 1 General Information 1.1

1.1.0 Introduction 1.1
1.1.1 Project to do checklist (electrical) 1.2
1.1.2 Drawing design checklist (electrical) 1.5
1.1.3 Site design checklist (electrical) 1.8
1.1.4 Existing condition service and distribution checklist 1.10
1.1.5 Design coordination checklist (electrical) 1.13
1.1.6 Fire alarm system checklist 1.16
1.2.0 Electrical symbols 1.20
1.3.0 Mounting heights for electrical devices 1.31
1.4.0 NEMA configuration chart for general-purpose nonlocking plugs and receptacles 1.34
1.4.1 NEMA configuration chart for specific-purpose locking plugs and receptacles 1.35
1.5.0 IEEE standard protective device numbers 1.36
1.6.0 Comparison of specific applications of NEMA standard enclosures for indoor nonhazardous locations 1.42
1.6.1 Comparison of specific applications of NEMA standard enclosures for outdoor nonhazardous locations 1.42
1.6.2 Comparison of specific applications of NEMA standard enclosures for indoor hazardous locations 1.42
1.6.3 Knockout dimensions for NEMA standard enclosures 1.43
1.7.0 Formulas and terms 1.44
1.8.0 Introduction: typical equipment sizes, weights, and ratings 1.45
1.8.1 Typical equipment sizes: 600-V class 1.45
1.8.2 Transformer weight (lb) by kVA 1.46
1.8.3 Generator Weight (lb) by kW 1.46
1.8.4 Weight (lb/lf) of four-pole aluminum and copper bus duct by ampere rating 1.47
1.8.5 Conduit weight comparisons (lb per 100 ft) empty 1.47
1.8.6 Conduit weight comparisons (lb per 100 ft) with maximum cable fill 1.47
1.9.0 Seismic requirements 1.48

Section 2 Requirements for electrical installations 2.1

2.1.1 Description of fuse class designations 2.2
2.1.2 Maximum peak let-through current (I_p-amperes) and clearing I^2t (ampere-squared-seconds) 2.3
2.2.1 Working spaces 2.4
2.2.2 Examples of conditions 1, 2, and 3 (working spaces) 2.5
2.2.3 Example of exception 1 (working spaces) 2.6
2.2.4 Example of exception 3 (working spaces) 2.6
2.2.5 Required 30-in-wide front working space (working spaces) 2.7
2.2.6 Required full 90-degree opening of equipment doors (working spaces) 2.7
2.3.1 NEC Section 110.26(C), basic rule, first paragraph (access to working space) 2.8
2.3.2 NEC Section 110.26(C), basic rule, second paragraph (access to working space) 2.8
2.3.3 Example of an unacceptable arrangement of a large switchboard (access to working space) 2.9
2.3.4 Example of exception no. 1 (access to working space) 2.9
2.3.5 Example of exception no. 2 (access to working space) 2.10
2.4.1 Working space and dedicated electrical space 2.10
2.4.2 Working space in front of a panelboard as required by NEC Section 110.26 2.11
2.4.3 Dedicated electrical space over and under a panelboard 2.11
2.5.1 Minimum depth of clear working space at electrical equipment 2.12
2.5.2 Elevation of unguarded live parts above working space 2.12

Section 3 Overcurrent protection 3.1

3.1.0 Introduction 3.3
3.1.1 NEC Section 90.2 Scope of the NEC 3.3
3.2.1 NEC Section 110.3(A)(5), (6) and (8) Requirements for equipment selection 3.3
3.2.2 NEC Section 110.3(B) Requirements for proper installation of listed and labeled equipment 3.4
3.2.3 NEC Section 110.9 Requirements for proper interrupting rating of overcurrent protective devices 3.6
3.2.4 NEC Section 110.10 Proper protection of system components from short-circuits 3.13
3.2.5 NEC Section 110.22 Proper marking and identification of disconnecting means 3.16
3.3.1 NEC Section 210.20(A) Ratings of overcurrent devices on branch circuits serving continuous and noncontinuous loads 3.16
3.4.1 NEC Section 215.10 Requirements for ground-fault protection of equipment on feeders 3.16
3.5.1 NEC Section 230.82 Equipment allowed to be connected on the line side of the service disconnect 3.17
3.5.2 NEC Section 230.95 Ground-fault protection for services 3.17
3.6.1 NEC Section 240.1 Scope of Article 240 on overcurrent protection 3.18
3.6.2 NEC Section 240.3 Protection of conductors other than flexible cords and fixture wires 3.19

3.6.3	NEC Section 240.4 Proper protection of fixture wires and flexible cords	3.20
3.6.4	NEC Section 240.6 Standard ampere ratings	3.21
3.6.5	NEC Sections 240.8 and 380.17 Protective devices used in parallel	3.21
3.6.6	NEC Section 240.9 Thermal devices	3.21
3.6.7	NEC Section 240.10 Requirements for supplementary overcurrent protection	3.22
3.6.8	NEC Section 240.11 Definition of current-limiting overcurrent protective devices	3.23
3.6.9	NEC Section 240.12 System coordination or selectivity	3.24
3.6.10	NEC Section 240.13 Ground-fault protection of equipment on buildings or remote structures	3.25
3.6.11	NEC Section 240.21 Location requirements for overcurrent devices and tap conductors	3.25
3.6.12	NEC Section 240.40 Disconnecting means for fuses	3.27
3.6.13	NEC Section 240.50 Plug fuses, fuseholders, and adapters	3.28
3.6.14	NEC Section 240.51 Edison-base fuses	3.28
3.6.15	NEC Section 240.53 Type S fuses	3.28
3.6.16	NEC Section 240.54 Type S fuses, adapters, and fuseholders	3.29
3.6.17	NEC Section 240.60 Cartridge fuses and fuseholders	3.29
3.6.18	NEC Section 240.61 Classification of fuses and fuseholders	3.29
3.6.19	NEC Section 240.86 Series ratings	3.30
3.6.20	NEC Sections 240.90 and 240.91 Supervised industrial installations	3.30
3.6.21	NEC Section 240.92(B) Transformer secondary conductors of separately derived systems (supervised industrial installations only)	3.31
3.6.22	NEC Section 240.92(B)(1) Short-circuit and ground-fault protection (supervised industrial installations only)	3.31
3.6.23	NEC Section 240.92(B)(2) Overload protection (supervised industrial installations only)	3.31
3.6.24	NEC Section 240.92(C) Outside feeder taps (supervised industrial installations only)	3.31
3.6.25	NEC Section 240.100 Feeder and branch-circuit protection over 600 V nominal	3.32
3.6.26	NEC Section 240.100(C) Conductor protection	3.32
3.7.1	NEC Section 250.2(D) Performance of fault-current path	3.32
3.7.2	NEC Section 250.90 Bonding requirements and short-circuit current rating	3.32
3.7.3	NEC Section 250.96(A) Bonding other enclosures and short-circuit current requirements	3.32
3.7.4	NEC Section 250.122 Sizing of equipment grounding conductors	3.33
3.8.1	NEC Section 310.10 Temperature limitation of conductors	3.34
3.9.1	NEC Section 364.11 Protection at a busway reduction	3.34
3.10.1	NEC Section 384.16 Panelboard overcurrent protection	3.34
3.11.1	NEC Section 430.1 Scope of motor article	3.35
3.11.2	NEC Section 430.6 Ampacity of conductors for motor branch circuits and feeders	3.35
3.11.3	NEC Section 430.8 Marking on controllers	3.35
3.11.4	NEC Section 430.32 Motor overload protection	3.36

3.11.5 NEC Section 430.36 Fuses used to provide overload and single-phasing protection 3.36
3.11.6 NEC Section 430.52 Sizing of various overcurrent devices for motor branch-circuit protection 3.37
3.11.7 NEC Section 430.53 Connecting several motors or loads on one branch circuit 3.38
3.11.8 NEC Section 430.71 Motor control-circuit protection 3.38
3.11.9 NEC Section 430.72(A) Motor control-circuit overcurrent protection 3.39
3.11.10 NEC Section 430.72(B) Motor control-circuit conductor protection 3.39
3.11.11 NEC Section 430.72(C) Motor control-circuit transformer protection 3.41
3.11.12 NEC Section 430.94 Motor control-center protection 3.42
3.11.13 NEC Section 430.109(A)(6) Manual motor controller as a motor disconnect 3.42
3.12.1 NEC Section 440.5 Marking requirements on HVAC controllers 3.42
3.12.2 NEC Section 440.22 Application and selection of the branch-circuit protection for HVAC equipment 3.43
3.13.1 NEC Section 450.3 Protection requirements for transformers 3.43
3.13.2 NEC Section 450.3(A) Protection requirements for transformers over 600 V 3.45
3.13.3 NEC Section 450.3(B) Protection requirements for transformers 600 V or less 3.45
3.13.4 NEC Section 450.6(A)(3) Tie-circuit protection 3.45
3.14.1 NEC Section 455.7 Overcurrent protection requirements for phase converters 3.46
3.15.1 NEC Section 460.8(B) Overcurrent protection of capacitors 3.46
3.16.1 NEC Section 501.6(B) Fuses for Class 1, Division 2 locations 3.46
3.17.1 NEC Section 517.17 Requirements for ground-fault protection and coordination in health care facilities 3.47
3.18.1 NEC Section 520.53(F)(2) Protection of portable switchboards on stage 3.47
3.19.1 NEC Section 550.6(B) Overcurrent protection requirements for mobile homes and parks 3.48
3.20.1 NEC Section 610.14(C) Conductor sizes and protection for cranes and hoists 3.48
3.21.1 NEC Section 620.62 Selective coordination of overcurrent protective devices for elevators 3.48
3.22.1 NEC Section 670.3 Industrial machinery 3.49
3.23.1 NEC Section 700.5 Emergency systems: their capacity and rating 3.50
3.23.2 NEC Section 700.16 Emergency illumination 3.50
3.23.3 NEC Section 700.25 Emergency system overcurrent protection requirements (FPN) 3.51
3.24.1 NEC Section 705.16 Interconnected electric power production sources: interrupting and short-circuit current rating 3.51
3.25.1 NEC Section 725.23 Overcurrent protection for Class 1 circuits 3.52
3.26.1 NEC Section 760.23 Requirements for non-power-limited fire alarm signaling circuits 3.52

Section 4 Wiring methods and materials 4.1

4.1.0 NEC Table 300.1 (C), Metric Designator and Trade Sizes 4.4
4.1.1 NEC Table 300.5. Minimum cover requirements, 0 to 600 V, nominal, burial in millimeters (inches) 4.5
4.2.1 NEC Table 300.19(A) Spacings for conductor supports 4.6
4.2.2 Examples of installed support bushings and cleats 4.7
4.3.1 NEC Table 300.50. Minimum cover requirements 4.8
4.4.1 NEC Table 310.5. Minimum size of conductors 4.9
4.4.2 NEC Table 310.13. Conductor application and insulations 4.10
4.4.3 Conductor characteristics 4.15
4.4.4 NEC Table 310.15(B)(2)(a). Adjustment factors for more than three current-carrying conductors in a raceway or cable 4.15
4.4.5 NEC Table 310.16. Allowable ampacities of insulated conductors rated 0 through 2000 V, 60°C through 90°C (140°F through 194°F) not more than three current-carrying conductors in a raceway, cable, or earth (directly buried), based on ambient air temperature of 30°C (86°F) 4.16
4.4.6 NEC Table 310.17. Allowable ampacities of single-insulated conductors rated 0 through 2000 V in free air, based on ambient air temperature of 30°C (86°F) 4.17
4.4.7 NEC Table 310.18. Allowable ampacities of insulated conductors, rated 0 through 2000 V, 150°C through 250°C (302°F through 482°F), in raceway or cable, based on ambient air temperature of 40°C (104°F) 4.20
4.4.8 NEC Table 310.19. Allowable ampacities of single-insulated conductors, rated 0 through 2000 V, 150°C through 250°C (302°F through 482°F), in free air, based on ambient air temperature of 40°C (104°F) 4.21
4.4.9 NEC Table 310.20. Ampacities of not more than three single insulated conductors, rated 0 through 2000 V, supported on a messenger, based on ambient air temperature of 40°C (104°F) 4.22
4.4.10 NEC Table 310.21. Ampacities of bare or covered conductors in free air, based on 40°C (104°F) ambient, 80°C (176°F) total conductor temperature, 610 mm/sec (2 ft/sec) wind velocity 4.23
4.4.11 NEC Table 310.61. Conductor application and insulation 4.23
4.4.12 NEC Table 310.62. Thickness of insulation for 601- to 2000-V nonshielded types RHH and RHW 4.24
4.4.13 NEC Table 310.63. Thickness of insulation and jacket for nonshielded solid-dielectric insulated conductors rated 2001 to 8000 V 4.24
4.4.14 NEC Table 310.64. Thickness of insulation for shielded solid-dielectric insulated conductors rated 2001 to 35,000 V 4.25
4.4.15 NEC Table 310.67. Ampacities of insulated single copper conductor cables triplexed in air based on conductor temperatures of 90°C (194°F) and 105°C (221°F) and ambient air temperature of 40°C (104°F) 4.25
4.4.16 NEC Table 310.68. Ampacities of insulated single aluminum conductor cables triplexed in air based on conductor temperatures of 90°C (194°F) and 105°C (221°F) and ambient air temperature of 40°C (104°F) 4.26

4.4.17 NEC Table 310.69. Ampacities of insulated single copper conductor isolated in air based on conductor temperatures of 90°C (194°F) and 105°C (221°F) and ambient air temperature of 40°C (104°F) 4.26
4.4.18 NEC Table 310.70. Ampacities of insulated single aluminum conductor isolated in air based on conductor temperatures of 90°C (194°F) and 105°C (221°F) and ambient air temperature of 40°F (104°F) 4.27
4.4.19 NEC Table 310.71. Ampacities of an insulated three-conductor copper cable isolated in air based on conductor temperatures of 90°C (194°F) and 105°C (221°F) and ambient air temperature of 40°C (104°F) 4.27
4.4.20 NEC Table 310.72. Ampacities of insulated three-conductor aluminum cable isolated in air based on conductor temperatures of 90°C (194°F) and 105°C (221°F) and ambient air temperature of 40°C (104°F) 4.28
4.4.21 NEC Table 310.73. Ampacities of an insulated triplexed or three single-conductor copper cables in isolated conduit in air based on conductor temperatures of 90°C (194°F) and 105°C (221°F) and ambient air temperature of 40°C (104°F) 4.28
4.4.22 NEC Table 310.74. Ampacities of an insulated triplexed or three single-conductor aluminum cables in isolated conduit in air based on conductor temperatures of 90°C (194°F) and 105°C (221°F) and ambient air temperature of 40°C (104°F) 4.29
4.4.23 NEC Table 310.75. Ampacities of an insulated three-conductor copper cable in isolated conduit in air based on conductor temperatures of 90°C (194°F) and 105°C (221°F) and ambient air temperature of 40°C (104°F) 4.29
4.4.24 NEC Table 310.76. Ampacities of an insulated three-conductor aluminum cable in isolated conduit in air based on conductor temperatures of 90°C (194°F) and 105°C (221°F) and ambient air temperature of 40°C (104°F) 4.30
4.4.25 NEC Figure 310.60. Cable installation dimensions for use with Tables 4.4.26 through 4.4.35 (NEC Tables 310.77 through 310.86) 4.31
4.4.26 NEC Table 310.77. Ampacities of three single-insulated copper conductors in underground electrical ducts (three conductors per electrical duct) based on ambient earth temperature of 20°C (68°F), electrical duct arrangement per Figure 4.4.25 (NEC Figure 310.60), 100 percent load factor, thermal resistance (RHO) of 90, conductor temperatures of 90°C (194°F) and 105°C (221°F) 4.32
4.4.27 NEC Table 310.78. Ampacities of three single-insulated aluminum conductors in underground electrical ducts (three conductors per electrical duct) based on ambient earth temperature of 20°C (68°F), electrical duct arrangement per Figure 4.4.25 (NEC Figure 310.60), 100 percent load factor, thermal resistance (RHO) of 90, conductor temperatures of 90°C (194°F) and 105°C (221°F) 4.33
4.4.28 NEC Table 310.79. Ampacities of three insulated copper conductors cabled within an overall covering (three-conductor cable) in underground electrical ducts (one cable per electrical duct) based on ambient earth temperature of 20°C (68°F),

electrical duct arrangement per Figure 4.4.25 (NEC Figure 310.60), 100 percent load factor, thermal resistance (RHO) of 90, conductor temperatures of 90°C (194°F) and 105°C (221°F) 4.34

4.4.29 NEC Table 310.80. Ampacities of three insulated aluminum conductors cabled within an overall covering (three-conductor cable) in underground electrical ducts (one cable per electrical duct) based on ambient earth temperature of 20°C (68°F), electrical duct arrangement per Figure 4.4.25 (NEC Figure 310.60), 100 percent load factor, thermal resistance (RHO) of 90, conductor temperatures of 90°C (194°F) and 105°C (221°F) 4.36

4.4.30 NEC Table 310.81. Ampacities of single-insulated copper conductors directly buried in earth based on ambient earth temperature of 20°C (68°F), arrangement per Figure 4.4.25 (NEC Figure 310.60), 100 percent load factor, thermal resistance (RHO) of 90, conductor temperatures of 90°C (194°F) and 105°C (221°F) 4.37

4.4.31 NEC Table 310.82. Ampacities of single-insulated aluminum conductors directly buried in earth based on ambient earth temperature of 20°C (68°F), arrangement per Figure 4.4.25 (NEC Figure 310.60), 100 percent load factor, thermal resistance (RHO) of 90, conductor temperatures of 90°C (194°F) and 105°C (221°F) 4.38

4.4.32 NEC Table 310.83. Ampacities of three insulated copper conductors cabled within an overall covering (three-conductor cable), directly buried in earth based on ambient earth temperature of 20°C (68°F), arrangement per Figure 4.4.25 (NEC Figure 310.60), 100 percent load factor, thermal resistance (RHO) of 90, conductor temperatures of 90°C (194°F) and 105°C (221°F) 4.39

4.4.33 NEC Table 310.84. Ampacities of three insulated aluminum conductors cabled within an overall covering (three-conductor cable), directly buried in earth based on ambient earth temperature of 20°C (68°F), arrangement per Figure 4.4.25 (NEC Figure 310.60), 100 percent load factor, thermal resistance (RHO) of 90, conductor temperatures of 90°C (194°F) and 105°C (221°F) 4.40

4.4.34 NEC Table 310.85. Ampacities of three triplexed single-insulated copper conductors directly buried in earth based on ambient earth temperature of 20°C (68°F), arrangement per Figure 4.4.25 (NEC Figure 310.60), 100 percent load factor, thermal resistance (RHO) of 90, conductor temperatures of 90°C (194°F) and 105°C (221°F) 4.41

4.4.35 NEC Table 310.86. Ampacities of three triplexed single-insulated aluminum conductors directly buried in earth based on ambient earth temperature of 20°C (68°F), arrangement per Figure 4.4.25 (NEC Figure 310.60), 100 percent load factor, thermal resistance (RHO) of 90, conductor temperatures of 90°C (194°F) and 105°C (221°F) 4.42

4.5.1 NEC Table 392.7(B)(2). Metal area requirements for cable trays used as equipment grounding conductor 4.43

4.5.2 An example of multiconductor cables in cable trays with conduit runs to power equipment where bonding is provided 4.44

4.5.3 NEC Table 392.9. Allowable cable fill area for multiconductor cables in ladder, ventilated-trough, or solid-bottom cable trays for cables rated 2000 V or less 4.45

4.5.4 NEC Table 392-9(E). Allowable cable fill area for multiconductor cables in ventilated channel cable trays for cables rated 2000 V or less 4.45

4.5.4.1 NEC Table 392-9(E). Allowable cable fill area for multiconductor cables in solid channel cable trays for cables rated 2000 V or less 4.46

4.5.5 NEC Table 392-10(A). Allowable cable fill area for single-conductor cables in ladder or ventilated-trough cable trays for cables rated 2000 V or less 4.46

4.5.6 An illustration of Section 392.11(A)(3) for multiconductor cables, 2000 V or less, with not more than three conductors per cable (ampacity to be determined from Table B.310.3 in annex B 4.46

4.5.7 An illustration of Section 392.11(B)(4), for three single conductors installed in a triangular configuration with spacing between groups of not less than 2.15 times the conductor diameter (ampacities to be determined from table 310.20) 4.46

4.6.1 An illustration of Section 332.24, for bends in type MI cable 4.47

4.6.2 600-V MI power cable: size and ampacities 4.49

4.6.3 300-V MI twisted-pair and shielded twisted-pair cable sizes 4.51

4.6.4 MI cable versus conventional construction in hazardous (classified) locations 4.51

4.6.5 Engineering data: calculating voltage drop and feeder sizing (MI cable) 4.52

4.7.1 NEC Table 344.24, radius of conduit bends for IMC, RMC, RNC, and EMT 4.53

4.7.2 Minimum support required for IMC, RMC, and EMT 4.53

4.7.3 NEC Table 344.30(B)(2). Supports for rigid metal conduit 4.54

4.7.4 NEC Table 352.30(B). Support of rigid nonmetallic conduit 4.54

4.7.5 NEC Table 352.44(A). Expansion characteristics of PVC rigid nonmetallic conduit coefficient of thermal expansion = 6.085×10^{-5} mm/mm/°C (3.38×10^{-5} in./in./°F) 4.55

4.7.6 NEC Table 352.44(B). Expansion characteristics of reinforced thermosetting resin conduit (RTRC) coefficient of thermal expansion = 2.7×10^{-5} mm/mm/°C (1.5×10^{-5} in./in./°F) 4.56

4.7.7 NEC Table 348.22. Maximum number of insulated conductors in metric designator 12 (3/8-in.) flexible metal conduit 4.56

4.7.8 Conductor fill table for various surface raceways 4.57

4.7.9 NEC Table 384.22. Channel size and inside diameter area 4.58

4.8.1 NEC Table 314.16(A). Metal boxes 4.59

4.8.2 NEC Table 314.16(B). Volume allowance required per conductor 4.59

4.9.1 NEC Table 400.4. Flexible cords and cables 4.60

4.9.2 NEC Table 400.5(A). Allowable ampacity for flexible cords and cables [based on ambient temperature of 30°C (86°F). See Section 400.13 and Table 400.4] 4.66

4.9.3 NEC Table 400.5(B). Ampacity of cable types SC, SCE, SCT, PPE, G, G-GC, and W [based on ambient temperature of 30°C (86°F). See Table 400.4] temperature rating of cable 4.67

4.9.4 NEC Table 400.5(B). Adjustment factors for more than three current-carrying conductors in a flexible card or cable 4.67
4.9.5 NEC Table 402.3. Fixture wires 4.68
4.9.6 NEC Table 402.5. Allowable ampacity for fixture wires 4.71

Section 5 Primary and secondary service and system configurations 5.1

5.1.0 Introduction 5.1
5.1.1 Radial circuit arrangements in commercial buildings 5.2
5.1.2 Radial circuit arrangement: common primary feeder to secondary unit substations 5.3
5.1.3 Radial circuit arrangement: individual primary feeder to secondary unit substations 5.4
5.1.4 Primary radial-selective circuit arrangements 5.5
5.1.5 Secondary-selective circuit arrangement (double-ended substation with single tie) 5.6
5.1.6 Secondary-selective circuit arrangement (individual substations with interconnecting ties) 5.7
5.1.7 Primary- and secondary-selective circuit arrangement (double-ended substation with selective primary) 5.8
5.1.8 Looped primary circuit arrangement 5.9
5.1.9 Distributed secondary network 5.10
5.1.10 Basic spot network 5.11

Section 6 Preliminary load calculations 6.1

6.1.0 Introduction 6.1
6.1.1 Prescriptive unit lighting power allowance (ULPA) (W/ft^2), gross lighted area of total building 6.2
6.1.2 Typical appliance/general-purpose receptacle loads (excluding plug-in-type A/C and heating equipment) 6.2
6.1.3 Typical apartment loads 6.3
6.1.4 Typical connected electrical load for air conditioning only 6.3
6.1.5 Central air conditioning watts per SF, BTUs per hour per SF of floor area, and SF per ton of air conditioning 6.4
6.1.6 All-weather comfort standard recommended heat-loss values 6.4
6.1.7 Typical power requirement (kW) for high-rise building water pressure—boosting systems 6.5
6.1.8 Typical power requirement (kW) for electric hot water—heating system 6.5
6.1.9 Typical power requirement (kW) for fire pumps in commercial buildings (light hazard) 6.5
6.1.10 Typical loads in commercial kitchens 6.6
6.1.11 Comparison of maximum demand 6.6
6.1.12 Connected load and maximum demand by tenant classification 6.7
6.1.13 Factors used in sizing distribution-system components 6.7
6.1.14 Factors used to establish major elements of the electrical system serving HVAC systems 6.8
6.1.15 Service entrance peak demand (Veterans Administration) 6.8
6.1.16 Service entrance peak demand (Hospital Corporation of America) 6.9

Section 7 Short-circuit calculations 7.1

7.1.0 Introduction 7.1
7.1.1 Point-to-point method, three-phase short-circuit calculations, basic calculation procedure and formulas 7.2
7.1.2 System A and system B circuit diagrams for sample calculations using point-to-point method 7.3
7.1.3 Point-to-point calculations for system A, to faults X_1 and X_2 7.4
7.1.4 Point-to-point calculations for system B, to faults X_1 and X_2 7.5
7.1.5 C Values for conductors and busway 7.6
7.1.6 Short cut method 1: adding Zs 7.7
7.1.7 Average characteristics of 600-V conductors (ohms per 100 ft): two or three single conductors 7.7
7.1.8 Average characteristics of 600-V conductors (ohms per 100 ft): three conductor cables (and interlocked armored cable) 7.8
7.1.9 LV busway, R, X, and Z (ohms per 100 ft) 7.8
7.1.10 Short cut method 2: chart approximate method 7.9
7.1.11 Conductor conversion (based on using copper conductor) 7.10
7.1.12 Charts 1 through 13 for calculating short-circuit currents using chart approximate method 7.10
7.1.13 Assumptions for motor contributions to fault currents 7.13
7.1.14 Secondary short-circuit capacity of typical power transformers 7.14

Section 8 Selective coordination of protective devices 8.1

8.1.0 Introduction 8.1
8.1.1 Recommended procedure for conducting a selective coordination study 8.1
8.1.2 Example system one-line diagram for selective coordination study 8.3
8.1.3 Time-current curve no. 1 for system shown in Figure 8.1.2 with analysis notes and comments 8.3
8.1.4 Time-current curve no. 2 for system shown in Figure 8.1.2 with analysis notes and comments 8.3
8.1.5 Time-current curve no. 3 for system shown in Figure 8.1.2 with analysis notes and comments 8.3
8.1.6 Short-cut ratio method selectivity guide 8.3

Section 9 Component short-circuit protection 9.1

9.1.0 Introduction 9.2
9.1.1 Short-circuit current withstand chart for copper cables with paper, rubber, or varnished-cloth insulation 9.2
9.1.2 Short-circuit current withstand chart for copper cables with thermoplastic insulation 9.4
9.1.3 Short-circuit current withstand chart for copper cables with cross-linked polyethylene and ethylene-propylene-rubber insulation 9.5
9.1.4 Short-circuit current withstand chart for aluminum cables with paper, rubber, or varnished-cloth insulation 9.6
9.1.5 Short-circuit current withstand chart for aluminum cables with thermoplastic insulation 9.7

9.1.6	Short-circuit current withstand chart for aluminum cables with cross-linked polyethylene and ethylene-propylene-rubber insulation	9.8
9.1.7	Comparison of equipment grounding conductor short-circuit withstand ratings	9.9
9.1.8	NEMA (Standard short-circuit ratings of busway)	9.10
9.1.9	U.L. no. 508 motor controller short-circuit test ratings	9.10
9.1.10	Molded-case circuit breaker interrupting capacities	9.11
9.1.11	NEC Table 450.3(A). Maximum rating or setting of overcurrent protection for transformers over 600 V (as a percentage of transformer-rated current)	9.18
9.1.12	NEC Table 450.3(B). Maximum rating or setting of overcurrent protection for transformers 600 V and less (as a percentage of transformer-rated current)	9.19
9.1.13	U.L. 1008 minimum withstand test requirement (for automatic transfer switches)	9.19
9.1.14	HVAC equipment short-circuit test currents,Table 55.1 of U.L. Standard 1995	9.20
9.2.1	Protection through current limitation	9.20
9.2.2	Current-limiting effect of fuses	9.21
9.2.3	Analysis of a current-limiting fuse	9.21
9.2.4	Let-thru data pertinent to equipment withstand	9.22
9.2.5	How to use the let-thru charts	9.22
9.2.6	Current-limitation curves: Bussmann low-peak time-delay fuse KRP-C800SP	9.23

Section 10 Motor feeders and starters 10.1

10.1.0	Introduction	10.1
10.1.1	Sizing motor-circuit feeders and their overcurrent protection	10.1
10.1.2	NEC Table 430.7(B). Locked-rotor-indicating code letters	10.3
10.1.3	Motor circuit data sheets	10.3
10.1.4	480-V system (460-V motors) three-phase motor-circuit feeders	10.4
10.1.5	208-V system (200-V motors) three-phase motor-circuit feeders	10.5
10.1.6	115-V single-phase motor-circuit feeders	10.6
10.1.7	200-V single-phase motor-circuit feeders	10.6
10.1.8	230-V single-phase motor-circuit feeders	10.7
10.1.9	Motor starter characteristics (for squirrel-cage motors)	10.7
10.1.10	Reduced-voltage starter characteristics	10.8
10.1.11	Reduced-voltage starter selection table	10.8

Section 11 Standard voltages and voltage drop 11.1

11.1.0	Introduction	11.2
11.1.1	System voltage classes	11.2
11.1.2	Standard nominal system voltages in the United States	11.2
11.1.3	Standard nominal system voltages and voltage ranges	11.3
11.1.4	Principal transformer connections to supply the system voltages of Table 11.1.3.	11.4
11.1.5	Application of voltage classes	11.4

11.1.6 Voltage systems outside of the United States 11.4
11.1 7 System voltage tolerance limits 11.5
11.1.8 Standard voltage profile for a regulated power-distribution system, 120-V base 11.6
11.1.9 Voltage profile of the limits of Range A, ANSI C84.1-1989 11.6
11.1.10 Voltage ratings of standard motors 11.6
11.1.11 General effect of voltage variations on induction motor characteristics 11.7
11.1.12 Voltage-drop calculations 11.7
11.1.13 Voltage-drop tables 11.8
11.1.14 Voltage drop for Al conductor, direct current 11.9
11.1.15 Voltage drop for Al conductor in magnetic conduit, 70 percent PF 11.9
11.1.16 Voltage drop for Al conductor in magnetic conduit, 80 percent PF 11.9
11.1.17 Voltage drop for Al conductor in magnetic conduit, 90 percent PF 11.9
11.1.18 Voltage drop for Al conductor in magnetic conduit, 95 percent PF 11.9
11.1.19 Voltage drop for Al conductor in magnetic conduit, 100 percent PF 11.9
11.1.20 Voltage drop for Al conductor in nonmagnetic conduit, 70 percent PF 11.9
11.1.21 Voltage drop for Al conductor in nonmagnetic conduit, 80 percent PF 11.9
11.1.22 Voltage drop for Al conductor in nonmagnetic conduit, 90 percent PF 11.9
11.1 23 Voltage drop for Al conductor in nonmagnetic conduit, 95 percent PF 11.9
11.1.24 Voltage drop for Al conductor in nonmagnetic conduit, 100 percent PF 11.9
11.1.25 Voltage drop for Cu conductor, direct current 11.21
11.1.26 Voltage drop for Cu conductor in magnetic conduit, 70 percent PF 11.21
11.1.27 Voltage drop for Cu conductor in magnetic conduit, 80 percent PF 11.21
11.1.28 Voltage drop for Cu conductor in magnetic conduit, 90 percent PF 11.21
11.1.29 Voltage drop for Cu conductor in magnetic conduit, 95 percent PF 11.21
11.1.30 Voltage drop for Cu conductor in magnetic conduit, 100 percent PF 11.21
11.1.31 Voltage drop for Cu conductor in nonmagnetic conduit, 70 percent PF 11.21
11.1.32 Voltage drop for Cu conductor in nonmagnetic conduit, 80 percent PF 11.21
11.1.33 Voltage drop for Cu conductor in nonmagnetic conduit, 90 percent PF 11.21
11.1.34 Voltage drop for Cu conductor in nonmagnetic conduit, 95 percent PF 11.21
11.1.35 Voltage drop for Cu conductor in nonmagnetic conduit, 100 percent PF 11.21

11.1.36 Voltage-drop curves for typical interleaved construction of copper busway at rated load, assuming 70°C (158°F) as the operating temperature 11.21
11.1.37 Voltage-drop values for three-phase busways with copper bus bars, in volts per 100 ft, line-to-line at rated current with balanced entire load at end 11.33
11.1.38 Voltage-drop values for three-phase busways with aluminum bus bars, in volts per 100 ft, line-to-line, at rated current with balanced entire load at end 11.34
11.1.39 Voltage-drop curves for typical plug-in-type Cu busway at balanced rated load, assuming 70°C (158°F) as the operating temperature 11.34
11.1.40 Voltage-drop curves for typical Cu feeder busways at balanced rated load mounted flat horizontally, assuming 70°C (158°F) as the operating temperature 11.35
11.1.41 Voltage-drop curve versus power factor for typical light-duty trolley busway carrying rated load, assuming 70°C (158°F) as the operating temperature 11.35
11.1.42 Voltage-drop curves for three-phase transformers, 225 to 10,000 kVA, 5 to 25 kV 11.35
11.1.43 Application tips 11.36
11.1.44 Flicker of incandescent lamps caused by recurrent voltage dips 11.37
11.1.45 Effect of voltage variations on incandescent lamps 11.37
11.1.46 General effect of voltage variations on induction motor characteristics 11.38
11.1.47 Calculation of voltage dips (momentary voltage variations) 11.38

Section 12 Transformers 12.1

12.1.0 Introduction 12.1
12.1.1 Typical transformer weights (lb) by kVA 12.2
12.1.2 Transformer full-load current, three-phase, self-cooled ratings 12.2
12.1.3 Typical impedances, three-phase, liquid-filled transformers 12.2
12.1.4 Approximate transformer loss and impedance data 12.2
12.1.5 Transformer primary (480-V, three-phase, delta) and secondary (208-Y/120-V, three-phase, four-wire) overcurrent protection, conductors and grounding 12.4
12.1.6 Maximum rating or setting of overcurrent protection for transformers over 600 V (as a percentage of transformer rated current) 12.4
12.1.7 Maximum rating or setting of overcurrent protection for transformers 600 V and less (as a percentage of transformer rated current) 12.5
12.2.1 Electrical connection diagrams 12.6
12.3.1 Auto zig-zag grounding transformers for deriving a neutral, schematic and wiring diagram 12.7
12.3.2 Auto zig-zag transformer ratings 12.7
12.4.1 Buck-boost transformer three-phase connection summary 12.8
12.4.2 Wiring diagrams for low-voltage single-phase buck-boost transformers 12.8
12.4.3 Connection diagrams for buck-boost transformers in autotransformer arrangement for single-phase system 12.9

12.4.4	Connection diagrams for buck-boost transformers in autotransformer arrangement for three-phase system	12.9
12.5.1	Maximum average sound levels for transformers	12.10
12.5.2	Typical building ambient sound levels	12.11
12.6.1	Transformer insulation system temperature ratings	12.11
12.7.1	k-rated transformers	12.11

Section 13 Grounding, ground-fault, and lightning protection 13.1

13.1.0	Introduction	13.1
13.1.1	NEC Table 250.122. Minimum size equipment grounding conductors for grounding raceway and equipment	13.1
13.2.1	Solidly grounded systems	13.2
13.2.2	Ungrounded systems	13.3
13.2.3	Resistence-grounded systems	13.3
13.2.4	Grounding-electrode system (NEC articles 250.50 and 250.52)	13.3
13.2.5	Grounding-electrode conductor for alternating-current systems (NEC table 250.66)	13.5
13.3.0	Ground-fault protection: introduction	13.5
13.3.1	Ground-return sensing method	13.6
13.3.2	Zero-sequence sensing method	13.6
13.3.3	Residual sensing method	13.7
13.3.4	Dual-source system: single-point grounding	13.8
13.4.0	Lightning protection	13.9
13.4.1	Annual isokeraunic map showing the average number of thunderstorm days per year (a) USA and (b) Canada	13.11
13.4.2	Rolling-ball theory	13.11
13.4.3	Cone of protection	13.13

Section 14 Emergency and standby power systems 14.1

14.1.0	Introduction	14.2
14.1.1	Summary of codes for emergency power in the United States by states and major cities (completed September 1984)	14.2
14.1.2	Condensed general criteria for preliminary consideration	14.2
14.1.3	Typical emergency/standby lighting recommendations	14.11
14.2.0	Emergency/standby power source options and arrangements	14.11
14.2.1	Two-utility-source system using one automatic transfer switch	14.12
14.2.2	Two-utility-source system where any two circuit breakers can be closed	14.12
14.2.3	Diagram illustrating multiple automatic double-throw transfer switches providing varying degrees of emergency and standby power	14.13
14.2.4	Typical transfer switching methods (a) total transfer and (b) critical-load transfer	14.13
14.2.5	Typical multiengine automatic paralleling system	14.15
14.2.6	Elevator emergency power transfer system	14.16
14.2.7	Typical hospital installation with a nonautomatic transfer switch and several automatic transfer switches	14.17
14.3.1	Generators and generator-set sizing: introduction	14.17
14.3.2	Engine—generator set load factor	14.19

14.3.3	Load management	14.21
14.3.4	Standards	14.21
14.3.5	Generator set sizing example	14.22
14.3.5.1	Generator sizing chart	14.22
14.3.5.2	Generator sizing chart (when using NEMA code letters)	14.23
14.3.6	Critical installation considerations	14.24
14.3.7	Illustration showing a typical emergency standby generator installation	14.25
14.4.0	Uninterruptible power supply (UPS) systems: introduction	14.26
14.4.1	Nonredundant UPS system configuration	14.27
14.4.2	"Cold" standby redundant UPS system	14.29
14.4.3	Parallel redundant UPS system	14.30
14.4.4	Isolated redundant UPS system	14.30
14.4.5	Application of UPS	14.31
14.5.0	Power-system configuration for 60-Hz distribution	14.31
14.5.1	Single-module UPS system	14.31
14.5.2	Parallel-capacity UPS system	14.32
14.5.3	Parallel redundant UPS system	14.32
14.5.4	Dual redundant UPS system	14.32
14.5.5	Isolated redundant UPS system	14.34
14.5.6	Parallel tandem UPS system	14.35
14.5.7	Hot tied-bus UPS system	14.35
14.5.8	Superredundant parallel system: hot tied-bus UPS system	14.36
14.5.9	Uninterruptible power with dual utility sources and static transfer switches	14.37
14.5.10	Power-system configuration with 60-Hz UPS	14.37
14.5.11	UPS distribution systems	14.37
14.6.1	Power-system configuration for 400-Hz distribution	14.39

Section 15 NEC Chapter 9 tables, and appendices B and C 15.1

15.1.1	NEC Chapter 9, Table 1, Percent of cross section of conduit and tubing for conductors	15.3
15.1.2	NEC Chapter 9, Table 4, Dimensions and percent area of conduit and tubing (areas of conduit or tubing for the combinations of wires permitted in Table 1, Chapter 9)	15.4
15.1.3	NEC Chapter 9, Table 5, Dimensions of insulated conductors and fixture wires	15.9
15.1.4	NEC Chapter 9, Table 5A, Compact aluminum building wire nominal dimensions* and areas	15.11
15.1.5	NEC Chapter 9, Table 8, Conductor properties	15.12
15.1.6	NEC Chapter 9, Table 9, Alternating-current resistance and reactance for 600-V cables, three-phase, 60 Hz, 75°C (167°F): three single conductors in conduit	15.13
15.1.7	NEC Chapter 9, Tables 11(A) and 11(B), Class 2 and Class 3, alternating-current and direct-current power-source limitations, respectively	15.14
15.1.8	NEC Chapter 9, Tables 12 (A) and 12 (B), PLFA alternating-current and direct-current power-source limitations, respectively	15.16
15.2.1	NEC (Annex B), Table B.310.1, Ampacities of two or three insulated conductors, rated 0 through 2000 V, within an overall covering (multiconductor cable), in raceway in free air	15.17

15.2.2	NEC (Annex B), Table B.310.3, Ampacities of multiconductor cables with not more than three insulated conductors, rated 0 through 2000 V, in free air (for type TC, MC, MI, UF, and USE cables)	15.18
15.2.3	NEC (Annex B), Table B.310.5, Ampacities of single-insulated conductors, rated 0 through 2000 V, in nonmagnetic underground electrical ducts (one conductor per electrical duct)	15.19
15.2.4	NEC (Annex B), Table B.310.6, Ampacities of three insulated conductors, rated 0 through 2000 V, within an overall covering (three-conductor cable) in underground electrical ducts (one cable per duct)	15.20
15.2.5	NEC (Annex B), Table B.310.7, Ampacities of three single-insulated conductors, rated 0 through 2000 V, in underground electrical ducts (three conductors per electrical duct)	15.21
15.2.6	NEC (Annex B), Table B.310.8, Ampacities of two or three insulated conductors, rated 0 through 2000 V cabled within an overall (two- or three-conductor) covering, directly buried in earth	15.22
15.2.7	NEC (Annex B), Table B.310.9, Ampacities of three triplexed single insulated conductors, rated 0 through 2000 V, directly buried in earth	15.23
15.2.8	NEC (Annex B), Table B.310.10, Ampacities of three single-insulated conductors, rated 0 through 2000 V, directly buried in earth	15.24
15.3.1	NEC (Annex B), Figure B.310.1, Interpolation chart for cables in a duct bank I_1 = ampacity for Rho = 60, 50 LF; I_2 = ampacity for Rho = 120, 100 LF (load factor); desired ampacity = $F \times I_1$	15.25
15.3.2	NEC (Annex B), Figure B.310.2, Cable installation dimensions for use with NEC Tables B.310.5 through B.310.10	15.26
15.3.3	NEC (Annex B), Figure B.310.3, Ampacities of single-insulated conductors rated 0 through 5000 V in underground electrical ducts (three conductors per electrical duct), nine single-conductor cables per phase	15.27
15.3.4	NEC (Annex B), Figure B.310.4, Ampacities of single-insulated conductors rated 0 through 5000 V in nonmagnetic underground electrical ducts (one conductor per electrical duct), four single-conductor cables per phase	15.28
15.3.5	NEC (Annex B), Table B.310.5, Ampacities of single-insulated conductors rated 0 through 5000 V in nonmagnetic underground electrical ducts (one conductor per electrical duct), five single-conductor cables per phase	15.29
15.3.6	NEC (Annex B), Table B.310.11, Adjustment factors for more than three current-carrying conductors in a raceway or cable with load diversity	15.30
15.4.0	NEC Annex C, conduit and tube fill tables for conductors and fixture wires of the same size	15.31
15.4.1	Table C1. Maximum number of conductors or fixture wires in electrical metallic tubing	15.31
15.4.2	Table C1(A). Maximum number of compact conductors in electrical metallic tubing	15.34

15.4.3	Table C2. Maximum number of conductors or fixture wires in electrical nonmetallic tubing	15.35
15.4.4	Table C2(A). Maximum number of compact conductors in electrical nonmetallic tubing	15.38
15.4.5	Table C3. Maximum number of conductors or fixture wires in flexible metal conduit	15.39
15.4.6	Table C3(A). Maximum number of compact conductors in flexible metal conduit	15.42
15.4.7	Table C4. Maximum number of conductors or fixture wires in intermediate metal conduit	15.43
15.4.8	Table C4(A). Maximum number of compact conductors in intermediate metal conduit	15.46
15.4.9	Table C5. Maximum number of conductors or fixture wires in liquidtight flexible nonmetallic conduit (Type LFNC-B)	15.47
15.4.10	Table C5(A). Maximum number of compact conductors in liquidtight flexible nonmetallic conduit (Type LFNC-B)	15.50
15.4.11	Table C6. Maximum number of conductors or fixture wires in liquidtight flexible nonmetallic conduit (Type LFNC-A)	15.51
15.4.12	Table C6(A). Maximum number of compact conductors in liquidtight flexible nonmetallic conduit (Type LFNC-A)	15.54
15.4.13	Table C7. Maximum number of conductors or fixture wires in liquidtight flexible metal conduit (LFMC)	15.55
15.4.14	Table C7(A). Maximum number of compact conductors in liquidtight flexible metal conduit	15.58
15.4.15	Table C8. Maximum number of conductors or fixture wires in rigid metal conduit	15.59
15.4.16	Table C8(A). Maximum number of compact conductors in rigid metal conduit	15.62
15.4.17	Table C9. Maximum number of conductors or fixture wires in rigid PVC conduit, Schedule 80	15.63
15.4.18	Table C9(A). Maximum number of compact conductors in rigid PVC conduit, Schedule 80	15.66
15.4.19	Table C10. Maximum number of conductors or fixture wires in rigid PVC conduit, Schedule 40 and HDPE conduit	15.67
15.4.20	Table C10(A). Maximum number of compact conductors in rigid PVC conduit, Schedule 40 and HDPE conduit	15.70
15.4.21	Table C11. Maximum number of conductors or fixture wires in Type A rigid PVC conduit	15.71
15.4.22	Table C11(A). Maximum number of compact conductors in Type A rigid PVC conduit	15.74
15.4.23	Table C12. Maximum number of conductors in Type EB PVC conduit	15.75
15.4.24	Table C12(A). Maximum number of compact conductors in Type EB PVC conduit	15.77

Section 16 Lighting 16.1

16.1.1	Conversion factors of units of illumination	16.2
16.1.2	U.S. and Canadian standards for ballast efficacy factor	16.2
16.1.3	Starting and restrike times among different HID lamps	16.3

16.1.4	How light affects color	16.3
16.1.5	Summary of light-source characteristics and effects on color	16.4
16.2.1	Determination of illuminance categories	16.5
16.3.1	Zonal cavity method of calculating illumination	16.5
16.3.2	Coefficients of utilization for typical luminaires	16.6
16.3.3	Light-loss factor (LLF)	16.6
16.3.4	Light-loss factors by groups	16.6
16.3.5	Light output change due to voltage change	16.19
16.3.6	Lumen output for HID lamps as a function of operating position	16.19
16.3.7	Lamp lumen depreciation	16.19
16.3.8	Procedure for determining luminaire maintenance categories	16.21
16.3.9	Evaluation of operating atmosphere	16.22
16.3.10	Five degrees of dirt conditions	16.22
16.3.11	Luminaire dirt depreciation (LDD) factors for six luminaire categories (I through VI) and for the five degrees of dirtiness as determined from Tables 16.3.8 or 16.3.9	16.22
16.3.12	Room surface dirt depreciation (RSDD) factors	16.22
16.3.13	Step-by-step calculations for the number of luminaires required for a particular room	16.24
16.3.14	Reflectance values of various materials and colors	16.26
16.3.15	Room cavity ratios	16.26
16 3.16	Percent effective ceiling or floor cavity reflectances for various reflectance combinations	16.27
16.3.17	Multiplying factors for effective floor cavity reflectances other than 20 percent (0.2)	16.30
16.3.18	Characteristics of typical lamps	16.31
16.3.19	Guide to lamp selection	16.32
16.3.20	Recommended reflectances of interior surfaces	16.35
16.3.21	Recommended luminance ratios	16.35
16.3.22	Average illuminance calculation sheet	16.36

Section 17 Hazardous (classified) locations 17.1

17.1.0	Introduction	17.1
17.1.1	Table summary classification of hazardous atmospheres (NEC Articles 500 through 504)	17.2
17.1.2	Classification of hazardous atmospheres	17.2
17.1.3	Prevention of external ignition and explosion	17.5
17.1.4	Equipment for hazardous areas	17.8
17.1.5	Wiring methods and materials	17.9
17.1.6	Maintenance principles	17.11
17.1.7	Gases and vapors: hazardous substances used in business and industry	17.12
17.1.8	Dusts: hazardous substances used in business and industry	17.16
17.1.9	NEC Table 500.8(B). Classification of maximum surface temperature	17.18
17.1.10	NEC Table 500.8(C)(2). Class II ignition temperatures	17.18
17.1.11	NEC Article 505, Class I, Zone 0, 1 and 2 locations	17.19
17.1.12	NEC Article 511, Commercial garages, repair and storage	17.20
17.1.13	NEC Article 513, Aircraft hangers	17.21

17.1.14	NEC Article 514, Motor fuel dispensing facilities	17.22
17.1.15	NEC Article 515, Bulk storage plants	17.28
17.1.16	NEC Article 516, Spray application, dipping, and coating processes	17.35
17.1.17	Installation diagram for sealing	17.37
17.1.18	Diagram for Class I, Zone 1 power and lighting installation	17.38
17.1.19	Diagram for Class I, Division 1 lighting installation	17.39
17.1.20	Diagram for Class I, Division 1 power installation	17.40
17.1.21	Diagram for Class I, Division 2 power and lighting installation	17.41
17.1.22	Diagram for Class II lighting installation	17.42
17.1.23	Diagram for Class II power installation	17.43
17.1.24	Crouse-Hinds "quick-selector": electrical equipment for hazardous locations	17.44
17.1.25	Worldwide explosion protection methods, codes, categories, classifications, and testing authorities	17.45

Section 18 Telecommunications structured cabling systems 18.1

18.1.0	Introduction	18.2
18.1.1	Important codes and standards	18.2
18.1.2	Comparison of ANSI/TIA/EIA, ISO/IEC, and CENELEC cabling standards	18.3
18.2.0	Major elements of a telecommunications structured cabling system	18.4
18.2.1	Typical ranges of cable diameter	18.5
18.2.2	Conduit sizing-number of cables	18.5
18.2.3	Bend radii guidelines for conduits	18.6
18.2.4	Guidelines for adapting designs to conduits with bends	18.6
18.2.5	Recommended pull box configurations	18.7
18.2.6	Minimum space requirements in pull boxes having one conduit each in opposite ends of the box	18.8
18.2.7	Cable tray dimensions (common types)	18.9
18.2.8	Topology	18.10
18.2.9	Horizontal cabling to two individual work areas	18.10
18.2.10	Cable lengths	18.11
18.2.11	Twisted-pair (balanced) cabling categories	18.12
18.2.12	Optical fiber cable performance	18.13
18.2.13	Twisted-pair work area cable	18.13
18.2.14	Eight-position jack pin/pair assignments (TIA-568A) (front view of connector)	18.14
18.2.15	Optional eight-position jack pin/pair assignments (TIA-568B)(front view of connector)	18.14
18.2.16	Termination hardware for category-rated cabling systems	18.15
18.2.17	Patch cord wire color codes	18.15
18.2.18	ANSI/TIA/EIA-568A categories of horizontal copper cables (twisted-pair media)	18.16
18.2.19	Work area copper cable lengths to a multi-user telecommunications outlet assembly (MUTOA)	18.17
18.2.20	U.S. twisted-pair cable standards	18.18
18.2.21	Optical fiber sample connector types	18.19
18.2.22	Duplex SC interface	18.19

18.2.23 Duplex SC adapter with simplex and duplex plugs 18.20
18.2.24 Duplex SC patch cord crossover orientation 18.20
18.2.25 Optical fibers 18.21
18.2.26 Backbone system components 18.21
18.2.27 Backbone star wiring topology 18.22
18.2.28 Example of combined copper/fiber backbone supporting voice and data traffic 18.23
18.2.29 Backbone distances 18.24
18.2.30 Determining 100 mm (4 in) floor sleeves 18.25
18.2.31 Determining size of floor slots 18.25
18.2.32 Conduit fill requirements for backbone cable 18.26
18.2.33 TR cross-connect field color codes 18.27
18.2.34 TR temperature ranges 18.27
18.2.35 TR size requirements 18.28
18.2.36 Allocating termination space in TRs 18.28
18.2.37 Typical telecommunications room (TR) layout 18.29
18.2.38 TR industry standards 18.30
18.2.39 TR regulatory and safety standards 18.30
18.2.40 Environmental control systems standards for equipment rooms (ERs) 18.31
18.2.41 Underground entrance conduits for entrance facilities (EFs) 18.31
18.2.42 Typical underground installation to EF 18.32
18.2.43 Equipment room (ER) floor space (special-use buildings) 18.32
18.2.44 Entrance facility (EF) wall space (minimum equipment and termination wall space) 18.33
18.2.45 Entrance facility (EF) floor space (minimum equipment and termination floor space) 18.33
18.2.46 Separation of telecommunications pathways from 480-Volt or less power lines 18.34
18.2.47 Cabling standards document summary 18.35
18.3.0 Blown optical fiber technology (BOFT) overview 18.36
18.3.1 Diagram showing key elements of BOFT system 18.36
18.3.2 BOFT indoor plenum 5-mm multiduct 18.38
18.3.3 BOFT outdoor 8-mm multiduct 18.39
18.3.4 BOFT installation equipment 18.40

Section 19 Miscellaneous special applications 19.1

19.1.0 Fire alarm systems: introduction 19.1
19.1.1 Fire alarm systems: common code requirements 19.2
19.1.2 Fire alarm system classifications 19.2
19.1.3 Fire alarm fundamentals: basic elements (typical local protective signaling system) 19.3
19.1.4 Fire alarm system circuit designations 19.4
19.1.5 Fire alarm system: class 19.4
19.1.6 Fire alarm system: style 19.4
19.1.7 Performance of initiating device circuits (IDCs) 19.4
19.1.8 Performance of signaling-line circuits (SLCs) 19.5
19.1.9 Notification-appliance circuits (NACs) 19.5
19.1.10 Installation of class A circuits 19.7
19.1.11 Secondary supply capacity and sources 19.7

19.1.12	Audible notification appliances to meet the requirements of ADA, NFPA 72 (1993), and BOCA	19.7
19.1.13	Visual notification appliances to meet the requirements of ADA, NFPA 72 (1993), and BOCA	19.7
19.1.14	ADA-complying mounting height for manual pull stations (high forward-reach limit)	19.10
19.1.15	ADA-complying mounting height for manual pull stations (high and low side-reach limits)	19.10
19.1.16	Application tips	19.10
19.2.1	Fire pump applications	19.11
19.2.2	Typical one-line diagram of fire pump system with separate ATS	19.13
19.2.3	Typical one-line diagram of fire pump system with ATS integrated with the fire pump controller	19.14
19.3.1	Wiring of packaged rooftop AHUs with remote VFDs	19.15
19.4.1	Wye-delta motor starter wiring	19.15
19.5.1	Elevator recall systems	19.17
19.5.2	Typical elevator recall/emergency shutdown schematic	19.18
19.5.3	Typical elevator hoistway/machine room device installation detail	19.18
19.6.0	Harmonic effects and mitigation	19.18

Section 20 Metrification 20.1

20.0.0	Introduction to the 1975 metric conversion act	20.2
20.1.0	What will change and what will remain the same	20.2
20.2.0	How metric units will apply in the construction industry	20.9
20.3.0	Metrification of pipe sizes	20.10
20.4.0	Metrification of standard lumber sizes	20.11
20.5.0	Metric rebar conversions	20.11
20.6.0	Metric conversion of ASTM diameter and wall-thickness designations	20.12
20.7.0	Metric conversion scales (temperature and measurements)	20.13
20.8.0	Approximate metric conversions	20.14
20.9.0	Quick imperial (metric equivalents)	20.16
20.10.0	Metric conversion factors	20.17

Index I.1

Introduction

The *Electrical Construction Databook* provides the electrical design consultant, project manager, contractor, field superintendent, facility owners, and operations and maintenance personnel with a one-source reference guide to the most commonly encountered (and needed) electrical design, installation, and application data. Valuable information ranging from *NEC®* installation requirements, wiring methods and materials, to lighting and telecommunications systems, with scores of topics in between, is included in this single easy-to-access volume.

Numerous carefully selected sections of the National Electric Code (*NEC®*) are included with critical data and tables for sizing conductors, conduits, overcurrent protection, pull-boxes, etc., and many illustrations to help clarify the Code's intent with regard to proper equipment installation, working clearances, acceptable installations under exceptions with certain conditions applied, and a plethora of others, including materials and methods.

The *Electrical Construction Databook* contains single-line diagrams of primary and secondary service and system configurations, emergency and standby generator system configurations, and uninterruptable power supply system configurations, each with their advantages, disadvantages, and operating characteristics concisely outlined for easy comparison in determining what's best for a given application. Even the sequence in which they are presented, in general, is from the least cost and reliability to the highest cost and reliability in order to broadly address the economic criteria.

In addition to recognized code and professional organizations, much of the material in this book has been gleaned from manufacturer's sources and trade association–supplied information; some of the manufacturer-supplied data may be proprietary in nature but generally is similar to products made by other vendors. And the reader should note that many manufacturers and related trade organizations are often eager to furnish additional and more specific information if requested.

The *Electrical Construction Databook* conforms to the newly published 2002 edition of the *NEC®*. There may be some minor subtext references that use the 1999 *NEC®* edition's section/paragraph nomenclature and a few illustrations that show English units only without the equivalent metric units, but they are still valid, to the best of the author's knowledge.

This one-source *Electrical Construction Databook* should prove invaluable for office- and field-based construction and design professionals, since it contains, in one volume, answers to so many of the design and application questions that arise before and during a construction project. As an electrical engineer who has worked in the trade as an electrician, I have tried, based on almost 40 years of experience in the construction industry, to blend together data and information that is useful and practical from both a design and construction installation perspective. I trust that I have met that goal.

I hope you find the *Electrical Construction Databook* a worthwhile addition to your construction library.

Bob Hickey

Credits

Reprinted with permission from NFPA 70 (2002), the *National Electrical Code®*, copyright © 2001, National Fire Protection Association, Quincy, MA 02269. This reprinted material is not the referenced subject which is represented only by the standard in its entirety.

Figures 4.4.25, 15.3.1–15.3.5, 17.1.14D, and 17.1.15A–D, and Tables 2.2.1, 2.5.1, 2.5.2, 4.1.0, 4.1.1, 4.2.1, 4.3.1, 4.4.1–4.4.24, 4.4.26–4.4.34, 4.5.1, 4.5.3–4.5.5, 4.7.1, 4.7.3–4.7.7, 4.7.9, 4.8.1, 4.8.2, 4.9.1–4.9.6, 9.1.11, 9.1.12, 10.1.2, 12.1.6, 12.1.7, 13.1.1, 13.2.5, 15.1.1–15.1.8, 15.2.1–15.2.8, 15.3.6, 15.4.1–15.4.24, 17.1.9, 17.1.10, 17.1.14, and 17.1.15

Reprinted with permission from the *National Electrical Code® Handbook*, copyright © 1999, National Fire Protection Association, Quincy, MA 02269. This reprinted material is not the referenced subject which is represented subject which is represented only by the standard in its entirety.

Figures 1.4.0, 1.4.1, 2.2.2–2.2.6, 2.3.1–2.3.5, 2.4.1–2.4.3, 4.2.2, 4.5.2, 4.5.6, 4.5.7, 4.6.1, 4.7.2, 4.7.8, 17.1.12A,B, 17.1.13, 17.1.14C–G, and 17.1.15E–H, and Tables 2.1.1, 2.1.2, and 4.4.3

Reprinted with permission from NFPA 72, *National Fire Alarm Code®*, copyright © 1996, National Fire Protection Association, Quincy, MA 02269. This reprinted material is not the referenced subject which is represented only by the standard in its entirety.

Tables 19.1.7, 19.1.8, 19.1.9 and excerpts of text on pp. 19.2–29.7

Tables 1.8.2–1.8.6, 6.1.5, and 12.1.1 are reprinted with permission from *1996 Means Electrical Cost Data*, copyright R.S. Means Co., Inc., Kingston, MA, 781-585-7880, all rights reserved.

ACME Electric, Caterpillar, Cooper Bussmann, Cooper Crouse-Hinds, Cutler-Hammer, Ferraz Shawmut, General Cable Corporation, OSRAM Sylvania, and Siemens Corporation.

Acknowledgments

Many thanks to the entire electrical staff at vanZelm, Heywood, & Shadford, Inc., for their valuable input, and to Kristine M. Buccino for her assistance in getting permission to reprint copyrighted material.

A special thanks to Larry Hager and Steve Melvin and their team at McGraw-Hill, whose wonderful collaborative spirit and many professional talents transformed the raw manuscript into a published reality.

And finally, a very special thanks to Chuck Durang of the National Fire Protection Association, whose invaluable cooperation and assistance made it possible to incorporate the new 2002 *NEC®* edition changes in time for printing of this book.

Section

1

General Information

1.1.0 Introduction
1.1.1 Project To-Do Checklist (Electrical)
1.1.2 Drawing Design Checklist (Electrical)
1.1.3 Site Design Checklist (Electrical)
1.1.4 Existing Condition Service and Distribution Checklist
1.1.5 Design Coordination Checklist (Electrical)
1.1.6 Fire Alarm System Checklist
1.2.0 Electrical Symbols
1.3.0 Mounting Heights for Electrical Devices
1.4.0 NEMA Configuration Chart for General-Purpose Nonlocking Plugs and Receptacles
1.4.1 NEMA Configuration Chart for Specific-Purpose Locking Plugs and Receptacles
1.5.0 IEEE Standard Protective Device Numbers
1.6.0 Comparison of Specific Applications of NEMA Standard Enclosures for Indoor Nonhazardous Locations
1.6.1 Comparison of Specific Applications of NEMA Standard Enclosures for Outdoor Nonhazardous Locations
1.6.2 Comparison of Specific Applications of NEMA Standard Enclosures for Indoor Hazardous Locations
1.6.3 Knockout Dimensions for NEMA Standard Enclosures
1.7.0 Formulas and Terms
1.8.0 Introduction: Typical Equipment Sizes, Weights, and Ratings
1.8.1 Typical Equipment Sizes: 600-V Class
1.8.2 Transformer Weight (lb) by kVA
1.8.3 Generator Weight (lb) by kW
1.8.4 Weight (lb/lf) of Four-Pole Aluminum and Copper Bus Duct by Ampere Rating
1.8.5 Conduit Weight Comparisons (lb per 100 ft) Empty
1.8.6 Conduit Weight Comparisons (lb per 100 ft) with Maximum Cable Fill
1.9.0 Seismic Requirements

1.1.0 Introduction

This section provides information of a general nature that is needed frequently by electrical design and construction professionals. Information that follows in subsequent sections is more specific in its applications.

1.1.1 Project To-Do Checklist (Electrical)

1.1.1

Page 1 of 3

Project Status
☐ SD
☐ DD
☐ CD

Project: ______________
Proj. No: ______________
PM/PE: ______________
Date: ______________

Project To Do List (Electrical)

PreDesign
- ☐ Review Contract Scope
- ☐ Review Design Budget with P.M.
- ☐ Establish design criteria
- ☐ Establish design schedule
- ☐ Schedule review meetings & team
- ☐ Setup project notebook
- ☐ Code review
- ☐ Obtain as-built drawings
- ☐ Site survey
- ☐ Start project data sheet
- ☐ Contact Power Company
- ☐ Contact Telephone Company
- ☐ Review client's design requirements
- ☐ ______________
- ☐ ______________
- ☐ ______________

Load Analysis
- ☐ Schematic, sq.foot basis
- ☐ Mechanical loads finalized
- ☐ Process equipment loads finalized
- ☐ Final design loads scheduled
- ☐ ______________

Fault Current Analysis
- ☐ Rough estimate pre-design
- ☐ Final analysis

Coordination Study
- ☐ Rough selection pre-design
- ☐ Final study

Design
- ☐ Main electric service
- ☐ Power Distribution system
- ☐ Branch circuits
- ☐ Building lighting
- ☐ Site lighting
- ☐ Main telephone service
- ☐ ______________
- ☐ ______________

Other Systems
- ☐ Communications Consultant
- ☐ AV Consultant
- ☐ Food Service Consultant
- ☐ Elevator Consultant
- ☐ Theatre Consultant
- ☐ Division 16 coordinated with Div. 15/13
- ☐ ______________

Special Systems
- ☐ Fire alarm & smoke detection system
- ☐ Telephone outlets
- ☐ TV outlets
- ☐ Elevator System
- ☐ Data outlets
- ☐ Intercom system
- ☐ Security system
- ☐ Standby generators & Automatic Transfer Switch
- ☐ Energy Management System
- ☐ Grounding systems
- ☐ Lightning Protection system
- ☐ ______________
- ☐ ______________

(continued)

1.1.1

Project Status
☐ SD
☐ DD
☐ CD

Project: ____________
Proj. No: ____________
PM/PE: ____________
Date: ____________

Project To Do List (Electrical)

Specification
☐ Cover
☐ Bidding forms
☐ General Conditions & Division 1
☐ Non-electrical sections
☐ Division 13 sections
☐ Division 15 sections
☐ Division 16 sections

Construction Estimates
☐ Schematic design
☐ Design development
☐ Construction documents

Drawings
☐ Title block & drawing size
☐ Site plans
☐ Demolition plans
☐ Symbol list
☐ Abbreviation list
☐ General notes
☐ Power plans
☐ Lighting plans
☐ Fixture schedule
☐ One-line power diagram
☐ Switchboard schedules
☐ MCC schedules
☐ Distribution panelboard schedules
☐ Lighting panelboard schedules
☐ Fire detection & alarm plans
☐ Fire detection & alarm one-line diagram
☐ Building grounding grid plan
☐ Lightning protection plan

Electrical Details
☐ Front Elevation Switchboards
☐ Front Elevation MCCs
☐ ____________
☐ ____________

Site Details
☐ Concrete Bases for Lighting Poles
☐ Transformer Concrete Pads & Grounding
☐ Equipment Concrete Pads & Grounding
☐ Manholes, Ductbanks, Grounding
☐ Trench, backfill & reseed
☐ Pavement
☐ ____________

Drawing Check
☐ Overlay electrical drawings
☐ Complete drawing checklists
☐ Complete site checklists
☐ ____________
☐ ____________

In House Review
☐ Conceptual review
☐ Schematic Design
☐ Design Development
☐ Construction Documents

Client Submission
☐ Schematic Design
☐ Design Development
☐ Construction Documents

(continued)

1.1.1

Project Status
☐ SD
☐ DD
☐ CD

Project: ______________
Proj. No: ______________
PM/PE: ______________
Date: ______________

Project To Do List (Electrical)

Design Closeout

☐ Complete project data sheet

☐ Project profile completed

☐ File the design calculations

☐ Complete the design notebook
Has Power Company Reviewed Designed Service? ☐ Yes ☐ No ☐ Not Required
Charges: $______ ☐ Unknown
Has Power Company Been Sent Electrical Loads, Drawings and Specs?
☐ Yes ☐ No ☐ Not Required

☐ Send client record documents

☐ ______________________________

☐ ______________________________

1.1.2 Drawing Design Checklist (Electrical)

1.1.2

Page 1 of 3

Project Status
☐ SD
☐ DD
☐ CD

Project: ______________
Proj. No: ______________
PM/PE: ______________
Date: ______________

Drawing Design Checklist (Electrical)

Items Included

☐ Power Plan
☐ Lighting Plan
☐ Site Plan
☐ Special System Plans
☐ Symbol List
☐ Abbreviation List
☐ One Line - Power Diagram
☐ One Line - Special Systems
☐ Switchboard Schedules
☐ Panelboard Schedules
☐ Fixture Schedules
☐ Site Details
☐ Electrical Details
☐ Building Grounding Plan
☐ Lightning Protection Plan
☐ General Notes
☐ ______________________
☐ ______________________
☐ ______________________

General Items to Check

☐ Title Blocks
  ☐ Firm Logo
  ☐ Job Number
  ☐ Drawing Title
  ☐ Drawing Numbers
  ☐ Date
☐ Plan Titles with Scale
☐ Detail Titles with Scale
☐ Detail Designation Symbols
☐ Symbol List Agrees with Drawing
☐ Abbreviation List Agrees with Drawings
☐ No Abbreviation Used Less than Five Times

☐ Openings and Floor Plans for Installation and Removal of Electrical and Generator Equipment
☐ Electrical equipment access and clearances
☐ Elevator Size Accommodates All Equipment
☐ Electrical Plans Overlayed on:
  ☐ Architectural Plans
  ☐ Reflected Ceiling Plans
  ☐ Mechanical Plans

One-Line Power Diagram

☐ Primary Distribution
  ☐ Voltage
  ☐ Fault Current Available
  ☐ Cables and Raceways
  ☐ Manholes and Pullboxes
  ☐ Terminations and Splices

☐ Primary Switchgear
  ☐ Enclosure
    ☐ Indoor ☐ Weatherproof ☐ Walk-In
  ☐ Selector Switches
    ☐ Non-fused ☐ Fuse Size
  ☐ Protective Devices
    ☐ Stationary ☐ Drawout
    ☐ Manual ☐ Electrical
    ☐ Active ☐ Space & Busing
    ☐ Breaker ☐ Trip Setting
    ☐ Relay ☐ Trip Setting
    ☐ Circuit Numbering
    ☐ Arresters
    ☐ Interlocks
    ☐ Fault Rating

(continued)

1.1.2

Project Status
☐ SD
☐ DD
☐ CD

Project: ____________
Proj. No: ____________
PM/PE: ____________
Date: ____________

Drawing Design Checklist (Electrical)

☐ Primary Metering
  ☐ Owner ☐ Power Co.
☐ Transformers
  ☐ Primary Voltage
  ☐ Primary Connection
    ☐ Delta ☐ Wye ☐ Double Bushing
  ☐ Secondary Voltage
  ☐ Secondary Connection
    ☐ Delta ☐ Wye
  ☐ Grounding
  ☐ KVA & Percent Impedance (Min.)
  ☐ Type:
  (Oil, Dry, Padmount, Open, WP, etc.)
  ☐ Secondary Compartment C/Bs
  ☐ Surge Arresters
  ☐ Power Company Supplied

☐ Secondary Distribution
  ☐ Voltage
  ☐ Fault Current Available
  ☐ Cables and Raceways
  ☐ Manholes and Pullboxes
  ☐ Termination and Splices

Secondary Switchboard
  ☐ Switchboard (NEMA PB-2 and UL 891)
  ☐ Switchgear (ANSI C37 and UL 1558)
  ☐ Rating ☐ Current ☐ Voltage
  ☐ Phase ☐ Wire
  ☐ Fault Rating
  ☐ Service Entrance?
☐ Enclosure
  ☐ Free-standing ☐ Non-freestanding
☐ Accessible
  ☐ Front ☐ Rear ☐ Side

☐ Main Protective Device
  ☐ Fuse/Sw ☐ Size & Class of Fuse
  ☐ Power Breaker ☐ Insulated Case
    ☐ Molded Case
    ☐ Indv. Mount ☐ Group Mount
  ☐ Stationary ☐ Drawout
  ☐ Manual ☐ Electrical
  ☐ Thermal/Magnetic ☐ Solid State
  ☐ Number of Poles & Trip/Frame Amps
  ☐ 100% Duty ☐ 80% Duty
  ☐ Shunt Trip
  ☐ Interlocks or Ties
☐ Ground Fault Protection
  ☐ Selective ☐ Time Delay
☐ Service Ground
  ☐ Water Service
  ☐ Building Steel
  ☐ Ground Rod
  ☐ Ground Grid - Substation
  ☐ Ground Grid - Building
☐ Revenue Metering
  ☐ Active ☐ Reactive
  ☐ CT's ☐ PT's
☐ Owner Metering
  ☐ Volt ☐ Amp ☐ Watt ☐ VA
  ☐ Watt Hr ☐ VARS
  ☐ AMSS ☐ VMSS
  ☐ Electronic
☐ Busing
  ☐ Full Neutral
  ☐ Ground Bus
  ☐ Equipment Ground
  ☐ Grounding Electrode Conductor
  ☐ Connection

(continued)

1.1.2

Page 3 of 3

Project Status
☐ SD
☐ DD
☐ CD

Project: ______________
Proj. No: ______________
PM/PE: ______________
Date: ______________

Drawing Design Checklist (Electrical)

- ☐ Main Feeder Cable and Raceways
- ☐ Transfer Switches
  - ☐ Type
    - ☐ Automatic ☐ Manual
  - ☐ Current Rating and # Poles
  - ☐ Control Connection
  - ☐ Load Feeder Cable and Raceway
  - ☐ 3 Pole or 4 Pole
  - ☐ Neutral and Ground Connection
- ☐ Standby Generator ☐ Emergency Generator
  - ☐ Line Circuit Breaker ☐ Main Lug
    - ☐ Thermal ☐ Magnetic
    - ☐ Solid State
  - ☐ Number of Poles & Trip/Frame Amps
  - ☐ GFP ☐ Sel. ☐ Timedelay
  - ☐ Load Feeder Cable and Raceway
  - ☐ Neutral and Ground Connections
- ☐ Power Distribution (Panelboard and MCC)
  - ☐ Bus Data
    - ☐ Current
    - ☐ Voltage
    - ☐ Phase ☐ Wire
    - ☐ Fault Current
    - ☐ Full Neutral
    - ☐ Equipment Ground
    - ☐ Insulated
    - ☐ Enclosure
    - ☐ Weatherproof ☐ Walk-in
  - ☐ Mounting
    - ☐ Individual ☐ Group (Panel Sched.)
    - ☐ Stationary ☐ Drawout
  - ☐ Operation
    - ☐ Manual ☐ Automatic
- ☐ Protective Devices
  - ☐ Circuit Numbering
  - ☐ Fuse/Switch
    - ☐ Fuse Size/Class
  - ☐ Combination Starter
    - ☐ Fuse/Switch & Fuses
    - ☐ Circuit Breaker
      - ☐ Mag. Only
    - ☐ Starter Size & Type
    - ☐ Overload Relays
  - ☐ Circuit Breaker
    - ☐ Power
    - ☐ Insulated Case
    - ☐ Molded Case
    - ☐ 100% Duty
      - ☐ Mixed Duty
    - ☐ Thermal/Magnetic
    - ☐ Magnetic
    - ☐ Solid State
    - ☐ Number of Poles
    - ☐ Trip/Frame Amps
    - ☐ Ground Fault Protection
    - ☐ Selective ☐ Time Delay
    - ☐ Interlocks
    - ☐ Key ☐ Electric

1.1.3 Site Design Checklist (Electrical)

1.1.3

Page 1 of 2

Project Status
- ☐ SD
- ☐ DD
- ☐ CD

Project: ____________
Proj. No: ____________
PM/PE: ____________
Date: ____________

Site Design Check List (Electrical)

Site Drawings - Plans

- ☐ Title
- ☐ Scale
- ☐ Benchmark
- ☐ Topo Lines

Top Elevation on:
- ☐ Transformer Pads
- ☐ Switchgear Pads
- ☐ Pole Bases for Site Lighting
- ☐ Standby Generator Pads
- ☐ Manholes
- ☐ Pullboxes

- ☐ Existing Utility Poles and Numbers
- ☐ New Utility Poles and Guys (by whom)
- ☐ Pole Transformers (by whom)
- ☐ Pad Mount Transformers (by whom)
- ☐ Revenue Meters
- ☐ Site Lighting Poles
- ☐ Generator (Outdoor)
- ☐ Switchgear (Outdoor)
- ☐ Manholes
- ☐ Pullboxes

Check Site Planting, Grades, Fences, Equipment for Truck Access to:
- ☐ Padmount Transformers
- ☐ Utility Poles
- ☐ Site Lighting Poles

Aerial Distribution
- ☐ Electric Primary
- ☐ Electric Secondary
- ☐ Telephone
- ☐ Site Lighting
- ☐ TV

Underground Distribution
- ☐ Electric Primary
- ☐ Electric Secondary
- ☐ Telephone
- ☐ TV
- ☐ Site Lighting
- ☐ Conduit Sleeves Under Pavement

Fuel Oil Systems
- ☐ Fuel Oil Tank
- ☐ Supply and Return Lines
- ☐ Fill Cap and Fill Lines
- ☐ Vent Cap and Vent Lines
- ☐ Tank Level Gauge Line
- ☐ Soil Condition - Anodes, FG
- ☐ Direction of Line Pitch

Check Truck Wheel Loading Cover:
- ☐ Fuel Oil Tanks
- ☐ Underground Lines
- ☐ Manholes
- ☐ Pullboxes

(continued)

1.1.3

Page 2 of 2

Project Status
☐ SD
☐ DD
☐ CD

Project: ______________
Proj. No: ______________
PM/PE: ______________
Date: ______________

Site Checklist (Electrical)

Site Drawings - Details

- ☐ Titles
- ☐ Scale
- ☐ Utility Pole Riser
- ☐ Revenue Meter Riser

Trench Cross Sections
- ☐ Electric, Telephone and TV Lines
- ☐ Duct Banks, Concrete and Grounding

- ☐ Padmount Transformer, Concrete Pad & Grounding
- ☐ Exterior Switchgear, Concrete Pad & Grounding
- ☐ Generator, Concrete Pad & Grounding
- ☐ Manholes, Concrete, Cable Racks & Grounding
- ☐ Pullboxes, Concrete, & Grounding
- ☐ Pole Bases for Site Lighting and Signs

Fuel Oil Systems
- ☐ Fuel Oil Tank, Concrete Pad
- ☐ Trench Cross Sections for Supply & Return Lines
- ☐ Fill, Vent and Level Gage Lines
- ☐ Fuel Fill Cap
- ☐ Fuel Vent Cap

1.1.4 Existing Condition Service and Distribution Checklist

1.1.4

Page 1 of 3
Project: ____________
Proj. No: ____________
PM/PE: ____________
Date: ____________

Existing Condition Service & Distribution Checklist

Power Company Service

Power Company: ______________________
Rep Name: ______________________
Telephone: ______________________

Type of Service:

☐ Primary ☐ Secondary ☐ Unknown

☐ Underground ☐ Overhead
☐ Combination ☐ Unknown
☐ ____________________

Transformation

☐ Pad ☐ Pole ☐ N/A ☐ Unknown
KVA: ____________ ☐ Unknown
% Impedance: ______ ☐Unknown
Primary Voltage _____ ☐ Unknown
Secondary Voltage: ____ ☐ Unknown

Short Circuit Fault Current Available

☐ Power Company ☐ Sym
☐ Primary ☐ MVA
☐ Secondary: _______ ☐ A
☐ Unknown

Power Company Pole #: ______ ☐ Unknown

New Poles: ☐ Street Line ☐ Private
☐ N/A ☐ Unknown

Primary Service

Raceway Size: _______ ☐ Unknown
Type: ☐ RSC ☐ PVC ☐ PVC/Conc.
☐ DB:_____ ☐ Unknown
Cable: ____________ ☐ Unknown
Ground Conductor: ______ ☐ Unknown

Secondary Service

Raceway Size: _______ ☐ Unknown
Type: ☐ RSC ☐ PVC ☐ PVC/Conc.
☐ DB:_____ ☐ Unknown
Cable: ____________ ☐ Unknown

Type of Power Available at Site Line

Primary ☐ 1PH ☐ 3PH ☐ Unknown
Sec ☐ 1PH ☐ 3 PH ☐ Unknown

Has Power Company Been Conducted for Existing Loads and Requirements for new services?

☐ Yes ☐ No ☐ Not Req.

Comments:

____
____
____

Main Electric Service

Main Entrance Capacity:

Size ____________ A ☐ Unknown
Total Load
________ KW ________ KVA
Power Factor ______ ☐ Unknown
Largest Connected Motor ☐ N/A
__________ HP ☐ Unknown
Starter Size & Type ____ ☐Unknown

(continued)

1.1.4

Page 2 of 3
Project: ______________
Proj. No: ______________
PM/PE: ______________
Date: ________________

Existing Condition Service & Distribution Checklist

Main Protective Device:

☐ Fuse/Switch ☐ MCCB ☐ ICCB
☐ Power breaker_______ ☐Unknown
Duty: ☐ 80% ☐ 100% ☐Unknown
Type of Trip: ☐ Thermal ☐Magnetic
☐ Solid State
GFI ☐Yes ☐No ☐Unknown
☐ Selective ☐Time Delay
Current Limiting
☐Yes ☐No ☐Unknown

CT's Required: ☐Yes ☐No ☐Unknown
PT's Required: ☐Yes ☐No ☐Unknown

Who Supplies CT's and PT's:
______________ ☐Unknown

Revenue Meters
☐ Active ☐ Reactive ☐Unknown
☐ Inside ☐ Outside ☐Unknown

Type of Construction
☐ Panelboard ☐ Switchboard
☐ Unitized ☐ MCC __________

Grounding Electrode Conductor
Size _____________
☐ Ground Rod ☐ Water Service

Rating of Gear
_____________ AIC Sym. ☐Unknown

Comments:
____
____
____

Power Distribution System ☐ N/A

Main Distribution Bus ________ A ☐Unknown
Rating ________ AIC Sym ☐Unknown
Distribution Devices
☐ Fuse/Switch ☐ MCCB ☐ ICCB
☐ Power breaker_______ ☐Unknown
Duty: ☐ 80% ☐ 100% ☐Unknown
Type of Trip: ☐ Thermal ☐Magnetic
☐ Solid State
GFI ☐Yes ☐No ☐Unknown
☐ Selective ☐Time Delay
Current Limit ☐Yes ☐No ☐Unknown

Raceways ☐ Aluminum ☐ RSC ☐ ISC
☐ EMT ☐ PVC ☐ Unknown

Conductor Type_____________ ☐ Unknown

Voltage Systems #1____________ ☐ Unknown
#2______________________

Raceway Location ☐ Exposed ☐ Unknown
Concealed in: ☐ Walls ☐ Ceilings
☐ Floors ☐ Unknown

Busway ☐ Aluminum ☐ Copper ☐ WP
☐ N/A ☐ Unknown
☐ Feeder ☐ Plug-in ☐ Standard
☐ LVD ☐ CL
Plug In Unit: ☐ Fuse/Switch ☐ N/A
☐Circuit Breaker ☐Unknown

Dry Type Transformer
☐ 1 PH ☐ 3 PH ☐ N/A ☐Unknown
Minimum Impedance _________%

(continued)

1.1.4

Project: ______________
Proj. No: ______________
PM/PE: ______________
Date: ______________

Existing Condition Service & Distribution Checklist

Sub Panels:

☐ 1 PH ☐ 3 PH ☐ N/A ☐Unknown

Rating: __________ AIC sym ☐Unknown

Branch Breakers: ☐ Standard

☐ Switching Duty ☐Unknown

Comments:

____

____

____

1.1.5 Design Coordination Checklist (Electrical)

1.1.5

Page 1 of 3

Project Status
☐ SD
☐ DD
☐ CD

Project: ______________
Proj. No: ______________
PM/PE: ______________
Date: ______________

Design Coordination Checklist (Electrical)

Electrical Drawings - Plans	**Coord./% Complete**		
Check that electrical floor plans match architectural and mechanical plans.	Y	N	N/A
Check that the location of floor mounted equipment is consistent between disciplines.	Y	N	N/A
Check that the location of light fixtures matches architectural reflected ceiling plan.	Y	N	N/A
Check that elevator power, telephone and recall systems are shown and coordinated with architectural and fire protection	Y	N	N/A
Check that light fixtures do not conflict with the structure or the mechanical HVAC system.	Y	N	N/A
Check electrical connections to major equipment. Check that horsepower rating, phase, voltage, starter and drive types are consistent with other trade schedules.	Y	N	N/A
Check that locations of panelboards are consistent with architectural floor plans, mechanical floor plans, plumbing & fire protection floor plans.	Y	N	N/A
Check that the panelboards are indicated on the electrical riser diagram.	Y	N	N/A
Check that HVAC control power needs are addressed.			
Check that notes are referenced.	Y	N	N/A
Check that locations of electrical conduit runs, floor trenches, and openings are coordinated with structural plans.	Y	N	N/A
Check that electrical panels are not recessed in fire rated walls.	Y	N	N/A
Check that locations of exterior electrical equipment are coordinated with site paving, grading and landscaping.	Y	N	N/A
Check that structural supports are provided for rooftop electrical equipment.	Y	N	N/A
Food Service Drawings			
Check that the equipment layout matches other trade floor plans.	Y	N	N/A
Check that there are no conflicts with columns.	Y	N	N/A
Check that equipment is connected to utility systems.	Y	N	N/A

(continued)

1.1.5

Project Status
☐ SD
☐ DD
☐ CD

Project: ____________
Proj. No: ____________
PM/PE: ____________
Date: ____________

Design Coordination Checklist (Electrical)	**Coord./% Complete**		
Check that equipment as scheduled on the drawings matches the kitchen floor plans and specifications.	Y	N	N/A
Check that floor depressions and floor troughs are coordinated.	Y	N	N/A
Check that kitchen equipment is scheduled and coordinated with floor plans.	Y	N	N/A
Communication Drawings			
Check that equipment layout matches Architect and Consultant Plans.	Y	N	N/A
Check for conflicts between equipment/device spacing, clearances and access.	Y	N	N/A
Check for Architect's or Consultant's typical elevations and details show special device location and mounting heights.	Y	N	N/A
Check empty raceway systems for coordination with Consultant's equipment and wiring.	Y	N	N/A
Check for coordination between Specialty Contractor responsibility and Electrical Contractor responsibility.	Y	N	N/A
A/V Drawings			
Check that equipment layout matches Architect and Consultant Plans.	Y	N	N/A
Check for conflicts between equipment/device spacing, clearances and access.	Y	N	N/A
Check for Architect's or Consultant's typical elevations and details show special device location and mounting heights.	Y	N	N/A
Check empty raceway systems for coordination with Consultant's equipment and wiring.	Y	N	N/A
Check for coordination between Specialty Contractor responsibility and Electrical Contractor responsibility.	Y	N	N/A
Theatre Drawings			
Check that equipment layout matches Architect and Consultant Plans.	Y	N	N/A
Check for conflicts between equipment/device spacing, clearances and access.	Y	N	N/A

(continued)

1.1.5

Project Status	
☐ SD	Project: ____________
☐ DD	Proj. No: ____________
☐ CD	PM/PE: ____________
	Date: ____________

Design Coordination Checklist (Electrical)	**Coord./% Complete**
Check for Architect's or Consultant's typical elevations and details show special device location and mounting heights.	Y N N/A
Check empty raceway systems for coordination with Consultant's equipment and wiring.	Y N N/A
Check for coordination between Specialty Contractor responsibility and Electrical Contractor responsibility.	Y N N/A
Specifications	
Check that bid items explicitly state what is intended.	Y N N/A
Check specifications for phasing of construction.	Y N N/A
Check that architectural finish schedule agrees with specification index.	Y N N/A
Check that major equipment items are coordinated with contract drawings.	Y N N/A
Check that items specified "as indicated" and "where indicated" in the specifications are in fact indicated on the contract drawings.	Y N N/A
Check that the table of contents matches the sections contained in the body of the specifications.	Y N N/A

1.1.6 Fire Alarm System Checklist

1.1.6

Page 1 of 4

Project: ______________
Proj. No: ______________
PM/PE: ______________
Date: ______________

Fire Alarm System Checklist

Part One - Central Reporting Requirements

		Y	N	N/A
Emergency Forces Notification		Y	N	N/A
Auxiliary Alarm System: (Alarms transmitted directly to municipal communication center)		Y	N	N/A
Central Station: (Alarms transmitted to a station location with 24-hour supervision)		Y	N	N/A
Central Station System: (Alarms automatically transmitted to, recorded in, maintained and supervised from an approved central supervising station)		Y	N	N/A
Proprietary Protective System: (Alarms automatically transmitted to a central supervising station on the Agency property with trained personnel and 24-hour supervision)		Y	N	N/A
Remote Station System: (Alarms transmitted to a location remote from the building where circuits are supervised and appropriate action is taken)		Y	N	N/A

Part Two - Fire Alarm System

		Y	N	N/A
Is there a building presently equipped with a Fire Alarm System? If yes: indicate Make/Model ______________ Type: ______________ Date Installed: ______________		Y	N	N/A
Will this project extend/expand the existing system?		Y	N	N/A
Does the existing system conform to current Codes?	NFPA	Y	N	N/A
	BOCA	Y	N	N/A
	ADA	Y	N	N/A
	NEC	Y	N	N/A
Is the existing system a conventional or an addressable system?		Y	N	N/A
Is all existing equipment of the same make and manufacturer?		Y	N	N/A
Is the "Fire Alarm Control Panel" located at the Primary Building Entrance or Main Lobby?		Y	N	N/A
Is the "Fire Alarm Control Panel" and "Annunciator" currently located at a location approved by the State or local Fire Marshal?		Y	N	N/A

(continued)

1.1.6

Project: ______________
Proj. No: ______________
PM/PE: ______________
Date: ______________

Are system components readily available?	Y	N	N/A
Have you inspected the existing Fire Alarm System?	Y	N	N/A
Have you received Agency information on the operational status of the existing system?	Y	N	N/A
Is the building equipped with adequate peripheral devices (i.e., pull stations, back up power, heat and smoke detectors, horn/speaker and strobe lights)?	Y	N	N/A
Is the existing panel and annunciator capable of accommodating the system expansion due to the new renovations?	Y	N	N/A
Have you requested copies of the latest State Fire Marshal citations?	Y	N	N/A
Are there smoke detectors at the elevator lobbies for the elevator recall system where required by Code?	Y	N	N/A
Are there smoke detectors in locations required by the Elevator Code (ASME/ANSI A17.1)?	Y	N	N/A
Are there adequate quantities of horn/speaker and strobe lights in the corridors?	Y	N	N/A
Is the building equipped with a Fire-Fighter's phone system at each stairwell and elevator lobby?	Y	N	N/A
Have you verified that smoke detectors in residential rooms have been located away from cooking stoves and shower stalls?	Y	N	N/A
Have you specified "single-station" and not "system" detectors in the sleeping residential areas?	Y	N	N/A
Have air handling units been equipped with duct-smoke detectors, as required by NFPA Codes?	Y	N	N/A
Are air handling units annunciated at the building annunciator for easy identification of alarm location?	Y	N	N/A
Is the existing system connected to a Fire Department or other answering service?	Y	N	N/A
If a new building, is the system specified compatible with the existing campus system?	Y	N	N/A
Is the system specified as a "Proprietary" system?	Y	N	N/A

(continued)

1.1.6

Project: ______________
Proj. No: ______________
PM/PE: ______________
Date: ______________

Does the Specification cite three manufacturers of equal quality meeting DPW and Agency requirements?	Y	N	N/A
If building is a high-rise, does the fire alarm system conform to BOCA high-rise requirements?	Y	N	N/A
Are stairwells required to have a pressurized smoke ventilation system?	Y	N	N/A
Is the building sprinkler system connected to the Fire Alarm Control Panel and "Annunciator" system?	Y	N	N/A
Is the building equipped with a fire pump?	Y	N	N/A
Is the fire alarm system backed up by a battery and standby generator system?	Y	N	N/A
Is the Fire Pump Electrical Service connected ahead of the Main Service Entrance switch?	Y	N	N/A

Part Three - Elevator Related Questions

Does BOCA or NFPA Code require this building to be fully sprinklered?	Y	N	N/A
When the building is fully sprinklered; are there sprinkler heads in the Elevator Machine Room, and at the top and bottom of each elevator shaft?	Y	N	N/A
Is the power to the elevator automatically shut off by a heat detector and shunt trip breaker; prior to sprinkler discharge?	Y	N	N/A
Are elevator recall smoke detectors isolated from the building's Fire Alarm System?	Y	N	N/A
Do the elevator detectors report to the main Fire Alarm Panel?	Y	N	N/A
Is the proposed elevator room steel fire proofing provided by a material acceptable to the State Elevator Inspector?	Y	N	N/A
Is there a sump pit and duplex outlet in each elevator pit?	Y	N	N/A
Is the elevator pit equipped with a guarded lighting fixture, light switch and duplex outlet?	Y	N	N/A
Does the electrical wiring, equipment, pipes, ducts, etc. in hoistways and machine rooms conform to Section 102 of the Elevator Code (ASME/ANSI A17.1 Code)?	Y	N	N/A

(continued)

1.1.6

Page 4 of 4

Project: ______________

Proj. No: ______________

PM/PE: ______________

Date: ______________

Is there any water piping in the elevator shaft or machine room which does not serve the shaft or machine room?	Y	N	N/A
If there is a standby generator in the building, is any elevator connected to the standby power?	Y	N	N/A
Does the design comply with ADA Section 4.10 requirements for elevators?	Y	N	N/A

1.2.0 Electrical Symbols

Electrical symbols can vary widely, but the following closely adhere to industry standards. Industry standard symbols often are modified to meet client- and/or project-specific requirements.

1.2.0

LIGHTING	
SYMBOL	DESCRIPTION
○X	CEILING MOUNTED LIGHT FIXTURE; SUBLETTER INDICATES FIXTURE TYPE
X○	WALL MOUNTED LIGHT FIXTURE; SUBLETTER INDICATES FIXTURE TYPE
X	2'x4' CEILING MOUNTED FLUORESCENT LIGHT FIXTURE; SUBLETTER INDICATES FIXTURE TYPE
X	DUAL BALLAST 2'x4' CEILING MOUNTED FLUORESCENT LIGHT FIXTURE; SUBLETTER INDICATES FIXTURE TYPE
X	1'x4' CEILING MOUNTED FLUORESCENT LIGHT FIXTURE; SUBLETTER INDICATES FIXTURE TYPE
X	2'x2' CEILING MOUNTED FLUORESCENT LIGHT FIXTURE; SUBLETTER INDICATES FIXTURE TYPE
X	TYPICAL CEILING MOUNTED FLUORESCENT FIXTURE– NORMAL/EMERGENCY
	CONTINUOUS FLUORESCENT LIGHT FIXTURE
X	WALL WASHER LIGHT FIXTURE
	LIGHT ON EMERGENCY CIRCUIT
X	FLUORESCENT STRIP LIGHT FIXTURE; SUBLETTER INDICATES FIXTURE TYPE
	POWER LIGHT TRACK WITH NUMBER OF FIXTURES AS INDICATED ON PLANS; SUBLETTER INDICATES FIXTURE TYPE
	SINGLE OR DUAL HEAD, WALL MOUNTED, REMOTE EMERGENCY LIGHT
	DOUBLE FACED CEILING OR WALL–MOUNTED, EXIT SIGN WITH EMERGENCY POWER BACK UP AND DIRECTIONAL ARROWS AS INDICATED ON PLANS
	SINGLE FACED CEILING OR WALL–MOUNTED EXIT SIGN WITH EMERGENCY POWER BACK UP AND DIRECTIONAL ARROWS AS INDICATED ON PLANS
	CEILING OR WALL–MOUNTED, SELF–CONTAINED EMERGENCY LIGHT UNIT; FIXTURE SHALL MONITOR LIGHTING CIRCUIT IN AREA.
	EMERGENCY LIGHTING BATTERY UNIT

(continued)

1.2.0

SWITCHES	
S	SINGLE-POLE SWITCH
S_2	DOUBLE-POLE SWITCH
S_3	3-WAY SWITCH
S_4	4-WAY SWITCH
S_P	SINGLE-POLE SWITCH AND PILOT LIGHT
S_{be}	BOILER EMERGENCY SWITCH
S_{DM}	SINGLE-POLE DIMMER SWITCH
S_{DM3}	3-WAY DIMMER SWITCH
S_T	SINGLE-POLE SWITCH WITH THERMAL OVERLOAD PROTECTION
S_K	SINGLE-POLE KEYED SWITCH
S_{K3}	KEYED, 3-WAY SWITCH
S_{K4}	KEYED, 4-WAY SWITCH
S_{MC}	MOMENTARY CONTACT SWITCH
S_{PROJ}	MOTORIZED PROJECTION SCREEN RAISE/LOWER SWITCH
S_O	OCCUPANCY SENSOR SWITCH
[S_O]	CEILING MOUNTED OCCUPANCY SENSOR
[C]	CONTACTOR, COMPLETE WITH NEMA ENCLOSURE
[TC]	TIME CLOCK, AS INDICATED ON PLANS
[PC]	PHOTOCELL
[•]	PUSHBUTTON SWITCH
E,G	EMERGENCY SHUT-OFF SWITCH. SUBLETTER "E" INDICATES ELECTRICAL. SUBLETTER "G" INDICATES GAS
KE,KG	MASTER EMERGENCY SHUT-OFF/KEYED RESET SWITCH. SUBLETTER "KE" INDICATES ELECTRICAL. SUBLETTER "KG" INDICATES GAS

(continued)

1.2.0

POWER

a,b	DUPLEX RECEPTACLE; SUBLETTER "a" INDICATES RECEPTACLE TO BE MOUNTED 6" ABOVE COUNTER TOP OR 48" AFF. SUBLETTER "b" INDICATES MOUNTED IN ARCHITECTURAL MILLWORK. COORDINATE INSTALLATION WITH ARCHITECT.
a,b	DOUBLE DUPLEX RECEPTACLE; SUBLETTER "a" INDICATES RECEPTACLE TO BE MOUNTED 6" ABOVE COUNTER TOP OR 48" AFF. SUBLETTER "b" INDICATES MOUNTED IN ARCHITECTURAL MILLWORK. COORDINATE INSTALLATION WITH ARCHITECT.
	SINGLE RECEPTACLE
R,F,S	FLOOR MOUNTED DUPLEX RECEPTACLE: SUBLETTER "R" INDICATES RECESSED BACKBOX. SUBLETTER "F" INDICATES FLUSH BACKBOX. SUBLETTER "S" INDICATED SURFACE BACKBOX (MONUMENT)
	DUPLEX RECEPTACLE-ONE OUTLET SWITCHED
c	DUPLEX RECEPTACLE. SUBLETTER "C" INDICATES CEILING MOUNTED
TV	DUPLEX RECEPTACLE FOR TELEVISION. MOUNTING HEIGHT AS NOTED ON PLANS
E	ELECTRICAL FLOOR MONUMENT WITH LFMC WHIP CONNECTION
	SPECIAL-PURPOSE OUTLET. AMPERAGE AND VOLTAGE AS INDICATED ON PLANS. VERIFY NEMA CONFIGURATION WITH EQUIPMENT MANUFACTURER
a,b	DUPLEX RECEPTACLE, EMERGENCY POWER; SUBLETTER "a" INDICATES RECEPTACLE TO BE MOUNTED 6" ABOVE COUNTER TOP OR 48" AFF. SUBLETTER "b" INDICATES MOUNTED IN ARCHITECTURAL MILLWORK. COORDINATE INSTALLATION WITH ARCHITECT.
a,b	DOUBLE DUPLEX RECEPTACLE, EMERG. POWER; SUBLETTER "a" INDICATES RECEPTACLE TO BE MOUNTED 6" ABOVE COUNTER TOP OR 48" AFF. SUBLETTER "b" INDICATES MOUNTED IN ARCHITECTURAL MILLWORK. COORDINATE INSTALLATION WITH ARCHITECT.
	SURFACE RACEWAY WITH OUTLETS AS INDICATED ON PLANS, MOUNTED AT 18" AFF, UNLESS OTHERWISE NOTED
T/P	TELEPHONE/POWER POLE

	ELECTRICAL PANEL 480/277 VOLT
	ELECTRICAL PANEL 120/208 VOLT
	SPECIAL-PURPOSE ELECTRICAL PANEL OR EQUIPMENT
T	ELECTRICAL POWER TRANSFORMER
	MAGNETIC STARTER
F XX XX	FUSED DISCONNECT SWITCH WITH SIZE/RATING
	NON-FUSED DISCONNECT SWITCH
	COMBINATION MAGNETIC STARTER AND DISCONNECT SWITCH
	ELECTRIC MOTOR
VFD	VARIABLE FREQUENCY DRIVE
J	FLOOR OR CEILING MOUNTED JUNCTION BOX
J	WALL MOUNTED JUNCTION BOX
	ELECTRIFIED BUS DUCT WITH FUSIBLE, PLUG-IN, BRANCH CIRCUIT DEVICE
	HARD-WIRED EQUIPMENT CONNECTION
R	RELAY
EDO	ELECTRIC DOOR OPENER
	ELECTRIC DOOR OPENER ACTUATOR PUSH PLATE

(continued)

1.2.0

SPECIAL SYSTEMS

SYMBOL	DESCRIPTION
R,F,S	FLOOR MOUNTED TEL/DATA OUTLET: SUBLETTER "R" INDICATES RECESSED BACKBOX. SUBLETTER "F" INDICATES FLUSH BACKBOX. SUBLETTER "S" INDICATES SURFACE BACKBOX (MONUMENT)
C	COMMUNICATIONS FLOOR MONUMENT WITH LFMC WHIP CONNECTION
	COMBINATION DATA/TELEPHONE OUTLET WITH BACKBOX AND EMPTY CONDUIT STUBBED UP TO ABOVE FINISHED CEILING, INCLUDING DRAG LINE
W	TELEPHONE OUTLET WITH BACKBOX AND EMPTY CONDUIT, STUBBED UP TO ABOVE FINISHED CEILING, INCLUDING DRAG LINE. SUBLETTER "W" INDICATES WALL-MOUNTED;
HP	HANDICAP PAY TELEPHONE OUTLET WITH BACKBOX AND CONDUIT STUBBED UP TO ABOVE FINISHED CEILING, INCLUDING DRAG LINE
	DATA OUTLET WITH BACKBOX AND EMPTY CONDUIT STUBBED UP TO ABOVE FINISHED CEILING, INCLUDING DRAGLINE
TV	TELEVISION CABLE OUTLET; MOUNT AT "18" AFF UNLESS OTHERWISE NOTED.
S	CEILING-MOUNTED, SOUND SYSTEM SPEAKER
S	WALL-MOUNTED, SOUND SYSTEM SPEAKER
VC	SOUND SYSTEM VOLUME CONTROLLER
M F,W	SOUND SYSTEM MICROPHONE JACK; SUBLETTER "F" INDICATES FLOOR-MOUNTED. SUBLETTER "W" INDICATES WALL-MOUNTED
	PA/SOUND SYSTEM HANDSET
S	PA/SOUND SYSTEM CLOCK AND SPEAKER MOUNTED IN COMMON ENCLOSURE
	WALL CLOCK WITH HANGER TYPE OUTLET
P	PROGRAM BELL
E	EMERGENCY CALL-FOR-AID AUDIO INDICATING UNIT
E PC	EMERGENCY CALL-FOR-AID SWITCH
E PB	EMERGENCY CALL-FOR-AID PUSHBUTTON
	EMERGENCY CALL-FOR-AID VISUAL INDICATING UNIT
E	EMERGENCY CALL-FOR-AID VISUAL/AUDIO INDICATING UNIT
AMP	AMPLIFIER
I M	INTERCOM STATION; SUBLETTER "M" INDICATES MASTER

(continued)

1.2.0

HOSPITAL SYMBOLS	
SYMBOL	DESCRIPTION
$N_{1/2}$	NURSE CALL BEDSIDE STATION – SUBNUMBER INDICATES SINGLE OR DOUBLE BED
N_E	EMERGENCY NURSE CALL STATION
$N_{M/S}$	NURSE CALL MICROPHONE/SPEAKER UNIT
N_{SR}	NURSE CALL STAFF REGISTER
N_{ACU}	NURSE CALL AREA CONTROL UNIT
N_{FCS}	NURSE CALL FLOOR CONTROL STATION
N_B	NURSE CALL CODE BLUE
N_{SS}	NURSE CALL STAFF STATION
N_{DS}	NURSE CALL DUTY STATION
FM	FETAL MONITORING STATION
PM	PATIENT MONITORING STATION
D_N D_N	NURSE CALL CORRIDOR DOME LIGHT – CEILING OR WALL MOUNTED.
D_Z D_Z	NURSE CALL CORRIDOR ZONE LIGHT – CEILING OR WALL MOUNTED.
TA	TELEMETRY RECEIVER
CTM	CENTRAL TELEMETRY UNIT
$PU_{NC/PM}$	PRINTER UNIT, SUBLETTER "NC" INDICATES NURSE CALL; SUBLETTER "PM" INDICATES PATIENT MONITOR
MGAP	MEDICAL GAS ALARM PANEL
CMS	CENTRAL PATIENT MONITOR STATION

(continued)

1.2.0

FIRE	
M	FIRE ALARM MAGNETIC DOOR HOLD DEVICE
FS	SPRINKLER FIRE ALARM FLOW SWITCH
SS	SPRINKLER FIRE ALARM SUPERVISORY SWITCH
PS	SPRINKLER FIRE ALARM PRESSURE SWITCH
F	MASTER FIRE ALARM PULL BOX
S E	SMOKE DETECTOR FOR ELEVATOR RECALL CONTROLS
	EXTERIOR REMOTE FIRE ALARM FLASHING STROBE LIGHT
FACP	FIRE ALARM CONTROL PANEL
RAP	REMOTE ANNUNCIATOR PANEL
F	MANUAL FIRE ALARM PULL STATION
	FIRE ALARM VISUAL INDICATING UNIT
	FIRE ALARM AUDIO/VISUAL INDICATING UNIT
S	FIRE ALARM SPEAKER/VISUAL INDICATING UNIT (VOICE EVAC. SYSTEM)
F	FIRE ALARM CEILING-MOUNTED SPEAKER
M	FIRE ALARM MINI SPEAKER
H B	AUTOMATIC FIRE ALARM HEAT DETECTOR. SUBLETTER "B" INDICATES 200 DEGREES F. HEAT DETECTOR
FP	FIREFIGHTERS TELEPHONE OUTLET
AOR	AREA OF REFUGE TELEPHONE OUTLET
EM	EMERGENCY TELEPHONE OUTLET
S	AUTOMATIC FIRE ALARM SMOKE DETECTOR
S S	AUTOMATIC FIRE ALARM SMOKE DETECTOR WITH SOUNDER BASE
DS	DUCT SMOKE FIRE ALARM DETECTOR
DH	DUCT HEAT FIRE ALARM DETECTOR
TS	SMOKE DETECTOR TEST SWITCH

(continued)

1.2.0

SECURITY	
ES	DOOR STRIKE
	DOOR/WINDOW CONTACT
	VIDEO CAMERA, WITH MOUNTING HARDWARE
VM	VIDEO MONITOR
VR	VIDEO RECORDER
CR	CARD READER
M M	CEILING OR WALL–MOUNTED MOTION DETECTOR

WIRING	
———	BRANCH CIRCUIT POWER WIRING
— — —	BRANCH CIRCUIT SWITCHED WIRING
—— - - ——	BRANCH CIRCUIT AC OR DC CONTROL WIRING
—EM—	BRANCH CIRCUIT EMERGENCY AC OR DC WIRING. 3/4" CONDUIT, 2#10 AND 1#10 GROUND, UNLESS OTHERWISE NOTED
—CT—	CABLETRAY
	CONDUIT DOWN
	CONDUIT UP
	HOME RUN. 3/4" CONDUIT, 2#12 AND 1#12 GROUND, UNLESS OTHERWISE NOTED. NOTE: HOME RUN SHALL BE FROM FIRST ELECTRICAL DEVICE BACKBOX IN CIRCUIT TO ELECTRICAL PANEL

(continued)

1.2.0

ONE-LINE	
	POTHEAD
	STRESSCONE
	CURRENT TRANSFORMER
	POTENTIAL TRANSFORMER
	FUSE
	FUSE CUT OUT
	FUSE & SWITCH
	SWITCH
	CIRCUIT BREAKER
	DRAWOUT CIRCUIT BREAKER
	MEDIUM VOLTAGE DRAWOUT CIRCUIT BREAKER
	BUSPLUG CIRCUIT BREAKER
	BUSPLUG FUSE & SWITCH
	GROUND
	THERMAL OVERLOAD
C	RELAY/COIL
	N/O CONTACT
	N.C. CONTACT
	PROTECTIVE RELAY
A	AMMETER
AS	AMMETER SWITCH
V	VOLTMETER
VS	VOLTMETER SWITCH
WHM	WATTHOUR METER
WM	WATTMETER

(continued)

1.2.0

ONE-LINE	
WHD	WATTHOUR DEMAND METER
	TRANSFORMER
	SHIELDED TRANSFORMER
	AUTO TRANSFORMER
	LIGHTNING ARRESTER
G	GENERATOR
Δ	DELTA
Y	WYE
K #	KEY INTERLOCK
N E L	AUTOMATIC TRANSFER SWITCH (A.T.S.)
	MAIN LUG ONLY PANELBOARD
	MAIN CIRCUIT BREAKER PANELBOARD
XXXAF / XXXAT	CIRCUIT BREAKER WITH AMP FRAME OVER AMP TRIP
XXX / XXX	FUSED DISCONNECT SWITCH, WITH SWITCH SIZE OVER FUSE SIZE

(continued)

1.2.0

ABBREVIATIONS	
SYMBOL	DESCRIPTION
A	AMPERE
C	CONDUIT
P	POLE
W	WIRE
T	TELEPHONE SERVICE
FA	FIRE ALARM
NF	NON–FUSED
WP	WEATHERPROOF
C/B	CIRCUIT BREAKER
AFF	ABOVE FINISHED FLOOR
AFG	ABOVE FINISHED GRADE
CIR	CIRCUIT
TX	TRANSFORMER
MD	MOTORIZED DAMPER
PE	PRIMARY ELECTRIC SERVICE
SE	SECONDARY ELECTRIC SERVICE
RTU	ROOFTOP UNIT
TCP	TEMPERATURE CONTROL PANEL
SD	SMOKE DAMPER
IG	ISOLATED GROUND
RMC	RIGID METALLIC CONDUIT
EMT	ELECTRIC METALLIC TUBING

(continued)

1.2.0

ABBREVIATIONS	
SYMBOL	DESCRIPTION
FMC	FLEXIBLE METALLIC TUBING
TV	TELEVISION
PVC	POLYVINYL CHLORIDE CONDUIT
EF	EXHAUST FAN
REF	ROOF EXHAUST FAN
AHU	AIR HANDLING UNIT
CUH	CABINET UNIT HEATER
EWC	ELECTRIC WATER COOLER
EWH	ELECTRIC WATER HEATER
GFI	GROUND FAULT INTERRUPTER
MAU	MAKE-UP AIR UNIT
WG	WIRE GUARD
S&P	SPACE AND PROVISION
E	EXISTING TO REMAIN
RE	REMOVE EXISTING
RL	RELOCATE EXISTING
NL	NEW LOCATION OF EXISTING RELOCATED
NR	NEW TO REPLACE EXISTING
RR	REMOVE AND REPLACE ON NEW SURFACE

1.3.0 Mounting Heights for Electrical Devices

Mounting heights for electrical devices are influenced by and must be closely coordinated with the architectural design. However, there are industry standard practices followed by architects as well as code and legal requirements, such as Americans with Disabilities Act (ADA) guidelines. The following recommended mounting heights for electrical devices provide a good guideline in the absence of any specific information and are ADA compliant.

TABLE 1.3.0

MOUNTING HEIGHTS FOR ELECTRICAL DEVICES AND/OR APPURTENANCES (ADA COMPLYING)

	DEVICE	MOUNTING HEIGHTS
1.	Light switches, wall-mounted occupancy sensors	48" to centerline of box Exception: 44" maximum to top above counters which are 20"-25"D.
2.	Wall-mounted exit signs	90" to centerline of sign or centered in wall area between top of door and ceiling.
2A.	Ceiling-mounted exit signs and pendant-mounted fixtures	80" to bottom of fixture
3.	Receptacles	16" to bottom of box Exception: 44" maximum to top above counters which are 20"-25"D.
4.	Special outlets or receptacles	16" to bottom of box or as noted on drawings Exception: 44" maximum to top above counters which are 20"-25"D.
5.	Plugmold or Wiremold	As noted on drawings, limited same as for receptacles.
6.	Clock outlets	12" from ceiling to centerline or 7'-0" to centerline if ceiling is over 8'-0"
7.	Data/communication or telephone outlets	16" to bottom of box Exception: 44" maximum to top above counters which are 20"-25"D.

(continued)

TABLE 1.3.0

8.	Telephone outlets - wall type	54" to Dial Center (non-accessible) 48" to highest operable part (accessible)
9.	Pay type telephone outlets	48" maximum to coin slot
10.	Fire alarm manual pull stations	48" to centerline of box - not more than 5'-0" from exit
11.	Combination Fire alarm audio/visual units	80" to bottom of backbox or 6" below ceiling to top of backbox, whichever is lower so that entire lens is within the 80" - 96" area required by ADA & NFPA72, spacing shall be such that no point is more than 50' away without obstruction
12.	Wall-mounted remote indicator light	80" to centerline of device or 6" below ceiling, whichever is lower
13.	Area of Refuge Telephone	Same as telephone - accessible
14.	Call-for-Aid switch with pull chain to floor	48" to centerline of box minimum (toilets) 66" to centerline of box maximum (showers - located out of spray area)
15.	Card reader	48" to highest operable part (side or forward access)
16.	Intercom station	54" to highest operable part (side access) 48" highest operable part (forward access)
17.	Sound system volume control	54" to highest operable part (side access) 48" to highest operable part (forward access)
18.	Microphone outlets	16" to bottom of box
19.	Thermostats	54" to highest operable part (side access) 48" to highest operable part (forward access)
20.	Temperature/Humidity Sensors	60" to centerline of box

NOTES:

1. All dimensions are considered from finished floor and, unless noted otherwise, shall not vary.
2. All dimensions shall be coordinated with architectural details and may be adjusted to conform with architectural requirements as long as no code restriction is violated.
3. Outlets installed lower than 15" AFF (forward reach) and 9" AFF (side reach) are in violation of ADA.

(continued)

TABLE 1.3.0

SPECIAL NOTES:

1. Exit signs shall NOT be installed so that it blocks fire alarm visual devices.

2. Where floor proximity exit signs are required by NFPA 101, the bottom shall be not less than 6" nor higher than 8" above floor.

3. Wall-mounted light fixtures:

   a. Bottom of fixture at 80" AFF or greater.
   b. Bottom of fixture at less than 80" AFF protrusion into space shall be not more than 4".

4. For fire alarm, if you can't make your installation work with these requirements or your just not sure if it's right or not, **REFER TO NFPA 72 AND/OR ADA!!**

1.4.0 NEMA Configuration Chart for General-Purpose Nonlocking Plugs and Receptacles

Reproduced from NEMO WD 6-88, Wiring Devices—Dimensional Requirements (revision and redesignation of ANSI C73-73).

1.4.0

		15 AMPERE		20 AMPERE		30 AMPERE		50 AMPERE		60 AMPERE	
		RECEPTACLE	PLUG	RECEPTACLE	PLUG	RECEPTACLE	PLUG	RECEPTACLE	PLUG	RECEPTACLE	PLUG
2-POLE 2-WIRE	1 125 V	1-15R	1-15P		1-20P						
2-POLE 2-WIRE	2 250 V		2-15P	2-20R	2-20P	2-30R	2-30P				
2-POLE 2-WIRE	3 277 V	(RESERVED FOR FUTURE CONFIGURATIONS)									
2-POLE 2-WIRE	4 600 V	(RESERVED FOR FUTURE CONFIGURATIONS)									
2-POLE 3-WIRE GROUNDING	5 125 V	5-15R	5-15P	5-20R	5-20P	5-30R	5-30P	5-50R	5-50P		
2-POLE 3-WIRE GROUNDING	6 250 V	6-15R	6-15P	6-20R	6-20P	6-30R	6-30P	6-50R	6-50P		
2-POLE 3-WIRE GROUNDING	7 277 V AC	7-15R	7-15P	7-20R	7-20P	7-30R	7-30P	7-50R	7-50P		
2-POLE 3-WIRE GROUNDING	24 347 V AC	24-15R	24-15P	24-20R	24-20P	24-30R	24-30P	24-50R	24-50P		
2-POLE 3-WIRE GROUNDING	8 480 V AC	(RESERVED FOR FUTURE CONFIGURATIONS)									
2-POLE 3-WIRE GROUNDING	9 600 V AC	(RESERVED FOR FUTURE CONFIGURATIONS)									
3-POLE 3-WIRE	10 125/250 V			10-20R	10-20P	10-30R	10-30P	10-50R	10-50P		
3-POLE 3-WIRE	11 3ø 250 V	11-15R	11-15P	11-20R	11-20P	11-30R	11-30P	11-50R	11-50P		
3-POLE 3-WIRE	12 3ø 480 V	(RESERVED FOR FUTURE CONFIGURATIONS)									
3-POLE 3-WIRE	13 3ø 600 V	(RESERVED FOR FUTURE CONFIGURATIONS)									
3-POLE 4-WIRE GROUNDING	14 125/250 V	14-15R	14-15P	14-20R	14-20P	14-30R	14-30P	14-50R	14-50P	14-60R	14-60P
3-POLE 4-WIRE GROUNDING	15 3ø 250 V	15-15R	15-15P	15-20R	15-20P	15-30R	15-30P	15-50R	15-50P	15-60R	15-60P
3-POLE 4-WIRE GROUNDING	16 3ø 480 V	(RESERVED FOR FUTURE CONFIGURATIONS)									
3-POLE 4-WIRE GROUNDING	17 3ø 600 V	(RESERVED FOR FUTURE CONFIGURATIONS)									
4-POLE 4-WIRE	18 3ø 208 Y 120 V	18-15R	18-15P	18-20R	18-20P	18-30R	18-30P	18-50R	18-50P	18-60R	18-60P
4-POLE 4-WIRE	19 3ø 408 Y 277 V	(RESERVED FOR FUTURE CONFIGURATIONS)									
4-POLE 4-WIRE	20 3ø 600 Y 347 V	(RESERVED FOR FUTURE CONFIGURATIONS)									
4-POLE 5-WIRE GROUNDING	21 3ø 208 Y 120 V	(RESERVED FOR FUTURE CONFIGURATIONS)									
4-POLE 5-WIRE GROUNDING	22 3ø 408 Y 277 V	(RESERVED FOR FUTURE CONFIGURATIONS)									
4-POLE 5-WIRE GROUNDING	23 3ø 600 Y 347 V	(RESERVED FOR FUTURE CONFIGURATIONS)									

1.4.1 NEMA Configuration Chart for Specific-Purpose Locking Plugs and Receptacles *Reproduced from NEMO WD 6-88, Wiring Devices—Dimensional Requirements (revision and redesignation of ANSI C73-73).*

1.4.1

		15 AMPERE		20 AMPERE		30 AMPERE		50 AMPERE		60 AMPERE	
		RECEPTACLE	PLUG	RECEPTACLE	PLUG	RECEPTACLE	PLUG	RECEPTACLE	PLUG	RECEPTACLE	PLUG
2-POLE 2-WIRE	1 125 V	L1-15R	L1-15P								
	2 250 V			L2-20R	L2-20P						
	3 277 V				(RESERVED FOR FUTURE CONFIGURATIONS)						
	4 600 V				(RESERVED FOR FUTURE CONFIGURATIONS)						
2-POLE 3-WIRE GROUNDING	5 125 V	L5-15R	L5-15P	L5-20R	L5-20P	L5-30R	L5-30P	L5-50R	L5-50P	L5-60R	L5-60P
	6 250 V	L6-15R	L6-15P	L6-20R	L6-20P	L6-30R	L6-30P	L6-50R	L6-50P	L6-60R	L6-60P
	7 277 V AC	L7-15R	L7-15P	L7-20R	L7-20P	L7-30R	L7-30P	L7-50R	L7-50P	L7-60R	L7-60P
	24 347 V AC			L24-20R	L24-20P						
	8 480 V AC			L8-20R	L8-20P	L8-30R	L8-30P	L8-50R	L7-50P	L8-60R	L8-60P
	9 600 V AC			L9-20R	L9-20P	L9-30R	L9-30P	L9-50R	L9-50P	L9-60R	L9-60P
3-POLE 3-WIRE	10 125/250 V			L10-20R	L10-20P	L10-30R	L10-30P				
	11 3ø 250 V	L11-15R	L11-15P	L11-20R	L11-20P	L11-30R	L11-30P				
	12 3ø 480 V			L12-20R	L12-20P	L12-30R	L12-30P				
	13 3ø 600 V					L13-30R	L13-30P				
3-POLE 4-WIRE GROUNDING	14 125/250 V			L14-20R	L14-20P	L14-30R	L14-30P	L14-50R	L14-50P	L14-60R	L14-60P
	15 3ø 250 V			L15-20R	L15-20P	L15-30R	L15-30P	L15-50R	L15-50P	L15-60R	L15-60P
	16 3ø 480 V			L16-20R	L16-20P	L16-30R	L16-30P	L16-50R	L16-50P	L16-60R	L16-60P
	17 3ø 600 V					L17-30R	L17-30P	L17-50R	L17-50P	L17-60R	L17-60P
4-POLE 4-WIRE	18 3ø 208 Y 120 V			L18-20R	L18-20P	L18-30R	L18-30P				
	19 3ø 408 Y 277 V			L19-20R	L19-20P	L19-30R	L19-30P				
	20 3ø 600 Y 347 V			L20-20R	L20-20P	L20-30R	L20-30P				
4-POLE 5-WIRE GROUNDING	21 3ø 208 Y 120 V			L21-20R	L21-20P	L21-30R	L21-30P	L21-50R	L21-50P	L21-60R	L21-60P
	22 3ø 408 Y 277 V			L22-20R	L22-20P	L22-30R	L22-30P	L22-50R	L22-50P	L22-60R	L22-60P
	23 3ø 600 Y 347 V			L23-20R	L23-20P	L23-30R	L23-30P	L23-50R	L23-50P	L23-60R	L23-60P

1.5.0 IEEE Standard Protective Device Numbers

TABLE 1.5.0

device number	definition and function
1	**master element** is the initiating device, such as a control switch, voltage relay, float switch, etc., which serves either directly, or through such permissive devices as protective and time-delay relays to place an equipment in or out of operation.
2	**time-delay starting, or closing, relay** is a device which functions to give a desired amount of time delay before or after any point or operation in a switching sequence or protective relay system, except as specifically provided by device functions 62 and 79 described later.
3	**checking or interlocking relay** is a device which operates in response to the position of a number of other devices, or to a number of predetermined conditions in an equipment to allow an operating sequence to proceed, to stop, or to provide a check of the position of these devices or of these conditions for any purpose.
4	**master contactor** is a device, generally controlled by device No. 1 or equivalent, and the necessary permissive and protective devices, which serves to make and break the necessary control circuits to place an equipment into operation under the desired conditions and to take it out of operation under other or abnormal conditions.
5	**stopping device** functions to place and hold an equipment out of operation.
6	**starting circuit breaker** is a device whose principal function is to connect a machine to its source of starting voltage.
7	**anode circuit breaker** is one used in the anode circuits of a power rectifier for the primary purpose of interrupting the rectifier circuit if an arc back should occur.
8	**control power disconnecting device** is a disconnecting device—such as a knife switch, circuit breaker or pullout fuse block—used for the purpose of connecting and disconnecting, respectively, the source of control power to and from the control bus or equipment. **note:** Control power is considered to include auxiliary power which supplies such apparatus as small motors and heaters.
9	**reversing device** is used for the purpose of reversing a machine field or for performing any other reversing functions.
10	**unit sequence switch** is used to change the sequence in which units may be placed in and out of service in multiple-unit equipments.
11	Reserved for future application.
12	**over-speed device** is usually a direct-connected speed switch which functions on machine overspeed.
13	**synchronous-speed device,** such as a centrifugal-speed switch, a slip-frequency relay, a voltage relay, an undercurrent relay or any type of device, operates at approximately synchronous speed of a machine.
14	**under-speed device** functions when the speed of a machine falls below a predetermined value.
15	**speed or frequency, matching device** functions to match and hold the speed or the frequency of a machine or of a system equal to, or approximately equal to, that of another machine, source or system.
16	Reserved for future application.
17	**shunting, or discharge, switch** serves to open or to close a shunting circuit around any piece of apparatus (except a resistor), such as a machine field, a machine armature, a capacitor or a reactor. **note:** This excludes devices which perform such shunting operations as may be necessary in the process of starting a machine by devices 6 or 42, or their equivalent, and also excludes device 73 function which serves for the switching of resistors.
18	**accelerating or decelerating device** is used to close or to cause the closing of circuits which are used to increase or to decrease the speed of a machine.
19	**starting-to-running transition contactor** is a device which operates to initiate or cause the automatic transfer of a machine from the starting to the running power connection.
20	**electrically operated valve** is a solenoid- or motor-operated valve which is used in a vacuum, air, gas, oil, water, or similar, lines. **note:** The function of the valve may be indicated by the insertion of descriptive words such as "Brake" or "Pressure Reducing" in the function name, such as "Electrically Operated *Brake* Valve".
21	**distance relay** is a device which functions when the circuit admittance, impedance or reactance increases or decreases beyond predetermined limits.
22	**equalizer circuit breaker** is a breaker which serves to control or to make and break the equalizer or the current-balancing connections for a machine field, or for regulating equipment, in a multiple-unit installation.
23	**temperature control device** functions to raise or to lower the temperature of a machine or other apparatus, or of any medium, when its temperature falls below, or rises above, a predetermined value. **note:** An example is a thermostat which switches on a space heater in a switchgear assembly when the temperature falls to a desired value as distinguished from a device which is used to provide automatic temperature regulation between close limits and would be designated as 90T.

(continued)

TABLE 1.5.0

device number	definition and function
24	Reserved for future application.
25	**synchronizing, or synchronism-check, device** operates when two a-c circuits are within the desired limits of frequency, phase angle or voltage, to permit or to cause the paralleling of these two circuits.
26	**apparatus thermal device** functions when the temperature of the shunt field or the armortisseur winding of a machine, or that of a load limiting or load shifting resistor or of a liquid or other medium exceeds a predetermined value; or if the temperature of the protected apparatus, such as a power rectifier, or of any medium decreases below a predetermined value.
27	**undervoltage relay** is a device which functions on a given value of undervoltage.
28	Reserved for future application.
29	**isolating contactor** is used expressly for disconnecting one circuit from another for the purposes of emergency operation, maintenance, or test.
30	**annunciator relay** is a nonautomatically reset device which gives a number of separate visual indications upon the functioning of protective devices, and which may also be arranged to perform a lockout function.
31	**separate excitation device** connects a circuit such as the shunt field of a synchronous converter to a source of separate excitation during the starting sequence; or one which energizes the excitation and ignition circuits of a power rectifier.
32	**directional power relay** is one which functions on a desired value of power flow in a given direction, or upon reverse power resulting from arc back in the anode or cathode circuits of a power rectifier.
33	**position switch** makes or breaks contact when the main device or piece of apparatus, which has no device function number, reaches a given position.
34	**motor-operated sequence switch** is a multi-contact switch which fixes the operating sequence of the major devices during starting and stopping, or during other sequential switching operations.
35	**brush-operating, or slip-ring short-circuiting, device** is used for raising, lowering, or shifting the brushes of a machine, or for short-circuiting its slip rings, or for engaging or disengaging the contacts of a mechanical rectifier.
36	**polarity device** operates or permits the operation of another device on a predetermined polarity only.
37	**undercurrent or underpower relay** is a device which functions when the current or power flow decreases below a predetermined value.
38	**bearing protective device** is one which functions on excessive bearing temperature, or on other abnormal mechanical conditions, such as undue wear, which may eventually result in excessive bearing temperature.
39	Reserved for future application.
40	**field relay** is a device that functions on a given or abnormally low value or failure of machine field current, or on an excessive value of the reactive component of armature current in an a-c machine indicating abnormally low field excitation.
41	**field circuit breaker** is a device which functions to apply, or to remove, the field excitation of a machine.
42	**running circuit breaker** is a device whose principal function is to connect a machine to its source of running voltage after having been brought up to the desired speed on the starting connection.
43	**manual transfer or selector device** transfers the control circuits so as to modify the plan of operation of the switching equipment or of some of the devices.
44	**unit sequence starting relay** is a device which functions to start the next available unit in a multiple-unit equipment on the failure or on the non-availability of the normally preceding unit.
45	Reserved for future application.
46	**reverse-phase, or phase-balance, current relay** is a device which functions when the polyphase currents are of reverse-phase sequence, or when the polyphase currents are unbalanced or contain negative phase-sequence components above a given amount.
47	**phase-sequence voltage relay** is a device which functions upon a predetermined value of polyphase voltage in the desired phase sequence.
48	**incomplete sequence relay** is a device which returns the equipment to the normal, or off, position and locks it out if the normal starting, operating or stopping sequence is not properly completed within a predetermined time.

(continued)

TABLE 1.5.0

device number	definition and function
49	**machine, or transformer, thermal relay** is a device which functions when the temperature of an a-c machine armature, or of the armature or other load carrying winding or element of a d-c machine, or converter or power rectifier or power transformer (including a power rectifier transformer) exceeds a predetermined value.
50	**instantaneous overcurrent, or rate-of-rise relay** is a device which functions instantaneously on an excessive value of current, or on an excessive rate of current rise, thus indicating a fault in the apparatus or circuit being protected.
51	**a-c time overcurrent relay** is a device with either a definite or inverse time characteristic which functions when the current in an a-c circuit exceeds a predetermined value.
52	**a-c circuit breaker** is a device which is used to close and interrupt an a-c power circuit under normal conditions or to interrupt this circuit under fault or emergency conditions.
53	**exciter or d-c generator relay** is a device which forces the d-c machine field excitation to build up during starting or which functions when the machine voltage has built up to a given value.
54	**high-speed d-c circuit breaker** is a circuit breaker which starts to reduce the current in the main circuit in 0.01 second or less, after the occurrence of the d-c overcurrent or the excessive rate of current rise.
55	**power factor relay** is a device which operates when the power factor in an a-c circuit becomes above or below a predetermined value.
56	**field application relay** is a device which automatically controls the application of the field excitation to an a-c motor at some predetermined point in the slip cycle.
57	**short-circuiting or grounding device** is a power or stored energy operated device which functions to short-circuit or to ground a circuit in response to automatic or manual means.
58	**power rectifier misfire relay** is a device which functions if one or more of the power rectifier anodes fails to fire.
59	**overvoltage relay** is a device which functions on a given value of overvoltage.
60	**voltage balance relay** is a device which operates on a given difference in voltage between two circuits.
61	**current balance relay** is a device which operates on a given difference in current input or output of two circuits.
62	**time-delay stopping, or opening, relay** is a time-delay device which serves in conjunction with the device which initiates the shutdown, stopping, or opening operation in an automatic sequence.
63	**liquid or gas pressure, level, or flow relay** is a device which operates on given values of liquid or gas pressure, flow or level, or on a given rate of change of these values.
64	**ground protective relay** is a device which functions on failure of the insulation of a machine, transformer or of other apparatus to ground, or on flashover of a d-c machine to ground. **note:** This function is assigned only to a relay which detects the flow of current from the frame of a machine or enclosing case or structure of a piece of apparatus to ground, or detects a ground on a normally ungrounded winding or circuit. It is not applied to a device connected in the secondary circuit or secondary neutral of a current transformer, or current transformers, connected in the power circuit of a normally grounded system.
65	**governor** is the equipment which controls the gate or valve opening of a prime mover.
66	**notching, or jogging, device** functions to allow only a specified number of operations of a given device, or equipment, or a specified number of successive operations within a given time of each other. It also functions to energize a circuit periodically, or which is used to permit intermittent acceleration or jogging of a machine at low speeds for mechanical positioning.
67	**a-c directional overcurrent relay** is a device which functions on a desired value of a-c overcurrent flowing in a predetermined direction.
68	**blocking relay** is a device which initiates a pilot signal for blocking of tripping on external faults in a transmission line or in other apparatus under predetermined conditions, or co-operates with other devices to block tripping or to block reclosing on an out-of-step condition or on power swings.
69	**permissive control device** is generally a two-position, manually operated switch which in one position permits the closing of a circuit breaker, or the placing of an equipment into operation, and in the other position prevents the circuit breaker or the equipment from being operated.
70	**electrically operated rheostat** is a rheostat which is used to vary the resistance of a circuit in response to some means of electrical control.

(continued)

TABLE 1.5.0

device number	definition and function
71	Reserved for future application.
72	**d-c circuit breaker** is used to close and interrupt a d-c power circuit under normal conditions or to interrupt this circuit under fault or emergency conditions.
73	**load-resistor contactor** is used to shunt or insert a step of load limiting, shifting, or indicating resistance in a power circuit, or to switch a space heater in circuit, or to switch a light, or regenerative, load resistor of a power rectifier or other machine in and out of circuit.
74	**alarm relay** is a device other than an annunciator, as covered under device No. 30, which is used to operate, or to operate in connection with, a visual or audible alarm.
75	**position changing mechanism** is the mechanism which is used for moving a removable circuit breaker unit to and from the connected, disconnected, and test positions.
76	**d-c overcurrent relay** is a device which functions when the current in a d-c circuit exceeds a given value.
77	**pulse transmitter** is used to generate and transmit pulses over a telemetering or pilot-wire circuit to the remote indicating or receiving device.
78	**phase angle measuring, or out-of-step protective relay** is a device which functions at a predetermined phase angle between two voltages or between two currents or between voltage and current.
79	**a-c reclosing relay** is a device which controls the automatic reclosing and locking out of an a-c circuit interrupter.
80	Reserved for future application.
81	**frequency relay** is a device which functions on a predetermined value of frequency—either under or over or on normal system frequency—or rate of change of frequency.
82	**d-c reclosing relay** is a device which controls the automatic closing and reclosing of a d-c circuit interrupter, generally in response to load circuit conditions.
83	**automatic selective control, or transfer, relay** is a device which operates to select automatically between certain sources or conditions in an equipment, or performs a transfer operation automatically.
84	**operating mechanism** is the complete electrical mechanism or servo-mechanism, including the operating motor, solenoids, position switches, etc., for a tap changer, induction regulator or any piece of apparatus which has no device function number.
85	**carrier, or pilot-wire, receiver relay** is a device which is operated or restrained by a signal used in connection with carrier-current or d-c pilot-wire fault directional relaying.
86	**locking-out relay** is an electrically operated hand or electrically reset device which functions to shut down and hold an equipment out of service on the occurrence of abnormal conditions.
87	**differential protective relay** is a protective device which functions on a percentage or phase angle or other quantitative difference of two currents or of some other electrical quantities.
88	**auxiliary motor, or motor generator** is one used for operating auxiliary equipment such as pumps, blowers, exciters, rotating magnetic amplifiers, etc.
89	**line switch** is used as a disconnecting or isolating switch in an a-c or d-c power circuit, when this device is electrically operated or has electrical accessories, such as an auxiliary switch, magnetic lock, etc.
90	**regulating device** functions to regulate a quantity, or quantities, such as voltage, current, power, speed, frequency, temperature, and load, at a certain value or between certain limits for machines, tie lines or other apparatus.
91	**voltage directional relay** is a device which operates when the voltage across an open circuit breaker or contactor exceeds a given value in a given direction.
92	**voltage and power directional relay** is a device which permits or causes the connection of two circuits when the voltage difference between them exceeds a given value in a predetermined direction and causes these two circuits to be disconnected from each other when the power flowing between them exceeds a given value in the opposite direction.
93	**field changing contactor** functions to increase or decrease in one step the value of field excitation on a machine.
94	**tripping, or trip-free, relay** is a device which functions to trip a circuit breaker, contactor, or equipment, or to permit immediate tripping by other devices; or to prevent immediate reclosure of a circuit interrupter, in case it should open automatically even though its closing circuit is maintained closed.
95 to **99**	Used only for specific applications on individual installations where none of the assigned numbered functions from 1 to 94 is suitable.

note: A similar series of numbers, starting with 201 instead of 1, shall be used for those device functions in a machine, feeder or other equipment when these are controlled directly from the supervisory system. Typical examples of such device functions are 201, 205, and 294.

(continued)

TABLE 1.5.0

suffix letters

Suffix letters are used with device function numbers for various purposes. In order to prevent possible conflict, any suffix letter used singly, or any combination of letters, denotes only one word or meaning in an individual equipment. All other words should use the abbreviations as contained in American Standard Z32.13-1950, or latest revision thereof, or should use some other distinctive abbreviation, or be written out in full each time they are used. Furthermore, the meaning of each single suffix letter, or combination of letters, should be clearly designated in the legend on the drawings or publications applying to the equipment.

The following suffix letters generally form part of the device function designation and thus are written directly behind the device number, such as 23X, 90V, or 52BT.

These letters denote **separate auxiliary devices,** such as

X, Y, Z —auxiliary relay*
R —raising relay
L —lowering relay
O —opening relay
C —closing relay
CS —control switch
CL —"a" auxiliary-switch relay
OP —"b" auxiliary-switch relay
U —"up" position-switch relay
D —"down" position-switch relay
PB —push button

*note: In the control of a circuit breaker with so-called X-Y relay control scheme, the X relay is the device whose main contacts are used to energize the closing coil and the contacts of the Y relay provide the anti-pump feature for the circuit breaker.

These letters indicate the **condition or electrical** quantity to which the device responds, or the medium in which it is located, such as:

A —air, or amperes
C —current
E —electrolyte
F —frequency, or flow
L —level, or liquid
P —power, or pressure
PF —power factor
Q —oil
S —speed
T —temperature
V —voltage, volts, or vacuum
VAR—reactive power
W —water, or watts

These letters denote the **location of the main device in the circuit,** or the type of circuit in which the device is used or the type of circuit or apparatus with which it is associated, when this is necessary, such as:

A —alarm or auxiliary power
AC —alternating current
AN —anode
B —battery, or blower, or bus
BK —brake
BP —bypass
BT —bus tie
C —capacitor, or condenser, compensator, or carrier current
CA —cathode
DC —direct current
E —exciter
F —feeder, or field, or filament
G —generator, or ground**
H —heater, or housing
L —line
M —motor, or metering
N —network, or neutral**
P —pump
R —reactor, or rectifier
S —synchronizing
T —transformer, or test, or thyratron
TH —transformer (high-voltage side)
TL —transformer (low-voltage side)
TM —telemeter
U —unit

**Suffix "N" is generally used in preference to "G" for devices connected in the secondary neutral of current transformers, or in the secondary of a current transformer whose primary winding is located in the neutral of a machine or power transformer, except in the case of transmission line relaying, where the suffix "G" is more commonly used for those relays which operate on ground faults.

These letters denote **parts of the main device,** divided in the two following categories:

all parts, except auxiliary contacts and limit switches as covered later.

Many of these do not form part of the device number, and should be written directly below the device number, such as $\frac{20}{\text{LS}}$ or $\frac{43.}{\text{A}}$

BB —bucking bar (for high speed d-c circuit breaker)
BK —brake
C —coil, or condenser, or capacitor
CC —closing coil
HC —holding coil
IS —inductive shunt
L —lower operating coil
M —operating motor
MF —fly-ball motor
ML —load-limit motor
MS —speed adjusting, or synchronizing, motor
S —solenoid
TC —trip coil
U —upper operating coil
V —valve

(continued)

TABLE 1.5.0

All auxiliary contacts and limit switches for such devices and equipment as circuit breakers, contactors, valves and rheostats. These are designated as follows:

a —Auxiliary switch, open when the main device is in the de-energized or non-operated position.
b —Auxiliary switch, closed when the main device is in the de-energized or non-operated position.
aa—Auxiliary switch, open when the operating mechanism of the main device is in the de-energized or non-operated position.
bb—Auxiliary switch, closed when the operating mechanism of the main device is in the de-energized or non-operated position.

e, f, h, etc., ab, ac, ad, etc., or ba, bc, bd, etc., are special auxiliary switches other than a, b, aa, and bb. Lower-case (small) letters are to be used for the above auxiliary switches.

note: If several similar auxiliary switches are present on the same device, they should be designated numerically 1, 2, 3, etc. when necessary.

LC—Latch-checking switch, closed when the circuit breaker-mechanism linkage is relatched after an opening operation of the circuit breaker.
LS—limit switch

These letters cover **all other distinguishing features or characteristics or conditions,** not specifically described in 2-9.4.1 to 2-9.4.4, which serve to describe the use of the device or its contacts in the equipment such as:

A —accelerating, or automatic
B —blocking, or backup
C —close, or cold
D —decelerating, detonate, or down
E —emergency
F —failure, or forward
H —hot, or high
HR —hand reset
HS —high speed
IT —inverse time
L —left, or local, or low, or lower, or leading
M —manual
OFF —off
ON —on
O —open
P —polarizing
R —right, or raise, or reclosing, or receiving, or remote, or reverse
S —sending, or swing
T —test, or trip, or trailing
TDC—time-delay closing
TDO—time-delay opening
U —up

suffix numbers

If two or more devices with the same function number and suffix letter (if used) are present in the same equipment, they may be distinguished by numbered suffixes as for example, 52X-1, 52X-2 and 52X-3, when necessary.

devices performing more than one function

If one device performs two relatively important functions in an equipment so that it is desirable to identify both of these functions, this may be done by using a double function number and name such as:

27-59 undervoltage and overvoltage relay.

1.6.0 Comparison of Specific Applications of NEMA Standard Enclosures for Indoor Nonhazardous Locations

TABLE 1.6.0

Provides a Degree of Protection Against the Following Environmental Conditions	Type of Enclosures									
	1[1]	2[1]	4	4X	5	6	6P	12	12K	13
Incidental contact with the enclosed equipment	X	X	X	X	X	X	X	X	X	X
Falling dirt	X	X	X	X	X	X	X	X	X	X
Falling liquids and light splashing	...	X	X	X	X	X	X	X	X	X
Circulating dust, lint, fibers, and flyings[2]	...	...	X	X	...	X	X	X	X	X
Settling airborne dust, lint, fibers, and flyings[2]	...	...	X	X	X	X	X	X	X	X
Hosedown and splashing water	...	...	X	X	...	X	X	...	...	...
Oil and coolant seepage	...	...	...	...	...	...	...	X	X	X
Oil or coolant spraying and splashing	...	...	...	...	...	...	...	...	...	X
Corrosive agents	...	...	...	X	...	...	X	...	...	...
Occasional temporary submersion	...	...	...	...	...	X	X	...	...	...
Occasional prolonged submersion	...	...	...	...	...	...	X	...	...	...

1.6.1 Comparison of Specific Applications of NEMA Standard Enclosures for Outdoor Nonhazardous Locations

TABLE 1.6.1

Provides a Degree of Protection Against the Following Environmental Conditions	Type of Enclosures						
	3	3R[3]	3S	4	4X	6	6P
Incidental contact with the enclosed equipment	X	X	X	X	X	X	X
Rain, snow, and sleet[4]	X	X	X	X	X	X	X
Sleet[5]	...	...	X	...	...	...	...
Windblown dust	X	...	X	X	X	X	X
Hosedown	...	...	...	X	X	X	X
Corrosive agents	...	...	...	...	X	...	X
Occasional temporary submersion	...	...	...	...	...	X	X
Occasional prolonged submersion	...	...	...	...	...	...	X

1.6.2 Comparison of Specific Applications of NEMA Standard Enclosures for Indoor Hazardous Locations

TABLE 1.6.2

(If the installation is outdoors and/or additional protection is required by Tables 1.6.0 and 1.6.1, a combination-type enclosure is required. See paragraph 3.2.)

Provides a Degree of Protection Against Atmospheres Typically Containing (For Complete Listing, See NFPA 497M-1986, *Classification of Gases, Vapors and Dusts for Electrical Equipment in Hazardous (Classified) Locations)*	Class	Type of Enclosure 7 and 8, Class I Groups[6]				Type of Enclosure 9, Class II Groups[6]			
		A	B	C	D	E	F	G	10
Acetylene	I	X	...	...	...	...	...	...	...
Hydrogen, manufactured gas	I	...	X	...	...	...	...	...	...
Diethel ether, ethylene, cyclopropane	I	...	...	X	...	...	...	...	...
Gasoline, hexane, butane, naphtha, propane, acetone, toluene, isoprene	I	...	...	...	X	...	...	...	...
Metal dust	II	...	...	...	...	X	...	...	...
Carbon black, coal dust, coke dust	II	...	...	...	...	...	X	...	...
Flour, starch, grain dust	II	...	...	...	...	...	...	X	...
Fibers, flyings[7]	III	...	...	...	...	...	...	X	...
Methane with or without coal dust	MSHA	...	...	...	...	...	...	...	X

1.6.3 Knockout Dimensions for NEMA Standard Enclosures

TABLE 1.6.3

Conduit Trade Size, Inches	Knockout Diameter, Inches		
	Minimum	Nominal	Maximum
1/2	0.859	0.875	0.906
3/4	1.094	1.109	1.141
1	1.359	1.375	1.406
1 1/4	1.719	1.734	1.766
1 1/2	1.958	1.984	2.016
2	2.433	2.469	2.500
2 1/2	2.938	2.969	3.000
3	3.563	3.594	3.625
3 1/2	4.063	4.125	4.156
4	4.563	4.641	4.672
5	5.625	5.719	5.750
6	6.700	6.813	6.844

① These enclosures may be ventilated. However, Type 1 may not provide protection against small particles of falling dirt when ventilation is provided in the enclosure top. Consult the manufacturer.
② These fibers and flying are nonhazardous materials and are not considered the Class III type ignitable fibers or combustible flyings. For Class III type ignitable fibers or combustible flyings see the National Electrical Code, Article 500.
③ External operating mechanisms are not required to be operable when the enclosure is ice covered.
④ External operating mechanisms are operable when the enclosure is ice covered.
⑤ These enclosures may be ventilated.
⑥ For Class III type ignitable fibers or combustible flyings see the National Electrical Code, Article 500.
⑦ Due to the characteristics of the gas, vapor, or dust, a product suitable for one Class or Group may not be suitable for another Class or Group unless so marked on the product.

1.7.0 Formulas and Terms

1.7.0

Formulas for Determining Amperes, hp, kW, and kVA

To Find	Direct Current	Alternating Current		
		Single-Phase	Two-Phase — 4-Wire①	Three-Phase
Amperes (I) When Horsepower is Known	$\frac{hp \times 746}{E \times \%\ eff}$	$\frac{hp \times 746}{E \times \%\ eff \times pf}$	$\frac{hp \times 746}{2 \times E \times \%\ eff \times pf}$	$\frac{hp \times 746}{\sqrt{3} \times E \times \%\ eff \times pf}$
Amperes (I) When Kilowatts is Known	$\frac{kW \times 1000}{E}$	$\frac{kW \times 1000}{E \times pf}$	$\frac{kW \times 1000}{2 \times E \times pf}$	$\frac{kW \times 1000}{\sqrt{3} \times E \times \%\ pf}$
Amperes (I) When kVA is Known		$\frac{kVA \times 1000}{E}$	$\frac{kVA \times 1000}{2 \times E}$	$\frac{kVA \times 1000}{\sqrt{3} \times E}$
Kilowatts	$\frac{I \times E}{1000}$	$\frac{I \times E \times pf}{1000}$	$\frac{I \times E \times 2 \times pf}{1000}$	$\frac{I \times E \times \sqrt{3} \times pf}{1000}$
kVA		$\frac{I \times E}{1000}$	$\frac{I \times E \times 2}{1000}$	$\frac{I \times E \times \sqrt{3}}{1000}$
Horsepower (Output)	$\frac{I \times E \times \%\ eff}{746}$	$\frac{I \times E \times \%\ eff \times pf}{746}$	$\frac{I \times E \times 2 \times \%\ eff \times pf}{746}$	$\frac{I \times E \times \sqrt{3} \times \%\ eff \times pf}{746}$

Common Electrical Terms

Ampere (I) = unit of current or rate of flow of electricity

Volt (E) = unit of electromotive force

Ohm (R) = unit of resistance

Ohms law: $I = \frac{E}{R}$ (DC or 100% pf)

Megohm = 1,000,000 ohms

Volt Amperes (VA) = unit of apparent power
= E × I (single-phase)
= $E \times I \times \sqrt{3}$

Kilovolt Amperes (kVA) = 1000 volt-amperes

Watt (W) = unit of true power
= VA × pf
= .00134 hp

Kilowatt (kW) = 1000 watts

Power Factor (pf) = ratio of true to apparent power
$= \frac{W}{VA} \quad \frac{kW}{kVA}$

Watt-hour (Wh) = unit of electrical work
= one watt for one hour
= 3,413 Btu
= 2,655 ft. lbs.

Kilowatt-hour (kWh) = 1000 watt-hours

Horsepower (hp) = measure of time rate of doing work
= equivalent of raising 33,000 lbs. one ft. in one minute
= 746 watts

Demand Factor = ratio of maximum demand to the total connected load

Diversity Factor = ratio of the sum of individual maximum demands of the various subdivisions of a system to the maximum demand of the whole system

Load Factor = ratio of the average load over a designated period of time to the peak load occurring in that period

How to Compute Power Factor

Determining watts: $pf = \frac{watts}{volts \times amperes}$

1. From watt-hour meter.
   Watts = rpm of disc × 60 × Kh

   Where Kh is meter constant printed on face or nameplate of meter.

   If metering transformers are used, above must be multiplied by the transformer ratios.

2. Directly from wattmeter reading.
   Where:

   Volts = line-to-line voltage as measured by voltmeter.

   Amps = current measured in line wire (not neutral) by ammeter.

Temperature Conversion

(F° to C°) (C° to F°)			C°=5/9 (F°)-32° F°=9/5(C°)+32°					
C°	-15	-10	-5	0	5	10	15	20
F°	5	14	23	32	41	50	59	68
C°	25	30	35	40	45	50	55	60
F°	77	86	95	104	113	122	131	140
C°	65	70	75	80	85	90	95	100
F°	149	158	167	176	185	194	203	212

1 inch = 2.54 centimeters
1 kilogram = 2.20 lbs.
1 square inch = 1,273,200 circular mills
1 circular mill = .785 square mil
1 Btu = 778 ft. lbs.
= 252 calories
1 year = 8,760 hours

① For 2-phase, 3-wire circuits the current in the common conductor is $\sqrt{2}$ times that in either of the two other conductors.

② Units of measurement and definitions for E (volts), I (amperes), and other abbreviations are given below under Common Electrical Terms.

1.8.0 Introduction: Typical Equipment Sizes, Weights, and Ratings

Tables 1.8.1 through 1.8.7 provide typical equipment sizes, weights, and ratings to assist in the preliminary design and layout of an electrical distribution system. The reader is cautioned that these data are only representative of industry manufacturers and should consult specific vendors for detailed information. This information could prove useful in determining initial space requirements and weight impacts for structural purposes.

1.8.1 Typical Equipment Sizes: 600-V Class

TABLE 1.8.1

Equipment	KVA Rating	Dimensions (inches)			Weight Lbs. (CU)	Weight Lbs. (AL)
		H	W	D		
Switchboards (per Section)	N/A	90	26 - 45	24 - 60	Varies	Varies
Motor Control Centers (per Section)	N/A	90	20	16 - 22	Varies	Varies
Power Panel	N/A	To 80	30 - 48	6 - 12	Varies	Varies
Lighting/Small Appliance Panels	N/A	30 - 50	22	6	Varies	Varies
Transformers 3-phase, Dry Type, General Purpose	30 45 75 112.5 150 300 500	30 30 40 40 46 56 75	20 20 26 26 26 32 45	15 15 20 20 21 24 36	300 370 550 675 850 1750 3100	230 310 480 600 760 1300 2400
Transformers 3-phase, Dry Type, K-Rated	30 45 75 112.5 150 300 500	31 40 40 56 56 75 90	21 26 26 31 31 45 69	15 20 20 24 24 36 42	370 575 675 850 1200 3100 see mfg.	310 480 600 760 1100 2400 4500

1.8.2 Transformer Weight (lb) by kVA

TABLE 1.8.2

Oil Filled 3 Phase 5/15 KV To 480/277			
KVA	Lbs.	KVA	Lbs.
150	1800	1000	6200
300	2900	1500	8400
500	4700	2000	9700
750	5300	3000	15000
Dry 240/480 To 120/240 Volt			
1 Phase		3 Phase	
KVA	Lbs.	KVA	Lbs.
1	23	3	90
2	36	6	135
3	59	9	170
5	73	15	220
7.5	131	30	310
10	149	45	400
15	205	75	600
25	255	112.5	950
37.5	295	150	1140
50	340	225	1575
75	550	300	1870
100	670	500	2850
167	900	750	4300

1.8.3 Generator Weight (lb) by kW

TABLE 1.8.3

3 Phase 4 Wire 277/480 Volt			
Gas		Diesel	
KW	Lbs.	KW	Lbs.
7.5	600	30	1800
10	630	50	2230
15	960	75	2250
30	1500	100	3840
65	2350	125	4030
85	2570	150	5500
115	4310	175	5650
170	6530	200	5930
		250	6320
		300	7840
		350	8220
		400	10750
		500	11900

1.8.4 Weight (lb/lf) of Four-Pole Aluminum and Copper Bus Duct by Ampere Rating

TABLE 1.8.4

Amperes	Aluminum Feeder	Copper Feeder	Aluminum Plug-In	Copper Plug-In
225			7	7
400			8	13
600	10	10	11	14
800	10	19	13	18
1000	11	19	16	22
1350	14	24	20	30
1600	17	26	25	39
2000	19	30	29	46
2500	27	43	36	56
3000	30	48	42	73
4000	39	67		
5000		78		

1.8.5 Conduit Weight Comparisons (lb per 100 ft) Empty

TABLE 1.8.5

Type	1/2"	3/4"	1"	1-1/4"	1-1/2"	2"	2-1/2"	3"	3-1/2"	4"	5"	6"
Rigid Aluminum	28	37	55	72	89	119	188	246	296	350	479	630
Rigid Steel	79	105	153	201	249	332	527	683	831	972	1314	1745
Intermediate Steel (IMC)	60	82	116	150	182	242	401	493	573	638		
Electrical Metallic Tubing (EMT)	29	45	65	96	111	141	215	260	365	390		
Polyvinyl Chloride, Schedule 40	16	22	32	43	52	69	109	142	170	202	271	350
Polyvinyl Chloride Encased Burial						38		67	88	105	149	202
Fibre Duct Encased Burial						127		164	180	206	400	511
Fibre Duct Direct Burial						150		251	300	354		
Transite Encased Burial						160		240	290	330	450	550
Transite Direct Burial						220		310		400	540	640

1.8.6 Conduit Weight Comparisons (lb per 100 ft) with Maximum Cable Fill

TABLE 1.8.6

Type	1/2"	3/4"	1"	1-1/4"	1-1/2"	2"	2-1/2"	3"	3-1/2"	4"	5"	6"
Rigid Galvanized Steel (RGS)	104	140	235	358	455	721	1022	1451	1749	2148	3083	4343
Intermediate Steel (IMC)	84	113	186	293	379	611	883	1263	1501	1830		
Electrical Metallic Tubing (EMT)	54	116	183	296	368	445	641	930	1215	1540		

1.9.0 Seismic Requirements

The design of seismic restraint systems for electrical distribution equipment and raceways is usually done by a structural engineer through performance specifications by the electrical design professional. It is therefore necessary for the electrical designer generally to be familiar with the seismic code requirements and the seismic zone that are applicable to a project. The following will serve as an introduction.

Seismic Requirements

Uniform Building Code (UBC)
The 1994 Uniform Building Code (UBC) includes Volume 2 for earthquake design requirements. Sections 1624-1633 of this reference specifically require that structures and portions of structures shall be designed to withstand the seismic ground motion specified in the code. The design engineer must evaluate the effect of lateral forces not only on the building structure but also on the equipment in determining whether the design will withstand those forces. In the code electrical equipment such as control panels, motors, switchgear, transformers, and associated conduit are specifically identified.

The criteria for selecting the seismic requirements are defined in Section 1627 of the code. Figure 16-2 of the code includes a seismic zone map of the United States. Figure 16-3 of the code includes the normalized response spectra shapes for different soil conditions. The damping value is 5% of the critical damping.

The seismic requirements in the UBC can be completely defined as the Zero Period Acceleration (ZPA) and Spectrum Accelerations are computed. In a test program, these values are computed conservatively to envelop the requirements of all seismic zones. The lateral force on elements of structures and nonstructural components are defined in Section 1630. The dynamic lateral forces are defined in Section 1629. These loads are converted to seismic accelerations according to the normalized response spectra shown in Figure 16-3 of the UBC.

The total design lateral force required is:

Force Fp = Z Ip Cp Wp

Dividing both sides by Wp, the acceleration requirement in g's is equal to:

Acceleration = Fp/Wp = Z Ip Cp

Where:

Z: is the seismic zone factor and is taken equal to 0.4. This is the maximum value provided in Table 16-I of the code.

Ip: is the importance factor and is taken equal to 1.5. This is the maximum value provided in Table 16-K of the code.

Cp: is the horizontal force factor and is taken equal to 0.75 for rigid equipment as defined in Table 16-O. For flexible equipment, this value is equal to twice the value for the rigid equipment: 2 x 0.75 = 1.5. This is the maximum value provided in the code.

Wp: is the weight of the equipment.

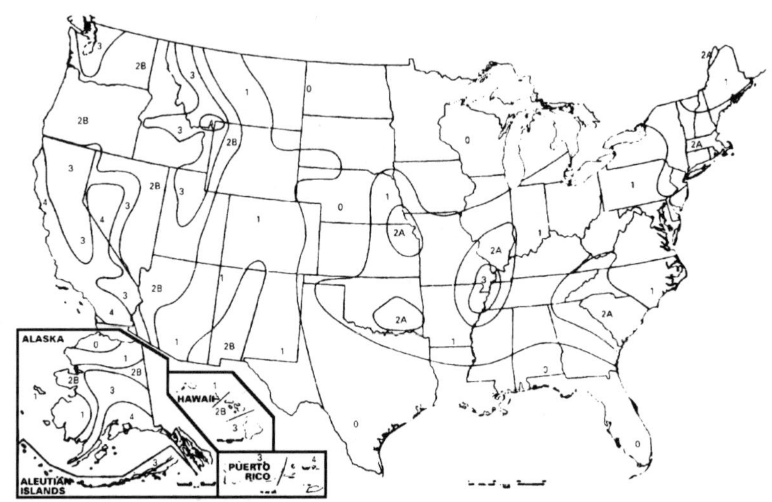

Figure 16-2. Seismic Zone Map of the United States

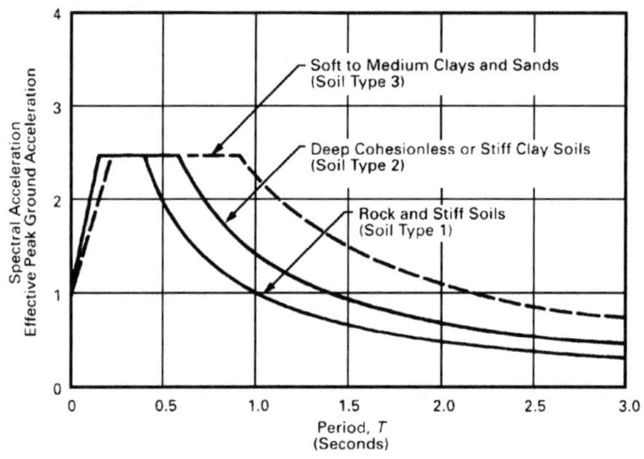

Figure 16-3. Normalized Response Spectra Shapes

Therefore, the maximum acceleration for rigid equipment is:

Acceleration = Fp/Wp
= Z Ip Cp
= 0.4 x 1.5 x 0.75
= 0.45g

The maximum acceleration for flexible equipment is:

Acceleration = Fp/Wp
= Z Ip Cp
= 0.4 x 1.5 x 1.5
= 0.9g

Flexible equipment is defined in the UBC as equipment with a period of vibration equal to or greater than 0.06 seconds. This period of vibration corresponds to a dominant frequency of vibration equal to 16.7 Hz.

Equipment must be designed and tested to the UBC requirements to determine that it will be functional following a seismic event. In addition, a structural or civil engineer must perform calculations based on data received from the equipment manufacturer specifying the size, weight, center of gravity, and mounting provisions of the equipment to determine its method of attachment so it will remain attached to its foundation during a seismic event. Finally, the contractor must properly install the equipment in accordance with the anchorage design.

Section

2

Requirements for Electrical Installations

2.1.1 Description of Fuse Class Designations
2.1.2 Maximum Peak Let-Through Current (I_p-Amperes) and Clearing I^2t (Amperes-Squared-Seconds)
2.2.1 Working Spaces [NEC Table 110.26(A)]
2.2.2 Examples of Conditions 1, 2, and 3 (Working Spaces)
2.2.3 Example of Exception 1 (Working Spaces)
2.2.4 Example of Exception 3 (Working Spaces)
2.2.5 Required 30-In-Wide Front Working Space (Working Spaces)
2.2.6 Required Full 90-Degree Opening of Equipment Doors (Working Spaces)
2.3.1 *NEC* Section 110.26(C), Basic Rule, First Paragraph (Access to Working Space)
2.3.2 *NEC* Section 110.26(C), Basic Rule, Second Paragraph (Access to Working Space)
2.3.3 Example of an Unacceptable Arrangement of a Large Switchboard (Access to Working Space)
2.3.4 Example of Exception No. 1 (Access to Working Space)
2.3.5 Example of Exception No. 2 (Access to Working Space)
2.4.1 Working Space and Dedicated Electrical Space
2.4.2 Working Space in Front of a Panelboard as Required by *NEC* Section 110.26
2.4.3 Dedicated Electrical Space Over and Under a Panelboard
2.5.1 Minimum Depth of Clear Working Space at Electrical Equipment
2.5.2 Elevation of Unguarded Live Parts Above Working Space

2.1.1 Description of Fuse Class Designations

TABLE 2.1.1

Class Designation	Subclasses	Voltage Rating, V_{ac}	Ampere Ratings	Interrupting Rating, rms Symmetrical Amperes	Current Limitation
CC		600	0–30	200,000	Current limiting
G		480	0–60	100,000	Current limiting
H		250 or 600	0–600	10,000	None
J		600	0–600	200,000	Current limiting
K	K1, K5, and K9	250 or 600	0–600	200,000	Not marked current limiting
L		600	601–6000	200,000	Current limiting
Plug Fuse	Edison Base	125	0–30	10,000	Not current limiting
	Type S	125	0–30	10,000	Not current limiting
R	RK1	250 or 600	0–600	200,000	High degree of current limitation
	RK5	250 or 600	0–600	200,000	Moderate degree of current limitation
T		300	0–1200	200,000	Current limiting
		600	0–1200	200,000	Current limiting

2.1.2 Maximum Peak Let-Through Current (I_p-Amperes) and Clearing I^2t (Amperes-Squared-Seconds)

TABLE 2.1.2

Fuse Class	Current Rating Range, A	Between Threshold and 50 kA		100 kA		200 kA	
		$I_p \times 10^3$	$I^2t \times 10^3$	$I_p \times 10^3$	$I^2t \times 10^3$	$I_p \times 10^3$	$I^2t \times 10^3$
CC	0–15	3	2	3	2	4	3
	16–20	3	2	4	3	5	3
	21–30	6	7	7.5	7	12	7
G	0–15	—	—	4	3.8	n/a	n/a
	16–20	—	—	5	5	n/a	n/a
	21–30	—	—	7	7	n/a	n/a
	31–60	—	—	10.5	25	n/a	n/a
J and 600 V Class T	0–30	6	7	7.5	7	12	7
	31–60	8	30	10	30	16	30
	61–100	12	60	14	80	20	80
	101–200	16	200	20	300	30	300
	201–400	25	1000	30	1100	45	1100
	401–600	35	2500	45	2500	70	2500
	601–800	50	4000	55	4000	75	4000
Class T only	801–1200	55	7500	70	7500	88	7500
300 V Class T	0–30	5	3.5	7	3.5	9	3.5
	31–60	7	15	9	15	12	15
	61–100	9	40	12	40	15	40
	101–200	13	150	16	150	150	20
	201–400	22	550	28	550	35	550
	401–600	29	1000	37	1000	46	1000
	601–800	37	1500	50	1500	65	1500
	801–1200	50	3500	65	3500	80	4000
L	601–800	80	10	80	10	80	10
	801–1200	80	12	80	12	120	15
	1201–1600	100	22	100	22	150	30
	1601–2000	110	35	120	35	165	40
	2001–2500	—	—	165	75	180	75
	2501–3000	—	—	175	100	200	100
	3001–4000	—	—	220	150	250	150
	4001–5000	—	—	—	350	300	350
	5001–6000	—	—	—	350	350	500
RK1	0–30	6	10	10	10	12	11
	31–60	10	40	12	40	16	50
	61–100	14	100	16	100	20	100
	101–200	18	400	22	400	30	400
	201–400	33	1200	35	1200	50	1600
	401–600	43	3000	50	3000	70	4000
RK5	0–30	11	50	11	50	14	50
	31–60	20	200	21	200	26	200
	61–100	22	500	25	500	32	500
	101–200	32	1600	40	1600	50	2000
	201–400	50	5000	60	5000	75	6000
	401–600	65	10,000	80	10,000	100	12,000

2.2.1 Working Spaces

TABLE 2.2.1

Nominal Voltage to Ground	Minimum Clear Distance		
	Condition 1	Condition 2	Condition 3
0–150	900 mm (3 ft)	900 mm (3 ft)	900 mm (3 ft)
151–600	900 mm (3 ft)	1 m ($3^1/_2$ ft)	1.2 m (4 ft)

Note: Where the conditions are as follows:
Condition 1 — Exposed live parts on one side and no live or grounded parts on the other side of the working space, or exposed live parts on both sides effectively guarded by suitable wood or other insulating materials. Insulated wire or insulated busbars operating at not over 300 volts to ground shall not be considered live parts.
Condition 2 — Exposed live parts on one side and grounded parts on the other side. Concrete, brick, or tile walls shall be considered as grounded.
Condition 3 — Exposed live parts on both sides of the work space (not guarded as provided in Condition 1) with the operator between.

Exception No. 1. Working space shall not be required in back or sides of assemblies, such as dead-front switchboards or motor control centers, where there are no renewable or adjustable parts, such as fuses or switches, on the back or sides and where all connections are accessible from locations other than the back or sides. Where rear access is required to work on deenergized parts on the back of enclosed equipment, a minimum working space of 30 in (762 mm) horizontally shall be provided.

Exception No. 2. By special permission, smaller spaces shall be permitted where all uninsulated parts are at a voltage no greater than 30 V rms, 42 V peak, or 60 V dc.

Exception No. 3. In existing buildings where electrical equipment is being replaced, condition 2 working clearance shall be permitted between dead-front switchboards, panelboards, or motor control centers located across the aisle from each other where conditions of maintenance and supervision ensure that written procedures have been adopted to prohibit equipment on both sides of the aisle from being open at the same time and qualified persons who are authorized will service the installation.

2.2.2 Examples of Conditions 1, 2, and 3 (Working Spaces)

Distances are measured from the live parts if the live parts are exposed, or from the enclosure front if live parts are enclosed. If any assemblies, such as switchboards or motor-control centers, are accessible from the back and expose live parts, the working clearance dimensions would be required at the rear of the equipment, as illustrated. Note that for Condition 3, where there is an enclosure on opposite sides of the working space, the clearance for only one working space is required.

2.2.2

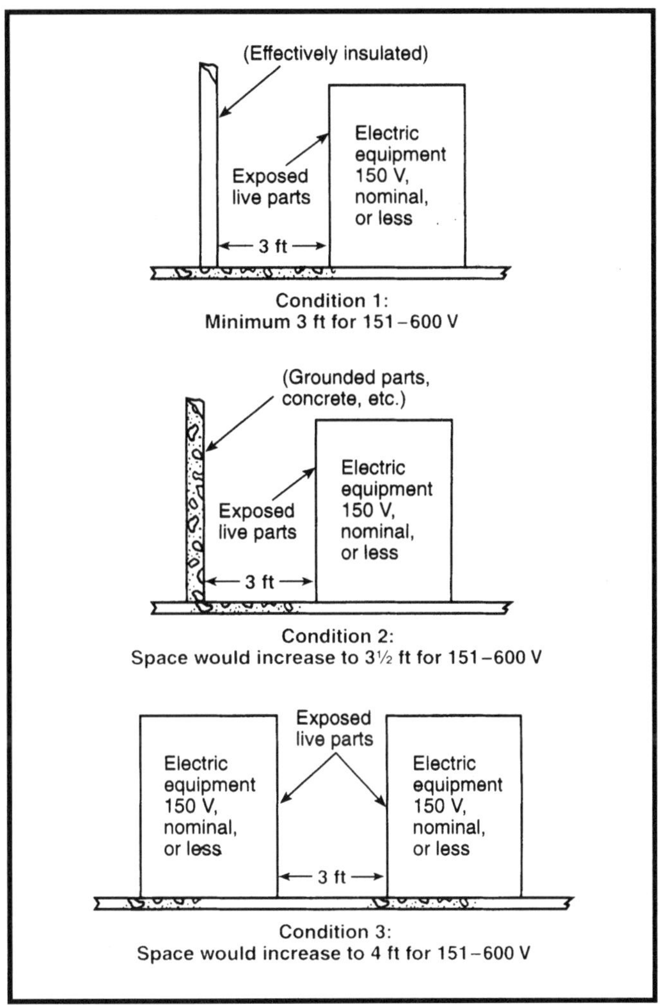

2.2.3 Example of Exception 1 (Working Spaces)

2.2.3

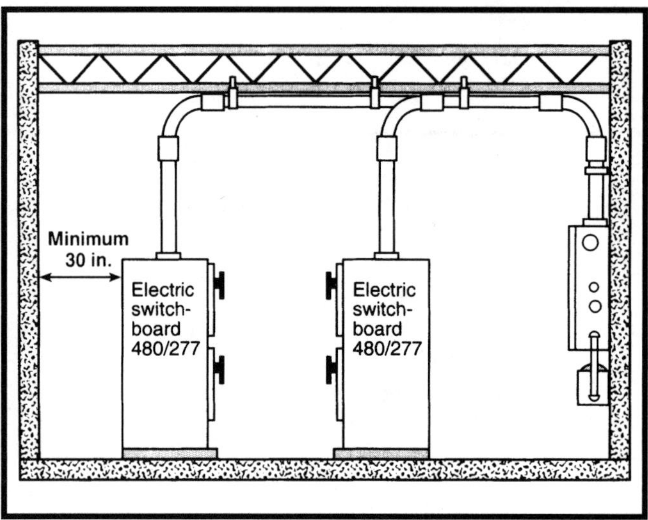

(© 1999, NFPA)

2.2.4 Example of Exception 3 (Working Spaces)

2.2.4

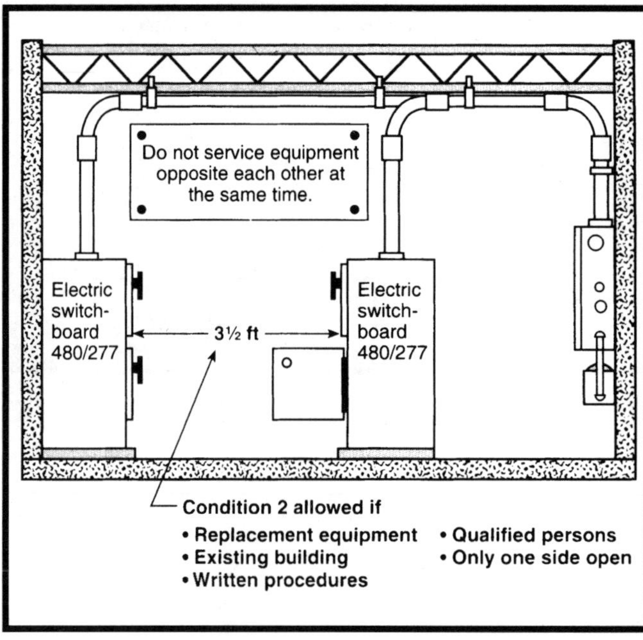

(© 1999, NFPA)

2.2.5 Required 30-In-Wide Front Working Space (Working Spaces)

The 30-inch-wide front working space is not required to be directly centered on the electrical equipment if it can be ensured that the space is sufficient for safe operation and maintenance of such equipment.

2.2.5

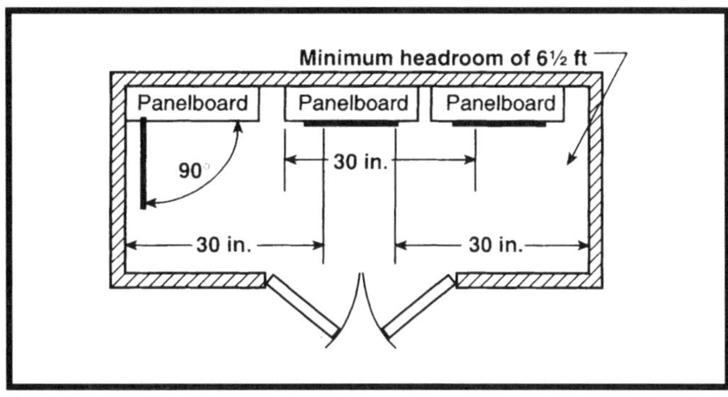

(© 1999, NFPA)

2.2.6 Required Full 90-Degree Opening of Equipment Doors (Working Spaces)

Equipment doors are required to open a full 90 degrees to ensure a safe working space.

2.2.6

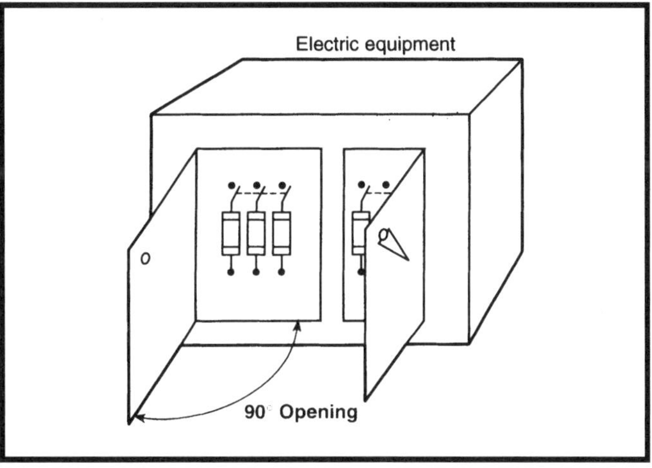

(© 1999, NFPA)

2.3.1 *NEC* Section 110.26(C), Basic Rule, First Paragraph (Access to Working Space)

Section 110.25 (C), Basic Rule, first paragraph. At least one entrance is required to provide access to the working space around electrical equipment. The installation shown on the bottom would not be acceptable if the electrical equipment was a switchboard over 6 feet wide and rated 1200 amperes or more.

2.3.1

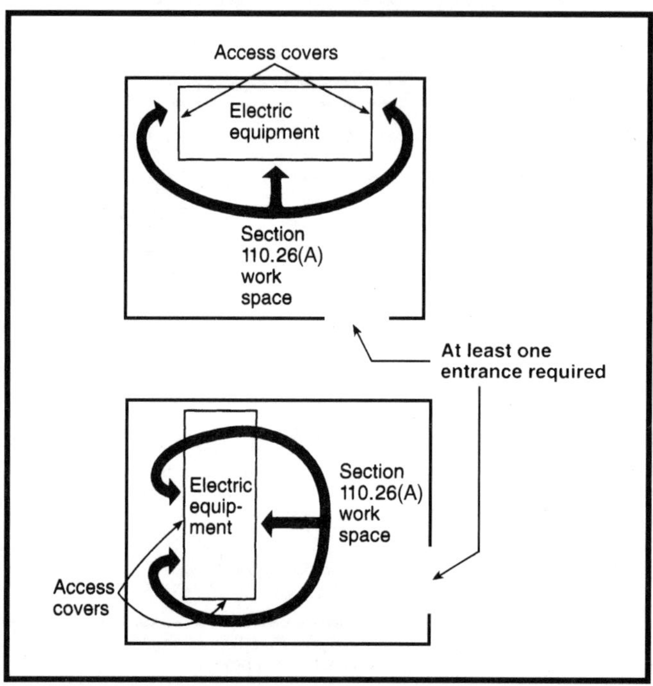

(© 1999, NFPA)

2.3.2 *NEC* Section 110.26(C), Basic Rule, Second Paragraph (Access to Working Space)

Section 110.26 (C), Basic Rule, second paragraph. For equipment rated 1200 amperes or more and over 6 feet wide, one entrance not less than 24 inches wide and 6-1/2 feet high is required at each end.

2.3.2

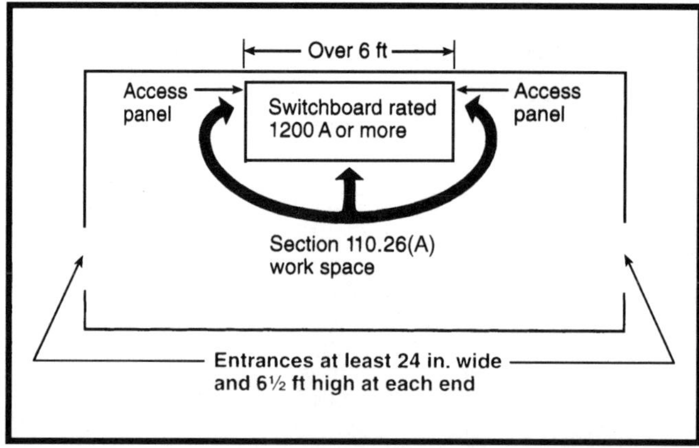

(© 1999, NFPA)

2.3.3 Example of an Unacceptable Arrangement of a Large Switchboard (Access to Working Space)

Unacceptable arrangement of large switchboard—a person could be trapped behind arcing electrical equipment.

2.3.3

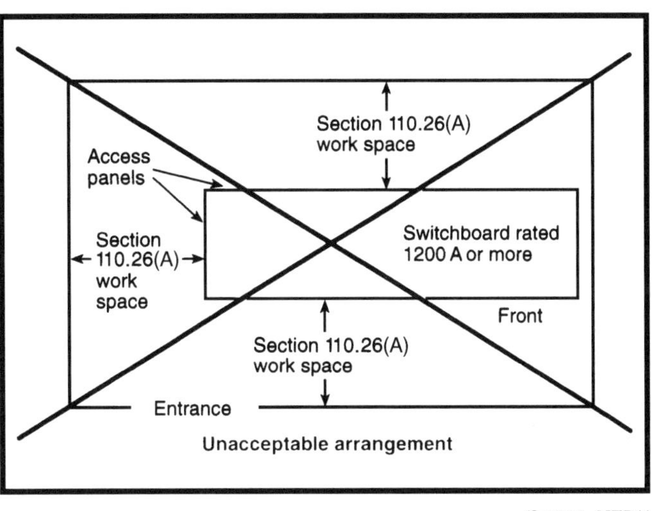

(© 1999, NFPA)

2.3.4 Example of Exception No. 1 (Access to Working Space)

The equipment location permits a continuous and unobstructed way of exit travel.

2.3.4

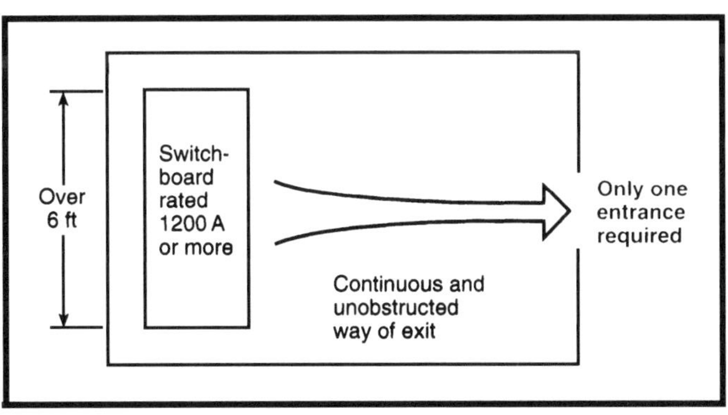

(© 1999, NFPA)

2.3.5 Example of Exception No. 2 (Access to Working Space)

If the working space required by Section 110.26 (A) is doubled, only one entrance to the working space is required.

2.3.5

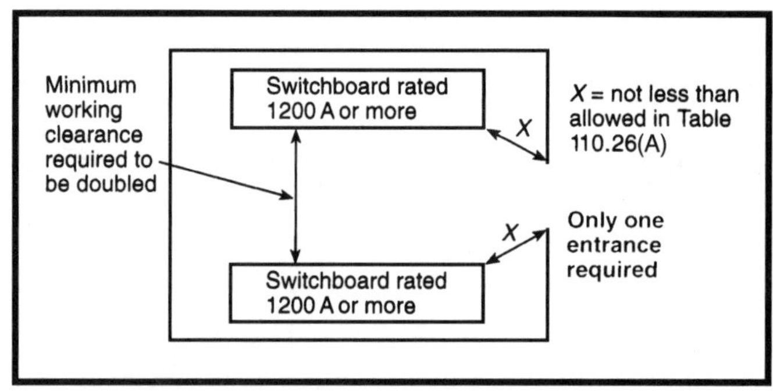

2.4.1 Working Space and Dedicated Electrical Space

The two distinct indoor installation spaces required by Section 110.26 (A) and 110.26 (F), that is, the "working space" and the "dedicated electrical space."

2.4.1

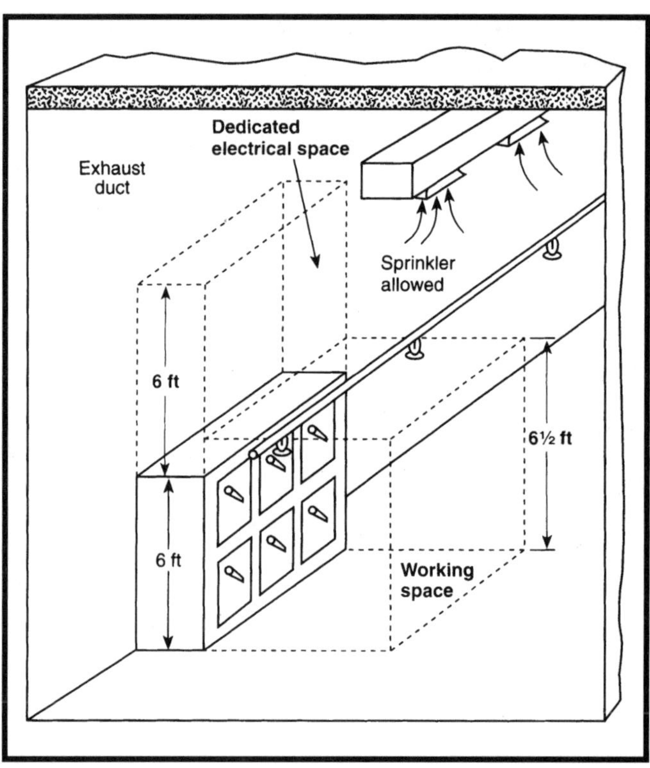

2.4.2 Working Space in Front of a Panelboard as Required by *NEC* Section 110.26

The working space in front of a panelboard as required by Section 110.26. This illustration supplements the dedicated equipment space shown in Figure 2.4.3.

2.4.2

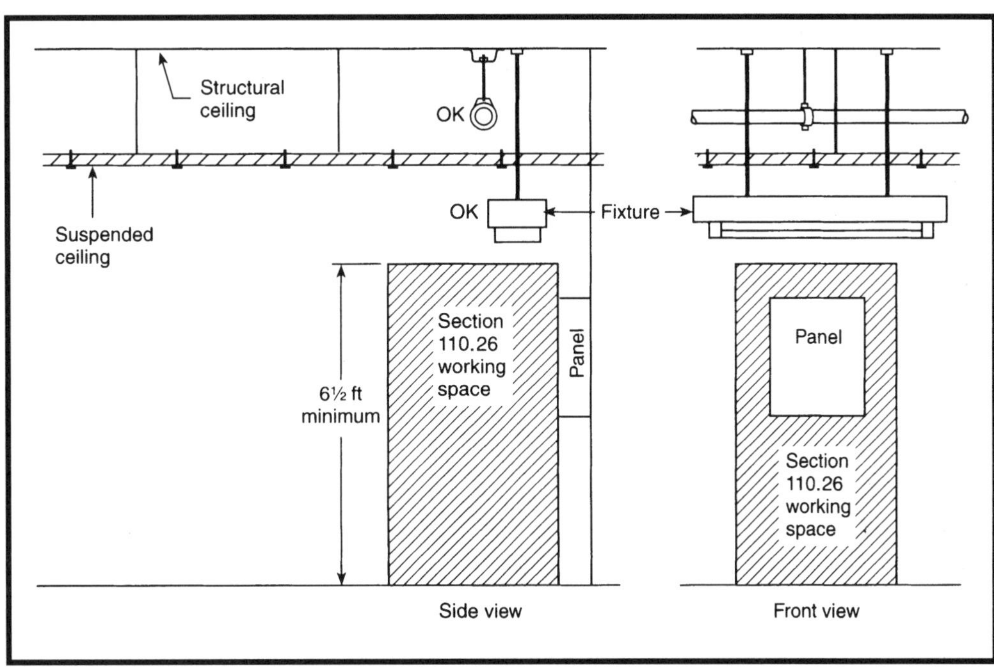

(© 1999, NFPA)

2.4.3 Dedicated Electrical Space Over and Under a Panelboard

The dedicated electrical space over and under a panelboard, as required by Section 110.26 (F) (1).

2.4.3

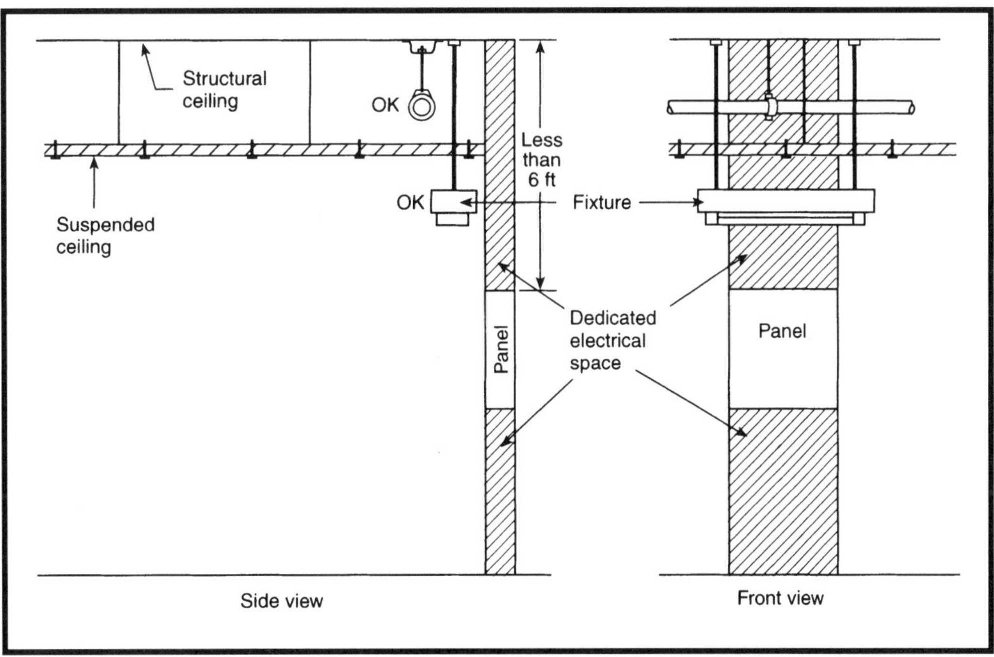

(© 1999, NFPA)

2.5.1 Minimum Depth of Clear Working Space at Electrical Equipment [*NEC* Table 110.34(A)]

TABLE 2.5.1

Nominal Voltage to Ground	Minimum Clear Distance		
	Condition 1	Condition 2	Condition 3
601–2500 V	900 mm (3 ft)	1.2 m (4 ft)	1.5 m (5 ft)
2501–9000 V	1.2 m (4 ft)	1.5 m (5 ft)	1.8 m (6 ft)
9001–25,000 V	1.5 m (5 ft)	1.8 m (6 ft)	2.8 m (9 ft)
25,001–75 kV	1.8 m (6 ft)	2.5 m (8 ft)	3.0 m (10 ft)
Above 75 kV	2.5 m (8 ft)	3.0 m (10 ft)	3.7 m (12 ft)

Note: Where the conditions are as follows:
Condition 1 — Exposed live parts on one side and no live or grounded parts on the other side of the working space, or exposed live parts on both sides effectively guarded by suitable wood or other insulating materials. Insulated wire or insulated busbars operating at not over 300 volts shall not be considered live parts.
Condition 2 — Exposed live parts on one side and grounded parts on the other side. Concrete, brick, or tile walls will be considered as grounded surfaces.
Condition 3 — Exposed live parts on both sides of the work space (not guarded as provided in Condition 1) with the operator between.

2.5.2 Elevation of Unguarded Live Parts Above Working Space [*NEC* Table 110.34(A)]

TABLE 2.5.2

Nominal Voltage Between Phases	Elevation
601–7500 V	2.8 m (9 ft)
7501–35,000 V	2.9 m ($9^1/_2$ ft)
Over 35 kV	2.9 m + 9.5 mm/kV above 35 ($9^1/_2$ ft + 0.37 in./kV above 35)

Section

3

Overcurrent Protection

3.1.0 Introduction
3.1.1 *NEC* Section 90.2, Scope of the *NEC*
3.2.1 *NEC* Section 110.3(A)(5), (6), and (8), Requirements for Equipment Selection
3.2.2 *NEC* Section 110.3(B), Requirements for Proper Installation of Listed and Labeled Equipment
3.2.3 *NEC* Section 110.9, Requirements for Proper Interrupting Rating of Overcurrent Protective Devices
3.2.4 *NEC* Section 110.10, Proper Protection of System Components from Short Circuits
3.2.5 *NEC* Section 110.22, Proper Marking and Identification of Disconnecting Means
3.3.1 *NEC* Section 210.20(A), Ratings of Overcurrent Devices on Branch Circuits Serving Continuous and Noncontinuous Loads
3.4.1 *NEC* Section 215.10, Requirements for Ground-Fault Protection of Equipment on Feeders
3.5.1 *NEC* Section 230.82, Equipment Allowed to Be Connected on the Line Side of the Service Disconnect
3.5.2 *NEC* Section 230.95, Ground-Fault Protection for Services
3.6.1 *NEC* Section 240.1, Scope of Article 240 on Overcurrent Protection
3.6.2 *NEC* Section 240.3, Protection of Conductors Other than Flexible Cords and Fixture Wires
3.6.3 *NEC* Section 240.4, Proper Protection of Fixture Wires and Flexible Cords
3.6.4 *NEC* Section 240.6, Standard Ampere Ratings
3.6.5 *NEC* Sections 240.8 and 380.17, Protective Devices Used in Parallel
3.6.6 *NEC* Section 240.9, Thermal Devices
3.6.7 *NEC* Section 240.10, Requirements for Supplementary Overcurrent Protection
3.6.8 *NEC* Section 240.11, Definition of Current-Limiting Overcurrent Protective Devices
3.6.9 *NEC* Section 240.12, System Coordination or Selectivity
3.6.10 *NEC* Section 240.13, Ground-Fault Protection of Equipment on Buildings or Remote Structures
3.6.11 *NEC* Section 240.21, Location Requirements for Overcurrent Devices and Tap Conductors
3.6.12 *NEC* Section 240.40, Disconnecting Means for Fuses
3.6.13 *NEC* Section 240.50, Plug Fuses, Fuseholders, and Adapters
3.6.14 *NEC* Section 240.51, Edison-Base Fuses
3.6.15 *NEC* Section 240.53, Type S Fuses
3.6.16 *NEC* Section 240.54, Type S Fuses, Adapters, and Fuseholders

3.6.17 *NEC* Section 240.60, Cartridge Fuses and Fuseholders
3.6.18 *NEC* Section 240.61, Classification of Fuses and Fuseholders
3.6.19 *NEC* Section 240.86, Series Ratings
3.6.20 *NEC* Section 240.90, Supervised Industrial Installations
3.6.21 *NEC* Section 240.92(B), Transformer Secondary Conductors of Separately Derived Systems (Supervised Industrial Installations Only)
3.6.22 *NEC* Section 240.92(B)(1), Short-Circuit and Ground-Fault Protection (Supervised Industrial Installations Only)
3.6.23 *NEC* Section 240.92(B)(2), Overload Protection (Supervised Industrial Installations Only)
3.6.24 *NEC* Section 240.92(C), Outside Feeder Taps (Supervised Industrial Installations Only)
3.6.25 *NEC* Section 240.100, Feeder and Branch-Circuit Protection Over 600 V Nominal
3.6.26 *NEC* Section 240.100(C), Conductor Protection
3.7.1 *NEC* Section 250.2(D), Performance of Fault-Current Path
3.7.2 *NEC* Section 250.90, Bonding Requirements and Short-Circuit Current Rating
3.7.3 *NEC* Section 250.96(A), Bonding Other Enclosures and Short-Circuit Current Requirements
3.7.4 *NEC* Section 250.122, Sizing of Equipment Grounding Conductors
3.8.1 *NEC* Section 310.10, Temperature Limitation of Conductors
3.9.1 *NEC* Section 364.11, Protection at a Busway Reduction
3.10.1 *NEC* Section 384.16, Panelboard Overcurrent Protection
3.11.1 *NEC* Section 430.1, Scope of Motor Article
3.11.2 *NEC* Section 430.6, Ampacity of Conductors for Motor Branch Circuits and Feeders
3.11.3 *NEC* Section 430.8, Marking on Controllers
3.11.4 *NEC* Section 430.32, Motor Overload Protection
3.11.5 *NEC* Section 430.36, Fuses Used to Provide Overload and Single-Phasing Protection
3.11.6 *NEC* Section 430.52, Sizing of Various Overcurrent Devices for Motor Branch-Circuit Protection
3.11.7 *NEC* Section 430.53, Connecting Several Motors or Loads on One Branch Circuit
3.11.8 *NEC* Section 430.71, Motor Control-Circuit Protection
3.11.9 *NEC* Section 430.72(A), Motor Control-Circuit Overcurrent Protection
3.11.10 *NEC* Section 430.72(B), Motor Control-Circuit Conductor Protection
3.11.11 *NEC* Section 430.72(C), Motor Control-Circuit Transformer Protection
3.11.12 *NEC* Section 430.94, Motor Control-Center Protection
3.11.13 *NEC* Section 430.109(A)(6), Manual Motor Controller as a Motor Disconnect
3.12.1 *NEC* Section 440.5, Marking Requirements on HVAC Controllers
3.12.2 *NEC* Section 440.22, Application and Selection of the Branch-Circuit Protection for HVAC Equipment
3.13.1 *NEC* Section 450.3, Protection Requirements for Transformers
3.13.2 *NEC* Section 450.3(A), Protection Requirements for Transformers Over 600 V
3.13.3 *NEC* Section 450.3(B), Protection Requirements for Transformers 600 V or Less
3.13.4 *NEC* Section 450.6(A)(3), Tie-Circuit Protection
3.14.1 *NEC* Section 455.7, Overcurrent Protection Requirements for Phase Converters
3.15.1 *NEC* Section 460.8(B), Overcurrent Protection of Capacitors

3.16.1 *NEC* Section 501.6(B), Fuses for Class 1, Division 2 Locations
3.17.1 *NEC* Section 517.17, Requirements for Ground-Fault Protection and Coordination in Health Care Facilities
3.18.1 *NEC* Section 520.53(F)(2), Protection of Portable Switchboards on Stage
3.19.1 *NEC* Section 550.6(B), Overcurrent Protection Requirements for Mobile Homes and Parks
3.20.1 *NEC* Section 610.14(C), Conductor Sizes and Protection for Cranes and Hoists
3.21.1 *NEC* Section 620.62, Selective Coordination of Overcurrent Protective Devices for Elevators
3.22.1 *NEC* Section 670.3, Industrial Machinery
3.23.1 *NEC* Section 700.5, Emergency Systems: Their Capacity and Rating
3.23.2 *NEC* Section 700.16, Emergency Illumination
3.23.3 *NEC* Section 700.25, Emergency System Overcurrent Protection Requirements (FPN)
3.24.1 *NEC* Section 705.16, Interconnected Electric Power Production Sources: Interrupting and Short-Circuit Current Rating
3.25.1 *NEC* Section 725.23, Overcurrent Protection for Class 1 Circuits
3.26.1 *NEC* Section 760.23, Requirements for Non-Power-Limited Fire Alarm Signaling Circuits

3.1.0 Introduction

In this section, a presentation format of questions and answers has been used to focus on the factors that are pertinent to a basic understanding and application of overcurrent protective devices. Relevant sections of the *National Electrical Code* (*NEC*) are referenced and analyzed in detail. Each section is translated into simple, easily understood language, complemented by one-line diagrams giving sound, practical means of applying overcurrent protection, as well as affording compliance with the *NEC*. The following is by permission of Cooper Bussmann, Inc.

3.1.1 *NEC* Section 90.2, Scope of the *NEC*

3.1.1

What does this Section mean?
90-2(b) covers installations that are not covered by requirements of the N.E.C. However, the fine print note states that it is the intent of this section that utility installed utilization equipment located on private property is subject to the National Electrical Code.

3.2.1 *NEC* Section 110.3(A)(5), (6), and (8) Requirements for Equipment Selection

3.2.1

What does 110-3(a)(5), (6) and (8) require?
When equipment is selected its arc-flash protection capability and finger-safe rating must be evaluated. When equipment is energized, and the door is open the possibility exists that an employee could accidentally create an arcing fault or come into contact with a live part. Equipment must be evaluated for both possibilities, and be chosen for minimum employee exposure to either danger.

3.2.2 *NEC* Section 110.3(B), Requirements for Proper Installation of Listed and Labeled Equipment

3.2.2

What is the importance of Section 110-3(b)?
Equipment that is listed is subject to specific conditions of installation or operation. The conditions must be followed for safe and proper operation.

What is the protection requirement of an air conditioner when its name plate specifies Maximum Fuse Size Amperes?

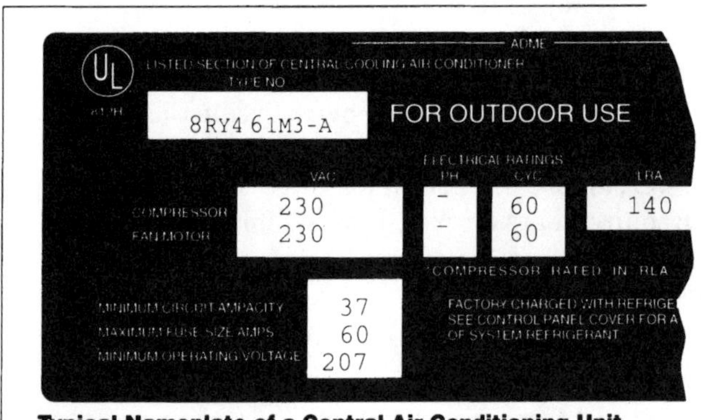

Typical Nameplate of a Central Air Conditioning Unit.

Fuse protection in the branch circuit is mandatory to meet the requirements of the U.L. Listings and the National Electrical Code®.

Note that the U.L. Orange Book "Electrical Appliance and Utilization Equipment Directory," April 1998, requires the following for central cooling, air conditioners: "Such multimotor and combination load equipment is to be connected only to a circuit protected by fuses or a circuit breaker with a rating which does not exceed the value marked on the data plate. This marked protective device rating is the maximum for which the equipment has been investigated and found acceptable. Where the marking specified fuses, or "HACR Type" circuit breakers, the circuit is intended to be protected only by the type of protective device specified." U.L. Standard 1995 also covers this subject.

What about a motor starter heater table (such as that shown below) which specifies Maximum Fuse?

Heater Code Marking	Full-Load Current of Motor (Amperes) (40°C Ambient)	Max. Fuse
XX03	.25- .27	1
XX04	.28- .31	3
XX05	.32- .34	3
XX06	.35- .38	3
XX14	.76- .83	6
XX15	.84- .91	6
XX16	.92-1.00	6
XX17	1.01-1.11	6
XX18	1.12-1.22	6

Above Heaters for use on Size 0

Like an air conditioner, use of fuse protection is mandatory. Also, the fuse must provide branch circuit protection and be no larger than the specified size [430-53(c)]. The chart shown, for example, is typical for starter manufacturers and may be found on the inside of the door of the starter enclosure. (See starter manufacturer for specific recommendations.)

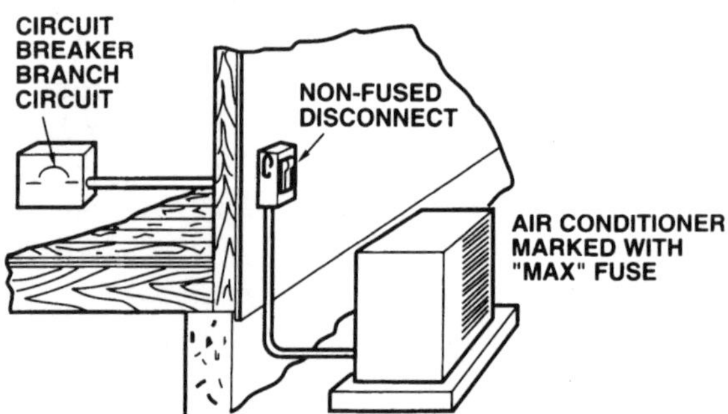

Violates N.E.C. & Listing Requirements

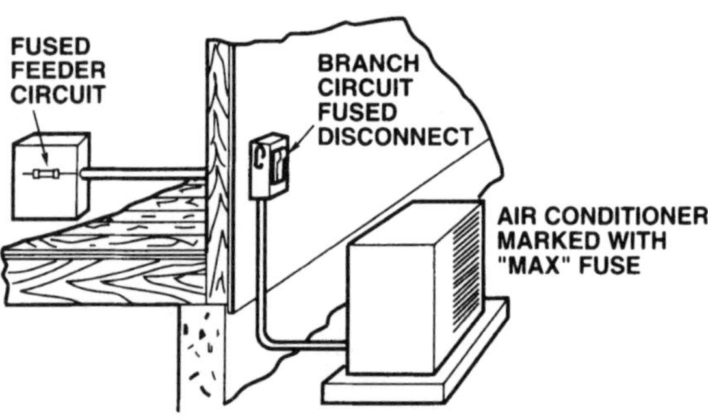

Conforms to N.E.C. & Listing Requirements

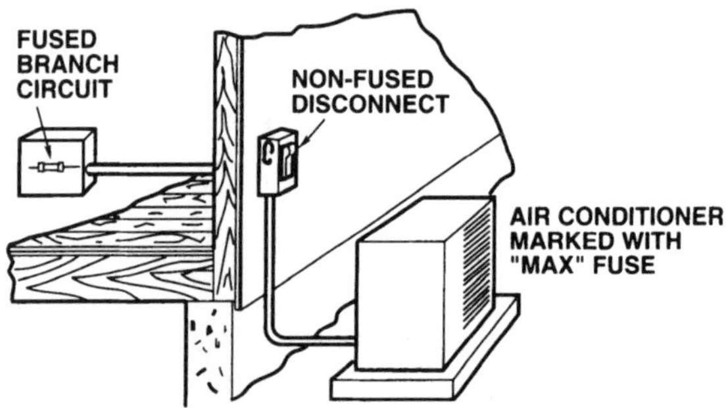

Conforms to N.E.C. & Listing Requirements

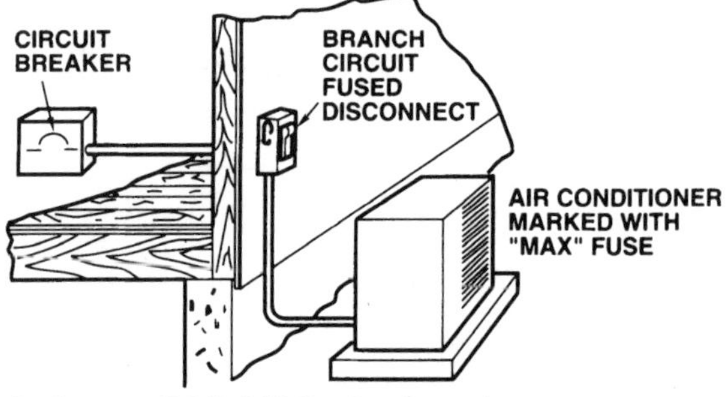

Conforms to N.E.C. & Listing Requirements

(continued)

3.2.2 (*Continued*)

What violation exists when a "series-rated" panelboard with a "42/10" system rating has the potential to see a fault current less than 4 ft. from the loadside circuit breaker?

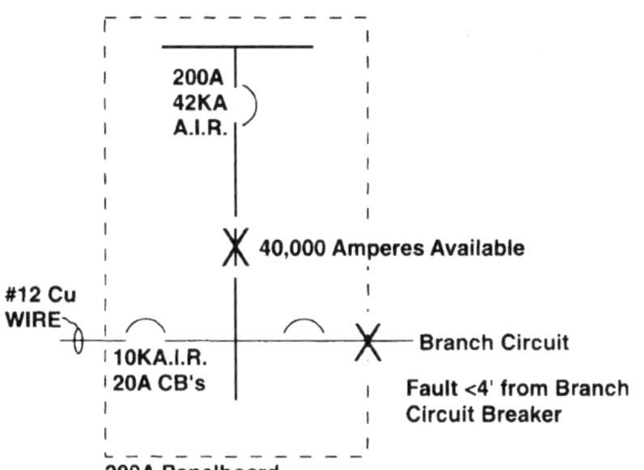

U.L. 489 Series Rating tests allow a maximum of 4 ft. of rated wire to be connected to the branch circuit breaker. Whenever the potential for a fault exists closer than 4 ft. from the circuit breaker, i.e., where the #12 wire leaves the enclosure, or a maintenance man is working on the equipment "hot", a violation of 110-3b exists, as does a potentially hazardous condition. In this situation, the interrupting capacity of the circuit breaker may not equal its marked interrupting rating!

3.2.3 *NEC* Section 110.9, Requirements for Proper Interrupting Rating of Overcurrent Protective Devices

3.2.3

What is the importance of Section 110-9?
Equipment designed to break fault or operating currents must have a rating sufficient to withstand such currents. This article emphasizes the difference between clearing fault level currents and clearing operating currents. Protective devices such as fuses and circuit breakers are designed to clear fault currents and, therefore, must have short-circuit interrupting ratings sufficient for fault levels. Equipment such as contactors and switches have interrupting ratings for currents at other than fault levels. Thus, the interrupting rating of electrical equipment is divided into two parts.

Most people are familiar with the normal current carrying ampere rating of a fuse or circuit breaker; however, what is a short-circuit interrupting rating?
It is the maximum short-circuit current that an overcurrent protective device can safely interrupt under specified test conditions.

What is a device's interrupting capacity?
The following definition of Interrupting Capacity is from the IEEE Standard Dictionary of Electrical and Electronic Terms:

Interrupting Capacity: The highest current at rated voltage that the device can interrupt.

Because of the way fuses are short-circuit tested (without additional cable impedance), their interrupting capacity is greater than or equal to their interrupting rating. Because of the way circuit breakers are short circuit tested (with additional cable impedance), their interrupting capacity can be less than, equal to, or greater than their interrupting rating.

How does Section 110-9 pertain to services?
Service equipment must be able to withstand available short-circuit currents. More specifically, the service switchboard, panelboard, etc., and the protective devices which they incorporate must have a short-circuit rating equal to or greater than the short-circuit current available at the line side of the equipment.

In this circuit, what must be the short-circuit rating of the switchboard?

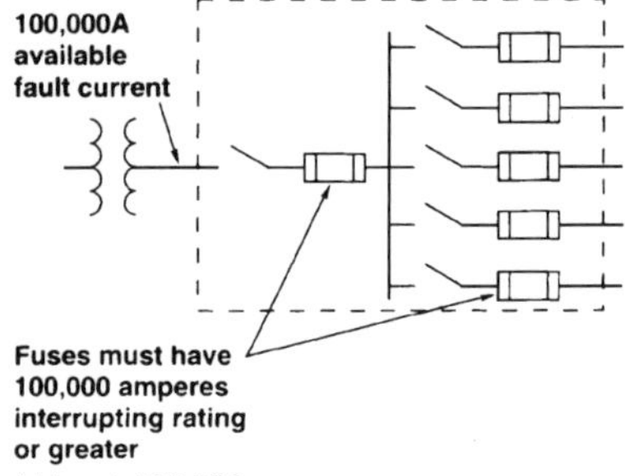

At least 100,000 amperes.

What must be the interrupting rating of the fuses?
100,000 amperes or greater. (Most current-limiting fuses have an interrupting rating of 200,000 or 300,000 amperes.)

In this circuit, what must be the interrupting capacity of the main circuit breaker, and the short-circuit rating of the switchboard?

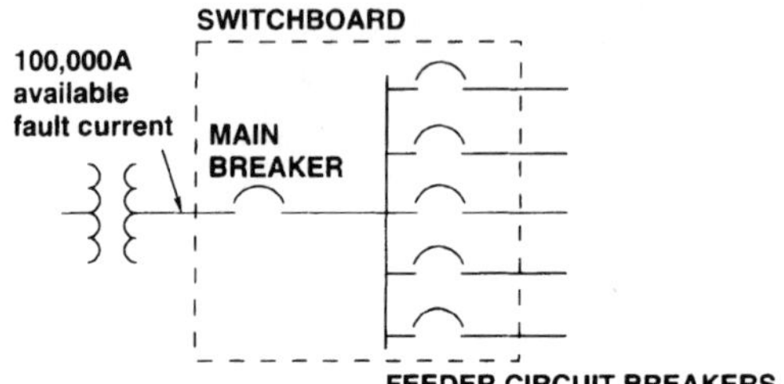

At least 100,000 amperes.

As shown in the circuit, can fuses be used to protect circuit breakers with a low interrupting rating.

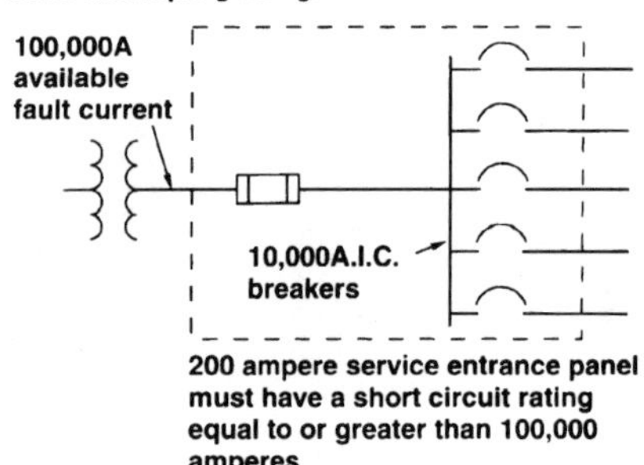

Yes. Properly selected fuses can protect circuit breakers as well as branch circuit conductors by limiting short-circuit currents to a low level even though available short-circuit current is as high as 100,000 amperes. (Buss® LOW-PEAK® YELLOW™ or T-TRON® fuses give optimum protection.)

(continued)

3.2.3 (*Continued*)

Can cable limiters protect service entrance equipment from short-circuit currents?

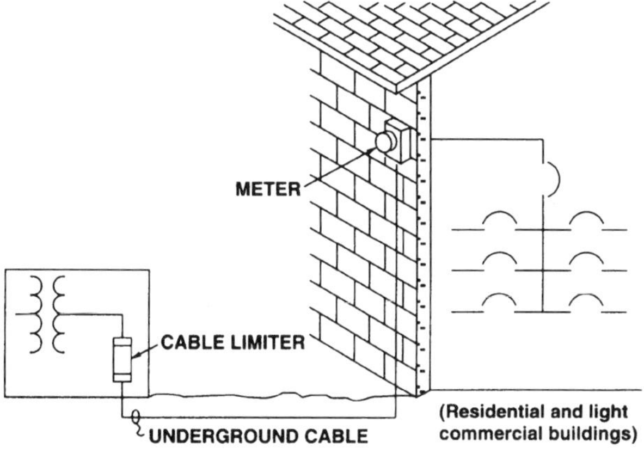

Current-limiting cable limiters not only can be used to isolate a "faulted" service cable, but also can help to protect utility meters with low withstand ratings against high short-circuit currents. (See Section 230-82).

Application Note:
Residential—100 ampere and 200 ampere fused main-branch circuit breaker panels are commercially available. These load centers incorporate the small-sized T-TRON® JJN fuses which make it possible to obtain a 100,000 amperes short-circuit current rating. Mobile home meter pedestals are also available incorporating the T-TRON® JJN fuses in a Fuse Pullout Unit.

Apartment Complexes—Have high densities of current and, therefore, high short-circuit currents for the typical meters.

Grouped meter stacks are commercially available using the T-TRON® JJN fuses (up to 1200 amperes) to give the proper short-circuit protection. Meter stacks are also available with Class T fuse pullouts on the load side of each meter.

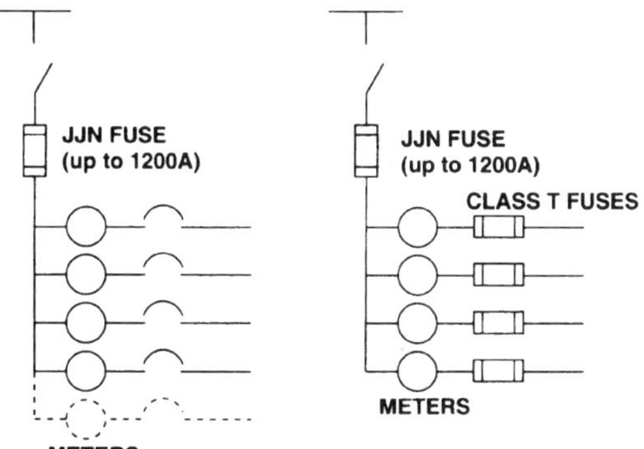

What happens if a fault current exceeds the interrupting rating of a fuse or the interrupting capacity of a circuit breaker?
It can be damaged or destroyed. Severe equipment damage and personnel injury can result.

In this circuit, what interrupting rating must the fuse have?

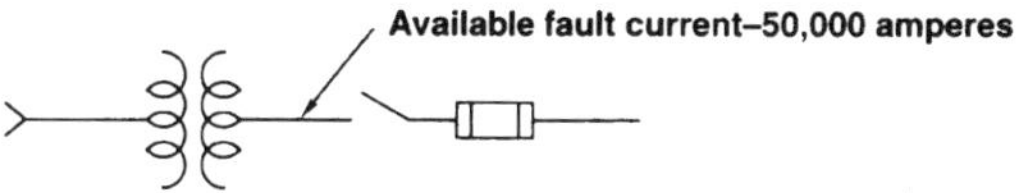

At least 50,000 amperes. (Class R, J, T, L and CC fuses have an Interrupting Rating of at least 200,000 amperes. The interrupting rating of a fuse and switch combination may also be 200,000 amperes. . .well above the available short-circuit current of 50,000 amperes. The interrupting rating of Class G fuses is 100,000 amperes; K1 and K5 fuses can be 50,000, 100,000, or 200,000 amperes.)

In this circuit, what interrupting rating must the circuit breaker have?

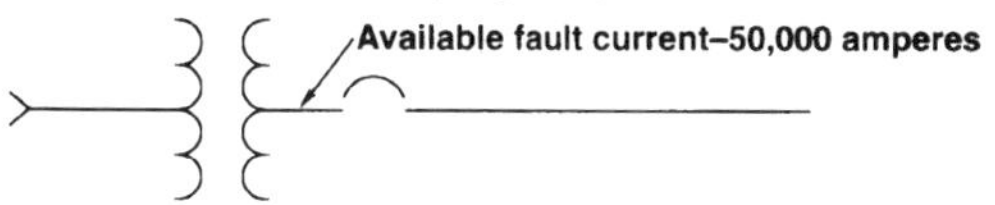

Some value greater than or equal to 50,000 amperes. See discussion on circuit breaker interrupting rating in Section 110-10 for a further evaluation. (Faults within four feet of the breaker could cause complete destruction of the breaker if it is applied where the available fault current approaches the tested interrupting capacity of the breaker.)

Section 110-9 also requires the overcurrent device to have a sufficient interrupting rating for both phase voltage and phase-to-ground voltage.

What is the significance of this requirement?
Certain molded case circuit breakers have lower single-pole interrupting ratings than their multi-pole interrupting rating.

What are the single-pole interrupting ratings for overcurrent devices?
What are the single-pole interrupting ratings for overcurrent devices? Modern current limiting fuses such as Class RK1, J and L have single-pole interrupting ratings of at least 200,000 amperes RMS symmetrical. For example, per UL/CSA 248-8, a 600 volt Class J fuse is tested at a minimum of 200,000 amperes at 600 volts across one pole. Bussmann® has recently introduced the above fuse types with 300,000 ampere single-pole interrupting ratings. Per ANSI C37.13 and C37.16, an airframe/power circuit breaker has a single-pole rating of 87% of its three-pole rating. Listed three- pole molded case circuit breakers have minimum single-pole interrupting ratings according to Table 7.1.7.2 of U.L. 489. The following table indicates the single-pole ratings of various three-pole molded-case circuit breakers taken from Table 7.1.7.2 of U.L. 489. A similar table is shown on page 54 of the IEEE "Blue Book" (Std 1015-1997). Molded-case circuit breakers may or may not be able to safely interrupt single-pole faults above these values since they are typically not tested beyond these values.

If the ratings shown in this table are too low for the application, the actual single-pole rating for the breaker must be ascertained to insure proper application. Or, modern current limiting fuses or airframe/power circuit breakers can be utilized.

(*continued*)

3.2.3 (*Continued*)

As an example of single-pole interrupting ratings in a typical installation, consider a common three-pole, 20 amp, 480 volt circuit breaker with a three-pole interrupting rating of 65,000 amperes. Referring to the table, this breaker has an 8,660 ampere single-pole interrupting rating for faults across one pole. If the available line-to-ground fault current exceeds 8660 amps, the MCCB may be misapplied. In this case, the breaker manufacturer must be consulted to verify interrupting ratings and proper application.

Single-Pole Interrupting Ratings for Three Pole Molded Case Circuit Breakers (ANY I.R.)

FRAME RATING	240V	480/277V	480V	600/347V	600V
100A Maximum 250V Max	4,330	—	—	—	—
100A Maximum 251-600V	—	10,000	8,660	10,000	8,660
101 - 800	8,660	10,000	8,660	10,000	8,660
801 - 1200	12,120	14,000	12,120	14,000	12,120
1201 - 2000	14,000	14,000	14,000	14,000	14,000
2001 - 2500	20,000	20,000	20,000	20,000	20,000
2501 - 3000	25,000	25,000	25,000	25,000	25,000
3001 - 4000	30,000	30,000	30,000	30,000	30,000
4001 - 5000	40,000	40,000	40,000	40,000	40,000
5001 - 6000	50,000	50,000	50,000	50,000	50,000

How much short-circuit current will flow in a ground fault condition?
The answer is dependent upon the location of the fault with respect to the transformer secondary. Referring to Figures 3 and 4, the ground fault current flows through one coil of the wye transformer secondary and through the phase conductor to the point of the fault. The return path is through the enclosure and conduit to the bonding jumper and back to the secondary through the grounded neutral. Unlike three-phase faults, the impedance of the return path must be used in determining the magnitude of ground fault current. This ground return impedance is usually difficult to calculate. If the ground return path is relatively short (i.e. close to the center tap of the transformer), the ground fault current will approach the three-phase short-circuit current.

Theoretically, a bolted line-to-ground fault may be higher than a three-phase bolted fault since the zero-sequence impedance can be less than the positive sequence impedance. The ground fault location will determine the level of short-circuit current available. However, to insure a safe system, the prudent design engineer should assume that the ground fault current equals at least the three-phase current and should assure that the overcurrent devices are rated accordingly.

How does a solidly grounded wye system affect the requirements for single-pole interrupting ratings?
The Solidly Grounded Wye system shown in Figures 1 and 2 is by far the most common type of electrical system. This system is typically delta connected on the primary and has an intentional solid connection between the ground and the center of the wye connected secondary (neutral). The grounded neutral conductor carries single-phase or unbalanced three-phase current. This system lends itself well to industrial applications where 480V(L-L-L) three-phase motor loads and 277V(L-N) lighting is required.

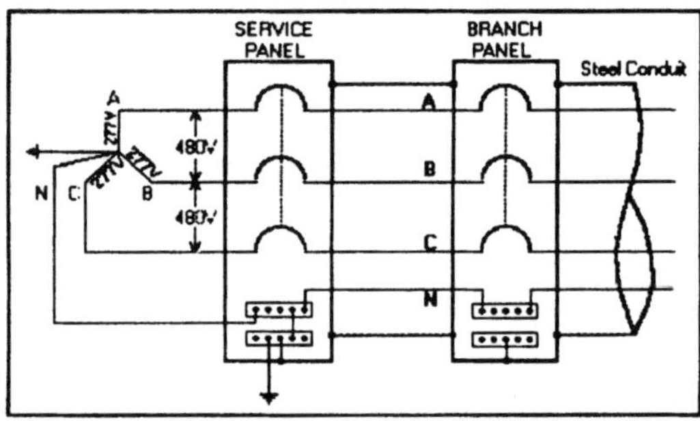

Figure 1 - Solidly Grounded WYE System - Circuit Breakers

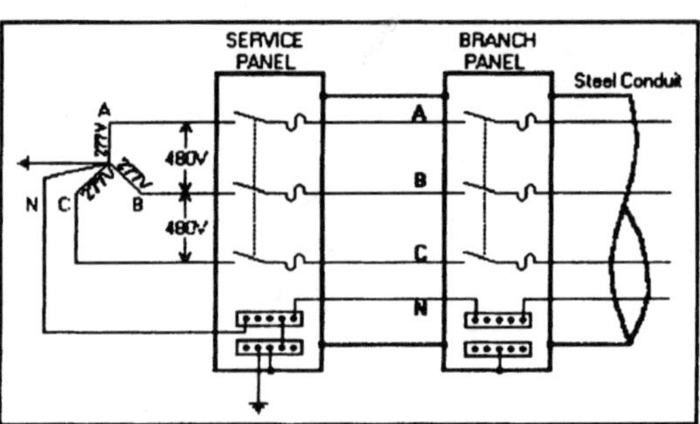

Figure 2 - Solidly Grounded WYE System - Fuses

If a fault occurs between any phase conductor and ground (Figures 3 and 4), the available short-circuit current is limited only by the combined impedance of the transformer winding, the phase conductor and the equipment ground path from the point of the fault back to the source. [Some current (typically 5%) will flow in the parallel earth ground path. Since the earth impedance is typically much greater than the equipment ground path, current flow through earth ground is generally negligible.]

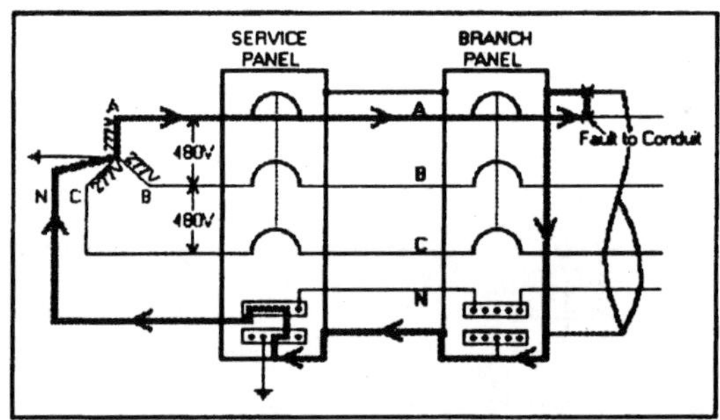

Figure 3 - Single-Pole Fault to Ground - Circuit Breakers

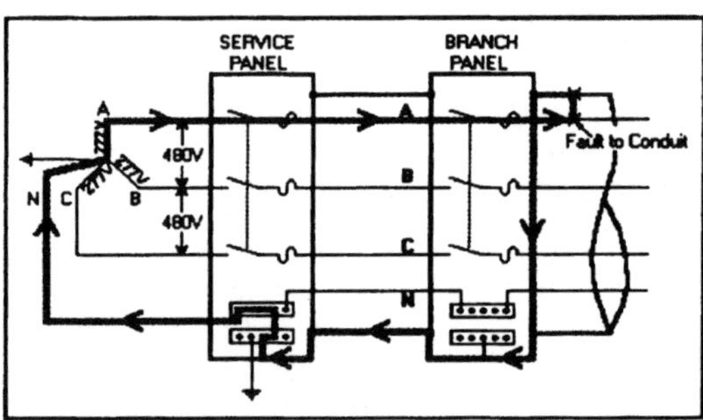

Figure 4 - Single-Pole Fault to Ground - Fuses

(*continued*)

3.2.3 *(Continued)*

In solidly grounded wye systems, the first low impedance fault to ground is generally sufficient to open the overcurrent device on the faulted leg. In Figures 3 and 4, this fault current causes the branch circuit overcurrent device to clear the 277 volt fault. This system requires compliance with single-pole interrupting ratings for 277 volt faults on one pole. If the overcurrent devices have a single-pole interrupting rating adequate for the available short-circuit current, then the system meets Section 110-9 of the National Electrical Code®.

Although not as common as the solidly grounded wye connection, the following systems are typically found in industrial installations where continuous operation is essential. Whenever these systems are encountered, it is absolutely essential that the single-pole ratings of overcurrent devices be investigated. This is due to the fact that full phase-to-phase voltage can appear across just one pole. Phase-to-phase voltage across one pole is much more difficult for an overcurrent device to clear than the line-to-neutral voltage associated with the solidly grounded wye systems.

How does a corner-grounded-delta system affect the requirements for single-pole interrupting ratings?
The systems of Figures 5 and 6 have a delta-connected secondary and are solidly grounded on the B-phase. If the B-phase should short to ground, no fault current will flow because it is already solidly grounded.

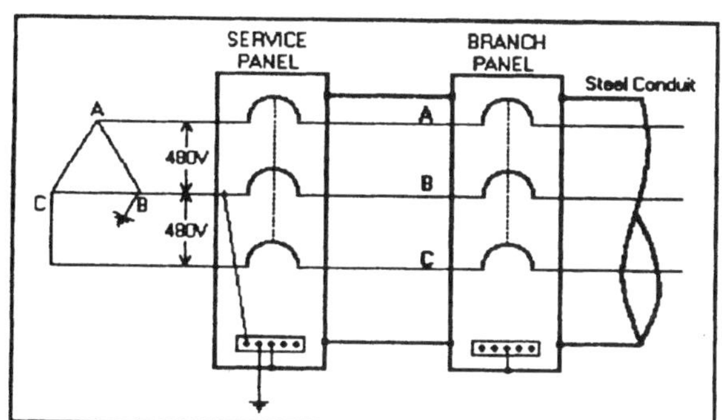

Figure 5 - B-Phase Grounded (Solidly) System - Circuit Breakers

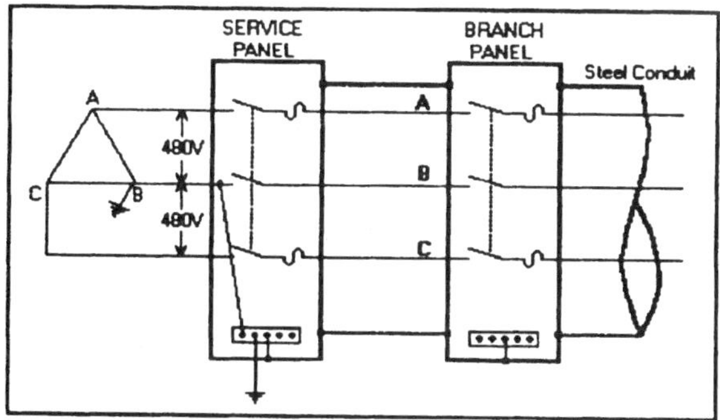

Figure 6 - B-Phase Grounded (Solidly) System - Fuses

If either Phase A or C is shorted to ground, only one pole of the overcurrent device will see the 480V fault as shown in Figures 7 and 8. This system requires compliance with single-pole interrupting ratings for 480 volt faults on one pole.

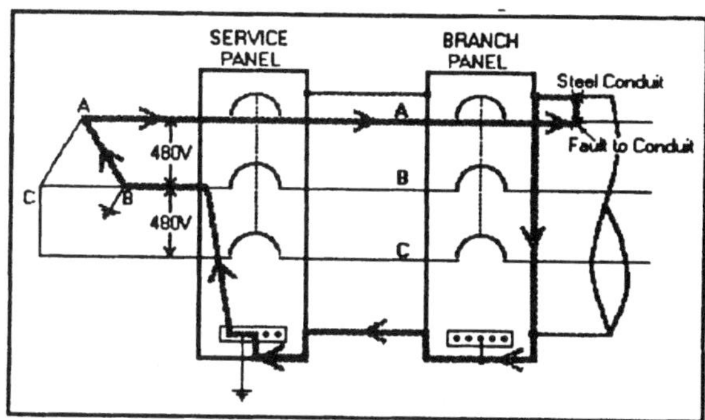

Figure 7 - B-Phase Solidly Grounded System - Circuit Breakers

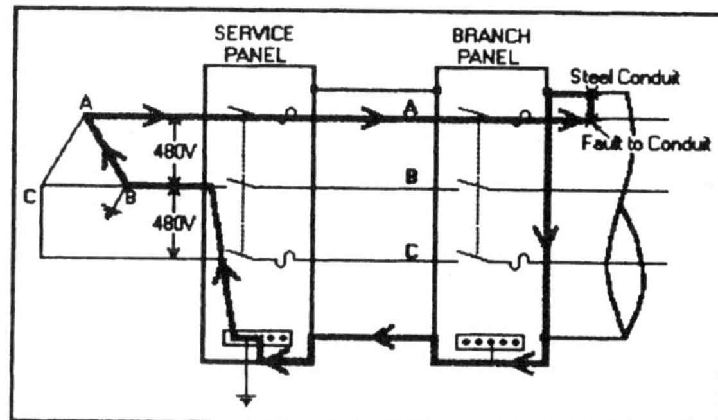

Figure 8 - B-Phase Solidly Grounded System - Fuses

A disadvantage of B-phase solidly grounded systems is the inability to readily supply voltage levels for fluorescent or HID lighting (277V). Installations with this system require a 480-120V transformer to supply 120V lighting. Another disadvantage, as given on page 33 of IEEE Std 142-1991, Section 1.5.1(4) (Green Book) is **"the possibility of exceeding interrupting capabilities of marginally applied circuit breakers, because for a ground fault, the interrupting duty on the affected circuit breaker pole exceeds the three-phase fault duty."**

How does a resistance-grounded system affect the requirements for single-pole interrupting ratings?
"Low or High" resistance grounding schemes are found primarily in industrial installations. These systems are used to limit, to varying degrees, the amount of current that will flow in a phase to ground fault.
"Low" resistance grounding is used to limit ground fault current to values acceptable for relaying schemes. This type of grounding is used mainly in medium voltage systems and is not widely installed in low voltage applications (600V or below).
The "High" Resistance Grounded System offers the advantage that the first fault to ground will not draw enough current to cause the overcurrent device to open. This system will reduce the stresses, voltage dips, heating effects, etc. normally associated with high short-circuit current. Referring to Figures 9 and 10, High Resistance Grounded Systems have a resistor between the center tap of the wye transformer and ground.

(continued)

3.2.3 (*Continued*)

With high resistance grounded systems, line-to-neutral loads are not permitted per the (2002) National Electrical Code, Section 250-36(4)

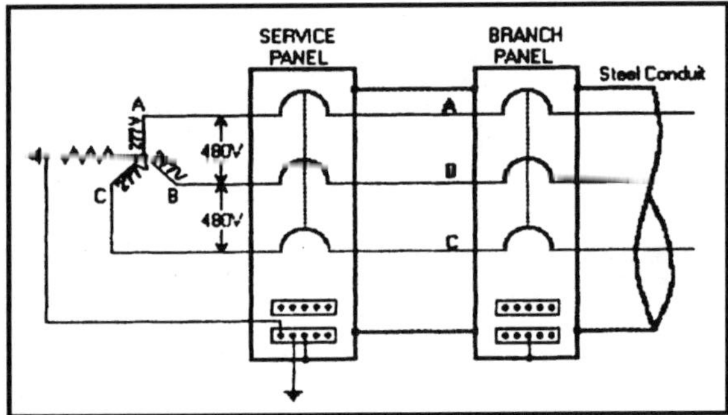

Figure 9 - Resistance Grounded System - Circuit Breakers

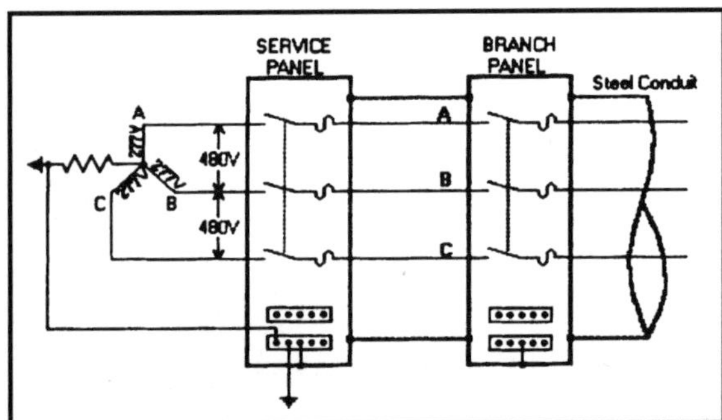

Figure 10 - Resistance Grounded System - Fuses

When the first fault occurs from phase to ground as shown in Figures 11 and 12, the current path is through the grounding resistor. Because of this inserted resistance, the fault current is not high enough to open protective devices. This allows the plant to continue "on line". NEC&RM 250-36(3) requires ground detectors to be installed on these systems, so that the first fault can be found and fixed before a second fault occurs on another phase.

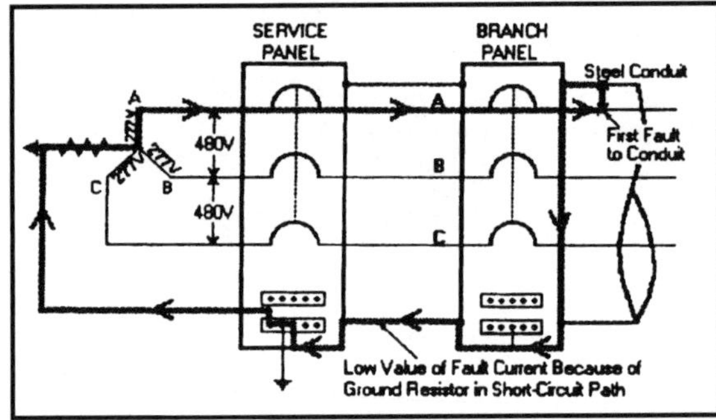

Figure 11 - First Fault in Resistance Grounded System - Circuit Breakers

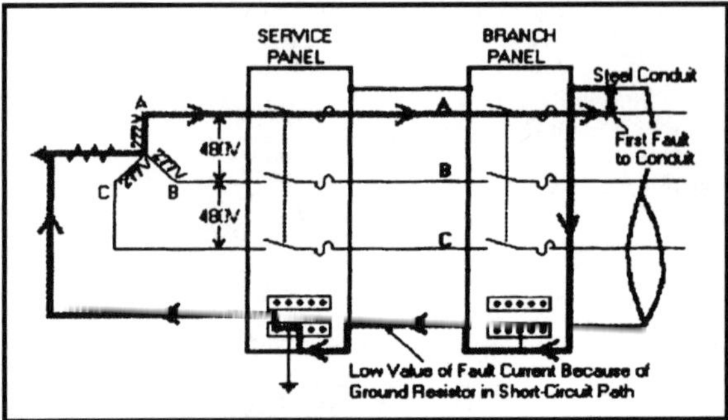

Figure 12 - First Fault in Resistance Grounded System - Fuses

Even though the system is equipped with a ground alarm, the exact location of the ground fault may be difficult to determine. The first fault to ground MUST be removed before a second phase goes to ground, creating a 480 volt fault across only one pole of the affected branch circuit device. Figures 13 and 14 show how the 480 volt fault can occur across the branch circuit device.

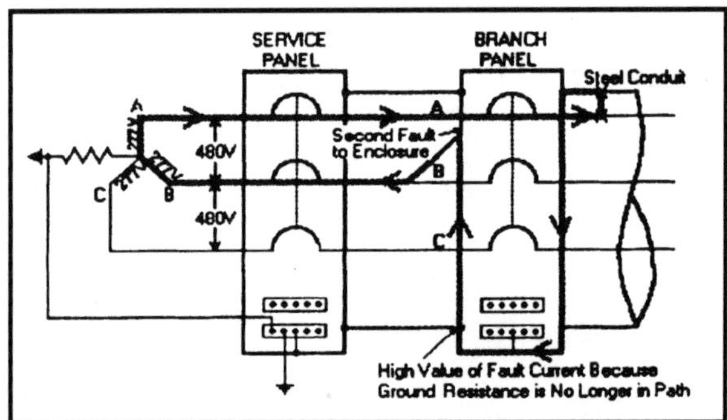

Figure 13 - Second fault in Resistance Grounded System - Circuit Breakers

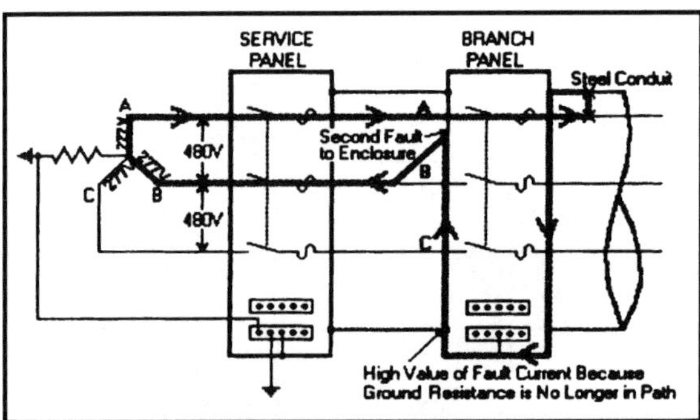

Figure 14 - Second fault in Resistance Grounded System - Fuses

The magnitude of this fault current can approach 87% of the L-L-L short-circuit current. Because of the possibility that a second fault will occur, single-pole ratings must be investigated. The IEEE "Red Book", Std 141-1993, page 367, supports this requirement, "One final consideration for resistance-grounded systems is the necessity to apply overcurrent devices based upon their "single-pole" short-circuit interrupting rating, which can be equal to or in some cases less than their 'normal rating'."

(continued)

3.2.3 *(Continued)*

How does an ungrounded system affect the requirements for single-pole interrupting ratings?
The Ungrounded Systems of Figures 15 and 16 offer the same advantage for continuity of service that are characteristic of high resistance grounded systems.

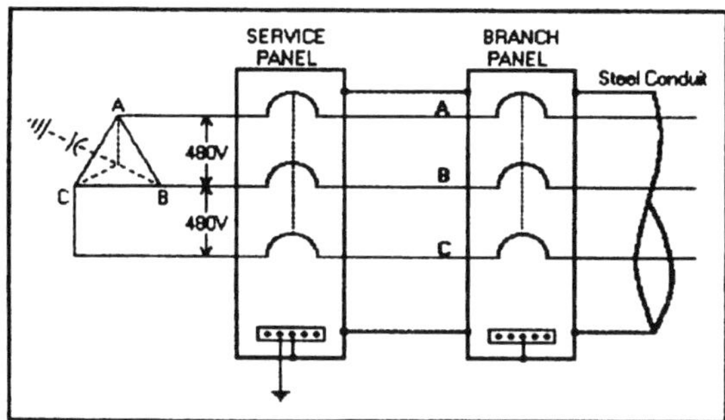

Figure 15 - Ungrounded System Circuit Breakers

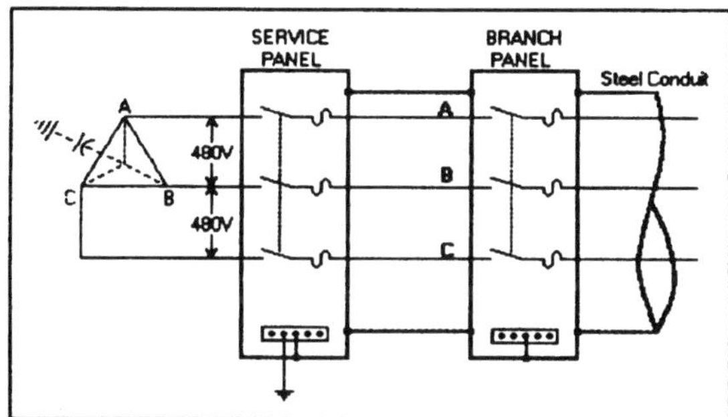

Figure 16 - Ungrounded System - Fuses

Although not physically connected, the phase conductors are capacitively coupled to ground. The first fault to ground is limited by the large impedance through which the current has to flow (Figures 17 and 18). Since the fault current is reduced to such a low level, the overcurrent devices do not open and the plant continues to "run".

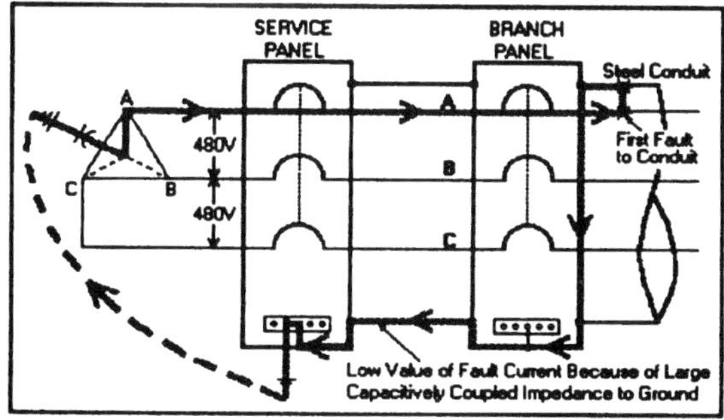

Figure 17 - First Fault to Conduit in Ungrounded System Circuit Breakers

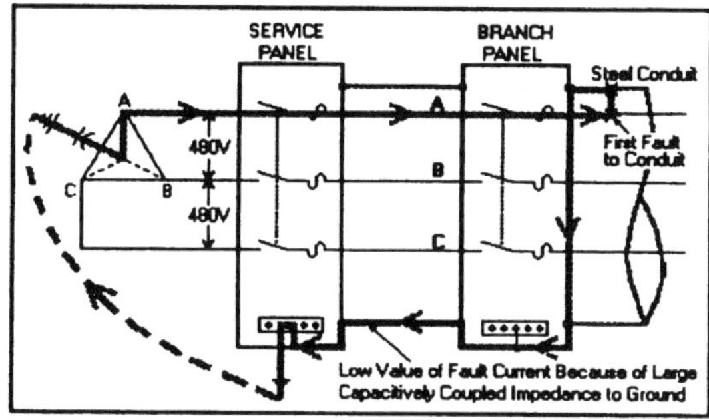

Figure 18 - First Fault to Conduit in Ungrounded System - Fuses

As with High Resistance Grounded Systems, ground detectors should warn the maintenance crew to find and fix the fault before a second fault from another phase also goes to ground (Figures 19 and 20).

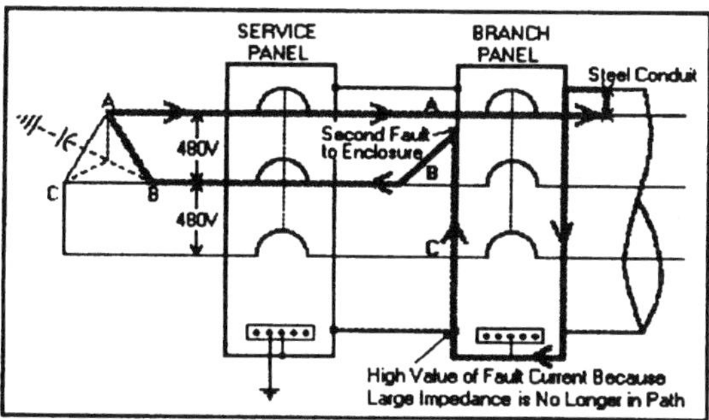

Figure 19 - Second Fault to Conduit in Ungrounded System - Circuit Breakers

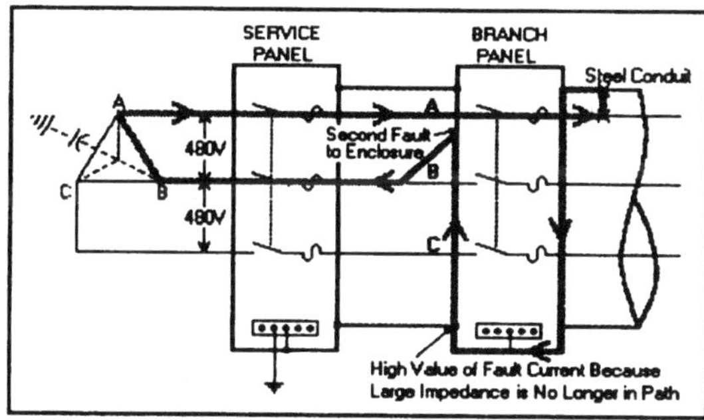

Figure 20 - Second Fault to Conduit in Ungrounded System - Fuses

The second fault from Phase B to ground (in Figures 19 and 20) will create a 480 volt fault across only one pole at the branch circuit overcurrent device. Again, the values from Table 1 must be used for molded case circuit breaker systems as the tradeoff for the increased continuity of service. Or, properly rated current limiting fuses and air frame/power circuit breakers can be utilized to meet the interrupting rating requirements. The IEEE "Red Book", Std 141-1993, page 366, supports this requirement, "One final consideration for ungrounded systems is the necessity to apply overcurrent devices based upon their "single-pole" short-circuit interrupting rating, which can be equal to or in some cases less than their normal rating."

A simple solution exists to insure adequate interrupting ratings both in present installations and in future upgrades. Modern current-limiting fuses are available that have tested single-pole interrupting ratings of 300,000 amps. Air frame/power circuit breakers are also available that have tested single-pole interrupting ratings that are 87% of the published three-pole rating.

(continued)

3.2.3 (*Continued*)

Does an overcurrent protective device with a high interrupting rating assure circuit component protection?

No. Choosing overcurrent protective devices strictly on the basis of voltage, current, and interrupting rating alone will not assure component protection from short-circuit currents. High interrupting capacity electro-mechanical overcurrent protective devices, (circuit breakers) especially those that are not current-limiting, may not be capable of protecting wire, cable, starters, or other components within the higher short-circuit ranges. See discussion of Sections 110-10 and 240-1 for the requirements that overcurrent protective devices must meet to protect components such as motor starters, contactors, relays, switches, conductors, and bus structures.

Note: Breaking current at other than fault levels.

The rating of contactors, motor starters, switches, circuit breakers and other devices for closing in and/or disconnecting loads at operating current levels must be sufficient for the current to be interrupted, including inrush currents of transformers, tungsten lamps, capacitors, etc. In addition to handling the full-load current of a motor, a switch and motor starter must also be capable of handling its locked rotor current. If the switch or motor starter has a horsepower rating at least as great as that of the motor, they will adequately disconnect even the locked rotor current of the motor.

It is necessary to calculate available short-circuit currents at various points in a system to determine whether the equipment meets the requirements of Sections 110-9 and 110-10. How does one calculate the values of short-circuit currents at various points throughout a distribution system?

There are a number of methods. Some give approximate values; some require extensive computations and are quite exacting. A simple, usually adequate method is the Buss® Point-To-Point procedure presented in Buss® bulletin SPD, Selecting Protective Devices. The point-to-point method is based on computation of the two main circuit impedance parameters: those of transformers and cables. Of these two components, the transformer is generally the major short-circuit current factor for faults near the service entrance. The percent impedance of the transformer can vary considerably. Thus, the transformer specification should always be checked. As shown in the illustration of a typical transformer nameplate, "%" impedance is specifically designated.

Given the full-load transformer secondary amperage and percent impedance of a transformer, how can you compute the level of short-circuit amperes that can be delivered at the secondary terminals (Assuming an infinite, unlimited, short-circuit current at the primary)?

$$I_{SCA} = (F.L.A.) \times \left[\frac{100}{\%Z \times .9^{\dagger\dagger}}\right]$$

Given: 1.3% impedance from nameplate of 500 KVA transformer with a 480V secondary
601 Full-Load Amperes (from Table below)

$$I_{SCA} = \frac{601 \times 100}{1.3 \times .9} = 51{,}368 \text{ Amperes}$$

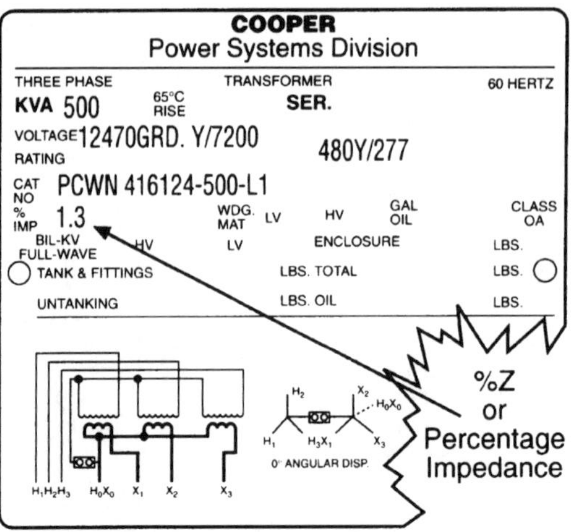

What are typical values of transformer short-circuit currents?

Short-Circuit Currents Available from Various Size Transformers

Voltage+ and Phase	KVA	Full-Load Amperes	% Impedance †† (Name plate)	† Short-Circuit Amperes
120/240 1 ph.*	25	104	1.58	11,574
	37½	156	1.56	17,351
	50	209	1.54	23,122
	75	313	1.6	32,637
	100	417	1.6	42,478
	167	695	1.8	60,255
120/208 3 ph.	150	416	1.07	43,198
	225	625	1.12	62,004
	300	833	1.11	83,383
	500	1388	1.24	124,373
	750	2082	3.5	66,095
	1000	2776	3.5	88,127
	1500	4164	3.5	132,190
	2000	5552	5.0	123,377
	2500	6950	5.0	154,444
277/480 3 ph.	112½	135	1.0	15,000
	150	181	1.2	16,759
	225	271	1.2	25,082
	300	361	1.2	33,426
	500	601	1.3	51,368
	750	902	3.5	28,410
	1000	1203	3.5	38,180
	1500	1804	3.5	57,261
	2000	2406	5.0	53,461
	2500	3007	5.0	66,822

† Three-phase short-circuit currents based on "infinite" primary.
* Single-phase values are L-N values at transformer terminals. These figures are based on change in turns ratio between primary and secondary, 100,000 KVA primary, zero feet from terminals of transformer, 1.2 (%X) and 1.5 (%R) multipliers for L-N vs. L-L reactance and resistance values, and transformer X/R ratio = 3.
†† U.L. listed transformers 25KVA or greater have a ±10% impedance tolerance. "Short-Circuit Amperes" reflect a worst case scenario.
\+ Fluctuations in system voltage will affect the available short-circuit current. For example, a 10% increase in system voltage will result in a 10% increase in the available short-circuit currents shown in the table.

3.2.4 *NEC* Section 110.10, Proper Protection of System Components from Short Circuits

3.2.4

What is the importance of Section 110-10?
The design of a system must be such that short-circuit currents cannot exceed the short-circuit current ratings of the components selected as part of the system. Given specific system components and level of "available" short-circuit currents which could occur, overcurrent protective devices (mainly fuses and/or circuit breakers) must be used which will limit the energy let-through of fault currents to levels within the short-circuit current ratings of the system components. (Current- limitation is treated under 240-11 of this bulletin). The last sentence of Section 110-10 emphasizes the requirement to thoroughly review the product standards and to apply components within the short-circuit current ratings in those standards.

What is component short-circuit current rating?
It is a current rating given to conductors, switches, circuit breakers and other electrical components, which, if exceeded by fault currents, will result in "extensive" damage to the component. The rating is expressed in terms of time intervals and/or current values. Short-circuit damage can be heat generated or the the result of electro-mechanical force of high-intensity, magnetic fields.

Conductor Protection

How is the component withstand rating of conductors expressed?
As shown in the table below, component withstand of conductors is expressed in terms of maximum short-circuit current vs. cycles (or time).

Table—Copper, 75° Thermoplastic Insulated Cable Damage Table* (Based on 60 HZ).

Copper Wire Size 75° Thermoplastic	Maximum Short-Circuit Withstand Current in Amperes For 1/2 Cycle**	For 1 Cycle	For 2 Cycles	For 3 Cycles**
#14	2,400	1,700**	1,200**	1,000
#12	3,800	2,700**	1,900**	1,550
#10	6,020	4,300	3,000	2,450
#8	9,600	6,800	4,800	3,900
#6	15,200	10,800	7,600	6,200
#4	24,200	17,100	12,100	9,900

Footnotes—*Reprinted from ICEA. **From ICEA formula

In this circuit, what is the maximum permissible available short-circuit current?

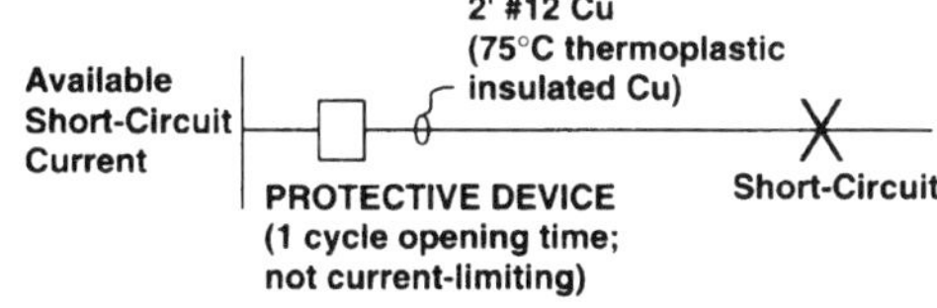

2700 amperes. Since the protective device is not current-limiting, the short-circuit current must not exceed the one cycle withstand of the #12 conductor, or 2700 amperes.

In this 20 ampere circuit with a non-current-limiting protective device, what would be the smallest size conductor that would have to be used?

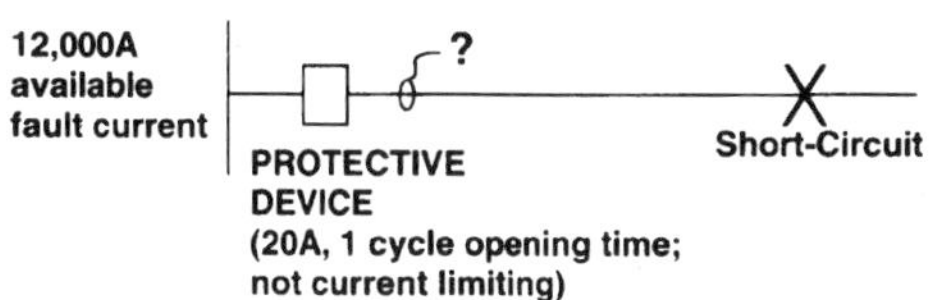

No. 4 wire. Since the protective device is not current-limiting, the wire selected must withstand 12,000 amperes for one cycle.

In this circuit, what type of protective device must be used?

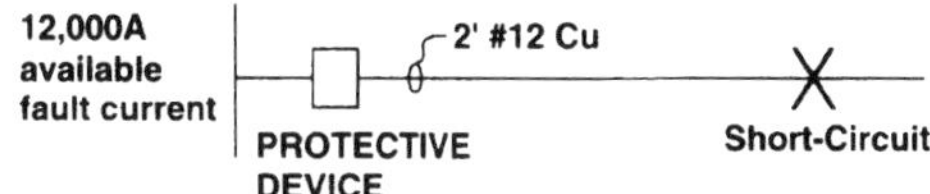

It must be current-limiting. When the available short-circuit current exceeds the short-circuit current rating of the wire, a protective device such as a current-limiting fuse, properly selected, will limit fault current to a level lower than the wire short-circuit current rating (3,800 amperes for 1/2 cycle). (See Section 240-1 FPN.) For instance, a LOW-PEAK® YELLOW™ LPN-RK20SP fuse will limit the 12,000 amperes available short-circuit to less than 1000 amperes and clear in less than 1/2 cycle.

Protection of Motor Controllers, Contacts and Relays

In this circuit, what kind of fuse must be used to provide adequate protection of the starter?

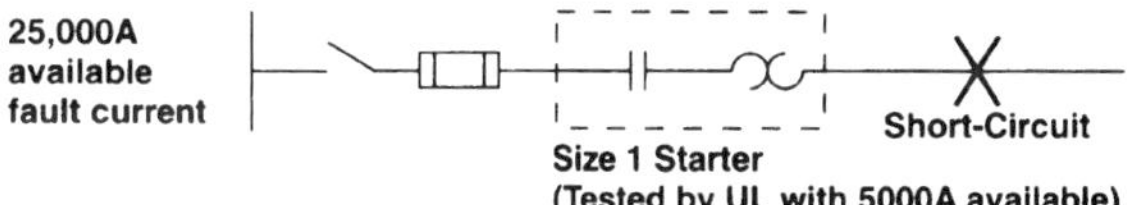

A current-limiting fuse, such as the Buss® LOW-PEAK® YELLOW™ or FUSETRON® dual-element fuse. Such a fuse must limit fault currents to a value below the withstand rating of the starter and clear the fault in less than 1/2 cycle.

(*continued*)

3.2.4 (*Continued*)

What is Type 2, motor starter protection?
UL 508E and IEC 947-4-1 have test procedures designed to verify that motor controllers will not be a safety hazard and will not cause a fire.

These standards offer guidance in evaluating the level of damage likely to occur during a short-circuit with various branch circuit protective devices. They address the coordination between the branch circuit protective device and the motor starter. They also provide a method to measure the performance of these devices should a short-circuit occur. They define two levels of protection (coordination) for the motor starter:

Type 1. Considerable damage to the contactor and overload relay is acceptable. Replacement of components or a completely new starter may be needed. There must be no discharge of parts beyond the enclosure.

Type 2. No damage is allowed to either the contactor or overload relay. Light contact welding is allowed, but must be easily separable.

Where Type 2 protection is desired, the controller manufacturer must verify that Type 2 protection can be achieved by using a specified protective device. Many U.S. manufacturers have both their NEMA and IEC motor controllers verified to meet the Type 2 requirements. Only current-limiting devices have been able to provide the current-limitation necessary to provide verified Type 2 protection. In many cases, Class J, Class RK1, or Class CC fuses are required, because most Class RK5 fuses and circuit breakers aren't fast enough under short-circuit conditions to provide Type 2 protection.

Type 2 protection is defined and suggested in the notes to Table 1 of NFPA 79 (Industrial Machinery).

Protection of Circuit Breakers

There are several key concepts about the protection of circuit breakers that need to be understood.

1. The user should be aware of the potential problems associated with series-rated circuit breakers. The engineer can not always "engineer" the installation as before because,
2. A molded case circuit breaker's interrupting capacity may be substantially less than its interrupting rating, and
3. Some molded case circuit breakers exhibit "dynamic" operation that begins in less than 1/2 cycle. This makes them more difficult to protect than other static electrical circuit components.

The most practical and reliable solution is to specify a fully-rated fusible system.

Molded Case Circuit Breakers—U.L. 489 and CSA5 Test Procedures
U.L. 489 requires a unique test set-up for testing circuit breaker interrupting ratings. Figure F illustrates a typical calibrated test circuit waveform for a 20 ampere, 240 volt, 2-pole molded case circuit breaker, with a marked interrupting rating of 22,000 amperes, RMS symmetrical.

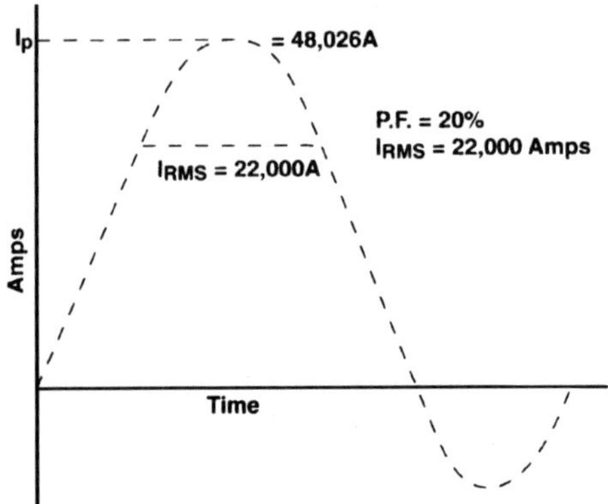

Figure F

Figure G illustrates the test circuit as allowed by U.L. 489.

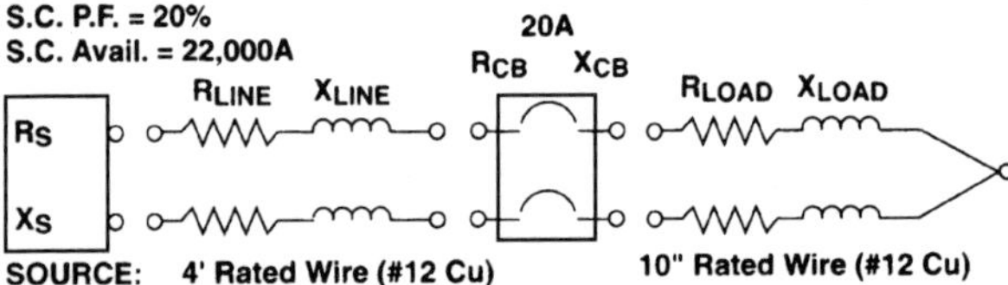

Note: For calculations, R_{CB} and X_{CB} are assumed negligible.

Figure G

Standard interrupting rating tests will allow for a maximum 4 ft. rated wire on the line side, and 10 in. rated wire on the load side of the circuit breaker. Performing a short-circuit analysis of this test circuit results in the following short-circuit parameters, as seen by the circuit breaker.

- Actual short-circuit RMS current = 9900 amperes RMS symmetrical
- Actual short-circuit power factor = 88%
- Actual short-circuit peak current = 14,001 amperes

(continued)

3.2.4 (*Continued*)

Following is an example of a partial table showing the actual I_P and I_{RMS} values to which the circuit breaker is tested.

240V–2-Pole MCCB INTERRUPTING CAPACITIES (KA)

CB	10KA		14KA		18KA		22KA	
RATING	I_p	I_{rms}	I_p	I_{rms}	I_p	I_{rms}	I_p	I_{rms}
15A	7.2	5.1	8.7	6.1	9.3	6.6	9.9	7.0
20A	8.9	6.3	11.4	8.1	12.6	8.9	14.0	9.9
25A	10.7	7.5	14.2	10.1	16.5	11.7	19.9	13.5
30A	10.7	7.5	14.2	10.1	16.5	11.7	19.9	13.5
40A	11.7	8.3	16.0	11.3	19.2	13.6	22.7	16.1
50A	11.7	8.3	16.0	11.3	19.2	13.6	22.7	16.1
60A	12.5	8.8	17.3	12.2	21.3	15.1	25.6	18.1
70A	13.0	9.2	18.1	12.8	22.6	16.0	27.4	19.4
80A	13.0	9.2	18.1	12.8	22.6	16.0	27.4	19.4
90A	13.2	9.3	18.3	12.9	23.0	16.3	27.9	19.7
100A	13.2	9.3	18.3	12.9	23.0	16.3	27.9	19.7

These values are known as the circuit breaker's interrupting capacities.

What about the "bus shot" tests? Won't those prove that circuit breakers can safely and properly interrupt their marked interrupting rating?
No! Beginning 10/31/2000, UL 489 required circuit breakers rated 100A and less to additionally be tested under "bus bar conditions." In this test, line and load terminals will be connected to 10" of rated conductor. For single pole circuit breakers, these 10" leads are then connected to 4' of #1 for connection to the test station. For multipole circuit breakers, the 10" line side leads are connected to the test station through 4' of #1. The load side is shorted by 10' leads of rated conductor per pole. These "bus shots" still do not fully address the situation where a fault can occur less than 4'10" from the circuit breaker. For example, 7.1.11.6.3.1 of UL 489 states "The inability to relatch, reclose, or otherwise reestablish continuity ... shall be considered acceptable for circuit breakers which are tested under "bus bar conditions." This says the circuit breaker doesn't have to work after a close-in fault occurs, and is in violation of the 1999 NEC requirement for a circuit breaker which is found in the definition. The NEC defines a circuit breaker as "A device designed to open and close a circuit by nonautomatic means and to open the circuit automatically on a predetermined overcurrent without damage to itself when properly applied within its rating."

Protection of Bus Structures

In the circuit below, what must be the busway short-circuit bracing?

100,000 amperes, because the overcurrent device is not current-limiting.

In this circuit, what would the busway short-circuit bracing have to be?

36,000 amperes (as shown in the Minimum Bracing Table). With an available short-circuit current of 100,000 amperes, the LOW-PEAK® YELLOW™ KRP-C1600SP fuse will only let-through an equivalent of 36,000 amperes, RMS symmetrical.

Minimum Bracing Required for Bus Structures at 480V. (Amperes RMS Symmetrical)

Rating*						
Busway	Fuse	Available Short-Circuit Amperes RMS Sym.				
		25,000	50,000	75,000	100,000	200,000
100	100	3,400	4,200	4,800	5,200	6,500
225	225	6,000	7,000	8,000	9,000	12,000
400	400	9,200	11,00	13,000	14,000	17,000
600	600	12,000	15,000	17,000	19,000	24,000
601	601	11,000	14,500	17,000	18,000	24,000
800	800	14,200	17,500	20,000	23,000	29,000
1200	1200	16,000	22,500	26,000	28,000	39,000
1600	1600	22,500	28,500	33,000	36,000	46,000
2000	2000	25,000	32,000	37,000	40,000	52,000
3000	3000	25,000	43,000	50,000	58,000	73,000
4000	4000	25,000	48,000	58,000	68,000	94,000

*Fuses are: 100-600 Ampere—LOW-PEAK® YELLOW™ Dual-Element Fuses—LPS-RK_SP (Class RK1) or LPJ_SP (Class J); 800-4000 Ampere—LOW-PEAK® YELLOW™ Time-Delay Fuses—KRP-C_SP (Class L). (LOW-PEAK® YELLOW™ fuses are current-limiting fuses.)

3.2.5 *NEC* Section 110.22, Proper Marking and Identification of Disconnecting Means

3.2.5

What does this new Section require?

Labeling Considerations
N.E.C. Sections 110-22 and 240-83 require special marking for a testing agency listed series-rated systems.

On listed series-rated systems, the downstream equipment will be marked by the manufacturer per applicable standards [240-86(a)]. The N.E.C. requires that the main or upstream protective device be marked with a field installed label per N.E.C. Section 110-22. This is the responsibility of the electrical contractor.

Short-circuit calculations **must be performed** at panel locations where series-rated systems are specified.

3.3.1 *NEC* Section 210.20(A), Ratings of Overcurrent Devices on Branch Circuits Serving Continuous and Noncontinuous Loads

3.3.1

What is the importance of this Section?
The overcurrent protective device provided for branch circuits, must not be less than the total non-continuous load, plus 125% of the continuous load (defined as a load that continues for 3 hours or more).

Rating not less than = [(10A) x 1.0] + [(8A) x 1.25]
= 20A

EXAMPLE

20A Rating

Non-Continuous 10A

Continuous Load 8A

The branch circuit rating shall not be less than 20 amperes.

3.4.1 *NEC* Section 215.10, Requirements for Ground-Fault Protection of Equipment on Feeders

3.4.1

What is the importance of this Section?

Equipment classified as a feeder disconnect, as shown in these examples, must have ground fault protection as specified in Section 230-95.

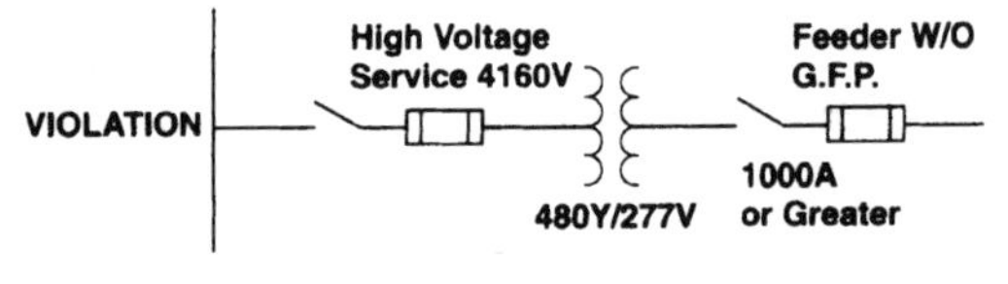

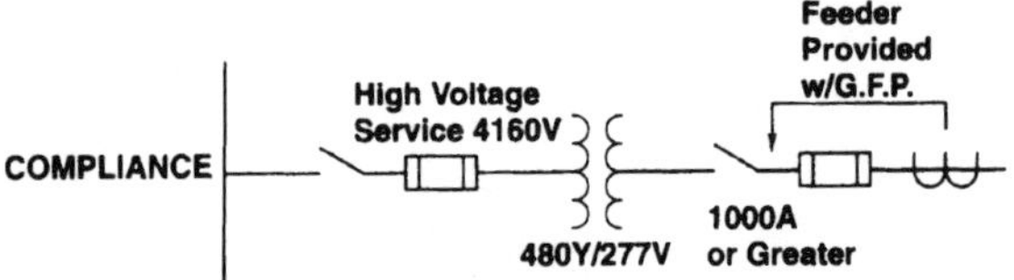

G.F.P. is not required on feeder equipment when it is provided on the supply side of the feeder (except for certain Health Care Facilities requirements, Article 517).

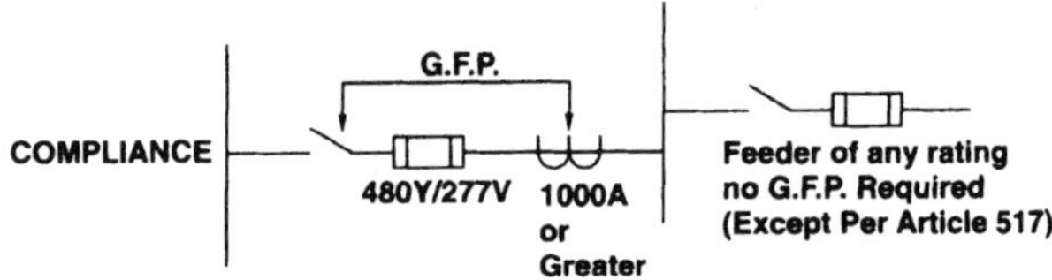

Additionally, the requirements of this section do not apply to fire pumps or to a continuous industrial process where a nonorderly shutdown will introduce additional or increased hazards.

See Section 230-95 for an in-depth discussion of Ground Fault Protection.

Ground fault protection without current-limitation may not protect system components. See Section 110-10.

3.5.1 *NEC* Section 230.82, Equipment Allowed to Be Connected on the Line Side of the Service Disconnect

3.5.1

What are the advantages of using cable limiters on the supply side of the service disconnect.
Typical cable installations are shown in the illustration below. The benefits of cable limiters are several:
1. The isolation of a faulted cable permits the convenient scheduling of repair service.
2. Continuity of service is sustained even though one or more cables are faulted.
3. The possibility of severe equipment damage or burn down as a result of a fault is greatly reduced. (Typically, without cable limiters the circuit from the transformer to the service equipment is afforded little or no protection.)
4. Their current-limiting feature can be used to provide protection against high short-circuit currents for utility meters and provide compliance with Section 110-10.

COMMERCIAL/INDUSTRIAL SERVICE ENTRANCE
(Multiple cables per phase)

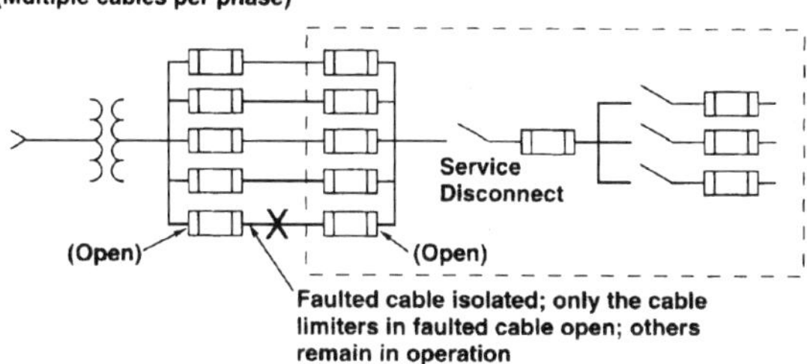

RESIDENTIAL SERVICE ENTRANCE
(Single cable per phase)

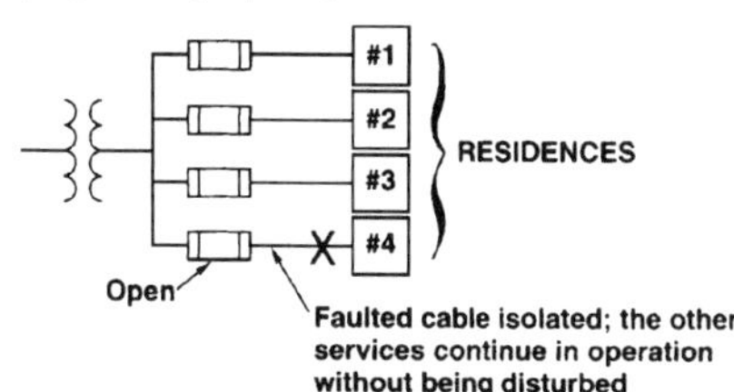

What do (6) and (7) mean?
The control circuit for power operable service disconnecting means and ground fault protection must have a means for disconnection and adequate overcurrent protection–interrupting rating and component protection.

3.5.2 *NEC* Section 230.95, Ground-Fault Protection for Services

3.5.2

What is the importance of this section?
This section means that 480Y/277 volt, solidly grounded "wye" only connected service disconnects, 1000 amperes and larger, must have ground fault protection in addition to conventional overcurrent protection. Ground fault protection, however, is not required on a fire pump or a service disconnect for a continuous process where its opening will increase hazards. All delta connected services are not required to have ground fault protection. The maximum setting for the ground fault relay (or sensor) must be set to pick up ground-faults which are 1200 amperes or more and actuate the main switch or circuit breaker to disconnect all phase conductors. A ground fault relay with a deliberate time delay characteristic of up to 3000 amperes for 1 second can be used. (The use of such a relay greatly enhances system coordination and minimizes power outages).

Under short-circuit conditions, unlike current-limiting fuses, ground fault protection in itself **will not** limit the line-to-ground or phase-to-phase short-circuit current. When mechanical protective devices such as conventional circuit breakers are used with G.F.P., all of the available short-circuit current will flow to the point of fault limited only by circuit impedance. Therefore, it is recommended that current-limiting overcurrent protective devices be used in conjunction with G.F.P. relays.

In this circuit, what protection does the fuse provide in addition to that provided by the ground fault equipment?

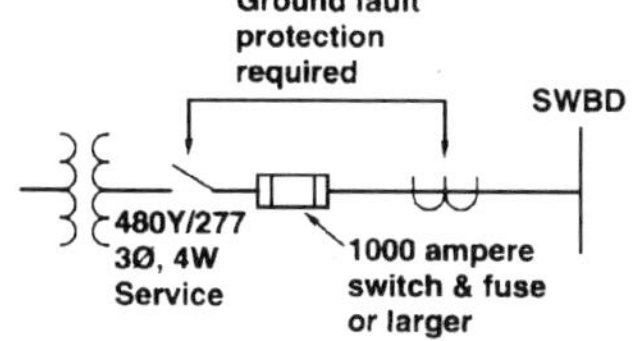

Current limitation under short-circuit conditions and high-level ground-faults.

In this circuit, is protection provided against high magnitude ground-faults as well as low level faults?

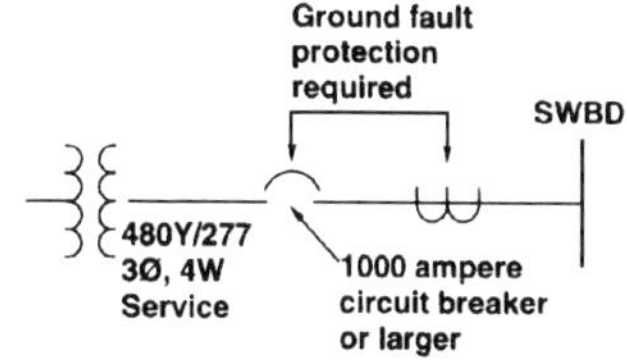

No, it is not. There is no current-limitation.

Is G.F.P. required on all services?
No. The following do not require G.F.P.:
1. Continuous industrial process where non-orderly shutdown would increase hazard.
2. All services where disconnect is less than 1000 amperes.
3. All 120/208 volts, 3Ø, 4W (wye) services.
4. All single-phase services including 120/240 volt, 1Ø, 3W.
5. High or medium voltage services. (See N.E.C. Sections 240-13 and 215-10 for equipment and feeder requirements.)
6. All services on delta systems (grounded or ungrounded) such as: 240 volt, 3Ø, 3W Delta, 480 volt, 3Ø, 3W Delta, or 240 volt, 3Ø, 4W Delta with midpoint tap.
7. Service with 6 disconnects or less (Section 230-71) where each disconnect is less than 1000 amperes. A 4000 ampere service could be split into five 800 ampere switches.
8. Resistance or impedance grounded systems.

(continued)

3.5.2 *(Continued)*

What are some of the problems associated with G.F.P.?
Incorrect settings, false tripping and, eventually, disconnection. (The knocking-out of the total building service or large feeders as a result of minor faults or nuisance tripping cannot be tolerated in many facilities). Unnecessary plant down time is often more critical, or even more dangerous, than a minor ground fault.

Note: G.F.P. without current limitation may not protect system components. See Section 110-10 and 250-2(d).

How can ground faults be minimized?
1. To prevent blackouts, make sure that all overcurrent protective devices throughout the overall system are selectively coordinated. When maximum continuity of electrical service is necessary, ground fault protective equipment should be incorporated in feeders and branch circuits. [Per Section 230-95 (FPN No. 2).]
2. Insulating bus structures can greatly minimize the possibility of faults. The hazard of personnel exposure to energized electrical equipment is also reduced with insulated bus structures.
3. Specify switchboards and other equipment with adequate clearance between phase conductors and ground. Ground faults are rare on 120/208 volt systems because equipment manufacturers provide ample spacing for this voltage. Insist on greater spacing for 277/480 volt equipment and the likelihood of ground faults will be greatly reduced.
4. Avoid unusually large services; split the service whenever possible.
5. Adequately bond all metallic parts of the system to enhance ground fault current flow. Then, if a ground fault does occur, it is more likely to be sensed by fuses or circuit breakers.

To respond properly to a line-to-ground type fault, what should be the setting of a ground fault relay located on the main disconnect?
The setting should allow the feeder circuit (or preferably the branch) overcurrent protective devices to function without disturbing the G.F.P. relay.

How is a G.F.P. setting determined?
By making a coordination study. Such a study requires the plotting of the time-current curves of the protective devices.

A simple solution to the problem of coordinating ground fault relays with overcurrent protective devices is shown in the system represented in the graph at right. The G.F.P. relay coordinates with the feeder fuses KTS-R 250. The G.F.P. relay with a degree of inverse time characteristics provides coordination with feeder fuses in order to avoid outages. (Section 230-95 permits an inverse time-delay relay with a delay of up to 1 second at 3000 amperes.)

Conventional mechanical tripping overcurrent protective devices often do not permit a selectively coordinated system* and BLACKOUTS can occur. For ground faults (and short-circuit current as well) of current magnitude above the instantaneous trip setting on the main circuit breaker's overcurrent element, the main will nuisance trip (open) causing a blackout even though the fault is on a feeder or branch circuit. Appropriate selection of current-limiting fuses with proper G.F.P. settings can provide the highest degree of coordination and prevent blackouts.

* A system wherein only the protective device nearest the fault operates and none of the other protective devices in the system are disturbed.

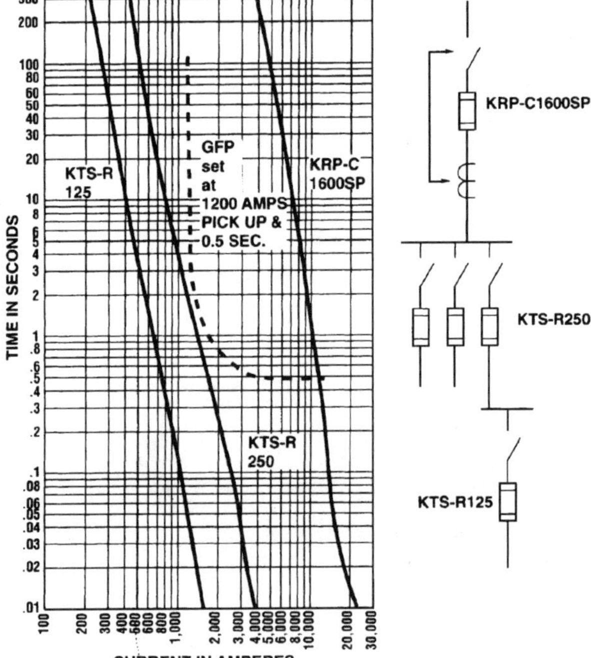

3.6.1 *NEC* Section 240.1, Scope of Article 240 on Overcurrent Protection

3.6.1

What is the importance of this Section?
The basic purpose of overcurrent protection is to open a circuit before conductors or conductor insulation are damaged when an overcurrent condition exists. An overcurrent condition can be the result of an overload or a short-circuit. It must be removed before the damage point of conductor insulation is reached. Conductor insulation damage points can be established from available engineering information, i.e., Publication P-32-382, "Short-Circuit Characteristics of Cable", ICEA, (Insulated Cable Engineers Association, Inc.), IEEE Color Books, Canadian Electrical Code, and IEC Wiring Regulations.

When selecting an overcurrent protective device to protect a conductor, is it adequate to simply match the ampere rating of the device to the ampacity of the conductor?
No. Although conductors do have maximum allowable ampacity ratings, they also have maximum allowable short-circuit current withstand rating. Damage ranging from slight degradation of insulation to violent vaporization of the conductor metal can result if the short-circuit withstand is exceeded. (See Section 110-10.)

Why, in the circuit below, is the #10 wire protected even though the available short-circuit current exceeds the wire withstand? The #10 conductor can withstand 4300 amperes for one cycle and 6020 amperes for one-half cycle.**

**Footnote—From ICEA tables and formula.

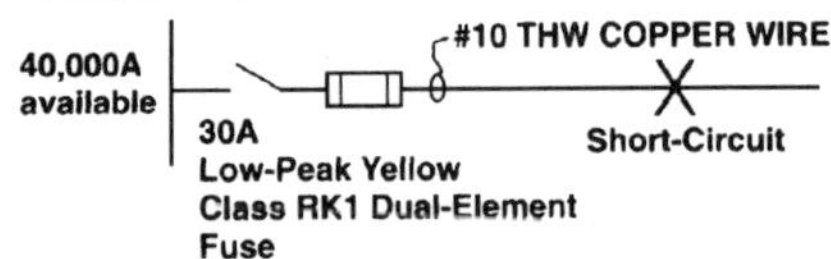

Under short-circuits, the LOW-PEAK® YELLOW™ Dual-Element fuse (30 ampere) is fast acting. It will clear and limit (cut off) short-circuit current before it can build up to a level higher than the wire withstand. The opening time of the fuse is less than one-half cycle (less than 0.008 seconds). In this particular example, the prospective current let-thru by the fuse is less than 1850 amperes. Thus, opening time and current let-through of the fuse is far lower

(continued)

3.6.1 *(Continued)*

than the wire withstand. (Conductor protection is not a problem when the conductor is protected by current-limiting fuses which have an ampere rating that is the same as the conductor. In the case of short-circuit protection only, fuses can often be sized many times higher than the wire current rating, depending upon the current-limiting characteristics of the fuse.)

Does the circuit below represent a misapplication? (#10 THW insulated copper wire can withstand 4300 amperes for one cycle and 6020 amperes for one-half cycle).

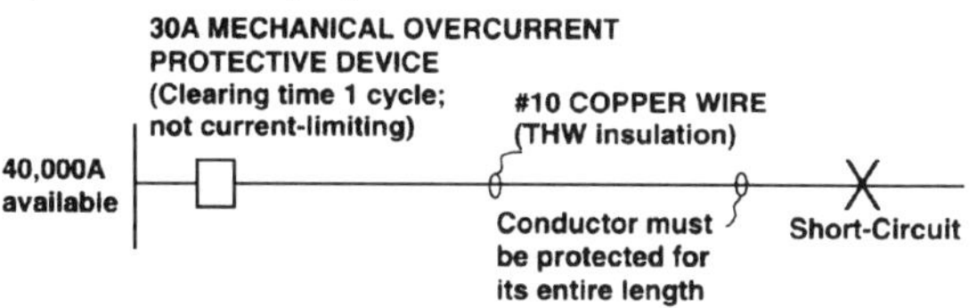

Yes. The 40,000 ampere short-circuit current far exceeds the withstand of the #10 THW wire. Note the table and chart which follow.

What can be done to correct the above misapplication?
There are two possible solutions:
1. Use a larger size conductor (i.e., 1/0), one with a withstand greater than the short-circuit for one cycle (see chart below).
2. Use an overcurrent protective device which is current-limiting such as that shown in the previous question.

The following table is based on Insulated Cable Engineers Association, Inc. (ICEA) insulated cable damage charts in Publication 32-382. This table assumes that the conductor is preloaded to its ampacity before a short-circuit is incurred. The formula that was used to develop the ICEA Damage Charts is given following the table. This formula can be used to extrapolate withstand data for wire sizes or time durations not furnished in the ICEA Publication 32-382 charts. A sample chart is shown at right.

The mechanical overcurrent protective device opening time and any impedance (choking) effect should be known along with the available short-circuit current and cable withstand data to determine the proper conductor that must be used.

Insulated Cable Damage Table (60Hz)†

Wire Size (THW Cu)	Maximum Short-Circuit Withstand Current Amperes) at Various Withstand Times			
	1 Cycle	1/2 Cycle	1/4 Cycle	1/8 Cycle
#14	1,700*	2,400*	3,400*	4,800*
#12	2,700*	3,800*	5,400*	7,600*
#10	4,300	6,020*	8,500*	12,000*
#8	6,800	9,600*	13,500*	19,200*
#6	10,800	15,200*	21,500*	30,400*
#4	17,100	24,200*	34,200*	48,400*

† See Insulated Cable Engineers Association, Inc., "Short-Circuit Characteristics of Cable", Pub. P-32-382, and circuit breaker manufactures' published opening times for various types of circuit breakers.

Copper—Thermoplastic Conductor Insulation

$$\left[\frac{I}{A}\right]^2 t = 0.0297 \log \left[\frac{T_2 + 234}{T_1 + 234}\right]$$

Aluminum—Thermoplastic Conductor Insulation

$$\left[\frac{I}{A}\right]^2 t = 0.0125 \log \left[\frac{T_2 + 228}{T_1 + 228}\right]$$

Where:
I = Short-Circuit Current—Amperes
A = Conductor Area—Circular Mils
t = Time of Short-Circuit—Seconds
T_1 = Maximum Operating Temperature—75°C
T_2 = Maximum Short-Circuit Temperature—150°C

Note: ICEA (Insulated Cable Engineers Association) is the most widely accepted authority on conductor short-circuit withstand ratings.

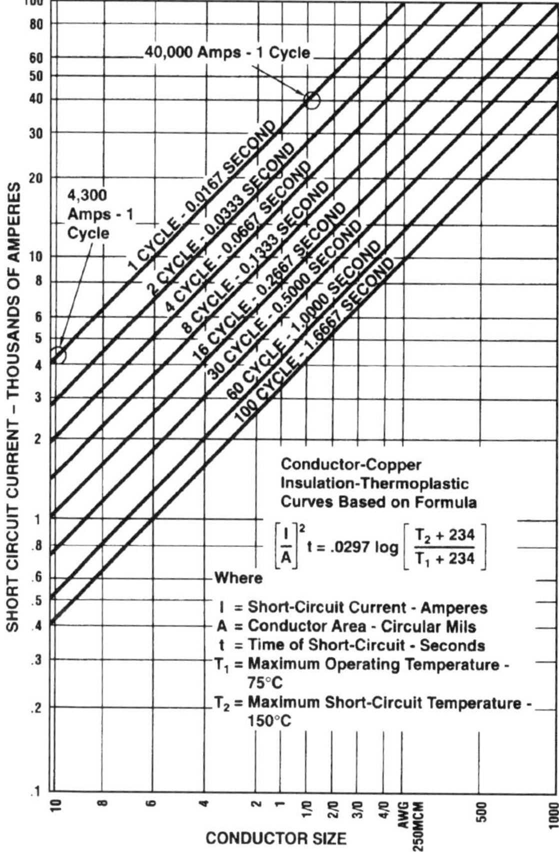

3.6.2 *NEC* Section 240.3, Protection of Conductors Other than Flexible Cords and Fixture Wires

3.6.2

What is the meaning of 240-3(b) and 240-3(c)?
Where the ampacity of a conductor does not correspond with a standard rating (240-6) of a fuse, the next standard rating may be used as long as the fuse is not above 800 amps and the conductors are not part of a multi-outlet branch circuit supplying receptacles for cord and plug-connected portable loads.

What does 240-3(f) mean?
Conductors fed from single-phase, 2-wire secondary transformers and three phase, delta-delta connected transformers with three-wire (single-voltage) secondaries can be considered protected by the primary side fuses if the transformer is properly protected in accordance with Section 450-3. The primary fuse must be less than or equal to the secondary conductor ampacity times the secondary-to-primary transformer voltage ratio.

What is the definition of a "tap" conductor?
A tap conductor is defined in 240-3(e) as a conductor, other than a service conductor that has overcurrent protection ahead of its point of supply, that exceeds the value permitted for similar conductors that are protected as described elsewhere in this section.

3.6.3 *NEC* Section 240.4, Proper Protection of Fixture Wires and Flexible Cords

3.6.3

What is the importance of this section?
Flexible cords and extension cords shall have overcurrent protection rated at their ampacities. Supplementary fuse protection is an acceptable method of protection. For #18 fixture wire 50 feet or over, a 6 ampere fuse would provide necessary protection, and for #16 100 feet or over, an 8 ampere fuse would provide the necessary protection. #18 extension cords must be protected by a 7 ampere fuse.

Also, Section 760-23, covering special non-power-limited fire alarm circuits, requires 7 ampere protection for #18 conductors and 10 ampere protection for #16 conductors.

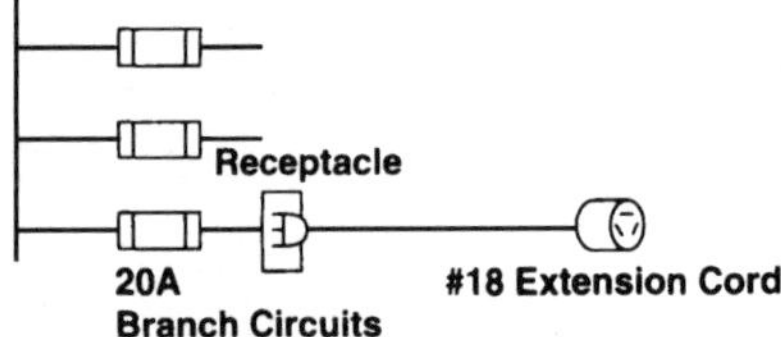

Compliance
(EXTENSION CORD)

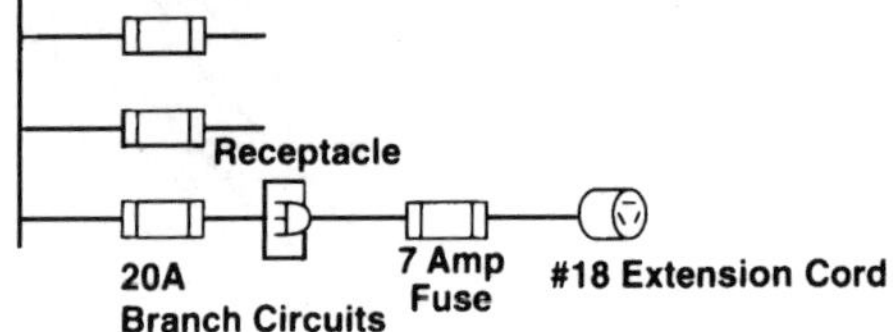

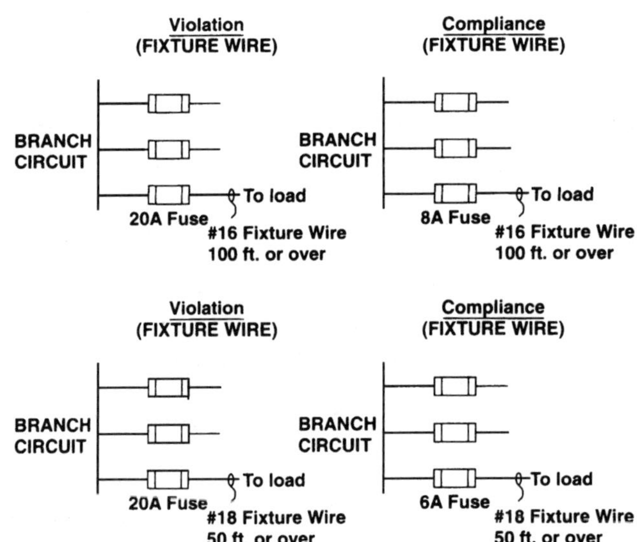

3.6.4 *NEC* Section 240.6, Standard Ampere Ratings

3.6.4

What is the importance of this section?
In addition to the standard ratings of fuses and circuit breakers, this section states that the rating of an adjustable trip circuit breaker is considered to be the highest possible setting. This becomes important when protecting conductors or motor circuits. For example, if a copper 75°C conductor is required to carry 200 amperes continuously, a 250 kcmil cable might be chosen. If a circuit breaker were chosen to protect this cable with an external adjustable trip from 225 through 400 amperes, the rating of the breaker would be 400 amperes, and 500 kcmil cable would therefore be required, increasing costs significantly. However, if this adjusting means is "restricted", such as behind a bolted equipment door, behind locked doors accessible only by a qualified person, or if a removable and sealable cover is over the adjusting means, then the rating can be considered to be equal to the adjusted setting.

Note: Standard ampere ratings for fuses and inverse time circuit breakers are 15, 20, 25, 30, 35, 40, 45, 50, 60, 70, 80, 90, 100, 110, 125, 150, 175, 200, 225, 250, 300, 350, 400, 450, 500, 600, 700, 800, 1000, 1200, 1600, 2000, 2500, 3000, 4000, 5000 and 6000 amperes. In addition, standard fuse ratings are 1, 3, 6, 10 and 601. The use of fuses and inverse time circuit breakers with non-standard ampere ratings shall be permitted.

3.6.5 *NEC* Sections 240.8 and 380.17, Protective Devices Used in Parallel

3.6.5

What do these sections mean?
There are cases in which an original equipment manufacturer, for various reasons, must parallel fuses and receive an appropriate equipment listing. For example, this would be the case of some solid-state power conversion equipment. However, for the standard safety switch, conventional branch circuit applications, switch-boards, and panelboards, the use of parallel fuses is not allowed.

3.6.6 *NEC* Section 240.9, Thermal Devices

3.6.6

What does this section mean?
Thermal overload devices generally can neither withstand opening a circuit under short-circuit conditions nor even carry short-circuit currents of higher magnitudes. When using thermal overload protective devices, the use of a current-limiting fuse will not only provide short circuit protection for the circuit but for the thermal overload device as well.

3.6.7 *NEC* Section 240.10, Requirements for Supplementary Overcurrent Protection

3.6.7

What is the importance of this section?
Supplementary fuses, often used to provide protection for lighting fixtures, cannot be used where branch circuit protection is required.

What are the advantages of supplementary protection?
The use of supplementary protection for many types of appliances, fixtures, cords, decorator lighting (Christmas tree lights. . .)*, etc., is often well advised. There are several advantages:

1. Provides superior protection of the individual equipment by permitting close fuse sizing.
2. With an occurrence of an overcurrent, the equipment protected by the supplementary protected device is isolated; the branch circuit overcurrent device is not disturbed. For instance, the in-line-fuse and holder combination, such as the Type HLR fuseholder with Type GLR or GMF fuses, protects and isolates fluorescent lighting fixtures in the event of an overcurrent.
3. It is easier to locate equipment in which a malfunction has occurred. Also, direct access to the fuse of the equipment is possible.

*Footnote–Supplementary protection for series connected decorator lighting sets and parallel sets (Christmas tree string lights) was required in 1982. Manufacturers have implemented this requirement.

3.6.8 *NEC* Section 240.11, Definition of Current-Limiting Overcurrent Protective Devices

(This article is not a part of the 2002 NEC, but is included for reference from the 1999 NEC.)

3.6.8

What is the importance of this Section?

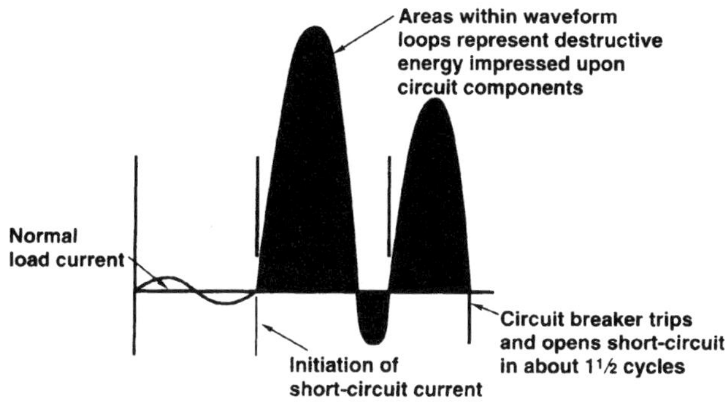

ACTION OF NON-CURRENT-LIMITING CIRCUIT BREAKER

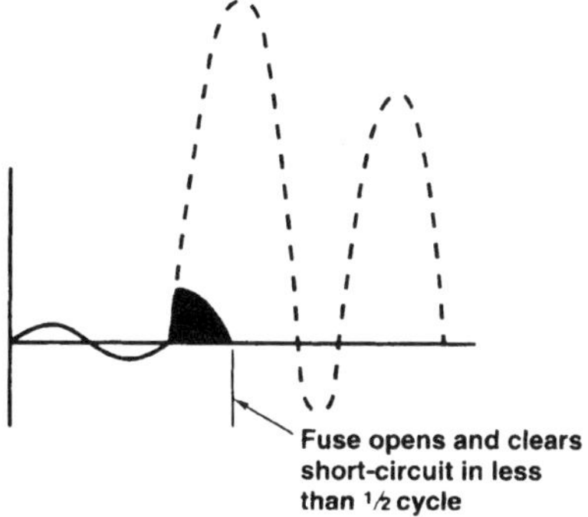

ACTION OF CURRENT-LIMITING FUSE.

Simply stated, a current-limiting protective device is one which cuts off a fault current in less than one-half cycle†. It thus prevents short-circuit currents from building up to their full available values.

The greatest damage done to components by a fault current occurs in the first half-cycle (or more precisely, "the first major loop" of the sinewave). Heating of components to very high temperatures can cause deterioration of insulation, or even explosion. Tremendous magnetic forces between conductors can crack insulators and loosen or rupture bracing structures.

The levels of both thermal energy and magnetic forces are proportionate to the square of current. Thermal energy is proportionate to the square of "RMS" current; maximum magnetic fields to the square of "peak" current. If a fault current is 100 times higher than normal current, its increased heating effects equals $(100)^2$ or 10,000 times higher than that of the normal current. Thus, to prevent circuit component damage, the use of current-limiting protective devices is extremely important, particularly since present-day distribution systems are capable of delivering high level fault currents.

† Footnote: The more technical definition of a current-limiting protective device is expressed by 240-11.

To further appreciate current-limitation, assume for example, that the available prospective short-circuit current in a circuit is 50,000 amperes. If a 200 ampere LOW-PEAK® YELLOW™ fuse is used to protect the circuit, the current let-through by the fuse will be only 6500 amperes instead of 50,000 amperes. Peak current will be only 15,000 amperes instead of a possible 115,000 amperes. Thus, in this particular example, currents are limited to only 13% of the available short-circuit values.

As is true of fuse application in general, the application of current-limiting fuses in respect to current-limitation and component protection (110-10) is quite simple. Graphs or tables such as the one shown below permit easy determination of the "let-thru" currents that a fuse will pass for various levels of prospective short-circuit currents. For example, the table below shows that the 200 ampere LOW-PEAK® YELLOW™ fuse will let-through 6500 amperes when prospective short-circuit current is 50,000 amperes.

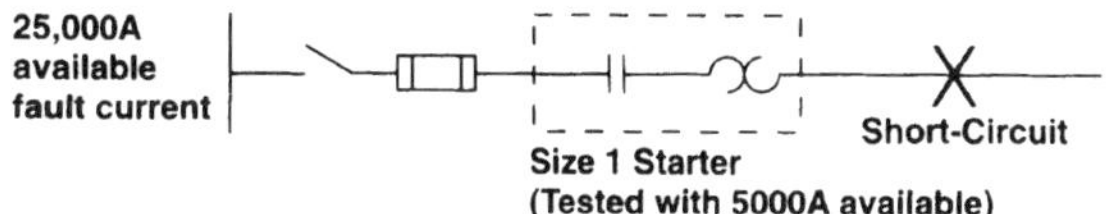

For the above circuit, the Size 1 Starter has a short-circuit withstand rating of 5000 amperes.* The question is, with the 25,000 ampere available short-circuit current, will a LOW-PEAK® YELLOW™ fuse provide adequate protection of the starter? By referring to the table below, it can easily be seen that for a prospective short-circuit current of 25,000 amperes, fuses with ratings of 100 amperes or less will limit fault currents to below the 5000 ampere withstand of the starter and, thus, provide adequate protection.

Current-Limiting Effects of RK1 LOW-PEAK® YELLOW™ Fuses.

Prospective Short-Circuit Current	Let-Through Current (Apparent RMS Symmetrical) LPS-RK_SP (600V) Fuse Ratings					
	30A	60A	100A	200A	400A	600A
5,000	980	1,600	2,100	3,200	5,000	5,000
10,000	1,200	2,000	2,550	4,000	6,750	9,150
15,000	1,400	2,300	2,900	4,800	7,850	10,200
20,000	1,500	2,500	3,150	5,200	8,250	11,300
25,000	1,600	2,650	3,400	5,450	9,150	12,200
30,000	1,650	2,850	3,550	5,650	9,550	12,800
35,000	1,750	2,950	3,750	5,850	10,000	13,500
40,000	1,850	3,100	3,900	6,100	10,450	13,900
50,000	1,950	3,300	4,150	6,500	11,300	15,000
60,000	2,050	3,500	4,350	6,950	11,950	16,100
80,000	2,250	3,850	4,800	7,850	13,000	17,400
100,000	2,450	4,050	5,200	8,250	13,900	18,700
150,000	2,750	4,800	6,100	9,550	15,900	21,300
200,000	3,000	5,200	6,500	10,000	17,400	23,500

RMS Symmetrical Amperes

*Footnote: See discussion on Section 110-10 in this bulletin.

The reader should note that much of the current-limitation claimed by small ampere circuits breakers is actually the result of the significant impedance added to the circuit breaker test circuit after the circuit has been calibrated. Refer to the circuit breaker protection portion of Section 110-10 for further information on circuit breaker test circuits.

3.6.9 *NEC* Section 240.12, System Coordination or Selectivity

3.6.9

What is the importance of this section?
Whenever a partial or total building blackout could cause hazard(s) to personnel or equipment, the fuses and/or circuit breakers must be coordinated in the short-circuit range. It is acceptable for a monitoring system to be used to indicate an overload condition, if the overcurrent protective devices cannot be coordinated in the overload region. However, in the vast majority of cases, both circuit breakers and fuses will be able to be coordinated in the overload range, so the monitoring systems will seldom be required. Typical installations where selective coordination would be required include hospitals, industrial plants, office buildings, schools, government buildings, military installations, high-rise buildings, or any installation where continuity of service is essential.*

*Footnote: See also Section 4-5.1 of NFPA 110 (Emergency and Standby Power Systems) and Sections 3-3.2.1.2(4) & 3-4.1.1.1 of NFPA 99 (Health Care Facilities) for additional information on selective coordination.

VIOLATION

1000A I.T.=10x Opens

225A I.T.=8x Opens

20A I.T.=8x Opens

22,000 Amp Short-Circuit

Fault exceeding the instantaneous trip setting of all 3 circuit breakers in series will open all 3. This will blackout the entire system.

COMPLIANCE

1000A Not Open

225A Not Open

20A Opens

22,000 Amp Short-Circuit

Fault opens the nearest upstream fuse, localizing the fault to the equipment affected. Service to the rest of the system remains energized.

If the ampere rating of a feeder overcurrent device is larger than the rating of the branch circuit device, are the two selectively coordinated?
No. A difference in rating does not in itself assure coordination. For example, a feeder circuit breaker may have a rating of 400 amperes and the branch breaker 90 amperes. Under overload conditions in the branch circuit, the 90 ampere breaker will open before, and without, the 400 ampere breaker opening. However, under short-circuit conditions, not only will the 90 ampere device open, the 400 ampere may also open. In order to determine whether the two devices will coordinate, it is necessary to plot their time-current curves as shown. For a short-circuit of 4000 amperes:

1. The 90 ampere breaker will unlatch (Point A) and free the breaker mechanism to start the actual opening process.
2. The 400 ampere breaker will unlatch (Point B) and it, too, would begin the opening process. Once a breaker unlatches, it will open. The process at the unlatching point is irreversible.
3. At Point C, the contacts of the 90 ampere breaker finally open and interrupt the fault current.
4. At Point D, the contacts of the 400 ampere breaker open. . .the entire feeder is "blacked out"!

Example of Non-Selective System.

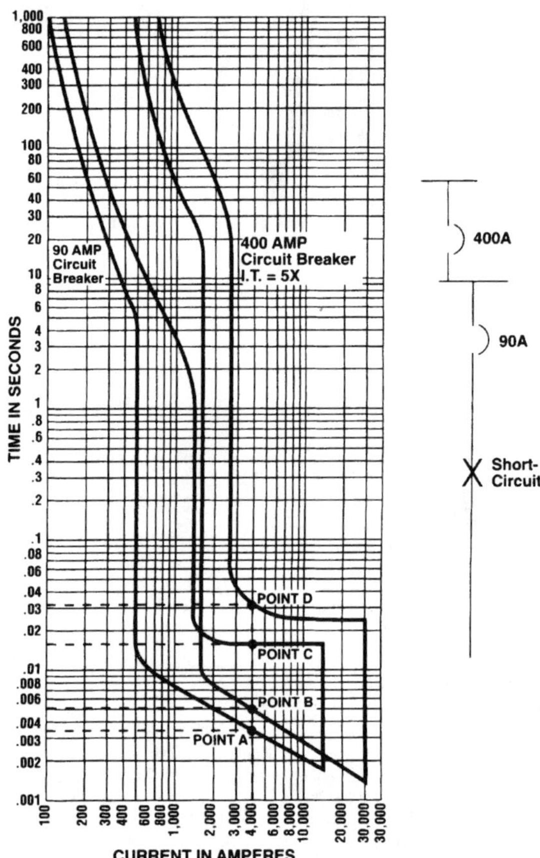

Now, let's take the case of fuse coordination. When selective coordination of current-limiting fuses is desired, the Selectivity Ratio Guide (next page) provides the sizing information necessary. In other words, it is not necessary to draw and compare curves. Current-limiting fuses can be selectively coordinated by maintaining at least a minimum ampere rating ratio between the main fuse and feeder fuses and between the feeder fuse and branch circuit fuses.

These ratios are based on the fact that the smaller downstream fuses will clear the overcurrent before the larger upstream fuses melt. An example of ratios of fuse ampere ratings which provide selective coordination is shown in the one-line circuit diagram.

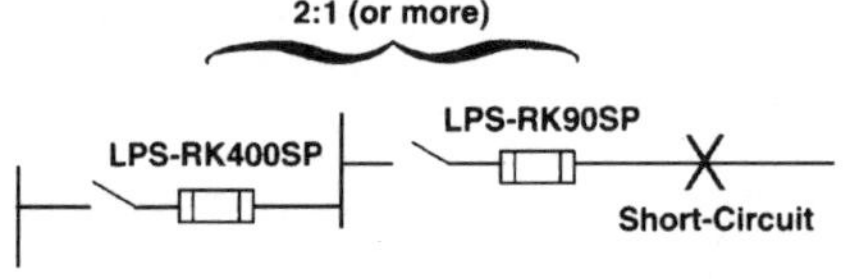

(continued)

3.6.9 (*Continued*)

***Selectivity Ratio Guide (Line-Side to Load-Side) for Blackout Prevention.**

Circuit				Load-Side Fuse									
Current Rating				601-6000A	601-4000A	0-600A			601-6000A	0-600A	0-1200A	0-600A	0-60A
Type				Time-Delay	Time-Delay	Dual-Element Time-Delay			Fast-Acting	Fast-Acting			Time-Delay
Trade Name &				LOW-PEAK® YELLOW™	LIMITRON®	LOW-PEAK® YELLOW™		FUSETRON®	LIMITRON®	LIMITRON®	T-TRON®	LIMITRON®	SC
Class				(L)	(L)	(RK1)	(J)**	(RK5)	(L)	(RK1)	(T)	(J)	(G)
			Buss Symbol	**KRP-C_SP**	**KLU**	**LPN-RK_SP LPS-RK_SP**	**LPJ_SP**	**FRN-R FRS-R**	**KTU**	**KTN-R KTS-R**	**JJN JJS**	**JKS**	**SC**
Line-Side Fuse: 601 to 6000A	Time-Delay	LOW-PEAK® YELLOW™ (L)	**KRP-C_SP**	2:1	2.5:1	2:1	2:1	4:1	2:1	2:1	2:1	2:1	N/A
601 to 4000A	Time-Delay	LIMITRON® (L)	**KLU**	2:1	2:1	2:1	2:1	4:1	2:1	2:1	2:1	2:1	N/A
0 to 600A	Dual Element	LOW-PEAK® YELLOW™ (RK1)	**LPN-RK_SP LPS-RK_SP**	–	–	2:1	2:1	8:1	–	3:1	3:1	3:1	4:1
		(J)	**LPJ_SP****	–	–	2:1	2:1	8:1	–	3:1	3:1	3:1	4:1
		FUSETRON® (RK5)	**FRN-R FRS-R**	–	–	1.5:1	1.5:1	2:1	–	1.5:1	1.5:1	1.5:1	1.5:1
601 to 6000A		LIMITRON® (L)	**KTU**	2:1	2.5:1	2:1	2:1	6:1	2:1	2:1	2:1	2:1	N/A
0 to 600A	Fast-Acting	LIMITRON® (RK1)	**KTN-R KTS-R**	–	–	3:1	3:1	8:1	–	3:1	3:1	3:1	4:1
0 to 1200A		T-TRON® (T)	**JJN JJS**	–	–	3:1	3:1	8:1	–	3:1	3:1	3:1	4:1
0 to 600A		LIMITRON® (J)	**JKS**	–	–	2:1	2:1	8:1	–	3:1	3:1	3:1	4:1
0 to 60A	Time-Delay	SC (G)	**SC**	–	–	3:1	3:1	4:1	–	2:1	2:1	2:1	2:1

* **Note:** At some values of fault current, specified ratios may be lowered to permit closer fuse sizing. Plot fuse curves or consult with Cooper Bussmann.
General Notes: Ratios given in this table apply only to Buss® fuses. When fuses are within the same case size, consult Cooper Bussmann.
** Consult Cooper Bussmann for latest LPJ__SP ratios.

3.6.10 *NEC* Section 240.13, Ground-Fault Protection of Equipment on Buildings or Remote Structures

3.6.10

What does this section require?
Equipment ground fault protection of the type required in Section 230-95 is required for each disconnect rated 1000 amperes or more in 480Y/277V systems that will serve as a main disconnect for a separate building or structure. Refer to Sections 215-10 and 230-95.

Note: G.F.P. that is not current-limiting may not protect system components. See Section 110-10 and 250-1 (FPN).

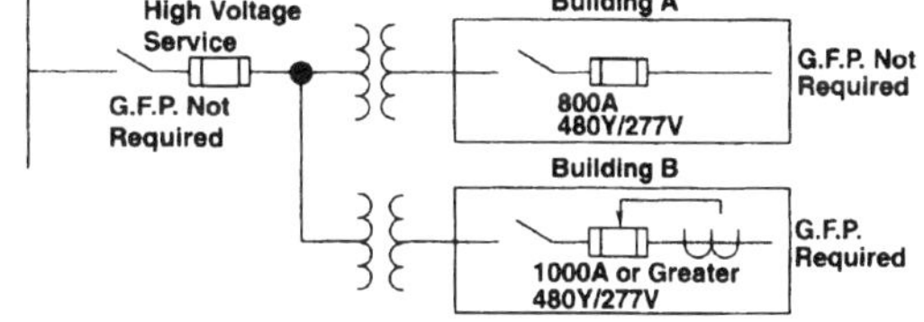

3.6.11 *NEC* Section 240.21, Location Requirements for Overcurrent Devices and Tap Conductors

3.6.11

What are the requirements of 240-21(b)(1)?
The basic content of this section remains unchanged. However, it has been rewritten to improve readability and comprehension. Typically, fuses must be installed at point where the conductor receives its supply, i.e., at the beginning or line side of a branch circuit or feeder. There are installations where this basic rule may not have to be followed.

Fuses are not required at the conductor supply if a feeder tap conductor is not over ten feet long; is enclosed in raceway; does not extend beyond the switchboard, panelboard or control device which it supplies; and has an ampacity not less than the combined computed loads supplied, and not less than the rating of the device supplied, unless the tap conductors are terminated in a fuse not exceeding the tap conductor's ampacity [240-21(b)]. For field installed taps, where the tap conductors leave the enclosure, the ampacity of the tap conductor must be at least 10% of the overcurrent device rating. See the following example.

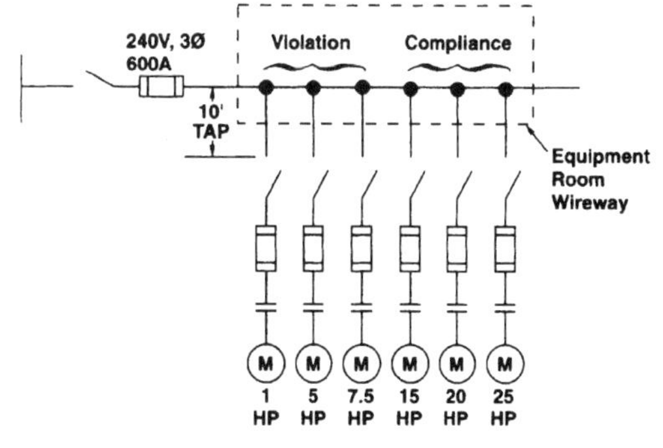

(continued)

3.6.11 (*Continued*)

In the previous diagram, the feeder overcurrent devices are sized per the N.E.C for the load served.

All taps to the motors are 10 foot taps.
Three of the motors are smaller motors:
one motor is a 1 HP motor,
one motor is a 5 HP motor,
and one motor is a 7½ HP motor.

The 1, 5 and 7½ HP motors will require a minimum of #14, #12 and #10 75°C conductors, respectively. For field wiring, these 10 foot taps are not permitted since the line side overcurrent device is 600 amperes. Section 240-21(b)(1)(d) requires that the maximum overcurrent protection for field installations shall not exceed 1000%, or 10 times the ampacity of the tap conductor, for example:

#14 conductor, 20 amperes ampacity, maximum line side overcurrent protection is 200 amperes.
#12 conductor, 25 amperes ampacity, maximum line side overcurrent protection is 250 amperes.
#10 conductor, 35 amperes ampacity, maximum line side overcurrent protection is 350 amperes.

To tap the above conductors to a 600 amperes feeder overcurrent device would be a violation of Section 240-21(b)(1)(d) of the Code.

The solution is to feed the smaller motors from a branch circuit panel or from a smaller feeder where the feeder overcurrent protection does not exceed the 10 times rating of the tap conductor's ampacity.

The smallest of the three larger motors is a 15 HP motor which requires a branch circuit conductor with a minimum ampacity of 52.5 amperes and which could be tapped to the 600 ampere feeder since a No. 6 75° conductor has an ampacity of 65 amperes and 10 x 65 = 650. In other words the No. 6 75° conductors could be tapped to an overcurrent device as high as 650 amperes.

Motor tap conductors that have a 60 ampere ampacity or greater could be tapped to a 600 ampere feeder overcurrent protective device.

What are the requirements of 240-21(b)(2)?

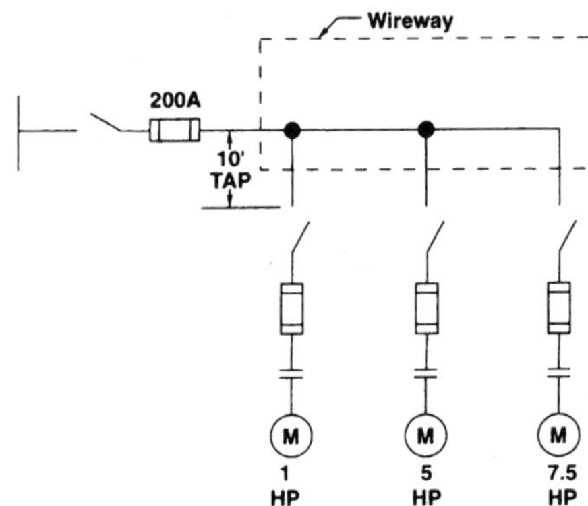

Fuses are not required at the conductor supply if a feeder tap conductor is not over 25 feet long, is suitably protected from physical damage; has an ampacity not less than ⅓ that of the feeder conductors or fuses from which the tap conductors receive their supply; and terminate in a single set of fuses sized not more than the tap conductor ampacity. See "Note".

What are the requirements of 240-21(b)(3)?

Fuses are not required at the conductor supply where the conductor feeds a transformer and the primary plus secondary is not over 25 ft. long and where all of the following conditions are met. (Any portion of the primary conductor that is protected at its ampacity is not included in the 25 feet)

(1) The primary conductor ampacity must be at least 1/3 of the rating of the fuse protecting the feeder

(2) The secondary conductor ampacity when multiplied by the secondary to primary voltage ratio must be at least 1/3 of the rating of the fuse protecting the feeder

(3) The primary and secondary conductors must be protected from physical damage

(4) The secondary conductors terminate in a single set of fuses that will limit the load to the ampacity of the secondary conductors.

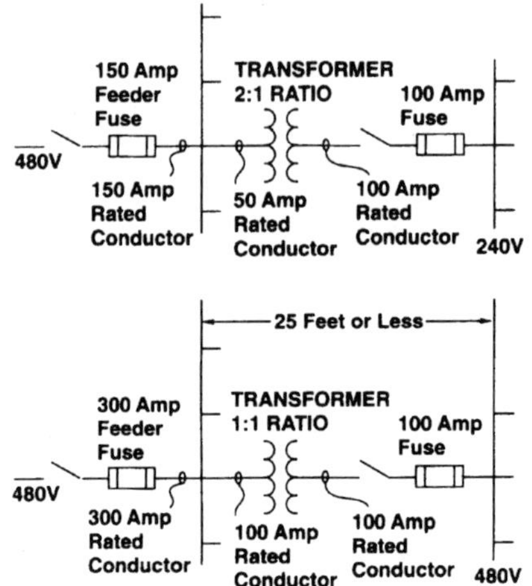

What are the requirements of 240-21(b)(4)?

Fuses are not required at the conductor supply if a feeder tap is not over 25 feet long horizontally and not over 100 feet long, total length, in high bay manufacturing buildings where only qualified persons will service such a system. Also, the ampacity of the tap conductors is not less than ⅓ of the fuse rating from which they are supplied, the size of the tap conductors must be at least No. 6 AWG copper or No. 4 AWG aluminum. They may not penetrate walls, floors, or ceilings, may not be spliced, and the taps are made no less than 30 feet from the floor.

What are the requirements of 240-21(b)(5)?

Fuses are not required at the supply for an outside tap of unlimited length where all of the following are met:

(1) The conductors are outdoors except at the point of termination
(2) The conductors are protected from physical damage
(3) The conductors terminate in a single set of fuses that limit the load to the ampacity of the conductors.
(4) The fuses are a part of or immediately adjacent to the disconnecting means.
(5) The disconnecting means is readily accessible and is installed outside or inside nearest the point of entrance.

Note: Smaller conductors tapped to larger conductors can be a serious hazard. If not adequately protected against short-circuit conditions (as required in sections 110-10 and 240-1), these unprotected conductors can vaporize or incur severe insulation damage. Molten metal and ionized gas created by a vaporized conductor can envelop other conductors (such as bare bus), causing equipment burndown. Adequate short-circuit protection is recommended for all conductors. When a tap is made to a switchboard bus for an adjacent panel, such as an emergency panel, the use of BUSS® Cable Limiters is recommended as supplementary protection of the tapped conductor. These current-limiting cable limiters are available in sizes designed for short-circuit protection of conductors from #12 to 1000 kcmil. BUSS® Cable Limiters are available in a variety of terminations to make adaption to bus structures or conductors relatively simple.

(*continued*)

3.6.11 (*Continued*)

What are the requirements of 240-21(c)(1)?
Fuses are not required on the secondary of a single phase 2-wire or three phase, three wire, delta-delta transformer to provide conductor protecting where all of the following are met:
(1) The transformer is protected in accordance with 450-3
(2) The overcurrent protective device on the primary of the transformer does not exceed the ampacity of the secondary conductor multiplied by the secondary to primary voltage ratio.

Note: Refer to 384-16 for panelboard protection requirements.

What are the requirements of 240-21(c)(2)?
Fuses are not required on the secondary of a transformer to provide conductor protection where all of the following are met:
(1) The secondary conductors are not over 10 ft long.
(2) The secondary conductor ampacity is not less than the combined computed loads
(3) The secondary conductor ampacity is not less than the rating of the device they supply or the rating of the overcurrent device at their termination
(4) The secondary conductors do not extend beyond the enclosure(s) of the equipment they supply and they are enclosed in a raceway

Note: Refer to 450-3 for transformer protection and 384-16 for panelboard protection requirements

What are the requirements of 240-21(c)(3)?
Transformer secondary conductors of separately derived systems do not require fuses at the transformer terminals when all of the following conditions are met.

1. Must be an industrial location.
2. Secondary conductors must be less than 25 feet long.
3. Secondary conductor ampacity must at least equal to the secondary full load current of transformer and sum of terminating, grouped, overcurrent devices.
4. Secondary conductors must be protected from physical damage.

Note: Switchboard and panelboard protection (384-16) and transformer protection (450-3) must still be observed.

What are the requirements of 240-21(c)(4)?
Fuses are not required on the secondary of a transformer to provide conductor protection where all of the following are met:
(1) The secondary conductors are located outdoors except at the point of termination and are protected from physical damage
(2) The secondary conductors terminate in one overcurrent device that limits the load to the ampacity of the conductors
(3) The overcurrent device is a part of or immediately adjacent to the disconnecting means.
(4) The disconnecting means for the conductors is readily accessible and outside or inside nearest the point of entrance of the conductors.

Note: Refer to 450-3 for transformer protection and 384-16 for panelboard protection requirements

3.6.12 *NEC* Section 240.40, Disconnecting Means for Fuses

3.6.12

What does the section require?
A line side disconnecting means must be provided for all cartridge fuses where accessible to other than qualified persons and for any fuse in circuits over 150 volts to ground. This section does not require a disconnecting means for the typical 120/240V single phase residential plug fuse application.

There is no requirement for a disconnecting means ahead of a current-limiting cable limiter or other current-limiting fuse ahead of the service disconnecting means.

One disconnect is allowed for multiple sets of fuses as provided in 430.112 for group motor applications and 424.22(C) for fixed electric space-heating equipment.

3.6.13 *NEC* Section 240.50, Plug Fuses, Fuseholders, and Adapters

3.6.13

What does this section mean?
Normally, plug fuses are applied in 125 volt circuits for appliances, small motors, machines, etc. They may be used on 240/120 volts single-phase circuits, and 208/120 volt three-phase circuits, where the neutral is solidly grounded.

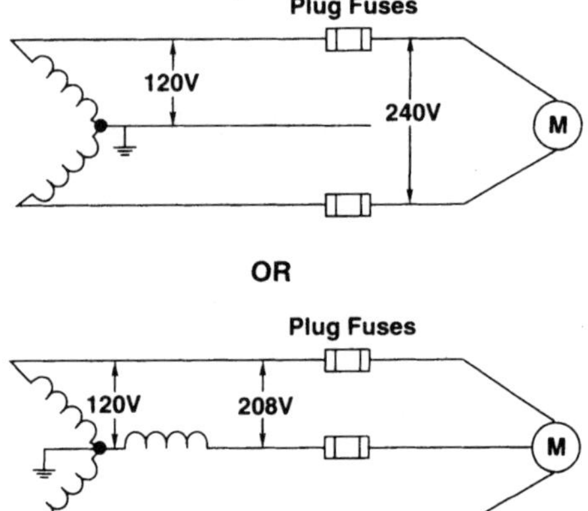

3.6.14 *NEC* Section 240.51, Edison-Base Fuses

3.6.14

What are these fuse types?
These are generally referred to as branch circuit listed fuses which are NOT size rejecting. They can provide protection for appliances and small motors in residential, commercial, and industrial applications. For branch circuit protection, these fuses may only be used for replacements in existing installations where there is no evidence of overfusing or tampering.

Edison-base fuses can be used for supplementary overcurrent protection in new installations.

3.6.15 *NEC* Section 240.53, Type S Fuses

3.6.15

What are these fuse types?
These are branch circuit listed fuses that are size (ampere) rejecting. They become size rejecting when a special Type S holder or Type S adapter is used. For example, when a 20 ampere adaptor is installed, it is very difficult to insert a 25 or 30 ampere fuses.

Type S fuses are required for new installation where plug fuses are to be used as the branch circuit protection.

3.6.16 *NEC* Section 240.54, Type S Fuses, Adapters, and Fuseholders

3.6.16

What are the advantages of Type S Fuses?
Type S fuses are size rejecting to prevent overfusing. They are used with special adapters that cannot easily be removed.

3.6.17 *NEC* Section 240.60, Cartridge Fuses and Fuseholders

3.6.17

What does this section mean?
300 volt rated fuses can be used to protect single-phase line-neutral loads when supplied from three-phase solidly grounded 480/277 volt circuits, where the single-phase line-to-neutral voltage is 277 volts.

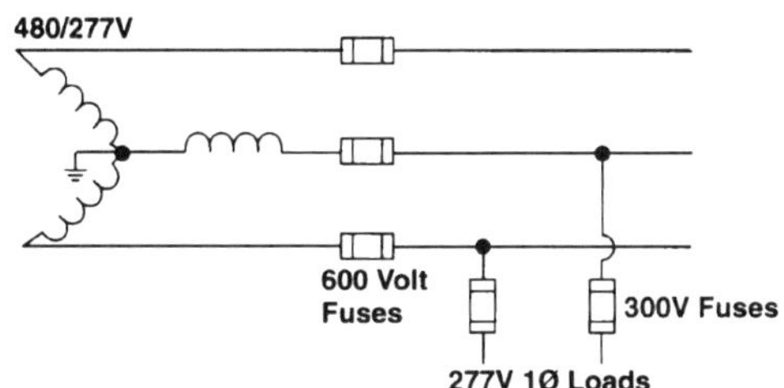

Branch circuit listed fuses are designed so that it is very difficult to replace an installed fuse with one of lesser capability. This is based on voltage, current, or current-limiting vs. non-current-limiting ratings.

The interrupting rating must be marked on all branch circuit fuses with interrupting ratings other than 10,000 amperes.

3.6.18 *NEC* Section 240.61, Classification of Fuses and Fuseholders

3.6.18

What does this section mean?
All low voltage branch circuit fuses have a voltage rating associated with them. They can be properly applied at system voltages up to that rating.

3.6.19 *NEC* Section 240.86, Series Ratings

3.6.19

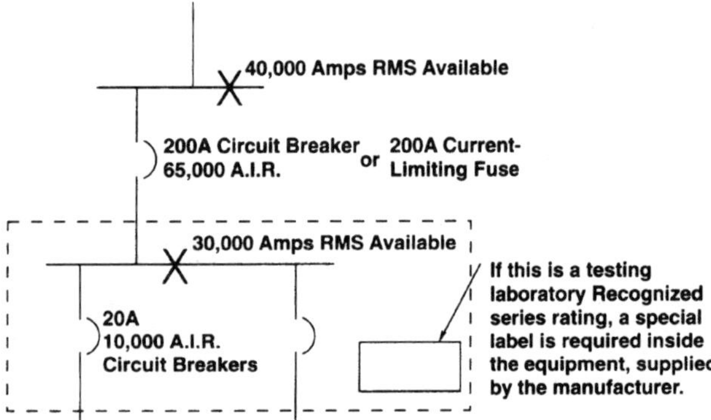

Note: Refer to Section 110-22 For marking requirements for the main or upstream protective device.

Can a series-rated system recognized by a testing laboratory be used in a facility with motors on the load side of the line side overcurrent device? (Assume an available short circuit current of 20,000 amperes)

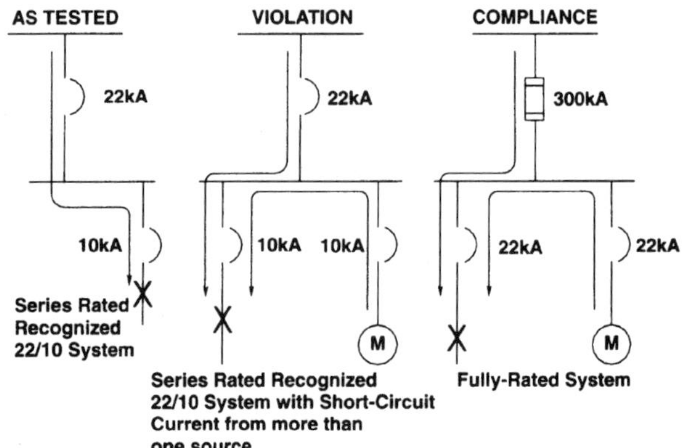

What does the section require?
Special marking requirements exist for series-rated systems which are tested and recognized. (One source is the U.L. Yellow Books.) This special marking on the equipment must state that the specific circuit breakers in the equipment have been series tested with special upstream devices. Any substitution of series-rated circuit breakers with a non-series-rated device will void the recognition and create a potentially dangerous situation. This labeling is supplied by the manufacturer.

The requirement of (b) is meant to assure that series rated systems are installed in the same way that they are tested. They are not tested with the infusion of current (motor contribution) between the series rated devices. Therefor, if the motor load exceeds 1% of the interrupting rating of the downstream breaker, the series rating cannot be used. In the series rated tests, all of the current seen by the downstream device is also seen by the upstream device. If there are motors fed by several downstream devices, that motor contribution (under a fault condition) will not be seen by the upstream device. Since the series rating is a fine tuned "combination" of action by both the upstream and downstream devices, any significant amount of current which is not seen by the upstream device would throw off the fine tuning.

Note: Refer to Section 110-22 for marking requirements for the main or upstream protective device.

Maybe, it could be a violation of 110-3b and 240-86(b), since the series-rated system is tested only with a short-circuit source on the line side of the main circuit breaker. Should motors be used on the load side of the main circuit breaker, the main circuit breaker will not "see" this contribution, but the branch circuit breaker will, thus changing the opening characteristics of the combination, and violating the recognition. Section 240-86(b) allows for a connected motor load of up to 1% of the interrupting rating of the load side circuit breaker.

The recommended solution would be to specify a fully-rated system, with all devices meeting the requirements of Section 110-9.

3.6.20 *NEC* Section 240.90, Supervised Industrial Installations

3.6.20

What is the intent of Sections 240-90 and 240-91?
This is the new Part H. The special provisions of this Part apply only to the process and manufacturing portions of an industrial installation.

The intent of Part VIII is to limit its use to large industrial locations. The maintenance crew must be qualified and under engineering supervision. Total load must be 2500 KVA or greater as calculated in accordance with Article 220. And, there must be at least one service at 277/480 or 480 volts or higher.

3.6.21 *NEC* Section 240.92(B), Transformer Secondary Conductors of Separately Derived Systems (Supervised Industrial Installations Only)

3.6.21

What does this section mean?
Conductors may be connected directly to the secondary terminals of a transformer of a separately derived system, without overcurrent protection at the connection if the conductors meet special requirements for short-circuit, overload, and physical protection.

3.6.22 *NEC* Section 240.92(B)(1), Short-Circuit and Ground-Fault Protection (Supervised Industrial Installations Only)

3.6.22

What are the requirements of 240-92(b)(1)?
The cable can be 50' or less as long as the primary overcurrent device is no larger than 150% of the ampacity of the secondary conductor multiplied by the secondary to primary voltage ratio.

The cable can be 75' or less as long as the conductors are protected by a differential relay with a trip setting not greater than the secondary conductor ampacity.

The cable can be 75' or less as long as it is determined that it will be protected under short-circuit conditions by engineering calculations. Typical methods are found in IEEE Color Books, Canadial Electrical Code, IEC Wiring Regulations, ICEA (Insulated Cable Engineers Association), and manufacturers' literature.

3.6.23 *NEC* Section 240.92(B)(2), Overload Protection (Supervised Industrial Installations Only)

3.6.23

What are the requirements of 240-92(b)(2)?
Overload protection can be achieved by terminating in one overcurrent device, or in up to six overcurrent devices, grouped in one location, that add up to no more than the ampacity of the conductor. Engineering calculations can be used to demonstrate overload protection. Finally, relays can be used to limit the load to the ampacity by opening devices on the line or load side.

3.6.24 *NEC* Section 240.92(C), Outside Feeder Taps (Supervised Industrial Installations Only)

3.6.24

What are the requirements for Outside Feeder Taps?
Outdoor conductors may be tapped to a feeder or connected to the secondary of a transformer without overcurrent protection at the tap or connection if all 5 of the following conditions are met.

1) The conductors must be protected from physical damage.
2) The sum of the one to six grouped overcurrent devices at the termination of the outdoor conductor must limit the load to the ampacity of the conductor.
3) The conductors are outside except at the point of termination.
4) The overcurrent device must be a part of the disconnecting means or immediately adjacent to it.
5) The disconnecting means are readily accessible and located outside or inside nearest the point of entrance.

3.6.25 *NEC* Section 240.100, Feeder and Branch-Circuit Protection over 600 V Nominal

3.6.25

What are the requirements of this section?
Part I has been rewritten to include the overcurrent protection requirements that were previously in Article 710.

This major change now specifies the location of the overcurrent protective device for circuits of over 600 volts. It requires that protection be provided at the beginning of the feeder or branch-circuit unless another location is determined under engineering supervision. Previous editions of the Code did not specify the location of the protective device.

3.6.26 *NEC* Section 240.100(C), Conductor Protection

3.6.26

What is the meaning of 240-100(c)?
This was moved from a Fine Print Note in Article 710. It requires that the short-circuit ratings of the cable not be exceeded. These ratings can be found in the IEEE Color books, ICEA (Insulated Cable Engineers Association), IEC Wiring Regulations, Canadian Electrical Code, and manufacturers' literature.

3.7.1 *NEC* Section 250.2(D), Performance of Fault-Current Path

3.7.1

What does this section mean?
The effective grounding path shall have a low enough impedance to limit the voltage to ground and to facilitate the opening of the overcurrent protective device. It must be permanent and continuos and be able to safely conduct available fault current.

3.7.2 *NEC* Section 250.90, Bonding Requirements and Short-Circuit Current Rating

3.7.2

What does this section mean?
All bonding provided must have the capacity to conduct safely any fault current it is likely to see.

3.7.3 *NEC* Section 250.96(A), Bonding Other Enclosures and Short-Circuit Current Requirements

3.7.3

What do these sections require?
All materials used in the grounding and bonding of equipment shall be capable of safely carrying the short-circuit current that could flow through the ground path. This will, in many cases, require the use of a current-limiting fuse to protect the equipment from damage. See Section 110-10 for more on component protection.

3.7.4 *NEC* Section 250.122, Sizing of Equipment Grounding Conductors

3.7.4

What are the ramifications of 250-122 and especially the note at the bottom of Table 250-122?

The integrity of the grounding path is essential for safety; it facilitates the operation of the overcurrent protective devices. Improper sizing of the grounding conductors can result in their annealing, melting or vaporizing before the protective device clears the circuit. Generally, the grounding electrode conductor and the equipment grounding conductors are smaller than the circuit conductors and their ampere rating is less than that of the overcurrent protective device. The protective device may be too slow to protect an undersized conductor against high fault currents (see Section 240-1 of this Bulletin). Consideration must be given to the size of the grounding conductors, their withstand, the magnitude of ground fault currents, and the operating characteristics of circuit overcurrent devices. Where the protective device is not fast enough to protect the undersized equipment grounding conductor, the conductor size may need to be increased, or a different overcurrent device could be chosen which could provided adequate protection for the conductor. This section of the N.E.C. now **requires** this analysis.

Caution, Table 250-122 in the N.E.C. gives the "Minimum Size Equipment Grounding Conductors for Grounding Raceway and Equipment."

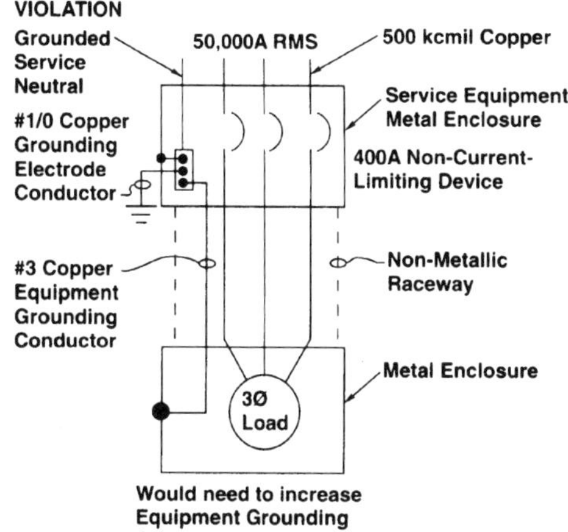

Would need to increase Equipment Grounding Conductor to 2/0.

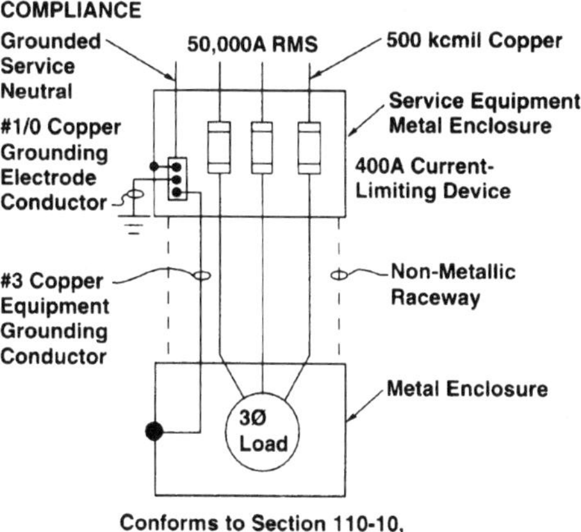

Conforms to Section 110-10, 250-2(d), and Table 250-122.

For example, Table 250-122 allows a circuit protected by a 400 ampere overcurrent device to have a #3 copper equipment grounding conductor. If the 400 ampere overcurrent device takes one cycle to open in a circuit where 50,000 amperes are available, typical cable manufacturer's withstand charts show that the #3 conductor would be damaged. One solution would be to install a #2/0 copper equipment grounding conductor which would be able to withstand the 50,000 amperes for one cycle. The other alternative is to limit the 50,000 amperes to within the 22,000 ampere for one cycle limit of the #3 conductor. This can be accomplished easily with the use of current-limiting fuses.

What is the importance of 250-122(d)?

Since instantaneous only circuit breakers (MCP's) can be set as high as 1700% of motor full-load current, the equipment grounding conductor shall be sized based on the motor overload relay.

What is the problem with 250-122(f)(2)?

This allows for protection of a parallelled equipment grounding conductor in a multiconductor cable with equipment ground fault protection.

However, ground fault protection is not a substitute for overcurrent protection. It is designed to prevent the burn down of switchboards. It was not designed for, nor is it fast enough to protect equipment grounding conductors from annealing under short-circuit conditions.

Take a 4,000 ampere circuit with 50,000 amperes available, as an example. Nine 750 kcmil/phase with one 500 kcmil as an EGC could be used. However, if 9 conduits are utilized, the previous code would require a 500 kcmil EGC in each conduit. The new requirements would allow for a #2 EGC in each conduit. (75°C, 750 kcmil is rated for 475 amps, and EGC associated with a 500 ampere overcurrent device is a #2.) The I^2t required to anneal a #2 copper equipment grounding conductor (Soares' validity rating) is 24.5 x 10^6 ampere squared seconds. The I^2t let-through for GFP, set at 475 amperes for a typical delay of .3 seconds at 50,000 amperes, is 50,000 x 50,000 x .3 = 750 x 10^6. That's more than 30 times the I^2t needed to anneal the copper. After a fault, the equipment grounding conductor would not be "tight" under the lug. In other words, there would no longer be an adequate ground return path. The 500 kcmil required by the previous NEC® has an I^2t rating of 1,389 x 10^6 ampere squared seconds, more than enough to stay tight under a lug after a fault occurs. For more detailed explanation of these concepts, review the latest edition of Soares Book on Grounding, now published by IAEI.

3.8.1 *NEC* Section 310.10, Temperature Limitation of Conductors

3.8.1

What is the purpose of the fine print note in this section?
The fine print note is intended to point out the need for conductor derating at high ambient temperatures. It also directs the user to be aware of other information, such as conductor size and number, to assure proper application.

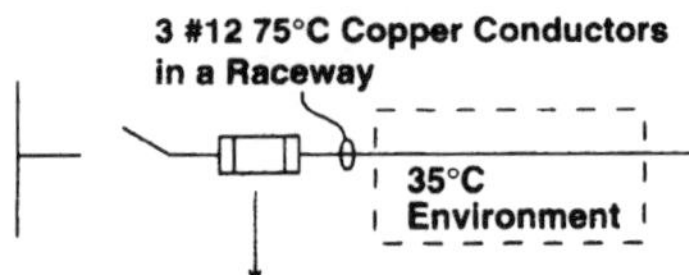

This fuse is sized at 25 (amperes) x .94 (temperature derating factor) = 23.5 amperes. The next standard size is 25 amperes, but the obelisk directs the reader to Section 240-3 where the maximum overcurrent device is given as 20 amperes.

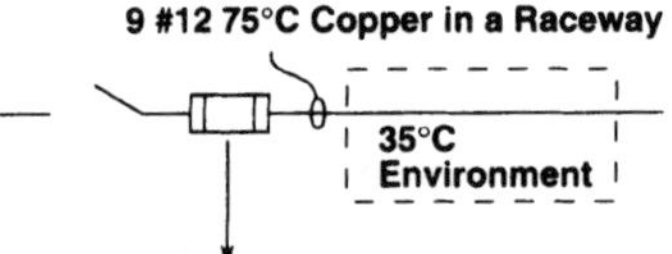

This fuse is sized at 25 (amperes) x .94 (temperature derating factor) x .70 (9 conductors in a raceway derating factor from Table 310-15(b)(2)(a) to ampacity tables) = 16.45 amperes. The next standard size is a 20 ampere Fuse.

3.9.1 *NEC* Section 364.11, Protection at a Busway Reduction

3.9.1

What does this section mean?
Overcurrent protection is required whenever busway is reduced in ampacity unless all of the following conditions are met:

1. Industrial establishment only.
2. Length of smaller bus does not exceed 50 feet.
3. Ampacity of smaller bus must be at least 1/3 that of the upstream overcurrent device.
4. Smaller bus must not contact combustible material.

3.10.1 *NEC* Section 384.16, Panelboard Overcurrent Protection

3.10.1

What is the meaning of 384-16(a)?
Lighting and appliance branch circuit panelboards must be protected by a main overcurrent device (up to two sets of fuses, as long as their combined ratings do not exceed that of the panelboard), unless the feeder has overcurrent protection not greater than the rating of the panelboard.

What is the meaning of 384-16(b)?
A Power Panelboard having supply conductors which include a neutral and having more than 10% of its overcurrent devices protecting branch circuits of 30 amperes or less, shall have individual protection on the line side not greater than the rating of the panelboard. Individual protection is not required when the power panel is used as service equipment in accordance with 230-71.

General Comment—The service entrance split bus load center or panelboard having up to 6 main disconnects is no longer permitted on new installations.

The tap rules found in Section 240-21 do not remove these requirements for lighting and appliance branch circuit panelboard protection, nor do they remove the requirements for transformer protection found in Section 450-3.

3.11.1 *NEC* Section 430.1, Scope of Motor Article

3.11.1

What is the importance of this section?
It offers an overview of protection for motors, motor circuits, motor controllers, and motor control centers.

3.11.2 *NEC* Section 430.6, Ampacity of Conductors for Motor Branch Circuits and Feeders

3.11.2

What is the importance of this section?
It states that conductors supplying motors shall be selected from applicable tables in Article 310 and Section 400-5. The determination of conductor ampacity, or ampere rating of switches, branch circuit protection, etc., should be taken from the motor F.L.A. tables in Article 430, Tables 430-147 through 430-150.

There is an exception for listed motor-operated appliances with both a horsepower rating and a full load current rating marked on the nameplate. In this case, the ampere rating on the nameplate should be used to determine the ampacity or rating of the motor circuit conductors, disconnecting means, controller, and the branch circuit short-circuit and ground fault protection. Similar exceptions exist for multispeed motors (Exc. 1) and equipment employing shaded pole or permanent-split capacitor-type fan or blower motor.

The separate overload device should always be based on the nameplate current rating.

3.11.3 *NEC* Section 430.8, Marking on Controllers

3.11.3

What is the purpose of the new FPN added to this section?
This new FPN was added to warn the user about the delicate nature of small contacts and overload relays which can easily be damaged under short circuit conditions unless properly protected by current-limiting protective devices.

3.11.4 *NEC* Section 430.32, Motor Overload Protection

3.11.4

What are the typical ways of providing motor overload protection external to the motor?
Generally, motor starters with overload relays and/or Class RK1 and RK5 dual-element fuses are used to provide motor running protection.

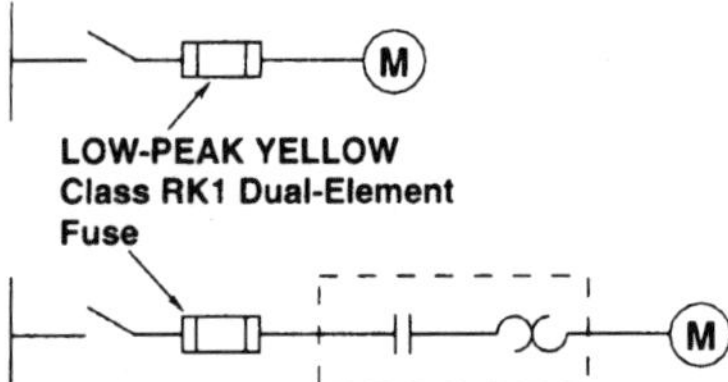

Typically, how are the devices selected for protection of motors?
With starters and overload relays, the proper heater element is selected from manufacturers' tables based on the motor full-load current rating. The level of protection reached in this selection process complies with Article 430.

When employing only dual-element Class RK1 and RK5 fuses for motor running overload protection, the rating of the fuse should be as follows:

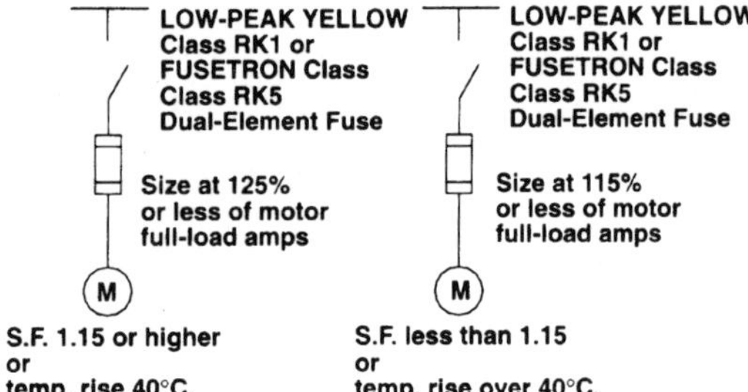

Do fuses sized as above also provide branch circuit protection requirements?
Yes. Sizing FUSETRON® Class RK5 and LOW-PEAK® YELLOW™ Class RK1 Dual-Element fuses for motor running overload protection also provides the necessary short-circuit protection per 430-52. The use of these dual-element fuses permits close sizing. Thus, fuse case sizes often can be smaller, thereby permitting the use of smaller switches.

Can circuit breakers and fuses other than Class RK1 and RK5 dual-element fuses be used to give motor overload protection?
Not generally. The conventional circuit breakers usually must be sized at 250% of the motor full-load amperes to avoid tripping on motor starting current, and thus cannot provide overload protection. Instantaneous only circuit breakers or motor short-circuit protectors are only equipped with a short-circuit tripping element and, therefore, are incapable of providing overload protection. For motor applications, the non-time-delay fuses such as the LIMITRON® KTS-R fuses normally have to be sized at 300% of a motor full-load current rating to avoid opening on motor start-up and, therefore, do not provide overload protection.

When single-phasing occurs on a 3-phase motor circuit, unbalanced currents flow through the motor, which can damage the motor if not taken off-line. Class RK1 and RK5 dual-element, time-delay fuses, sized for motor overload protection, can provide single-phase damage protection . See Section 430-36.

Footnote–Abnormal Motor Operation: The application of motors under certain abnormal operating conditions often requires the use of larger size fuses than would normally be required. The use of oversize fuses limits protection to short-circuit or branch circuit protection only. The types of abnormal motor installations that may be encountered include the following: (a) Fuses in high ambient temperature locations. (b) Motors having a high Code Letter (or possibly no Code Letter) with full-voltage start. (c) Motors driving high inertial loads or motors which must be frequently cycled off-and-on. Typical high inertial loads are machines such as punch presses having large mass flywheels, or machines such as centrifugal extractors and pulverizes, or large fans which cannot be brought up to speed quickly. (d) High efficiency motors with high inrush currents.

3.11.5 *NEC* Section 430.36, Fuses Used to Provide Overload and Single-Phasing Protection

3.11.5

What does this section require?
This section clarifies the need for overload protection in all three phases of a 3-phase, 3-wire system, where one phase also serves as the grounded conductor.

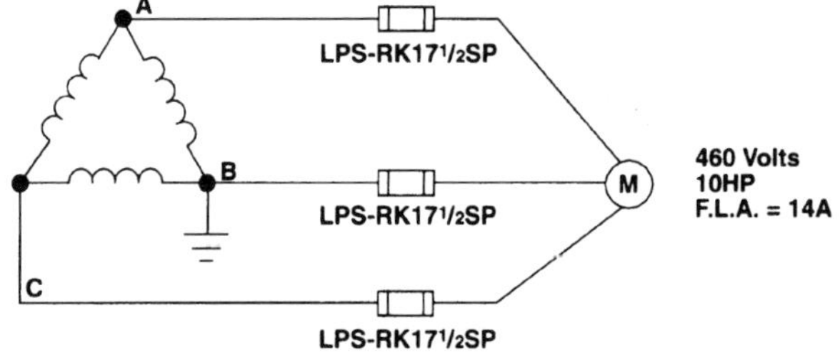

3.11.6 *NEC* Section 430.52, Sizing of Various Overcurrent Devices for Motor Branch-Circuit Protection

3.11.6

What is the basic content of this section?

This Section deals with the protection of motor branch circuits against short-circuit damage. It establishes the maximum permissible settings for overcurrent protective devices. (Branch circuits include all the circuit components–wire, switches, motor starters, etc.) As is apparent in Code Table 430-152, maximum settings vary with different types of motors, each type having unique starting characteristics. Motors to which the maximum permissible settings or ratings apply (shown in the condensed Table following) include all types of single-phase, three-phase squirrel cage and three-phase synchronous motors. The table below does not apply to Design E, Wound Rotor, and dc motors.

These maximum values do not preclude the application of lower sizes. Also, compliance with Section 110-10 must be analyzed. Motor starters have relatively low short-circuit current withstands. Refer to Buss® bulletin SPD for specific fuse recommendations.

Maximum Rating or Setting of Protective Devices†

Fuse		Circuit Breaker*	
Non-Time-Delay All Class CC	Dual-Element Time-Delay	Instantaneous Type Only	Inverse Time Type
300%	175%	800%	250%

†See Article 430, Section 430-52.
*For latest information, check manufacturer's data and/or Underwriters' Laboratories U.L. Standard #508 for damage and warning label requirements.

What about starter withstandability and Section 110-10 requirements for component protection?

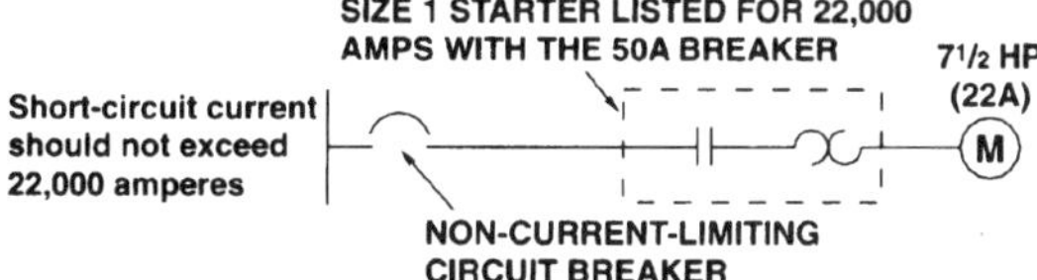

Under short-circuit conditions, the branch circuit protective device must protect the circuit components from extensive damage. Therefore, the following factors should be analyzed: available short-circuit current, let-through characteristics of the overcurrent protective device, and starter withstandability.

As an Example, this Size 1 Starter has been tested by U.L. with 22,000 ampere available short-circuit current per U.L. Standard 508. Thus, in the example above, the available short-circuit currents should not exceed 22,000 amperes since the circuit breaker is not current-limiting.

Additionally an MCP, if used in a combination controller, must be listed for that specific combination. The MCP cannot be used as a separate motor branch circuit short-circuit protective device to protect a motor controller. Applications of MCP's on many motors, i.e., high efficiency or high Code Letter, may cause the MCP to operate needlessly, even when sized at 1700% of motor current.

In the circuit below using a Buss® LOW-PEAK® YELLOW™ dual-element time-delay fuse, can available short-circuit current exceed 22000 amperes?

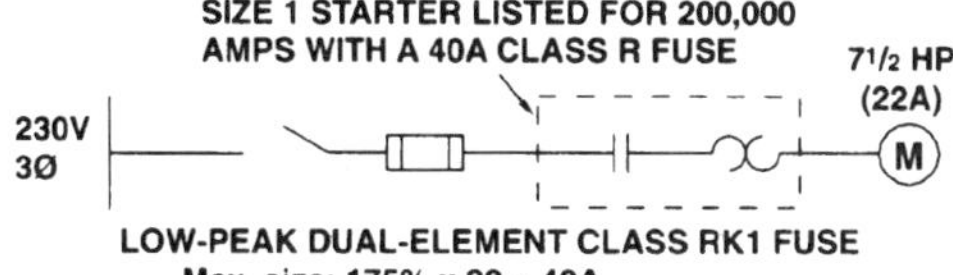

Yes. Because the LOW-PEAK® YELLOW™ fuse is "current-limiting," excellent short-circuit protection is provided, even though available short-circuit current greatly exceeds 22,000 amperes. (Specifically, the LOW-PEAK® YELLOW™ fuse would give protection against fault currents through 200,000 amperes.) It is also significant to note that because the Class RK1 LOW-PEAK® YELLOW™ fuse is a time-delay fuse, it actually could be sized at 125% of full-load current or the next larger size (30 amperes) with the advantage of permitting the use of a smaller disconnect switch, and providing backup overload protection and even better short-circuit protection.

These maximum sizing allowances are all overridden if a manufacturer's label shows overcurrent protection values lower than what 430-52 allows.

The overload relay heater elements of a motor controller often have relatively low short-circuit current withstand ratings. The maximum ratings of protective devices given in Table 430-152, thus, do not necessarily apply since they may be too large to provide adequate protection. Consequently, the starter manufacturer often includes an overload relay table within the starter enclosure. If the table states the maximum fuse size ratings to be used which will adequately protect the overload relays, the protective device must be a fuse.

TYPICAL EXAMPLE: The chart shown below is typical for starter manufacturers and may be found on the inside of the door of the starter enclosure. (See starter manufacturer for specific recommendation.)

Heater Code Marking	Full Load Current of Motor (Amperes) (40°C Ambient)	Max. Fuse
XX03	.25-.27	1
XX04	.28-.31	3
XX05	.32-.34	3
XX06	.35-.38	3
XX14	.76-.83	6
XX15	.84-.91	6
XX16	.92-1.00	6
XX17	1.01-1.11	6
XX18	1.12-1.22	6

Section 240-6 has an exception listing additional standard fuse ampere ratings of 1, 3, 6 and 10 amperes. The lower ratings were added to provide more effective protection for circuits with small motors, in accordance with Sections 430-52 and 430-40 and requirements for protecting the overload relays in controllers for very small motors. Fuse manufacturers have available other intermediate fuse ampere ratings to provide closer circuit protection (such as sizing Class RK1 and RK5 dual-element fuses at 125% of motor current) or to comply with "Maximum Fuse" sizes specified in controller manufacturer's overload relay tables.

430-52(c)(5) allows other fuses to be used in place of those allowed in Table 430-152. Why is this Code provision necessary?

Some "solid-state" motor starters and drives require fuses specifically designed to protect semiconductor components. The Code provision was necessary in order to give branch circuit, short-circuit and ground fault "status" to these fuses.

What is the significance of 430-52(c)(3) Exc. 1, (c)(6) & (c)(7)?

Design B energy efficient motors are included with Design E motors as far as protection with instantaneous trip circuit breakers (MCP's), self-protected combination controllers, and motor short-circuit protectors (MSCP's) are concerned. These branch-circuit devices may be set as high as 1700% of the motor full load current as shown in Tables 430-147 through 430-150. Motor controllers may have difficulty opening at current levels just below the 1700% rating.

3.11.7 *NEC* Section 430.53, Connecting Several Motors or Loads on One Branch Circuit

3.11.7

What does this section mean?
Simply stated, branch circuit protection for group motor installations must be testing agency and factory listed for such installations. This listing can be accomplished as a factory installed assembly with specified marking, or field installed as separate assemblies listed for use with each other, with instructions provided by the manufacturer. For the best protection of group motor installations, the branch circuit protective device must be current-limiting. The Fine Print Note reference to Section 110-10 emphasizes the necessity to comply with the component short-circuit withstand ratings.

If the equipment nameplate specifies "MAX" fuse for a multimotor circuit, what must the branch circuit device be?
It must be a fuse, rated at not more than what is specified on the nameplate. The best type of fuse to use is a current-limiting fuse.

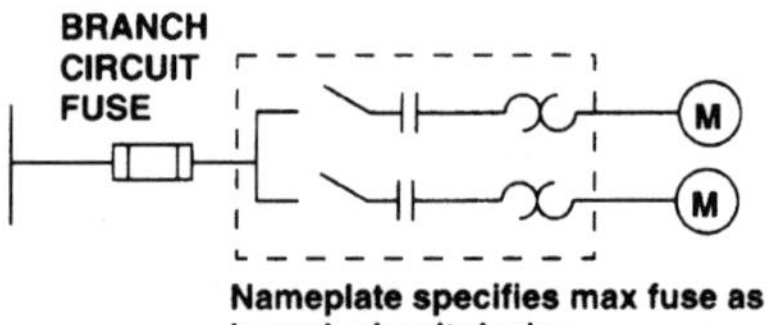

Nameplate specifies max fuse as branch circuit device.

If the equipment nameplate specifies "MAX" circuit breaker of a certain manufacturer and part number, what must be used?
Only that specific type and manufacturer may be used. In other words, that controller has been tested and listed with a certain circuit breaker, with certain short-circuit characteristics. Although breakers of other manufacturers and interrupting ratings may be interchangeable, that substitution is not allowed by 430-53(c)(3). This is due in part to the fact that there is no standardization of short-circuit performance of circuit breakers. Also, some circuit breakers exhibit current-limitation, to a degree, while not being marked current-limiting. This could prove to be a hazard if a non-current-limiting breaker of the same form and fit were to be installed.

3.11.8 *NEC* Section 430.71, Motor Control-Circuit Protection

3.11.8

What does this section mean?

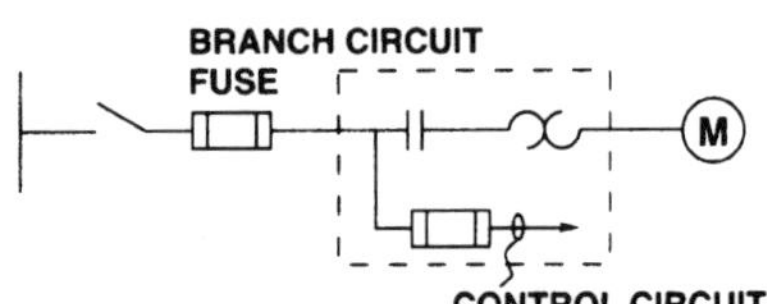

Section 430-71 defines the control circuit of a motor controller (control apparatus). The relationship of a control circuit to the circuit carrying the main power current is illustrated in the circuit diagram at left.

3.11.9 *NEC* Section 430.72(A), Motor Control-Circuit Overcurrent Protection

3.11.9

What does this section mean:

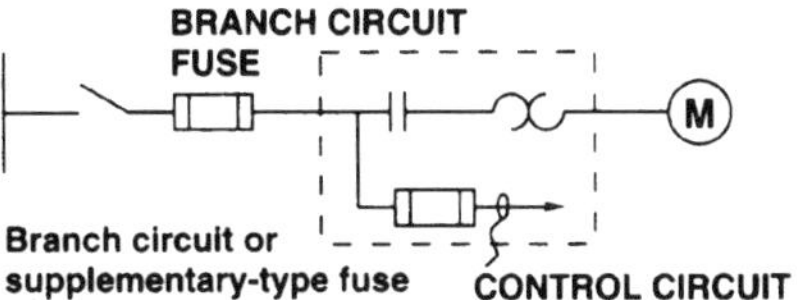

As shown in the above circuit, the motor control circuit tapped on the load side of the motor branch circuit protective device can be protected by either a branch or supplementary-type protective device (such a control circuit is not to be considered a branch circuit).

For motor controllers listed for available fault currents greater than 10,000 amperes, the control circuit fuse must be a branch circuit fuse with a sufficient interrupting rating. (The use of Buss® FNQ-R, KTK-R, LP-CC, LPJ_SP, JJS, or JJN fuses is recommended; these fuses have branch circuit listing status, high interrupting rating, current-limitation, and small size.)

3.11.10 *NEC* Section 430.72(B), Motor Control-Circuit Conductor Protection

3.11.10

What does this section mean?
Control circuit conductors must be protected by a fuse rated at not more than those values shown in Column "A" of Table 430-72(b).

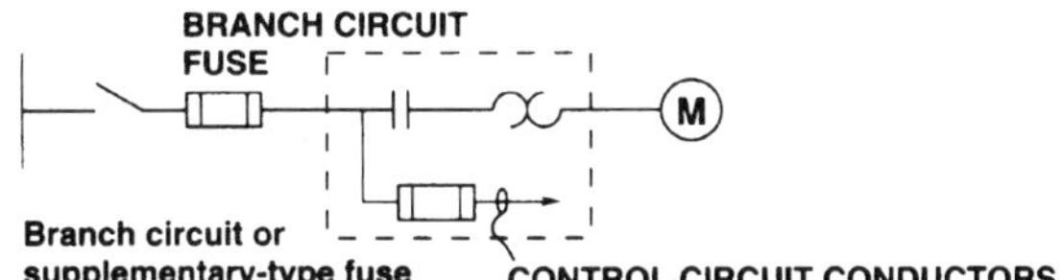

What if the control conductors remain within the enclosure?
If the control conductors do not leave the enclosure, they can be considered to be protected by the branch circuit fuse, if that fuse does not exceed the values of Table 430-72(b) Column B.

(430-72(b)(2))

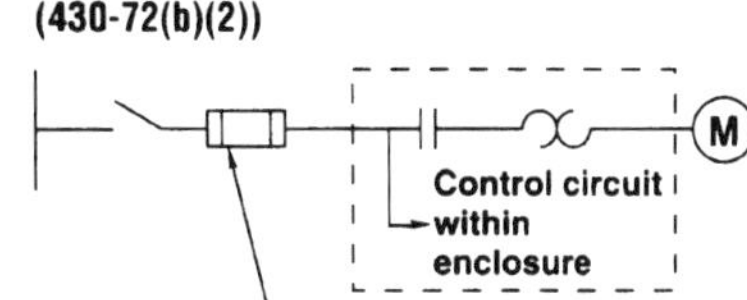

The motor branch protective device is considered to also protect the control conductors if the conductors do not extend beyond the enclosure and the maximum rating of the protective device is not greater than Table 430-72(b) Column B.

(continued)

3.11.10 (*Continued*)

Do the circuits shown below require individual control circuit protection?

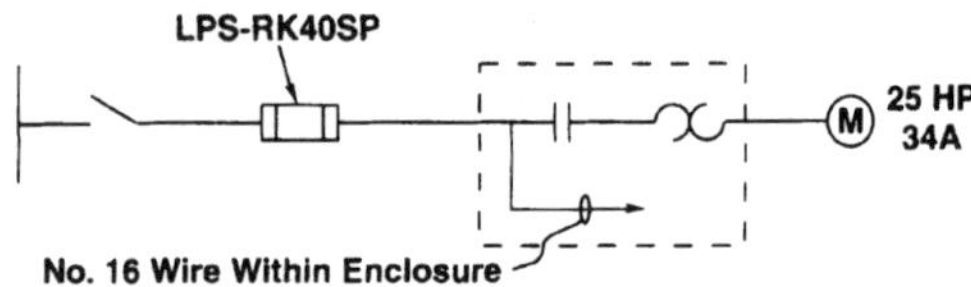

No. The LPS-RK40SP ampere fuses are sized within the 40 ampere requirement for #16 conductor within an enclosure. (See Table 430-72(b).)

80A
25 HP
34A
10A Required
No. 16 Wire Within Enclosure

Yes. Individual control circuit fuses are required since the 80 ampere circuit breaker has a rating in excess of the 40 ampere requirement for #16 conductor within an enclosure. (See Table 430-72(b).

Note: Sections 110-10 and 240-1 require that component withstand not be exceeded. Not all overcurrent devices sized per 430-72(b) can actually protect small conductors.

What if the control conductors leave the enclosure?
If the control conductors leave the enclosure, they can be considered to be protected by the branch circuit fuse, if that fuse does not exceed the values of Table 430-72(b) Column C.

(430-72(b)(2))

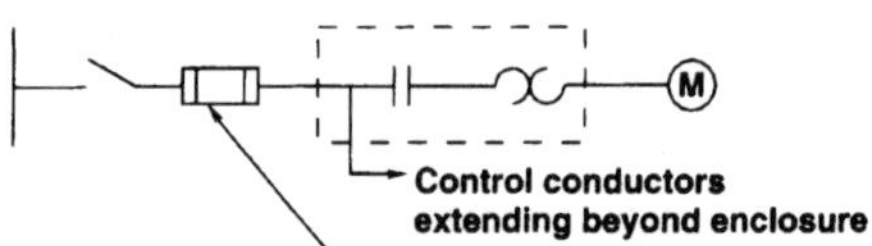

The motor branch circuit protective device is considered also to protect the control conductors if it does not exceed the values of Column C.

What does Exception No. 2 mean?
Primary fusing of a control transformer can be considered to protect the 2-wire, secondary conductors if the fuse rating does not exceed the value of multiplying the appropriate rating from Table 430-72(b) with the secondary-to-primary voltage ratio.

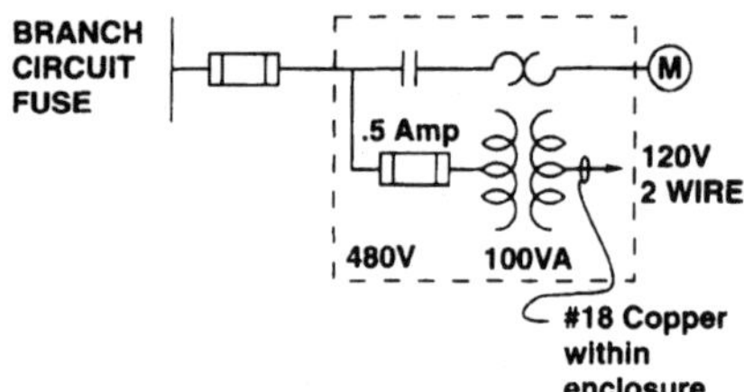

From Table 430-72(b)

Wire Size	**Max. Protection**
#18 Copper	7 Ampere Fuse

Maximum primary fuse shall not exceed 1.75A as determined by—

$$\frac{120V}{480V} \times 7A = 1.75A$$

Comparison By Largest HP Motor (460V) Circuit Where Branch Circuit Protective Device Is Considered To Protect The Control Conductors Per 430-72(b) (2).

Protective Device	**Approx. Size As Percent Motor F.L.A.**	**Level Of Protection**	**Control Circuit Within Enclosure**			**Control Circuit Extending Beyond Enclosure**		
			#18	**#16**	**#14**	**#18**	**#16**	**#14**
LOW-PEAK® YELLOW™ Class RK1 or FUSETRON® Class RK5 dual-element Fuse	125%	Overload and Branch Circuit	15HP	25HP	60HP	3HP	5HP	25HP
	175%		10HP	15HP	40HP	2HP	3HP	15HP
Non-Time-Delay Fuse	300%		5HP	7½HP	20HP	1HP	1½HP	10HP
Thermal Magnetic Circuit Breaker	250%	Branch Circuit	5HP	10HP	30HP	1½HP	2HP	10HP
Instantaneous Only Circuit Breaker	1000%*	Only	1HP	2HP	5HP	¼HP	½HP	2HP

*Instantaneous only circuit breakers cannot provide any overload protection. Typically to hold starting currents, instantaneous trip is set at 1000% to 1700% of motor full-load amperes.

Even though a fuse or circuit breaker can be sized at 300% or 400% of the conductor ampacity, what level of control conductor protection can be expected?
The protective device would respond only to high level conductor overcurrents; the control conductors would not be protected against lower overcurrent levels. This lack of protection could result in a prolonged 200% control circuit overcurrent and eventual insulation breakdown and melting of the conductors. For example, if the control circuit run were of considerable length, the conductor impedance might be sufficiently high to limit fault currents to 200% to 400% of the conductor ampacity. Thus, oversized overcurrent devices would provide inadequate protection. In contrast, fuses sized to the conductors ampacity would provide full-range overcurrent protection; their use is to be recommended.

3.11.11 *NEC* Section 430.72(C), Motor Control-Circuit Transformer Protection

3.11.11

What does this section mean?†

Primary Fuse Protection Only.

Transformer Primary Current	Primary Fuse Ampacity Must Not Exceed
Less than 2 amperes	500% (430-72(c)(4))
2 to 9 amperes	167%
9 amperes or more	125%

Primary and Secondary Fuse Protection.

Primary Fuse Does Not Exceed	Secondary Current	Secondary Fuse
250%	9 amperes or more	125%
250%	Less than 9 amperes	167%

The conditions of 430-72(c)(3), permit the use of a control transformer rated less than 50 VA* without the inclusion of individual protection on the primary side of the transformer in the control circuit proper. Thus, protection of the transformer primary against short-circuit currents is dependent upon the device used for branch circuit protection. However, consideration should be given to protecting the control transformer on the primary side with individual fuses specifically sized for control transformer protection.

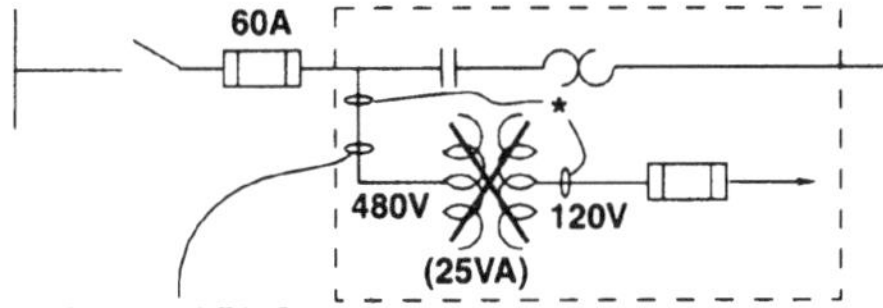

.05A normal F.L.C.
(breakdown of transformer windings could cause current to increase many times over normal level but less than 60A) *Conductor protection is still required per Section 430-72(b)

Take the case, for instance, in which a short occurs in a control transformer (such as would result from insulation deterioration and breakdown). (See diagram above in which a 60 ampere branch circuit fuse is shown.) Now, if the overcurrent drawn by the control circuit as a result of the shorted control transformer is relatively low (actually could be less than 60 amperes) compared to the response time of the 60 ampere branch circuit fuse or circuit breaker, the transformer could become so hot that extensive damage could be done to the insulation of the control conductors . . . the transformer itself could burst into flames.

*Control Transformers rated less than 50 VA are usually impedance protected or have other types of protection, such as inherent protection.

†Refer to Section 8.12 of NFPA 79 for the requirements of control transformers in Industrial Machinery.

However, inclusion of fuse protection in the primary of the control transformer would minimize this type of hazard. Buss® FNQ or FNQ-R Time-Delay fuses are excellent choices. When applying fuses, the time-current characteristics should be checked to determine if the fuse can hold the inrush magnetizing current of the transformer.

Fuses Commonly Used in Control Circuits.

There are several fuse types which have small dimensions that are ideally suited for control circuit protection. The KTK-R, FNQ-R and LP-CC fuses are listed as Class CC fuses, and JJN (JJS) fuses are listed as Class T fuses. When used for control transformer, coil, or solenoid protection, the fuse should be selected to withstand the inrush current for the required time.

Symbol	Voltage Rating	Ampere Rating	Class	Interrupting Rating	Comment
Branch Circuit Rejection Fuses					
FNQ-R	600V	15/100 thru 10	CC*	200KA	
LP-CC	600V	1/2 thru 30	CC*	200KA	Time-delay in
LPJ	600V	1 thru 600	J*	300KA	overload region
SC	600V	6 thru 20	G*	100KA	
SC	480V	25 thru 60	G*	100KA	
KTK-R	600V	1/10 thru 30	CC*	200KA	No intentional
JJN	300V	1 thru 1200	T*	200KA	time-delay
JJS	600V	1 thru 800	T*	200KA	in the overload
SC	600V	1/2 thru 5	G*	100KA	region
Supplementary Fuses					
FNQ	500V	1/10 thru 30	SUP.*	10KA	
FNW	250V	12 thru 30	SUP.*	10KA	
FNM	250V	0 thru 1	SUP.*	35A	
FNM	250V	1.1 thru 3.5	SUP.*	100A	
FNM	250V	3.6 thru 10	SUP.*	200A	Time-delay in the
FNM	125V	10.1 thru 15	SUP.*	10KA	overload region
FNM	32V	15.1 thru 30	SUP.	1KA	
FNA	250V	1/10 thru 8/10	SUP.*	35A	
FNA	125V	1 thru 15	SUP.*	10KA	
FNA	32V	15.1 thru 30	SUP.	1KA	
KTK	600V	1/10 thru 30	SUP.*	100KA	
BAF	250V	1/2 thru 1	SUP.*	35A	
BAF	250V	1.1 thru 3.5	SUP.*	100A	
BAF	250V	3.6 thru 10	SUP.*	200A	No intentional
BAF	250V	10.1 thru 15	SUP.*	750A	time-delay
BAF	125V	15.1 thru 30	SUP.	10KA	in the overload
BAN	250V	2/10 thru 1	SUP.	35A	region
BAN	250V	1.1 thru 3.5	SUP.	100A	
BAN	250V	3.6 thru 10	SUP.	200A	
BAN	250V	10.1 thru 15	SUP.	750A	
BAN	250V	15.1 thru 30	SUP.	1500A	

* U.L. Listed

3.11.12 *NEC* Section 430.94, Motor Control-Center Protection

3.11.12

What are the requirements of this section?
Where motor control centers (MCC) are specified, proper overcurrent protection shall be supplied in the MCC as an integral main, or remote main. These devices should be rated based on the common power bus rating.

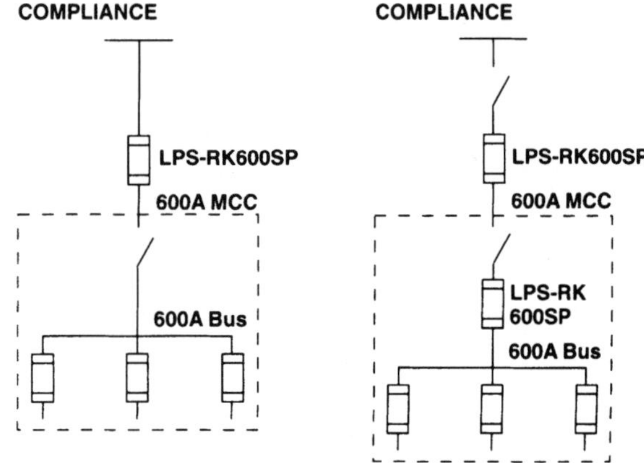

3.11.13 *NEC* Section 430.109(A)(6), Manual Motor Controller as a Motor Disconnect

3.11.13

What are the requirements of 430-109(a)(6)?
Manual motor protectors or manual motor controllers can be used as a motor disconnecting means if they are marked "Suitable as Motor Disconnect" and located between the final branch-circuit overcurrent device and the motor. The required location would preclude their use as the branch circuit disconnecting means. Note that these devices can not be used as the branch-circuit overcurrent device even though some of them have the ability to open short-circuit currents.

3.12.1 *NEC* Section 440.5, Marking Requirements on HVAC Controllers

3.12.1

How does this section affect the overcurrent protection requirements?
If the nameplate on the equipment controller is marked with "MAX FUSE", that means a fuse must be used to protect the equipment. See Section 110-3(b) for proper installation and protection.

3.12.2 *NEC* Section 440.22, Application and Selection of the Branch-Circuit Protection for HVAC Equipment

3.12.2

What are the requirements of 440-22(a)?
The branch circuit protective device may be sized at the maximum value of 175% of the motor-compressor rated load current. If the motor cannot start due to high inrush currents, this value may be increased to, but cannot exceed, 225% of motor rated current.

What are the requirements of 440-22(c)?
440-22(c) states that if the manufacturer's heater table shows a maximum protective device less than that allowed above, the protective device rating shall not exceed the manufacturer's values (refer to Section 430-52 also).

3.13.1 *NEC* Section 450.3, Protection Requirements for Transformers

3.13.1

What is the importance of Note 2 found on Table 450-3(a) and 450-3(b)?
The required secondary protection may be satisfied with multiple overcurrent devices that protect feeders fed from the transformer secondary. The total ampere rating of these multiple devices cannot exceed the allowed value of a single secondary overcurrent device. If this method is chosen, dual-element, time-delay fuse protection offers much greater flexibility.

Note the following examples:

This design utilized a single secondary overcurrent device. It provides the greatest degree of selectively coordinated transformer protection, secondary cable protection, and switchboard/panelboard/load center protection. The transformer cannot be overloaded to a significant degree if future loads are added (improperly) in the future.

If the single secondary overcurrent device is eliminated, much of the protection will be reduced.

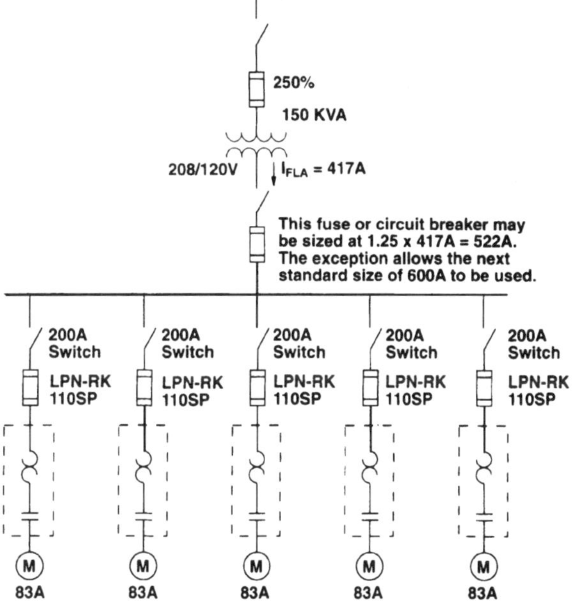

(continued)

3.13.1 (*Continued*)

Using the same logic, if the single secondary main is eliminated and thermal magnetic circuit breakers are utilized as branch circuit protection, only three of the motors can be connected because the thermal-magnetic breakers will have been sized at approximately 250% of motor F.L.A. (83 x 250% = 207.5A)

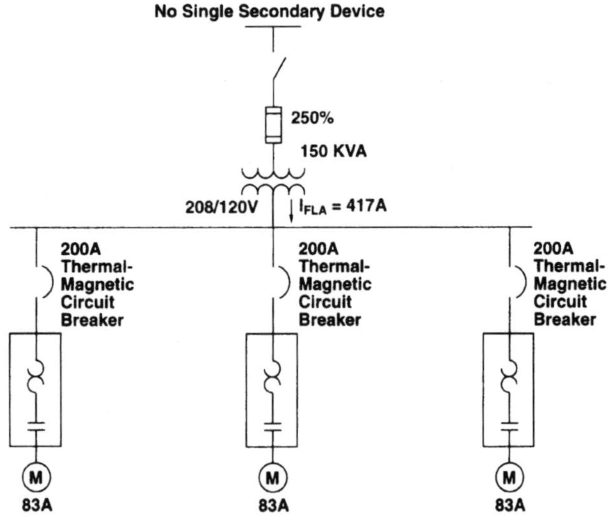

Using a 200 ampere circuit breaker would allow three (600 ÷ 200) motors to be connected.

If the single secondary main is eliminated and MCP's are utilized as branch circuit protection, the transformer will be seriously underutilized because only one motor can be connected. For one motor, 1 x 700% of 83 = 581 amperes. For two motors, 2 x 700% of 83 = 1162 amperes. Since the sum of the devices cannot exceed 600 amperes, only one motor can be connected when the motor circuit is protected by an MCP.

If the MCP will not hold at the 700% setting due to a high starting current, it cannot be adjusted beyond 722% (600÷83) and therefor it may not be able to be used.

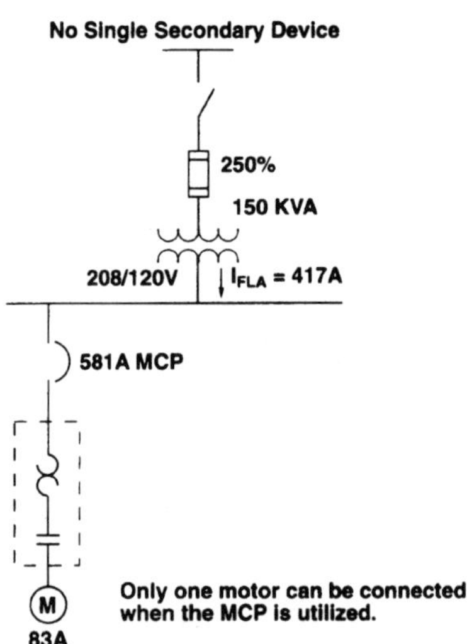

If the single secondary main is eliminated, and dual-element fuses are utilized as branch circuit protection, the transformer can continue to be loaded with the five 83 ampere motors because 5 x 110 = 550 amperes (which is less than the maximum of 600 amperes).

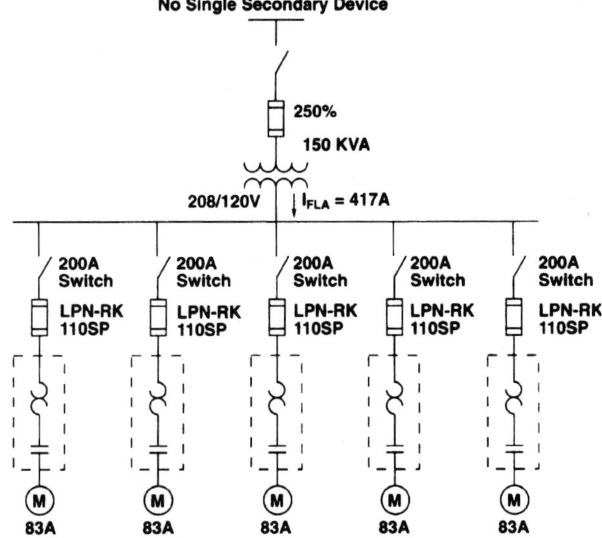

3.13.2 *NEC* Section 450.3(A), Protection Requirements for Transformers over 600 V

3.13.2

What is the general content of this section?

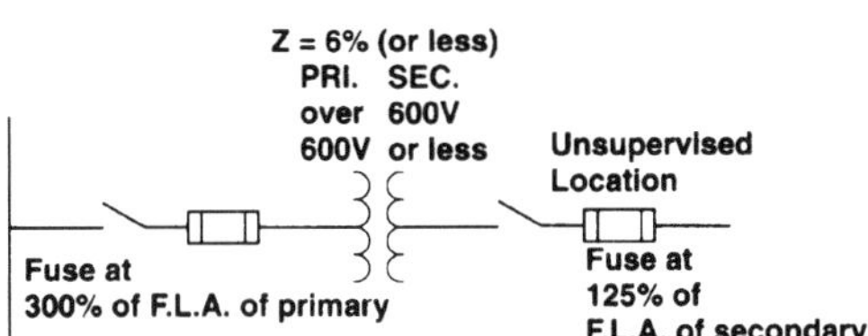

This part of the section sets the overcurrent protection requirements of transformers (over 600 volts): For primary and secondary protection, the primary should be protected by an individual protective device with fuse rating not in excess of 300% of the primary's rated current. Secondary sizing (600V and below) is at 125%* for any location or up to 250% for supervised installations. Secondary sizing (over 600V) can be up to 250%* for % Z ≤ 6% and 225% for % Z > 6%.

For supervised installations, secondary protection is not required (above, at, or below 600 volts) if the primary is at 250% (or the next standard size if 250% does not correspond to a standard fuse size) of the primary full load amps. Note that conductor protection and panelboard protection may still be required.

*Where this does not correspond to a standard fuse size, the next higher standard size shall be permitted.

3.13.3 *NEC* Section 450.3(B), Protection Requirements for Transformers 600 V or Less

3.13.3

What is the general content of this section?
This section covers protection requirements of transformers, 600 volts or less. Fusing requirements are shown in the illustrated example below.

Where the primary FLA is ≥ 2 amps, but < 9 amps, the primary fuse may be sized at 167% or less. If the primary FLA < 2 amps, the primary fuse may be sized at 300% or less.

PRIMARY PROTECTION ONLY

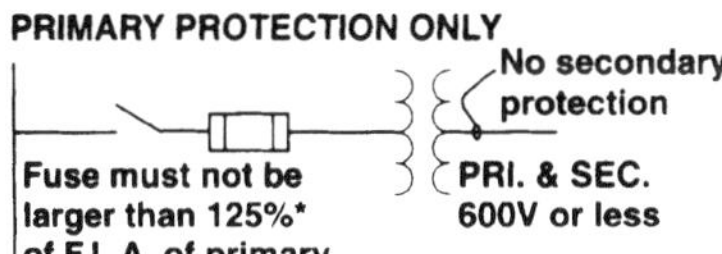

PRIMARY AND SECONDARY PROTECTION

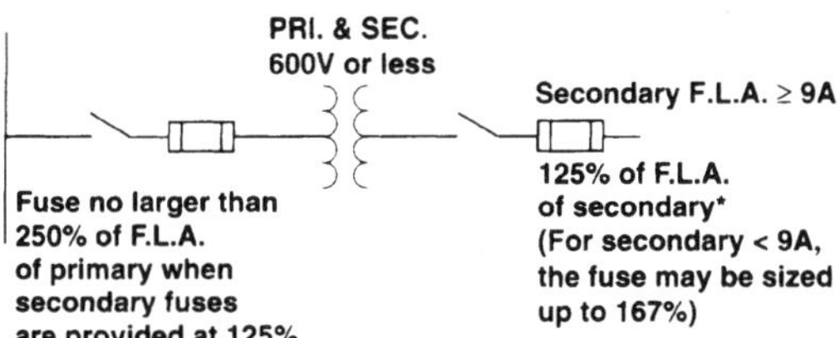

*Where this does not correspond to a standard fuse size, the next standard size may be used.

Protection of circuit conductors is required per Articles 240 and 310; protection of panelboards per Article 384. Specific sections which should be referenced are Sections 240-3, 240-21 and Section 384-16.

Note: Transformer overload protection will not be provided by using overcurrent protective devices sized much greater than the transformer F.L.A. The limits of 167%, 250% and 300% will not adequately protect transformers. It is suggested that for transformer overload protection, the fuse size should be within 125% of the transformer full-load amperes.

There is a wide range of fuse ampere ratings available to properly protect transformers. FUSETRON® (Class RK5) and LOW-PEAK® YELLOW™ (Class RK1) dual-element fuses can often be sized on the transformer primary and/or secondary, rated as low as 125% of the transformer F.L.A. These dual-element fuses have time delay to withstand the high magnetizing inrush currents of transformers. There is a wide ampere rating selection in the 0 to 15 ampere range for these dual-element fuses to provide protection for even small control transformers.

3.13.4 *NEC* Section 450.6(A)(3), Tie-Circuit Protection

3.13.4

What does this section require?
Current-limiting cable limiters shall be used on each end of the tie conductors, specified per the size of the conductors.

3.14.1 *NEC* Section 455.7, Overcurrent Protection Requirements for Phase Converters

3.14.1

What does this section mean?
Phase converters supplying variable loads must be protected at not more than 125% of the nameplate single-phase input full-load current.

For converters supplying fixed loads, the conductors shall be protected at their ampacity, but in no case can the overcurrent protection exceed 125% of the phase converter nameplate single phase current.

Where the required rating does not correspond to a standard rating, sizes up to the next standard rating may be used.

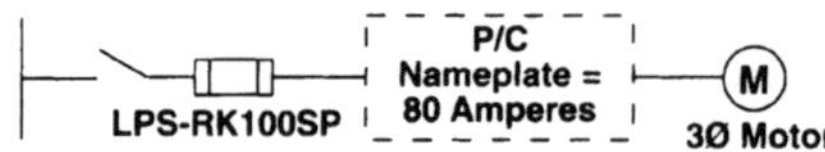

A maximum fuse rating LPS-RK100SP will meet the 125% requirements.

3.15.1 *NEC* Section 460.8(B), Overcurrent Protection of Capacitors

3.15.1

What are the requirements of this section?
Overcurrent protection must be provided in each ungrounded conductor supplying a capacitor bank, except for a capacitor located on the load side of a motor overload protective device.

The rating of this overcurrent protective device shall be as low as practical. Generally, dual-element time-delay fuses can be sized at 150% to 175% of the capacitor rated current.

3.16.1 *NEC* Section 501.6(B), Fuses for Class 1, Division 2 Locations

3.16.1

What is the meaning of 501-6(b)(3)?
The intent of this reference is to suggest the use of non-indicating, filled, current-limiting fuses. The following is a partial list of filled, non-indicating fuses which are current-limiting:

Class CC	LP-CC 1/2 - 30, KTK-R 3 - 30, FNQ-R 2 - 10
Class J	LPJ_SP 15 - 600, JKS 0 - 600
Class L	KRP-C_SP 601 - 6000, KTU 601 - 6000, KLU 601 - 4000
Class RK1	KTN-R 1 - 600, KTS-R 35 - 600

What is the importance of 501-6(b)(4)?
General Comment–These fuses are used to isolate a faulted fixture ballast and maintain continuity of service. Listed or recognized branch circuit or supplementary fuses may be used. Additionally, the GLR fuse is used on ballasts that have a 200 ampere short-circuit withstand rating such as Class P ballasts.

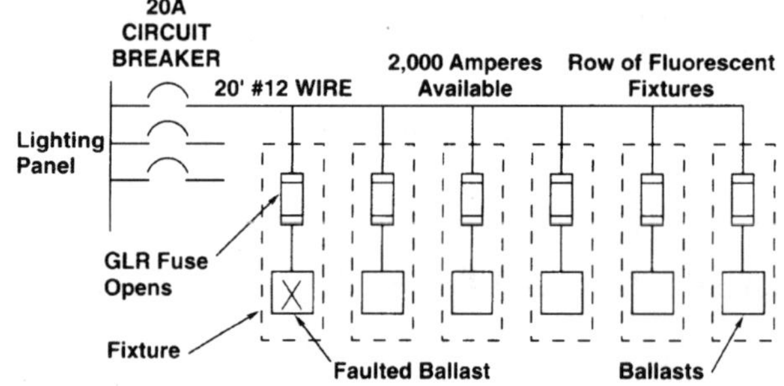

3.17.1 *NEC* Section 517.17, Requirements for Ground-Fault Protection and Coordination in Health Care Facilities

3.17.1

What does this section mean?
If ground fault protection is placed on the main service or feeder of a health care facility, ground fault protection must also be placed on the next level of feeders. The separation between ground fault relay time bands for any feeder and main ground fault relay must be at least 6 cycles in order to achieve coordination between these two ground fault relays. In health care facilities where no ground fault protection is placed on the main or feeder, no ground fault protection is necessary at the next level down. Therefore, if the requirements of Sections 230-95 and 215-10 do not require ground fault protection, then no ground fault protection is required on the downstream feeders either.

If the ground fault protection of the feeder coordinates with the main ground faul' protection, will complete coordination between the main and feeder be assured for all ground faults?
No, not necessarily! Merely providing coordinated ground fault relays does not prevent a main service blackout caused by feeder ground faults. The overcurrent protective devices must also be selectively coordinated. The intent of Section 517-17 is to achieve "100 percent selectivity" for all magnitudes of ground fault current and overcurrents. 100% selectivity requires that the overcurrent protective devices be selectively coordinated for medium and high magnitude ground fault currents because the conventional overcurrent devices may operate at these levels. (See discussion of Section 240-12, System Coordination, for a more detailed explanation of selective coordination).

What is one of the most important design parameters of the power distribution system of a health care facility?
Selective coordination. To minimize the disruption of power and blackouts in a distribution system, it is absolutely mandatory that the overcurrent protective devices be selectively coordinated.

What is selective coordination?
A selectively coordinated system is one in which the overcurrent protective devices have been selected so that only the overcurrent device protecting that circuit in which a fault occurs opens; other circuits in the system are not disturbed. The danger of a major power failure in a health care facility such as a hospital is self evident. In any facility, a power failure is at least inconvenient, if not quite costly; in a hospital, it can easily give rise to panic and endanger lives. Continuity of electrical service by selective coordination of the protection devices is a must. (See Section 240-12, System Coordination, of this bulletin for a more detailed explanation of selective coordination).

3.18.1 *NEC* Section 520.53(F)(2), Protection of Portable Switchboards on Stage

3.18.1

What does this section require?
Compliance with Sections 110-9 and 110-10 is mandatory. Short-circuit ratings must be marked on the switchboard.

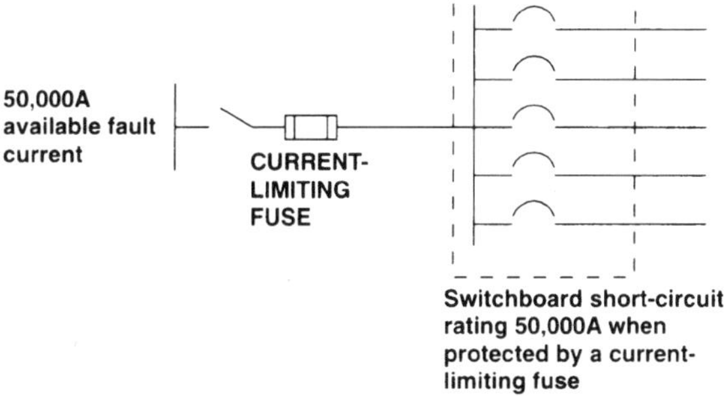

3.19.1 *NEC* Section 550.6(B), Overcurrent Protection Requirements for Mobile Homes and Parks

3.19.1

What does this section mean?
Branch circuit fuses installed in a mobile home should not exceed the rating of the conductors they supply. These fuses should not be more than 1.5 times the rating of an appliance rated 13.3 amperes or more on a single branch, and not more than the fuse size marked on the air conditioner or other motor operated appliance.
Do these branch circuit fuses conform to the requirements of 550-6(b)?
Yes.

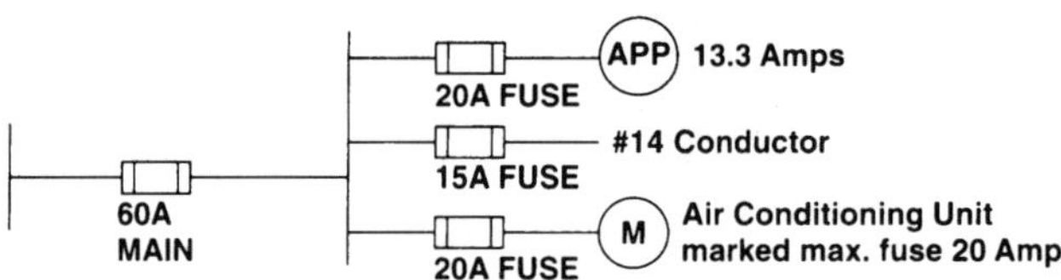

Note: If the nameplate on a device states "Maximum Fuse Size", then fuses that size or smaller must be used somewhere in the circuit.

3.20.1 *NEC* Section 610.14(C), Conductor Sizes and Protection for Cranes and Hoists

3.20.1

What does this section mean?
#18 conductors can be used in control circuits of cranes and hoists if they are fused at not greater than 7 amperes.

3.21.1 *NEC* Section 620.62, Selective Coordination of Overcurrent Protective Devices for Elevators

3.21.1

What does this section require?
When a feeder supplies more than one elevator, the elevator overcurrent protective devices must selectively coordinate with all upstream feeders and mains.

To be "selectively coordinated" means that should a fault (L-G, L-N, L-L, L-L-L) occur anywhere in a system, ONLY the first overcurrent device upstream of the fault will open. Thus, power is maintained on all other branches (feeders) in the system.

In the following diagram, a fault at X would trip both the 90 ampere breaker and the 400 ampere breaker. This non-selectivity would be present for all short-circuit current values higher than the instant trip setting of the 400 ampere breaker. In the example. . . 400 x 5 = 2,000 amperes. This results in total loss of power to the other elevators. The aftermath can be PANIC. This installation is a clear VIOLATION of Section 620-62.

If LOW-PEAK® YELLOW™ fuses were used (see the Selectivity Ratio Guide), a fault at X clears the 90 ampere fuses ONLY. The 400 ampere fuses remain intact, thereby maintaining power to the other elevators. This installation is in COMPLIANCE with Section 620-62.

(continued)

3.21.1 (*Continued*)

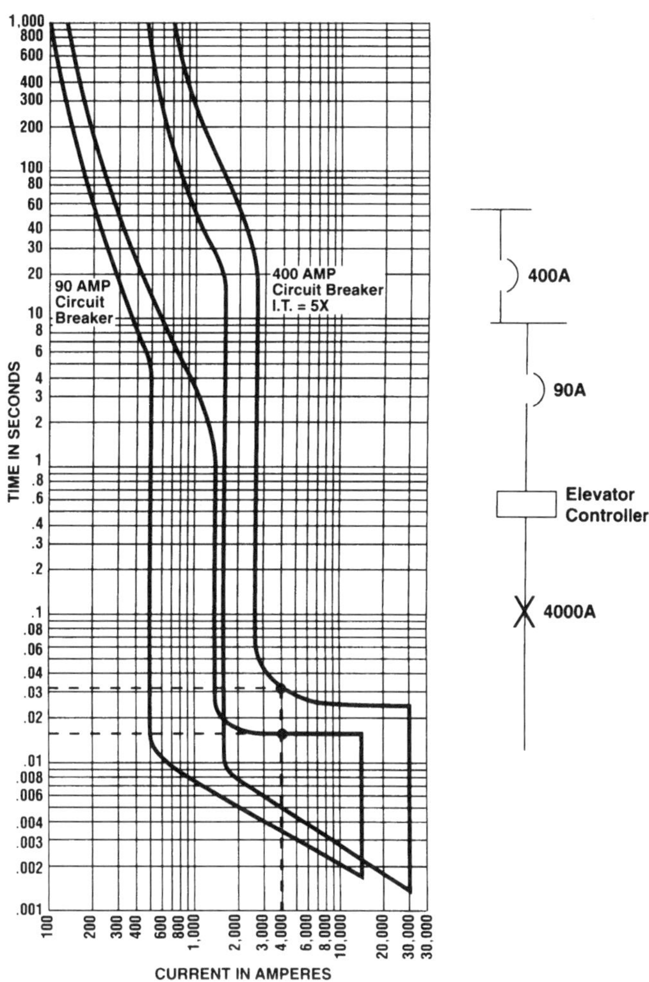

3.22.1 *NEC* Section 670.3, Industrial Machinery

3.22.1

What does this section mean?

If a main overcurrent protective device is provided on an industrial machine, the nameplate shall state, among other things, the interrupting capacity of the device. The machine shall also be marked "overcurrent protection provided at machine supply terminals".

3.23.1 *NEC* Section 700.5, Emergency Systems: Their Capacity and Rating

3.23.1

What does 700-5(a) require?
Emergency systems and equipment must be able to handle the available short-circuit current at their line side. If the equipment cannot, it may be damaged, causing additional hazards to personnel. The use of current-limiting fuses can be a solution to this high fault current problem.

3.23.2 *NEC* Section 700.16, Emergency Illumination

3.23.2

What does this section require?
Emergency lighting systems cannot allow a blackout in any area requiring emergency illumination due to the failure of any one element of the lighting system. Such failures could be caused by the burning out of a light bulb or the opening of a branch circuit protective device due to a faulted ballast. The solution to the burnt out light bulb is to have additional bulb(s) in the area. The solution to the open branch circuit protective device is to install listed supplementary fuses on each ballast. In that way, a faulted ballast would be taken off the line by the supplementary fuse, not by the branch circuit protective device, allowing the rest of the emergency system to remain energized.

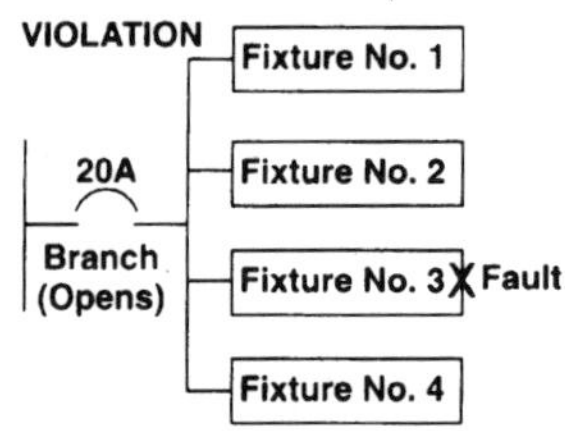

The fault in Fixture #3 causes the 20 ampere branch circuit overcurrent device to open, causing a blackout in the entire area.

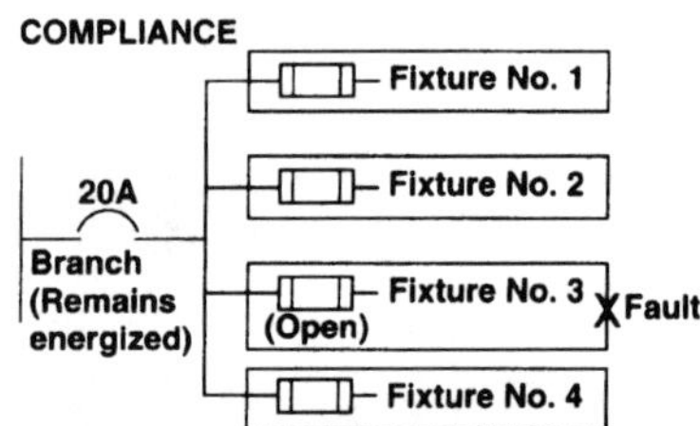

The fault in Fixture #3 will open just the supplementary fuse. The 20 ampere branch circuit device does not open and Fixtures 1, 2 and 4 remain energized, preventing a blackout.

3.23.3 *NEC* Section 700.25, Emergency System Overcurrent Protection Requirements (FPN)

3.23.3

What is the meaning of this fine print note?
In order to maximize the reliability of emergency systems, the overcurrent devices must be selectively coordinated. Time-current curves of both fuses and circuit breakers must be examined to determine whether or not only the overcurrent device closest to a fault opens. If additional upstream devices open, the system is not selectively coordinated, causing additional sections of the emergency system to black out and therefore, reducing the reliability of that system.*

BLACKOUT!
Reduced Reliability
VIOLATION

1000A
I.T.=10x
Opens

225A
I.T.=8x.
Opens

20A
I.T.=8A
Opens
22,000 Amp
Short-Circuit

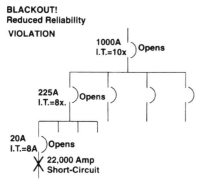

Fault exceeding the instantaneous trip setting of all three circuit breakers in series will open all three. This will blackout the entire emergency system.

*See also Section 4-5.1 of NFPA 110 (Emergency and Standby Power Systems) and Sections 3-3.2.1.2(4) & 3-4.1.1.1 of NFPA 99 (Health Care Facilities) for additional information on selective coordination.

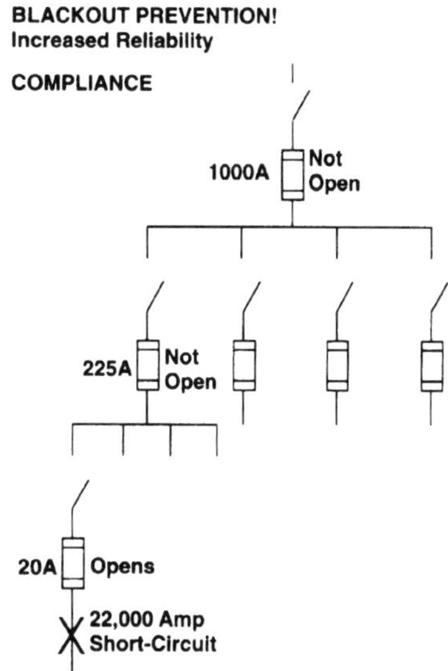

Fault opens the nearest upstream fuse, allowing other circuits to remain energized. Reliability of the emergency system is increased.

3.24.1 *NEC* Section 705.16, Interconnected Electric Power Production Sources: Interrupting and Short-Circuit Current Rating

3.24.1

What do these three sections mean?
The N.E.C. requires that emergency and standby systems shall have the capability of safely interrupting the available short-circuit current available at the line terminals of the equipment. Refer to Sections 110-9 and 110-10.

3.25.1 *NEC* Section 725.23, Overcurrent Protection for Class 1 Circuits

3.25.1

What does this Section mean?
Class 1 Control Circuit Conductors shall be protected by fuses at their ampacities. In addition, #18 and #16 shall be protected at 7 amperes and 10 amperes, respectively.

What must be added to this Control Circuit to comply with 725-23?

VIOLATION

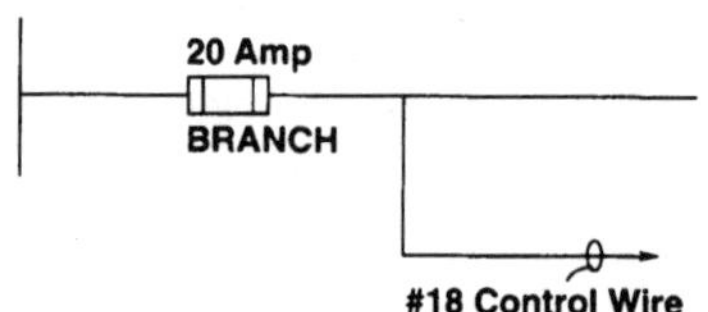

A 7 ampere fuse must be added to protect the #18 control wire.

COMPLIANCE

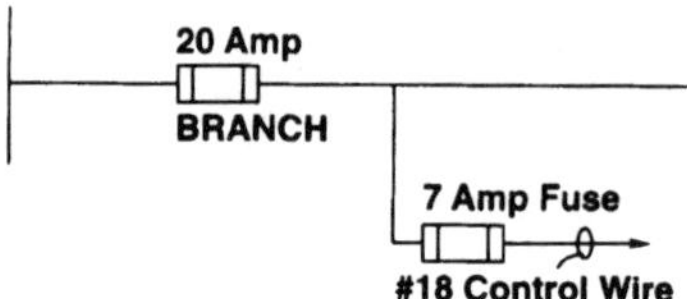

3.26.1 *NEC* Section 760.23, Requirements for Non-Power-Limited Fire Alarm Signaling Circuits

3.26.1

What does this provision require?
Fire protective signaling circuits with conductors #18 and larger must be protected at their ampacities as shown:

#18 ...7 ampere fuse maximum
#16 ...10 ampere fuse maximum
#14 (and larger). . .Max. fuse size as dictated in Section 310-15.

Fuses shall be located at the supply terminals of the conductor.

Section

4

Wiring Methods and Materials

4.1.0 NEC Table300.1(C), Metric Designator and Trade Sizes
4.1.1 *NEC* Table 300.5, Minimum Cover Requirements, 0 to 600 V, Nominal, Burial in Millimeters (Inches)
4.2.1 *NEC* Table 300.19(A), Spacings for Conductor Supports
4.2.2 Examples of Installed Support Bushings and Cleats
4.3.1 *NEC* Table 300.50, Minimum Cover Requirements
4.4.1 *NEC* Table 310.5, Minimum Size of Conductors
4.4.2 *NEC* Table 310.13, Conductor Application and Insulations
4.4.3 Conductor Characteristics
4.4.4 *NEC* Table 310.15(B)(2)(a), Adjustment Factors for More Than Three Current.Carrying Conductors in a Raceway or Cable
4.4.5 *NEC* Table 310.16, Allowable Ampacities of Insulated Conductors Rated 0 through 2000 V, 60°C through 90°C (140°F through 194°F), Not More Than Three Current.Carrying Conductors in a Raceway, Cable, or Earth (Directly Buried), Based on Ambient Air Temperature of 30°C (86°F)
4.4.6 *NEC* Table 310.17, Allowable Ampacities of Single Insulated Conductors Rated 0 through 2000 V in Free Air, Based on Ambient Air Temperature of 30°C (86°F)
4.4.7 *NEC* Table 310.18, Allowable Ampacities of Insulated Conductors, Rated 0 through 2000 V, 150°C through 250°C (302°F through 482°F), Not More Than Three Current-Carrying Conductors in Raceway or Cable, Based on Ambient Air Temperature of 40°C (104°F)
4.4.8 *NEC* Table 310.19, Allowable Ampacities of Single Insulated Conductors, Rated 0 through 2000 V, 150°C through 250°C (302°F through 482°F), in Free Air, Based on Ambient Air Temperature of 40°C (104°F)
4.4.9 *NEC* Table 310.20, Ampacities of Not More Than Three Single Insulated Conductors, Rated 0 through 2000 V, Supported on a Messenger, Based on Ambient Air Temperature of 40°C (104°F)
4.4.10 *NEC* Table 310.21, Ampacities of Bare or Covered Conductors in Free Air, Based on 40°C (104°F) Ambient, 80°C (176°F) Total Conductor Temperature, 610 mm/s (2 ft/s) Wind Velocity
4.4.11 *NEC* Table 310.61, Conductor Application and Insulation
4.4.12 *NEC* Table 310.62, Thickness of Insulation for 601. to 2000.V Nonshielded Types RHH and RHW
4.4.13 *NEC* Table 310.63, Thickness of Insulation and Jacket for Nonshielded Solid-Dielectric Insulated Conductors Rated 2001 to 8000 V
4.4.14 *NEC* Table 310.64, Thickness of Insulation for Shielded Solid-Dielectric Insulated Conductors Rated 2001 to 35,000 V
4.4.15 *NEC* Table 310.67, Ampacities of Insulated Single Copper Conductor Cables Triplexed in Air Based on Conductor Temperatures of 90°C (194°F)

and 105°C (221°F) and Ambient Air Temperature of 40°C (104°F)
4.4.16 *NEC* Table 310.68, Ampacities of Insulated Single Aluminum Conductor Cables Triplexed in Air Based on Conductor Temperatures of 90°C (194°F) and 105°C (221°F) and Ambient Air Temperature of 40°C (104°F)
4.4.17 *NEC* Table 310.69, Ampacities of Insulated Single Copper Conductor Isolated in Air Based on Conductor Temperatures of 90°C (194°F) and 105°C (221°F) and Ambient Air Temperature of 40°C (104°F)
4.4.18 *NEC* Table 310.70, Ampacities of Insulated Single Aluminum Conductor Isolated in Air Based on Conductor Temperatures of 90°C (194°F) and 105°C (221°F) and Ambient Air Temperature of 40°F (104°F)
4.4.19 *NEC* Table 310.71, Ampacities of an Insulated Three-Conductor Copper Cable Isolated in Air Based on Conductor Temperatures of 90°C (194°F) and 105°C (221°F) and Ambient Air Temperature of 40°C (104°F)
4.4.20 *NEC* Table 310.72, Ampacities of Insulated Three-Conductor Aluminum Cable Isolated in Air Based on Conductor Temperatures of 90°C (194°F) and 105°C (221°F) and Ambient Air Temperature of 40°C (104°F)
4.4.21 *NEC* Table 310.73, Ampacities of an Insulated Triplexed or Three Single-Conductor Copper Cables in Isolated Conduit in Air Based on Conductor Temperatures of 90°C (194°F) and 105°C (221°F) and Ambient Air Temperature of 40°C (104°F)
4.4.22 *NEC* Table 310.74, Ampacities of an Insulated Triplexed or Three Single-Conductor Aluminum Cables in Isolated Conduit in Air Based on Conductor Temperatures of 90°C (194°F) and 105°C (221°F) and Ambient Air Temperature of 40°C (104°F)
4.4.23 *NEC* Table 310.75, Ampacities of an Insulated Three-Conductor Copper Cable in Isolated Conduit in Air Based on Conductor Temperatures of 90°C (194°F) and 105°C (221°F) and Ambient Air Temperature of 40°C (104°F)
4.4.24 *NEC* Table 310.76, Ampacities of an Insulated Three-Conductor Aluminum Cable in Isolated Conduit in Air Based on Conductor Temperatures of 90°C (194°F) and 105°C (221°F) and Ambient Air Temperature of 40°C (104°F)
4.4.25 *NEC* Figure 310.60, Cable Installation Dimensions for Use with Tables 4.4.26 through 4.4.35 (*NEC* Tables 310.77 through 310.86)
4.4.26 *NEC* Table 310.77, Ampacities of Three Single-Insulated Copper Conductors in Underground Electrical Ducts (Three Conductors Per Electrical Duct) Based on Ambient Earth Temperature of 20°C (68°F), Electrical Duct Arrangement per Figure 4.4.25 (*NEC* Figure 310.60), 100 Percent Load Factor, Thermal Resistance (RHO) of 90, Conductor Temperatures of 90°C (194°F) and 105°C (221°F)
4.4.27 *NEC* Table 310.78, Ampacities of Three Single-Insulated Aluminum Conductors in Underground Electrical Ducts (Three Conductors per Electrical Duct) Based on Ambient Earth Temperature of 20°C (68°F), Electrical Duct Arrangement per Figure 4.4.25 (*NEC* Figure 310.60), 100 Percent Load Factor, Thermal Resistance (RHO) of 90, Conductor Temperatures of 90°C (194°F) and 105°C (221°F)
4.4.28 *NEC* Table 310.79, Ampacities of Three Insulated Copper Conductors Cabled Within an Overall Covering (Three-Conductor Cable) in Underground Electrical Ducts (One Cable per Electrical Duct) Based on Ambient Earth Temperature of 20°C (68°F), Electrical Duct Arrangement per Figure 4.4.25 (*NEC* Figure 310.60), 100 Percent Load Factor, Thermal Resistance (RHO) of 90, Conductor Temperatures of 90°C (194°F) and 105°C (221°F)
4.4.29 *NEC* Table 310.80, Ampacities of Three Insulated Aluminum Conductors Cabled Within an Overall Covering (Three-Conductor Cable) in Underground Electrical Ducts (One Cable per Electrical Duct) Based on Ambient Earth Temperature of 20°C (68°F), Electrical Duct Arrangement per Figure 4.4.25 (*NEC* Figure 310.60), 100 Percent Load Factor, Thermal Resistance (RHO)

of 90, Conductor Temperatures of 90°C (194°F) and 105°C (221°F)
4.4.30 *NEC* Table 310.81, Ampacities of Single-Insulated Copper Conductors Directly Buried in Earth Based on Ambient Earth Temperature of 20°C (68°F), Arrangement per Figure 4.4.25 (*NEC* Figure 310.60), 100 Percent Load Factor, Thermal Resistance (RHO) of 90, Conductor Temperatures of 90°C (194°F) and 105°C (221°F)
4.4.31 *NEC* Table 310.82, Ampacities of Single-Insulated Aluminum Conductors Directly Buried in Earth Based on Ambient Earth Temperature of 20°C (68°F), Arrangement per Figure 4.4.25 (*NEC* Figure 310.60), 100 Percent Load Factor, Thermal Resistance (RHO) of 90, Conductor Temperatures of 90°C (194°F) and 105°C (221°F)
4.4.32 *NEC* Table 310.83, Ampacities of Three Insulated Copper Conductors Cabled Within an Overall Covering (Three-Conductor Cable) Directly Buried in Earth Based on Ambient Earth Temperature of 20°C (68°F), Arrangement per Figure 4.4.25 (*NEC* Figure 310.60), 100 Percent Load Factor, Thermal Resistance (RHO) of 90, Conductor Temperatures of 90°C (194°F) and 105°C (221°F)
4.4.33 *NEC* Table 310.84, Ampacities of Three Insulated Aluminum Conductors Cabled Within an Overall Covering (Three-Conductor Cable) Directly Buried in Earth Based on Ambient Earth Temperature of 20°C (68°F), Arrangement per Figure 4.4.25 (*NEC* Figure 310.60), 100 Percent Load Factor, Thermal Resistance (RHO) of 90, Conductor Temperatures of 90°C (194°F) and 105°C (221°F)
4.4.34 *NEC* Table 310.85, Ampacities of Three Triplexed Single-Insulated Copper Conductors Directly Buried in Earth Based on Ambient Earth Temperature of 20°C (68°F), Arrangement per Figure 4.4.25 (*NEC* Figure 310.60), 100 Percent Load Factor, Thermal Resistance (RHO) of 90, Conductor Temperatures of 90°C (194°F) and 105°C (221°F)
4.4.35 *NEC* Table 310.86, Ampacities of Three Triplexed Single-Insulated Aluminum Conductors Directly Buried in Earth Based on Ambient Earth Temperature of 20°C (68°F), Arrangement per Figure 4.4.25 (*NEC* Figure 310.60), 100 Percent Load Factor, Thermal Resistance (RHO) of 90, Conductor Temperatures of 90°C (194°F) and 105°C (221°F)
4.5.1 *NEC* Table 392.7(B), Metal Area Requirements for Cable Trays Used as Equipment Grounding Conductor
4.5.2 An Example of Multiconductor Cables in Cable Trays with Conduit Runs to Power Equipment Where Bonding Is Provided
4.5.3 *NEC* Table 392.9, Allowable Cable Fill Area for Multiconductor Cables in Ladder, Ventilated-Trough, or Solid-Bottom Cable Trays for Cables Rated 2000 V or Less
4.5.4 *NEC* Table 392.9(E), Allowable Cable Fill Area for Multiconductor Cables in Ventilated-Channel Cable Trays for Cables Rated 2000 V or Less
4.5.4.1 *NEC* Table 393.9(F), Allowable Cable Fill Area for Multiconductor Cables in Solid Channel Cable Trays for Cables Rated 2000 V or Less
4.5.5 *NEC* Table 392.10(A), Allowable Cable Fill Area for Single-Conductor Cables in Ladder or Ventilated-Trough Cable Trays for Cables Rated 2000 V or Less
4.5.6 An Illustration of Section 392.11(A)(3) for Multiconductor Cables, 2000 V or Less, with Not More Than Three Conductors per Cable (Ampacity to be Determined from Table B.310.3 in Annex B)
4.5.7 An Illustration of Section 392.11(B)(4) for Three Single Conductors Installed in a Triangular Configuration with Spacing Between Groups of Not Less Than 2.15 Times the Conductor Diameter (Ampacities to Be Determined from Table 310.20)
4.6.1 An Illustration of Section 332.24 for Bends in Type MI Cable
4.6.2 600-V MI Power Cable: Size and Ampacities

4.6.3 300-V MI Twisted-Pair and Shielded Twisted-Pair Cable Sizes
4.6.4 MI Cable Versus Conventional Construction in Hazardous (Classified) Locations
4.6.5 Engineering Data: Calculating Voltage Drop and Feeder Sizing (MI Cable)
4.7.1 *NEC* Table 344.24, Radius of Conduit Bends for IMC, RMC, RNC, and EMT
4.7.2 Minimum Support Required for IMC, RMC, and EMT
4.7.3 *NEC* Table 344.30(B)(2), Supports for Rigid Metal Conduit
4.7.4 *NEC* Table 352.30(B), Support of Rigid Nonmetallic Conduit (RNC)
4.7.5 *NEC* Table 352.44(A), Expansion Characteristics of PVC Rigid Nonmetallic Conduit, Coefficient of Thermal Expansion = 6.084×10^{-5} mm/mm/°C (3.38×10^{-5} in/in/°F)
4.7.6 *NEC* Table 352.44(B), Expansion Characteristics of Reinforced Thermosetting Resin Conduit (RTRC), Coefficient of Thermal Expansion = 2.7×10^{-5} mm/mm/°C (1.5×10^{-5} in/in/°F)
4.7.7 *NEC* Table 348.22, Maximum Number of Insulated Conductors in Metric Designator 12 (Trade Size $^3/_8$ in) Flexible Metal Conduit
4.7.8 Conductor Fill Table for Various Surface Raceways (Based on Wiremold)
4.7.9 *NEC* Table 384.22, Channel Size and Inside Diameter Area
4.8.1 *NEC* Table 314.16(A), Metal Boxes
4.8.2 *NEC* Table 314.16(B), Volume Allowance Required per Conductor
4.9.1 *NEC* Table 400.4, Flexible Cords and Cables
4.9.2 *NEC* Table 400.5(A), Allowable Ampacity for Flexible Cords and Cables [Based on Ambient Temperature of 30°C (86°F); See Section 400.13 and Table 400.4]
4.9.3 *NEC* Table 400.5(B), Ampacity of Cable Types SC, SCE, SCT, PPE, G, G-GC, and W [Based on Ambient Temperature of 30°C (86°F); See Table 400-4] Temperature Rating of Cable
4.9.4 NEC Table 400.5, Adjustment Factors for More Than Three Current-Carrying Conductors in a Flexible Cord or Cable
4.9.5 *NEC* Table 402.3, Fixture Wires
4.9.6 *NEC* Table 402.5, Allowable Ampacities for Fixture Wires

4.1.0 *NEC* Table 300.1(C), Metric Designator and

TABLE 4.1.0

Metric Designator	Trade Size
12	⅜
16	½
21	¾
27	1
35	1¼
41	1½
53	2
63	2½
78	3
91	3½
103	4
129	5
155	6

Note: The metric designators and trade sizes are for identification

Trade Sizes

4.1.1 *NEC* Table 300 5, Minimum Cover

TABLE 4.1.1

Location of Wiring Method or Circuit	Type of Wiring Method or Circuit									
	Column 1 Direct Burial Cables or Conductors		Column 2 Rigid Metal Conduit or Intermediate Metal Conduit		Column 3 Nonmetallic Raceways Listed for Direct Burial Without Concrete Encasement or Other Approved Raceways		Column 4 Residential Branch Circuits Rated 120 Volts or Less with GFCI Protection and Maximum Overcurrent Protection of 20 Amperes		Column 5 Circuits for Control of Irrigation and Landscape Lighting Limited to Not More Than 30 Volts and Installed with Type UF or in Other Identified Cable or Raceway	
	mm	in.	mm	in.	mm	in.	mm	in.	mm	in.
All locations not specified below	600	24	150	6	450	18	300	12	150	6
In trench below 50-mm (2-in.) thick concrete or equivalent	450	18	150	6	300	12	50	6	50	6
Under a building	0 (in raceway only)	0	0	0	0	0	0 (in raceway only)	0	0 (in raceway only)	0
Under minimum of 102-mm (4-in.) thick concrete exterior slab with no vehicular traffic and the slab extending not less than 152 mm (6 in.) beyond the underground installation	450	18	100	4	100	4	150 (direct burial) 100 (in raceway)	6 4	150	6
Under streets, highways, roads, alleys, driveways, and parking lots	600	24	600	24	600	24	600	24	600	24
One- and two-family dwelling driveways and outdoor parking areas, and used only for dwelling-related purposes	450	18	450	18	450	18	300	12	450	18
In or under airport runways, including adjacent areas where trespassing prohibited	450	18	450	18	450	18	450	18	450	18

Notes:

1. Cover is defined as the shortest distance in millimeters (inches) measured between a point on the top surface of any direct-buried conductor, cable, conduit, or other raceway and the top surface of finished grade, concrete, or similar cover.
2. Raceways approved for burial only where concrete encased shall require concrete envelope not less than 50 mm (2 in.) thick.
3. Lesser depths shall be permitted where cables and conductors rise for terminations or splices or where access is otherwise required.
4. Where one of the wiring method types listed in Columns 1–3 is used for one of the circuit types in Columns 4 and 5, the shallower depth of burial shall be permitted.
5. Where solid rock prevents compliance with the cover depths specified in this table, the wiring shall be installed in metal or nonmetallic raceway permitted for direct burial. The raceways shall be covered by a minimum of 50 mm (2 in.) of concrete extending down to rock.

Requirements, 0 to 600 V, Nominal, Burial in Millimeters (Inches)

Burial is in inches (*cover* is defined as the shortest distance in inches measured between a point on the top surface of any direct-buried conductor, cable, conduit, or other raceway and the top surface of finished grade, concrete, or similar cover).

4.2.1 *NEC* Table 300.19(A), Spacings for Conductor

TABLE 4.2.1

		Conductors			
		Aluminum or Copper-Clad Aluminum		Copper	
Size of Wire	Support of Conductors in Vertical Raceways	m	ft	m	ft
18 AWG through 8 AWG	Not greater than	30	100	30	100
6 AWG through 1/0 AWG	Not greater than	60	200	30	100
2/0 AWG through 4/0 AWG	Not greater than	55	180	25	80
Over 4/0 AWG through 350 kcmil	Not greater than	41	135	18	60
Over 350 kcmil through 500 kcmil	Not greater than	36	120	15	50
Over 500 kcmil through 750 kcmil	Not greater than	28	95	12	40
Over 750 kcmil	Not greater than	26	85	11	35

Supports (Maximum Spacing Intervals in Vertical Raceways)

Exception. Steel wire-armored cable shall be supported at the top of the riser with a cable support that clamps the steel wire armor. A safety device shall be permitted at the lower end of the riser to hold the cable in the event there is slippage of the cable in the wire-armored cable support. Additional wedge-type supports shall be permitted to relieve the strain on the equipment terminals caused by expansion of the cable under load.

4.2.2 (a)

4.2.2 (b)

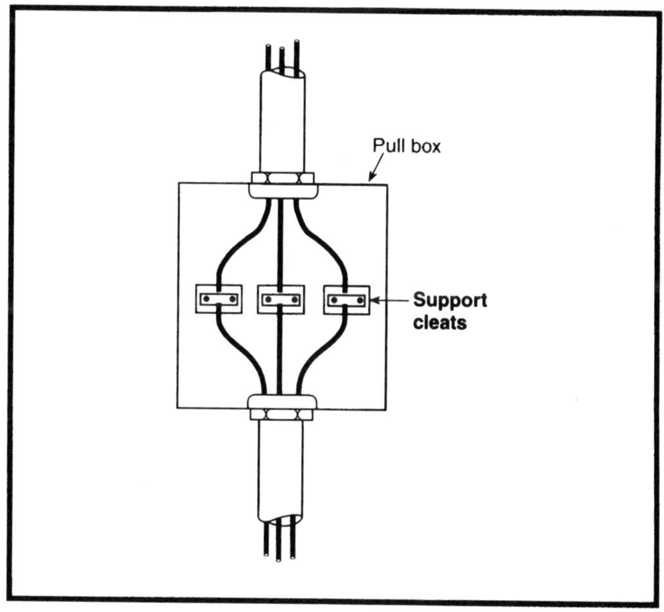

4.2.2 Examples of Installed Support Bushings and Cleats

TABLE 4.3.1

Circuit Voltage	Direct-Buried Cables		Rigid Nonmetallic Conduit Approved for Direct Burial*		Rigid Metal Conduit and Intermediate Metal Conduit	
	mm	in.	mm	in.	mm	in.
Over 600 V through 22 kV	750	30	450	18	150	6
Over 22 kV through 40 kV	900	36	600	24	150	6
Over 40 kV	1000	42	750	30	150	6

Note: *Cover* is defined as the shortest distance in millimeters measured between a point on the top surface of any direct-buried conductor, cable, conduit, or other raceway and the top surface of finished grade, concrete, or similar cover.

*Listed by a qualified testing agency as suitable for direct burial without encasement. All other nonmetallic systems shall require 50 mm (2 in.) of concrete or equivalent above conduit in addition to above depth.

4.3.1 *NEC* Table 300.50, Minimum Cover Requirements (Over 600 V)

Cover is defined as the shortest distance in inches measured between a point on the top surface of any direct-buried conductor, cable, conduit, or other raceway and the top surface of finished grade, concrete, or similar cover.

Exception No. 1. Areas subject to vehicular traffic, such as thoroughfares or commercial parking areas, shall have a minimum cover of 600 mm (24 in).

Exception No. 2. The minimum cover requirements for other than rigid metal conduit and intermediate metal conduit shall be permitted to be reduced 150 mm (6 in) for each 50 mm (2 in) of concrete or equivalent protection placed in the trench over the underground installation.

Exception No. 3. The minimum cover requirements for conduits shall not apply if the installation meets either one of the following conditions. A warning ribbon or other effective means suitable for the conditions shall be placed above the underground installation. (a) The minimum cover requirements shall not apply to conduits or other raceways that are located under a building: or (b) The minimum cover requirements shall not apply to conduits or other raceways that are located under an exterior concrete slab not less than 100 mm (4 in) in thickness and extending not less than 150 mm (6 in) beyond the underground installation.

Exception No. 4. Lesser depths shall be permitted where cables and conductors rise for terminations or splices or where access is otherwise required.

Exception No. 5. In or under airport runways, including adjacent defined areas where trespass is prohibited, cable shall be permitted to be buried not less than 450 mm (18 in) deep and without raceways, concrete enclosement, or equivalent.

Exception No. 6. In or under airport runways, including adjacent defined areas where trespass is prohibited, conduit shall be permitted to be buried not less than 450 mm (18 in) deep and without concrete enclosement or equivalent.

Exception No. 7. Raceways installed in solid rock shall be permitted to be buried at lesser depth where covered by 50 mm (2 in) of concrete, which shall be permitted to extend to the rock surface.

TABLE 4.4.1

Conductor Voltage Rating (Volts)	Minimum Conductor Size (AWG)	
	Copper	Aluminum or Copper-Clad Aluminum
0–2000	14	12
2001–8000	8	8
8001–15,000	2	2
15,001–28,000	1	1
28,001–35,000	1/0	1/0

4.4.1 *NEC* Table 310.5, Minimum Size of Conductors

Exception No. 1. For flexible cords as permitted by Section 400.12.

Exception No. 2. For fixture wire as permitted by Section 402.6.

Exception No. 3. For motors rated 1 hp or less as permitted by Section 430.22(C).

Exception No. 4. For cranes and hoists as permitted by Section 610.14.

Exception No. 5. For elevator control and signaling circuits as permitted by Section 620.12.

Exception No. 6. For class 1, class 2, and class 3 circuits as permitted by Sections 725.27(A) and 725.51, Exception.

Exception No. 7. For fire alarm circuits as permitted by Sections 760.27(A), 760.51, Exception, and 760.71(B).

Exception No. 8. For motor-control circuits as permitted by Section 430.72.

Exception No. 9. For control and instrumentation circuits as permitted by Section 727.6.

Exception No. 10. For electric signs and outline lighting as permitted in Sections 600.31(B) and 600.32(B).

TABLE 4.4.2

Trade Name	Type Letter	Maximum Operating Tempera-ture	Application Provisions	Insulation	Thickness of Insulation			Outer Covering[1]
					AWG or kcmil	mm	Mils	
Fluorinated ethylene propylene	FEP or FEPB	90°C 194°F	Dry and damp locations	Fluorinated ethylene propylene	14–10 8–2	0.51 0.76	20 30	None
					14–8	0.36	14	Glass braid
		200°C 392°F	Dry locations — special applications[2]	Fluorinated ethylene propylene	6–2	0.36	14	Glass or other suitable braid material
Mineral insulation (metal sheathed)	MI	90°C 194°F 250°C 482°F	Dry and wet locations For special applications[2]	Magnesium oxide	18–16[3] 16–10 9–4 3–500	0.58 0.91 1.27 1.40	23 36 50 55	Copper or alloy steel
Moisture-, heat-, and oil-resistant thermoplastic	MTW	60°C 140°F 90°C 194°F	Machine tool wiring in wet locations as permitted in NFPA 79 (See Article 670.) Machine tool wiring in dry locations as permitted in NFPA 79 (See Article 670.)	Flame-retardant moisture-, heat-, and oil-resistant thermo-plastic	 22–12 10 8 6 4–2 1–4/0 213–500 501–1000	(A) (B) 0.76 0.38 0.76 0.51 1.14 0.76 1.52 0.76 1.52 1.02 2.03 1.27 2.41 1.52 2.79 1.78	(A) (B) 30 15 30 20 45 30 60 30 60 40 80 50 95 60 110 70	(A) None (B) Nylon jacket or equivaler
Paper		85°C 185°F	For underground service conductors, or by special permission	Paper				Lead sheath
Perfluoro-alkoxy	PFA	90°C 194°F 200°C 392°F	Dry and damp locations Dry locations — special applications[2]	Perfluoro-alkoxy	14–10 8–2 1–4/0	0.51 0.76 1.14	20 30 45	None

(continued)

TABLE 4.4.2

Trade Name	Type Letter	Maximum Operating Tempera-ture	Application Provisions	Insulation	Thickness of Insulation			Outer Covering[1]
					AWG or kcmil	mm	Mils	
Perfluoro-alkoxy	PFAH	250°C 482°F	Dry locations only. Only for leads within apparatus or within raceways connected to apparatus (nickel or nickel-coated copper only)	Perfluoro-alkoxy	14–10 8–2 1–4/0	0.51 0.76 1.14	20 30 45	None
Thermoset	RHH	90°C 194°F	Dry and damp locations		14-10 8–2 1–4/0 213–500 501–1000 1001–2000 For 601–2000, *see* Table 310.62.	1.14 1.52 2.03 2.41 2.79 3.18	45 60 80 95 110 125	Moisture-resistant, flame-retardant, nonmetallic covering[1]
Moisture-resistant thermoset	RHW[4]	75°C 167°F	Dry and wet locations	Flame-retardant, moisture-resistant thermo-set	14–10 8–2 1–4/0 213–500 501–1000 1001–2000 For 601–2000, *see* Table 310.62.	1.14 1.52 2.03 2.41 2.79 3.18	45 60 80 95 110 125	Moisture-resistant, flame-retardant, nonmetallic covering[5]
Moisture-resistant thermoset	RHW-2	90°C 194°F	Dry and wet locations	Flame-retardant moisture-resistant thermo-set	14–10 8–2 1–4/0 213–500 501–1000 1001–2000 For 601–2000, see Table 310.62.	1.14 1.52 2.03 2.41 2.79 3.18	45 60 80 95 110 125	Moisture-resistant, flame-retardant, nonmetallic covering[5]
Silicone	SA	90°C 194°F 200°C 392°F	Dry and damp locations For special application[2]	 Silicone rubber	14–10 8–2 1–4/0 213–500 501–1000 1001–2000	1.14 1.52 2.03 2.41 3.18 3.18	45 60 80 95 110 125	Glass or other suitable braid material
Thermoset	SIS	90°C 194°F	Switchboard wiring only	Flame-retardant thermoset	14–10 8–2 1–4/0	0.76 1.14 2.41	30 45 95	None

(continued)

TABLE 4.4.2

Trade Name	Type Letter	Maximum Operating Tempera-ture	Application Provisions	Insulation	Thickness of Insulation			Outer Covering[1]
					AWG or kcmil	mm	Mils	
Thermoplastic and fibrous outer braid	TBS	90°C 194°F	Switchboard wiring only	Thermo-plastic	14–10 8 6–2 1–4/0	0.76 1.14 1.52 2.03	30 45 60 80	Flame-retardant, nonmetallic covering
Extended polytetra-fluoro-ethylene	TFE	250°C 482°F	Dry locations only. Only for leads within apparatus or within raceways connected to apparatus, or as open wiring (nickel or nickel-coated copper only)	Extruded polytetra-fluoro-ethylene	14–10 8–2 1–4/0	0.51 0.76 1.14	20 30 45	None
Heatresistant thermoplastic	THHN	90°C 194°F	Dry and damp locations	Flame-retardant, heat-resistant thermo-plastic	14–12 10 8–6 4–2 1–4/0 250–500 501–1000	0.38 0.51 0.76 1.02 1.27 1.52 1.78	15 20 30 40 50 60 70	Nylon jacket or equivalent
Moisture- and heat-resistant thermoplastic	THHW	75°C 167°F 90°C 194°F	Wet location Dry location	Flame-retardant, moisture- and heat-resistant thermo-plastic	14–10 8 6–2 1–4/0 213–500 501–1000	0.76 1.14 1.52 2.03 2.41 2.79	30 45 60 80 95 110	None
Moisture- and heat-resistant thermoplastic	THW [4]	75°C 167°F 90°C 194°F	Dry and wet locations Special applications within electric discharge lighting equipment. Limited to 1000 open-circuit volts or less. (size 14-8 only as permitted in 410.33)	Flame-retardant, moisture- and heat-resistant thermo-plastic	14–10 8 6–2 1–4/0 213–500 501–1000 1001–2000	0.76 1.14 1.52 2.03 2.41 2.79 3.18	30 45 60 80 95 110 125	None
Moisture- and heat-resistant thermoplastic	THWN[4]	75°C 167°F	Dry and wet locations	Flame-retardant, moisture- and heat-resistant thermo-plastic	14–12 10 8–6 4–2 1–4/0 250–500 501–1000	0.38 0.51 0.76 1.02 1.27 1.52 1.78	15 20 30 40 50 60 70	Nylon jacket or equivalent

(continued)

TABLE 4.4.2

Trade Name	Type Letter	Maximum Operating Tempera- ture	Application Provisions	Insulation	Thickness of Insulation			Outer Covering[1]
					AWG or kcmil	mm	Mils	
Moisture-resistant thermo-plastic	TW	60°C 140°F	Dry and wet locations	Flame-retardant, moisture-resistant thermo-plastic	14–10 8 6–2 1–4/0 213–500 501–1000 1001–2000	0.76 1.14 1.52 2.03 2.41 2.79 3.18	30 45 60 80 95 110 125	None
Underground feeder and branch-circuit cable — single conductor (For Type UF cable employing more than one conductor, *see* Articles 339, 340.)	UF	60°C 140°F 75°C 167°F[7]	See Article 340.	Moisture-resistant Moisture- and heat-resistant	14–10 8–2 1–4/0	1.52 2.03 2.41	60[6] 80[6] 95[6]	Integral with insulation
Underground service-entrance cable — single conductor (For Type USE cable employing more than one conductor, *see* Article 338.)	USE[4]	75°C 167°F	See Article 338.	Heat- and moisture-resistant	14–10 8–2 1–4/0 213–500 501–1000 1001–2000	1.14 1.52 2.03 2.41 2.79 3.18	45 60 80 95[8] 110 125	Moisture-resistant nonmetallic covering (See 338.2.)
Thermoset	XHH	90°C 194°F	Dry and damp locations	Flame-retardant thermoset	14–10 8–2 1–4/0 213–500 501–1000 1001–2000	0.76 1.14 1.40 1.65 2.03 2.41	30 45 55 65 80 95	None
Moisture-resistant thermoset	XHHW[4]	90°C 194°F 75°C 167°F	Dry and damp locations Wet locations	Flame-retardant, moisture-resistant thermoset	14–10 8–2 1–4/0 213–500 501–1000 1001–2000	0.76 1.14 1.40 1.65 2.03 2.41	30 45 55 65 80 95	None

(continued)

TABLE 4.4.2

Trade Name	Type Letter	Maximum Operating Temperature	Application Provisions	Insulation	Thickness of Insulation			Outer Covering[1]
					AWG or kcmil	mm	Mils	
Moisture-resistant thermoset	XHHW-2	90°C 194°F	Dry and wet locations	Flame-retardant, moisture-resistant thermoset	14–10 8–2 1–4/0 213–500 501–1000 1001–2000	0.76 1.14 1.40 1.65 2.03 2.41	30 45 55 65 80 95	None
Modified ethylene tetra-fluoro-ethylene	Z	90°C 194°F 150°C 302°F	Dry and damp locations Dry locations — special applications[2]	Modified ethylene tetra-fluoro-ethylene	14–12 10 8–4 3–1 1/0–4/0	0.38 0.51 0.64 0.89 1.14	15 20 25 35 45	None
Modified ethylene tetra-fluoro-ethylene	ZW[4]	75°C 167°F 90°C 194°F 150°C 302°F	Wet locations Dry and damp locations Dry locations — special applications[2]	Modified ethylene tetra-fluoro-ethylene	14–10 8–2	0.76 1.14	30 45	None

[1] Some insulations do not require an outer covering.

[2] Where design conditions require maximum conductor operating temperatures above 90°C (194°F).

[3] For signaling circuits permitting 300-volt insulation.

[4] Listed wire types designated with the suffix "2," such as RHW-2, shall be permitted to be used at a continuous 90°C (194°F) operating temperature, wet or dry.

[5] Some rubber insulations do not require an outer covering.

[6] Includes integral jacket.

[7] For ampacity limitation, see 340.80.

[8] Insulation thickness shall be permitted to be 2.03 mm (80 mils) for listed Type USE conductors that have been subjected to special investigations. The nonmetallic covering over individual rubber-covered conductors of aluminum-sheathed cable and of lead-sheathed or multiconductor cable shall not be required to be flame retardant.
For Type MC cable, see 330.104. For nonmetallic-sheathed cable, *see* Article 334, Part III. For Type UF cable, *see* Article 340, Part III.

4.4.2 *NEC* Table 310.13, Conductor Application and Insulations

TABLE 4.4.3

Characteristic	Copper	Copper-Clad Aluminum	Aluminum
Density (lb/in.^{3})	0.323	0.121	0.098
Density (g/cm^3)	8.91	3.34	2.71
Resistivity ohms/CMF	10.37	16.08	16.78
Resistivity Microhm — CM	1.724	2.673	2.790
Conductivity (IACS %)	100	61–63	61.0
Weight % Copper	100	26.8	—
Tensile K psi — Hard	65.0	30.0	27.0
Tensile kg/mm^2 — Hard	45.7	21.1	19.0
Tensile K psi — Annealed	35.0	17.0	17.0*
Tensile kg/mm^2 — Annealed	24.6	12.0	12.0
Specific Gravity	8.91	3.34	2.71

*Semi-annealed

4.4.3 Conductor Characteristics

TABLE 4.4.4

Number of Current-Carrying Conductors	Percent of Values in Tables 310.16 through 310.19 as Adjusted for Ambient Temperature if Necessary
4–6	80
7–9	70
10–20	50
21–30	45
31–40	40
41 and above	35

FPN: See Annex B, Table B.310.11, for adjustment factors for more than three current-carrying conductors in a raceway or cable with load diversity.

4.4.4 *NEC* Table 310.15(B)(2)(a), Adjustment Factors for More Than Three Current-Carrying Conductors in a Raceway or Cable

Exception No. 1. Where conductors of different systems, as provided in Section 300.3, are installed in a common raceway or cable, the derating factors shown in Table 310.15(B)(2)(a) shall apply to the number of power and lighting conductors only (Articles 210, 215, 220, and 230).

Exception No. 2. For conductors installed in cable trays, the provisions of Section 318.11 shall apply.

Exception No. 3. Derating factors shall not apply to conductors in nipples having a length not exceeding 600 mm (24 in).

Exception No. 4. Derating factors shall not apply to underground conductors entering or leaving an outdoor trench if those conductors have physical protection in the form of rigid metal conduit, intermediate metal conduit, or rigid nonmetallic conduit having a length not exceeding 3.05 m (10 ft) and the number of conductors does not exceed four.

Exception No. 6. Adjustment factors shall not apply to Type AC cable or to Type MC cable without an overall outer jacket under the following conditions: (a) Each cable has not more than three current-carrying conductors; (b) The conductors are 12 AWG copper; (c) Not more than 20 current-carrying conductors are bundled, stacked, or supported on "bridle rings." A 60-percent adjustment factor shall be applied where the current-carrying conductors in these cables that are stacked or bundled longer than 600 mm (24 in) without maintaining spacing exceeds 20.

TABLE 4.4.5

Size AWG or kcmil	Temperature Rating of Conductor (See Table 310.13.)						Size AWG or kcmil
	60°C (140°F)	75°C (167°F)	90°C (194°F)	60°C (140°F)	75°C (167°F)	90°C (194°F)	
	Types TW, UF	Types RHW, THHW, THW, THWN, XHHW, USE, ZW	Types TBS, SA, SIS, FEP, FEPB, MI, RHH, RHW-2, THHN, THHW, THW-2, THWN-2, USE-2, XHH, XHHW, XHHW-2, ZW-2	Types TW, UF	Types RHW, THHW, THW, THWN, XHHW, USE	Types TBS, SA, SIS, THHN, THHW, THW-2, THWN-2, RHH, RHW-2, USE-2, XHH, XHHW, XHHW-2, ZW-2	
	COPPER			ALUMINUM OR COPPER-CLAD ALUMINUM			
18	—	—	14	—	—	—	—
16	—	—	18	—	—	—	—
14*	20	20	25	—	—	—	—
12*	25	25	30	20	20	25	12*
10*	30	35	40	25	30	35	10*
8	40	50	55	30	40	45	8
6	55	65	75	40	50	60	6
4	70	85	95	55	65	75	4
3	85	100	110	65	75	85	3
2	95	115	130	75	90	100	2
1	110	130	150	85	100	115	1
1/0	125	150	170	100	120	135	1/0
2/0	145	175	195	115	135	150	2/0
3/0	165	200	225	130	155	175	3/0
4/0	195	230	260	150	180	205	4/0
250	215	255	290	170	205	230	250
300	240	285	320	190	230	255	300
350	260	310	350	210	250	280	350
400	280	335	380	225	270	305	400
500	320	380	430	260	310	350	500
600	355	420	475	285	340	385	600
700	385	460	520	310	375	420	700
750	400	475	535	320	385	435	750
800	410	490	555	330	395	450	800
900	435	520	585	355	425	480	900
1000	455	545	615	375	445	500	1000
1250	495	590	665	405	485	545	1250
1500	520	625	705	435	520	585	1500
1750	545	650	735	455	545	615	1750
2000	560	665	750	470	560	630	2000
	CORRECTION FACTORS						
Ambient Temp. (°C)	For ambient temperatures other than 30°C (86°F), multiply the allowable ampacities shown above by the appropriate factor shown below.						Ambient Temp. (°F)
21–25	1.08	1.05	1.04	1.08	1.05	1.04	70–77
26–30	1.00	1.00	1.00	1.00	1.00	1.00	78–86
31–35	0.91	0.94	0.96	0.91	0.94	0.96	87–95
36–40	0.82	0.88	0.91	0.82	0.88	0.91	96–104
41–45	0.71	0.82	0.87	0.71	0.82	0.87	105–113
46–50	0.58	0.75	0.82	0.58	0.75	0.82	114–122
51–55	0.41	0.67	0.76	0.41	0.67	0.76	123–131
56–60	—	0.58	0.71	—	0.58	0.71	132–140
61–70	—	0.33	0.58	—	0.33	0.58	141–158
71–80	—	—	0.41	—	—	0.41	159–176

* See 240.4(D).

4.4.5 *NEC* Table 310.16, Allowable Ampacities of Insulated Conductors Rated 0 through 2000 V, 60°C through 90°C (140°F through 194°F), Not More Than Three Current-Carrying Conductors in a Raceway, Cable, or Earth (Directly Buried), Based on Ambient Air Temperature of 30°C (86°F) (see *page 4.17*)

TABLE 4.4.6

	Temperature Rating of Conductor (See Table 310.13.)						
	60°C (140°F)	75°C (167°F)	90°C (194°F)	60°C (140°F)	75°C (167°F)	90°C (194°F)	
	Types TW, UF	Types RHW, THHW, THW, THWN, XHHW, ZW	Types TBS, SA, SIS, FEP, FEPB, MI, RHH, RHW-2, THHN, THHW, THW-2, THWN-2, USE-2, XHH, XHHW, XHHW-2, ZW-2	Types TW, UF	Types RHW, THHW, THW, THWN, XHHW	Types TBS, SA, SIS, THHN, THHW, THW-2, THWN-2, RHH, RHW-2, USE-2, XHH, XHHW, XHHW-2, ZW-2	
Size AWG or kcmil	COPPER			ALUMINUM OR COPPER-CLAD ALUMINUM			Size AWG or kcmil
18	—	—	18	—	—	—	—
16	—	—	24	—	—	—	—
14*	25	30	35	—	—	—	—
12*	30	35	40	25	30	35	12*
10*	40	50	55	35	40	40	10*
8	60	70	80	45	55	60	8
6	80	95	105	60	75	80	6
4	105	125	140	80	100	110	4
3	120	145	165	95	115	130	3
2	140	170	190	110	135	150	2
1	165	195	220	130	155	175	1
1/0	195	230	260	150	180	205	1/0
2/0	225	265	300	175	210	235	2/0
3/0	260	310	350	200	240	275	3/0
4/0	300	360	405	235	280	315	4/0
250	340	405	455	265	315	355	250
300	375	445	505	290	350	395	300
350	420	505	570	330	395	445	350
400	455	545	615	355	425	480	400
500	515	620	700	405	485	545	500
600	575	690	780	455	540	615	600
700	630	755	855	500	595	675	700
750	655	785	885	515	620	700	750
800	680	815	920	535	645	725	800
900	730	870	985	580	700	785	900
1000	780	935	1055	625	750	845	1000
1250	890	1065	1200	710	855	960	1250
1500	980	1175	1325	795	950	1075	1500
1750	1070	1280	1445	875	1050	1185	1750
2000	1155	1385	1560	960	1150	1335	2000

(continued)

TABLE 4.4.6

Size AWG or kcmil	Temperature Rating of Conductor (See Table 310.13.)						Size AWG or kcmil
	60°C (140°F)	75°C (167°F)	90°C (194°F)	60°C (140°F)	75°C (167°F)	90°C (194°F)	
	Types TW, UF	Types RHW, THHW, THW, THWN, XHHW, ZW	Types TBS, SA, SIS, FEP, FEPB, MI, RHH, RHW-2, THHN, THHW, THW-2, THWN-2, USE-2, XHH, XHHW, XHHW-2, ZW-2	Types TW, UF	Types RHW, THHW, THW, THWN, XHHW	Types TBS, SA, SIS, THHN, THHW, THW-2, THWN-2, RHH, RHW-2, USE-2, XHH, XHHW, XHHW-2, ZW-2	
	COPPER			ALUMINUM OR COPPER-CLAD ALUMINUM			
	CORRECTION FACTORS						
Ambient Temp. (°C)	For ambient temperatures other than 30°C (86°F), multiply the allowable ampacities shown above by the appropriate factor shown below.						Ambient Temp. (°F)
21–25	1.08	1.05	1.04	1.08	1.05	1.04	70–77
26–30	1.00	1.00	1.00	1.00	1.00	1.00	78–86
31–35	0.91	0.94	0.96	0.91	0.94	0.96	87–95
36–40	0.82	0.88	0.91	0.82	0.88	0.91	96–104
41–45	0.71	0.82	0.87	0.71	0.82	0.87	105–113
46–50	0.58	0.75	0.82	0.58	0.75	0.82	114–122
51–55	0.41	0.67	0.76	0.41	0.67	0.76	123–131
56–60	—	0.58	0.71	—	0.58	0.71	132–140
61–70	—	0.33	0.58	—	0.33	0.58	141–158
71–80	—	—	0.41	—	—	0.41	159–176

* See 240.4(D).

4.4.6 *NEC* Table 310.17, Allowable Ampacities of Single-Insulated Conductors Rated 0 through 2000 V in Free Air, Based on Ambient Air Temperature of 30°C (86°F)

TABLE 4.4.7

	Temperature Rating of Conductor (See Table 310.13.)				
	150°C (302°F)	200°C (392°F)	250°C (482°F)	150°C (302°F)	
	Type Z	Types FEP, FEPB, PFA	Types PFAH, TFE	Type Z	
Size AWG or kcmil	COPPER		NICKEL OR NICKEL-COATED COPPER	ALUMINUM OR COPPER-CLAD ALUMINUM	Size AWG or kcmil
14	34	36	39	—	14
12	43	45	54	30	12
10	55	60	73	44	10
8	76	83	93	57	8
6	96	110	117	75	6
4	120	125	148	94	4
3	143	152	166	109	3
2	160	171	191	124	2
1	186	197	215	145	1
1/0	215	229	244	169	1/0
2/0	251	260	273	198	2/0
3/0	288	297	308	227	3/0
4/0	332	346	361	260	4/0
CORRECTION FACTORS					
Ambient Temp. (°C)	For ambient temperatures other than 40°C (104°F), multiply the allowable ampacities shown above by the appropriate factor shown below.				Ambient Temp. (°F)
41–50	0.95	0.97	0.98	0.95	105–122
51–60	0.90	0.94	0.95	0.90	123–140
61–70	0.85	0.90	0.93	0.85	141–158
71–80	0.80	0.87	0.90	0.80	159–176
81–90	0.74	0.83	0.87	0.74	177–194
91–100	0.67	0.79	0.85	0.67	195–212
101–120	0.52	0.71	0.79	0.52	213–248
121–140	0.30	0.61	0.72	0.30	249–284
141–160	—	0.50	0.65	—	285–320
161–180	—	0.35	0.58	—	321–356
181–200	—	—	0.49	—	357–392
201–225	—	—	0.35	—	393–437

4.4.7 *NEC* Table 310.18, Allowable Ampacities of Insulated Conductors, Rated 0 through 2000 V, 150°C through 250°C (302°F through 482°F), Not More Than Three Current-Carrying Conductors in Raceway or Cable, Based on Ambient Air Temperature of 40°C (104°F)

TABLE 4.4.8

	Temperature Rating of Conductor (See Table 310.13.)				
	150°C (302°F)	200°C (392°F)	250°C (482°F)	150°C (302°F)	
	Type Z	Types FEP, FEPB, PFA	Types PFAH, TFE	Type Z	
Size AWG or kcmil	COPPER		NICKEL, OR NICKEL-COATED COPPER	ALUMINUM OR COPPER-CLAD ALUMINUM	Size AWG or kcmil
14	46	54	59	—	14
12	60	68	78	47	12
10	80	90	107	63	10
8	106	124	142	83	8
6	155	165	205	112	6
4	190	220	278	148	4
3	214	252	327	170	3
2	255	293	381	198	2
1	293	344	440	228	1
1/0	339	399	532	263	1/0
2/0	390	467	591	305	2/0
3/0	451	546	708	351	3/0
4/0	529	629	830	411	4/0
CORRECTION FACTORS					
Ambient Temp. (°C)	For ambient temperatures other than 40°C (104°F), multiply the allowable ampacities shown above by the appropriate factor shown below.				Ambient Temp. (°F)
41–50	0.95	0.97	0.98	0.95	105–122
51–60	0.90	0.94	0.95	0.90	123–140
61–70	0.85	0.90	0.93	0.85	141–158
71–80	0.80	0.87	0.90	0.80	159–176
81–90	0.74	0.83	0.87	0.74	177–194
91–100	0.67	0.79	0.85	0.67	195–212
101–120	0.52	0.71	0.79	0.52	213–248
121–140	0.30	0.61	0.72	0.30	249–284
141–160	—	0.50	0.65	—	285–320
161–180	—	0.35	0.58	—	321–356
181–200	—	—	0.49	—	357–392
201–225	—	—	0.35	—	393–437

4.4.8 *NEC* Table 310.19, Allowable Ampacities of Single-Insulated Conductors, Rated 0 through 2000 V, 150°C through 250°C (302°F through 482°F), in Free Air, Based on Ambient Air Temperature of

TABLE 4.4.9

	Temperature Rating of Conductor (See Table 310.13.)				
	75°C (167°F)	90°C (194°F)	75°C (167°F)	90°C (194°F)	
	Types RHW, THHW, THW, THWN, XHHW, ZW	Types MI, THHN, THHW, THW-2, THWN-2, RHH, RHW-2, USE-2, XHHW, XHHW-2, ZW-2	Types RHW, THW, THWN, THHW, XHHW	Types THHN, THHW, RHH, XHHW, RHW-2, XHHW-2, THW-2, THWN-2, USE-2, ZW-2	
Size AWG or kcmil	COPPER		ALUMINUM OR COPPER-CLAD ALUMINUM		Size AWG or kcmil
8	57	66	44	51	8
6	76	89	59	69	6
4	101	117	78	91	4
3	118	138	92	107	3
2	135	158	106	123	2
1	158	185	123	144	1
1/0	183	214	143	167	1/0
2/0	212	247	165	193	2/0
3/0	245	287	192	224	3/0
4/0	287	335	224	262	4/0
250	320	374	251	292	250
300	359	419	282	328	300
350	397	464	312	364	350
400	430	503	339	395	400
500	496	580	392	458	500
600	553	647	440	514	600
700	610	714	488	570	700
750	638	747	512	598	750
800	660	773	532	622	800
900	704	826	572	669	900
1000	748	879	612	716	1000
	CORRECTION FACTORS				
Ambient Temp. (°C)	For ambient temperatures other than 40°C (104°F), multiply the allowable ampacities shown above by the appropriate factor shown below.				Ambient Temp. (°F)
21–25	1.20	1.14	1.20	1.14	70–77
26–30	1.13	1.10	1.13	1.10	79–86
31–35	1.07	1.05	1.07	1.05	88–95
36–40	1.00	1.00	1.00	1.00	97–104
41–45	0.93	0.95	0.93	0.95	106–113
46–50	0.85	0.89	0.85	0.89	115–122
51–55	0.76	0.84	0.76	0.84	124–131
56–60	0.65	0.77	0.65	0.77	133–140
61–70	0.38	0.63	0.38	0.63	142–158
71–80	—	0.45	—	0.45	160–176

40°C (104°F)

4.4.9 *NEC* Table 310.20, Ampacities of Not More Than Three Single Insulated Conductors, Rated 0

TABLE 4.4.10

Copper Conductors				AAC Aluminum Conductors			
Bare		Covered		Bare		Covered	
AWG or kcmil	Amperes	AWG or kcmil	Amperes	AWG or kcmil	Amperes	AWG or kcmil	Amperes
8	98	8	103	8	76	8	80
6	124	6	130	6	96	6	101
4	155	4	163	4	121	4	127
2	209	2	219	2	163	2	171
1/0	282	1/0	297	1/0	220	1/0	231
2/0	329	2/0	344	2/0	255	2/0	268
3/0	382	3/0	401	3/0	297	3/0	312
4/0	444	4/0	466	4/0	346	4/0	364
250	494	250	519	266.8	403	266.8	423
300	556	300	584	336.4	468	336.4	492
500	773	500	812	397.5	522	397.5	548
750	1000	750	1050	477.0	588	477.0	617
1000	1193	1000	1253	556.5	650	556.5	682
—	—	—	—	636.0	709	636.0	744
—	—	—	—	795.0	819	795.0	860
—	—	—	—	954.0	920	—	—
—	—	—	—	1033.5	968	1033.5	1017
—	—	—	—	1272	1103	1272	1201
—	—	—	—	1590	1267	1590	1381
—	—	—	—	2000	1454	2000	1527

through 2000 V, Supported on a Messenger, Based on Ambient Air Temperature of 40°C (104°F)

TABLE 4.4.11

Trade Name	Type Letter	Maximum Operating Temperature	Application Provision	Insulation	Outer Covering
Medium voltage solid dielectric	MV-90 MV–105*	90°C 105°C	Dry or wet locations rated 2001 volts and higher	Thermoplastic or thermosetting	Jacket, sheath, or armor

*Where design conditions require maximum conductor temperatures above 90°C.

4.4.10 *NEC* Table 310.21, Ampacities of Bare or Covered Conductors in Free Air, Based on 40°C

TABLE 4.4.12

Conductor Size (AWG or kcmil)	Column A[1]		Column B[2]	
	mm	mils	mm	mils
14–10	2.03	80	1.52	60
8	2.03	80	1.78	70
6–2	2.41	95	1.78	70
1–2/0	2.79	110	2.29	90
3/0–4/0	2.79	110	2.29	90
213–500	3.18	125	2.67	105
501–1000	3.56	140	3.05	120

[1]Column A insulations are limited to natural, SBR, and butyl rubbers.
[2]Column B insulations are materials such as cross-linked polyethylene, ethylene propylene rubber, and composites thereof.

(104°F) Ambient, 80°C (176°F) Total Conductor Temperature, 610 mm/s (2 ft/s) Wind Velocity

TABLE 4.4.13

Conductor Size (AWG or kcmil)	2001–5000 Volts												5001–8000 Volts 100 Percent Insulation Level Wet or Dry Locations					
	Dry Locations, Single Conductor						Wet or Dry Locations											
	Without Jacket Insulation		With Jacket				Single Conductor				Multi-conductor Insulation*		Single Conductor				Multi-conductor Insulation*	
			Insulation		Jacket		Insulation		Jacket				Insulation		Jacket			
	mm	mils	mm	mils	mm	mils	mm	mils	mm	mils	mm	mils	mm	mils	mm	mils	mm	mils
8	2.79	110	2.29	90	0.76	30	3.18	125	2.03	80	2.29	90	4.57	180	2.03	80	4.57	180
6	2.79	110	2.29	90	0.76	30	3.18	125	2.03	80	2.29	90	4.57	180	2.03	80	4.57	180
4–2	2.79	110	2.29	90	1.14	45	3.18	125	2.03	80	2.29	90	4.57	180	2.41	95	4.57	180
1–2/0	2.79	110	2.29	90	1.14	45	3.18	125	2.03	80	2.29	90	4.57	180	2.41	95	4.57	180
3/0–4/0	2.79	110	2.29	90	1.65	65	3.18	125	2.41	95	2.29	90	4.57	180	2.79	110	4.57	180
213–500	3.05	120	2.29	90	1.65	65	3.56	140	2.79	110	2.29	90	5.33	210	2.79	110	5.33	210
501–750	3.30	130	2.29	90	1.65	65	3.94	155	3.18	125	2.29	90	5.97	235	3.18	125	5.97	235
751–1000	3.30	130	2.29	90	1.65	65	3.94	155	3.18	125	2.29	90	6.35	250	3.56	140	6.35	250

*Under a common overall covering such as a jacket, sheath, or armor.

4.4.11 *NEC* Table 310.61, Conductor Application and Insulation

TABLE 4.4.14

Conductor Size (AWG or kcmil)	2001–5000 Volts		5001–8000 Volts				8001–15,000 Volts				15,001–25,000 Volts				25,001–28,000 Volts				28,001–35,000 Volts			
			100 Percent Insulation Level 1		133 Percent Insulation Level 2		100 Percent Insulation Level 1		133 Percent Insulation Level 2		100 Percent Insulation Level 1		133 Percent Insulation Level 2		100 Percent Insulation Level 1		133 Percent Insulation Level 2		100 Percent Insulation Level 1		133 Percent Insulation Level 2	
	mm	mils	mm	mils	mm	mils	mm	mils	mm	mils	mm	mils	mm	mils	mm	mils	mm	mils	mm	mils	mm	mils
8	2.29	90	—	—	—	—	—	—	—	—	—	—	—	—	—	—	—	—	—	—	—	—
6–4	2.29	90	2.92	115	3.56	140	—	—	—	—	—	—	—	—	—	—	—	—	—	—	—	—
2	2.29	90	2.92	115	3.56	140	4.45	175	5.46	215	—	—	—	—	—	—	—	—	—	—	—	—
1	2.29	90	2.92	115	3.56	140	4.45	175	5.46	215	6.60	260	8.76	345	7.11	280	8.76	345	—	—	—	—
1/0–2000	2.29	90	2.92	115	3.56	140	4.45	175	5.46	215	6.60	260	8.76	345	7.11	280	8.76	345	8.76	345	10.67	420

[1]**100 Percent Insulation Level.** Cables in this category shall be permitted to be applied where the system is provided with relay protection such that ground faults will be cleared as rapidly as possible but, in any case, within 1 minute. While these cables are applicable to the great majority of cable installations that are on grounded systems, they shall be permitted to be used also on other systems for which the application of cables is acceptable, provided the above clearing requirements are met in completely de-energizing the faulted section.

[2]**133 Percent Insulation Level.** This insulation level corresponds to that formerly designated for ungrounded systems. Cables in this category shall be permitted to be applied in situations where the clearing time requirements of the 100 percent level category cannot be met and yet there is adequate assurance that the faulted section will be de-energized in a time not exceeding 1 hour. Also, they shall be permitted to be used where additional insulation strength over the 100 percent level category is desirable.

4.4.12 *NEC* Table 310.62, Thickness of Insulation for 601- to 2000-V Nonshielded Types RHH and RHW

4.4.13 *NEC* Table 310.63, Thickness of Insulation

TABLE 4.4.15

Conductor Size (AWG or kcmil)	Temperature Rating of Conductor (See Table 310-61)			
	2001–5000 Volts Ampacity		5001–35,000 Volts Ampacity	
	90°C (194°F) Type MV-90	105°C (221°F) Type MV-105	90°C (194°F) Type MV-90	105°C (221°F) Type MV-105
8	65	74	—	—
6	90	99	100	110
4	120	130	130	140
2	160	175	170	195
1	185	205	195	225
1/0	215	240	225	255
2/0	250	275	260	295
3/0	290	320	300	340
4/0	335	375	345	390
250	375	415	380	430
350	465	515	470	525
500	580	645	580	650
750	750	835	730	820
1000	880	980	850	950

and Jacket for Nonshielded Solid Dielectric Insulated Conductors Rated 2001 to 8000 V

4.4.14 *NEC* Table 310.64, Thickness of Insulation

TABLE 4.4.16

Conductor Size (AWG or kcmil)	Temperature Rating of Conductor (See Table 310-61)			
	2001–5000 Volts Ampacity		5001–35,000 Volts Ampacity	
	90°C (194°F) Type MV-90	105°C (221°F) Type MV-105	90°C (194°F) Type MV-90	105°C (221°F) Type MV-105
8	50	57	—	—
6	70	77	75	84
4	90	100	100	110
2	125	135	130	150
1	145	160	150	175
1/0	170	185	175	200
2/0	195	215	200	230
3/0	225	250	230	265
4/0	265	290	270	305
250	295	325	300	335
350	365	405	370	415
500	460	510	460	515
750	600	665	590	660
1000	715	800	700	780

(© 2001, NFPA)

for Shielded Solid Dielectric Insulated Conductors Rated 2001 to 35,000 V

TABLE 4.4.17

Conductor Size (AWG or kcmil)	Temperature Rating of Conductor (See Table 310-61)					
	2001–5000 Volts Ampacity		5001–15,000 Volts Ampacity		15,001–35,000 Volts Ampacity	
	90°C (194°F) Type MV-90	105°C (221°F) Type MV-105	90°C (194°F) Type MV-90	105°C (221°F) Type MV-105	90°C (194°F) Type MV-90	105°C (221°F) Type MV-105
8	83	93	—	—	—	—
6	110	120	110	125	—	—
4	145	160	150	165	—	—
2	190	215	195	215	—	—
1	225	250	225	250	225	250
1/0	260	290	260	290	260	290
2/0	300	330	300	335	300	330
3/0	345	385	345	385	345	380
4/0	400	445	400	445	395	445
250	445	495	445	495	440	490
350	550	615	550	610	545	605
500	695	775	685	765	680	755
750	900	1000	885	990	870	970
1000	1075	1200	1060	1185	1040	1160
1250	1230	1370	1210	1350	1185	1320
1500	1365	1525	1345	1500	1315	1465
1750	1495	1665	1470	1640	1430	1595
2000	1605	1790	1575	1755	1535	1710

(© 2001, NFPA)

4.4.15 *NEC* Table 310.67, Ampacities of Insulated Single Copper Conductor Cables Triplexed in Air Based on Conductor Temperatures of 90°C (194°F) and 105°C (221°F) and Ambient Air Temperature of 40°C (104°F)

TABLE 4.4.18

	Temperature Rating of Conductor (See Table 310-61)					
	2001–5000 Volts Ampacity		5001–15,000 Volts Ampacity		15,001–35,000 Volts Ampacity	
Conductor Size (AWG or kcmil)	90°C (194°F) Type MV-90	105°C (221°F) Type MV-105	90°C (194°F) Type MV-90	105°C (221°F) Type MV-105	90°C (194°F) Type MV-90	105°C (221°F) Type MV-105
8	64	71	—	—	—	—
6	85	95	87	97	—	—
4	115	125	115	130	—	—
2	150	165	150	170	—	—
1	175	195	175	195	175	195
1/0	200	225	200	225	200	225
2/0	230	260	235	260	230	260
3/0	270	300	270	300	270	300
4/0	310	350	310	350	310	345
250	345	385	345	385	345	380
350	430	480	430	480	430	475
500	545	605	535	600	530	590
750	710	790	700	780	685	765
1000	855	950	840	940	825	920
1250	980	1095	970	1080	950	1055
1500	1105	1230	1085	1215	1060	1180
1750	1215	1355	1195	1335	1165	1300
2000	1320	1475	1295	1445	1265	1410

4.4.16 *NEC* Table 310.68, Ampacities of Insulated Single Aluminum Conductor Cables Triplexed in Air Based on Conductor Temperatures of 90°C (194°F) and 105°C (221°F) and Ambient Air Temperature of 40°C (104°F)

TABLE 4.4.19

	Temperature Rating of Conductor (See Table 310-61)			
	2001–5000 Volts Ampacity		5001–35,000 Volts Ampacity	
Conductor Size (AWG or kcmil)	90°C (194°F) Type MV-90	105°C (221°F) Type MV-105	90°C (194°F) Type MV-90	105°C (221°F) Type MV-105
8	59	66	—	—
6	79	88	93	105
4	105	115	120	135
2	140	154	165	185
1	160	180	185	210
1/0	185	205	215	240
2/0	215	240	245	275
3/0	250	280	285	315
4/0	285	320	325	360
250	320	355	360	400
350	395	440	435	490
500	485	545	535	600
750	615	685	670	745
1000	705	790	770	860

4.4.17 *NEC* Table 310.69, Ampacities of Insulated Single Copper Conductor Isolated in Air Based on Conductor Temperatures of 90°C (194°F) and 105°C (221°F) and Ambient Air Temperature of 40°C (104°F)

TABLE 4.4.20

	Temperature Rating of Conductor (See Table 310-61)			
	2001–5000 Volts Ampacity		5001–35,000 Volts Ampacity	
Conductor Size (AWG or kcmil)	90°C (194°F) Type MV-90	105°C (221°F) Type MV-105	90°C (194°F) Type MV-90	105°C (221°F) Type MV-105
8	46	51	—	—
6	61	68	72	80
4	81	90	95	105
2	110	120	125	145
1	125	140	145	165
1/0	145	160	170	185
2/0	170	185	190	215
3/0	195	215	220	245
4/0	225	250	255	285
250	250	280	280	315
350	310	345	345	385
500	385	430	425	475
750	495	550	540	600
1000	585	650	635	705

4.4.18 *NEC* Table 310.70, Ampacities of Insulated Single Aluminum Conductor Isolated in Air Based on Conductor Temperatures of 90°C (194°F) and 105°C (221°F) and Ambient Air Temperature of 40°F (104°F)

TABLE 4.4.21

	Temperature Rating of Conductor (See Table 310-61)			
	2001–5000 Volts Ampacity		5001–35,000 Volts Ampacity	
Conductor Size (AWG or kcmil)	90°C (194°F) Type MV-90	105°C (221°F) Type MV-105	90°C (194°F) Type MV-90	105°C (221°F) Type MV-105
8	55	61	—	—
6	75	84	83	93
4	97	110	110	120
2	130	145	150	165
1	155	175	170	190
1/0	180	200	195	215
2/0	205	225	225	255
3/0	240	270	260	290
4/0	280	305	295	330
250	315	355	330	365
350	385	430	395	440
500	475	530	480	535
750	600	665	585	655
1000	690	770	675	755

4.4.19 *NEC* Table 310.71, Ampacities of an Insulated Three-Conductor Copper Cable Isolated in Air Based on Conductor Temperatures of 90°C (194°F) and 105°C (221°F) and Ambient Air Temperature of 40°C (104°F)

TABLE 4.4.22

	Temperature Rating of Conductor (See Table 310-61)			
	2001–5000 Volts Ampacity		5001–35,000 Volts Ampacity	
Conductor Size (AWG or kcmil)	90°C (194°F) Type MV-90	105°C (221°F) Type MV-105	90°C (194°F) Type MV-90	105°C (221°F) Type MV-105
8	43	48	—	—
6	58	65	65	72
4	76	85	84	94
2	100	115	115	130
1	120	135	130	150
1/0	140	155	150	170
2/0	160	175	175	200
3/0	190	210	200	225
4/0	215	240	230	260
250	250	280	255	290
350	305	340	310	350
500	380	425	385	430
750	490	545	485	540
1000	580	645	565	640

(© 2001, NFPA)

4.4.20 *NEC* Table 310.72, Ampacities of Insulated Three-Conductor Aluminum Cable Isolated in Air Based on Conductor Temperatures of 90°C (194°F) and 105°C (221°F) and Ambient Air Temperature of 40°C (104°F)

TABLE 4.4.23

	Temperature Rating of Conductor (See Table 310-61)			
	2001–5000 Volts Ampacity		5001–35,000 Volts Ampacity	
Conductor Size (AWG or kcmil)	90°C (194°F) Type MV-90	105°C (221°F) Type MV-105	90°C (194°F) Type MV-90	105°C (221°F) Type MV-105
8	52	58	—	—
6	69	77	83	92
4	91	100	105	120
2	125	135	145	165
1	140	155	165	185
1/0	165	185	195	215
2/0	190	210	220	245
3/0	220	245	250	280
4/0	255	285	290	320
250	280	315	315	350
350	350	390	385	430
500	425	475	470	525
750	525	585	570	635
1000	590	660	650	725

(© 2001, NFPA)

4.4.21 *NEC* Table 310.73, Ampacities of an Insulated Triplexed or Three Single-Conductor Copper Cables in Isolated Conduit in Air Based on Conductor Temperatures of 90°C (194°F) and 105°C (221°F) and Ambient Air Temperature of 40°C (104°F)

TABLE 4.4.24

	Temperature Rating of Conductor (See Table 310-61)			
	2001–5000 Volts Ampacity		5001–35,000 Volts Ampacity	
Conductor Size (AWG or kcmil)	90°C (194°F) Type MV-90	105°C (221°F) Type MV-105	90°C (194°F) Type MV-90	105°C (221°F) Type MV-105
8	41	46	—	—
6	53	59	64	71
4	71	79	84	94
2	96	105	115	125
1	110	125	130	145
1/0	130	145	150	170
2/0	150	165	170	190
3/0	170	190	195	220
4/0	200	225	225	255
250	220	245	250	280
350	275	305	305	340
500	340	380	380	425
750	430	480	470	520
1000	505	560	550	615

4.4.22 *NEC* Table 310.74, Ampacities of an Insulated Triplexed or Three Single-Conductor Aluminum Cables in Isolated Conduit in Air Based on Conductor

4.4.25

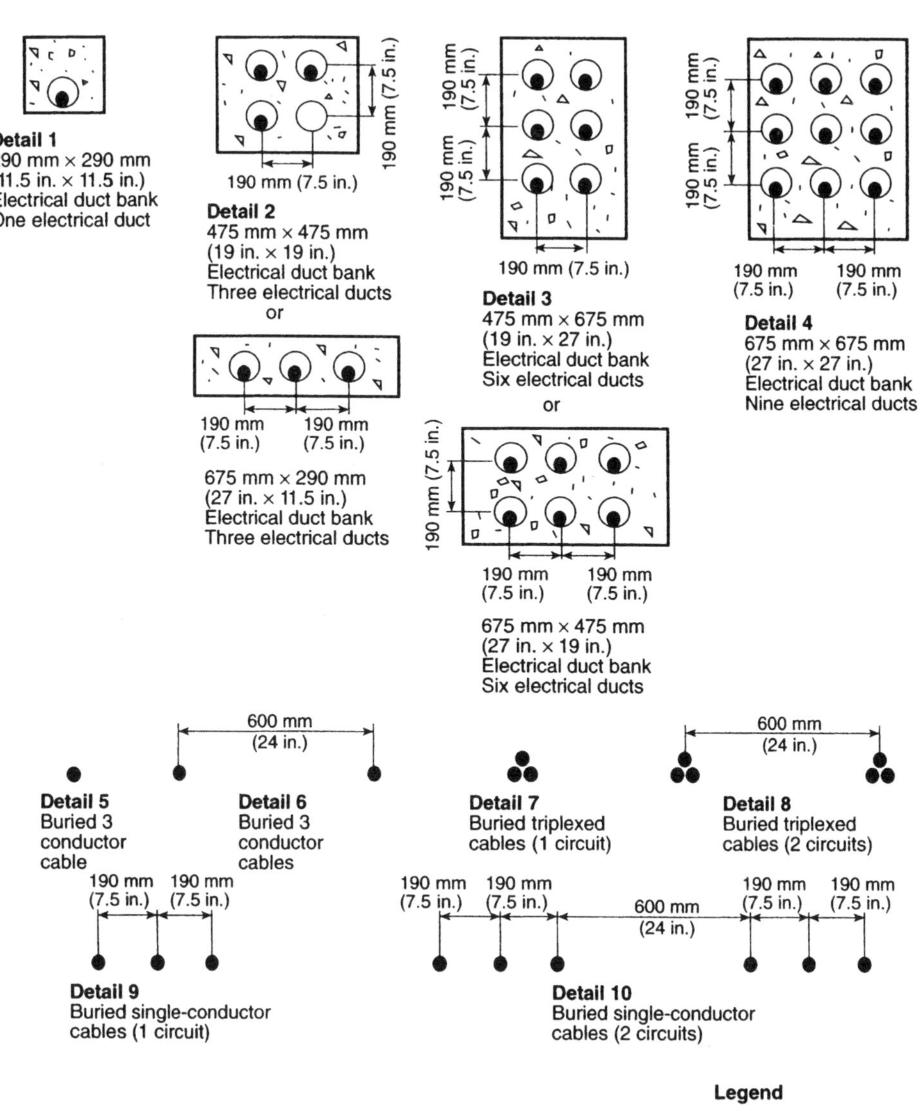

Note: Minimum burial depths to top electrical ducts or cables shall be in accordance with Section 300.50. Maximum depth to the top of electrical duct banks shall be 750 mm (30 in.) and maximum depth to the top of direct buried cables shall be 900 mm (36 in.).

Temperatures of 90°C (194°F) and 105°C (221°F) and Ambient Air Temperature of 40°C (104°F)

4.4.23 *NEC* Table 310.75, Ampacities of an Insulated Three-Conductor Copper Cable in Isolated Conduit in Air Based on Conductor Temperatures of 90°C (194°F) and 105°C (221°F) and Ambient Air

TABLE 4.4.26

	Temperature Rating of Conductor (See Table 310-61)			
	2001–5000 Volts Ampacity		5001–35,000 Volts Ampacity	
Conductor Size (AWG or kcmil)	90°C (194°F) Type MV-90	105°C (221°F) Type MV-105	90°C (194°F) Type MV-90	105°C (221°F) Type MV-105
One Circuit (See Figure 310-60, Detail 1)				
8	64	69	—	—
6	85	92	90	97
4	110	120	115	125
2	145	155	155	165
1	170	180	175	185
1/0	195	210	200	215
2/0	220	235	230	245
3/0	250	270	260	275
4/0	290	310	295	315
250	320	345	325	345
350	385	415	390	415
500	470	505	465	500
750	585	630	565	610
1000	670	720	640	690
Three Circuits (See Figure 310-60, Detail 2)				
8	56	60	—	—
6	73	79	77	83
4	95	100	99	105
2	125	130	130	135
1	140	150	145	155
1/0	160	175	165	175
2/0	185	195	185	200
3/0	210	225	210	225
4/0	235	255	240	255
250	260	280	260	280
350	315	335	310	330
500	375	405	370	395
750	460	495	440	475
1000	525	565	495	535
Six Circuits (See Figure 310-60, Detail 3)				
8	48	52	—	—
6	62	67	64	68
4	80	86	82	88
2	105	110	105	115
1	115	125	120	125
1/0	135	145	135	145
2/0	150	160	150	165
3/0	170	185	170	185
4/0	195	210	190	205
250	210	225	210	225
350	250	270	245	265
500	300	325	290	310
750	365	395	350	375
1000	410	445	390	415

Temperature of 40°C (104°F)

4.4.24 *NEC* Table 310.76, Ampacities of an Insulated Three-Conductor Aluminum Cable in Isolated Conduit in Air Based on Conductor Temperatures of 90°C (194°F) and 105°C (221°F) and Ambient Air Temperature of 40°C (104°F)

TABLE 4.4.27

	Temperature Rating of Conductor (See Table 310-61)			
	2001–5000 Volts Ampacity		5001–35,000 Volts Ampacity	
Conductor Size (AWG or kcmil)	90°C (194°F) Type MV-90	105°C (221°F) Type MV-105	90°C (194°F) Type MV-90	105°C (221°F) Type MV-105
One Circuit (See Figure 310-60, Detail 1)				
8	50	54	—	—
6	66	71	70	75
4	86	93	91	98
2	115	125	120	130
1	130	140	135	145
1/0	150	160	155	165
2/0	170	185	175	190
3/0	195	210	200	215
4/0	225	245	230	245
250	250	270	250	270
350	305	325	305	330
500	370	400	370	400
750	470	505	455	490
1000	545	590	525	565
Three Circuits (See Figure 310-60, Detail 2)				
8	44	47	—	—
6	57	61	60	65
4	74	80	77	83
2	96	105	100	105
1	110	120	110	120
1/0	125	135	125	140
2/0	145	155	145	155
3/0	160	175	165	175
4/0	185	200	185	200
250	205	220	200	220
350	245	265	245	260
500	295	320	290	315
750	370	395	355	385
1000	425	460	405	440
Six Circuits (See Figure 310-60, Detail 3)				
8	38	41	—	—
6	48	52	50	54
4	62	67	64	69
2	80	86	80	88
1	91	98	90	99
1/0	105	110	105	110
2/0	115	125	115	125
3/0	135	145	130	145
4/0	150	165	150	160
250	165	180	165	175
350	195	210	195	210
500	240	255	230	250
750	290	315	280	305
1000	335	360	320	345

4.4.25 *NEC* Figure 310.60, Cable Installation Dimensions for Use with Tables 4.4.26 through 4.4.35 (*NEC* Tables 310.77 through 310.86)

4.4.26 *NEC* Table 310.77, Ampacities of Three Single-Insulated Copper Conductors in Underground Electrical Ducts (Three Conductors Per Electrical Duct) Based on Ambient Earth Temperature of 20°C

TABLE 4.4.28

Conductor Size (AWG or kcmil)	Temperature Rating of Conductor (See Table 310-61)			
	2001–5000 Volts Ampacity		5001–35,000 Volts Ampacity	
	90°C (194°F) Type MV-90	105°C (221°F) Type MV-105	90°C (194°F) Type MV-90	105°C (221°F) Type MV-105
One Circuit (See Figure 310-60, Detail 1)				
8	59	64	—	—
6	78	84	88	95
4	100	110	115	125
2	135	145	150	160
1	155	165	170	185
1/0	175	190	195	210
2/0	200	220	220	235
3/0	230	250	250	270
4/0	265	285	285	305
250	290	315	310	335
350	355	380	375	400
500	430	460	450	485
750	530	570	545	585
1000	600	645	615	660
Three Circuits (See Figure 310-60, Detail 2)				
8	53	57	—	—
6	69	74	75	81
4	89	96	97	105
2	115	125	125	135
1	135	145	140	155
1/0	150	165	160	175
2/0	170	185	185	195
3/0	195	210	205	220
4/0	225	240	230	250
250	245	265	255	270
350	295	315	305	325
500	355	380	360	385
750	430	465	430	465
1000	485	520	485	515
Six Circuits (See Figure 310-60, Detail 3)				
8	46	50	—	—
6	60	65	63	68
4	77	83	81	87
2	98	105	105	110
1	110	120	115	125
1/0	125	135	130	145
2/0	145	155	150	160
3/0	165	175	170	180
4/0	185	200	190	200
250	200	220	205	220
350	240	270	245	275
500	290	310	290	305
750	350	375	340	365
1000	390	420	380	405

(68°F), Electrical Duct Arrangement per Figure 4.4.25 (*NEC* Figure 310-60), 100 Percent Load Factor, Thermal Resistance (RHO) of 90, Conductor Temperatures of 90°C (194°F) and 105°C (221°F)

4.4.27 *NEC* Table 310.78, Ampacities of Three Single-Insulated Aluminum Conductors in Underground Electrical Ducts (Three Conductors per Electrical

TABLE 4.4.29

	Temperature Rating of Conductor (See Table 310-61)			
	2001–5000 Volts Ampacity		5001–35,000 Volts Ampacity	
Conductor Size (AWG or kcmil)	90°C (194°F) Type MV-90	105°C (221°F) Type MV-105	90°C (194°F) Type MV-90	105°C (221°F) Type MV-105
One Circuit (See Figure 310-60, Detail 1)				
8	46	50	—	—
6	61	66	69	74
4	80	86	89	96
2	105	110	115	125
1	120	130	135	145
1/0	140	150	150	165
2/0	160	170	170	185
3/0	180	195	195	210
4/0	205	220	220	240
250	230	245	245	265
350	280	310	295	315
500	340	365	355	385
750	425	460	440	475
1000	495	535	510	545
Three Circuits (See Figure 310-60, Detail 2)				
8	41	44	—	—
6	54	58	59	64
4	70	75	75	81
2	90	97	100	105
1	105	110	110	120
1/0	120	125	125	135
2/0	135	145	140	155
3/0	155	165	160	175
4/0	175	185	180	195
250	190	205	200	215
350	230	250	240	255
500	280	300	285	305
750	345	375	350	375
1000	400	430	400	430
Six Circuits (See Figure 310-60, Detail 3)				
8	36	39	—	—
6	46	50	49	53
4	60	65	63	68
2	77	83	80	86
1	87	94	90	98
1/0	99	105	105	110
2/0	110	120	115	125
3/0	130	140	130	140
4/0	145	155	150	160
250	160	170	160	170
350	190	205	190	205
500	230	245	230	245
750	280	305	275	295
1000	320	345	315	335

Duct) Based on Ambient Earth Temperature of 20°C (68°F), Electrical Duct Arrangement per Figure 4.4.25 (*NEC* Figure 310.60), 100 Percent Load Factor, Thermal Resistance (RHO) of 90, Conductor Temperatures of 90°C (194°F) and 105°C (221°F)

TABLE 4.4.30

	Temperature Rating of Conductor (See Table 310-61)			
	2001–5000 Volts Ampacity		5001–35,000 Volts Ampacity	
Conductor Size (AWG or kcmil)	90°C (194°F) Type MV-90	105°C (221°F) Type MV-105	90°C (194°F) Type MV-90	105°C (221°F) Type MV-105
One Circuit, Three Conductors (See Figure 310-60, Detail 9)				
8	110	115	—	—
6	140	150	130	140
4	180	195	170	180
2	230	250	210	225
1	260	280	240	260
1/0	295	320	275	295
2/0	335	365	310	335
3/0	385	415	355	380
4/0	435	465	405	435
250	470	510	440	475
350	570	615	535	575
500	690	745	650	700
750	845	910	805	865
1000	980	1055	930	1005
Two Circuits, Six Conductors (See Figure 310-60, Detail 10)				
8	100	110	—	—
6	130	140	120	130
4	165	180	160	170
2	215	230	195	210
1	240	260	225	240
1/0	275	295	255	275
2/0	310	335	290	315
3/0	355	380	330	355
4/0	400	430	375	405
250	435	470	410	440
350	520	560	495	530
500	630	680	600	645
750	775	835	740	795
1000	890	960	855	920

4.4.28 *NEC* Table 310.79, Ampacities of Three Insulated Copper Conductors Cabled Within an Overall Covering (Three-Conductor Cable) in Underground Electrical Ducts (One Cable per Electrical Duct) Based on Ambient Earth Temperature of 20°C (68°F), Electrical Duct Arrangement per Figure 4.4.25 (*NEC* Figure 310.60), 100 Percent Load Factor, Thermal Resistance (RHO) of 90, Conductor

TABLE 4.4.31

	Temperature Rating of Conductor (See Table 310-61)			
	2001–5000 Volts Ampacity		5001–35,000 Volts Ampacity	
Conductor Size (AWG or kcmil)	90°C (194°F) Type MV-90	105°C (221°F) Type MV-105	90°C (194°F) Type MV-90	105°C (221°F) Type MV-105
One Circuit, Three Conductors (See Figure 310-60, Detail 9)				
8	85	90	—	—
6	110	115	100	110
4	140	150	130	140
2	180	195	165	175
1	205	220	185	200
1/0	230	250	215	230
2/0	265	285	245	260
3/0	300	320	275	295
4/0	340	365	315	340
250	370	395	345	370
350	445	480	415	450
500	540	580	510	545
750	665	720	635	680
1000	780	840	740	795
Two Circuits, Six Conductors (See Figure 310-60, Detail 10)				
8	80	85	—	—
6	100	110	95	100
4	130	140	125	130
2	165	180	155	165
1	190	200	175	190
1/0	215	230	200	215
2/0	245	260	225	245
3/0	275	295	255	275
4/0	310	335	290	315
250	340	365	320	345
350	410	440	385	415
500	495	530	470	505
750	610	655	580	625
1000	710	765	680	730

Temperatures of 90°C (194°F) and 105°C (221°F)

4.4.29 *NEC* Table 310.80, Ampacities of Three Insulated Aluminum Conductors Cabled Within an Overall Covering (Three-Conductor Cable) in Underground Electrical Ducts (One Cable per Electrical Duct) Based on Ambient Earth

TABLE 4.4.32

	Temperature Rating of Conductor (See Table 310-61)			
	2001–5000 Volts Ampacity		5001–35,000 Volts Ampacity	
Conductor Size (AWG or kcmil)	90°C (194°F) Type MV-90	105°C (221°F) Type MV-105	90°C (194°F) Type MV-90	105°C (221°F) Type MV-105
One Circuit (See Figure 310-60, Detail 5)				
8	85	89	—	—
6	105	115	115	120
4	135	150	145	155
2	180	190	185	200
1	200	215	210	225
1/0	230	245	240	255
2/0	260	280	270	290
3/0	295	320	305	330
4/0	335	360	350	375
250	365	395	380	410
350	440	475	460	495
500	530	570	550	590
750	650	700	665	720
1000	730	785	750	810
Two Circuits (See Figure 310-60, Detail 10)				
8	80	84	—	—
6	100	105	105	115
4	130	140	135	145
2	165	180	170	185
1	185	200	195	210
1/0	215	230	220	235
2/0	240	260	250	270
3/0	275	295	280	305
4/0	310	335	320	345
250	340	365	350	375
350	410	440	420	450
500	490	525	500	535
750	595	640	605	650
1000	665	715	675	730

Temperature of 20°C (68°F), Electrical Duct Arrangement per Figure 4.4.25 (*NEC* Figure 310.60), 100 Percent Load Factor, Thermal Resistance (RHO) of 90, Conductor Temperatures of 90°C (194°F) and 105°C (221°F)

4.4.30 *NEC* Table 310.81, Ampacities of Single-

TABLE 4.4.33

	Temperature Rating of Conductor (See Table 310-61)			
	2001–5000 Volts Ampacity		5001–35,000 Volts Ampacity	
Conductor Size (AWG or kcmil)	90°C (194°F) Type MV-90	105°C (221°F) Type MV-105	90°C (194°F) Type MV-90	105°C (221°F) Type MV-105
One Circuit (See Figure 310-60, Detail 5)				
8	65	70	—	—
6	80	88	90	95
4	105	115	115	125
2	140	150	145	155
1	155	170	165	175
1/0	180	190	185	200
2/0	205	220	210	225
3/0	230	250	240	260
4/0	260	280	270	295
250	285	310	300	320
350	345	375	360	390
500	420	450	435	470
750	520	560	540	580
1000	600	650	620	665
Two Circuits (See Figure 310-60, Detail 6)				
8	60	66	—	—
6	75	83	80	95
4	100	110	105	115
2	130	140	135	145
1	145	155	150	165
1/0	165	180	170	185
2/0	190	205	195	210
3/0	215	230	220	240
4/0	245	260	250	270
250	265	285	275	295
350	320	345	330	355
500	385	415	395	425
750	480	515	485	525
1000	550	590	560	600

Insulated Copper Conductors Directly Buried in Earth Based on Ambient Earth Temperature of 20°C (68°F), Arrangement per Figure 4.4.25 (*NEC* Figure 310.60), 100 Percent Load Factor, Thermal Resistance (RHO) of 90, Conductor Temperatures of 90°C (194°F) and 105°C (221°F)

TABLE 4.4.34

	Temperature Rating of Conductor (See Table 310-61)			
	2001–5000 Volts Ampacity		5001–35,000 Volts Ampacity	
Conductor Size (AWG or kcmil)	90°C (194°F) Type MV-90	105°C (221°F) Type MV-105	90°C (194°F) Type MV-90	105°C (221°F) Type MV-105
One Circuit, Three Conductors (See Figure 310-60, Detail 7)				
8	90	95	—	—
6	120	130	115	120
4	150	165	150	160
2	195	205	190	205
1	225	240	215	230
1/0	255	270	245	260
2/0	290	310	275	295
3/0	330	360	315	340
4/0	375	405	360	385
250	410	445	390	410
350	490	580	470	505
500	590	635	565	605
750	725	780	685	740
1000	825	885	770	830
Two Circuits, Six Conductors (See Figure 310-60, Detail 8)				
8	85	90	—	—
6	110	115	105	115
4	140	150	140	150
2	180	195	175	190
1	205	220	200	215
1/0	235	250	225	240
2/0	265	285	255	275
3/0	300	320	290	315
4/0	340	365	325	350
250	370	395	355	380
350	445	480	425	455
500	535	575	510	545
750	650	700	615	660
1000	740	795	690	745

4.4.31 *NEC* Table 310.82, Ampacities of Single-Insulated Aluminum Conductors Directly Buried in Earth Based on Ambient Earth Temperature of 20°C (68°F), Arrangement per Figure 4.4.25 (*NEC* Figure 310.60), 100 Percent Load Factor, Thermal Resistance (RHO) of 90, Conductor Temperatures of 90°C (194°F) and 105°C (221°F)

TABLE 4.4.35

	Temperature Rating of Conductor (See Table 310-61)			
	2001–5000 Volts Ampacity		5001–35,000 Volts Ampacity	
Conductor Size (AWG or kcmil)	90°C (194°F) Type MV-90	105°C (221°F) Type MV-105	90°C (194°F) Type MV-90	105°C (221°F) Type MV-105
One Circuit, Three Conductors (See Figure 310-60, Detail 7)				
8	70	75	—	—
6	90	100	90	95
4	120	130	115	125
2	155	165	145	155
1	175	190	165	175
1/0	200	210	190	205
2/0	225	240	215	230
3/0	255	275	245	265
4/0	290	310	280	305
250	320	350	305	325
350	385	420	370	400
500	465	500	445	480
750	580	625	550	590
1000	670	725	635	680
Two Circuits, Six Conductors (See Figure 310-60, Detail 8)				
8	65	70	—	—
6	85	95	85	90
4	110	120	105	115
2	140	150	135	145
1	160	170	155	170
1/0	180	195	175	190
2/0	205	220	200	215
3/0	235	250	225	245
4/0	265	285	255	275
250	290	310	280	300
350	350	375	335	360
500	420	455	405	435
750	520	560	485	525
1000	600	645	565	605

4.4.32 *NEC* Table 310.83, Ampacities of Three Insulated Copper Conductors Cabled Within an Overall Covering (Three-Conductor Cable) Directly

TABLE 4.5.1

Maximum Fuse Ampere Rating, Circuit Breaker Ampere Trip Setting, or Circuit Breaker Protective Relay Ampere Trip Setting for Ground-Fault Protection of Any Cable Circuit in the Cable Tray System	Minimum Cross-Sectional Area of Metal[a]			
	Steel Cable Trays		Aluminum Cable Trays	
	mm^2	in.^{2}	mm^2	in.^{2}
60	129	0.20	129	0.20
100	258	0.40	129	0.20
200	451.5	0.70	129	0.20
400	645	1.00	258	0.40
600	967.5	1.50[b]	258	0.40
1000	—	—	387	0.60
1200	—	—	645	1.00
1600	—	—	967.5	1.50
2000	—	—	1290	2.00[b]

[a]Total cross-sectional area of both side rails for ladder or trough cable trays; or the minimum cross-sectional area of metal in channel cable trays or cable trays of one-piece construction.

[b]Steel cable trays shall not be used as equipment grounding conductors for circuits with ground-fault protection above 600 amperes. Aluminum cable trays shall not be used as equipment grounding conductors for circuits with ground-fault protection above 2000 amperes.

Buried in Earth Based on Ambient Earth Temperature of 20°C (68°F), Arrangement per Figure 4.4.25 (*NEC* Figure 310.60), 100 Percent Load Factor, Thermal Resistance (RHO) of 90, Conductor

4.5.2

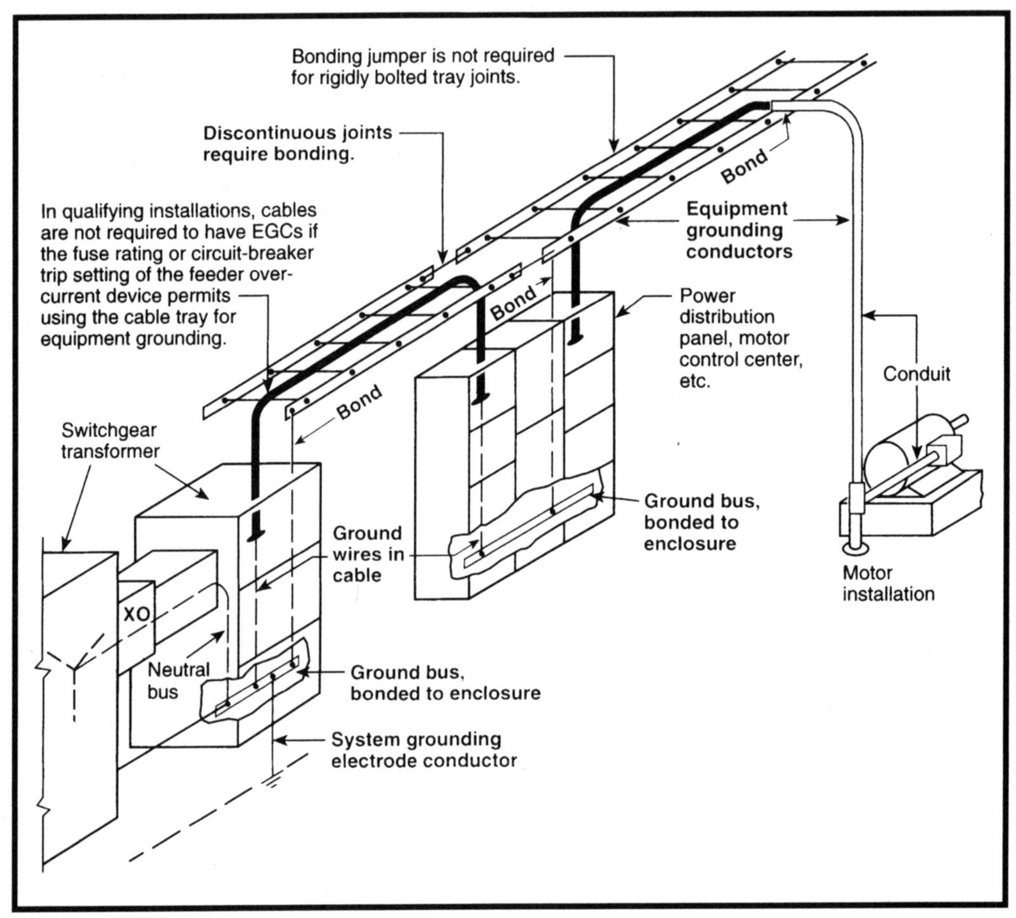

Temperatures of 90°C (194°F) and 105°C (221°F)

4.4.33 *NEC* Table 310.84, Ampacities of Three

TABLE 4.5.3

Inside Width of Cable Tray		Maximum Allowable Fill Area for Multiconductor Cables							
		Ladder or Ventilated Trough Cable Trays, 392.9(A)				Solid Bottom Cable Trays, 392.9(C)			
		Column 1 Applicable for 392.9(A)(2) Only		Column 2[a] Applicable for 392.9(A)(3) Only		Column 3 Applicable for 392.9(C)(2) Only		Column 4[a] Applicable for 392.9(C)(3) Only	
mm	in.	mm^2	in.^{2}	mm^2	in.^{2}	mm^2	in.^{2}	mm^2	in.^{2}
150	6.0	4,500	7.0	4,500 – (1.2 Sd)[b]	7 – (1.2 Sd)[b]	3,500	5.5	3,500 – Sd[b]	5.5 – Sdb
225	9.0	6,800	10.5	6,800 – (1.2 Sd)	10.5 – (1.2 Sd)	5,100	8.0	5,100 – Sd	8.0 – Sd
300	12.0	9,000	14.0	9,000 – (1.2 Sd)	14 – (1.2 Sd)	7,100	11.0	7,100 – Sd	11.0 – Sd
450	18.0	13,500	21.0	13,500 – (1.2 Sd)	21 – (1.2 Sd)	10,600	16.5	10,600 – Sd	16.5 – Sd
600	24.0	18,000	28.0	18,000 – (1.2 Sd)	28 – (1.2 Sd)	14,200	22.0	14,200 – Sd	22.0 – Sd
750	30.0	22,500	35.0	22,500 – (1.2 Sd)	35 – (1.2 Sd)	17,700	27.5	17,700 – Sd	27.5 – Sd
900	36.0	27,000	42.0	27,000 – (1.2 Sd)	42 – (1.2 Sd)	21,300	33.0	21,300 – Sd	33.0 – Sd

[a]The maximum allowable fill areas in Columns 2 and 4 shall be computed. For example, the maximum allowable fill in mm^2 for a 150-mm wide cable tray in Column 2 shall be 4500 minus (1.2 multiplied by Sd) [the maximum allowable fill, in square inches, for a 6-in. wide cable tray in Column 2 shall be 7 minus (1.2 multiplied by Sd)].
[b]The term *Sd* in Columns 2 and 4 is equal to the sum of the diameters, in mm, of all cables 107.2 mm (in inches, of all 4/0 AWG) and larger multiconductor cables in the same cable tray with smaller cables.

Insulated Aluminum Conductors Cabled Within an Overall Covering (Three-Conductor Cable) Directly Buried in Earth Based on Ambient Earth

TABLE 4.5.4

Inside Width of Cable Tray		Maximum Allowable Fill Area for Multiconductor Cables			
		Column 1 One Cable		Column 2 More Than One Cable	
mm	in.	mm^2	in.^{2}	mm^2	in.^{2}
75	3	1500	2.3	850	1.3
100	4	2900	4.5	1600	2.5
150	6	4500	7.0	2450	3.8

Temperature of 20°C (68°F), Arrangement per Figure 4.4.25 (*NEC* Figure 310.60), 100 Percent Load Factor, Thermal Resistance (RHO) of 90, Conductor

TABLE 4.5.4.1

Inside Width of Cable Tray		Column 1 One Cable		Column 2 More Than One Cable	
mm	in.	mm²	in.²	mm²	in.²
50	2	850	1.3	500	0.8
75	3	1300	2.0	700	1.1
100	4	2400	3.7	1400	2.1
150	6	3600	5.5	2100	3.2

(© 2001, NFPA)

Temperatures of 90°C (194°F) and 105°C (221°F)

4.4.34 *NEC* Table 310.85, Ampacities of Three

TABLE 4.5.5

Inside Width of Cable Tray		Maximum Allowable Fill Area for Single-Conductor Cables in Ladder or Ventilated-Trough Cable Trays: Column 1 Applicable for 392.10(A)(2) Only		Column 2[a] Applicable for 392.10(A)(3) Only	
mm	in.	mm²	in.²	mm²	in.²
150	6	4,200	6.5	4,200–(1.1Sd)[b]	6.5–(1.1Sd)[b]
225	9	6,100	9.5	6,100–(1.1 Sd)	9.5–(1.1 Sd)
300	12	8,400	13.0	8,400–(1.1 Sd)	13.0–(1.1 Sd)
450	18	12,600	19.5	12,600–(1.1 Sd)	19.5–(1.1 Sd)
600	24	16,800	26.0	16,800–(1.1 Sd)	26.0–(1.1 Sd)
750	30	21,000	32.5	21,000–(1.1 Sd)	32.5–(1.1 Sd)
900	36	25,200	39.0	25,200–(1.1 Sd)	39.0–(1.1 Sd)

[a]The maximum allowable fill areas in Column 2 shall be computed. For example, the maximum allowable fill, in mm² for a 150 mm wide cable tray in Column 2 shall be 4192.5 minus (1.1 multiplied by Sd) [the maximum allowable fill, in square inches, for a 6-in. wide cable tray in Column 2 shall be 6.5 minus (1.1 multiplied by Sd)].

[b]The term *Sd* in Column 2 is equal to the sum of the diameters, in mm, of all cables 507 mm² (in inches, of all 1000 kcmil) and larger single-conductor cables in the same ladder or ventilated-trough cable tray with small cables.

(© 2001, NFPA)

Triplexed Single-Insulated Copper Conductors Directly Buried in Earth Based on Ambient Earth Temperature of 20°C (68°F), Arrangement per Figure 4.4.25 (*NEC* Figure 310.60), 100 Percent Load Factor, Thermal Resistance (RHO) of 90, Conductor

4.5.6

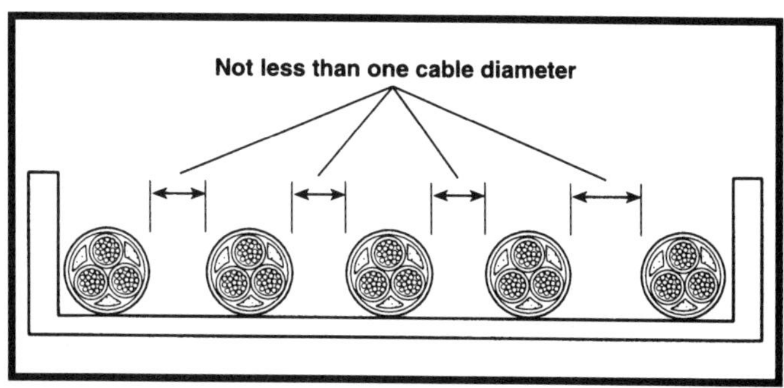

(© 1999, NFPA)

Temperatures of 90°C (194°F) and 105°C (221°F)

4.4.35 *NEC* Table 310.86, Ampacities of Three Triplexed Single-Insulated Aluminum Conductors Directly Buried in Earth Based on Ambient Earth

4.5.7

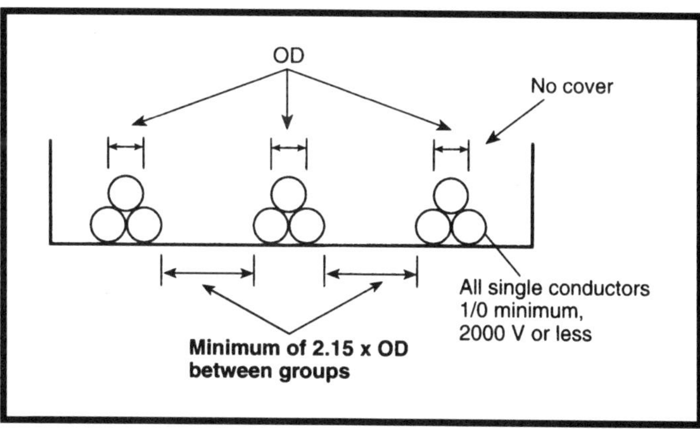

(© 1999, NFPA)

Temperature of 20°C (68°F), Arrangement per Figure 4.4.25 (*NEC* Figure 310.60), 100 Percent Load

4.6.1

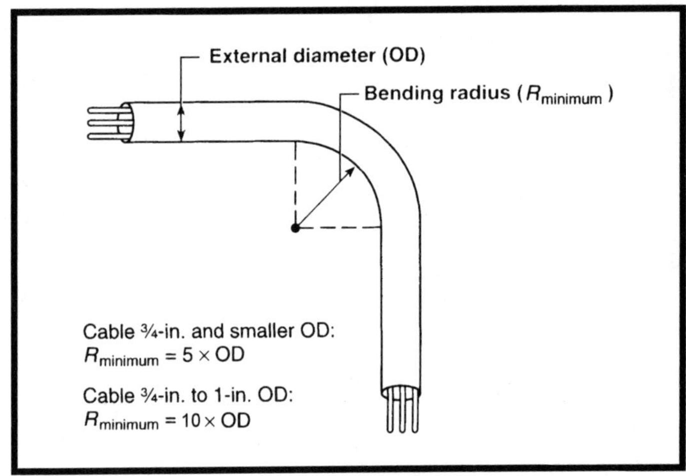

TABLE 4.6.2

16 AWG CURRENT RATING (75°C/90°C)*	– /24	– /18	– /18	– /14.4	– /12.6
TERMINATION SIZE	1/2"	1/2"	1/2"	1/2"	3/4"
CABLE REFERENCE	1850/215/1	1850/340/2	1850/355/3	1850/387/4	1850/449/7
14 AWG CURRENT RATING (75°C/90°C)*	30/35	20/25	20/25	16/20	14/17.5
TERMINATION SIZE	1/2"	1/2"	1/2"	3/4"	3/4"
CABLE REFERENCE	1850/230/1	1850/371/2	1850/387/3	1850/418/4	1850/496/7
12 AWG CURRENT RATING (75°C/90°C)*	35/40	25/30	25/30	20/24	17.5/21
TERMINATION SIZE	1/2"	1/2"	1/2"	3/4"	3/4"
CABLE REFERENCE	1850/246/1	1850/402/2	1850/434/3	1850/465/4	1850/543/7
10 AWG CURRENT RATING (75°C/90°C)*	50/55	35/40	35/40	28/32	24.5/28
TERMINATION SIZE	1/2"	3/4"	3/4"	3/4"	1"
CABLE REFERENCE	1850/277/1	1850/449/2	1850/480/3	1850/527/4	1850/621/7
8 AWG CURRENT RATING (75°C/90°C)*	70/80	50/55	50/55	40/44	
TERMINATION SIZE	1/2"	3/4"	3/4"	3/4"	
CABLE REFERENCE	1850/309/1	1850/512/2	1850/543/3	1850/590/4	
6 AWG CURRENT RATING (75°C/90°C)*	95/105	65/75	65/75	52/60	
TERMINATION SIZE	1/2"	3/4"	3/4"	1"	
CABLE REFERENCE	1850/340/1	1850/590/2	1850/621/3	1850/684/4	

(continued)

TABLE 4.6.2

4 AWG

CURRENT RATING (75°C/90°C)*	125/140	85/95	85/95		
TERMINATION SIZE	1/2"	1"	1"		
CABLE REFERENCE	1850/402/1	1850/684/2	1850/730/3		

	3 AWG	2 AWG	1 AWG	1/0 AWG	2/0AWG
CURRENT RATING (75°C/90°C)*	145/165	170/190	195/220	230/260	265/300
TERMINATION SIZE	1/2"	3/4"	3/4"	3/4"	3/4"
CABLE REFERENCE	1850/434/1	1850/465/1	1850/496/1	1850/543/1	1850/590/1

	3/0 AWG	4/0 AWG	250 kcmil	350 kcmil	500 kcmil
CURRENT RATING (75°C/90°C)*	310/350	360/405	405/455	505/570	620/700
TERMINATION SIZE	3/4"	1"	1"	1 1/4"	1 1/4"
CABLE REFERENCE	1850/637/1	1850/699/1	1850/746/1	1850/834/1	1850/1000/1

(BICC Pyrotenax)

Temperatures of 90°C (194°F) and 105°C (221°F)

TABLE 4.6.3

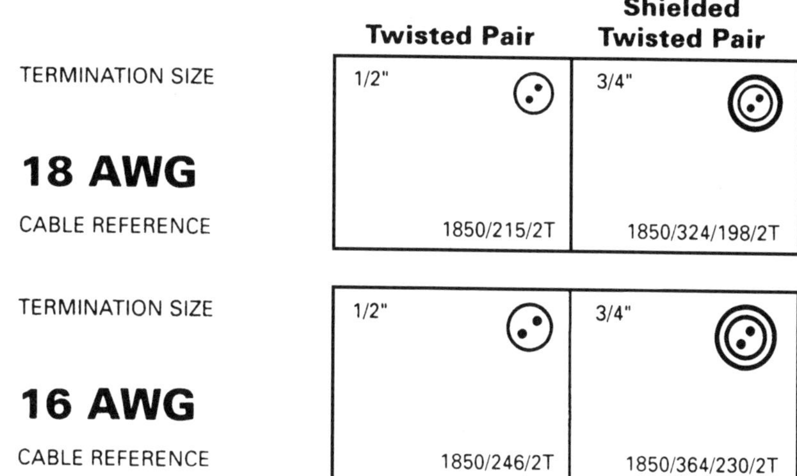

		Twisted Pair	Shielded Twisted Pair
18 AWG	TERMINATION SIZE	1/2"	3/4"
	CABLE REFERENCE	1850/215/2T	1850/324/198/2T
16 AWG	TERMINATION SIZE	1/2"	3/4"
	CABLE REFERENCE	1850/246/2T	1850/364/230/2T

(BICC Pyrotenax)

4.5.1 *NEC* Table 392.7(B), Metal Area Requirements for Cable Trays Used as Equipment

4.6.4

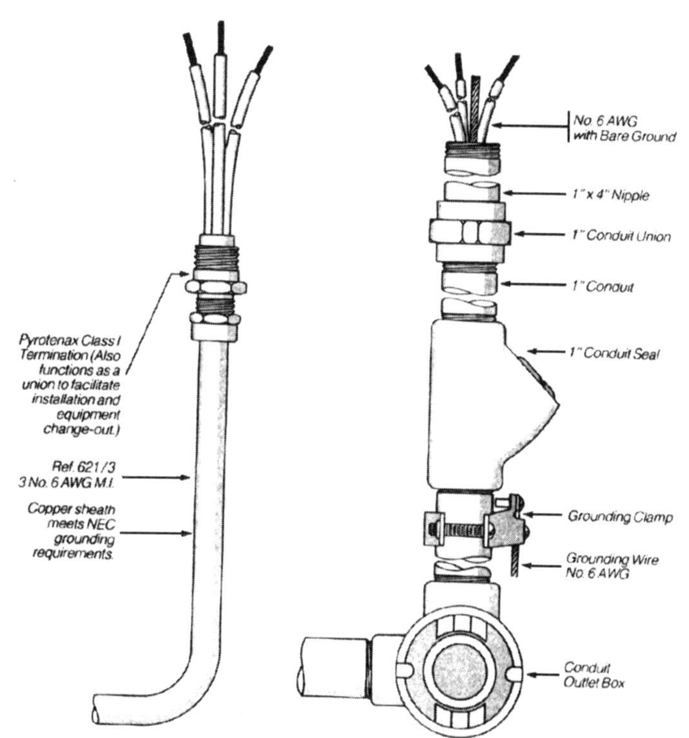

(BICC Pyrotenax)

Grounding Conductor

TABLE 4.6.5

Step I Determine Feeder Size

Estimate feeder size using the Voltage Drop Chart at right as in the following example:

Run Length = 100'
Circuit Voltage = 208 volts
Circuit Amps = 400 amps
Required Voltage Drop = 2% or 4.16 volts

Step II Verify Feeder Size

Using the formula and tables below, verify choice from Step I.

1. Voltage Drop = $\frac{\text{(Run Length) X (Circuit Current) X (Temperature Constant) X (Factor from Voltage Drop Calculations Chart) X .87*}}{1000}$

* .87 is multiplyer for 3-phase. Omit if making single phase calculation

2. Using the values of the example:

   $\frac{100' \times 400 \times 1.0 \times .1112 \times .87}{1000}$ = 3.87 Volts Voltage Drop

3. Percentage Voltage Drop = $\frac{\text{Voltage Drop}}{\text{Circuit Voltage}}$ X 100%

4. Values from example:

   $\frac{3.87}{208}$ X 100% = 1.86% Percent Voltage Drop

5. Conclusion: Since 1.86% is better than the 2% voltage drop required, the choice of 250 MCM Pyrotenax MI Cable (746/1) is confirmed.

Temperature Constant Chart

Cable at full rated current	1.00
Cable at 3/4 rated current	0.95
Cable at 1/2 rated current	0.91
Cable at 1/4 rated current	0.88

Factors For Calculating Voltage Drop Using Pyrotenax MI Cable

AWG	Single Conductor	2 Conductor	3 Conductor	4 Conductor	7 Conductor
18		15.06	15.57	15.16	15.60
16	9.2	9.40	9.48	9.63	9.63
14	5.7	5.46	5.67	5.50	5.86
12	3.46	3.43	3.49	3.49	3.62
10	2.24	2.20	2.24	2.20	2.32
8	1.492	1.470	1.512	1.480	
6	.954	.928	.968	.944	
4	.602	.580	.608		
3	.478				
2	.406				
1	.314				
1/0	.254				
2/0	.202				
3/0	.1626				
4/0	.1296				
250 MCM	.1112				
350 MCM	.086				
500 MCM	.064				

Shaded area figures include an allowance for the effect of sheath loss (assuming the cables are run close together).

(BICC Pyrotenax)

4.5.2 An Example of Multiconductor Cables in Cable Trays with Conduit Runs to Power Equipment

TABLE 4.7.1

Conduit Size		One Shot and Full Shoe Benders		Other Bends	
Metric Designator	Trade Size	mm	in.	mm	in.
16	½	101.6	4	101.6	4
21	¾	114.3	4½	127	5
27	1	146.05	5¾	152.4	6
35	1¼	184.15	7¼	203.2	8
41	1½	209.55	8¼	254	10
53	2	241.3	9½	304.8	12
63	2½	266.7	10½	381	15
78	3	330.2	13	457.2	18
91	3½	381	15	533.4	21
103	4	406.4	16	609.6	24
129	5	609.6	24	762	30
155	6	762	30	914.4	36

(© 2001, NFPA)

Where Bonding Is Provided

4.7.2

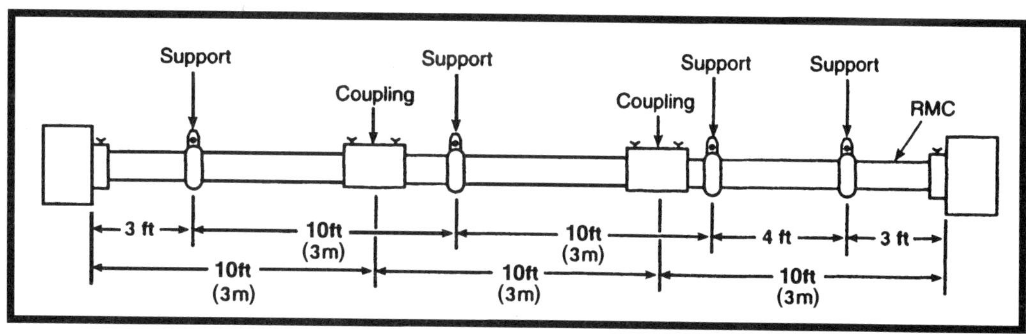

(© 1999, NFPA)

4.5.3 *NEC* Table 392.9, Allowable Cable Fill Area for Multiconductor Cables in Ladder, Ventilated-Trough,

TABLE 4.7.3

Conduit Size		Maximum Distance Between Rigid Metal Conduit Supports	
Metric Designator	Trade Size	m	ft
16 – 21	½ – ¾	3.0	10
27	1	3.7	12
35 – 41	1¼ – 1½	4.3	14
53 – 63	2 – 2½	4.9	16
78 and larger	3 and larger	6.1	20

(© 2001, NFPA)

or Solid-Bottom Cable Trays for Cables Rated 2000 V or Less

TABLE 4.7.4

Conduit Size		Maximum Spacing Between Supports	
Metric Designator	Trade Size	mm or m	ft
16–27	½–1	900 mm	3
35–53	1¼–2	1.5 m	5
63–78	2½–3	1.8 m	6
91–129	3½–5	2.1 m	7
155	6	2.5 m	8

(© 2001, NFPA)

4.5.4 *NEC* Table 392.9(E), Allowable Cable Fill Area for Multiconductor Cables in Ventilated Channel Cable Trays for Cables Rated 2000 V or Less

TABLE 4.7.5

Temperature Change (°C)	Length Change of PVC Conduit (mm/m)	Temperature Change (°F)	Length Change of PVC Conduit (in./100 ft)	Temperature Change (°F)	Length Change of PVC Conduit (in./100 ft)
5	0.30	5	0.20	105	4.26
10	0.61	10	0.41	110	4.46
15	0.91	15	0.61	115	4.66
20	1.22	20	0.81	120	4.87
25	1.52	25	1.01	125	5.07
30	1.83	30	1.22	130	5.27
35	2.13	35	1.42	135	5.48
40	2.43	40	1.62	140	5.68
45	2.74	45	1.83	145	5.88
50	3.04	50	2.03	150	6.08
55	3.35	55	2.23	155	6.29
60	3.65	60	2.43	160	6.49
65	3.95	65	2.64	165	6.69
70	4.26	70	2.84	170	6.90
75	4.56	75	3.04	175	7.10
80	4.87	80	3.24	180	7.30
85	5.17	85	3.45	185	7.50
90	5.48	90	3.65	190	7.71
95	5.78	95	3.85	195	7.91
100	6.08	100	4.06	200	8.11

4.5.4.1 *NEC* Table 392.9(E), Allowable Cable Fill Area for Multiconductor Cables in Solid Channel Cable Trays for Cables Rated 2000 V or Less

TABLE 4.7.6

Temperature Change (°C)	Length Change of RTRC Conduit (mm/m)	Temperature Change (°F)	Length Change of RTRC Conduit (in./100 ft)	Temperature Change (°F)	Length Change of RTRC Conduit (in./100 ft)
5	0.14	5	0.09	105	1.89
10	0.27	10	0.18	110	1.98
15	0.41	15	0.27	115	2.07
20	0.54	20	0.36	120	2.16
25	0.68	25	0.45	125	2.25
30	0.81	30	0.54	130	2.34
35	0.95	35	0.63	135	2.43
40	1.08	40	0.72	140	2.52
45	1.22	45	0.81	145	2.61
50	1.35	50	0.90	150	2.70
55	1.49	55	0.99	155	2.79
60	1.62	60	1.08	160	2.88
65	1.76	65	1.17	165	2.97
70	1.89	70	1.26	170	3.06
75	2.03	75	1.35	175	3.15
80	2.16	80	1.44	180	3.24
85	2.30	85	1.53	185	3.33
90	2.43	90	1.62	190	3.42
95	2.57	95	1.71	195	3.51
100	2.70	100	1.80	200	3.60

4.5.5 *NEC* Table 392.10(A), Allowable Cable Fill Area for Single-Conductor Cables in Ladder or Ventilated-Trough Cable Trays for Cables Rated

TABLE 4.7.7

	Types RFH-2, SF-2		Types TF, XHHW, TW		Types TFN, THHN, THWN		Types FEP, FEBP, PF, PGF	
Size (AWG)	Fittings Inside Conduit	Fittings Outside Conduit	Fittings Inside Conduit	Fittings Outside Conduit	Fittings Inside Conduit	Fittings Outside Conduit	Fittings Inside Conduit	Fittings Outside Conduit
18	2	3	3	5	5	8	5	8
16	1	2	3	4	4	6	4	6
14	1	2	2	3	3	4	3	4
12	—	—	1	2	2	3	2	3
10	—	—	1	1	1	1	1	2

*In addition, one covered or bare equipment grounding conductor of the same size shall be permitted.

2000 V or Less

TABLE 4.7.8

Type of Raceway	Wire Size Gauge No.	Number of Wires: Types RHH, RHW	Type THW	Type TW	Types THHN,THWN
No. 200 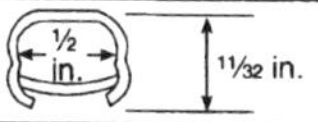 (½ in.; 11/32 in.)	12	—	2	3	3
	14	—	2	3	5
No. 500 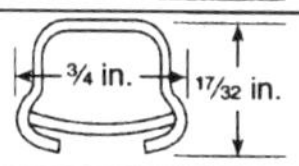 (¾ in.; 17/32 in.)	8	—	—	2	2
	10	2	2	3	4
	12	2	3	4	7
	14	2	4	6	9
No. 700 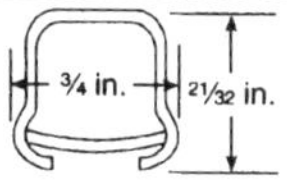 (¾ in.; 21/32 in.)	6	—	—	—	2
	8	—	2	2	3
	10	2	3	4	5
	12	2	4	6	8
	14	3	5	7	11
No. 1500 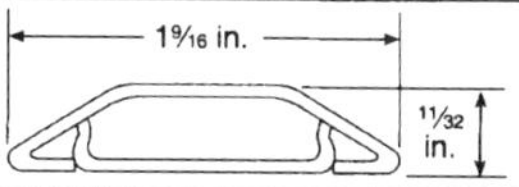 (1 9/16 in.; 11/32 in.)	6	—	—	—	2
	8	—	—	2	3
	10	2	3	4	5
	12	2	3	5	7
	14	2	4	6	10
No. 2000[a] 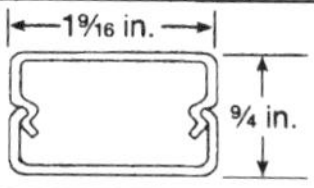 (1 9/16 in.; ¾ in.)	12	—	—	7	7
	14	—	—	7	7
No. 2100[a] 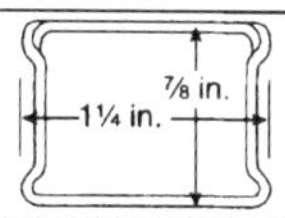 (7/8 in.; 1¼ in.)	6	2	4	4	6
	8	4	6	8	10
	10	7	10	14	17
	12	8	13	19	28
	14	10	15	24	37
No. 2200[a] 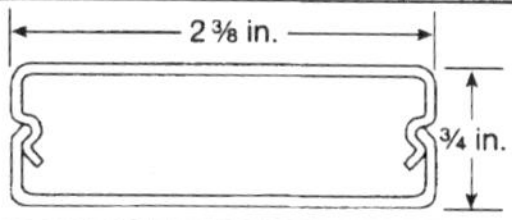 (2 3/8 in.; ¾ in.)	6	5	7	3[b] 7	11
	8	8	11	7[b] 14	19
	10	13	19	10[b] 26	32
	12	15	23	10[b] 34	51
	14	18	29	10[b] 44	69
No. 2600 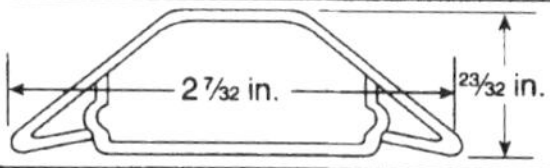 (2 7/32 in.; 23/32 in.)	6	2	3	3	5
	8	4	5	7	9
	10	6	9	12	15
	12	7	11	16	24
	14	9	14	21	33
G-3000 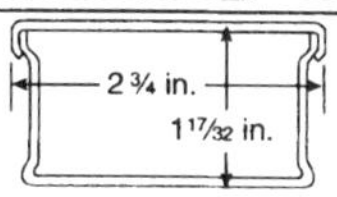 (2¾ in.; 1 17/32 in.)	6	4[b] 11	6[b] 19	6[b] 17	6[b] 27
	8	6[b] 18	8[b] 26	8[b] 34	8[b] 44
	10	10[b] 30	10[b] 45	10[b] 62	10[b] 76
	12	14[b] 36	18[b] 55	18[b] 81	18[b] 119
	14	16[b] 42	26[b] 67	26[b] 103	26[b] 160
G-4000, with divider 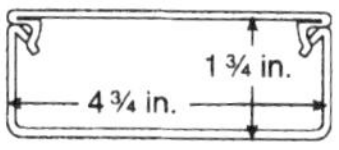 (1¾ in.; 4¾ in.)	2	— 7	— 10	— 10	— 12
	3	— 8	— 11	— 11	— 15
	4	— 9	— 13	— 13	— 17
	6	4[b] 12	4[b] 18	4[b] 18	7[b] 28
	8	7[b] 19	7[b] 28	7[b] 36	8[b] 47
	10	11[b] 32	11[b] 48	11[b] 66	15[b] 81
	12	15[b] 39	15[b] 59	15[b] 86	24[b] 128
	14	17[b] 45	17[b] 72	17[b] 110	32[b] 171
G-4000, without divider 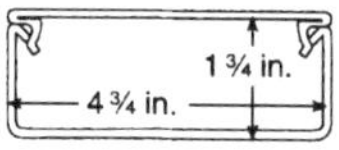 (1¾ in.; 4¾ in.)	2	— 14	— 20	— 20	— 25
	3	— 16	— 23	— 23	— 30
	4	— 18	— 27	— 27	— 35
	6	8[b] 24	8[b] 36	8[b] 36	10[b] 57
	8	10[b] 39	10[b] 57	10[b] 78	15[b] 94
	10	15[b] 65	15[b] 96	12[b] 133	18[b] 163
	12	21[b] 78	21[b] 119	16[b] 174	34[b] 256
	14	21[b] 91	21[b] 145	17[b] 222	34[b] 344
G-6000 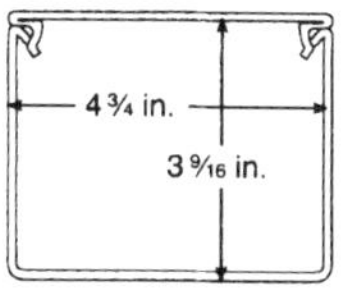 (4¾ in.; 3 9/16 in.)	2/0	10[b] 17	12[b] 22	12[b] 22	15[b] 27
	1/0	11[b] 20	14[b] 26	14[b] 26	18[b] 33
	1	12[b] 23	17[b] 31	17[b] 31	21[b] 39
	2	16[b] 30	23[b] 43	23[b] 43	29[b] 53
	3	19[b] 34	27[b] 50	27[b] 50	34[b] 63
	4	21[b] 39	32[b] 58	32[b] 58	40[b] 74
	6	27[b] 51	42[b] 77	42[b] 77	66[b] 122
	8	40[b] 74	57[b] 106	73[b] 134	92[b] 169
	10	75[b] 137	111[b] 203	154[b] 282	187[b] 343
	12	90[b] 164	137[b] 252	200[b] 386	295[b] 540
	14	105[b] 193	167[b] 307	255[b] 469	396[b] 726

[a]Figures for Nos. 2000, 2100, 2200, G-3000, G-4000, and G-6000 are without receptacles, except where noted.
[b]With receptacles.

4.5.6 An Illustration of Section 392.11(A)(3) for Multiconductor Cables, 2000 V or Less, with Not

TABLE 4.7.9

Size Channel	Area		40% Area*		25% Area**	
	in.²	mm²	in.²	mm²	in.²	mm²
1⅝ × 13/16	0.887	572	0.355	229	0.222	143
1⅝ × 1	1.151	743	0.460	297	0.288	186
1⅝ × 1⅜	1.677	1076	0.671	433	0.419	270
1⅝ × 1⅝	2.028	1308	0.811	523	0.507	327
1⅝ × 27/16	3.169	2045	1.267	817	0.792	511
1⅝ × 3¼	4.308	2780	1.723	1112	1.077	695
1½ × ¾	0.849	548	0.340	219	0.212	137
1½ × 1½	1.828	1179	0.731	472	0.457	295
1½ × 1⅞	2.301	1485	0.920	594	0.575	371
1½ × 3	3.854	2487	1.542	995	0.964	622

*Raceways with external joiners shall use a 40 percent wire fill calculation to determine the number of conductors permitted.

**Raceways with internal joiners shall use a 25 percent wire fill calculation to determine the number of conductors permitted.

More Than Three Conductors per Cable (Ampacity

TABLE 4.8.1

Box Trade Size			Minimum Volume		Maximum Number of Conductors*						
mm	in.		cm³	in.³	18	16	14	12	10	8	6
100 × 32	(4 × 1¼)	round/octagonal	205	12.5	8	7	6	5	5	5	2
100 × 38	(4 × 1½)	round/octagonal	254	15.5	10	8	7	6	6	5	3
100 × 54	(4 × 2⅛)	round/octagonal	353	21.5	14	12	10	9	8	7	4
100 × 32	(4 × 1¼)	square	295	18.0	12	10	9	8	7	6	3
100 × 38	(4 × 1½)	square	344	21.0	14	12	10	9	8	7	4
100 × 54	(4 × 2⅛)	square	497	30.3	20	17	15	13	12	10	6
120 × 32	(4¹¹⁄₁₆ × 1¼)	square	418	25.5	17	14	12	11	10	8	5
120 × 38	(4¹¹⁄₁₆ × 1½)	square	484	29.5	19	16	14	13	11	9	5
120 × 54	(4¹¹⁄₁₆ × 2⅛)	square	689	42.0	28	24	21	18	16	14	8
75 × 50 × 38	(3 × 2 × 1½)	device	123	7.5	5	4	3	3	3	2	1
75 × 50 × 50	(3 × 2 × 2)	device	164	10.0	6	5	5	4	4	3	2
75 × 50 × 57	(3 × 2 × 2¼)	device	172	10.5	7	6	5	4	4	3	2
75 × 50 × 65	(3 × 2 × 2½)	device	205	12.5	8	7	6	5	5	4	2
75 × 50 × 70	(3 × 2 × 2¾)	device	230	14.0	9	8	7	6	5	4	2
75 × 50 × 90	(3 × 2 × 3½)	device	295	18.0	12	10	9	8	7	6	3
100 × 54 × 38	(4 × 2⅛ × 1½)	device	169	10.3	6	5	5	4	4	3	2
100 × 54 × 48	(4 × 2⅛ × 1⅞)	device	213	13.0	8	7	6	5	5	4	2
100 × 54 × 54	(4 × 2⅛ × 2⅛)	device	238	14.5	9	8	7	6	5	4	2
95 × 50 × 65	(3¾ × 2 × 2½)	masonry box/gang	230	14.0	9	8	7	6	5	4	2
95 × 50 × 90	(3¾ × 2 × 3½)	masonry box/gang	344	21.0	14	12	10	9	8	7	2
min. 44.5 depth	FS — single cover/gang (1¾)		221	13.5	9	7	6	6	5	4	2
min. 60.3 depth	FD — single cover/gang (2⅜)		295	18.0	12	10	9	8	7	6	3
min. 44.5 depth	FS — multiple cover/gang (1¾)		295	18.0	12	10	9	8	7	6	3
min. 60.3 depth	FD — multiple cover/gang (2⅜)		395	24.0	16	13	12	10	9	8	4

*Where no volume allowances are required by 314.16(B)(2) through 314.16(B)(5).

to Be Determined from Table B.310.3 in Annex B)

TABLE 4.8.2

Size of Conductor (AWG)	Free Space Within Box for Each Conductor	
	cm³	in.³
18	24.6	1.50
16	28.7	1.75
14	32.8	2.00
12	36.9	2.25
10	41.0	2.50
8	49.2	3.00
6	81.9	5.00

4.5.7 An Illustration of Section 392.11(B)(4) for

TABLE 4.9.1

Trade Name	Type Letter	Voltage	AWG or kcmil	Number of Conductors	Insulation	Nominal Insulation Thickness[1]			Braid on Each Conductor	Outer Covering	Use		
						AWG or kcmil	mm	mils					
Lamp cord	C	300 600	18–16 14–10	2 or more	Thermoset or thermo-plastic	18–16 14–10	0.76 1.14	30 45	Cotton	None	Pendant or portable	Dry locations	Not hard usage
Elevator cable	E See Note 5. See Note 9. See Note 10.	300 or 600	20–2	2 or more	Thermoset	20–16 14–12 12–10 8–2	0.51 0.76 1.14 1.52	20 30 45 60	Cotton	Three cotton, Outer one flame-retardant & moisture-resistant. See Note 3.	Elevator lighting and control	Unclassified locations	
						20–16 14–12 12–10 8–2	0.51 0.76 1.14 1.52	20 30 45 60	Flexible nylon jacket				
Elevator cable	EO See Note 5. See Note 10.	300 or 600	20–2	2 or more	Thermoset	20–16 14–12 12–10 8–2	0.51 0.76 1.14 1.52	20 30 45 60	Cotton	Outer one Three cotton, flame-retardant & moisture-resistant. See Note 3.	Elevator lighting and control	Unclassified locations	
										One cotton and a neoprene jacket. See Note 3.		Hazardous (classified) locations	
Elevator cable	ET See Note 5. See Note 10.	300 or 600	20–2	2 or more	Thermo-plastic	20–16 14–12 12–10 8–2	0.51 0.76 1.14 1.52	20 30 45 60	Rayon	Three cotton or equivalent. Outer one flame-retardant & moisture-resistant. See Note 3.	Unclassified locations		
	ETLB See Note 5. See Note 10.	300 or 600							None				
	ETP See Note 5. See Note 10.	300 or 600							Rayon	Thermoplastic	Hazardous (classified) locations		
	ETT See Note 5. See Note 10.	300 or 600							None	One cotton or equivalent and a thermo-plastic jacket			
Portable power cable	G	2000	12–500	2–6 plus grounding conductor(s)	Thermoset	12–2 1–4/0 250–500	1.52 2.03 2.41	60 80 95		Oil-resistant thermoset	Portable and extra hard usage		

TABLE 4.9.1

Trade Name	Type Letter	Voltage	AWG or kcmil	Number of Conductors	Insulation	Nominal Insulation Thickness[1]			Braid on Each Conductor	Outer Covering	Use		
						AWG or kcmil	mm	mils					
	G-GC	2000	12–500	3–6 plus grounding conductors and 1 ground check conductor	Thermoset	12–2 1–4/0 250–500	1.52 2.03 2.41	60 80 95		Oil-resistant thermoset			
Heater cord	HPD	300	18–12	2, 3, or 4	Thermoset	18–16 14–12	0.38 0.76	15 30	None	Cotton or rayon	Portable heaters	Dry locations	Not hard usage
Parallel heater cord	HPN See Note 6.	300	18–12	2 or 3	Oil-resistant thermoset	18–16 14–12	1.14 1.52 2.41	45 60 95	None	Oil-resistant thermoset	Portable	Damp locations	Not hard usage
Thermoset jacketed heater cords	HSJ	300	18–12	2, 3, or 4	Thermoset	18–16	0.76	30	None	Cotton and Thermoset	Portable or portable heater	Damp	Hard usage
	HSJO	300	18–12		Oil-resistant thermoset	14–12	1.14	45		Cotton and oil-resistant thermoset			
	HSJOO	300	18–12										
Non-integral parallel cords	NISP-1 See Note 6.	300	20-18	2 or 3	Thermoset	20-18	0.38	15	None	Thermoset	Pendant or portable	Damp locations	Not hard usage
	NISP-2 See Note 6.	300	18-16			18-16	0.76	30					
	NISPE-1 See Note 6.	300	20-18		Thermo-plastic elastomer	20-18	0.38	15		Thermoplastic elastomer			
	NISPE-2 See Note 6.	300	18-16			18-16	0.76	30					
	NISPT-1 See Note 6.	300	20-18		Thermo-plastic	20-18	0.38	15		Thermoplastic			
	NISPT-2 See Note 6.	300	18-16			18-16	0.76	30					
Twisted portable cord	PD	300 600	18–16 14-10	2 or more	Thermoset or thermo-plastic	18–16 14–10	0.76 1.14	30 45	Cotton	Cotton or rayon	Pendant or portable	Dry locations	Not hard usage
Portable power cable	PPE	2000	12–500	1–6 plus optional grounding conductor(s)	Thermo-plastic elastomer	12–2 1–4/0 250–500	1.52 2.03 2.41	60 80 95		Oil-resistant thermoplastic elastomer	Portable, extra hard usage		
Hard service cord	S See Note 4.	600	18–12	2 or more	Thermoset	18–16 14–10 8–2	0.76 1.14 1.52	30 45 60	None	Thermoset	Pendant or portable	Damp locations	Extra hard usage
Flexible stage and lighting power cable	SC	600	8–250	1 or more		8–2 1–4/0 250	1.52 2.03 2.41	60 80 95		Thermoset[2]	Portable, extra hard usage		
	SCE	600			Thermo-plastic elastomer					Thermoplastic elastomer[2]			
	SCT	600			Thermo-plastic					Thermoplastic[2]			
Hard service cord	SE See Note 4.	600	18–2	2 or more	Thermo-plastic elastomer	18–16 14–10 8–2	0.76 1.14 1.52	30 45 60	None	Thermoplastic elastomer	Pendant or portable	Damp locations	Extra hard usage
	SEW See Note 4.	600											

(continued)

TABLE 4.9.1

Trade Name	Type Letter	Voltage	AWG or kcmil	Number of Conductors	Insulation	Nominal Insulation Thickness[1]			Braid on Each Conductor	Outer Covering	Use		
						AWG or kcmil	mm	mils					
	SEO See Note 4.	600								Oil-resistant thermoplastic elastomer			
	SEOW See Note 4.	600											
	SEOO See Note 4.	600			Oil-resistant thermo-plastic elastomer								
	SEOOW See Note 4.	600											
Junior hard service cord	SJ	300	18–10	2–6	Thermoset	18–12	0.76	30	None	Thermoset	Pendant or portable	Damp locations	Hard usage
	SJE	300			Thermo-plastic elastomer					Thermoplastic elastomer			
	SJEW	300											
	SJEO	300								Oil-resistant thermoplastic elastomer			
	SJEOW	300											
	SJEOO	300			Oil-resistant thermo-plastic elastomer								
	SJEOOW	300											
	SJO	300			Thermoset					Oil-resistant thermoset			
	SJOW	300											
	SJOO	300			Oil-resistant thermoset								
	SJOOW	300											
	SJT	300			Thermo-plastic	10	1.14	45		Thermoplastic			
	SJTW	300											
	SJTO	300			Thermo-plastic	18–12	0.76	30		Oil-resistant thermoplastic			
	SJTOW	300											
	SJTOO	300			Oil-resistant thermo-plastic								
	SJTOOW	300											
Hard service cord	SO See Note 4.	600	18–2	2 or more	Thermoset	18–16	0.76	30		Oil-resistant thermoset	Pendant or portable	Damp locations	Extra hard usage
	SOW See Note 4.	600											
	SOO See Note 4.	600			Oil-resistant thermoset	14–10 8–2	1.14 1.52	45 60					
	SOOW See Note 4.	600											

(continued)

TABLE 4.9.1

Trade Name	Type Letter	Voltage	AWG or kcmil	Number of Conductors	Insulation	Nominal Insulation Thickness[1]			Braid on Each Conductor	Outer Covering	Use		
						AWG or kcmil	mm	mils					
All thermoset parallel cord	SP-1 See Note 6.	300	20–18	2 or 3	Thermoset	20–18	0.76	30	None	None	Pendant or portable	Damp locations	Not hard usage
	SP-2 See Note 6.	300	18–16			18-16	1.14	45					
	SP-3 See Note 6.	300	18–10			18–16 14 12 10	1.52 2.03 2.41 2.80	60 80 95 110			Refrigerators, room air conditioners, and as permitted in 422.16(B)		
All elastomer (thermoplastic) parallel cord	SPE-1 See Note 6.	300	20-18	2 or 3	Thermoplastic elastomer	20–18	0.76	30	None	None	Pendant or portable	Damp locations	Not Hard usage
	SPE-2 See Note 6.	300	18–16			18–16	1.14	45					
	SPE-3 See Note 6.	300	18–10			18–16 14 12 10	1.52 2.03 2.41 2.80	60 80 95 110			Refrigerators, room air conditioners, and as permitted in 422.16(B)		
All plastic parallel cord	SPT-1 See Note 6.	300	20–18	2 or 3	Thermoplastic	20–18	0.76	30	None	None	Pendant or portable	Damp locations	Not hard usage
	SPT-1W See Note 6.	300											
	SPT-2 See Note 6.	300	18–16			18–16	1.14	45					
	SPT-2W See Note 6.	300											
	SPT-3 See Note 6.	300	18–10			18–16 14 12 10	1.52 2.03 2.41 2.80	60 80 95 110			Refrigerators, room air conditioners, and as permitted in 422.16(B)	Damp locations	Not hard usage
Range, dryer cable	SRD	300	10–4	3 or 4	Thermoset	10–4	1.14	45	None	Thermoset	Portable	Damp locations	Ranges, dryers
	SRDE	300	10–4	3 or 4	Thermoplastic elastomer				None	Thermoplastic elastomer			
	SRDT	300	10–4	3 or 4	Thermoplastic				None	Thermoplastic			
Hard service cord	ST See Note 4.	600	18–2	2 or more	Thermoplastic	18–16 14–10 8–2	0.76 1.14 1.52	30 45 60	None	Thermoplastic	Pendant or portable	Damp locations	Extra hard usage
	STW See Note 4.	600								Oil-resistant thermoplastic			
	STO See Note 4.	600											
	STOW See Note 4.	600											

(continued)

TABLE 4.9.1

Trade Name	Type Letter	Voltage	AWG or kcmil	Number of Conductors	Insulation	Nominal Insulation Thickness[1]			Braid on Each Conductor	Outer Covering	Use		
						AWG or kcmil	mm	mils					
	STOO See Note 4.	600			Oil-resistant thermo-plastic								
	STOOW See Note 4.	600											
Vacuum cleaner cord	SV See Note 6.	300	18–16	2 or 3	Thermoset	18–16	0.38	15	None	Thermoset	Pendant or portable	Damp locations	Not hard usage
	SVE See Note 6.	300			Thermo-plastic elastomer					Thermoplastic elastomer			
	SVEO See Note 6.	300								Oil-resistant thermoplastic elastomer			
	SVEOO See Note 6.	300			Oil-resistant thermo-plas tic elastomer								
	SVO	300			Thermoset					Oil-resistant thermoset			
	SVOO	300			Oil-resistant thermoset					Oil-resistant thermoset			
	SVT See Note 6.	300			Thermo-plastic					Thermoplastic			
	SVTO See Note 6.	300			Thermo-plastic					Oil-resistant thermoplastic			
	SVTOO	300			Oil-resistant thermo-plastic								
Parallel tinsel cord	TPT See Note 2.	300	27	2	Thermo-plastic	27	0.76	30	None	Thermoplastic	Attached to an appliance	Damp locations	Not hard usage
Jacketed tinsel cord	TST See Note 2.	300	27	2	Thermo-plastic	27	0.38	15	None	Thermoplastic	Attached to an appliance	Damp locations	Not Hard Usage
Portable power-cable	W	2000	12–500 501–1000	1–6 1	Thermoset	12–2 1–4/0 250–500 501–1000	1.52 2.03 2.41 2.80	60 80 95 110		Oil-resistant thermoset	Portable, extra hard usage		
Electric vehicle cable	EV	600	18–500 See Note 11.	2 or more plus grounding conduc-tor(s), plus optional hybrid data, signal com-munica-tions, and optical fiber cables	Thermoset with optional nylon See Note 12.	18–16 14–10 8–2 1–4/0 250–500	0.76 (0.51) 1.14 (0.76) 1.52 (1.14) 2.03 (1.52) 2.41 (1.90)	30 (20) 45 (30) 60 (45) 80 (60) 95 (75) See Note 12.	Optional	Thermoset	Electric vehicle charging	Wet locations	Extra hard usage
	EVJ	300	18-12 See Note 11.			18–12	0.76 (0.51)	30 (20) See Note 12.					Hard usage

(continued)

TABLE 4.9.1

Trade Name	Type Letter	Voltage	AWG or kcmil	Number of Conductors	Insulation	Nominal Insulation Thickness[1]			Braid on Each Conductor	Outer Covering	Use		
						AWG or kcmil	mm	mils					
	EVE	600	18-500 See Note 11.	2 or more plus grounding conduc-tor(s), plus optional hybrid data, signal com-munica-tions, and optical fiber cables	Thermo-plastic elastomer with optional nylon See Note 12.	18–16 14–10 8–2 1–4/0 250–500	0.76 (0.51) 1.14 (0.76) 1.52 (1.14) 2.03 (1.52) 2.41 (1.90)	30 (20) 45 (30) 60 (45) 80 (60) 95 (75) See Note 12.		Thermoplastic elastomer			Extra hard usage
	EVJE	300	18–12 See Note 11.			18–12	0.76 (0.51)	30 (20) See Note 12.					Hard usage
	EVT	600	18–500 See Note 11.	2 or more plus grounding conduc-tor(s), plus optional hybrid data, signal com-munica-tions, and optical fiber cables	Thermo-plastic with optional nylon See Note 12.	18–16 14–10 8–2 1–4/0 250–500	0.76 (0.51) 1.14 (0.76) 1.52 (1.14) 2.03 (1.52) 2.41 (1.90)	30 (20) 45 (30) 60 (45) 80 (60) 95 (75) See Note 12.	Optional	Thermoplastic	Electric vehicle charging	Wet Locations	Extra hard usage
	EVJT	300	18–12 See Note 11.			18–12	0.76 (0.51)	30 (20) See Note 12.					Hard usage

*See Note 8.

**The required outer covering on some single-conductor cables may be integral with the insulation.

Notes:

1. All types listed in Table 400.4 shall have individual conductors twisted together except for Types HPN, SP-1, SP-2, SP-3, SPE-1, SPE-2, SPE-3, SPT-1, SPT-2, SPT-3, TPT, NISP-1, NISP-2, NISPT-1, NISPT-2, NISPE-1, NISPE-2, and three-conductor parallel versions of SRD, SRDE, and SRDT.

2. Types TPT and TST shall be permitted in lengths not exceeding 2.5 m (8 ft) where attached directly, or by means of a special type of plug, to a portable appliance rated at 50 watts or less and of such nature that extreme flexibility of the cord is essential.

3. Rubber-filled or varnished cambric tapes shall be permitted as a substitute for the inner braids.

4. Types G, G-GC, S, SC, SCE, SCT, SE, SEO, SEOO, SO, SOO, ST, STO, STOO, PPE, and W shall be permitted for use on theater stages, in garages, and elsewhere where flexible cords are permitted by this *Code*.

5. Elevator traveling cables for operating control and signal circuits shall contain nonmetallic fillers as necessary to maintain concentricity. Cables shall have steel supporting members as required for suspension by 620.41. In locations subject to excessive moisture or corrosive vapors or gases, supporting members of other materials shall be permitted. Where steel supporting members are used, they shall run straight through the center of the cable assembly and shall not be cabled with the copper strands of any conductor.

In addition to conductors used for control and signaling circuits, Types E, EO, ET, ETLB, ETP, and ETT elevator cables shall be permitted to incorporate in the construction, one or more 20 AWG telephone conductor pairs, one or more coaxial cables, or one or more optical fibers. The 20 AWG conductor pairs shall be permitted to be covered with suitable shielding for telephone, audio, or higher frequency communications circuits; the coaxial cables consist of a center conductor, insulation, and shield for use in video or other radio frequency communications circuits. The optical fiber shall be suitably covered with flame-retardant thermoplastic. The insulation of the conductors shall be rubber or thermoplastic of thickness not less than specified for the other conductors of the particular type of cable. Metallic shields shall have their own protective covering. Where used, these components shall be permitted to be incorporated in any layer of the cable assembly but shall not run straight through the center.

6.The third conductor in these cables shall be used for equipment grounding purpose only. The insulation of the grounding conductor for Types SPE-1, SPE-2, SPE-3, SPT-1, SPT-2, SPT-3, NISPT-1, NISPT-2, NISPE-1, and NISPE-2 shall be permitted to be thermoset polymer.

7. The individual conductors of all cords, except those of heat-resistant cords, shall have a thermoset or thermoplastic insulation, except that the equipment grounding conductor where used shall be in accordance with 400.23(B).

8. Where the voltage between any two conductors exceeds 300, but does not exceed 600, flexible cord of 10 AWG and smaller shall have thermoset or thermoplastic insulation on the individual conductors at least 1.14 mm (45 mils) in thickness, unless Type S, SE, SEO, SEOO, SO, SOO, ST, STO, or STOO cord is used.

9. Insulations and outer coverings that meet the requirements as flame retardant, limited smoke, and are so listed, shall be permitted to be marked for limited smoke after the code type designation.

10. Elevator cables in sizes 20 AWG through 14 AWG are rated 300 volts, and sizes 10 through 2 are rated 600 volts. 12 AWG is rated 300 volts with a 0.76-mm (30-mil) insulation thickness and 600 volts with a 1.14-mm (45-mil) insulation thickness.

11. Conductor size for Types EV, EVJ, EVE, EVJE, EVT, and EVJT cables apply to nonpower-limited circuits only. Conductors of power-limited (data, signal, or communications) circuits may extend beyond the stated AWG size range. All conductors shall be insulated for the same cable voltage rating.

12. Insulation thickness for Types EV, EVJ, EVEJE, EVT, and EVJT cables of nylon construction is indicated in parentheses.

Three Single Conductors Installed in a Triangular Configuration with Spacing Between Groups of Not Less Than 2.15 Times the Conductor Diameter (Ampacities to Be Determined from Table 310.20)

TABLE 4.9.2

Size (AWG)	Thermo-plastic Types TPT, TST	Thermoset Types C, E, EO, PD, S, SJ, SJO, SJOW, SJOO, SJOOW, SO, SOW, SOO, SOOW, SP-1, SP-2, SP-3, SRD, SV, SVO, SVOO / Thermoplastic Types ET, ETLB, ETP, ETT, SE, SEW, SEO, SEOW, SEOOW, SJE, SJEW, SJEO, SJEOW, SJEOOW, SJT, SJTW, SJTO, SJTOW, SJTOO, SJTOOW, SPE-1, SPE-2, SPE-3, SPT-1, SPT-1W, SPT-2, SPT-2W, SPT-3, ST, SRDE, SRDT, STO, STOW, STOO, STOOW, SVE, SVEO, SVT, SVTO, SVTOO		Types HPD, HPN, HSJ, HSJO, HSJOO
		A+	B+	
27*	0.5	—	—	—
20	—	5**	***	—
18	—	7	10	10
17	—	—	12	13
16	—	10	13	15
15	—	—	—	17
14	—	15	18	20
12	—	20	25	30
10	—	25	30	35
8	—	35	40	—
6	—	45	55	—
4	—	60	70	—
2	—	80	95	—

*Tinsel cord.

**Elevator cables only.

***7 amperes for elevator cables only; 2 amperes for other types.

+The allowable currents under subheading A apply to 3-conductor cords and other multiconductor cords connected to utilization equipment so that only 3 conductors are current carrying. The allowable currents under subheading B apply to 2-conductor cords and other multiconductor cords connected to utilization equipment so that only 2 conductors are current carrying.

4.6.1 An Illustration of Section 332.24 for Bends in Type MI Cable

TABLE 4.9.3

Size (AWG or kcmil)	60°C (140°F)			75°C (167°F)			90°C (194°F)		
	D[1]	E[2]	F[3]	D[1]	E[2]	F[3]	D[1]	E[2]	F[3]
12	—	31	26	—	37	31	—	42	35
10	—	44	37	—	52	43	—	59	49
8	60	55	48	70	65	57	80	74	65
6	80	72	63	95	88	77	105	99	87
4	105	96	84	125	115	101	140	130	114
3	120	113	99	145	135	118	165	152	133
2	140	128	112	170	152	133	190	174	152
1	165	150	131	195	178	156	220	202	177
1/0	195	173	151	230	207	181	260	234	205
2/0	225	199	174	265	238	208	300	271	237
3/0	260	230	201	310	275	241	350	313	274
4/0	300	265	232	360	317	277	405	361	316
250	340	296	259	405	354	310	455	402	352
300	375	330	289	445	395	346	505	449	393
350	420	363	318	505	435	381	570	495	433
400	455	392	343	545	469	410	615	535	468
500	515	448	392	620	537	470	700	613	536
600	575	—	—	690	—	—	780	—	—
700	630	—	—	755	—	—	855	—	—
750	655	—	—	785	—	—	885	—	—
800	680	—	—	815	—	—	920	—	—
900	730	—	—	870	—	—	985	—	—
1000	780	—	—	935	—	—	1055	—	—

[1]The ampacities under subheading D shall be permitted for single-conductor Types SC, SCE, SCT, PPE, and W cable only where the individual conductors are not installed in raceways and are not in physical contact with each other except in lengths not to exceed 600 mm (24 in.) where passing through the wall of an enclosure.
[2]The ampacities under subheading E apply to two-conductor cables and other multiconductor cables connected to utilization equipment so that only two conductors are current carrying.
[3]The ampacities under subheading F apply to three-conductor cables and other multiconductor cables connected to utilization equipment so that only three conductors are current carrying.

4.6.2 600-V MI Power Cable: Size and Ampacities

TABLE 4.9.4

Number of Conductors	Percent of Value in Tables 400.5(A) and 400.5(B)
4 – 6	80
7 – 9	70
10 – 20	50
21 – 30	45
31 – 40	40
41 and above	35

4.6.3 300-V MI Twisted-Pair and Shielded

TABLE 4.9.5

Name	Type Letter	Insulation	Thickness of Insulation			Outer Covering	Maximum Operating Temperature	Application Provisions
			AWG	mm	mils			
Heat-resistant rubber-covered fixture wire — flexible stranding	FFH-2	Heat-resistant rubber Cross-linked synthetic polymer	18–16 18–16	0.76 0.76	30 30	Nonmetallic covering	75°C 167°F	Fixture wiring
ECTFE — solid or 7-strand	HF	Ethylene chlorotrifluoroethylene	18–14	0.38	15	None	150°C 302°F	Fixture wiring
ECTFE — flexible stranding	HFF	Ethylene chlorotrifluoroethylene	18–14	0.38	15	None	150°C 302°F	Fixture wiring
Tape insulated fixture wire — solid or 7-strand	KF-1	Aromatic polyimide tape	18–10	0.14	5.5	None	200°C 392°F	Fixture wiring — limited to 300 volts
	KF-2	Aromatic polyimide tape	18–10	0.21	8.4	None	200°C 392°F	Fixture wiring
Tape insulated fixture wire — flexible stranding	KFF-1	Aromatic polyimide tape	18–10	0.14	5.5	None	200°C 392°F	Fixture wiring — limited to 300 volts
	KFF-2	Aromatic polyimide tape	18–10	0.21	8.4	None	200°C 392°F	Fixture wiring
Perfluoro-alkoxy — solid or 7-strand (nickel or nickel-coated copper)	PAF	Perfluoro-alkoxy	18–14	0.51	20	None	250°C 482°F	Fixture wiring (nickel or nickel-coated copper)
Perfluoro-alkoxy — flexible stranding	PAFF	Perfluoro-alkoxy	18–14	0.51	20	None	150°C 302°F	Fixture wiring

(continued)

TABLE 4.9.5

Name	Type Letter	Insulation	Thickness of Insulation AWG	mm	mils	Outer Covering	Maximum Operating Temperature	Application Provisions
Fluorinated ethylene propylene fixture wire — solid or 7-strand	PF	Fluorinated ethylene propylene	18–14	0.51	20	None	200°C 392°F	Fixture wiring
Fluorinated ethylene propylene fixture wire — flexible stranding	PFF	Fluorinated ethylene propylene	18–14	0.51	20	None	150°C 302°F	Fixture wiring
Fluorinated ethylene propylene fixture wire — solid or 7-strand	PGF	Fluorinated ethylene propylene	18–14	0.36	14	Glass braid	200°C 392°F	Fixture wiring
Fluorinated ethylene propylene fixture wire — flexible stranding	PGFF	Fluorinated ethylene propylene	18–14	0.36	14	Glass braid	150°C 302°F	Fixture wiring
Extruded polytetra-fluoroethylene — solid or 7-strand (nickel or nickel-coated copper)	PTF	Extruded polytetra-fluoroethyl-ene	18–14	0.51	20	None	250°C 482°F	Fixture wiring (nickel or nickel-coated copper)
Extruded polytetra-fluoroethylene — flexible stranding 26-36 (AWG silver or nickel-coated copper)	PTFF	Extruded polytetra-fluoroethyl-ene	18–14	0.51	20	None	150°C 302°F	Fixture wiring (silver or nickel-coated copper)
Heat-resistant rubber-covered fixture wire — solid or 7-strand	RFH-1	Heat-resis-tant rubber	18	0.38	15	Nonmetallic covering	75°C 167°F	Fixture wiring — limited to 300 volts
	RFH-2	Heat-resis-tant rubber Cross-linked synthetic polymer	18–16	0.76	30	None or nonmetallic covering	75°C 167°F	Fixture wiring
Heat-resistant cross-linked synthetic polymer-insulated fixture wire — solid or stranded	RFHH-2*	Cross-linked synthetic polymer	18–16	0.76	30	None or nonmetallic covering	90°C 194°F	Fixture wiring — multi-conductor cable
	RFHH-3*		18–16	1.14	45			
Silicone insulated fixture wire — solid or 7-strand	SF-1	Silicone rubber	18	0.38	15	Nonmetallic covering	200°C 392°F	Fixture wiring — limited to 300 volts
	SF-2	Silicone rubber	18–12 10	0.76 1.14	30 45	Nonmetallic covering	200°C 392°F	Fixture wiring

(continued)

TABLE 4.9.5

Name	Type Letter	Insulation	Thickness of Insulation			Outer Covering	Maximum Operating Temperature	Application Provisions
			AWG	mm	mils			
Silicone insulated fixture wire — flexible stranding	SFF-1	Silicone rubber	18	0.38	15	Nonmetallic covering	150°C 302°F	Fixture wiring — limited to 300 volts
	SFF-2	Silicone rubber	18–12 10	0.76 1.14	30 45	Nonmetallic covering	150°C 302°F	Fixture wiring
Thermoplastic covered fixture wire — solid or 7-strand	TF*	Thermoplastic	18–16	0.76	30	None	60°C 140°F	Fixture wiring
Thermoplastic covered fixture wire — flexible stranding	TFF*	Thermoplastic	18–16	0.76	30	None	60°C 140°F	Fixture wiring
Heat-resistant thermoplastic covered fixture wire — solid or 7-strand	TFN*	Thermoplastic	18–16	0.38	15	Nylon-jacketed or equivalent	90°C 194°F	Fixture wiring
Heat-resistant thermoplastic covered fixture wire — flexible stranded	TFFN*	Thermoplastic	18–16	0.38	15	Nylon-jacketed or equivalent	90°C 194°F	Fixture wiring
Cross-linked polyolefin insulated fixture wire — solid or 7-strand	XF*	Cross-linked polyolefin	18–14 12-10	0.76 1.14	30 45	None	150°C 302°F	Fixture wiring — limited to 300 volts
Cross-linked polyolefin insulated fixture wire — flexible stranded	XFF*	Cross-linked polyolefin	18–14 12–10	0.76 1.14	30 45	None	150°C 302°F	Fixture wiring — limited to 300 volts
Modified ETFE — solid or 7-strand	ZF	Modified ethylene tetrafluoro-ethylene	18–14	0.38	15	None	150°C 302°F	Fixture wiring
Flexible stranding	ZFF	Modified ethylene tetrafluoro-ethylene	18–14	0.38	15	None	150°C 302°F	Fixture wiring
High temp. modified ETFE — solid or 7-strand	ZHF	Modified ethylene tetrafluoro-ethylene	18–14	0.38	15	None	200°C 392°F	Fixture wiring

*Insulations and outer coverings that meet the requirements of flame retardant, limited smoke, and are so listed shall be permitted to be marked for limited smoke after the *Code* type designation.

Twisted-Pair Cable Sizes

TABLE 4.9.6

Size (AWG)	Allowable Ampacity
18	6
16	8
14	17
12	23
10	28

(© 2001, NFPA)

Section

5

Primary and Secondary Service and System Configurations

5.1.0 Introduction
5.1.1 Radial Circuit Arrangements in Commercial Buildings
5.1.2 Radial Circuit Arrangement: Common Primary Feeder to Secondary Unit Substations
5.1.3 Radial Circuit Arrangement: Individual Primary Feeder to Secondary Unit Substations
5.1.4 Primary Radial-Selective Circuit Arrangements
5.1.5 Secondary-Selective Circuit Arrangement (Double-Ended Substation with Single Tie)
5.1.6 Secondary-Selective Circuit Arrangement (Individual Substations with Interconnecting Ties)
5.1.7 Primary- and Secondary-Selective Circuit Arrangement (Double-Ended Substation with Selective Primary)
5.1.8 Looped Primary Circuit Arrangement
5.1.9 Distributed Secondary Network
5.1.10 Basic Spot Network

5.1.0 Introduction

In order to provide electrical service to a building or buildings, you must first determine what type of system is available from the utility company or from a privately owned and operated system such as might be found on a college or university campus or industrial or commercial complex, as the case may be. Once this is known, it is important to understand the characteristics of the system—not only voltage, capacity, and available fault current but also the operational, reliability, and relative cost characteristics inherent to the system by virtue of its configuration or arrangement. Knowing the characteristics associated with the system arrangement, the most appropriate service and distribution system for the application at hand can be determined.

5.1.1 Radial Circuit Arrangements in Commercial Buildings

5.1.1

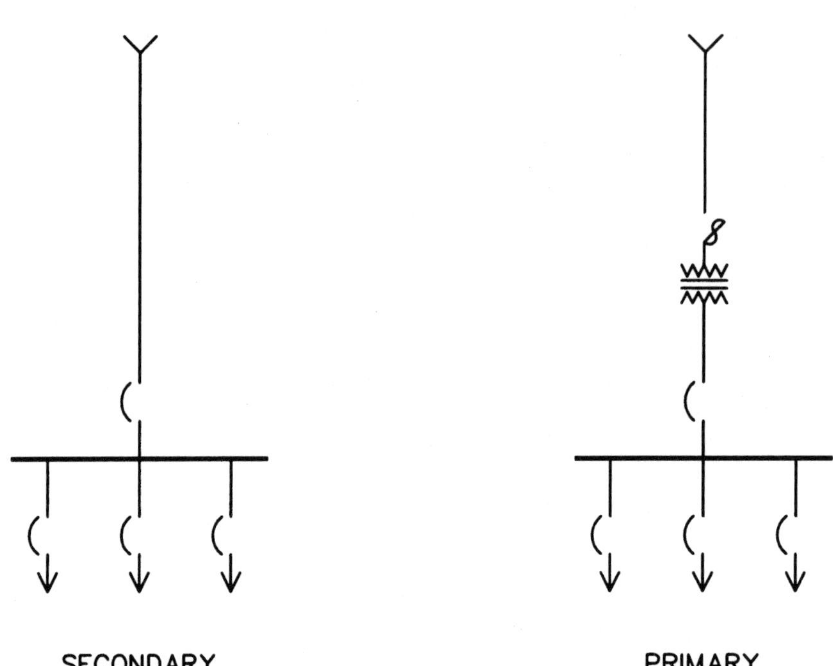

Characteristics:

- Simplest and lowest cost way of distributing power.
- Lowest reliability. A fault in the supply circuit, transformer, or the main bus will cause interruption of service to all loads.
- Modern distribution equipment has demonstrated sufficient reliability to justify use of the radial circuit arrangement in many applications.
- Most commonly used circuit arrangement.

5.1.2 Radial Circuit Arrangement: Common Primary Feeder to Secondary Unit Substations

5.1.2

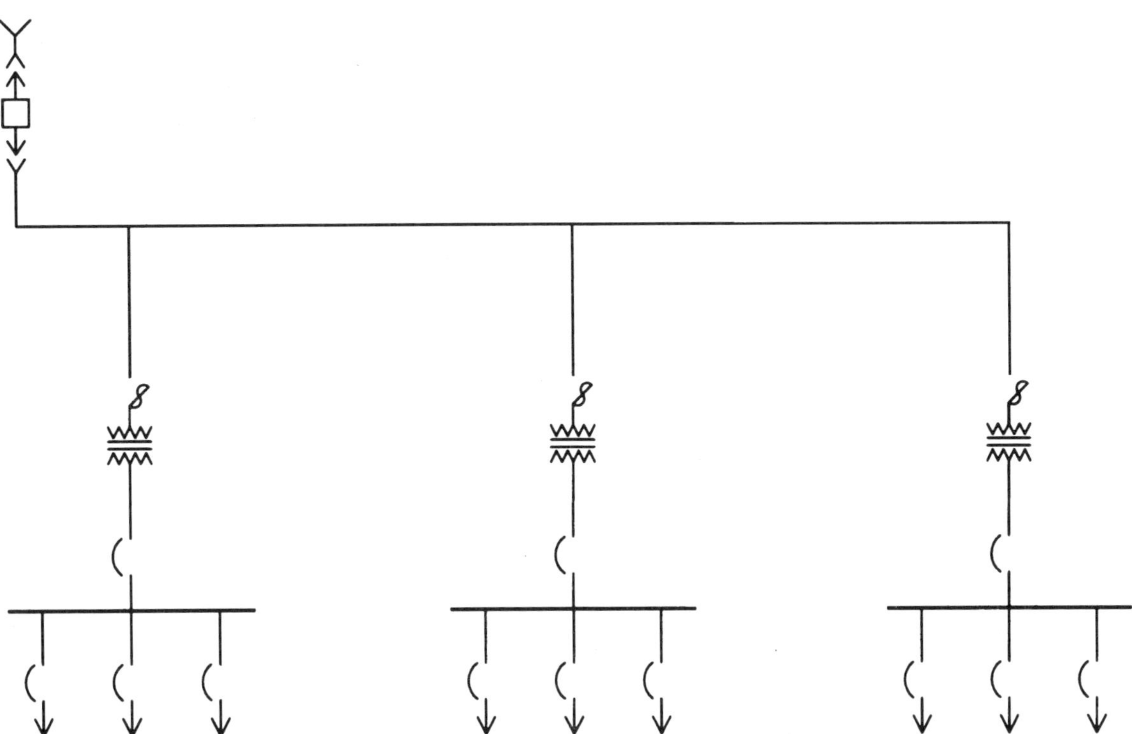

Characteristics:

- Multiple small rather than single large secondary substation.
- Used when demand, size of building, or both may be required to maintain adequate voltage at the utilization equipment.
- Smaller substations located close to center of load area.
- Provides better voltage conditions, lower system losses, less expensive installation cost than using relatively long, high–amperage, low–voltage feeder circuits.
- A primary feeder fault will cause the main protective device to operate and interrupt service to all loads. Service cannot be restored until the source of trouble has been eliminated.
- If a fault were in a transformer, service could be restored to all loads except those served by that transformer.

5.1.3 Radial Circuit Arrangement: Individual Primary Feeder to Secondary Unit Substations

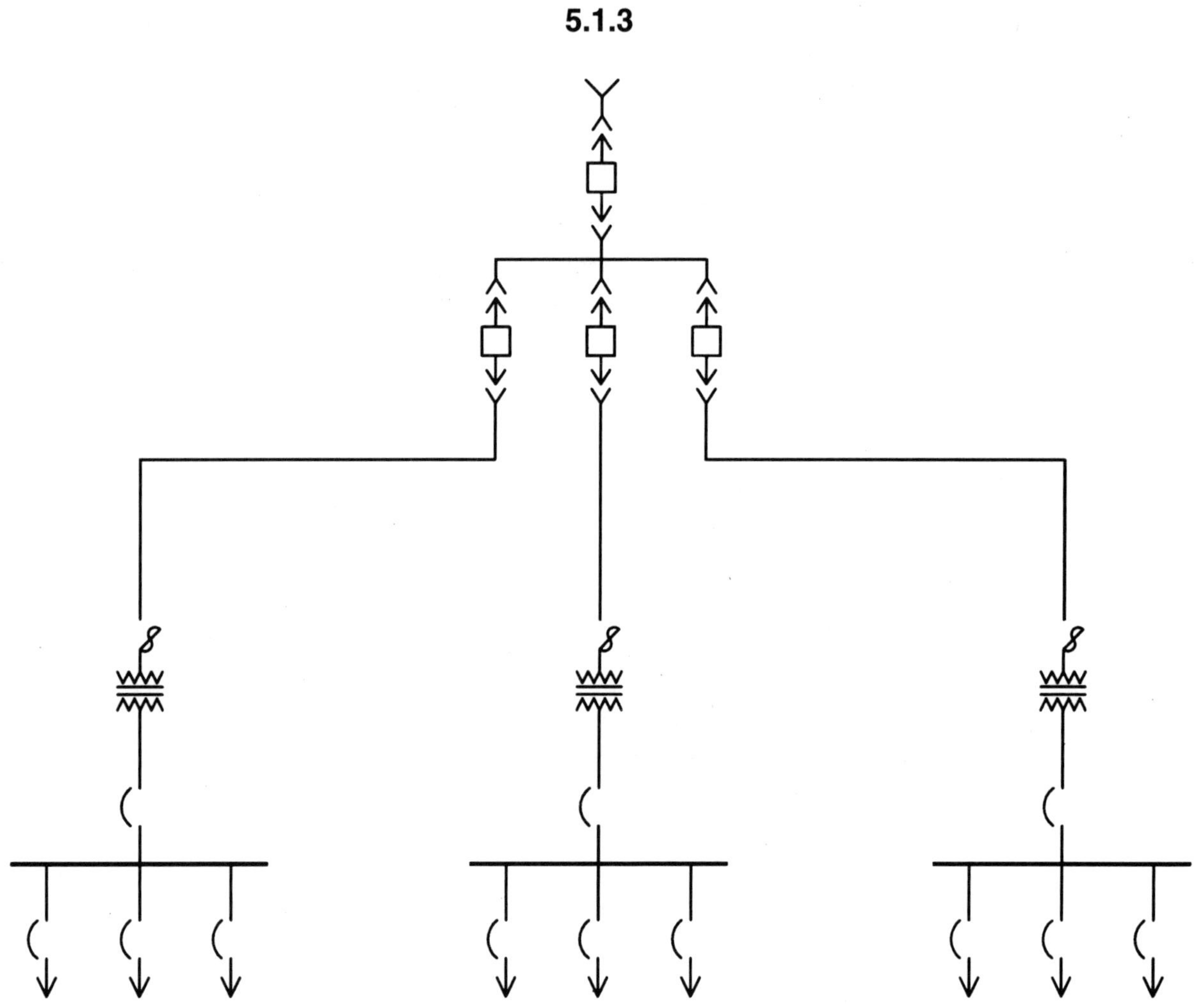

Characteristics:

- First three characteristics are the same as Figure 5.1.2.
- This arrangement has the advantage of limiting outages, due to a feeder or transformer, to the loads associated with the faulted equipment.
- The cost is usually higher than the arrangement shown in Figure 5.1.2.

5.1.4 Primary Radial-Selective Circuit Arrangements

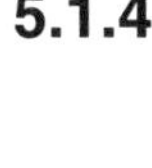

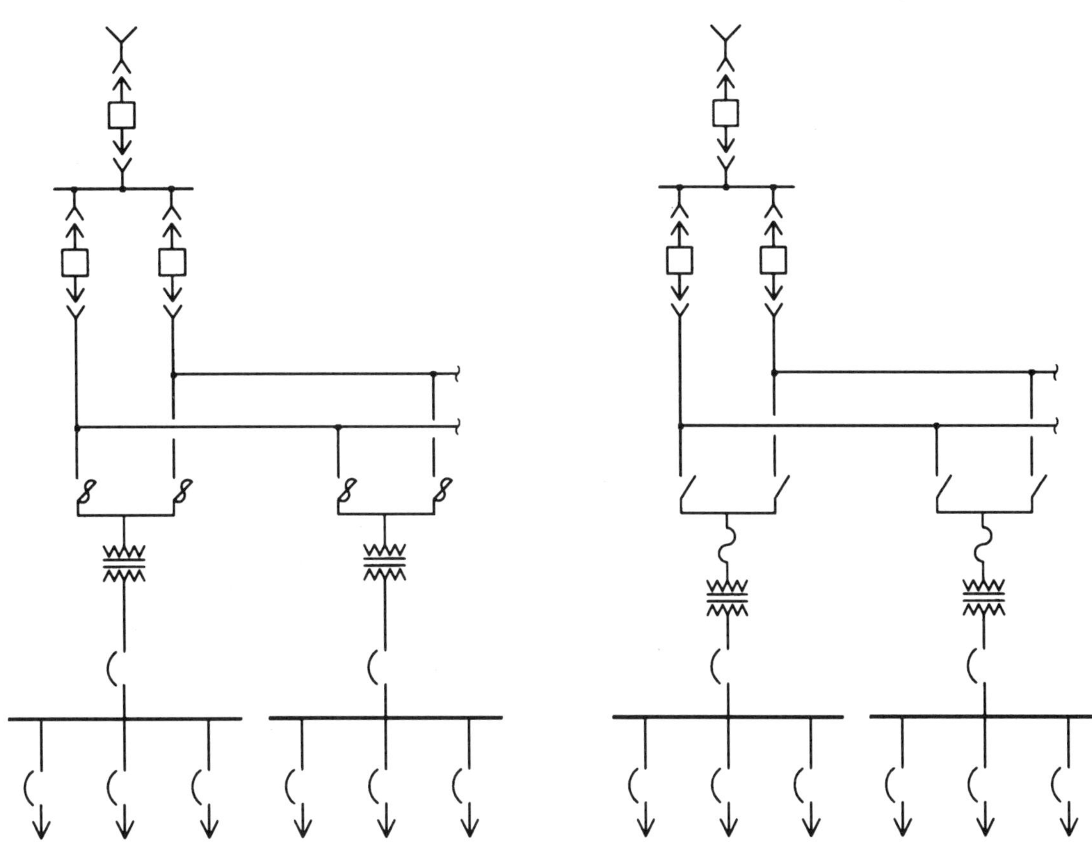

(a) Dual Fused Switches

(b) Duplex Load Interrupter Switches with Transformer Primary Fuses

Characteristics:

- These circuit arrangements reduce both the extent and duration of an outage caused by a primary feeder fault.
- Operating feature — duplicate primary feeder circuits and load interrupter switches, permit connection to either primary feeder circuit.
- Each feeder must be capable of saving the entire load.
- Suitable safety interlocks usually required to prevent closing of both switches at the same time.
- Under normal operating conditions, appropriate switches are closed to balance loads between two primary feeder circuits.
- Primary-selective switches are usually manually operated, but can be automated for quicker restoration of service. Automated switching is more costly but may be justified in many applications.
- If a fault occurs in a secondary substation transformer, service can be restored to all loads except those served from the faulted transformer.
- The higher degree of service continuity afforded by the primary-selective arrangement is realized at a cost that is usually 10%–20% higher than the circuit arrangement of Figure 5.1.3.

5.1.5 Secondary-Selective Circuit Arrangement (Double-Ended Substation with Single Tie)

5.1.5

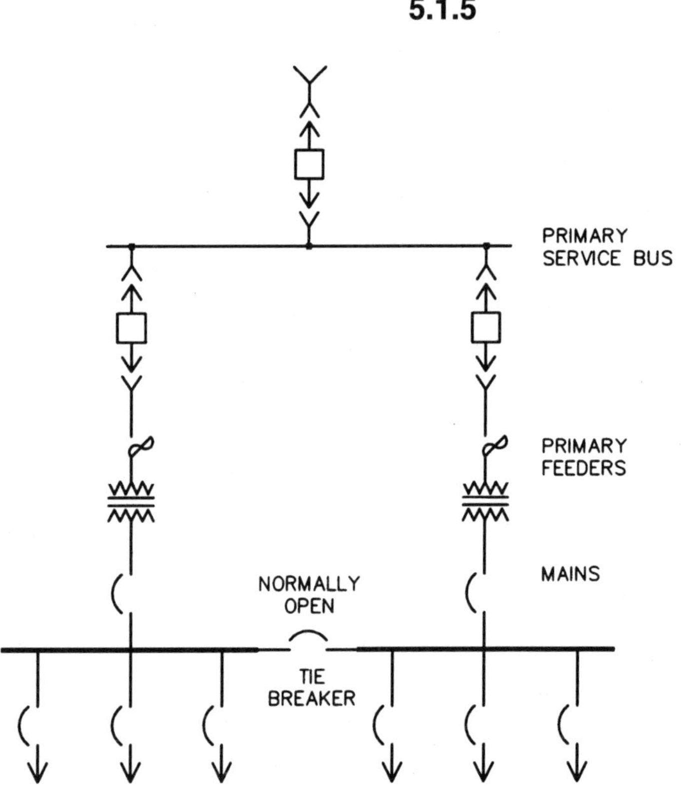

Characteristics:

- Under normal conditions, operates as two separate radial systems with the secondary bus–tie circuit breaker normally open.
- Loads should be divided equally between the two bus sections.
- If a fault occurs on a primary feeder or in a transformer, service is interrupted to all loads served from that half of the double–ended arrangement. Service can be restored to all secondary buses by opening the secondary main on the faulted side and closing the tie breaker.
- The main–tie –main breakers are normally interlocked to prevent paralleling the transformers and to prevent closing into a secondary bus fault. They can also be automated to transfer to standby operation and retransfer to normal operation.
- Cost of this arrangement will depend upon the spare capacity in the transformers and primary feeders. The minimum will be determined by the essential loads that need to be served under standby operating conditions. If service is to be provided for all loads under standby conditions, then the primary feeders and transformers must be capable of carrying the total load on both substation buses.
- This circuit arrangement is more expensive than either the radial or primary selective circuit configuration. This is primarily due to the redundant transformers.

5.1.6 Secondary-Selective Circuit Arrangement (Individual Substations with Interconnecting Ties)

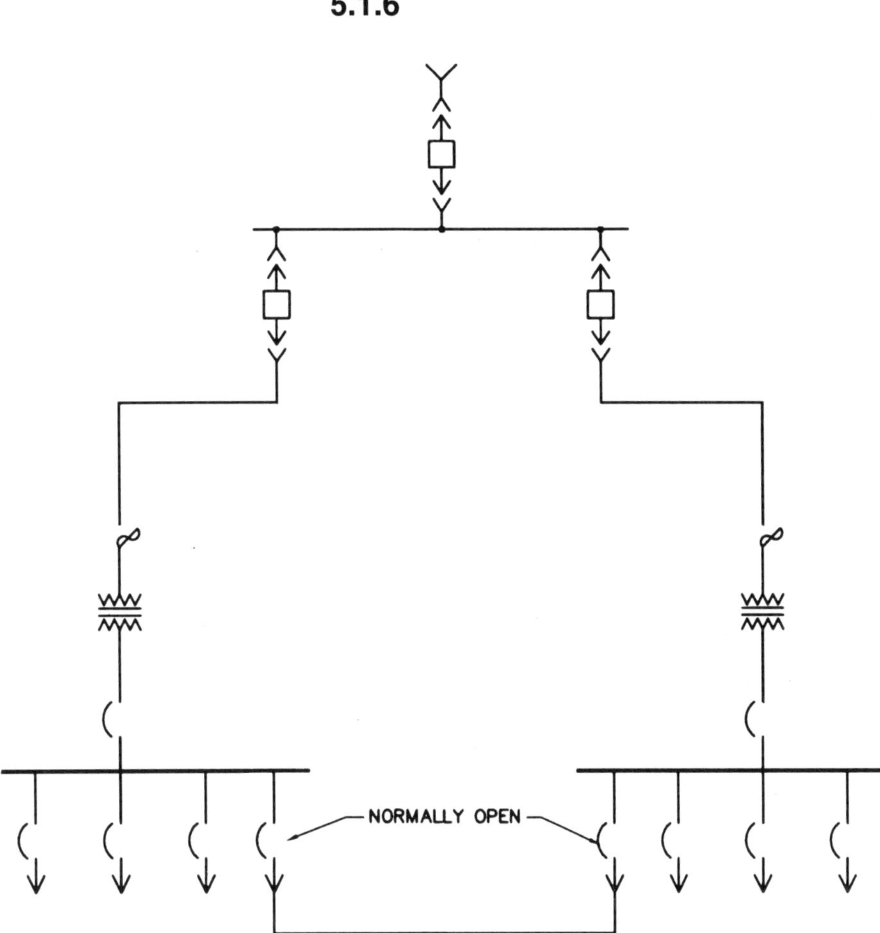

Characteristics:

- In this modification of the secondary–selective circuit arrangement shown if Figure 5.1.5, there is only one transformer in each secondary substation; but adjacent substations are interconnected in pairs by a normally open low–voltage tie circuit.

- When the primary feeder or transformer supplying one secondary substation bus is out of service, essential loads on that substation bus can be supplied over the tie circuit.

- Operating aspects of this system are somewhat complicated if the two substations are separated by distance.

- This would not be a desirable choice in a new building service design because a multiple key interlock system would be required to avoid tying the two substations together while they were both energized.

5.1.7 Primary- and Secondary-Selective Circuit Arrangement (Double-Ended Substation with Selective Primary)

5.1.7

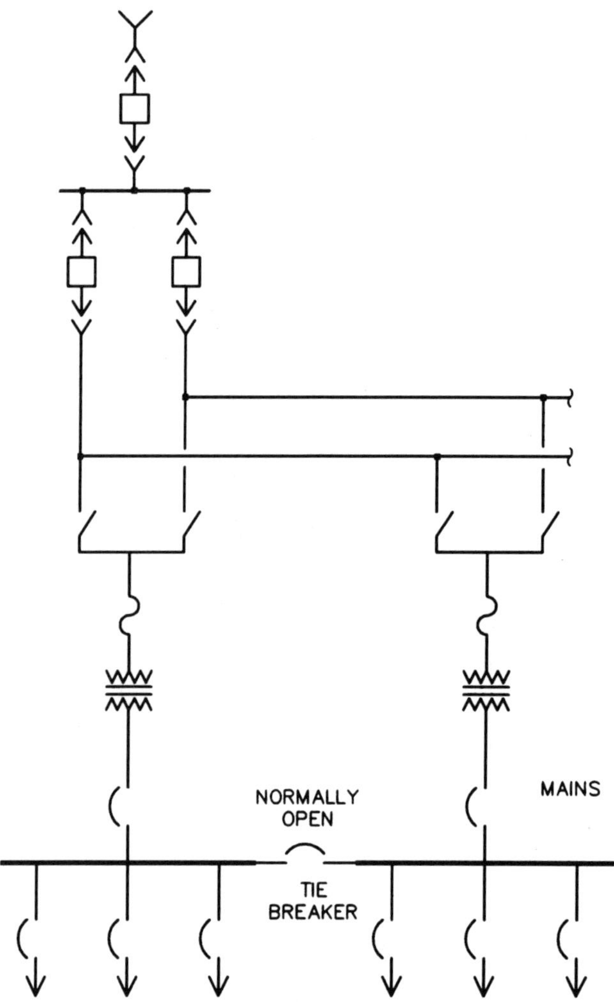

Characteristics:

- Used when highly reliable service is needed, such as hospital or data center loads.
- Has the combined benefits and characteristics of the arrangements shown in Figure 5.1.4 and 5.1.5.
- Small premium cost over configuration shown in Figure 5.1.5 for primary selector switches.

5.1.8 Looped Primary Circuit Arrangement

5.1.8

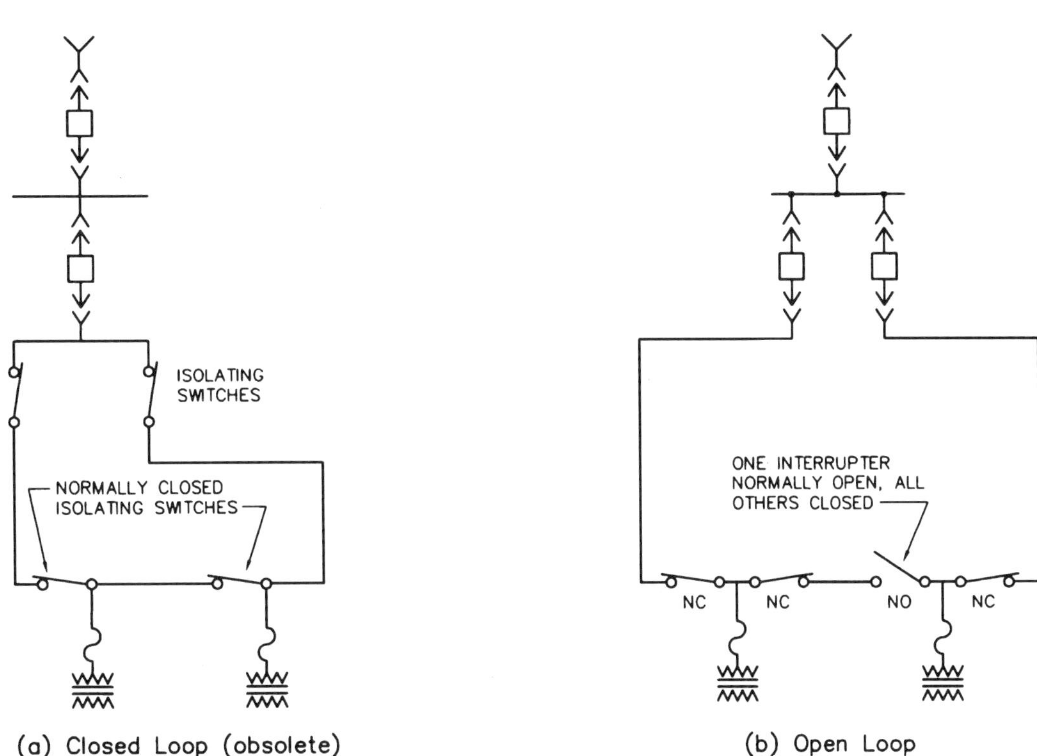

(a) Closed Loop (obsolete)

(b) Open Loop

Characteristics:

- Basically a two-circuit radial system with the ends connected together forming a continuous loop.

- Early versions of the closed loop in (a) above, although relatively inexpensive, fell into disfavor because of its apparent reliability advantages are offset by interruption of all service from a fault occurring anywhere in the loop, by the difficulty of locating primary faults, and by safety problems associated with the nonload break, or "dead break", isolating switches.

- Newer open-loop versions as shown in (b) above, designed for modern underground commercial and residential distribution systems, utilize fully rated air, oil, vacuum, and SF6 interrupter switches. Equipment available up to 34.5KV with interrupting ratings for both continuous load and fault currents to meet most system requirements. Certain equipment can close and latch on fault currents, equal to the equipment interrupting values, and still be operational without maintenance.

- Major advantages of the open-loop primary system over the simple radial system is the isolation of cable or transformer faults or both, while maintaining continuity of service to the remaining loads. With coordinated transformer fusing provided in the loop-tap position, transformer faults can be isolated without any interruption of primary service. Primary cable faults will temporarily drop service to half of the connected loads until the fault is located; then, by selective switching the unfaulted sections can be restored to service, leaving only the faulted section to be repaired.

- Disadvantages; increased costs to fully size cables, protective devices and interrupters to total capacity of the load, and the time delay necessary to locate the fault, isolate the section, and restore service. Safety considerations in maintaining a loop system are more complex than for a radial or a primary-selective system.

5.1.9 Distributed Secondary Network

5.1.9

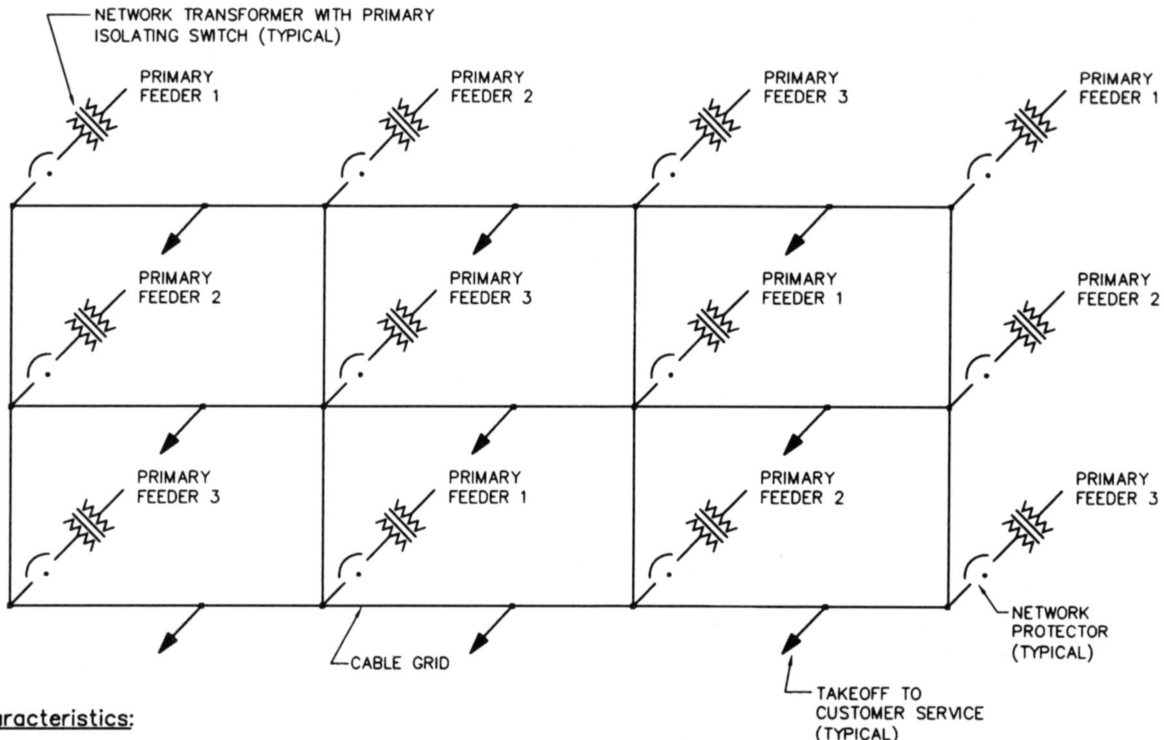

Characteristics:

- A secondary network is formed when two or more transformers having the same characteristics are supplied from separate feeders, and are connected to a common bus through network protectors.
- This arrangement is usually found in high load density urban areas where the highest level of service reliability is required. The cable grid shown typically represents a city block (for each square). Additional transformers are added as needed at locations where exceptionally high load customer service take-offs occur.
- Transformer and cable grid capacities are sized initially and added to as needed, to maintain voltage regulation, and load capacity.
- Transformer capacity and impedance characteristics are the same for equal load sharing. Likewise, the cable grid is sized for balanced load flow and to maintain voltage regulation.
- A typical grid voltage is 216Y/125-volts to provide preferred nominal 208Y/120-volt service and utilization. This helps to provide better voltage regulation.
- These systems are designed for 1st contingency operation, i.e., to provide full capacity with no interruption of service with the loss of one of the primary network feeders.
- Operational experience has shown that three primary feeders provides optimum reliability. Four or more primary network feeders provide virtually no additional reliability.
- 216V secondary network cable grids are designed to operate so that faults at the grid are allowed to burn clear rather than incur a disruption of service. This is accomplished by providing cable limiters in each end of each conductor in a parallel cable grouping that forms the secondary grid.
- Network transformers have a higher impedance than conventional transformers (typically 7% vs 5.75% for a 500KVA unit) to help limit available fault current. Secondary networks typically have an available fault current in the order of 200,000 amperes RMS symmetrical.

5.1.10 Basic Spot Network

5.1.10

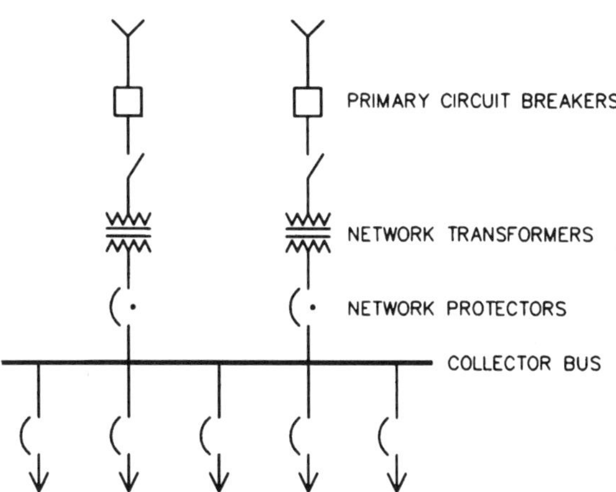

Characteristics:

- The spot network is a localized distribution center consisting of two or more transformer/network protector units connected to a common bus called a "collector bus". A building may have one or more spot network services.

- Spot networks are employed to provide a reliable source of power to important electrical loads. Spare capacity is built in to allow for at least one contingency, i.e., loss of a transformer or primary network feeder will cause no interruption of service.

- Planning for service continuity should be extended beyond the consideration of using a utility primary feeder or transformer. The consequences of severe equipment damage including the resulting system downtime, should also be considered.

- The primary side of a spot network transformer usually contains an isolating/load interrupter switch, primary fuses, and a non–load break grounding switch located within the same enclosure. Although the grounding switch has a fault closing rating, it cannot be operated until the safety requirements of a key interlock scheme have been satisfied. The key interlocks prevent closing the grounding switch until all possible sources of supply to the feeder have been isolated.

- Conventional automatic network protectors are sophisticated devices. They are self–contained units consisting of an electrically operated circuit breaker, special network relays, control transformers, instrument transformers, and open–type fuse links. The protector will automatically close when the oncoming transformer voltage is greater than the collector bus voltage and will open when reverse current flows from the collector bus into the transformer. Reverse current flow can be the result of a fault beyond the line side of the protector, supplying load current back into the primary distribution system when the collector bus voltage is higher than the individual transformer voltage, or the opening of the transformer primary feeder breaker, which causes the collector bus to supply transformer magnetizing current via the transformer secondary winding.

- Most spot network applications for commercial buildings provide 480Y/277–volt utilization, thus requiring ground fault protection. Relay protection is the most common method of ground–fault protection. The fault current may be sensed by the ground return, residual, or zero–sequence method. Each of the methods have proved successful where appropriately applied; but they share a common limitation in that they cannot distinguish between in–zone and thru–zone ground faults unless incorporated in a complex protection scheme. One particular method of ground–fault detection that is not prone to unnecessary tripping is enclosure monitoring. This method offers the distinct advantage of not requiring coordination with other protective devices.

Section

6

Preliminary Load Calculations

6.1.0 Introduction
6.1.1 Prescriptive Unit Lighting Power Allowance (ULPA) (W/ft^2), Gross Lighted Area of Total Building
6.1.2 Typical Appliance/General-Purpose Receptacle Loads (Excluding Plug-In-Type A/C and Heating Equipment)
6.1.3 Typical Apartment Loads
6.1.4 Typical Connected Electrical Load for Air Conditioning Only
6.1.5 Central Air Conditioning Watts per SF, BTUs per Hour per SF of Floor Area, and SF per Ton of Air Conditioning
6.1.6 All-Weather Comfort Standard Recommended Heat-Loss Values
6.1.7 Typical Power Requirement (kW) for High-Rise Building Water Pressure–Boosting Systems
6.1.8 Typical Power Requirement (kW) for Electric Hot Water–Heating System
6.1.9 Typical Power Requirement (kW) for Fire Pumps in Commercial Buildings (Light Hazard)
6.1.10 Typical Loads in Commercial Kitchens
6.1.11 Comparison of Maximum Demand
6.1.12 Connected Load and Maximum Demand by Tenant Classification
6.1.13 Factors Used in Sizing Distribution-System Components
6.1.14 Factors Used to Establish Major Elements of the Electrical System Serving HVAC Systems
6.1.15 Service Entrance Peak Demand (Veterans Administration)
6.1.16 Service Entrance Peak Demand (Hospital Corporation of America)

6.1.0 Introduction

The following tables are provided to assist the user in estimating preliminary loads for various building types. Considerable judgment should be used in the application of these data. Power densities typically are given in watts per square foot (W/ft^2) or volt-amperes per square foot (VA/ft^2) and are used interchangeably because unity power factor is assumed for preliminary load calculations.

These tables give estimated *connected* loads. To these the user must apply a demand factor to estimate the actual demand load. Demand factors for buildings typically range between 50 and 80 percent of the connected load. For most building types, the demand factor at the service where the maximum diversity is experienced is usually 60 to 75 percent of the connected load. Specific portions of the system may have much higher demand factors, even approaching 100 percent.

6.1.1 Prescriptive Unit Lighting Power Allowance (ULPA) (W/ft²), Gross Lighted Area of Total Building

TABLE 6.1.1

Building Type or Space Activity	0 to 2000 ft²	2001 to 10 000 ft²	10 001 to 25 000 ft²	25 001 to 50 000 ft²	50 001 to 250 000 ft²	>250 000 ft²
Food Service						
Fast Food/Cafeteria	1.50	1.38	1.34	1.32	1.31	1.30
Leisure Dining/Bar	2.20	1.91	1.71	1.56	1.46	1.40
Offices	1.90	1.81	1.72	1.65	1.57	1.50
Retail*	3.30	3.08	2.83	2.50	2.28	2.10
Mall Concourse	1.60	1.58	1.52	1.46	1.43	1.40
Multiple-Store Service						
Service Establishment	2.70	2.37	2.08	1.92	1.80	1.70
Garages	0.30	0.28	0.24	0.22	0.21	0.20
Schools						
Preschool/Elementary	1.80	1.80	1.72	1.65	1.57	1.50
Jr. High/High School	1.90	1.90	1.88	1.83	1.76	1.70
Technical/Vocational	2.40	2.33	2.17	2.01	1.84	1.70
Warehouse/Storage	0.80	0.66	0.56	0.48	0.43	0.40

NOTE: *Includes general, merchandising, and display lighting.

This prescriptive table is intended primarily for core-and-shell (i.e., speculative) buildings or for use during the preliminary design phase (i.e., when the space uses are less than 80% defined). The values in this table are not intended to represent the needs of all buildings within the types listed.

Reprinted from ASHRAE/IES 90.1-1989, Energy Efficient Design of New Buildings Except New Low-Rise Residential Buildings. Used with the permission of ASHRAE.

6.1.2 Typical Appliance/General-Purpose Receptacle Loads (Excluding Plug-In-Type A/C and Heating Equipment)

TABLE 6.1.2

Type of Occupancy	Unit Load (VA/ft²) Low	High	Average
Auditoriums	0.1	0.3	0.2
Cafeterias	0.1	0.3	0.2
Churches	0.1	0.3	0.2
Drafting rooms	0.4	1.0	0.7
Gymnasiums	0.1	0.2	0.15
Hospitals	0.5	1.5	1.0
Hospitals, large	0.4	1.0	0.7
Machine shops	0.5	2.5	1.5
Office buildings	0.5	1.5	1.0
Schools, large	0.2	1.0	0.6
Schools, medium	0.25	1.2	0.7
Schools, small	0.3	1.5	0.9

Other Unit Loads:
- Specific appliances — ampere rating of appliance
- Supplying heavy-duty lampholders — 5 A/outlet

6.1.3 Typical Apartment Loads

TABLE 6.1.3

Type	Load
Lighting and convenience outlets (except appliance)	3 VA/ft^2
Kitchen, dining appliance circuits	1.5 kVA each
Range	8 to 12 kW
Microwave oven	1.5 kW
Refrigerator	0.3 to 0.6 kW
Freezer	0.3 to 0.6 kW
Dishwasher	1.0 to 2.0 kW
Garbage disposal	0.33 to 0.5 hp
Clothes washer	0.33 to 0.5 hp
Clothes dryer	1.5 to 6.5 kW
Water heater	1.5 to 9.0 kW
Air conditioner (0.5 hp/room)	0.8 to 4.6 kW

6.1.4 Typical Connected Electrical Load for Air Conditioning Only

TABLE 6.1.4

Type of Building	Conditioned Area (VA/ft^2)
Bank	7
Department store	3 to 5
Hotel	6
Office building	6
Telephone equipment building	7 to 8
Small store (shoe, dress, etc.)	4 to 12
Restaurant (not including kitchen)	8

6.1.5 Central Air Conditioning Watts per SF, BTUs per Hour per SF of Floor Area, and SF per Ton of Air Conditioning

TABLE 6.1.5

Type Building	Watts per S.F.	BTUH per S.F.	S.F. per Ton	Type Building	Watts per S.F.	BTUH per S.F.	S.F. per Ton	Type Building	Watts per S.F.	BTUH per S.F.	S.F. per Ton
Apartments, Individual	3	26	450	Dormitory, Rooms	4.5	40	300	Libraries	5.7	50	240
Corridors	2.5	22	550	Corridors	3.4	30	400	Low Rise Office, Ext.	4.3	38	320
Auditoriums & Theaters	3.3	40	300/18*	Dress Shops	4.9	43	280	Interior	3.8	33	360
Banks	5.7	50	240	Drug Stores	9	80	150	Medical Centers	3.2	28	425
Barber Shops	5.5	48	250	Factories	4.5	40	300	Motels	3.2	28	425
Bars & Taverns	15	133	90	High Rise Off.-Ext. Rms.	5.2	46	263	Office (small suite)	4.9	43	280
Beauty Parlors	7.6	66	180	Interior Rooms	4.2	37	325	Post Office, Int. Office	4.9	42	285
Bowling Alleys	7.8	68	175	Hospitals, Core	4.9	43	280	Central Area	5.3	46	260
Churches	3.3	36	330/20*	Perimeter	5.3	46	260	Residences	2.3	20	600
Cocktail Lounges	7.8	68	175	Hotels, Guest Rooms	5	44	275	Restaurants	6.8	60	200
Computer Rooms	16	141	85	Public Spaces	6.2	55	220	Schools & Colleges	5.3	46	260
Dental Offices	6	52	230	Corridors	3.4	30	400	Shoe Stores	6.2	55	220
Dept. Stores, Basement	4	34	350	Industrial Plants, Offices	4.3	38	320	Shop'g. Ctrs., Sup. Mkts.	4	34	350
Main Floor	4.5	40	300	General Offices	4	34	350	Retail Stores	5.5	48	250
Upper Floor	3.4	30	400	Plant Areas	4.5	40	300	Specialty Shops	6.8	60	200

*Persons per ton

12,000 BTUH = 1 ton of air conditioning

6.1.6 All-Weather Comfort Standard Recommended Heat-Loss Values

TABLE 6.1.6

Degree Days	Design Heat Loss per Square Foot of Floor Area (Btu/h)	(watts)
Over 8000	40	11.7
7001 to 8000	38	11.3
6001 to 7000	35	10.3
5001 to 6000	32	9.4
3001 to 5000	30	8.8
Under 3001	28	8.2

6.1.7 Typical Power Requirement (kW) for High-Rise Building Water Pressure–Boosting Systems

TABLE 6.1.7

Building Type	Unit Quantity	Number of Stories			
		5	10	25	50
Apartments	10 apt./floor	—	15	90	350
Hospitals	30 patients/floor	10	45	250	—
Hotels/Motels	40 rooms/floor	7	35	175	450
Offices	10 000 ft^2/floor	—	15	75	250

6.1.8 Typical Power Requirement (kW) for Electric Hot Water–Heating System

TABLE 6.1.8

Building Type	Unit Quantity	Load
Apartments/Condominiums	20 apt/condo	30
Dormitories	100 residents	75
Elementary schools	100 students	6
High schools	100 students	12
Restaurant (full service)	100 servings/h	30
Restaurant (fast service)	100 servings/h	15
Nursing homes	100 residents	60
Hospitals	100 patient beds	200
Office buildings	10 000 ft^2	5

6.1.9 Typical Power Requirement (kW) for Fire Pumps in Commercial Buildings (Light Hazard)

TABLE 6.1.9

Area/Floor (ft^2)	Number of Stories			
	5	10	25	50
5000	40	65	150	250
10 000	60	100	200	400
25 000	75	150	275	550
50 000	120	200	400	800

*Based on zero pressure at floor 1.

6.1.10 Typical Loads in Commercial Kitchens

TABLE 6.1.10

	Number Served	Connected Load (kW)
Lunch counter (gas ranges, with 40 seats)		30
Cafeteria	800	150
Restaurant (gas cooking)		90
Restaurant (electric cooking)		180
Hospital (electric cooking)	1200	300
Diet kitchen (gas cooking)		200
Hotel (typical)		75
Hotel (modern, gas ranges, three kitchens)		150
Penitentiary (gas cooking)		175

Note: As an alternative to the preceding table, you may use 25 W/ft^2 for commercial kitchens using natural gas for cooking or 125 W/ft^2 for electric cooking. The applicable square footage in calculating kitchen floor area should include cooking and preparation, dishwashing, storage, walk-in refrigerators and freezers, food serving lines, tray assembly, and offices.

6.1.11 Comparison of Maximum Demand

TABLE 6.1.11

	Shopping Center A, New Jersey No Refrigeration*		Shopping Center B, New Jersey Refrigeration		Shopping Center C, New York Refrigeration	
Type of Store	Gross Area (ft^2)	(W/ft^2)	Gross Area (ft^2)	(W/ft^2)	Gross Area (ft^2)	(W/ft^2)
Bank					4000	9.0
Book	3700	6.0	2500	6.7		
Candy	1600	6.9			2000	10.8
Department	343 500	4.7	222 000	7.3	226 900	8.0
	84 000	3.1	114 000	5.6		
Drug	7000	6.1	6000	7.7		
Men's wear	17 000	5.5	17 000	9.9	2000	10.8
	28 000	4.9	9100	8.8		
Paint					15 600	8.5
Pet					2000	12.1
Restaurant					4000	9.0
Shoe	11 000	6.3	7000	12.5	3300	15.4
	4000	8.0	4400	12.9	2100	9.0
Supermarket	32 000	5.7	25 000	8.6	37 600	11.5
Variety	31 000	4.6	24 000	6.8	37 400	7.1
	30 000	4.4			30 000	7.0
Women's wear	20 400	4.7	19 300	8.9	1360	13.0
	1000	5.8	4500	9.6	1000	11.7

*Loads include all lighting and power, but no power for air-conditioning refrigeration (chilled water), which is supplied from a central plant.

6.1.12 Connected Load and Maximum Demand by Tenant Classification

TABLE 6.1.12

Classification	Connected Load (W/ft^2)	Maximum Demand (W/ft^2)	Demand Factor
10 Women's wear	7.7	5.9	0.75
3 Men's wear	7.2	5.6	0.78
6 Shoe store	8.5	6.9	0.79
2 Department store	6.0	4.7	0.74
2 Variety store	10.5	4.5	0.45
2 Drug store	11.7	6.7	0.57
5 Household goods	5.4	3.9	0.76
10 Specialty shop	8.1	6.8	0.79
4 Bakery and candy	17.1	12.1	0.71
3 Food store (supermarkets)	9.9	5.9	0.60
5 Restaurant	15.9	7.1	0.45

NOTE: Connected load includes an allowance for spares.

6.1.13 Factors Used in Sizing Distribution-System Components

TABLE 6.1.13

Distribution System Component	Lighting Demand Factor
Lighting panelboard buss and main overcurrent device	1.0
Lighting panelboard feeder and feeder overcurrent device	1.0
Distribution panelboard buss and main overcurrent device	
First 50 000 W or less	0.5
All over 50 000 W	0.4
Remaining components	0.4

6.1.14 Factors Used to Establish Major Elements of the Electrical System Serving HVAC Systems

TABLE 6.1.14

Item	Unit
Refrigeration Machines:	kVA/Ton of Chiller Capacity
Absorption	
Centrifugal	1.00
Reciprocating	
Auxiliary Pumps & Fans:	
Chilled Water Pumps	0.08
Condenser Water Pumps	
Absorption	
Centrifugal/Reciprocating	0.07
Cooling Tower Fans	
Absorption	
Centrifugal/Reciprocating	0.07
Boilers:	kVA/Boiler Horsepower
Natural Gas/Fuel Oil	0.07
Coal	
Boiler Auxiliary Pumps:	kVA/Boiler Horsepower
Deaerator	0.10
Auxiliary Equipment:	kVA/Bed
Clinical Vacuum Pumps	0.18
Clinical Air Compressors	0.10

Note: For a primary cooling system comprised of electrical centrifugal chillers, chilled water pumps, condenser water pumps, and cooling tower fans, a factor of 1.7 kVA/ton provides a good estimate.

6.1.15 Service Entrance Peak Demand (Veterans Administration)

TABLE 6.1.15

Hospital	Floor Area Square Feet	Beds*	Degree Days† Cooling	Degree Days† Heating	Principal‡ Fuel-HVAC	Watts Per Sq ft§ Maximum	Watts Per Sq ft§ Average
V.A. Hospital #1	821 000	922	234	3536	NG/FO	4.5	3.5
V.A. Hospital #2	334 000	500	863	5713	NG/FO	5.2	3.9
V.A. Hospital #3	645 995	670	3488	1488	NG/FO	3.8	2.8
V.A. Hospital #4	681 000	600	1016	654	NG/FO	6.1	4.0
V.A. Hospital #5	503 500	697	3495	841	NG/FO	7.2	5.5
V.A. Hospital #6	800 000	1050	600	7400	NG/FO	5.9	4.2

* Total beds shown. Beds actually occupied could affect values shown for watts per square foot.

† Degree Days: Normals, Base 65 °F, based on 1941-70 period. From *Local Climatological Data* Series, 1974, NOAA.

‡ NG/FO = Natural Gas/Fuel Oil; E = Electricity. Principal fuel is defined as that used for heating. In all cases, electricity was the fuel used for refrigeration.

§ Watts per square foot based on measured values at service entrance during metering periods ranging from 9 to 17 days, during cooling season in all instances, 1981.

6.1.16 Service Entrance Peak Demand (Hospital Corporation of America)

TABLE 6.1.16

Hospital and Location	Floor Area Square Feet	Beds*	Degree Days† Cooling	Degree Days† Heating	Principal‡ Fuel-HVAC	Watts Per Sq ft§ Maximum
#1 — East	273 000	458	1353	3939	NG/FO	6.8
#2 — Southeast	278 000	250	2294	2240	NG/FO	6.3
#3 — Central	123 000	157			NG/FO	7.5
#4 — Central	36 365	62	2029	3227‖	E	13.7
#5 — Central	318 000	300	1107	4306	NG/FO	4.6
#6 — Southeast	182 000	225	3786	299‖	NG/FO	5.3
#7 — East	283 523	320	1030	4307	NG/FO	6.8
#8 — Southwest	135 396	150	2250	2621‖	NG/FO	6.6
#9 — West	190 000	97	927	5983	NG/FO	2.8
#10 — Southeast	161 000	170	3226	733‖	NG/FO	6.3
#11 — Southeast	157 639	214	2078	2146	NG/FO	7.3
#12 — Southeast	162 187	222	2143	2378‖	NG/FO	4.3
#13 — East	109 617	146	1030	4307‖	NG/FO	5.7
#14 — East	76 000	153	1030	4307‖	E	8.8
#15 — Southeast	135 150	190	1995	2547‖	NG/FO	5.9
#16 — Southwest	75 769	131	2587	2382‖	NG/FO	7.4
#17 — Central	75 769	128	1636	3505‖	NG/FO	6.3
#18 — Northwest	129 000	150	714	5833‖	NG/FO	4.4
#19 — Central	54 938	108	1694	3696‖	E	13.3
#20 — West	144 000	160	2814	1752	NG/FO	4.5
#21 — Southeast	149 000	123	2078	2146‖	NG/FO	4.5
#22 — Central	89 000	128	2029	3227‖	E	8.4
#23 — Central	128 500	150	1197	4729‖	NG/FO	6.2
#24 — West	135 169	170	927	5983‖	NG/FO	4.7
#25 — Southeast	80 000	124	1722	2975‖	NG/FO	6.2
#26 — Southeast	83 117	126	3226	733‖	NG/FO	8.5
#27 — Central	51 000	97	1569	3478	E	8.8
#28 — Southeast	66 528	120	2929	902‖	E	9.7
#29 — East	112 000	140	1394	3514	NG/FO	4.3
#30 — Central	202 000	223	1636	3505	NG/FO	4.8
#31 — Southeast	56 000	51	3786	299‖	NG/FO	7.4
#32 — West	47 434	50	927	5983‖	NG/E	7.0
#33 — Central	23 835	32	1694	3696‖	E	10.8
#34 — Southeast	105 000	95	2706	1465‖	NG/FO	8.3
#35 — West	48 575	60	3042	108‖	NG/E	7.7
#36 — Southwest	133 000	185	2587	2382‖	NG/FO	6.3
#37 — Central	42 879	66	1694	3696‖	E	15.7

* Total beds shown. Beds actually occupied could affect values shown for watts per square foot.

† Degree Days: Normals, Base 65 °F, based on 1941-70 period. From *Local Climatological Data* Series, 1974, NOAA.

‡ NG/FO = Natural Gas/Fuel Oil; E = Electricity. Principal fuel is defined as that used for heating. In all cases, electricity was the fuel used for refrigeration.

§ Watts per square foot based on measured values by utility company meter at service entrance, 1977.

‖ Data shown for nearest recorded location.

(Each facility was self-contained, in that refrigeration and air conditioning equipment loads are included in power demands shown.)

Section

7

Short-Circuit Calculations

7.1.0 Introduction
7.1.1 Point-to-Point Method, Three-Phase Short-Circuit Calculations, Basic Calculation Procedure and Formulas
7.1.2 System A and System B Circuit Diagrams for Sample Calculations Using Point-to-Point Method
7.1.3 Point-to-Point Calculations for System A to Faults X_1 and X_2
7.1.4 Point-to-Point Calculations for System B to Faults X_1 and X_2
7.1.5 *C* Values for Conductors and Busway
7.1.6 Shortcut Method 1: Adding *Z*s
7.1.7 Average Characteristics of 600-V Conductors (Ohms per 100 ft): Two or Three Single Conductors
7.1.8 Average Characteristics of 600-V Conductors (Ohms per 100 ft): Three Conductor Cables (and Interlocked Armored Cable)
7.1.9 LV Busway, *R, X,* and *Z* (Ohms per 100 ft)
7.1.10 Shortcut Method 2: Chart Approximate Method
7.1.11 Conductor Conversion (Based on Using Copper Conductor)
7.1.12 Charts 1 through 13 for Calculating Short-Circuit Currents Using Chart Approximate Method
7.1.13 Assumptions for Motor Contributions to Fault Currents
7.1.14 Secondary Short-Circuit Capacity of Typical Power Transformers

7.1.0 Introduction

Of the four basic methods used to calculate short-circuit currents, the point-to-point method offers a simple, effective, and quick way to determine available short-circuit levels in simple to medium-complexity three-phase and single-phase electrical distribution systems with a reasonable degree of accuracy. This method is best illustrated by the figures and table that follow. Figure 7.1.1 shows the steps and equations needed in the point-to-point method. Figure 7.1.2 shows one-line diagrams of two systems (A and B) to be used as illustrative examples. Figures 7.1.3 and 7.1.4 show the calculations for these two examples. And Table 7.1.5 provides the circuit constants needed in the equations for the point-to-point method.

The point-to-point method is followed by two shortcut methods for determining short-circuit currents at ends of conductors, specifically, adding *Z*s and the chart approximate method. These two methods make use of simplifications that are reasonable under most circumstances and almost certainly will yield answers that are on the safe side.

7.1.1 Point-to-Point Method, Three-Phase Short-Circuit Calculations, Basic Calculation Procedure and Formulas

7.1.1

The application of the point-to-point method permits the determination of available short-circuit currents with a reasonable degree of accuracy at various points for either 3Ø or 1Ø electrical distribution systems. This method can assume unlimited primary short-circuit current (infinite bus).

Basic Point-to-Point Calculation Procedure

Step 1. Determine the transformer full load amperes from either the nameplate or the following formulas:

3Ø Transformer $I_{f.l.} = \dfrac{KVA \times 1000}{E_{L-L} \times 1.732}$

1Ø Transformer $I_{f.l.} = \dfrac{KVA \times 1000}{E_{L-L}}$

Step 2. Find the transformer multiplier.

$$\text{Multiplier} = \frac{100}{{}^{*}\%Z_{trans}}$$

* **Note.** Transformer impedance (Z) helps to determine what the short circuit current will be at the transformer secondary. Transformer impedance is determined as follows: The transformer secondary is short circuited. Voltage is applied to the primary which causes full load current to flow in the secondary. This applied voltage divided by the rated primary voltage is the impedance of the transformer.
Example: For a 480 volt rated primary, if 9.6 volts causes secondary full load current to flow through the shorted secondary, the transformer impedance is 9.6/480 = .02 = 2%Z.
In addition, UL listed transformer 25KVA and larger have a ± 10% impedance tolerance. Short circuit amperes can be affected by this tolerance.

Step 3. Determine the transformer let-thru short-circuit current**.

$$I_{S.C.} = I_{f.l.} \times \text{Multiplier}$$

** **Note.** Motor short-circuit contribution, if significant, may be added to the transformer secondary short-circuit current value as determined in Step 3. Proceed with this adjusted figure through Steps 4, 5 and 6. A practical estimate of motor short-circuit contribution is to multiply the total motor current in amperes by 4.

Step 4. Calculate the "f" factor.

3Ø Faults $f = \dfrac{1.732 \times L \times I}{C \times E_{L-L}}$

1Ø Line-to-Line (L-L) Faults on 1Ø Center Tapped Transformer $f = \dfrac{2 \times L \times I}{C \times E_{L-L}}$

1Ø Line-to-Neutral (L-N) Faults on 1Ø Center Tapped Transformer $f = \dfrac{2 \times L \times I^{\dagger}}{C \times E_{L-N}}$

Where:

L = length (feet) of circuit to the fault.

C = constant from Table 7.1.5. For parallel runs, multiply C values by the number of conductors per phase.

I = available short-circuit current in amperes at beginning of circuit.

† **Note.** The L-N fault current is higher than the L-L fault current at the secondary terminals of a single-phase center-tapped transformer. The short-circuit current available (I) for this case in Step 4 should be adjusted at the transformer terminals as follows:
At L-N center tapped transformer terminals,
I = 1.5 x L-L Short-Circuit Amperes at Transformer Terminals

At some distance from the terminals, depending upon wire size, the L-N fault current is lower than the L-L fault current. The 1.5 multiplier is an approximation and will theoretically vary from 1.33 to 1.67. These figures are based on change in turns ratio between primary and secondary, infinite source available, zero feet from terminals of transformer, and 1.2 x %X and 1.5 x %R for L-N vs. L-L resistance and reactance values. Begin L-N calculations at transformer secondary terminals, then proceed point-to-point.

Step 5. Calculate "M" (multiplier).

$$M = \frac{1}{1 + f}$$

Step 6. Calculate the available short-circuit symmetrical RMS current at the point of fault.

$$I_{S.C.\ sym\ RMS} = I_{S.C.} \times M$$

Calculation of Short-Circuit Currents at Second Transformer in System

Use the following procedure to calculate the level of fault current at the secondary of a second, downstream transformer in a system when the level of fault current at the transformer primary is known.

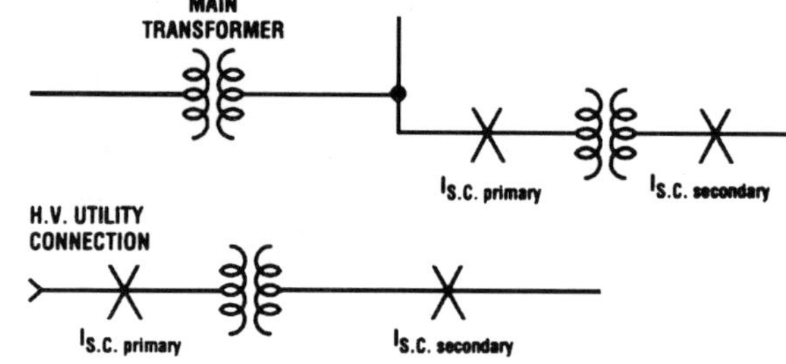

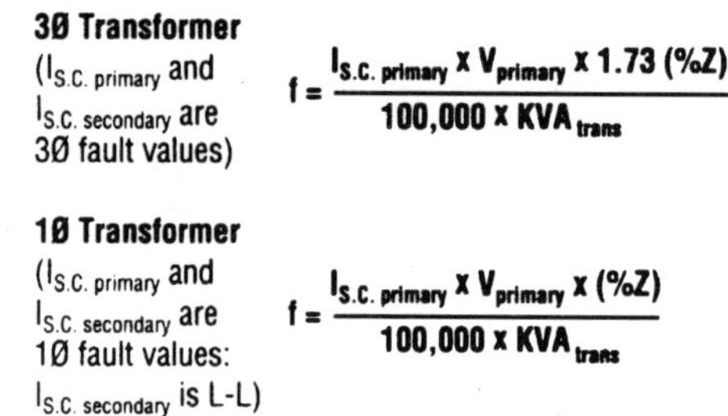

Procedure for Second Transformer in System

Step 1. Calculate the "f" factor ($I_{S.C.\ primary}$ known)

3Ø Transformer ($I_{S.C.\ primary}$ and $I_{S.C.\ secondary}$ are 3Ø fault values) $f = \dfrac{I_{S.C.\ primary} \times V_{primary} \times 1.73\ (\%Z)}{100{,}000 \times KVA_{trans}}$

1Ø Transformer ($I_{S.C.\ primary}$ and $I_{S.C.\ secondary}$ are 1Ø fault values: $I_{S.C.\ secondary}$ is L-L) $f = \dfrac{I_{S.C.\ primary} \times V_{primary} \times (\%Z)}{100{,}000 \times KVA_{trans}}$

Step 2. Calculate "M" (multiplier).

$$M = \frac{1}{1 + f}$$

Step 3. Calculate the short-circuit current at the secondary of the transformer. (See Note under Step 3 of "Basic Point-to-Point Calculation Procedure".)

$$I_{S.C.\ secondary} = \frac{V_{primary}}{V_{secondary}} \times M \times I_{S.C.\ primary}$$

7.1.2 System A and System B Circuit Diagrams for Sample Calculations Using Point-to-Point Method

7.1.2

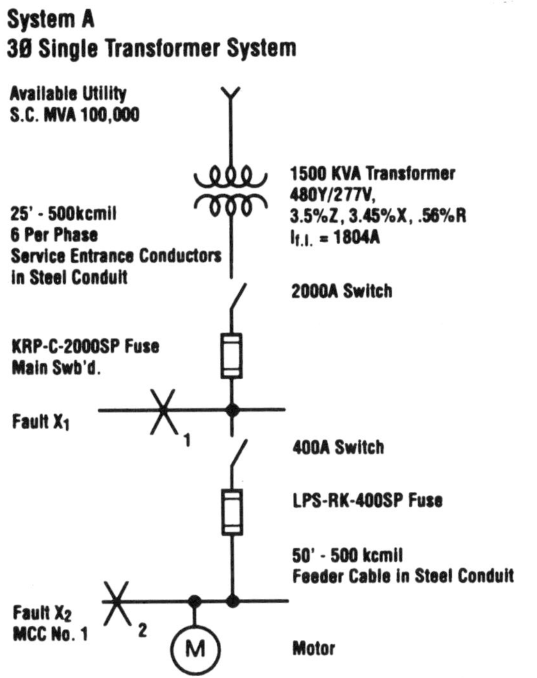

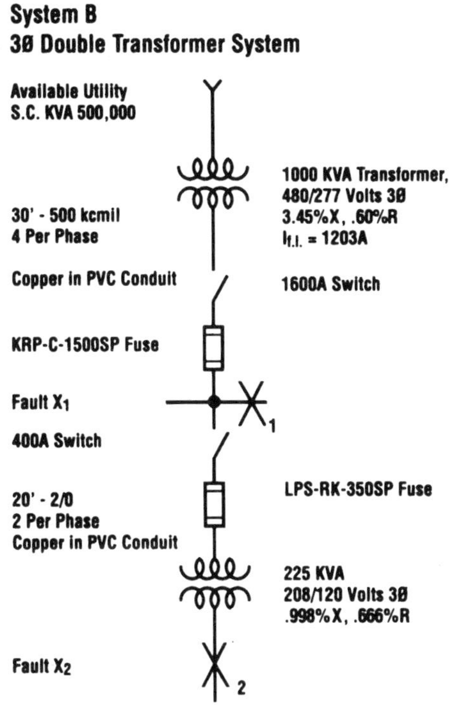

7.1.3 Point-to-Point Calculations for System A to Faults X_1 and X_2

7.1.3

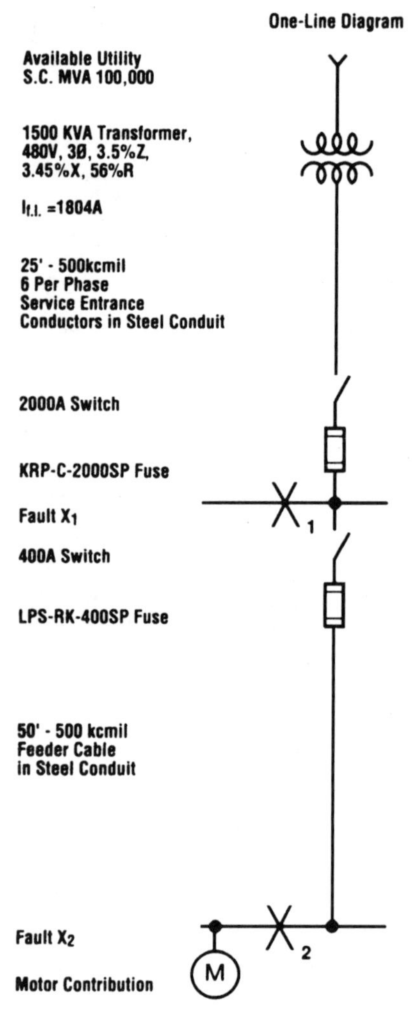

Fault X_1

Step 1. $I_{f.l.} = \dfrac{1500 \times 1000}{480 \times 1.732} = 1804A$

Step 2. $\text{Multiplier} = \dfrac{100}{3.5} = 28.57$

Step 3. $I_{S.C.} = 1804 \times 28.57 = 51{,}540A$

Step 4. $f = \dfrac{1.732 \times 25 \times 51{,}540}{6 \times 22{,}185 \times 480} = 0.0349$

Step 5. $M = \dfrac{1}{1 + .0349} = .9663$

Step 6. $I_{S.C.sym\ RMS} = 51{,}540 \times .9663 = 49{,}803A$

$I_{S.C.motor\ contrib} = 4 \times 1{,}804 = 7{,}216A$

$I_{total S.C.\ sym\ RMS\ (fault\ X_1)} = 49{,}803 + 7{,}216 = 57{,}019A$

Fault X_2

Step 4. Use $I_{S.C.sym\ RMS}$ @ Fault X_1 to calculate "f"

$f = \dfrac{1.732 \times 50 \times 49{,}803}{22{,}185 \times 480} = .4050$

Step 5. $M = \dfrac{1}{1 + .4050} = .7117$

Step 6. $I_{S.C.sym\ RMS} = 49{,}803 \times .7117 = 35{,}445A$

$I_{sym\ motor\ contrib} = 4 \times 1{,}804 = 7{,}216A$

$I_{total\ S.C.\ sym\ RMS\ (fault\ X_2)} = 35{,}445 + 7{,}216 = 42{,}661A$

7.1.4 Point-to-Point Calculations for System B to Faults X_1 and X_2

7.1.4

One-Line Diagram

Available Utility
500,000 S.C KVA

1000 KVA Transformer,
480V, 3Ø,
3.5%Z

$I_{f.l.}$ = 1203A

30' - 500 kcmil
4 Per Phase
Copper in PVC Conduit

1600A Switch

KRP-C-1500SP Fuse

Fault X_1

400A Switch

LPS-RK-350SP Fuse

20' - 2/0
2 Per Phase
Copper in PVC Conduit

225 KVA transformer,
208V, 3Ø
1.2%Z

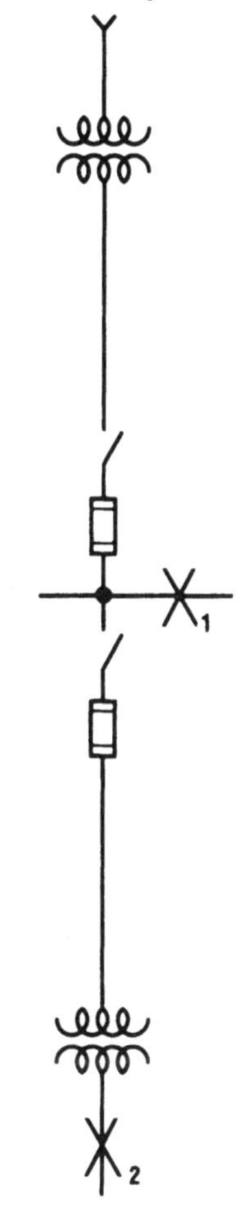

Fault X_1

Step 1. $I_{f.l.} = \dfrac{1000 \times 1000}{480 \times 1.732} = 1203A$

Step 2. $\text{Multiplier} = \dfrac{100}{3.5} = 28.57$

Step 3. $I_{S.C.} = 1203 \times 28.57 = 34{,}370A$

Step 4. $f = \dfrac{1.732 \times 30 \times 34{,}370}{4 \times 26{,}706 \times 480} = .0348$

Step 5. $M = \dfrac{1}{1 + .0348} = .9664$

Step 6. $I_{S.C.\,sym\,RMS} = 34{,}370 \times .9664 = 33{,}215A$

Fault X_2

Step 4. $f = \dfrac{1.732 \times 20 \times 33{,}215}{2 \times 11{,}423 \times 480} = .1049$

Step 5. $M = \dfrac{1}{1 + .1049} = .905$

Step 6. $I_{S.C.\,sym\,RMS} = 33{,}215 \times .905 = 30{,}059A$

Fault X_2

$$f = \frac{30{,}059 \times 480 \times 1.732 \times 1.2}{100{,}000 \times 225} = 1.333$$

$$M = \frac{1}{1 + 1.333} = .4286$$

$$I_{S.C.\,sym\,RMS} = \frac{480 \times .4286 \times 30{,}059}{208} = 29{,}731A$$

7.1.5 *C* Values for Conductors and Busway

TABLE 7.1.5

" C" Values for Conductors and Busway

Copper												
AWG or kcmil	Three Single Conductors						Three-Conductor Cable					
	Conduit						Conduit					
	Steel			Nonmagnetic			Steel			Nonmagnetic		
	600V	5KV	15KV	600V	5KV	15KV	600V	5KV	15KV	600V	5KV	15KV
14	389	389	389	389	389	389	389	389	389	389	389	389
12	617	617	617	617	617	617	617	617	617	617	617	617
10	981	981	981	981	981	981	981	981	981	981	981	981
8	1557	1551	1557	1558	1555	1558	1559	1557	1559	1559	1558	1559
6	2425	2406	2389	2430	2417	2406	2431	2424	2414	2433	2428	2420
4	3806	3750	3695	3825	3789	3752	3830	3811	3778	3837	3823	3798
3	4760	4760	4760	4802	4802	4802	4760	4790	4760	4802	4802	4802
2	5906	5736	5574	6044	5926	5809	5989	5929	5827	6087	6022	5957
1	7292	7029	6758	7493	7306	7108	7454	7364	7188	7579	7507	7364
1/0	8924	8543	7973	9317	9033	8590	9209	9086	8707	9472	9372	9052
2/0	10755	10061	9389	11423	10877	10318	11244	11045	10500	11703	11528	11052
3/0	12843	11804	11021	13923	13048	12360	13656	13333	12613	14410	14118	13461
4/0	15082	13605	12542	16673	15351	14347	16391	15890	14813	17482	17019	16012
250	16483	14924	13643	18593	17120	15865	18310	17850	16465	19779	19352	18001
300	18176	16292	14768	20867	18975	17408	20617	20051	18318	22524	21938	20163
350	19703	17385	15678	22736	20526	18672	19557	21914	19821	22736	24126	21982
400	20565	18235	16365	24296	21786	19731	24253	23371	21042	26915	26044	23517
500	22185	19172	17492	26706	23277	21329	26980	25449	23125	30028	28712	25916
600	22965	20567	47962	28033	25203	22097	28752	27974	24896	32236	31258	27766
750	24136	21386	18888	28303	25430	22690	31050	30024	26932	32404	31338	28303
1000	25278	22539	19923	31490	28083	24887	33864	32688	29320	37197	35748	31959
Aluminum												
14	236	236	236	236	236	236	236	236	236	236	236	236
12	375	375	375	375	375	375	375	375	375	375	375	375
10	598	598	598	598	598	598	598	598	598	598	598	598
8	951	950	951	951	950	951	951	951	951	951	951	951
6	1480	1476	1472	1481	1478	1476	1481	1480	1478	1482	1481	1479
4	2345	2332	2319	2350	2341	2333	2351	2347	2339	2353	2349	2344
3	2948	2948	2948	2958	2958	2958	2948	2956	2948	2958	2958	2958
2	3713	3669	3626	3729	3701	3672	3733	3719	3693	3739	3724	3709
1	4645	4574	4497	4678	4631	4580	4686	4663	4617	4699	4681	4646
1/0	5777	5669	5493	5838	5766	5645	5852	5820	5717	5875	5851	5771
2/0	7186	6968	6733	7301	7152	6986	7327	7271	7109	7372	7328	7201
3/0	8826	8466	8163	9110	8851	8627	9077	8980	8750	9242	9164	8977
4/0	10740	10167	9700	11174	10749	10386	11184	11021	10642	11408	11277	10968
250	12122	11460	10848	12862	12343	11847	12796	12636	12115	13236	13105	12661
300	13909	13009	12192	14922	14182	13491	14916	14698	13973	15494	15299	14658
350	15484	14280	13288	16812	15857	14954	15413	16490	15540	16812	17351	16500
400	16670	15355	14188	18505	17321	16233	18461	18063	16921	19587	19243	18154
500	18755	16827	15657	21390	19503	18314	21394	20606	19314	22987	22381	20978
600	20093	18427	16484	23451	21718	19635	23633	23195	21348	25750	25243	23294
750	21766	19685	17686	23491	21769	19976	26431	25789	23750	25682	25141	23491
1000	23477	21235	19005	28778	26109	23482	29864	29049	26608	32938	31919	29135

Ampacity	Busway				
	Plug-In	Feeder		High Impedance	
	Copper	Aluminum	Copper	Aluminum	Copper
225	28700	23000	18700	12000	—
400	38900	34700	23900	21300	—
600	41000	38300	36500	31300	—
800	46100	57500	49300	44100	—
1000	69400	89300	62900	56200	15600
1200	94300	97100	76900	69900	16100
1350	119000	104200	90100	84000	17500
1600	129900	120500	101000	90900	19200
2000	142900	135100	134200	125000	20400
2500	143800	156300	180500	166700	21700
3000	144900	175400	204100	188700	23800
4000	—	—	277800	256400	—

7.1.6 Shortcut Method 1: Adding *Z*s

This method uses the approximation of adding *Z*s instead of the accurate method of *R*s and *X*s (in complex form).

Example

1. For a 480/277-V system with 30,000 A symmetrical available at the line side of a conductor run of 100 ft of two 500-kcmil per phase and neutral, the approximate fault current at the load side end of the conductors can be calculated as follows:
2. 277 V/30,000 A = 0.00923 Ω (source impedance).
3. Conductor ohms for 500-kcmil conductor from Table 7.1.7 in magnetic conduit is 0.00546 Ω per 100 ft. For 100 ft and two conductors per phase, we have
4. 0.00546/2 = 0.00273 Ω (conductor impedance).
5. Add source and conductor impedance, or 0.00923 + 0.00273 = 0.01196 Ω total.
6. Next, 277 V/0.001196 Ω = 23,160 A rms at load side of conductors.

For impedance values, refer to Tables 7.1.7, 7.1.8, and 7.1.9.

7.1.7 Average Characteristics of 600-V Conductors (Ohms per 100 ft): Two or Three Single Conductors

TABLE 7.1.7

Wire Size, AWG or kcmil	Copper Conductors						Aluminum Conductors					
	Magnetic Conduit			Nonmagnetic Conduit			Magnetic Conduit			Nonmagnetic Conduit		
	R	X	Z	R	X	Z	R	X	Z	R	X	Z
14	.3130	.00780	.3131	.3130	.00624	.3131	–	–	–	–	–	–
12	.1968	.00730	.1969	.1968	.00584	.1969	–	–	–	–	–	–
10	.1230	.00705	.1232	.1230	.00564	.1231	–	–	–	–	–	–
8	.0789	.00691	.0792	.0789	.00553	.0791	–	–	–	–	–	–
6	.0490	.00640	.0494	.0490	.00512	.0493	.0833	.00509	.0835	.0833	.00407	.0834
4	.0318	.00591	.0323	.0318	.00473	.0321	.0530	.00490	.0532	.0530	.00392	.0531
2	.0203	.00548	.0210	.0203	.00438	.0208	.0335	.00457	.0338	.0335	.00366	.0337
1	.0162	.00533	.0171	.0162	.00426	.0168	.0267	.00440	.0271	.0267	.00352	.0269
1/0	.0130	.00519	.01340	.0129	.00415	.01360	.0212	.00410	.0216	.0212	.00328	.0215
2/0	.0104	.00511	.01159	.0103	.00409	.01108	.0170	.00396	.0175	.0170	.00317	.0173
3/0	.00843	.00502	.00981	.00803	.00402	.00898	.01380	.00386	.0143	.01380	.00309	.01414
4/0	.00696	.00489	.00851	.00666	.00391	.00772	.01103	.00381	.0117	.01097	.00305	.01139
250	.00588	.00487	.00763	.00578	.00390	.00697	.00936	.00375	.01008	.00933	.00300	.00980
300	.00512	.00484	.00705	.00501	.00387	.00633	.00810	.00366	.00899	.00797	.00293	.00849
350	.00391	.00480	.00619	.00380	.00384	.00540	.00694	.00360	.00782	.00688	.00288	.00746
400	.00369	.00476	.00602	.00356	.00381	.00521	.00618	.00355	.00713	.00610	.00284	.00673
450	.00330	.00467	.00595	.00310	.00374	.00486	.00548	.00350	.00650	.00536	.00280	.00605
500	.00297	.00458	.00546	.00275	.00366	.00458	.00482	.00346	.00593	.00470	.00277	.00546
600	.00261	.00455	.00525	.00241	.00364	.00437	.00409	.00355	.00542	.00395	.00284	.00486
700	.00247	.00448	.00512	.00247	.00358	.00435	.00346	.00340	.00485	.00330	.00272	.00428
750	.00220	.00441	.00493	.00198	.00353	.00405	.00308	.00331	.00452	.00278	.00265	.00384
1000	–	–	–	–	–	–	.00250	.00330	.00414	.00230	.00264	.00350

① Resistance and reactance are phase-to-neutral values, based on 60 Hertz ac, 3-phase, 4-wire distribution, in ohms per 100 feet of circuit length (not total conductor lengths).

② Based upon conductivity of 100% for copper, 61% for aluminum.

③ Based on conductor temperatures of 75°C. Reactance values will have negligible variation with temperature. Resistance of both copper and aluminum conductors will be approximately 5% lower at 60°C or 5% higher at 90°C. Data shown in tables may be used without significant error between 60°C and 90°C.

④ For interlocked armored cable, use magnetic conduit data for steel armor and non-magnetic conduit data for aluminum armor.

⑤ $= \sqrt{X^2} + R$

7.1.8 Average Characteristics of 600-V Conductors (Ohms per 100 ft): Three Conductor Cables (and Interlocked Armored Cable)

TABLE 7.1.8

Wire Size, AWG or kcmil	Copper Conductors						Aluminum Conductors					
	Magnetic Conduit			Nonmagnetic Conduit			Magnetic Conduit			Nonmagnetic Conduit		
	R	X	Z	R	X	Z	R	X	Z	R	X	Z
14	.3130	.00597	.3131	.3130	.00521	.3130	–	–	–	–	–	–
12	.1968	.00558	.1969	.1968	.00487	.1969	–	–	–	–	–	–
10	.1230	.00539	.1231	.1230	.00470	.1231	–	–	–	–	–	–
8	.0789	.00529	.0790	.0789	.00461	.0790	–	–	–	–	–	–
6	.0490	.00491	.0492	.0490	.00427	.0492	.0833	.00509	.0834	.0833	.00407	.0834
4	.0318	.00452	.0321	.0318	.00394	.0320	.0530	.00490	.0532	.0530	.00392	.0531
2	.0203	.00420	.0207	.0203	.00366	.0206	.0335	.00457	.0338	.0335	.00366	.0337
1	.0162	.00408	.0167	.0162	.00355	.0166	.0267	.00440	.0271	.0267	.00352	.0269
1/0	.0130	.00398	.0136	.0129	.00346	.0134	.0212	.00410	.0216	.0212	.00328	.0215
2/0	.0104	.00390	.0111	.0103	.00341	.0108	.0170	.00396	.0175	.0170	.00317	.0173
3/0	.00843	.00384	.00926	.00803	.00335	.00870	.01380	.00389	.0143	.01380	.00309	.01414
4/0	.00696	.00375	.00791	.00666	.00326	.00742	.01103	.00381	.0117	.01097	.00305	.01139
250	.00588	.00373	.00696	.00578	.00325	.00663	.00936	.00375	.01006	.00933	.00300	.00980
300	.00512	.00370	.00632	.00501	.00323	.00596	.00810	.00366	.00889	.00797	.00293	.00849
350	.00391	.00365	.00535	.00380	.00320	.00497	.00694	.00360	.00782	.00688	.00288	.00746
400	.00369	.00360	.00516	.00356	.00318	.00477	.00618	.00355	.00713	.00610	.00284	.00673
450	.00360	.00351	.00503	.00310	.00312	.00440	.00548	.00350	.00650	.00536	.00280	.00605
500	.00297	.00343	.00454	.00275	.00305	.00411	.00482	.00346	.00593	.00470	.00277	.00546
600	.00261	.00337	.00426	.00241	.00303	.00387	.00409	.00355	.00542	.00395	.00284	.00486
700	.00247	.00330	.00412	.00227	.00298	.00375	.00346	.00341	.00486	.00330	.00272	.00428
750	.00220	.00323	.00391	.00198	.00294	.00354	.00308	.00331	.00452	.00278	.00265	.00384
1000	–	–	–	–	–	–	.00250	.00330	.00414	.00230	.00264	.00350

① Resistance and reactance are phase-to-neutral values, based on 60 Hertz ac, 3-phase, 4-wire distribution, in ohms per 100 feet of circuit length (not total conductor lengths).

② Based upon conductivity of 100% for copper, 61% for aluminum.

③ Based on conductor temperatures of 75°C. Reactance values will have negligible variation with temperature. Resistance of both copper and aluminum conductors will be approximately 5% lower at 60°C or 5% higher at 90°C. Data shown in tables may be used without significant error between 60°C and 90°C.

④ For interlocked armored cable, use magnetic conduit data for steel armor and non-magnetic conduit data for aluminum armor.

⑤ $= \sqrt{X^2} + R$

7.1.9 LV Busway, *R*, *X*, and *Z* (Ohms per 100 ft)

TABLE 7.1.9

Ampere Rating	Plug-in			Feeder		
	Resistance	Reactance	Impedance	Resistance	Reactance	Impedance
Aluminum						
225	.00737	.00323	.00805	.00737	.00323	.00805
400	.00371	.00280	.00465	.00371	.00280	.00465
600	.00291	.00212	.00360	.00289	.00127	.00316
800	.00248	.00114	.00273	.00244	.000660	.00253
1000	.00188	.00100	.00213	.00197	.000552	.00205
1200	.00155	.000755	.00172	.00159	.000490	.00166
1350	.00130	.000600	.00143	.00134	.000385	.00139
1600	.00106	.000480	.00116	.00112	.000350	.00117
2000	.000841	.000449	.000953	.000864	.000310	.000918
2500	.000648	.000290	.000710	.000664	.000250	.000710
3000	.000521	.000183	.000552	.000558	.000197	.000592
4000	.000397	.000175	.000434	.000409	.000135	.000431
Copper						
225	.00425	.00323	.00534	.00425	.00323	.00534
400	.00291	.00301	.00419	.00291	.00301	.00419
600	.00212	.00234	.00316	.00202	.00170	.00264
800	.00169	.00212	.00271	.00188	.00149	.00240
1000	.00144	.00114	.00184	.00158	.000965	.00185
1200	.00112	.00100	.00150	.00120	.000552	.00132
1350	.00101	.000960	.00139	.00108	.000510	.00119
1600	.000898	.000716	.00115	.000920	.000480	.00104
2000	.000667	.000562	.000872	.000724	.000434	.000844
2500	.000494	.000449	.000668	.000520	.000305	.000603
3000	.000465	.000355	.000585	.000488	.000290	.000568
4000	.000336	.000242	.000414	.000378	.000203	.000429
5000	......	......	......	.000264	.000139	.000298

7.1.10 Shortcut Method 2: Chart Approximate Method

The chart method is based on the following:

Motor Contribution Assumptions

120/208-V systems	50 percent motor load
	4 times motor FLA contribution
240/480-V systems	100 percent motor load
	4 times motor FLA contribution

Feeder Conductors. The conductor sizes most commonly used for feeders from molded-case circuit breakers are shown. For conductor sizes not shown, Table 7.1.11 has been included for conversion to equivalent arrangements. In some cases, it may be necessary to interpolate for unusual feeder ratings. Table 7.1.11 is based on using copper conductor.

Short-Circuit Current Readout. The readout obtained from the charts is the rms symmetrical amperes available at the given distance from the transformer. The circuit breaker should have an interrupting capacity at least as large as this value.

How to Use the Short-Circuit Charts

Step 1. Obtain the following data:

- System voltage
- Transformer kVA rating
- Transformer impedance
- Primary source fault energy available in kVA

Step 2. Select the applicable chart from Figure 7.1.12 (Charts 1–13). The charts are grouped by secondary system voltage, which is listed with each transformer. Within each group, the chart for the lowest-kVA transformer is shown first, followed in ascending order to the highest-rated transformer.

Step 3. Select the family of curves that is closest to the "available source kVA." The upper-value-line family of curves is for a source of 500,000 kVA. The lower-value-line family of curves is for a source of 50,000 kVA. You may interpolate between curves if necessary, but for values above 100,000 kVA, it is appropriate to use the 500,000-kVA curves.

Step 4. Select the specific curve for the conductor size being used. If your conductor size is something other than the sizes shown on the chart, refer to the conductor conversion table (Table 7.1.11).

Step 5. Enter the chart along the bottom horizontal scale with the distance (in feet) from the transformer to the fault point. Draw a vertical line up the chart to the point where it intersects the selected curve. Then draw a horizontal line to the left from this point to the scale along the left side of the chart.

Step 6. The value obtained from the left-hand vertical scale is the fault current (in thousands of amperes) available at the fault point.

7.1.11 Conductor Conversion (Based on Using Copper Conductor)

TABLE 7.1.11

Conductor Conversion
(Based on Using Copper Conductor)

If Your Conductor is:	Use Equivalent Arrangement
3 – No. 4/0 cables	2 – 500 MCM
4 – No. 2/0 cables	2 – 500 MCM
3 – 2000 MCM cables	4 – 750 MCM
5 – 400 MCM cables	4 – 750 MCM
6 – 300 MCM cables	4 – 750 MCM
800 Amp busway	2 – 500 MCM
1000 Amp busway	2 – 500 MCM
1600 Amp busway	4 – 750 MCM

7.1.12 Charts 1 through 13 for Calculating Short-Circuit Currents Using Chart Approximate Method

7.1.12

Chart 1 – 225 kVA Transformer/4.5% Impedance/208 Volts

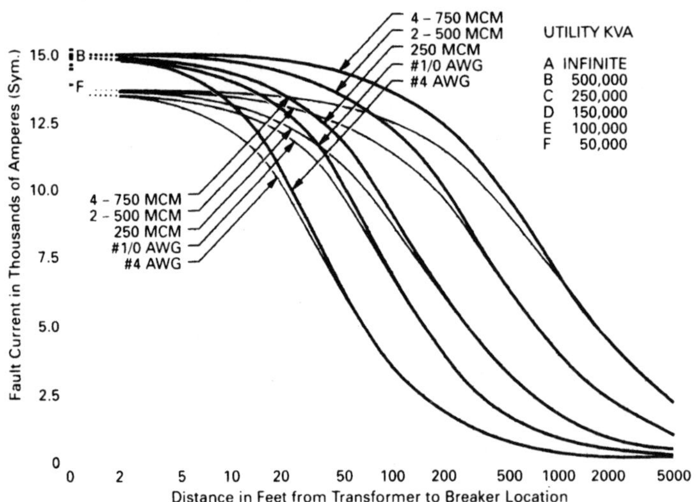

(continued)

7.1.12

Chart 2 – 300 kVA Transformer/4.5% Impedance/208 Volts

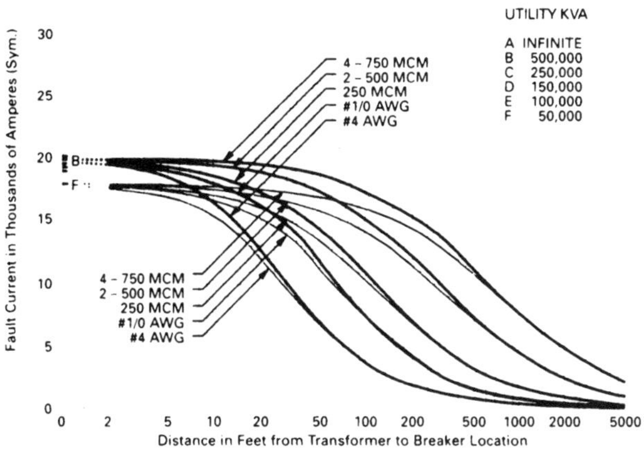

Chart 5 – 1000 kVA Transformer/5.5% Impedance/208 Volts

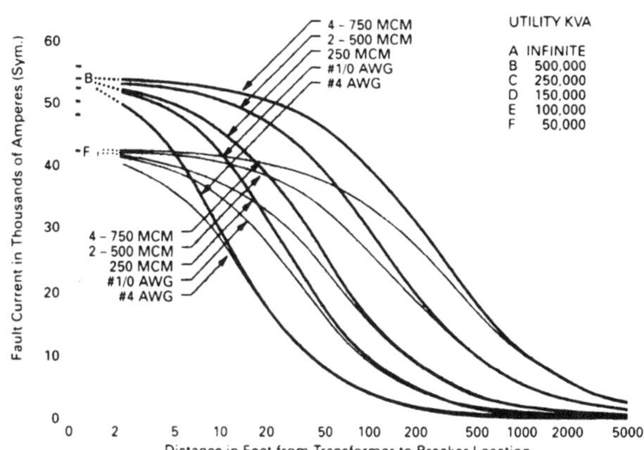

Chart 3 – 500 kVA Transformer/4.5% Impedance/208 Volts

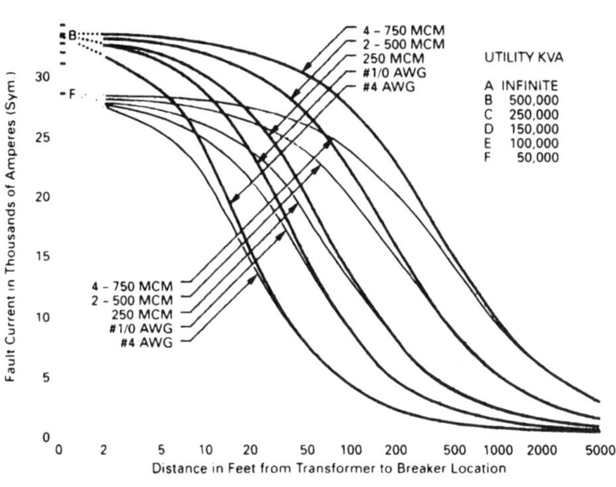

Chart 6 – 1500 kVA Transformer/5.5% Impedance/208 Volts

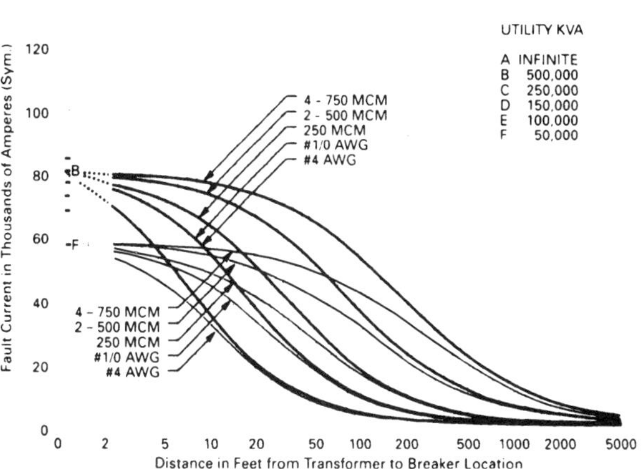

Chart 4 – 750 kVA Transformer/5.5% Impedance/208 Volts

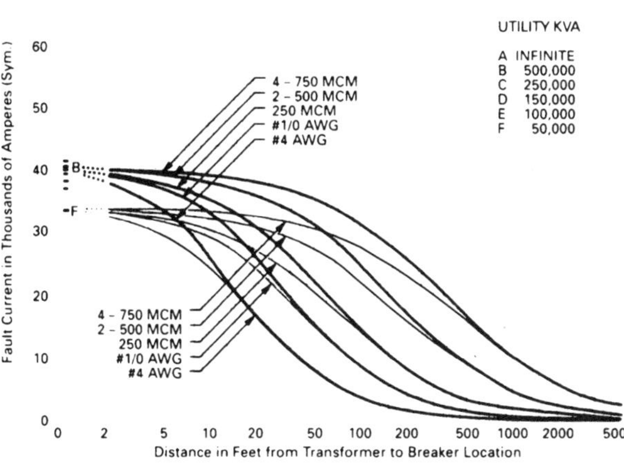

Chart 7 – 2000 kVA Transformer/5.5% Impedance/208 Volts

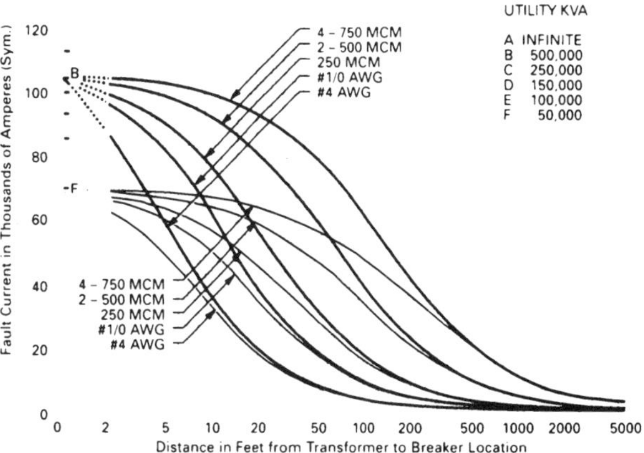

(continued)

7.1.12

Chart 8 – 300 kVA Transformer/4.5% Impedance/480 Volts

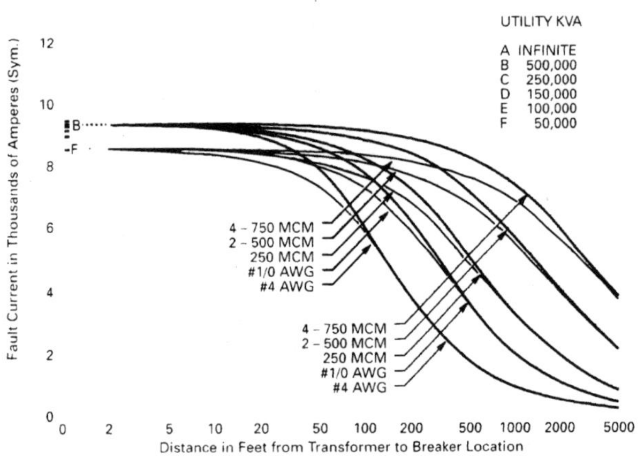

Chart 11 – 1000 kVA Transformer/5.5% Impedance/480 Volts

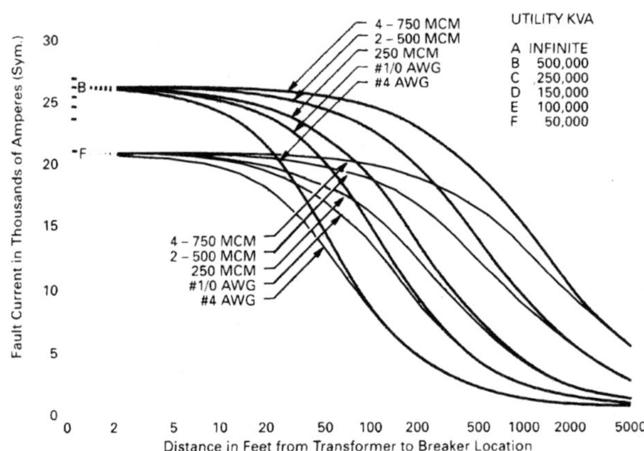

Chart 9 – 500 kVA Transformer/4.5% Impedance/480 Volts

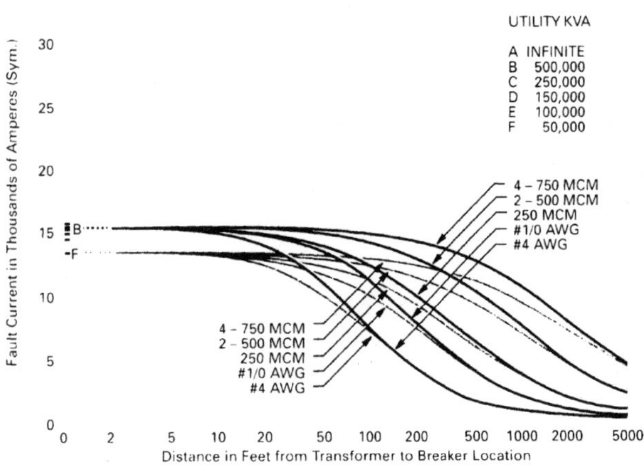

Chart 12 – 1500 kVA Transformer/5.5% Impedance/480 Volts

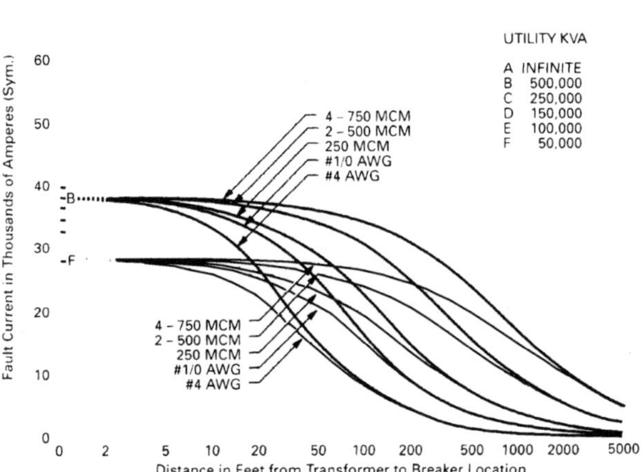

Chart 10 – 750 kVA Transformer/5.5% Impedance/480 Volts

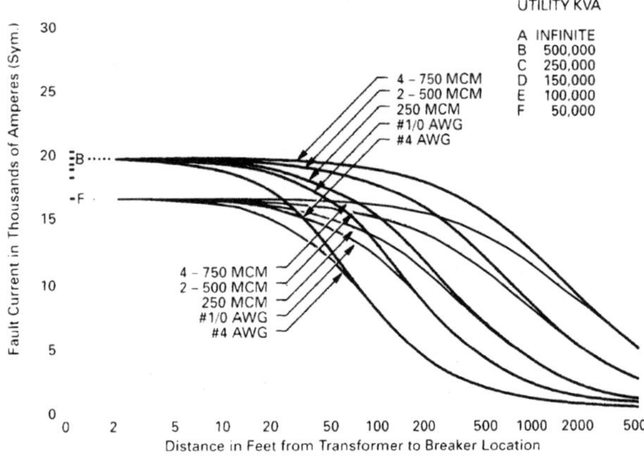

Chart 13 – 2000 kVA Transformer/5.5% Impedance/480 Volts

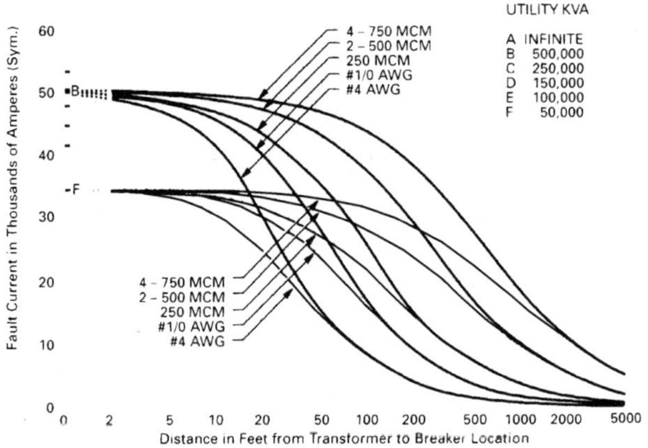

7.1.13 Assumptions for Motor Contributions to Fault Currents

To determine the motor contribution to the first half-cycle fault current when the system motor load is known, the following assumptions generally are made:

Induction motors: Use 4.0 times motor full-load current (impedance value of 25 percent).

Synchronous motors: Use 5.0 times motor full-load current (impedance value of 20 percent).

When the motor load is not known, the following assumptions generally are made:

208Y/120-V systems

- Assume 50 percent lighting and 50 percent motor load.
- Assume motor feedback contribution of 2.0 times full-load current of transformer.

240-480-600-V three-phase, three-wire systems

- Assume 100 percent motor load.
- Assume motors 25 percent synchronous and 75 percent induction.
- Assume motor feedback contribution of 4.0 times full-load current of transformer.

480Y/277-V systems in commercial buildings

- Assume 50 percent induction motor load.
- Assume motor feedback contribution of 2.0 times full-load current of transformer or source.
- For industrial plants, make same assumptions as for three-phase, three-wire systems (above).

Medium-voltage motors

- If known, use actual values. Otherwise, use the values indicated in the preceding for the same type of motor.

Note on asymmetrical currents. The calculation of asymmetrical currents is a laborious procedure because the degree of asymmetry is not the same on all three phases. It is common practice to calculate the rms symmetrical fault current, with the assumption being made that the dc component has decayed to zero, and then apply a multiplying factor to obtain the first half-cycle rms asymmetrical current, which is called the *momentary current.* For medium-voltage systems (defined by the IEEE as greater than 1000 V up to 69,000 V), the multiplying factor is established by NEMA and ANSI standards depending on the operating speed of the breaker; for low-voltage systems, 600 V and below, the multiplying factor is usually 1.17 (based on generally accepted use of an X/R ratio of 6.6 representing a source short-circuit power factor of 15 percent). These values take into account that medium-voltage breakers are rated on maximum asymmetry and low-voltage breakers are rated on average asymmetry.

7.1.14 Secondary Short-Circuit Capacity of Typical Power Transformers

TABLE 7.1.14

Trans-Former Rating 3-Phase kVA and Imped-ance Percent	Maximum Short Circuit kVA Available From Primary System	208 Volts, 3-Phase				240 Volts, 3-Phase				480 Volts, 3-Phase				600 Volts, 3-Phase			
		Rated Load Contin-uous Current, Amps	Short-Circuit Current RMS Symmetrical Amps			Rated Load Contin-uous Current, Amps	Short-Circuit Current RMS Symmetrical Amps			Rated Load Contin-uous Current, Amps	Short-Circuit Current RMS Symmetrical Amps			Rated Load Contin-uous Current, Amps	Short-Circuit Current RMS Symmetrical Amps		
			Trans-former Alone ①	50% Motor Load ②	Com-bined		Trans-former Alone ①	100% Motor Load ②	Com-bined		Trans-former Alone ②	100% Motor Load ①	Com-bined		Trans-former Alone ②	100% Motor Load ①	Com-bined
300 5%	50000	834	14900	1700	16600	722	12900	2900	15800	361	6400	1400	7800	289	5200	1200	6400
	100000		15700		17400		13600		16500		6800		8200		5500		6700
	150000		16000		17700		13900		16800		6900		8300		5600		6800
	250000		16300		18000		14100		17000		7000		8400		5600		6800
	500000		16500		18200		14300		17200		7100		8500		5700		6900
	Unlimited		16700		18400		14400		17300		7200		8600		5800		7000
500 5%	50000	1388	21300	2800	25900	1203	20000	4800	24800	601	10000	2400	12400	481	8000	1900	9900
	100000		25200		28000		21900		26700		10900		13300		8700		10600
	150000		26000		28800		22500		27300		11300		13700		9000		10900
	250000		26700		29500		23100		27900		11600		14000		9300		11200
	500000		27200		30000		23600		28400		11800		14200		9400		11300
	Unlimited		27800		30600		24100		28900		12000		14400		9600		11500
750 5.75%	50000	2080	28700	4200	32900	1804	24900	7200	32100	902	12400	3600	16000	722	10000	2900	12900
	100000		32000		36200		27800		35000		13900		17500		11100		14000
	150000		33300		37500		28900		36100		14400		18000		11600		14500
	250000		34400		38600		29800		37000		14900		18500		11900		14800
	500000		35200		39400		30600		37800		15300		18900		12200		15100
	Unlimited		36200		40400		31400		38600		15700		19300		12600		15500
1000 5.75%	50000	2776	35900	5600	41500	2406	31000	9600	40600	1203	15500	4800	20300	962	12400	3900	16300
	100000		41200		46800		35600		45200		17800		22600		14300		18200
	150000		43300		48900		37500		47100		18700		23500		15000		18900
	250000		45200		50800		39100		48700		19600		24400		15600		19500
	500000		46700		52300		40400		50000		20200		25000		16200		20100
	Unlimited		48300		53900		41800		51400		20900		25700		16700		20600
1500 5.75%	50000	4164	47600	8300	55900	3609	41200	14400	55600	1804	20600	7200	27800	1444	16500	5800	22300
	100000		57500		65800		49800		64200		24900		32100		20000		25800
	150000		61800		70100		53500		57900		26700		33900		21400		27200
	250000		65600		73900		56800		71200		28400		35600		22700		28500
	500000		68800		77100		59600		74000		29800		37000		23900		29700
	Unlimited		72500		80800		62800		77200		31400		38600		25100		30900
2000 5.75%	50000									2406	24700	9600	34300	1924	19700	7800	27500
	100000										31000		40600		24800		32600
	150000										34000		43600		27200		35000
	250000										36700		46300		29400		37200
	500000										39100		48700		31300		39100
	Unlimited										41800		51400		33500		41300
2500 5.75%	50000									3008	28000	12000	40000	2405	22400	9600	32000
	100000										36500		48500		29200		38800
	150000										40500		52500		32400		42000
	250000										44600		56600		35600		45200
	500000										48100		60100		38500		48100
	Unlimited										52300		64300		41800		51400

① Short-circuit capacity values shown correspond to kVA and impedances shown in this table. For impedances other than these, short-circuit currents are inversely proportional to impedance.

② The motor's short-circuit current contributions are computed on the basis of motor characteristics that will give four times normal current. For 208 volts, 50% motor load is assumed while for other voltages 100% motor load is assumed. For other percentages, the motor short-circuit current will be in direct proportion.

Section

8

Selective Coordination of Protective Devices

8.1.0 Introduction
8.1.1 Recommended Procedure for Conducting a Selective Coordination Study
8.1.2 Example System One-Line Diagram for Selective Coordination Study
8.1.3 Time-Current Curve No. 1 for System Shown in Figure 8.1.2 with Analysis Notes and Comments
8.1.4 Time-Current Curve No. 2 for System Shown in Figure 8.1.2 with Analysis Notes and Comments
8.1.5 Time-Current Curve No. 3 for System Shown in Figure 8.1.2 with Analysis Notes and Comments
8.1.6 Shortcut Ratio Method Selectivity Guide

8.1.0 Introduction

It is not enough to select protective devices based solely on their ability to carry the system load current and interrupt the maximum fault current at their respective levels. A properly engineered system will allow *only* the protective device nearest the fault to open, leaving the remainder of the system undisturbed and preserving continuity of service.

We may then define *selective coordination* as "the act of isolating a faulted circuit from the remainder of the electrical system, thereby eliminating unnecessary power outages. The faulted circuit is isolated by the selective operation of only that overcurrent protective device closest to the overcurrent condition."

8.1.1 Recommended Procedure for Conducting a Selective Coordination Study

The following steps are recommended when conducting a selective coordination study:

1. *One-line diagram.* Obtain or develop the electrical system one-line diagram that identifies important system components, as given below.
   a. *Transformers.* Obtain the following data for protection of and coordination information about transformers:
      (1) kVA rating
      (2) Inrush points
      (3) Primary and secondary connections
      (4) Impedance
      (5) Damage curves

(6) Primary and secondary voltages
(7) Liquid or dry type

b. Conductors. Check phase, neutral, and equipment grounding. The one-line diagram should include information such as

(1) Conductor size
(2) Number of conductors per phase
(3) Material (copper or aluminum)
(4) Insulation
(5) Conduit (magnetic or nonmagnetic)

From this information, short-circuit withstand curves can be developed. This provides information on how overcurrent devices will protect conductors from overload *and* short-circuit damage.

c. Motors. The system one-line diagram should include motor information such as

(1) Full-load currents
(2) Horsepower
(3) Voltage
(4) Type of starting characteristic (across the line, etc.)
(5) Type of overload relay (class 10, 20, 30)

Overload protection of the motor and motor circuit can be determined from these data.

d. Fuse characteristics. Fuse types/classes should be identified on the one-line diagram.

e. Circuit breaker characteristics. Circuit breaker types should be identified on the one-line diagram.

f. Relay characteristics. Relay types should be identified on the one-line diagram.

2. *Short-circuit study.* Perform a short-circuit analysis, calculating maximum available short-circuit currents at critical points in the distribution system (such as transformers, main switchgear, panelboards, motor control centers, load centers, and large motors and generators). Refer to the preceding section.
3. *Helpful hints*
   a. Determine the ampere scale selection. It is most convenient to place the time-current curves in the center of log-log paper. This is accomplished by multiplying or dividing the ampere scale by a factor of 10.
   b. Determine the reference (base) voltage. The best reference voltage is the voltage level at which most of the devices being studied fall. On most low-voltage industrial and commercial studies, the reference voltage will be 208, 240, or 480 V. Devices at other voltage levels will be shifted by a multiplier based on the transformer turn ratio. The best reference voltage will require the least amount of manipulation. Most computer programs will make these adjustments automatically when the voltage levels of the devices are identified by the input data.
   c. Commencing the analysis. The starting point can be determined by the designer. Typically, studies begin with the main circuit devices and work down through the feeders and branches (right to left on your log-log paper).
   d. Multiple branches. If many branches are taken off one feeder and the branch loads are similar, the largest rated branch-circuit device should be checked for coordination with upstream devices. If the largest branch device will coordinate and the branch devices are similar, they generally will coordinate as well. (The designer may wish to verify other areas of protection on those branches, conductors, etc.)
   e. Do not overcrowd the study. Many computer-generated studies will allow a maximum of 10 device characteristics per page. It is good practice, however, to have a minimum of three devices in a coordination sequence so that there is always one step of overlap.

f. *Existing systems.* The designer should be aware that when conducting a coordination study on an existing system, optimal coordination cannot always be achieved, and compromise may be necessary. It is then necessary to exercise experience and judgment to achieve the best coordination possible to mitigate the effects of "blackout" conditions. The designer must set priorities within the constraints of the system under study.

g. *Conductor short-circuit protection.* In low-voltage (600 V or less) systems, it is generally safe to ignore possible damage to conductors from short circuits because the philosophy is to isolate a fault as quickly as possible; thus the I^2t energy damage curves do not have enough time to come into play (become a factor). In medium- and high-voltage systems, however, where the philosophy is to have the overcurrent protection "hang in" as long as possible, the contrary is true; thus it can be a significant factor.

h. *One-line diagram.* A one-line diagram of the study should be drawn for future reference.

8.1.2 Example System One-Line Diagram for Selective Coordination Study (*see page 8.4*)

The following example will analyze in detail the system shown. It is understood that a short-circuit study has been completed and that all protective devices have adequate interrupting ratings. A selective coordination analysis is the next step.

This simple radial system will involve three separate time-current studies, applicable to the three feeders/branches shown. The three time-current curves and their accompanying notes are self-explanatory (Figures 8.1.3 through 8.1.5).

8.1.3 Time-Current Curve No. 1 for System Shown in Figure 8.1.2 with Analysis Notes and Comments (*see page 8.5*)

8.1.4 Time-Current Curve No. 2 for System Shown in Figure 8.1.2 with Analysis Notes and Comments (*see page 8.7*)

8.1.5 Time-Current Curve No. 3 for System Shown in Figure 8.1.2 with Analysis Notes and Comments (*see page 8.9*)

8.1.6 Shortcut Ratio Method Selectivity Guide (*see page 8.11*)

This selectivity guide may be used for an easy check on fuse selectivity regardless of the short-circuit current levels involved. It also may be used for fixed thermal-magnetic trip circuit breakers (exercising good judgment) with a reasonable degree of accuracy. Where medium- and high-voltage primary fuses and relays are involved, the time-current characteristic curves should be plotted on standard log-log graph paper for proper study.

8.1.2

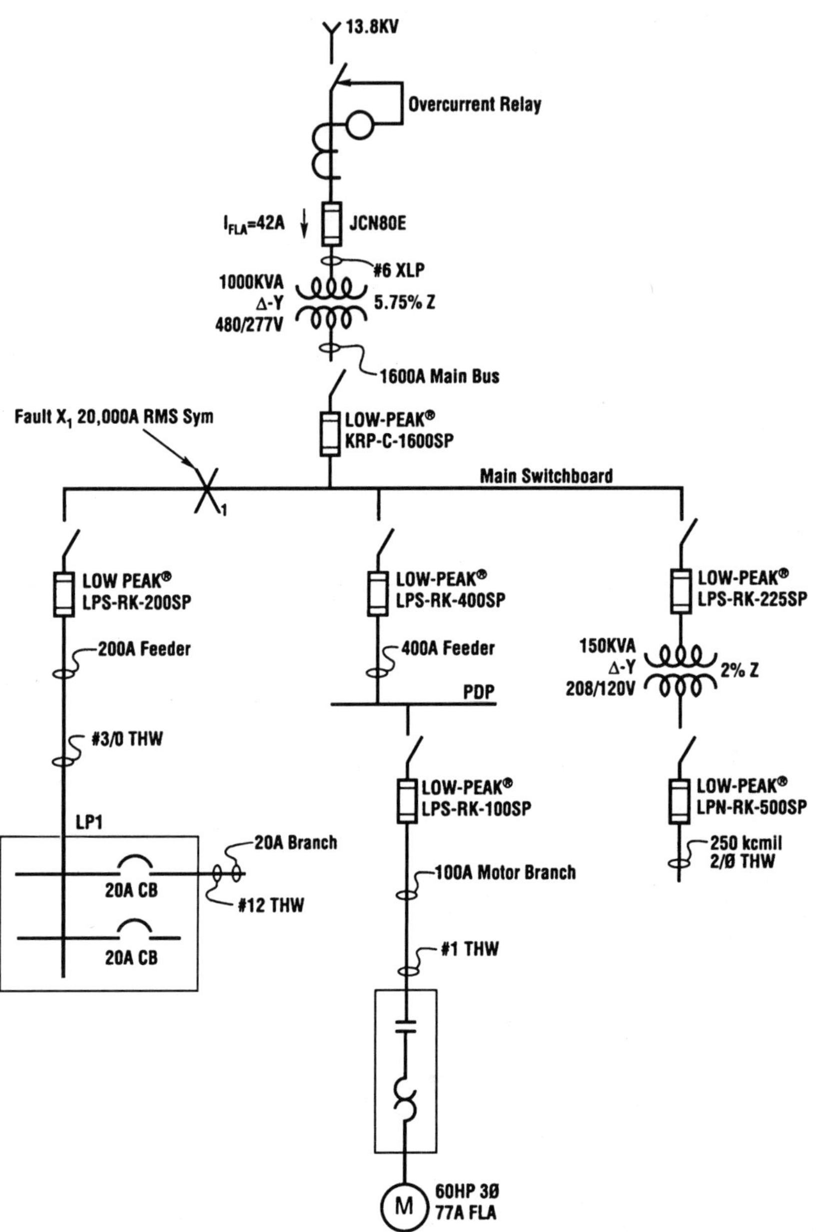
13.8KV
Overcurrent Relay
I_{FLA}=42A
JCN80E
#6 XLP
1000KVA
Δ-Y
480/277V
5.75% Z
1600A Main Bus
Fault X_1 20,000A RMS Sym
LOW-PEAK®
KRP-C-1600SP
Main Switchboard
X_1
LOW PEAK®
LPS-RK-200SP
200A Feeder
#3/0 THW
LP1
20A Branch
20A CB
#12 THW
20A CB
LOW-PEAK®
LPS-RK-400SP
400A Feeder
PDP
LOW-PEAK®
LPS-RK-100SP
100A Motor Branch
#1 THW
M
60HP 3Ø
77A FLA
LOW-PEAK®
LPS-RK-225SP
150KVA
Δ-Y
208/120V
2% Z
LOW-PEAK®
LPN-RK-500SP
250 kcmil
2/Ø THW

8.1.3

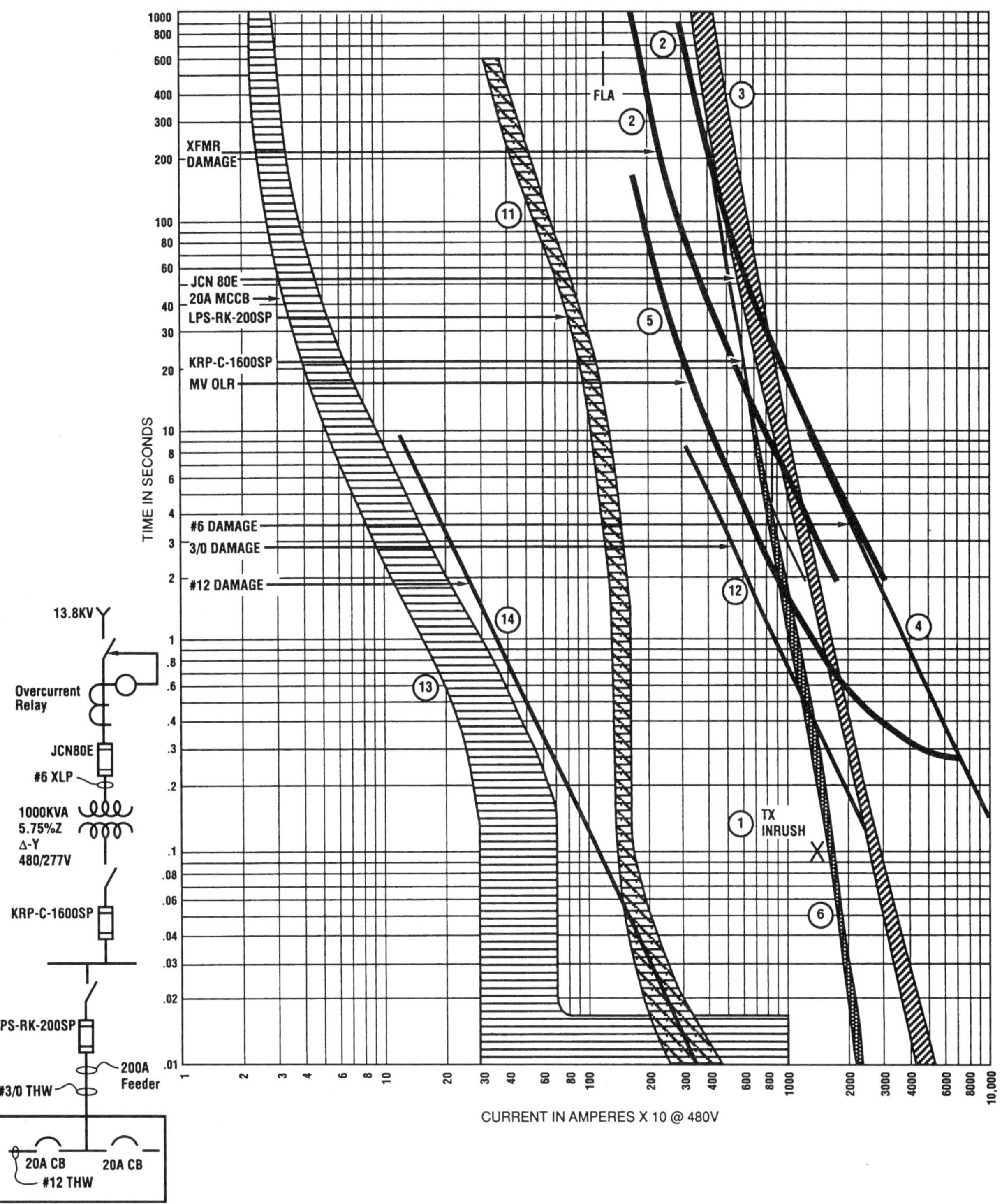

(continued)

8.1.3 (*Continued*)

Notes:

1. TCC1 includes the primary fuse, secondary main fuse, 200 ampere feeder fuse, and 20 ampere branch circuit breaker from LP1.

2. Analysis will begin at the main devices and proceed down through the system.

3. Reference (base) voltage will be 480 volts, arbitrarily chosen since most of the devices are at this level.

4. Selective coordination between the feeder and branch circuit is not attainable for faults above 2500 amperes that occur on the 20 amp branch circuit, from LP1. Notice the overlap of the 200 ampere fuse and 20 ampere circuit breaker.

5. The required minimum ratio of 2:1 is easily met between the KRP-C-1600SP and the LPS-RK-200SP.

Device ID	Description	Comments
①	1000KVA XFMR Inrush Point	12 x FLA @ .1 Seconds
②	1000KVA XFMR Damage Curves	5.75%Z, liquid filled (Footnote 1) (Footnote 2)
③	JCN 80E	E-Rated Fuse
④	#6 Conductor Damage Curve	Copper, XLP Insulation
⑤	Medium Voltage Relay	Needed for XFMR Primary Overload Protection
⑥	KRP-C-1600SP	Class L Fuse
⑪	LPS-RK-200SP	Class RK1 Fuse
⑫	3/0 Conductor Damage Curve	Copper THW Insulation
⑬	20A CB	Thermal Magnetic Circuit Breaker
⑭	#12 Conductor Damage Curve	Copper THW Insulation

Footnote 1: Transformer damage curves indicate when it will be damaged, thermally and/or mechanically, under overcurrent conditions.

Transformer impedance, as well as primary and secondary connections, and type, all will determine their damage characteristics.

Footnote 2: A Δ-Y transformer connection requires a 15% shift, to the right, of the L-L thermal damage curve. This is due to a L-L secondary fault condition, which will cause 1.0 p.u. to flow through one primary phase, and .866 p.u. through the two faulted secondary phases. (These currents are p.u. of 3-phase fault current.)

8.1.4

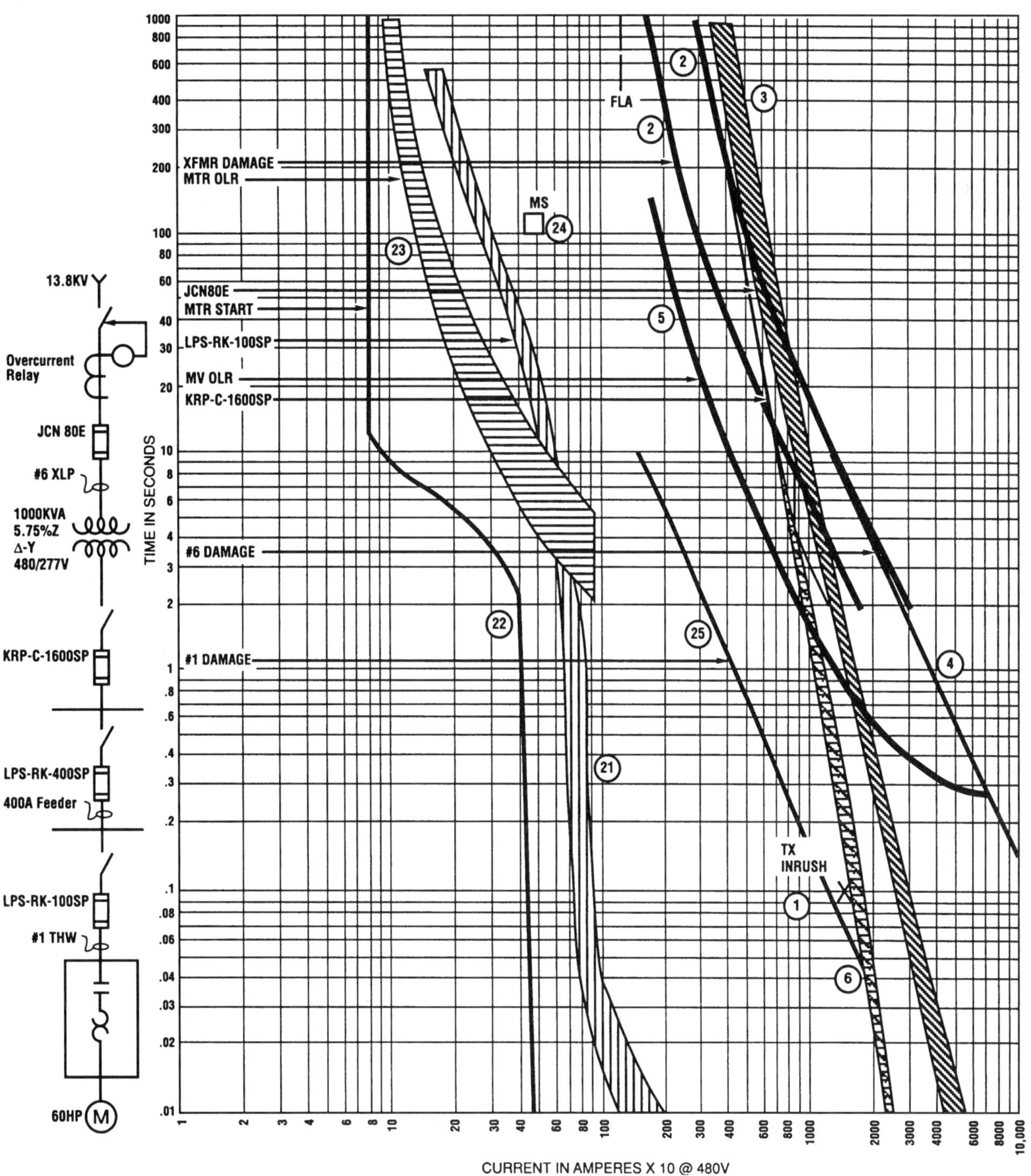

(continued)

8.1.4 (*Continued*)

Notes:

1. TCC2 includes the primary fuse, secondary main fuse, 400 ampere feeder fuse, 100 ampere motor branch fuse, 77 ampere motor and overload relaying.

2. Analysis will begin at the main devices and proceed down through the system.

3. Reference (base) voltage will be 480 volts, arbitrarily chosen since most of the devices are at this level.

Device ID	Description	Comment
①	1000KVA XFMR Inrush Point	12 x FLA @ .1 seconds
②	1000KVA XFMR Damage Curves	5.75%Z, liquid filled (Footnote 1) (Footnote 2)
③	JCN 80E	E-Rated Fuse
④	#6 Conductor Damage Curve	Copper, XLP Insulation
⑤	Medium Voltage Relay	Needed for XFMR Primary Overload Protection
⑥	KRP-C-1600SP	Class L Fuse
㉑	LPS-RK-100SP	Class RK1 Fuse
㉒	Motor Starting Curve	Across the Line Start
㉓	Motor Overload Relay	Class 10
㉔	Motor Stall Point	Part of a Motor Damage Curve
㉕	#1 Conductor Damage Curve	Copper THW Insulation

Footnote 1: Transformer damage curves indicate when it will be damaged, thermally and/or mechanically, under overcurrent conditions.

Transformer impedance, as well as primary and secondary connections, and type, all will determine their damage characteristics.

Footnote 2: A Δ-Y transformer connection requires a 15% shift, to the right, of the L-L thermal damage curve. This is due to a L-L secondary fault condition, which will cause 1.0 p.u. to flow through one primary phase, and .866 p.u. through the two faulted secondary phases. (These currents are p.u. of 3-phase fault current.)

8.1.5

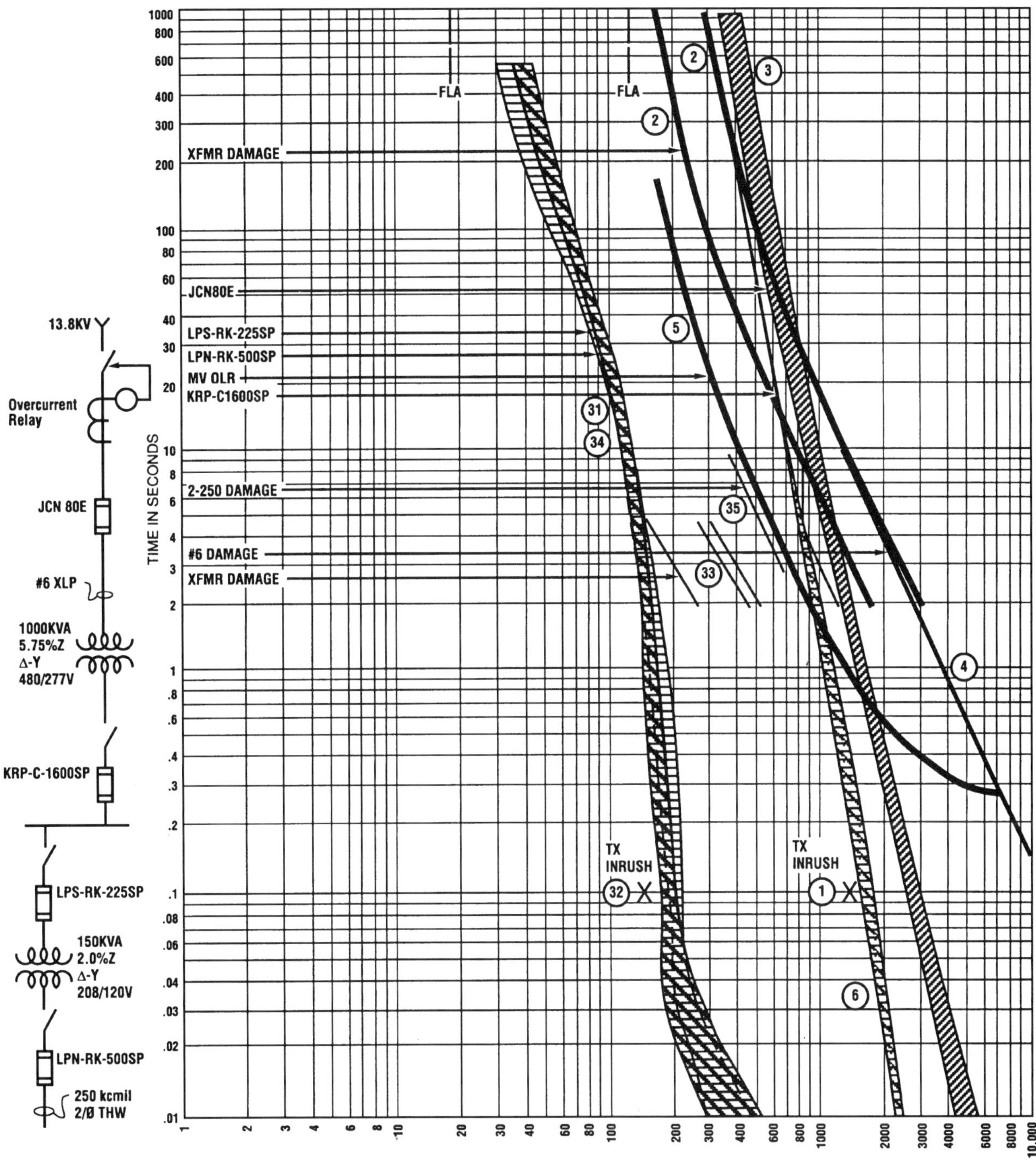

(continued)

8.1.5 (*Continued*)

Notes:

1. TCC3 includes the primary fuse, secondary main fuse, 225 ampere feeder/transformer primary and secondary fuses.

2. Analysis will begin at the main devices and proceed down through the system.

3. Reference (base) voltage will be 480 volts, arbitrarily chosen since most of the devices are at this level.

4. Relative to the 225 ampere feeder, coordination between primary and secondary fuses is not attainable, noted by overlap of curves.

5. Overload and short circuit protection for the 150 KVA transformer is afforded by the LPS-RK-225SP fuse.

Device ID	Description	Comment
(1)	1000KVA XFMR Inrush Point	12 x FLA @ .1 seconds
(2)	1000KVA XFMR Damage Curves	5.75%Z, liquid filled (Footnote 1) (Footnote 2)
(3)	JCN 80E	E-Rated Fuse
(4)	#6 Conductor Damage Curve	Copper, XLP Insulation
(5)	Medium Voltage Relay	Needed for XFMR Primary Overload Protection
(6)	KRP-C-1600SP	Class L Fuse
(31)	LPS-RK-225SP	Class RK1 Fuse
(32)	150 KVA XFMR Inrush Point	12 x FLA @.1 Seconds
(33)	150 KVA XFMR Damage Curves	2.00% Dry Type (Footnote 3)
(34)	LPN-RK-500SP	Class RK1 Fuse
(35)	2-250kcmil Conductors Damage Curve	Copper THW Insulation

Footnote 1: Transformer damage curves indicate when it will be damaged, thermally and/or mechanically, under overcurrent conditions.

Transformer impedance, as well as primary and secondary connections, and type, all will determine their damage characteristics.

Footnote 2: A Δ-Y transformer connection requires a 15% shift, to the right, of the L-L thermal damage curve. This is due to a L-L secondary fault condition, which will cause 1.0 p.u. to flow through one primary phase, and .866 p.u. through the two faulted secondary phases. (These currents are p.u. of 3-phase fault current.)

Footnote 3: Damage curves for a small KVA (<500KVA) transformer, illustrate thermal damage characteristics for Δ-Y connected. From right to left, these reflect damage characteristics, for a line-line fault, 3Ø fault, and L-G fault condition.

TABLE 8.1.6

*** Selectivity Ratio Guide (Line-Side to Load-Side) for Blackout Prevention**

Circuit				Load-Side Fuse									
Current Rating				601-6000A	601-4000A	0-600A			601-6000A	0-600A	0-1200A	0-600A	0-60A
	Type			Time-Delay	Time-Delay	Dual-Element Time-Delay			Fast-Acting	Fast-Acting			Time-Delay
		Trade Name		LOW-PEAK® YELLOW™	LIMITRON®	LOW-PEAK® YELLOW™		FUSETRON®	LIMITRON®	LIMITRON®	T-TRON®	LIMITRON®	SC
		Class		(L)	(L)	(RK1)	(J)**	(RK5)	(L)	(RK1)	(T)	(J)	(G)
		Buss Symbol		**KRP-CSP**	**KLU**	**LPN-RKSP LPS-RKSP**	**LPJ-SP**	**FRN-R FRS-R**	**KTU**	**KTN-R KTS-R**	**JJN JJS**	**JKS**	**SC**
Line-Side Fuse 601 to 6000A	Time-Delay	LOW-PEAK® YELLOW™ (L)	**KRP-CSP**	2:1	2.5:1	2:1	2:1	4:1	2:1	2:1	2:1	2:1	N/A
601 to 4000A	Time-Delay	LIMITRON® (L)	**KLU**	2:1	2:1	2:1	2:1	4:1	2:1	2:1	2:1	2:1	N/A
0 to 600A	Dual-Element	LOW-PEAK® YELLOW™ (RK1)	**LPN-RKSP LPS-RKSP**	–	–	2:1	2:1	8:1	–	3:1	3:1	3:1	4:1
		(J)	**LPJ-SP****	–	–	2:1	2:1	8:1	–	3:1	3:1	3:1	4:1
		FUSETRON® (RK5)	**FRN-R FRS-R**	–	–	1.5:1	1.5:1	2:1	–	1.5:1	1.5:1	1.5:1	1.5:1
601 to 6000A	Fast-Acting	LIMITRON® (L)	**KTU**	2:1	2.5:1	2:1	2:1	6:1	2:1	2:1	2:1	2:1	N/A
0 to 600A		LIMITRON® (RK1)	**KTN-R KTS-R**	–	–	3:1	3:1	8:1	–	3:1	3:1	3:1	4:1
0 to 1200A		T-TRON® (T)	**JJN JJS**	–	–	3:1	3:1	8:1	–	3:1	3:1	3:1	4:1
0 to 600A		LIMITRON® (J)	**JKS**	–	–	2:1	2:1	8:1	–	3:1	3:1	3:1	4:1
0 to 60A	Time-Delay	SC (G)	**SC**	–	–	3:1	3:1	4:1	–	2:1	2:1	2:1	2:1

* **Note:** At some values of fault current, specified ratios may be lowered to permit closer fuse sizing. Plot fuse curves or consult with Bussmann.
General Notes: Ratios given in this Table apply only to Buss fuses. When fuses are within the same case size, consult Bussmann.

** Consult Bussmann for latest LPJ-SP ratios.

Section

9

Component Short-Circuit Protection

9.1.0 Introduction
9.1.1 Short-Circuit Current Withstand Chart for Copper Cables with Paper, Rubber, or Varnished-Cloth Insulation
9.1.2 Short-Circuit Current Withstand Chart for Copper Cables with Thermoplastic Insulation
9.1.3 Short-Circuit Current Withstand Chart for Copper Cables with Cross-Linked Polyethylene and Ethylene-Propylene-Rubber Insulation
9.1.4 Short-Circuit Current Withstand Chart for Aluminum Cables with Paper, Rubber, or Varnished-Cloth Insulation
9.1.5 Short-Circuit Current Withstand Chart for Aluminum Cables with Thermoplastic Insulation
9.1.6 Short-Circuit Current Withstand Chart for Aluminum Cables with Cross-Linked Polyethylene and Ethylene-Propylene-Rubber Insulation
9.1.7 Comparison of Equipment Grounding Conductor Short-Circuit Withstand Ratings
9.1.8 NEMA (Standard Short-Circuit Ratings of Busway)
9.1.9 U.L. No. 508 Motor Controller Short-Circuit Test Ratings
9.1.10 Molded-Case Circuit Breaker Interrupting Capacities
9.1.11 *NEC* Table 450.3(A), Maximum Rating or Setting of Overcurrent Protection for Transformers over 600 V (as a Percentage of Transformer-Rated Current)
9.1.12 *NEC* Table 450.3(B), Maximum Rating or Setting of Overcurrent Protection for Transformers 600 V and Less (as a Percentage of Transformer-Rated Current)
9.1.13 U.L. 1008 Minimum Withstand Test Requirement (for Automatic Transfer Switches)
9.1.14 HVAC Equipment Short-Circuit Test Currents, Table 55.1 of U.L. Standard 1995
9.2.1 Protection through Current Limitation
9.2.2 Current-Limiting Effect of Fuses
9.2.3 Analysis of a Current-Limiting Fuse
9.2.4 Let-Thru Data Pertinent to Equipment Withstand
9.2.5 How to Use the Let-Thru Charts
9.2.6 Current-Limitation Curves: Bussmann Low-Peak Time-Delay Fuse KRP-C800SP

9.1.0 Introduction

Most electrical equipment has a withstand rating that is defined in terms of an rms symmetrical short-circuit current and, in some cases, peak let-thru current. These values have been established through short-circuit testing of that equipment according to an accepted industry standard. Or, as is the case with conductors, the withstand rating is based on a mathematical calculation and is also expressed as an rms symmetrical short-circuit current.

The following provides the short-circuit withstand data for each system component. Please note that where industry standards are given (e.g., NEMA), individual manufacturers of equipment often have withstand ratings that exceed industry standards.

9.1.1 Short-Circuit Current Withstand Chart for Copper Cables with Paper, Rubber, or Varnished-Cloth Insulation (*see page 9.3*)

9.1.1

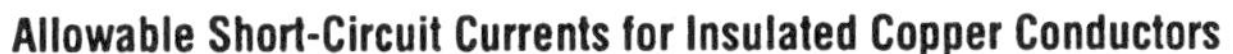

Allowable Short-Circuit Currents for Insulated Copper Conductors

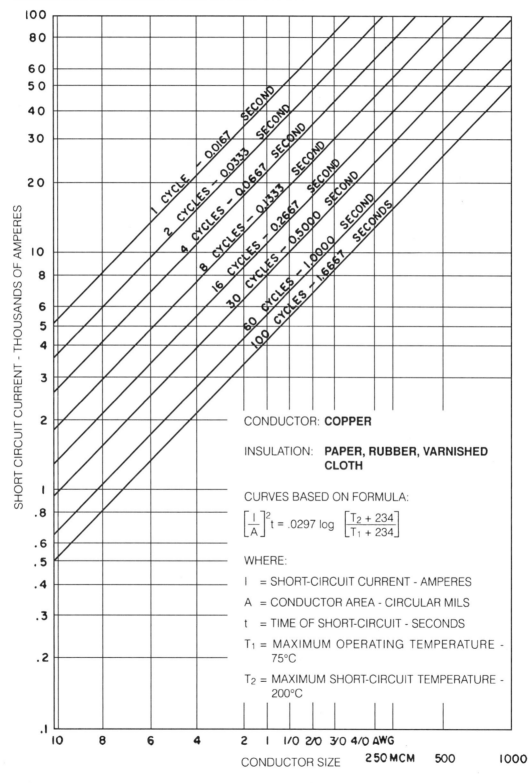

SHORT CIRCUIT CURRENT - THOUSANDS OF AMPERES

100, 80, 60, 50, 40, 30, 20, 10, 8, 6, 5, 4, 3, 2, 1, .8, .6, .5, .4, .3, .2, .1

1 CYCLE - 0.0167 SECOND

2 CYCLES - 0.0333 SECOND

4 CYCLES - 0.0667 SECOND

8 CYCLES - 0.1333 SECOND

16 CYCLES - 0.2667 SECOND

30 CYCLES - 0.5000 SECOND

60 CYCLES - 1.0000 SECOND

100 CYCLES - 1.6667 SECONDS

CONDUCTOR: **COPPER**

INSULATION: **PAPER, RUBBER, VARNISHED CLOTH**

CURVES BASED ON FORMULA:

$$\left[\frac{I}{A}\right]^2 t = .0297 \log \left[\frac{T_2 + 234}{T_1 + 234}\right]$$

WHERE:

I = SHORT-CIRCUIT CURRENT - AMPERES

A = CONDUCTOR AREA - CIRCULAR MILS

t = TIME OF SHORT-CIRCUIT - SECONDS

T_1 = MAXIMUM OPERATING TEMPERATURE - 75°C

T_2 = MAXIMUM SHORT-CIRCUIT TEMPERATURE - 200°C

10, 8, 6, 4, 2, 1, 1/0, 2/0, 3/0, 4/0 AWG

CONDUCTOR SIZE 250 MCM, 500, 1000

9.1.2 Short-Circuit Current Withstand Chart for Copper Cables with Thermoplastic Insulation

9.1.2

Allowable Short-Circuit Currents for Insulated Copper Conductors

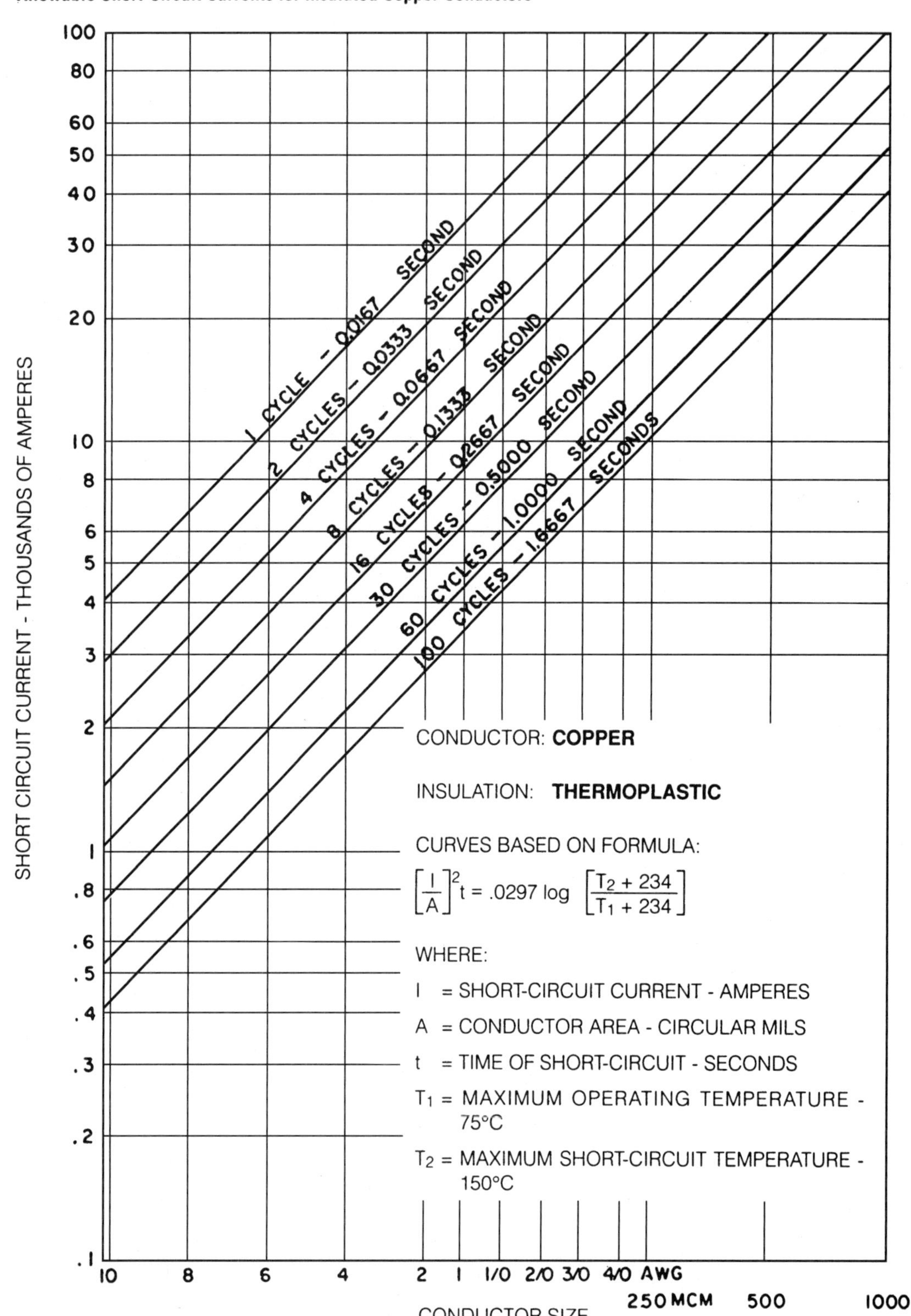

9.1.3 Short-Circuit Current Withstand Chart for Copper Cables with Cross-Linked Polyethylene and Ethylene-Propylene-Rubber Insulation

9.1.3

Allowable Short-Circuit Currents for Insulated Copper Conductors

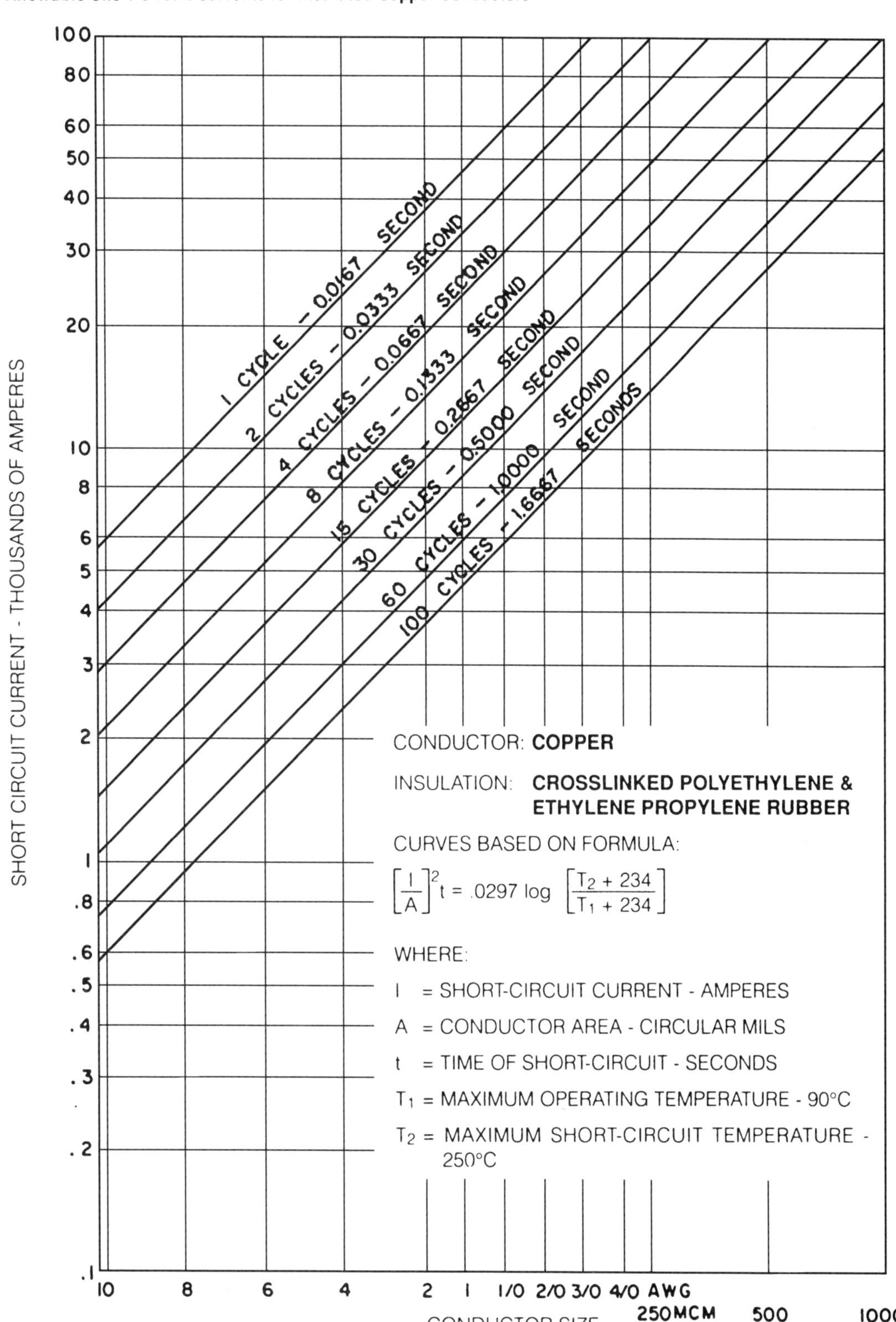

9.1.4 Short-Circuit Current Withstand Chart for Aluminum Cables with Paper, Rubber, or Varnished-Cloth Insulation

9.1.4

Allowable Short-Circuit Currents for Insulated Aluminum Conductors

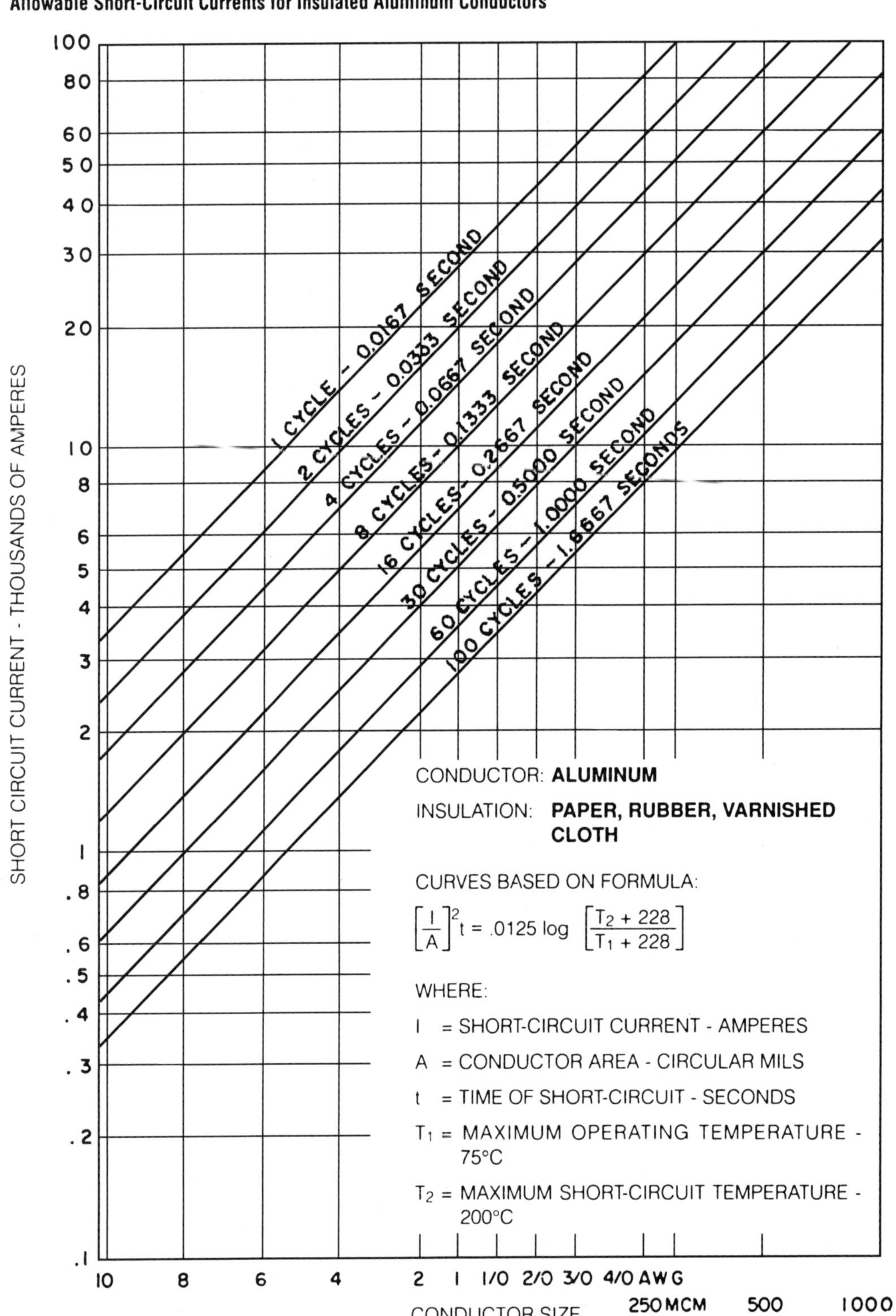

9.1.5 Short-Circuit Current Withstand Chart for Aluminum Cables with Thermoplastic Insulation

9.1.5

Allowable Short-Circuit Currents for Insulated Aluminum Conductors

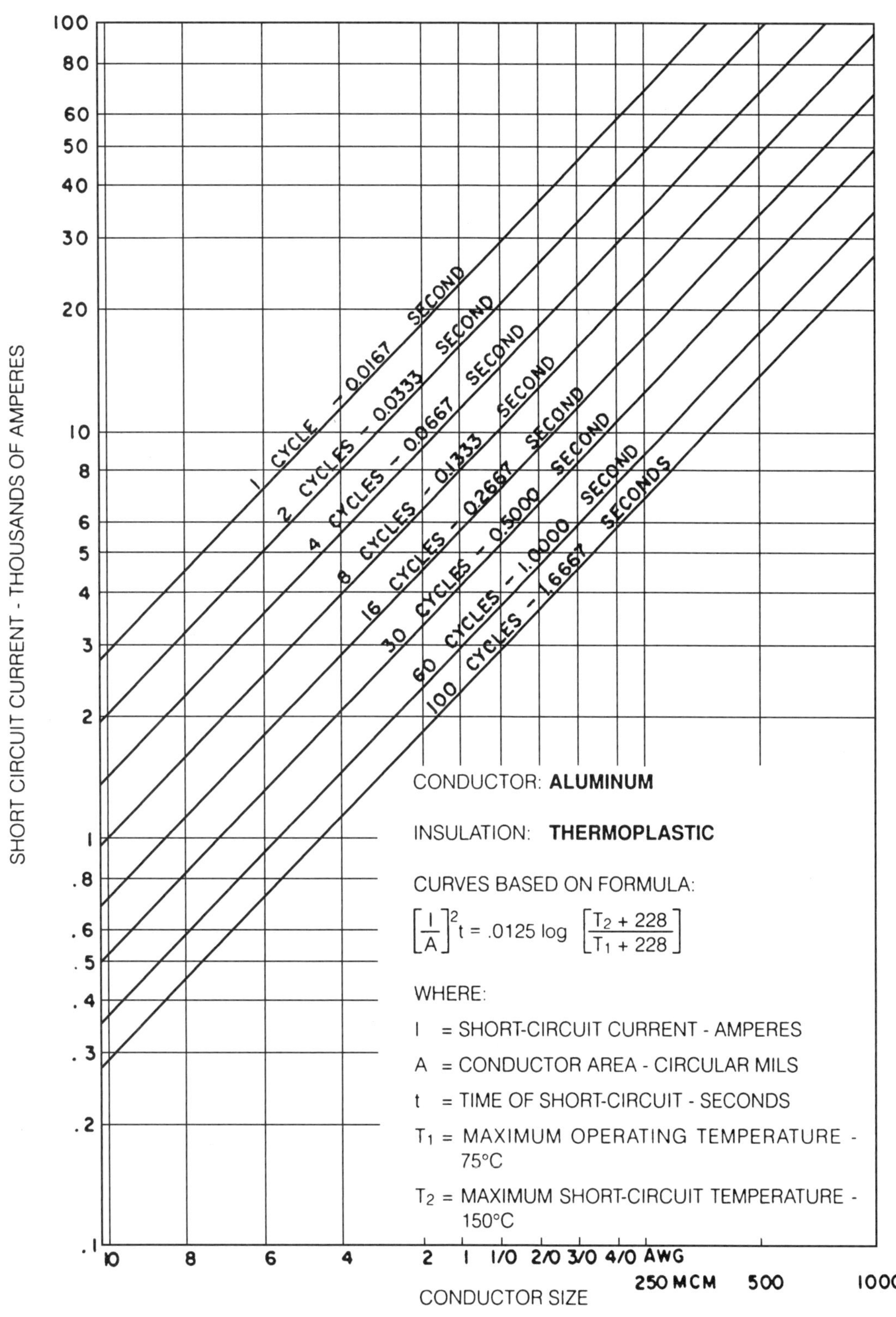

9.1.6 Short-Circuit Current Withstand Chart for Aluminum Cables with Cross-Linked Polyethylene and Ethylene-Propylene-Rubber Insulation

9.1.6

Allowable Short-Circuit Currents for Insulated Aluminum Conductors

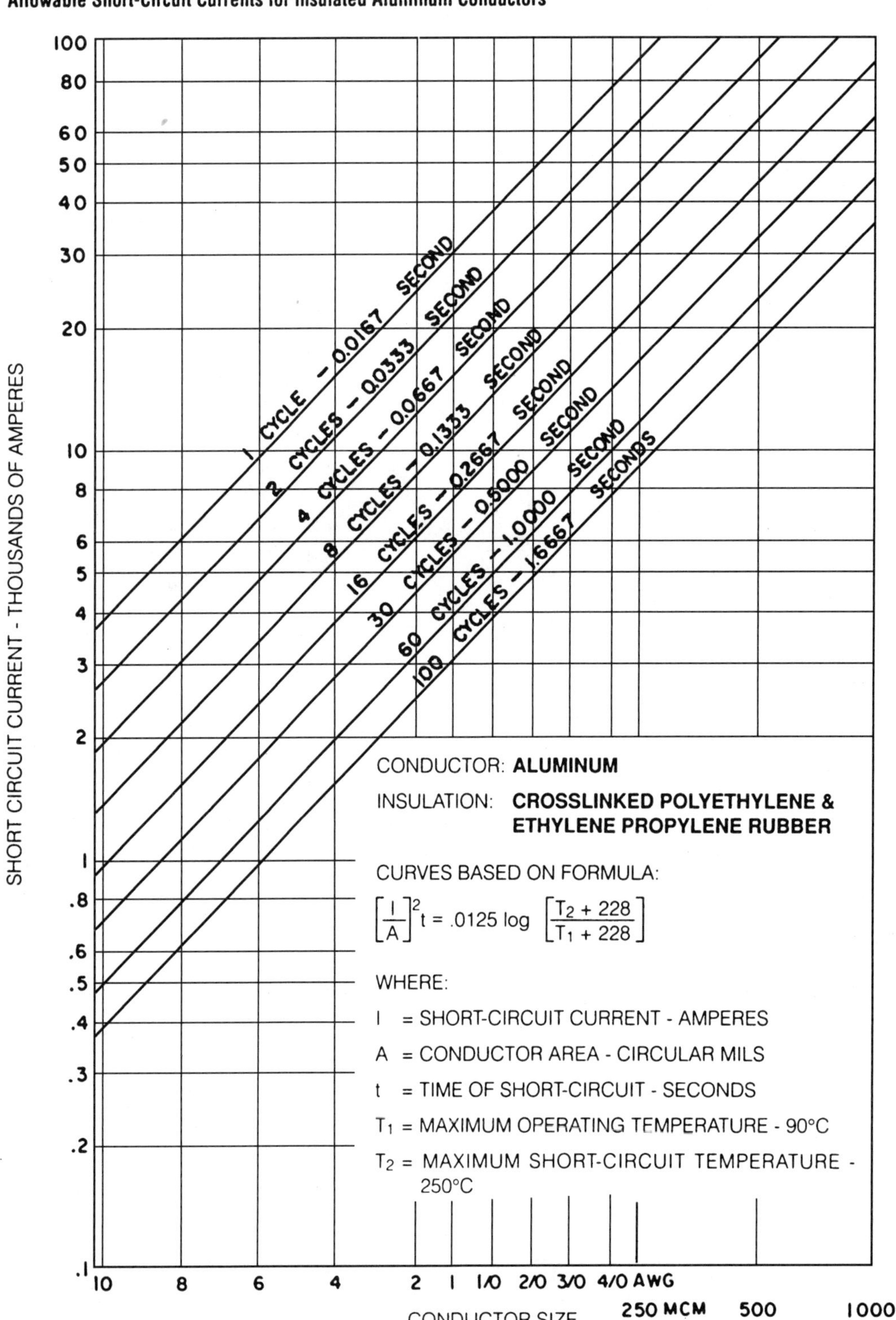

9.1.7 Comparison of Equipment Grounding Conductor Short-Circuit Withstand Ratings

TABLE 9.1.7

	5 Sec. Rating (Amps)			I^2t Rating x10^6 (Ampere Squared Seconds)		
Conductor Size	ICEA P32-382 Insulation Damage 150°C	Soares 1 Amp/30 cm Validity 250°C	Onderdonk Melting Point 1,083°C	ICEA P32-382 Insulation Damage 150°C	Soares 1 Amp/30 cm Validity 250°C	Onderdonk Melting Point 1,083°C
14	97	137	253	.047	.094	.320
12	155	218	401	.120	.238	.804
10	246	346	638	.303	.599	2.03
8	391	550	1,015	.764	1.51	5.15
6	621	875	1,613	1.93	3.83	13.0
4	988	1,391	2,565	4.88	9.67	32.9
3	1,246	1,754	3,234	7.76	15.4	52.3
2	1,571	2,212	4,078	12.3	24.5	83.1
1	1,981	2,790	5,144	19.6	38.9	132.0
1/0	2,500	3,520	6,490	31.2	61.9	210.0
2/0	3,150	4,437	8,180	49.6	98.4	331.0
3/0	3,972	5,593	10,313	78.9	156.0	532.0
4/0	5,009	7,053	13,005	125.0	248.0	845.0
250	5,918	8,333	15,365	175.0	347.0	1,180.0
300	7,101	10,000	18,438	252.0	500.0	1,700.0
350	8,285	11,667	21,511	343.0	680.0	2,314.0
400	9,468	13,333	24,584	448.0	889.0	3,022.0
500	11,835	16,667	30,730	700.0	1,389.0	4,721.0
600	14,202	20,000	36,876	1,008.0	2,000.0	6,799.0
700	16,569	23,333	43,022	1,372.0	2,722.0	9,254.0
750	17,753	25,000	46,095	1,576.0	3,125.0	10,623.0
800	18,936	26,667	49,168	1,793.0	3,556.0	12,087.0
900	21,303	30,000	55,314	2,269.0	4,500.0	15,298.0
1,000	23,670	33,333	61,460	2,801.0	5,555.0	18,867.0

9.1.8 NEMA (Standard Short-Circuit Ratings of Busway)

TABLE 9.1.8

Continuous Current Rating of Busway (Amperes)	Short-Circuit Current Ratings (Symmetrical Amperes)	
	Plug-In Duct	Feeder Duct
100	10,000	–
225	14,000	–
400	22,000	–
600	22,000	42,000
800	22,000	42,000
1000	42,000	75,000
1200	42,000	75,000
1350	42,000	75,000
1600	65,000	100,000
2000	65,000	100,000
2500	65,000	150,000
3000	85,000	150,000
4000	85,000	200,000
5000	–	200,000

This table pertains to feeder and plug-in busway. For switchboard and panelboard standard ratings refer to manufacturer.

U.L. Standard 891 details short-circuit durations for busway within switchboards for a minimum of three cycles, unless the main overcurrent device clears the short in less than three cycles.

9.1.9 U.L. No. 508 Motor Controller Short-Circuit Test Ratings

TABLE 9.1.9

Motor Controller HP Rating	Test Short Circuit Current Available
1HP or less and 300V or less	1,000A
50HP or less	5,000A
Greater than 50HP to 200HP	10,000A
201HP to 400HP	18,000A
401HP to 600HP	30,000A
601HP to 900HP	42,000A
901HP to 1600HP	85,000A

It should be noted that these are basic short-circuit requirements. Higher, combination ratings are attainable if tested to an applicable standard. However, damage is usually allowed.

9.1.10 Molded-Case Circuit Breaker Interrupting Capacities

TABLE 9.1.10

SIEMENS

Frame Size	Maximum Voltage Rating	Breaker Type	Ampere Rating	UL Interruption Capacity Symm. RMS (AC)	Dimensions (Inches)
100A	Standard Interrupting 240V AC 250V DC	ED2	15-100	120V AC 10 kA-1 Pole 240V AC 10 kA-2,3 Pole 125V DC 10 kA-1 Pole 250V DC 5 kA-2 Pole	W=1 (1 Pole) W=2 (2 Pole) W=3 (3 Pole) H=$6\frac{11}{32}$ D=4
125A	Standard Interrupting 480V AC 250V DC	ED4	15-125	120V AC 65 kA-1 Pole 240V AC 65 kA-2,3 Pole 277V AC 22 kA-1 Pole 480V AC 18 kA-2,3 Pole 125 V DC 5 kA-1 Pole 250V DC 30 kA-2 Pole	W=1 (1 Pole) W=2 (2 Pole) W=3 (3 Pole) H=$6\frac{11}{32}$ D=4
	Standard Interrupting 600V AC 500V DC	ED6	15-125	240V AC 65 kA-2,3 Pole 480V AC 25 kA-2,3 Pole 600V AC 18 kA-2,3 Pole 250V DC 30 kA-2 Pole 500V DC 18 kA-3 Pole	W=2,3 H=$6\frac{11}{32}$ D=4
	High Interrupting 480V AC 250V DC	HED4①	15-125	120V AC 100 kA-1 Pole 240V AC 100 kA-2,3 Pole 277V AC (15-30 Amperes) 65 kA-1 Pole 480V AC 42 kA-2,3 Pole 250V DC 30 kA-2 Pole	W=1,2,3 H=$6\frac{11}{32}$ D=4
	High Interrupting 600V AC 500V DC	HED6①	15-125	240V AC 100 kA-2,3 Pole 480V AC 30 kA-2,3 Pole 600V AC 18 kA-2,3 Pole 250V DC 30 kA-2 Pole 500V DC 25 kA-3 Pole	W=2,3 H=$6\frac{11}{32}$ D=4
	Current-Limiting 600V AC 500V DC	CED6	15-125	240V AC 200 kA-2,3 Pole 480V AC 200 kA-2,3 Pole 600V AC 30 kA-2,3 Pole 250V DC 30 kA-2 Pole 500V DC 50 kA-3 Pole	W=2,3 H=$9\frac{37}{64}$ D=4
225A	240V AC 2 or 3 Pole Construction	QJ2	60-225	240V AC 10 kA-2,3 Pole	W=3 (2 Pole) W=$4\frac{1}{2}$ (3 Pole) H=7 D=$2\frac{17}{32}$
		QJH2	60-225	240V AC 22 kA-2,3 Pole	
		QJ2-H	60-225	240V AC 42 kA-2,3 Pole	
250A	Standard Interrupting 600V AC 500V DC	FXD6 (Fix-Trip) FD6 (Interchangeable Trip)	70-250	240V AC 65 kA-2,3 Pole 480V AC 35 kA-2,3 Pole 600V AC 18 kA-2,3 Pole 250V DC 30 kA-2 Pole 500V DC 18 kA-3 Pole	W=$4\frac{1}{2}$ H=$9\frac{1}{2}$ D=4
	High Interrupting 600V AC 500V DC	HFD6② (Interchangeable Trip) HFXD6 (Fix-Trip)	70-250	240V AC 100 kA-2,3 Pole 480V AC 65 kA-2,3 Pole 600V AC 25 kA-2,3 Pole 250V DC 30 kA-2 Pole 500V DC 25 kA-3 Pole	
		HHFD6 (Interchangeable Trip) HHFXD6 (Fix-Trip)		240V AC 200 kA-3 Pole 480V AC 100 kA-3 Pole 600V AC 25 kA-3 Pole	
	Current-Limiting 600V AC 500V DC	CFD6 (Fix-Trip)	70-250	240V AC 200 kA-2,3 Pole 480V AC 200 kA-2,3 Pole 600V DC 100 kA-2,3 Pole 250V DC 30 kA-2 Pole 500V DC 50 kA-3 Pole	W=$4\frac{1}{2}$ H=$14\frac{1}{8}$ D=4

SQUARE D

Frame Size	Breaker Type	Ampere Rating	UL Interruption Capacity Symm. RMS (AC)	Dimensions (Inches)
100A	FAL	15-100	120V AC 10 kA-1 Pole 240V AC 10 kA-2,3 Pole 125V DC 5 kA-1 Pole 250V DC 5 kA-2,3 Pole	W=$1\frac{1}{2}$ (1Pole) W=3 (2 Pole) W=$4\frac{1}{2}$ (3 Pole) H=6 D=$3\frac{5}{32}$
125A	FAL	15-100	120V AC 25 kA-1 Pole 240V AC 25 kA-2,3 Pole 277V AC 18 kA-1 Pole 480V AC 18 kA-2,3 Pole 125V DC 18 kA-1 Pole 250V DC 10 kA-2,3 Pole	W=$1\frac{1}{2}$ (1Pole) W=3 (2 Pole) W=$4\frac{1}{2}$ (3 Pole) H=6 D=$3\frac{5}{32}$
	FAL	15-100	240V AC 25 kA-2,3 Pole 480V AC 18 kA-2,3 Pole 600V AC 14 kA-2,3 Pole 250V DC 10 kA-2,3 Pole	
	FCL	15-100	240V AC 100 kA-2,3 Pole 480V AC 65 kA-2,3 Pole	W=3 (2 Pole) W=$4\frac{1}{2}$ (3 Pole) H=6 D=$3\frac{5}{32}$
	FHL FHL-DC	15-100	240V AC 65 kA-2,3 Pole 480V AC 25 kA-2,3 Pole 600V AC 18 kA-2,3 Pole 250V DC 10 kA-2 Pole 500V DC 20 kA-3 Pole	
	FIL	15-100	240V AC 200 kA-2,3 Pole 480V AC 200 kA-2,3 Pole 600V DC 100 kA-2,3 Pole	W=$4\frac{1}{2}$ H=8 D=$2\frac{21}{32}$
225A	Q2L	100-225	240V AC 10 kA-2,3 Pole	W=3 (2 Pole) W=$4\frac{1}{2}$ (3 Pole) H=$6\frac{7}{16}$ D=$3\frac{59}{64}$
	Q2L-H	100-225	240V AC 22 kA-2,3 Pole	
	Q2LH	100-225	240V AC 42 kA-2,3 Pole	
250A	KAL (Fix-Trip)	70-250	240V AC 42 kA-2,3 Pole 480V AC 25 kA-2,3 Pole 600V AC 22 kA-2,3 Pole 250V DC 10 kA-2 Pole	W=$4\frac{1}{2}$ H=8 D=$3\frac{21}{32}$
	KHL (Fix-Trip) KHL-DC	70-250	240V AC 65 kA-2,3 Pole 480V AC 35 kA-2,3 Pole 600V AC 25 kA-2,3 Pole 250V DC 10 kA-2 Pole 500V DC 20 kA-3 Pole	
	Not Available			
	KIL (Fix-Trip)	110-250	240V AC 200 kA-2,3 Pole 480V AC 200 kA-2,3 Pole 600V AC 100 kA-2,3 Pole	W=$4\frac{1}{2}$ H=8 D=$3\frac{21}{32}$

WESTINGHOUSE

Frame Size	Breaker Type	Ampere Rating	UL Interruption Capacity Symm. RMS (AC)	Dimensions (Inches)
100A	EB	15-100	120V AC 10 kA-1 Pole 240V AC 10 kA-2,3 Pole 125V DC 5 kA-1 Pole 250V DC 5 kA-3 Pole	W=$1\frac{3}{8}$ (1 Pole) W=$2\frac{3}{4}$ (2 Pole) W=$4\frac{1}{8}$ (3 Pole) H=6 D=$3\frac{3}{8}$
125A	EHD	15-100	240V AC 18 kA-2,3 Pole 277V AC 14 kA-1 Pole 480V AC 14 kA-2,3 Pole 125 V DC 10 kA-1 Pole 250V DC 10 kA-2,3 Pole	W=$1\frac{3}{8}$ (1 Pole) W=$2\frac{3}{4}$ (2 Pole) W=$4\frac{1}{8}$ (3 Pole) H=6 D=$3\frac{3}{8}$
	FD	15-150	240V AC 65 kA-2,3 Pole 480V AC 25 kA-2,3 Pole 600V AC 18 kA-2,3 Pole 250V DC 10 kA-2,3 Pole	
	HFD	15-150	240V AC 100 kA-2,3 Pole 277V AC 65 kA-1 Pole 480V AC 65 kA-2,3 Pole 125V DC 10 kA-1 Pole 250V DC 10 kA-2,3 Pole	W=$1\frac{3}{8}$ (1 Pole) W=$2\frac{3}{4}$ (1 Pole) W=$4\frac{1}{8}$ (3 Pole) H=6
	HFD	15-150	240V AC 100 kA-2,3 Pole 480V AC 65 kA-2,3 Pole 600V AC 25 kA-2,3 Pole 250V DC 10 kA-2,3 Pole	
	FDC	15-150	240V AC 200 kA-2,3 Pole 480V AC 100 kA-2,3 Pole 600V AC 35 kA-2,3 Pole 250V DC 22 kA-2,3 Pole	
225A	CA	125-225	240V AC 10 kA-2,3 Pole	W=$2\frac{3}{4}$ (2 Pole) W=$4\frac{1}{8}$ (3 Pole) H=$6\frac{1}{2}$ D=$2\frac{11}{16}$
	CAH	125-225	240V AC 22 kA-2,3 Pole	
	HCA	125-225	240V AC 42 kA-2,3 Pole	
250A	JDB (Fix-Trip) JD (Interchangeable Trip)	70-250	240V AC 65 kA-2,3 Pole 480V AC 25 kA-2,3 Pole 600V AC 18 kA-2,3 Pole 250V DC 10 kA-2 Pole	W=$4\frac{1}{8}$ H=10 D=$4\frac{1}{16}$
	HJD (Interchangeable Trip)	70-250	240V AC 100 kA-2,3 Pole 480V AC 65 kA-2,3 Pole 600V AC 25 kA-2,3 Pole 250V DC 22 kA-2 Pole 500V DC 20 kA-3 Pole	
	Not Available			
	JDC (Interchangeable Trip)	70-250	240V AC 200 kA-2,3 Pole 480V AC 100 kA-2,3 Pole 600V AC 35 kA-2,3 Pole 250V DC 22 kA-2 Pole	

① Meets UL criteria for current limiting @ 240 VAC ② Meets UL criteria for current limiting @ 240 and 480 VAC ③ Current limiting @ 240 and 480 VAC

(continued)

TABLE 9.1.10 (*Continued*)

Frame Size	GENERAL ELECTRIC Breaker Type	Ampere Rating	UL Interruption Capacity Symm. RMS (AC)	Dimensions (inches)	CUTLER-HAMMER Breaker Type	Ampere Rating	UL Interruption Capacity Symm. RMS (AC)	Dimensions (inches)
100A	TEB	15-100	120V AC 10 kA-1 Pole 240V AC 10 kA-2,3 Pole 125V DC 5 kA-1 Pole 250V DC 5 kA-2,3 Pole	W=1 3/8 (1 Pole) W=2 3/4 (2 Pole) W=4 1/8 (3 Pole) H=6 9/16 D=3 3/8	FS	15-100	120V AC 65 kA-1 Pole (15-30 Amperes) 240V AC 10 kA-2,3 Pole 250V DC 120 kA-2 Pole	W=1 3/8 (1Pole) W=2 3/4 (2 Pole) W=4 1/8 (3 Pole) H=6 1/8 D=3 3/16
125A	TED	15-100	240V AC 18 kA-2,3 Pole 277V AC 14 kA-1 Pole 480V AC 14 kA-2,3 Pole 250V DC 10 kA-2,3 Pole	W=1 3/8 (1 Pole) W=2 3/4 (2 Pole) W=4 1/8 (3 Pole) H=6 9/16 D=3 3/8	FS	15-150	2120V AC 65 kA-1 Pole 277V AC 22 kA-1 Pole (15-30 Amperes) 240V AC 22 kA-2,3 Pole 480V AC 14 kA-2,3 Pole 250V DC 10 kA-2 Pole	W=1 3/8 (1Pole) W=2 3/4 (2Pole) W=4 1/8 (3Pole) H=6 1/8 D=3 3/16
	TED	15-100	240V AC 18 kA-2,3 Pole 480V AC 14 kA-2,3 Pole 600V AC 14 kA-2,3 Pole 250V DC 10 kA-2 Pole 500V DC 10 kA -3 Pole		FS	15-150	240V AC 22 kA-3 Pole 480V AC 14 kA-3 Pole 600V AC 14 kA-3 Pole 250V DC 10 kA-3 Pole	
	THED	15-150	240V AC (15-100 Amperes) 65 kA-2,3 Pole 240V AC (110-150 Amperes) 42 kA-2,3 Pole 277V AC (15-30 Amperes) 65 KA-1 Pole 480V AC 25 kA-2,3 Pole 250V DC 10 kA-2,3 Pole		FH	15-150	240V AC 100 kA-3 Pole 480V AC 30 kA-3 Pole	
	THED	15-150	240V AC (15-100 Amperes) 65 kA-2,3 Pole 240V AC (110-150 Amperes) 42 kA-2,3 Pole 480V AC 25 kA-2,3 Pole 600V AC 18 kA-2,3 Pole 250V DC 20 kA-2 Pole 500V DC 20 kA-3 Pole		FH	15-150	240V AC 100 kA-3 Pole 480V AC 30 kA,3 Pole 600V AC 18 kA,3 Pole 250V DC 10 kA-3 Pole	
	THLC1	15-150	240V AC 200 kA-3 Pole 480V AC (15-50 Amperes) 150 kA-3 Pole 480V AC (60-150 Amperes) 200 kA-3 Pole 600V AC 50 kA-3 Pole		FL Not UL Current Limiting	15-100	240V AC 200 kA-2,3 Pole 480V AC 100 kA-2,3 Pole 600V AC 25 kA-2,3 Pole 250V DC 10 kA-2,3 Pole	W=4 1/8 H=9 5/16 D=3 3/16
225A	TQD		240V AC 10 kA-2,3 Pole	W=2 3/4 (2 Pole) W=4 1/8 (3 Pole) H=6 9/16 D=2 5/8	CC	60-225	240V AC 10 KA-2,3 Pole	W=2 3/4 (2 Pole) W=4 1/8 (3Pole) H=6 1/8 D=2 57/64
	THQD		240V AC 22 kA-2,3 Pole		CCH	125-225	240V AC 25 KA-2,3 Pole	
	Not Available				CHH	60-225	240V AC 100 KA-2,3 Pole	
250A	TFJ (Fix-Trip)	70-250	240V AC 25 kA-2,3 Pole 480V AC 22 kA-2,3 Pole 600V AC 18 kA-3 Pole	W=4 1/8 H=10 1/8 D=3 3/16	JS (Fix-Trip)	100-250	240V AC 25 kA-3 Pole 480V AC 22 kA-3 Pole 600V AC 14 kA-3 Pole 250V DC 10 kA-3 Pole	W=4 1/8 H=12 D=3 13/16
	THFK (Interchangeable Trip)	70-250	240V AC 65 kA-2,3 Pole 480V AC 25 kA-2,3 Pole 600V AC 18 kA-3 Pole 250 V DC 10 kA-2 Pole		JH (Fix-Trip)	100-250	240V AC 100 kA-3 Pole 480V AC 30 kA-3 Pole 600V AC 18 kA-3 Pole 250V DC 10 kA-3 Pole	
	Current Limiting TFL (Fix-Trip)	70-250	240V AC 100 kA-3 Pole 480V AC 65 kA-3 Pole 600V AC 25 kA-3 Pole					
	Not Available				Not Available			
	THLC2 (Fix-Trip)	125-225	240V AC 200 kA-3 Pole 480V AC 200 kA-3 Pole 600V DC 50 kA-3 Pole	W=5 33/32 H=11 7/16 D=4 7/8	(Fix-Trip) JL Not UL Current Limiting	100-250	240V AC 200 kA,3 Pole 480V AC 100 kA-3 Pole 600V AC 25 kA-3 Pole 250V DC10 kA-3 Pole	

① Meets UL criteria for current limiting @ 240 VAC ② Meets UL criteria for current limiting @ 240 and 480 VAC ③ Current limiting @ 240 and 480 VAC

(*continued*)

TABLE 9.1.10 (*Continued*)

	SIEMENS				
400A	Standard Interrupting 240V AC	JXD2 (Fix-Trip)	200-400	240V AC 65 kA-2,3 Pole 250V DC 30 kA-2 Pole	$W=7\frac{1}{2}$ H=11 D=4
400A	Standard Interrupting 600V AC 500V DC	JXD6 (Fix-Trip) JD6 (Interchangeable Trip)	200-400	240V AC 65 kA-2,3 Pole 480V AC 35 kA-2,3 Pole 600V AC 25 kA-2,3 Pole 250V DC 30 kA-2 Pole 500V DC 25 kA-3 Pole	
400A	High Interrupting 600V AC 500V DC	HJD6 (Interchangeable Trip) HJXD6 (Fix-Trip)	200-400	240V AC 100 kA-2,3 Pole 480V AC 65 kA-2,3 Pole 600V AC 35 kA-2,3 Pole 250V DC 30 kA-2 Pole 500V DC 35 kA-3 Pole	
400A	High Interrupting① 600V AC	HHJXD6② (Fix-Trip) HHJD62② (Interchangeable Trip)	200-400	240V AC 200 kA-2,3 Pole 480V AC 100 kA-2,3 Pole 600V AC 50 kA-2,3 Pole	
400A	Current-Limiting 600V AC 500V DC	CJD6 (Fix-Trip)	200-400	240V AC 200 kA-3 Pole 480V AC 150 kA-3 Pole 600V AC 100 kA-3 Pole 250V DC 30 kA-2 Pole 500V DC 50 kA-3 Pole	$W=7\frac{1}{2}$ $H=17\frac{55}{64}$ D=4
400A	Standard Interrupting 600V AC	Solid State SJD6	200-400	240V AC 65 kA-3 Pole 480V AC 35 kA-3 Pole 600V AC 25 kA-3 Pole	$W=7\frac{1}{2}$ H=11 D=4
400A	High Interrupting 600V AC	Solid State SHJD6	200-400	240V AC 100 kA-3 Pole 480V AC 65 kA-3 Pole 600V AC 35 kA-3 Pole	
400A	Current-Limiting 600V AC	Solid State SCJD6	200-400	240V AC 200 kA-3 Pole 480V AC 150 kA-3 Pole 600V AC 100 kA-3 Pole	$W=7\frac{1}{2}$ $H=17\frac{55}{64}$ D=4
600A	Standard Interrupting 600V AC 500V DC	LXD6 (Fix-Trip) LD6 (Interchangeable Trip)	250-600	240V AC 65 kA-2,3 Pole 480V AC 35 kA-2,3 Pole 600V AC 25 kA-2,3 Pole 250V DC 30 kA-2 Pole 500V DC 25 kA-3 Pole	$W=7\frac{1}{2}$ H=11 D=4
600A	High Interrupting 600V AC 500V DC	HLXD6 (Fix-Trip) HLD6 (Interchangeable Trip)	250-600	240V AC 100 kA-2,3 Pole 480V AC 65 kA-2,3 Pole 600V AC 35 kA-2,3 Pole 250V DC 30 kA-2 Pole 500V DC 35 kA-3 Pole	
600A	High Interrupting① 600 Ampere 600V AC	HHLXD6② (Fix-Trip) HHLD6② (Interchangeable Trip)	250-600	240V AC 200 kA-2,3 Pole 480V AC 100 kA-2,3 Pole 600V AC 50 kA-2,3 Pole	
600A	Current Limiting 600V AC 500V DC	CLD6 (Fix-Trip)	450-600	240V AC 200 kA-3 Pole 480V AC 150 kA-3 Pole 600V AC 100 kA-3 Pole 250V DC 30 kA-2 Pole 500V DC 50 kA-3 Pole	$W=7\frac{1}{2}$ $H=17\frac{55}{64}$ D=4
600A	Standard Interrupting 600V AC	Solid State SLD6	300-600	240V AC 65 kA-3 Pole 480V AC 35 kA-3 Pole 600V AC 25 kA-3 Pole	$W=7\frac{1}{2}$ H=11 D=4
600A	High Interrupting 600V AC	Solid State SHLD6	300-600	240V AC 100 kA-3 Pole 480V AC 65 kA-3 Pole 600V AC 50 kA-3 Pole	
600A	Current Limiting 600V AC	Solid State SCLD6	300-600	240V AC 200 kA-3 Pole 480V AC 150 kA-3 Pole 600V AC 100 kA-3 Pole	$W=7\frac{1}{2}$ $H=17\frac{55}{64}$ D=4

SQUARE D			
Q4L	250-400	240V AC 25 kA-2,3 Pole	W=6 H=11 $D=4\frac{1}{16}$
LAL (Fix-Trip)	125-400	240V AC 42 kA-2,3 Pole 480V AC 30 kA-2,3 Pole 600V AC 22 kA-2,3 Pole 250V DC 10 kA-2 Pole	W=6 H=11 $D=4\frac{1}{16}$
LHL (Fix-Trip) LHL-DC	125-400	240V AC 65 kA-2,3 Pole 480V AC 35 kA-2,3 Pole 600V AC 25 kA-2,3 Pole 250V DC 10 kA-2 Pole 500V DC 20 kA-3 Pole	
Not Available			
LIL	300-400	240V AC 200 kA-2,3 Pole 480V AC 200 kA-2,3 Pole 600V AC 100 kA-2,3 Pole	$W=7\frac{1}{2}$ $H=11\frac{7}{8}$ $D=5\frac{1}{2}$
Not Available			
Solid State LXL	300-400	240V AC 100 kA-3 Pole 480V AC 65 kA-3 Pole 600V AC 35 kA-3 Pole	$W=7\frac{1}{2}$ $H=11\frac{7}{8}$ $D=5\frac{1}{2}$
Solid State LXIL	300-400	240V AC 200 kA,3 Pole 480V AC 200 kA-3 Pole 600V AC 100 kA-3 Pole	$W=7\frac{1}{2}$ $H=11\frac{7}{8}$ $D=5\frac{1}{2}$
Not Available			
LCL (Fix-Trip)	300-600	240V AC 100 kA-2,3 Pole 480V AC 65 kA-2,3 Pole 600V AC 35 kA-2,3 Pole	$W=7\frac{1}{2}$ $H=17\frac{55}{64}$ D=4
Not Available			
LIL (Fix-Trip)	450-600	240V AC 200 kA-2,3 Pole 480V AC 200 kA-2,3 Pole 600V AC 100 kA-2,3 Pole	$W=7\frac{1}{2}$ $H=17\frac{55}{64}$ D=4
Not Available			
Solid State LXL	400-600	240V AC 100 kA-3 Pole 480V AC 65 kA-3 Pole 600V AC 35 kA-3 Pole	$W=7\frac{1}{2}$ $H=11\frac{7}{8}$ $D=5\frac{1}{2}$
Solid State LXIL	400-600	240V AC 200 kA-3 Pole 480V AC 200 kA-3 Pole 600V AC 100 kA-3 Pole	

WESTINGHOUSE			
DK (Fix-Trip)	250-400	240V AC 65 kA-2,3 Pole 250V DC 10 kA-2,3 Pole	$W=5\frac{1}{2}$ $H=10\frac{1}{8}$ $D=4\frac{1}{16}$
KDB (Fix-Trip) KD (Interchangeable Trip)	100-400	240V AC 65 kA-2,3 Pole 480V AC 35 kA-2,3 Pole 600V AC 25 kA-2,3 Pole 250V DC 10 kA-2,3 Pole	$W=5\frac{1}{2}$ $H=10\frac{1}{8}$ $D=4\frac{1}{16}$
HKD (Interchangeable Trip)	100-400	240V AC 100 kA-2,3 Pole 480V AC 65 kA-2,3 Pole 600V AC 35 kA-2,3 Pole 250V DC 22 kA-2,3 Pole 500V DC 35 kA-3 Pole	
Current Limiting KDC (Interchangeable Trip)	100-400	240V AC 200 kA-2,3 Pole 480V DC 100 kA-2,3 Pole 600V AC 50 kA-2,3 Pole 250V DC 22 kA-2,3 Pole	
Current Limiting LCL (Fix-Trip)	125-400	240V AC 200 kA-2,3 Pole 480V AC 200 kA-2,3 Pole 600V AC 100 kA-2,3 Pole	$W=8\frac{1}{4}$ H=16 $D=4\frac{1}{16}$
Solid State KD	125-400	240V AC 65 kA-3 Pole 480V AC 35 kA-3 Pole 600V DC 25 kA-3 Pole	$W=5\frac{1}{2}$ $H=10\frac{1}{8}$ $D=4\frac{1}{16}$
Solid State HKD	125-400	240V AC 100 kA-3 Pole 480V AC 65 kA-3 Pole 600V AC 35 kA-3 Pole	
Solid State KDC	125-400	240V AC 200 kA,3 Pole 480V AC 100 kA-3 Pole 600V AC 50 kA-3 Pole	$W=5\frac{1}{2}$ $H=10\frac{1}{8}$ $D=4\frac{1}{16}$
LDB (Fix-Trip) LD (Interchangeable Trip)	300-600	240V AC 65 kA-2,3 Pole 480V AC 35 kA-2,3 Pole 600V AC 25 kA-2,3 Pole 250V DC 10 kA-2,3 Pole	$W=8\frac{1}{4}$ $H=10\frac{3}{4}$ $D=4\frac{1}{16}$
HLD (Interchangeable Trip)	250-600	240V AC 100 kA-2,3 Pole 480V AC 65 kA-2,3 Pole 600V AC 35 kA-2,3 Pole 250V DC 20 kA-2,3 Pole	
Current Limiting LDC (Interchangeable Trip)	300-600	240V AC 200 kA-2,3 Pole 480V AC 200 kA-2,3 Pole 600V AC 50 kA-2,3 Pole 250V DC 25 kA-2,3 Pole	
Not Available			
Solid State LD	600	240V AC 65 kA-3 Pole 480V AC 35 kA-3 Pole 600V AC 25 kA-3 Pole	$W=8\frac{1}{4}$ $H=10\frac{3}{4}$ $D=4\frac{1}{16}$
Solid State HLD	600	240V AC 100 kA-3 Pole 480V AC 65 kA-3 Pole 600V AC 35 kA-3 Pole	
Solid State LDC		240V AC 200 kA-3 Pole 480V AC 100 kA-3 Pole 600V AC 50 kA-3 Pole	

① Meets UL criteria for current limiting @ 240 VAC ② Meets UL criteria for current limiting @ 240 and 480 VAC ③ Current limiting @ 240 and 480 VAC

(*continued*)

TABLE 9.1.10 (*Continued*)

GENERAL ELECTRIC

Rating	Type	Amperes	Interrupting Rating	Dimensions
400A	TJD (Fix-Trip)	250-400	240V AC 22 kA-2,3 Pole 250V DC 10 kA-2 Pole	W=8¼ H=10⅛ D=3 13/16
400A	TJJ (Fix-Trip) TJK4 (Inchangeable Trip)	125-400	240V AC 42 kA-2,3 Pole 480V AC 30 kA-2,3 Pole 600V AC 22 kA-2,3 Pole 250V DC 10 kA-2 Pole 500V DC 20 kA-2 Pole	W=8¼ H=10⅛ D=3 13/16
400A	THJK4 (Interangeable Trip)	125-400	240V AC 65 kA-2,3 Pole 480V AC 35 kA-2,3 Pole 600V AC 25 kA-2,3 Pole 250V DC 10 kA-2 Pole	
400A	Current Limiting TLB4 (Fix-Trip)	250-400	240V AC 100 kA-3 Pole 480V DC 65 kA-3 Pole 600V AC 25 kA-3 Pole	
400A	THLC4 (Fix-Trip)	250-400	240V AC 200 kA-3 Pole 480V AC 200 kA-3 Pole 600V AC 50 kA-3 Pole	W=5 23/32 H=13 1/16 D=4⅞
400A	Solid State THJ4V	150-400	240V AC 65 kA-3 Pole 480V AC 35 kA-3 Pole 600V AC 25 kA-3 Pole	W=5½ H=10¾ D=3 13/16
400A	Solid State TJL4V	150-400	240V AC 100 kA-3 Pole 480V AC 65 kA-3 Pole 600V AC 30 kA-3 Pole	
400A	Not Available			
600A	TJK6 (Interchangeable Trip)	250-600	240V AC 42 kA-2,3 Pole 480V AC 30 kA-2,3 Pole 600V AC 22 kA-2,3 Pole 250V DC 10 kA-2 Pole 500V DC 20 kA-3 Pole	W=8¼ H=10⅛ D=3 13/16
600A	THJK6 (Interchangeable Trip)	250-600	240V AC 65 kA-2,3 Pole 480V AC 35 kA-2,3 Pole 600V AC 25 kA-2,3 Pole 250V DC 10 kA-2 Pole	
600A	Not Available			
600A	Not Available			
600A	Solid State THJ4V	150-600	240V AC 65 kA-3 Pole 480V AC 35 kA-3 Pole 600V AC 25 kA-3 Pole	W=8¼ H=10⅛ D=3 13/16
600A	Solid State TJL4V	150-600	240V AC 100 kA-3 Pole 480V AC 65 kA-3 Pole 600V AC 30 kA-3 Pole	
600A	Not Available			

CUTLER-HAMMER

Rating	Type	Amperes	Interrupting Rating	Dimensions
400A	KS (Fix-Trip)	250-400	240V AC 65 kA-2,3 Pole 250V DC 10 kA-2,3 Pole	W=5½ H=10⅛ 100-300 H=12 15/32 400 D=3 13/16
400A	KS (Fix-Trip)	100-400	240V AC 42 kA-3 Pole 480V AC 30 kA,3 Pole 600V AC 22 kA-3 Pole 250V DC 10 kA-3 Pole	
400A	KH (Fix-Trip)	100-400	240V AC 65 kA-3 Pole 480V AC 35 kA,3 Pole 600V DC 25 kA,3 Pole 250V DC 10 kA-3 Pole	
400A	Not Available			
400A	Not Available			
400A	Solid State KS	400	240V AC 42 kA-3 Pole 480V AC 30 kA-3 Pole 600V AC 22 kA-3 Pole	W=5½ H=10⅛ D=3 13/16
400A	Solid State KH	400	240V AC 65 kA-3 Pole 480V AC 35 kA-3 Pole 600V AC 22 kA-3 Pole	
400A	Not Available			
600A	LS—E (Fix-Trip) LS—E (Interchangeable Trip)	250-600	240V AC 65 kA-3 Pole 480V AC 35 kA-3 Pole 600V AC 25 kA-3 Pole 250V DC 10 kA-3 Pole	W=8¼ H=10¾ D=3 13/16
600A	LH—E (Interchangeable Trip)	250-600	240V AC 100 kA-3 Pole 480V AC 65 kA-3 Pole 600V AC 35 kA-3 Pole 250V DC 10 kA-3 Pole	
600A	LL—E Not UL Current Limiting	250-600	240V AC 200 kA-3 Pole 480V AC 150 kA-3 Pole 600V AC 100 kA-3 Pole 250V DC 50 kA-3 Pole	
600A	Not Available			
600A	Not Available			
600A	Not Available			
600A	Not Available			

① Meets UL criteria for current limiting @ 240 VAC ② Meets UL criteria for current limiting @ 240 and 480 VAC ③ Current limiting @ 240 and 480 VAC

(*continued*)

TABLE 9.1.10 (*Continued*)

	SIEMENS					SQUARE D				WESTINGHOUSE			
800A	Standard Interrupting 600V AC 500V DC	MXD6 (Fix-Trip) MD6 (Interchangeable Trip)	500-800	240V AC 65 kA-2,3 Pole 480V AC 50 kA-2,3 Pole 600V AC 25 kA-2,3 Pole 250V DC 30 kA-2 Pole 500V DC 25 kA-3 Pole	W=9 H=16 $D=6\frac{3}{16}$	MAL (Fix-Trip)	300-1000	240V AC 42 kA-2,3 Pole 480V AC 30 kA-2,3 Pole 600V AC 22 kA-2,3 Pole 250V DC 14 kA-2 Pole	W=9 H=14	MA (Interchangeable Trip)	125-800	240V AC 42 kA-2,3 Pole 480V AC 30 kA-2,3 Pole 600V AC 22 kA-2,3 Pole 250V DC 20 kA-2,3 Pole	$W=8\frac{1}{4}$ H=16 $D=4\frac{1}{6}$
	High Interrupting 600V AC 500V DC	HMXD6 (Fix-Trip) HMD6 (Interchangeable Trip)	500-800	240V AC 100 kA-2,3 Pole 480V AC 65 kA-2,3 Pole 600V AC 50 kA-2,3 Pole 250V DC 30 kA-2,3 Pole 500V DC 50 kA-2,3 Pole	W=9 H=16 $D=6\frac{3}{16}$	MHL (Fix-Trip) MHL-DC	300-1000	240V AC 65 kA-2,3 Pole 480V AC 65 kA-2,3 Pole 600V AC 25 kA-2,3 Pole 250V DC 14 kA-2 Pole 500V DC 20 kA-3 Pole	W=9 H=14 $D=4\frac{17}{32}$	HMA (Interchangeable Trip)	125-800	240V AC 65 kA-2,3 Pole 480V AC 35 kA-2,3 Pole 600V AC 25 kA-2,3 Pole 250V DC 20 kA-2,3 Pole	
	Current Limiting① 600V AC 500V DC	CMD6 (Fix-Trip)	500-800	240V AC 200 kA-3 Pole 480V AC 100 kA-3 Pole 600V AC 65 kA-3 Pole 250V DC 30 kA-2 Pole 500V DC 50 kA-3 Pole		Not Available				Not Available			
	Sandard Interrupting 600V AC	Solid State SMD6	600-800	240V AC 65 kA-3 Pole 480V AC 50 kA-3 Pole 600V AC 25 kA-3 Pole		Solid State MXL	450-800	240V AC 65 kA-3 Pole 480V AC 65 kA-3 Pole 600V AC 25 kA-3 Pole	W=9 $H=14\frac{3}{4}$ $D=4\frac{15}{32}$	Solid State ND	600-800	240V AC 65 kA-3 Pole 480V AC 50 kA-3 Pole 600V AC 25 kA-3 Pole	$W=8\frac{1}{4}$ H=16 $D=5\frac{1}{2}$
	High Interrupting 600V AC	Solid State SHMD6	600-800	240V AC 100 kA-3 Pole 480V AC 65 kA-3 Pole 600V AC 50 kA-3 Pole		Not Available				Solid State HND	600-800	240V AC 100 kA-3 Pole 480V AC 65 kA-3 Pole 600V AC 35 kA-3 Pole	
	Current Limiting 600V AC	Solid State SCMD6	600-800	240V AC 200 kA-3 Pole 480V AC 100 kA-3 Pole 600V AC 65 kA-3 Pole		Not Available				Solid State NDC	600-800	240V AC 200 kA-3 Pole 480V AC 100 kA-3 Pole 600V AC 50 kA-3 Pole	
1200A	Standard Interrupting 600V AC 500V DC	NXD6 (Fix-Trip) ND6 (Interchangeable Trip)	800-1200	240V AC 65 kA-2,3 Pole 480V AC 50 kA-2,3 Pole 600V AC 25 kA-2,3 Pole 250V DC 30 kA-3 Pole 500V DC 25 kA-3 Pole	W=9 H=16 $D=6\frac{3}{16}$	NAL (Fix-Trip)	600-1200	240V AC 100 kA-2,3 Pole 480V AC 50 kA-2,3 Pole 600V AC 25 kA-2,3 Pole	$W=14\frac{63}{64}$ $H=12\frac{1}{8}$ $D=6\frac{13}{32}$	NB (Interchangeable Trip)	700-1200	240V AC 42 kA-2,3 Pole 480V AC 30 kA-2,3 Pole 600V AC 22 kA-2,3 Pole	$W=8\frac{1}{4}$ H=16 $D=5\frac{1}{2}$
	High Interrupting 600V AC 500V DC	HNXD6 (Fix-Trip) HND6 (Interchangeable Trip)	800-1200	240V AC 100 kA-2,3 Pole 480V AC 65 kA-2,3 Pole 600V AC 50 kA-2,3 Pole 250V DC 30 kA-2 Pole 500V DC 50 kA-3 Pole		NCL (Fix-Trip)	600-1200	240V AC125 kA-2,3 Pole 480V AC 100 kA-2,3 Pole 600V AC 65 kA-2,3 Pole		HNB (Interchangeable Trip)	700-1200	240V AC 65 kA-2,3 Pole 480V AC 35 kA-2,3 Pole 600V AC 25 kA-2,3 Pole	
	Current Limiting③ 600V AC 500V DC	CND6 (Fix-Trip)	900-1200	240V AC 200 kA-3 Pole 480V AC 100 kA-3 Pole 600V AC 65 kA-3 Pole 250V DC 30 kA-2 Pole 500V DC 50 kA-3 Pole		Not Available				Not Available			
	Standard Interrupting 600V AC	Solid State SND6	800-1200	240V AC 65 kA-3 Pole 480V AC 50 kA-3 Pole 600V AC 25 kA-3 Pole		Not Available				Solid State ND	600-1200	240V AC 65 kA-3 Pole 480V AC 50 kA-3 Pole 600V AC 25 kA-3 Pole	$W=8\frac{1}{4}$ H=16 $D=5\frac{1}{2}$
	High Interrupting 600V AC	Solid State SHND6	800-1200	240V AC 100 kA-3 Pole 480V AC 65 kA-3 Pole 600V AC 50 kA-3 Pole		Solid State NXL	600-1200	240V AC 125 kA-3 Pole 480V AC 100 kA-3 Pole 600V AC 65 kA-3 Pole	$W=14\frac{63}{64}$ $H=12\frac{1}{8}$ $D=6\frac{13}{32}$	Solid State HND	600-1200	240V AC 100 kA-3 Pole 480V AC 65 kA-3 Pole 600V AC 35 kA-3 Pole	
	Current Limiting 600V AC	Solid State SCND6	800-1200	240V AC 200 kA-3 Pole 480V AC 100 kA-3 Pole 600V AC 65 kA-3 Pole		Not Available				Solid State NDC	600-1200	240V AC 200 kA-3 Pole 480V AC 100 kA-3 Pole 600V AC 50 kA-3 Pole	

① Meets UL criteria for current limiting @ 240 VAC ② Meets UL criteria for current limiting @ 240 and 480 VAC ③ Current limiting @ 240 and 480 VAC

(*continued*)

TABLE 9.1.10 *(Continued)*

Rating	GENERAL ELECTRIC				CUTLER-HAMMER			
800A	TKM8 (Interchangeable Trip)	300-800	240V AC 42 kA-2,3 Pole 480V AC 30 kA-2,3 Pole 600V AC 22 kA-2,3 Pole 250V DC 10 kA-2 Pole 500V DC 20 kA-3 Pole	W=$8\frac{1}{4}$ H=$15\frac{1}{2}$ D=$5\frac{1}{2}$	MS (Fix-Trip)	350-800	240V AC 42 kA-2,3 Pole 480V AC 30 kA-2,3 Pole 600V AC 22 kA-2,3 Pole 250V DC 10 kA-3 Pole (350-600 ONLY)	W=$8\frac{1}{4}$ H=16 D=$4\frac{1}{16}$
800A	THKM8 (Interchangeable Trip)	300-800	240V AC 65 kA-2,3 Pole 480V AC 35 kA-2,3 Pole 600V AC 25 kA-2,3 Pole 250V DC 10 kA-2 Pole 500V DC 10 kA-3 Pole		MH (Fix-Trip)	350-800	240V AC 65 kA-3 Pole 480V AC 35 kA-3 Pole 600V AC 25 kA-2,3 Pole 250V DC 10 kA-3 Pole (350-600 ONLY)	
800A	Not Available				Not Available			
800A	Solid State TK4V	800	240V AC 42 kA-3 Pole 480V AC 30 kA-3 Pole 600V AC 22 kA-3 Pole	W=$8\frac{1}{4}$ H=$15\frac{1}{2}$ D=$3\frac{13}{16}$	Not Available			
800A	Solid State TKL4V	800	240V AC 100 kA-3 Pole 480V AC 65 kA-3 Pole 600V AC 30 kA-3 Pole		Not Available			
800A	Not Available				Not Available			
1200A	TKM12 (Interchangeable Trip)	600-1200	240V AC 42 kA-2,3 Pole 480V AC 30 kA-2,3 Pole 600V AC 22 kA-2,3 Pole	W=$8\frac{1}{4}$ H=$15\frac{1}{2}$ D=$3\frac{13}{16}$	NS (Fix-Trip)	700-1200	240V AC 42 kA,3 Pole 480V AC 35 kA-3 Pole 600V AC 23 kA-3 Pole	W=$8\frac{1}{4}$ H=16 D=$5\frac{1}{2}$
1200A	THKM12 (Interchangeable Trip)	600-1200	240V AC 65 kA-2,3 Pole 480V AC 35 kA-2,3 Pole 600V AC 25 kA-2,3 Pole		NH (Fix-Trip)	700-1200	240V AC 65 kA-3 Pole 480V AC 30 kA-3 Pole 600V AC 22 kA-3 Pole	
1200A	Not Available				Not Available			
1200A	Solid State TKRV	800-1200	240V AC 42 kA-3 Pole 480V AC 30 kA-3 Pole 600V AC 25 kA-3 Pole	W=$8\frac{1}{4}$ H=$15\frac{1}{2}$ D=$5\frac{1}{2}$	Not Available			
1200A	Solid State TKL4V	800-1200	240V AC 100 kA-3 Pole 480V AC 65 kA-3 Pole 600V AC 30 kA-3 Pole		Not Available			
1200A	Not Available				Not Available			

① Meets UL criteria for current limiting @ 240 VAC ② Meets UL criteria for current limiting @ 240 and 480 VAC ③ Current limiting @ 240 and 480 VAC

(continued)

TABLE 9.1.10 *(Continued)*

SIEMENS

Frame	Type	Model	Ampere Range	Interrupting Ratings	Dimensions
1600A	Standard Interrupting 600V AC 500V DC	PXD6 (Fix-Trip) PD6 (Interchangeable Trip)	1200-1600	240V AC 65 kA-3 Pole 480V AC 50 kA-3 Pole 600V AC 25 kA-3 Pole 250V DC 30 kA-2 Pole 500V DC 25 kA-3 Pole	W=9 H=16 D=6 3/16
	High Interrupting 600V AC 500V DC	HPXD6 (Fix-Trip) HPD6 (Interchangeable Trip)		240V AC 100 kA-3 Pole 480V AC 65 kA-3 Pole 600V AC 50 kA-3 Pole 250V DC 30 kA-2 Pole 500V DC 50 kA-3 Pole	
	Current Limiting 600V AC 500V DC	CPD6 (Fix-Trip)	1200-1600	240V AC 200 kA-3 Pole 480V AC 100 kA-3 Pole 600V AC 65 kA-3 Pole 250V DC 30 kA-2 Pole 500V DC 50 kA-3 Pole	
	Standard Interrupting 600V AC	Solid State SPD6	1400-1600	240V AC 65 kA-3 Pole 480V AC 50 kA-3 Pole 600V AC 25 kA-3 Pole	
	High Interrupting 600V AC	Solid State SHPD6	1400-1600	240V AC 100 kA-3 Pole 480V AC 65 kA-3 Pole 600V AC 50 kA-3 Pole	
2000A	Standard Interrupting 600V AC 500V DC	RXD6 (Fix-Trip) RD6 (Interchangeable Trip)	1600-2000	240V AC 65 kA-3 Pole 480V AC 50 kA-3 Pole 600V AC 25 kA-3 Pole 250V DC 30 kA-2 Pole 500V DC 25 kA-3 Pole	W=9 H=16 D=6 3/16
	High Interrupting 600V AC 500V DC	HRXD6 (Fix-Trip) HRD6 (Interchangeable Trip)	1600-2000	240V AC 100 kA-3 Pole 480V AC 65 kA-3 Pole 600V AC 50 kA-3 Pole 250V DC 30 kA-2 Pole 500V DC 50 kA-3 Pole	

SQUARE D

Frame	Model	Ampere Range	Interrupting Ratings	Dimensions
1600A	Not Available			
	Not Available			
	Not Available			
	Not Available			
	Solid State PXF	1400-1600	240V AC 125 kA-3 Pole 480V AC 100 kA-3 Pole 600V AC 65 kA-3 Pole	W=23 3/16 H=26 1/16 D=13 21/64
2000A	PAF (Fix-Trip) PAF-DC	600-2000	240V AC 65 kA-2,3 Pole 480V AC 50 kA-2,3 Pole 600V AC 42 kA-2,3 Pole 500V DC 25 kA-3 Pole	W=13 3/4 H=20 1/16 D=7 1/4
	PHF (Fix-Trip)	600-2000	240V AC 125 kA-2,3 Pole 480V AC 100 kA-2,3 Pole 600V AC 65 kA-2,3 Pole	

WESTINGHOUSE

Frame	Model	Ampere Range	Interrupting Ratings	Dimensions
1600A	Not Available			
	Not Available			
	Not Available			
	Solid State RD	800-1600	240V AC 125 kA-3 Pole 480V AC 65 kA-3 Pole 600V AC 50 kA-3 Pole	W=15 1/2 H=16 D=9 3/4
	Solid State RDC	800-1600	240V AC 200 kA-3 Pole 480V AC 100 kA-3 Pole 600V AC 65 kA-3 Pole	
2000A	Not Available			
	Not Available			

GENERAL ELECTRIC

Frame	Model	Ampere Range	Interrupting Ratings	Dimensions
1600A	Not Available			
	Not Available			
	Not Available			
	Solid State TRLA	600-1600	240V AC 100 kA-3 Pole 480V AC 65 kA-3 Pole 600V AC 50 kA-3 Pole	W=13 1/2 H=17 1/2 D=8 5/16
	Solid State TRPA	600-1200	240V AC 125 kA-3 Pole 480V AC 100 kA-3 Pole 600V AC 65 kA-3 Pole	
2000A	Not Available			
	Not Available			

CUTLER-HAMMER

Frame	Model
1600A	Not Available
	Not Available
	Not Available
	Not Available
	Not Available
2000A	Not Available
	Not Available

① Meets UL criteria for current limiting @ 240 VAC ② Meets UL criteria for current limiting @ 240 and 480 VAC ③ Current limiting @ 240 and 480 VAC

(Courtesy of Siemens Corporation)

9.1.11 *NEC* Table 450.3(A), Maximum Rating or Setting of Overcurrent Protection for Transformers over 600 V (as a Percentage of Transformer-Rated Current)

TABLE 9.1.11

		Primary Protection Over 600 Volts		Secondary Protection (see Note 2)		
				Over 600 Volts		600 Volts or Below
Location Limitations	**Transformer Rated Impedance**	**Circuit Breaker (see Note 4)**	**Fuse Rating**	**Circuit Breaker (see Note 4)**	**Fuse Rating**	**Circuit Breaker or Fuse Rating**
Any location	Not more than 6%	600% (see Note 1)	300% (see Note 1)	300% (see Note 1)	250% (see Note 1)	125% (see Note 1)
	More than 6% and not more than 10%	400% (see Note 1)	300% (see Note 1)	250% (see Note 1)	225% (see Note 1)	125% (see Note 1)
Supervised locations only (see Note 3)	Any	300% (see Note 1)	250% (see Note 1)	Not required	Not required	Not required
	Not more than 6%	600%	300%	300% (See Note 5)	250% (See Note 5)	250% (See Note 5)
	More than 6% and not more than 10%	400%	300%	250% (See Note 5)	225% (See Note 5)	250% (See Note 5)

Notes:
1. Where the required fuse rating or circuit breaker setting does not correspond to a standard rating or setting, a higher rating or setting that does not exceed the next higher standard rating or setting shall be permitted. **(ROP 13-3)**
2. Where secondary overcurrent protection is required, the secondary overcurrent device shall be permitted to consist of not more than six circuit breakers or six sets of fuses grouped in one location. Where multiple overcurrent devices are utilized, the total of all the device ratings shall not exceed the allowed value of a single overcurrent device. If both circuit breakers and fuses are used as the overcurrent device, the total of the device ratings shall not exceed that allowed for fuses. **(ROP 13-4)**
3. A supervised location is a location where conditions of maintenance and supervision ensure that only qualified persons will monitor and service the transformer installation.
4. Electronically actuated fuses that may be set to open at a specific current shall be set in accordance with settings for circuit breakers.
5. A transformer equipped with a coordinated thermal overload protection by the manufacturer shall be permitted to have separate secondary protection omitted.

9.1.12 *NEC* Table 450.3(B), Maximum Rating or Setting of Overcurrent Protection for Transformers 600 V and Less (as a Percentage of Transformer-Rated Current)

TABLE 9.1.12

	Primary Protection			Secondary Protection (see Note 2)	
Protection Method	**Currents of 9 Amperes or More**	**Currents Less than 9 Amperes**	**Currents Less than 2 Amperes**	**Currents of 9 Amperes or More**	**Currents Less than 9 Amperes**
Primary only protection	125% (see Note 1)	167%	300%	Not required	Not required
Primary and secondary protection	250% (see Note 3)	250% (see Note 3)	250% (see Note 3)	125% (see Note 1)	167%

Notes:
1. Where 125 percent of this current does not correspond to a standard rating of a fuse or nonadjustable circuit breaker, a higher rating that does not exceed the next higher standard rating shall be permitted. **(ROP 13-3)**
2. Where secondary overcurrent protection is required, the secondary overcurrent device shall be permitted to consist of not more than six circuit breakers or six sets of fuses grouped in one location. Where multiple overcurrent devices are utilized, the total of all the device ratings shall not exceed the allowed value of a single overcurrent device. If both breakers and fuses are utilized as the overcurrent device, the total of the device ratings shall not exceed that allowed for fuses.
3. A transformer equipped with coordinated thermal overload protection by the manufacturer and arranged to interrupt the primary current, shall be permitted to have primary overcurrent protection rated or set at a current value that is not more than six times the rated current of the transformer for transformers having not more than 6 percent impedance, and not more than four times the rated current of the transformer for transformers having more than 6 percent but not more than 10 percent impedance.

(© 2001, NFPA)

9.1.13 U.L. 1008 Minimum Withstand Test Requirement (for Automatic Transfer Switches)

TABLE 9.1.13

Automatic Transfer Switch Rating	U.L. Minimum Current Amps	U.L. Test Current Power Factor
100 Amps or less	5,000	40% to 50%
101-400 Amps	10,000	40% to 50%
401 Amps and greater	20 times rating but not less than 10,000 Amps	40% to 50% for current of 10,000 Amps. OR 25% to 30% for currents of 20,000 Amps or less. OR 20% or less for current greater than 20,000 Amps.

9.1.14 HVAC Equipment Short-Circuit Test Currents, Table 55.1 of U.L. Standard 1995

TABLE 9.1.14

Product Ratings, A				
	Single-Phase			Circuit Capacity,
110-120V	200-208V	220-240V	254-277V	A
9.8 or less	5.4 or less	4.9 or less	–	200
9.9-16.0	5.5-8.8	5.0-8.0	6.65 or less	1000
16.1-34.0	8.9-18.6	8.1-17.0	–	2000
34.1-80.0	18.7-44.0	17.1-40.0	–	3500
Over 80.0	Over 44.0	Over 40.0	Over 6.65	5000
	3-Phase			Circuit Capacity,
200-208V	220-240V	440-480V	550-600V	A
2.12 or less	2.0 or less	–	–	200
2.13-3.7	2.1-3.5	1.8 or less	1.4 or less	1000
3.8-9.5	3.6-9.0	–	–	2000
9.6-23.3	9.1-22.0	–	–	3500
Over 23.3	Over 22.0	Over 1.8	Over 1.4	5000

9.2.1 Protection through Current Limitation

Today, most electrical distribution systems are capable of delivering very high short-circuit currents, some in excess of 200,000 A. If the components are not capable of handling these short-circuit currents, they could be easily damaged or destroyed. The current-limiting ability of today's modern fuses and current-limiting breakers (with current-limiting fuses) allows components with low short-circuit withstand ratings to be specified despite high available fault currents.

The concept of current limitation is pointed out and analyzed in Figures 9.2.2 and 9.2.3, respectively, where the prospective available fault current is shown in conjunction with the limited current resulting when a current-limiting fuse clears. The area under the current curve indicates the amount of short-circuit energy being dissipated in the circuit. Since both magnetic forces and thermal energy are directly proportional to the square of the current, it is important to limit the short-circuit current to as small a value as possible. Magnetic forces vary as the square of the *peak* current, and thermal energy varies as the square of the *rms* current.

Thus the current-limiting fuse in this example would limit the let-thru energy to a fraction of the value that is available from the system. In the first major loop of the fault current, standard non-current-limiting electromechanical devices would let through approximately 100 times as much destructive energy as the fuse would let through.

9.2.2 Current-Limiting Effect of Fuses

9.2.2

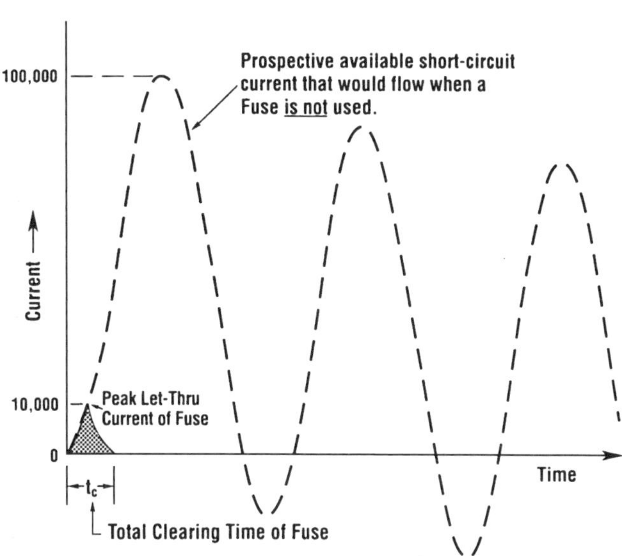

9.2.3 Analysis of a Current-Limiting Fuse

9.2.3

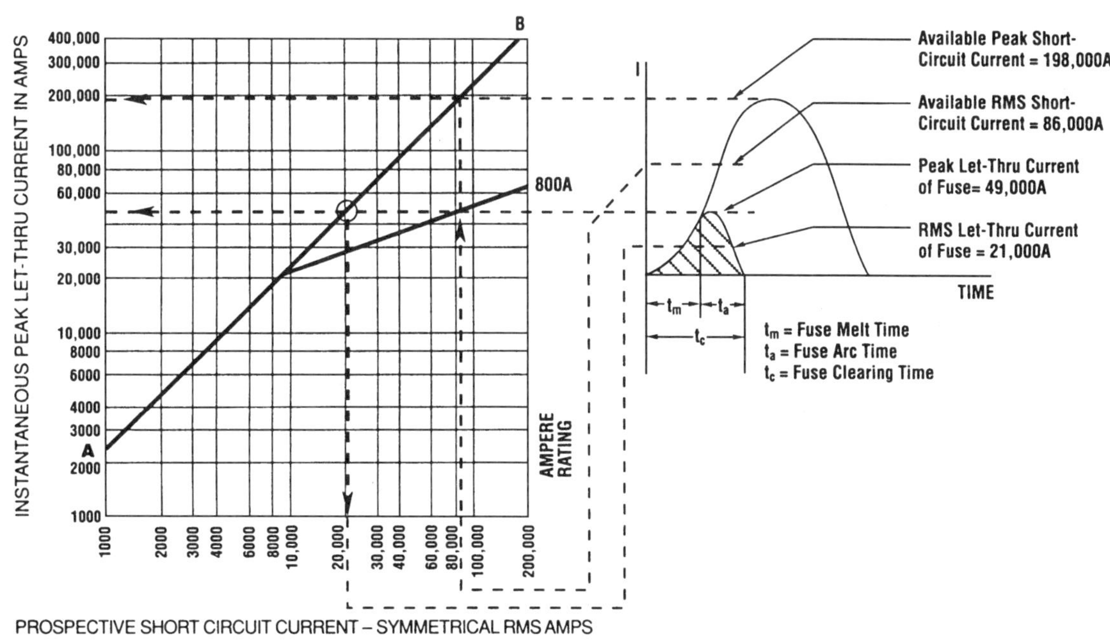

9.2.4 Let-Thru Data Pertinent to Equipment Withstand

Prior to using the fuse let-thru charts, it must be determined what let-thru data are pertinent to equipment withstand ratings.

Equipment withstand ratings can be described as, How much fault current can the equipment handle, and for how long? Based on standards currently available, the most important data that can be obtained from the fuse let-thru charts and their physical effects are the following:

- Peak let-thru current—mechanical forces
- Apparent prospective rms symmetrical let-thru current—heating effect

Figure 9.2.4 is a typical example showing the short-circuit current available to an 800-A circuit, an 800-A Bussmann Low-Peak current-limiting time-delay fuse, and the let-thru data of interest.

9.2.4

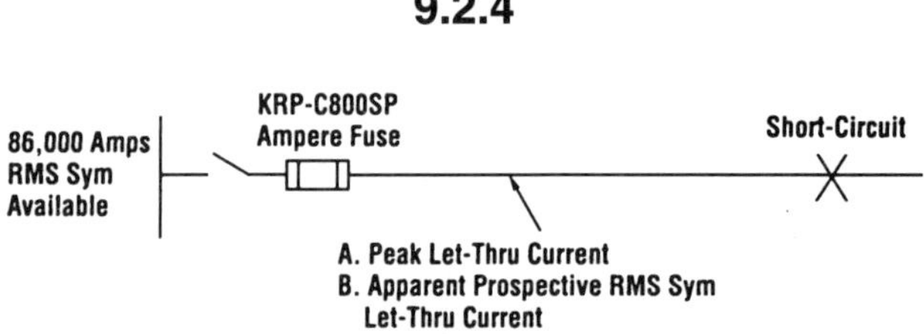

9.2.5 How to Use the Let-Thru Charts

Using the example given in Figure 9.2.4, one can determine the pertinent let-thru data for the Bussmann KRP-C800SP ampere Low-Peak fuse. The let-thru chart pertaining to the 800-A Low-Peak fuse is illustrated in Figure 9.2.6.

Determine the peak let-thru current

Step 1. Enter the chart on the prospective short-circuit current scale at 86,000 A, and proceed vertically until the 800-A fuse curve is intersected.

Step 2. Follow horizontally until the instantaneous peak let-thru current scale is intersected.

Step 3. Read the peak let-thru current as 49,000 A. (If a fuse had not been used, the peak current would have been 198,000 A.)

Determine the apparent prospective *rms* symmetrical let-thru current

Step 1. Enter the chart on the prospective short-circuit current scale at 86,000 A, and proceed vertically until the 800-A fuse curve is intersected.

Step 2. Follow horizontally until line *AB* is intersected.

Step 3. Proceed vertically down to the prospective short-circuit current.

Step 4. Read the apparent prospective rms symmetrical let-thru current as 21,000 A. (The rms symmetrical let-thru current would be 86,000 A if there were no fuse in the circuit.)

Refer to different fuse manufacturers' current-limitation characteristics for applications of different fuse types and sizes under various circuit conditions.

9.2.6 Current-Limitation Curves: Bussmann Low-Peak Time-Delay Fuse KRP-C800SP

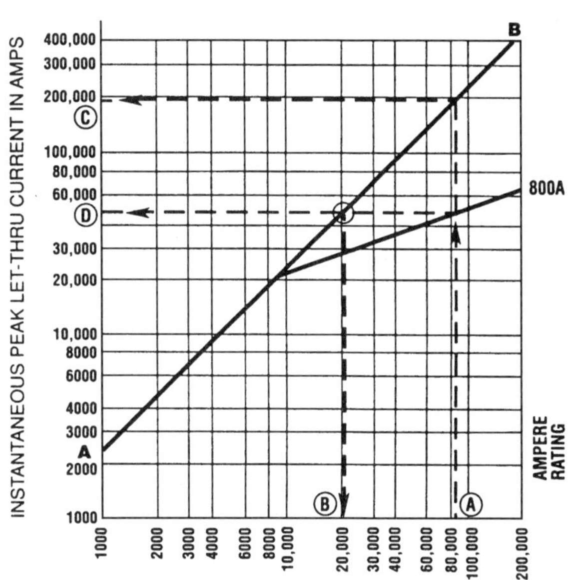

PROSPECTIVE SHORT CIRCUIT CURRENT – SYMMETRICAL RMS AMPS

Ⓐ I_{RMS} Available = 86,000 Amps

Ⓑ I_{RMS} Let-Thru = 21,000 Amps

Ⓒ I_p Available = 198,000 Amps

Ⓓ I_p Let-Thru = 49,000 Amps

Section

10

Motor Feeders and Starters

10.1.0 Introduction
10.1.1 Sizing Motor-Circuit Feeders and Their Overcurrent Protection
10.1.2 *NEC* Table 430.7(B), Locked-Rotor-Indicating Code Letters
10.1.3 Motor-Circuit Data Sheets
10.1.4 480-V System (460-V Motors) Three-Phase Motor-Circuit Feeders
10.1.5 208-V System (200-V Motors) Three-Phase Motor-Circuit Feeders
10.1.6 115-V Single-Phase Motor-Circuit Feeders
10.1.7 200-V Single-Phase Motor-Circuit Feeders
10.1.8 230-V Single-Phase Motor-Circuit Feeders
10.1.9 Motor Starter Characteristics (for Squirrel-Cage Motors)
10.1.10 Reduced-Voltage Starter Characteristics
10.1.11 Reduced-Voltage Starter Selection Table

10.1.0 Introduction

Motors comprise a significant portion of a building's electrical system loads. They are needed to power fans and pumps for basic mechanical building infrastructure such as heating, ventilation, air conditioning, plumbing, fire protection, elevators, and escalators. They are also needed to power equipment endemic to the occupancy, such as commercial kitchen equipment in an institutional facility, CT and MRI scanners in a hospital, and process equipment such as conveyors and machinery in an industrial plant or stone quarry. Consequently, designing motor-circuit feeders is very much in the mainstream of the electrical design professional's daily work. To save time in this process, the following information is provided.

10.1.1 Sizing Motor-Circuit Feeders and Their Overcurrent Protection

AC Single-Phase Motors, Polyphase Motors Other Than Wound-Rotor (Synchronous* and Induction Other Than Code E†)

1. Feeder wire size 125 percent of motor full-load current *minimum.*
2. Feeder breaker (thermal-magnetic fixed-trip type) 250 percent of full-load current *maximum.*
3. Feeder breaker (instantaneous magnetic-only type) 800 percent of full-load current *maximum.*

*Synchronous motors of the low-torque, low-speed type (usually 450 rpm or lower), such as those used to drive reciprocating compressors, pumps, etc., that start unloaded, do not require a fuse rating or circuit breaker setting in excess of 200 percent of full-load current.

†For code E induction motors, everything is the same as above except that if an instantaneous magnetic-only type circuit breaker is used, it shall have a maximum setting of 1100 percent.

4. Feeder fuse (dual-element time-delay type) 175 percent of full-load current *maximum.*
5. Feeder fuse (*NEC* non-time-delay type) 300 percent of full-load current *maximum.*

For Wound-Rotor Motors

1. Feeder wire size 125 percent of motor full-load current *minimum.*
2. Feeder breaker (thermal-magnetic fixed-trip type) 150 percent of full-load current *maximum.*
3. Feeder breaker (instantaneous magnetic-only type) 800 percent of full-load current *maximum.*
4. Feeder fuse (dual-element time-delay type) 150 percent of full-load current *maximum.*
5. Feeder fuse (*NEC* non-time-delay type) 150 percent of full-load current *maximum.*

For Hermetic Motors (Special Case). Hermetic motors are actually a combination consisting of a compressor and motor, both of which are enclosed in the same housing, with no external shaft or shaft seals, the motor operating in the refrigerant; thus their characteristics are different from those of standard induction motors. Calculating their feeder size and overcurrent protection is based on their nameplate branch-circuit selection current (BCSC) or their rated-load current (RLC), whichever is greater. The BCSC is always equal to or greater than the RLC. Hence, the following:

1. Feeder wire size 125 percent of BCSC/RLC *maximum.*
2. Feeder breaker (thermal-magnetic fixed-trip type) 175 to 225 percent of BCSC/RLC *maximum.*
3. Feeder breaker (instantaneous magnetic-only type) 800 percent of BCSC/RLC *maximum.*
4. Feeder fuse (dual-element time-delay type) 175 to 225 percent of BCSC/RLC *maximum.*
5. Feeder fuse (*NEC* non-time-delay type) *not recommended—Do not use.*

DC (Constant-Voltage) Motors

1. Feeder wire size 125 percent of motor full-load current *maximum.*
2. Feeder breaker (thermal-magnetic fixed-trip type) 150 percent of full-load current *maximum.*
3. Feeder breaker (instantaneous magnetic-only type) 250 percent of full-load current *maximum.*
4. Feeder fuse (dual-element time-delay type) 150 percent of full-load current *maximum.*
5. Feeder fuse (*NEC* non-time-delay type) 150 percent of full-load current *maximum.*

For Multiple Motors on One Feeder. First, size feeder and overcurrent protection for largest motor and add the full-load current of the remaining motors to size the overall feeder and overcurrent protection.

Application Tips

1. Refer to *NEC* Articles 430 and 440 for further details on sizing motor feeders and overcurrent protection.

2. For elevator motors, always try to get the full-load current because the nameplate horsepower on many machines is about 10 to 25 percent below the actual rating.
3. For packaged-type evaporative condensers with many small fans nominally rated 1 hp (for example), be sure to get the full-load current because these are really equivalent to about 2 hp (for example) each, and feeders sized on nominal horsepower ratings will be inadequate. Remember to size the feeder and overcurrent protection as a multiple motor load. Also refer to *NEC* Article 440.
4. Note that *maximum* and *minimum* have precise meanings and that feeder sizes shall not be less than the calculated minimum within 3 or 4 percent (e.g., 30-A-rated no. 10 wire is OK for a 31-A load) and breaker sizes shall not be more than the maximum indicated. In general, for larger motor sizes, the overcurrent protection needed decreases considerably from the maximum limit.
5. In sizing nonfused disconnects for motors, use the horsepower rating table in the manufacturer's catalog or realize that, in general, a nonfused disconnect switch should be rated the same as a switch fused with a dual-element time-delay fuse.
6. When sizing feeders for tape drives in mainframe data centers, it is usually necessary to oversize both the overcurrent protection and the feeder to accommodate the long acceleration time characteristic of this equipment.
7. Today's highly energy-efficient motors are characterized by low losses and high inrush currents, thus requiring overcurrent protection sized at or near the maximum limit prescribed by the *NEC* when these motors are used.
8. For *NEC* locked-rotor-indicating code letters, refer to Table 10.1.2 [*NEC* Table 430.7(B)].

10.1.2 *NEC* Table 430.7(B), Locked-Rotor-Indicating Code Letters

TABLE 10.1.2

Code Letter	Kilovolt-Amperes per Horsepower with Locked Rotor	Code Letter	Kilovolt-Amperes per Horsepower with Locked Rotor
A	0–3.14	L	9.0–9.99
B	3.15–3.54	M	10.0–11.19
C	3.55–3.99	N	11.2–12.49
D	4.0–4.49	P	12.5–13.99
E	4.5–4.99	R	14.0–15.99
F	5.0–5.59	S	16.0–17.99
G	5.6–6.29	T	18.0–19.99
H	6.3–7.09	U	29.0–22.39
J	7.1–7.99	V	22.4 and up
K	8.0–8.99		

10.1.3 Motor-Circuit Data Sheets

The following motor-circuit data sheets provide recommended design standards for branch-circuit protection and wiring of squirrel-cage induction motors of the sizes and voltages most frequently encountered in commercial, institutional, and industrial facilities. Experience has shown that most facilities of this type use copper wire and use no. 12 AWG wire and ¾-in conduit as minimum sizes for power distribution. These standards are reflected in the tables that follow. Refer also to the notes to these tables for assumptions and other criteria used.

10.1.4 480-V System (460-V Motors) Three-Phase Motor-Circuit Feeders

TABLE 10.1.4

HP	FLA	Safety Switch	Fuse	C/B	Wire	Ground (fuse)	Ground (C/B)	NEMA Starter Size
½	1.1	30	1.4	15	3 #12	#12	#12	00
3/4	1.6	30	2	15	3 #12	#12	#12	00
1	2.1	30	2.5	15	3 #12	#12	#12	00
1 ½	3.0	30	3.5	15	3 #12	#12	#12	00
2	3.4	30	4	15	3 #12	#12	#12	00
3	4.8	30	5.6	15	3 #12	#12	#12	0
5	7.6	30	9	20	3 #12	#12	#12	0
7 ½	11	30	15	30	3 #10	#12	#10	1
10	14	30	17.5	30	3 #10	#12	#10	1
15	21	30	25	50	3 #10	#10	#10	2
20	27	60	35	60	3 #10	#10	#10	2
25	34	60	40	70	3 #8	#10	#8	2
30	40	60	50	90	3 #8	#10	#8	3
40	52	100	60	125	3 #6	#10	#6	3
50	65	100	75	150	3 #4	#8	#6	3
60	77	100	90	175	3 #3	#8	#6	4
75	96	200	125	200	3 #1	#6	#6	4
100	124	200	150	250	3 #2/0	#6	#4	4
125	156	200	200	300	3 #3/0	#6	#4	5
150	180	400	225	350	3 #4/0	#4	#3	5
200	240	400	300	400	3-350	#4	#3	5
250	302	400	350	600	3-500	#3	#1	6
300	361	600	450	800	2 sets [3 #4/0]	#2	#1/0	6
350	414	600	500	800	2 sets [3-300]	#2	#1/0	--
400	477	800	600	1000	2 sets [3-350]	#1	#2/0	--
450	515	800	600	1000	2 sets [3-400]	#1	#2/0	--

Notes for Tables 10.1.4 through 10.1.8:

1. This table assumes that motor overload protection per 430 Part III is provided separately as part of the motor starter.
2. Horse powers are for atypical induction type squirrel cage motor and are taken from Table 430.150. Sizing characteristics are based on a FVNR, Design B, Code letter G motor. Multi speed, low power factor, high torque, high locked rotor or design E motors may have different FLA values and sizing requirements.
3. FLA (full load amperes) from Table 430.150.
4. Fuses and circuit breakers are sized for back up overload and short circuit protection per 430 Part IV.
5. Fuses are dual element time delay RK-1 type sized at a minimum 115% FLA based on service factors less than 1.15 or temperature greater than 40 degrees C.
6. Circuit breakers are general purpose thermal magnetic (inverse time) type sized at approximately 200% FLA.
7. Part 430.52 and Table 430.152 establish maximum sizes.
8. Wire sizes are based on 75 degrees C copper with THWN, THHN or XHHW insulation. Size may need to be increased because of voltage drop.
9. Maximum conduit fill depends on the wire insulation type and the type of conduit. These values have been calculated for conductors of the same size and are given in tabular form in Annex C of NFPA 70 [EMT - Table C1; IMC - Table C4; RMC - Table C8; Schedule 40 rigid PVC - Table C10; Schedule 80 rigid PVC - Table C9].

10.1.5 208-V System (200-V Motors) Three-Phase Motor-Circuit Feeders

TABLE 10.1.5

HP	FLA	Safety Switch	Fuse	C/B	Wire	Ground (fuse)	Ground (C/B)	NEMA Starter Size
½	2.5	30	3	15	3 #12	#12	#12	00
3/4	3.7	30	4.5	15	3 #12	#12	#12	00
1	4.8	30	5.6	15	3 #12	#12	#12	00
1 ½	6.9	30	8	20	3 #12	#12	#12	00
2	7.8	30	9	20	3 #12	#12	#12	0
3	11.0	30	15	25	3 #10	#12	#10	0
5	17.5	30	25	40	3 #10	#10	#10	1
7 ½	25.3	60	30	50	3 #10	#10	#10	1
10	32.2	60	40	70	3 #8	#10	#8	2
15	48.3	100	60	100	3 #6	#10	#8	3
20	62.1	100	75	150	3 #4	#8	#6	3
25	78.2	100	90	175	3 #3	#8	#6	3
30	92	200	110	200	3 #2	#6	#6	4
40	120	200	150	250	3 #1/0	#6	#4	4
50	150	200	175	300	3 #3/0	#6	#4	5
60	177	400	225	400	3 #4/0	#4	#3	5
75	221	400	300	400	3 #300	#4	#3	5
100	285	400	350	500	3 #500	#3	#2	6
125	359	600	450	600	2 sets [3 #4/0]	#2	#1	6
150	414	600	500	600	2 sets [3-300]	#2	#1	6

10.1.6 115-V Single-Phase Motor-Circuit Feeders

TABLE 10.1.6

HP	FLA	Safety Switch	Fuse	C/B	Wire	Ground (fuse)	Ground (C/B)	NEMA Starter Size
1/6	4.4	30	5.6/10	15	2 #12	#12	#12	00
1/4	5.8	30	7	15	2 #12	#12	#12	00
1/3	7.2	30	9	20	2 #12	#12	#12	0
1/2	9.8	30	12	20	2 #12	#12	#12	0
3/4	13.8	30	17.5	30	2 #10	#12	#10	0
1	16	30	20	40	2 #10	#12	#10	0
1-1/2	20	30	25	50	2 #12	#10	#10	1
2	24	30	30	60	2 #10	#10	#10	1

10.1.7 200-V Single-Phase Motor-Circuit Feeders

TABLE 10.1.7

HP	FLA	Safety Switch	Fuse	C/B	Wire	Ground (fuse)	Ground (C/B)	NEMA Starter Size
1/6	2.5	30	3	15	2 #12	#12	#12	00
1/4	3.3	30	4	15	2 #12	#12	#12	00
1/3	4.1	30	5	15	2 #12	#12	#12	00
1/2	5.6	30	7	15	2 #12	#12	#12	00
3/4	7.0	30	10	20	2 #12	#12	#12	00
1	9.2	30	12	20	2 #12	#12	#12	00
1-1/2	11.5	30	15	30	2 #10	#12	#10	0
2	13.8	30	17.5	30	2 #10	#12	#10	0
3	19.6	30	25	40	2 #10	#10	#10	1
5	32.2	60	40	70	2 #8	#10	#8	2
7-1/2	46	100	60	100	2 #6	#10	#8	2
10	57.5	100	70	125	2 #4	#8	#6	3

10.1.8 230-V Single-Phase Motor-Circuit Feeders

TABLE 10.1.8

HP	FLA	Safety Switch	Fuse	C/B	Wire	Ground (fuse)	Ground (C/B)	NEMA Starter Size
1/6	2.2	30	2.8	15	2 #12	#12	#12	00
1/4	2.9	30	3.5	15	2 #12	#12	#12	00
1/0	3.6	30	4.5	15	2 #12	#12	#12	00
1/2	4.9	30	6	15	2 #12	#12	#12	00
3/4	6.9	30	8	15	2 #12	#12	#12	00
1	8	30	10	20	2 #12	#12	#12	00
1-1/2	10	30	12	20	2 #12	#12	#12	0
2	12	30	15	30	2 #10	#12	#10	0
3	17	30	20	40	2 #10	#12	#10	1
5	28	60	35	60	2 #10	#10	#10	2
7-1/2	40	60	50	80	2 #8	#10	#8	2
10	50	100	60	100	2 #6	#10	#8	3

10.1.9 Motor Starter Characteristics (for Squirrel-Cage Motors)

There are fundamentally two types of motor starters, full voltage (both reversing and nonreversing) and reduced voltage. In the information that follows, their characteristics and selection criteria are briefly summarized.

Full-voltage starters. A squirrel-cage motor draws high starting current (inrush) and produces high starting torque when started at full voltage. While these values differ for different motor designs, for a typical NEMA design B motor, the inrush will be approximately 600 percent of the motor full-load amperage (FLA) rating, and the starting torque will be approximately 150 percent of full-load torque at full voltage. High current inrush and starting torque can cause problems in the electrical and mechanical systems and may even cause damage to utilization equipment or materials being processed.

Reduced-voltage starters. When a motor is started at reduced voltage, the current at the motor terminals is reduced in direct proportion to the voltage reduction, whereas the torque is reduced by the square of the voltage reduction. If the "typical" NEMA design B motor is started at 70 percent of line voltage, the starting current would be 70 percent of the full-voltage value (i.e., $0.70 \times 600\% = 420\%$ FLA). The torque would then be 0.70^2, or 49%, of the normal starting torque (i.e., $0.49 \times 150\% = 74\%$ full-load torque). Therefore, reduced-voltage starting provides an effective means of both reducing inrush current and starting torque.

If the motor has a high inertia, or if the motor rating is marginal for the applied load, reducing the starting torque may prevent the motor from reaching full speed before the thermal overloads trip. Applications that require high starting torque should be reviewed carefully to determine if reduced-voltage starting is suitable. As a rule, motors with a horsepower rating in excess of 15 percent of the kVA rating of the transformer feeding it should use reduced-voltage start.

There are several types of electromechanical as well as solid-state reduced-voltage starters that provide different starting characteristics. The following tables from Square D Company are a good representation of industry standard characteristics. Table 10.1.10 shows the starting characteristics for Square D's class 8600 series of reduced-voltage starters compared with full-voltage starting along with the advantages and disadvantages of each type. Table 10.1.11 provides an aid in the selection of the starter best suited for a particular application and desired starting characteristic.

10.1.10 Reduced-Voltage Starter Characteristics

TABLE 10.1.10

Characteristic	Full Voltage	Autotransformer Class 8606	Wye-Delta Class 8630	Part Winding Class 8640	Primary Resistance Class 8647	Solid State ATS23
Voltage at Motor	100%	50% / 65% / 80% (tap setting)	100%	100%	70%	Ramped Up
Line Current (% Full Load Current)	600%	150% / 250% / 380%	200%	390%	420%	200% to 500% (potentiometer adjustment)
Starting Torque (% Rated Torque)	150%	40% / 60% / 100%	50%	70%	75%	10% to 105% (function of i & V)
Start Time (Factory Setting)		6 - 7 sec	10 sec / 15 sec (open / closed transition)	1 - 1.5 sec	4 - 5 sec	10 sec (adjustable 5 to 30 sec)
Advantages	- Simple - Economical - High Starting Torque	- High torque/amp - High inertial loads - Flexibility	- High inertial loads - Long acceleration loads - Good torque/amp	- Simple - Small size	- Smooth acceleration Motor voltage increases with speed	- Greatest flexibility - Smooth ramp - Solid state O/L - Diagnostics
Disadvantages	- Abrupt starts - Large current inrush	- Large size	- Low torque - No flexibility	- Not suitable for: High inertial loads Frequent starting	- Low current limitation - Heat dissipation - Short start time	- SCR heat dissipation - Ambient limitations - Sensitive to power quality
Motor	Standard	Standard	Special	Special	Standard	Standard

(Courtesy of Square D / Schneider Electric)

10.1.11 Reduced-Voltage Starter Selection Table

TABLE 10.1.11

	Need		
Application	Smooth Acceleration	Minimum Line Current	**Comments**
High Inertial Loading	1. Solid State 2. Autotransformer 3. Primary Resistor 4. Wye Delta 5. Part Winding	1. Autotransformer 2. Solid State 3. Wye-Delta 4. Part Winding 5. Primary Resistor	
Long Acceleration Time	1. Solid State 2. Wye-Delta 3. Autotransformer 4. Primary Resistor	1. Solid State 2. Wye-Delta 3. Autotransformer 4. Primary Resistor	• For acceleration times greater than 5 sec primary resistor requires non-std resistors • Part winding not suitable for acceleration time greater than 2 seconds
Frequent Starting	1. Solid State 2. Wye-Delta 3. Primary Resistor 4. Autotransformer	1. Solid State 2. Wye-Delta 3. Primary Resistor 4. Autotransformer	• Part winding is unsuitable for frequent starts
Flexibility in Selecting Starter Characteristics	1. Solid State 2. Autotransformer 3. Primary Resistor 4. Part Winding	1. Solid State 2. Autotransformer 3. Primary Resistor 4. Part Winding	• For primary resistor, resistor change required to change starting characteristics • Starting characteristics cannot be changed for Wye-Delta starters

(Courtesy of Square D / Schneider Electric)

Section

11

Standard Voltages and Voltage Drop

11.1.0 Introduction
11.1.1 System Voltage Classes
11.1.2 Standard Nominal System Voltages in the United States
11.1.3 Standard Nominal System Voltages and Voltage Ranges
11.1.4 Principal Transformer Connections to Supply the System Voltages of Table 11.1.3
11.1.5 Application of Voltage Classes
11.1.6 Voltage Systems Outside the United States
11.1.7 System Voltage Tolerance Limits
11.1.8 Standard Voltage Profile for a Regulated Power-Distribution System, 120-V Base
11.1.9 Voltage Profile of the Limits of Range A, ANSI C84.1-1989
11.1.10 Voltage Ratings of Standard Motors
11.1.11 General Effect of Voltage Variations on Induction Motor Characteristics
11.1.12 Voltage-Drop Calculations
11.1.13 Voltage-Drop Tables
11.1.14 Voltage Drop for AL Conductor, Direct Current
11.1.15 Voltage Drop for AL Conductor in Magnetic Conduit, 70 Percent PF
11.1.16 Voltage Drop for AL Conductor in Magnetic Conduit, 80 Percent PF
11.1.17 Voltage Drop for AL Conductor in Magnetic Conduit, 90 Percent PF
11.1.18 Voltage Drop for AL Conductor in Magnetic Conduit, 95 Percent PF
11.1.19 Voltage Drop for AL Conductor in Magnetic Conduit, 100 Percent PF
11.1.20 Voltage Drop for AL Conductor in Nonmagnetic Conduit, 70 Percent PF
11.1.21 Voltage Drop for AL Conductor in Nonmagnetic Conduit, 80 Percent PF
11.1.22 Voltage Drop for AL Conductor in Nonmagnetic Conduit, 90 Percent PF
11.1 23 Voltage Drop for AL Conductor in Nonmagnetic Conduit, 95 Percent PF
11.1.24 Voltage Drop for AL Conductor in Nonmagnetic Conduit, 100 Percent PF
11.1.25 Voltage Drop for CU Conductor, Direct Current
11.1.26 Voltage Drop for CU Conductor in Magnetic Conduit, 70 Percent PF
11.1.27 Voltage Drop for CU Conductor in Magnetic Conduit, 80 Percent PF
11.1.28 Voltage Drop for CU Conductor in Magnetic Conduit, 90 Percent PF
11.1.29 Voltage Drop for CU Conductor in Magnetic Conduit, 95 Percent PF
11.1.30 Voltage Drop for CU Conductor in Magnetic Conduit, 100 Percent PF
11.1.31 Voltage Drop for CU Conductor in Nonmagnetic Conduit, 70 Percent PF
11.1.32 Voltage Drop for CU Conductor in Nonmagnetic Conduit, 80 Percent PF
11.1.33 Voltage Drop for CU Conductor in Nonmagnetic Conduit, 90 Percent PF
11.1.34 Voltage Drop for CU Conductor in Nonmagnetic Conduit, 95 Percent PF
11.1.35 Voltage Drop for CU Conductor in Nonmagnetic Conduit, 100 Percent PF
11.1.36 Voltage-Drop Curves for Typical Interleaved Construction of Copper Busway at Rated Load

11.1.37 Voltage-Drop Values for Three-Phase Busways with Copper Bus Bars, in Volts per 100 ft, Line-to-Line, at Rated Current with Balanced Entire Load at End
11.1.38 Voltage-Drop Values for Three-Phase Busways with Aluminum Bus Bars, in Volts per 100 ft, Line-to-Line, at Rated Current with Balanced Entire Load at End
11.1.39 Voltage-Drop Curves for Typical Plug-In-Type CU Busway at Balanced Rated Load
11.1.40 Voltage-Drop Curves for Typical CU Feeder Busways at Balanced Rated Load Mounted Flat Horizontally
11.1.41 Voltage-Drop Curve versus Power Factor for Typical Light-Duty Trolley Busway Carrying Rated Load
11.1.42 Voltage-Drop Curves for Three-Phase Transformers, 225 to 10,000 kVA, 5 to 25 kV
11.1.43 Application Tips
11.1.44 Flicker of Incandescent Lamps Caused by Recurrent Voltage Dips
11.1.45 Effect of Voltage Variations on Incandescent Lamps
11.1.46 General Effect of Voltage Variations on Induction Motor Characteristics
11.1.47 Calculation of Voltage Dips (Momentary Voltage Variations)

11.1.0 Introduction

An understanding of system voltage nomenclature and preferred voltage ratings of distribution apparatus and utilization equipment is essential to ensure the proper design and operation of a power-distribution system. The dynamic characteristics of the system should be recognized and the proper principles of voltage regulation applied so that satisfactory voltages will be supplied to utilization equipment under all normal conditions of operation.

11.1.1 System Voltage Classes

- *Low voltage.* A class of nominal system voltages 1000 V or less.
- *Medium voltage.* A class of nominal system voltages greater than 1000 V but less than 100,000 V.
- *High voltage.* A class of nominal system voltages equal to or greater than 100,000 V and equal to or less than 230,000 V.

11.1.2 Standard Nominal System Voltages in the United States

These voltages and their associated tolerance limits are listed in ANSI C84.1-1989 for voltages from 120 to 230,000 V and ANSI C92.2-1987 for alternating current (ac) electrical power systems and equipment operating at voltages above 230 kV nominal. The nominal system voltages and their associated tolerance limits and notes in the two standards have been combined in Table 11.1.3 to provide a single table listing all the nominal system voltages and their associated tolerance limits for the United States. Preferred nominal system voltages and voltage ranges are shown in boldface type, whereas other systems in substantial use that are recognized as standard voltages are shown in medium type. Other voltages may be encountered in older systems, but they are not recognized as standard voltages. The transformer connections from which these voltages are derived are shown in Figure 11.1.4.

11.1.3 Standard Nominal System Voltages and Voltage Ranges

TABLE 11.1.3 Standard Nominal System Voltages and Voltage Ranges (Preferred System Voltages in Boldface Type)

VOLTAGE CLASS	NOMINAL SYSTEM VOLTAGE (Note a)			Nominal Utilization Voltage (Note j)	VOLTAGE RANGE A (Note b)			VOLTAGE RANGE B (Note b)		
					Maximum	Minimum		Maximum	Minimum	
	Two-wire	Three-wire	Four-wire	Two-wire Three-wire Four-wire	Utilization and Service Voltage (Note c)	Service Voltage	Utilization Voltage	Utilization and Service Voltage	Service Voltage	Utilization Voltage
	Single-Phase Systems									
Low Voltage (Note 1)	120			**115**	126	114	110	127	110	106
		120/240		**115/230**	**126/252**	**114/228**	**110/220**	**127/254**	**110/220**	**106/212**
	Three-Phase Systems									
			208Y/120 (Note d)	**200**	**218Y/126**	**197Y/114**	**191Y/110**	**220Y/127**	**191Y/110** (Note 2)	**184Y/106** (Note 2)
			240/120	**230/115**	**252/126**	**228/114**	**220/110**	**254/127**	**220/110**	**212/106**
		240		**230**	252	228	220	254	220	212
			480Y/277	**460**	**504Y/291**	**456Y/263**	**440Y/254**	**508Y/293**	**440Y/254**	**424Y/245**
		480		**460**	**504**	**456**	**440**	**508**	**440**	**424**
		600 (Note e)		**575**	630 (Note e)	570	550	635 (Note e)	550	530
Medium Voltage		2 400			2 520	2 340	2 160	2 540	2 280	2 080
			4 160Y/2 400		4 370/2 520	4 050Y/2 340	3 740Y/2 160	4 400Y/2 540	3 950Y/2 280	3 600/2 080
		4 160			**4 370**	**4 050**	**3 740**	**4 400**	**3 950**	**3 600**
		4 800			5 040	4 680	4 320	5 080	4 560	4 160
		6 900			7 240	6 730	6 210	7 260	6 560	5 940
			8 320Y/4 800		8 730Y/5 040	8 110Y/4 680	(Note f)	8 800Y/5 080	7 900Y/4 560	(Note f)
			12 000Y/6 930		12 600Y/7 270	11 700Y/6 760		12 700Y/7 330	11 400Y/6 580	
			12 470Y/7 200		**13 090Y/7 560**	**12 160Y/7 020**		**13 200Y/7 620**	**11 850Y/6 840**	
			13 200Y/7 620		**13 860Y/8 000**	**12 870Y/7 430**		**13 970Y/8 070**	**12 504Y/7 240**	
			13 800Y/7 970		14 490Y/8 370	13 460Y/7 770		14 520Y/8 380	13 110Y/7 570	
		13 800			**14 490**	**13 460**	**12 420**	**14 520**	**13 110**	**11 880**
			20 780Y/12 000		21 820Y/12 600	20 260Y/11 700	(Note f)	22 000Y/12 700	19 740Y/11 400	(Note f)
			22 860Y/13 200		24 000Y/13 860	22 290Y/12 870		24 200Y/13 970	21 720Y/12 540	
		23 000			24 150	22 430		24 340	21 850	
			24 940Y/14 400		**26 190Y/15 120**	**24 320Y/14 040**		**26 400Y/15 240**	**23 690Y/13 680**	
			34 500Y/19 920		**36 230Y/20 920**	**33 640Y/19 420**		**36 510Y/21 080**	**32 780Y/18 930**	
		34 500			36 230	33 640		36 510	32 780	
					Maximum Voltage (Note g)					
		46 000			48 300					
		69 000			**72 500**					
High Voltage		**115 000**			**121 000**					
		138 000			**145 000**					
		161 000			169 000					
		230 000			**242 000**					
	(Note h)									
Extra-High Voltage		**345 000**			**362 000**					
		500 000			**550 000**					
		765 000			**800 000**					
Ultra-High Voltage		**1 100 000**			**1 200 000**					

NOTES: (1) Minimum utilization voltages for 120-600 volt circuits not supplying lighting loads are as follows

	Nominal System Voltage	Range A	Range B
	120	108	104
	120/240	**108/216**	**104/208**
(Note 2)	**208Y/120**	**187Y/108**	**180Y/104**
	240/120	**216/108**	**208/104**
	240	216	208
	480Y/277	**432Y/249**	**416Y/240**
	480	**432**	**416**
	600	540	520

(2) Many 220 volt motors were applied on existing 208 volt systems on the assumption that the utilization voltage would not be less than 187 volts. Caution should be exercised in applying the Range B minimum voltages of Table 17 and Note (1) to existing 208 volt systems supplying such motors

(continued)

TABLE 11.1.3 Standard Nominal System Voltages and Voltage Ranges (Preferred System Voltages in Boldface Type)

NOTES FOR TABLE 11.1.3:

(a) Three-phase, three-wire systems are systems in which only the three-phase conductors are carried out from the source for connection of loads. The source may be derived from any type of three-phase transformer connection, grounded or ungrounded. Three-phase, four-wire systems are systems in which a grounded neutral conductor is also carried out from the source for connection of loads. Four-wire systems in this table are designated by the phase-to-phase voltage, followed by the letter Y (except for the 240/120 V delta system), a slant line, and the phase-to-neutral voltage. Single-phase services and loads may be supplied from either single-phase or three-phase systems. The principal transformer connections that are used to supply single-phase and three-phase systems are illustrated in Fig 3.

(b) The voltage ranges in this table are illustrated in ANSI C84.1-1989, Appendix B [2].

(c) For 120-600 V nominal systems, voltages in this column are maximum service voltages. Maximum utilization voltages would not be expected to exceed 125 V for the nominal system voltage of 120, nor appropriate multiples thereof for other nominal system voltages through 600 V.

(d) A modification of this three-phase, four-wire system is available as a 120/208Y-volt service for single-phase, three-wire, open-wye applications.

(e) Certain kinds of control and protective equipment presently available have a maximum voltage limit of 600 V; the manufacturer or power supplier or both should be consulted to assure proper application.

(f) Utilization equipment does not generally operate directly at these voltages. For equipment supplied through transformers, refer to limits for nominal system voltage of transformer output.

(g) For these systems, Range A and Range B limits are not shown because, where they are used as service voltages, the operating voltage level on the user's system is normally adjusted by means of voltage regulation to suit their requirements.

(h) Standard voltages are reprinted from ANSI C92.2-1987 [3] for convenience only.

(i) Nominal utilization voltages are for low-voltage motors and control. See ANSI C84.1-1989, Appendix C [2] for other equipment nominal utilization voltages (or equipment nameplate voltage ratings).

11.1.4 Principal Transformer Connections to Supply the System Voltages of Table 11.1.3 (*see page 11.5*)

11.1.5 Application of Voltage Classes

1. Low-voltage-class voltages are used to supply utilization equipment.
2. Medium-voltage-class voltages are used as primary distribution voltages to supply distribution transformers that step the medium voltage down to a low voltage to supply utilization equipment. Medium voltages of 13,800 V and below are also used to supply utilization equipment, such as large motors.
3. High-voltage-class voltages are used to transmit large amounts of electric power over transmission lines that interconnect transmission substations.

11.1.6 Voltage Systems Outside the United States

Voltage systems in other countries (including Canada) generally differ from those in the United States. Also, the frequency in many countries is 50 Hz instead of 60 Hz, which affects the operation of some equipment, such as motors, which will run approximately 17 percent slower. Plugs and receptacles are generally different, which helps to prevent utilization equipment from the United States from being connected to the wrong voltage.

11.1.4

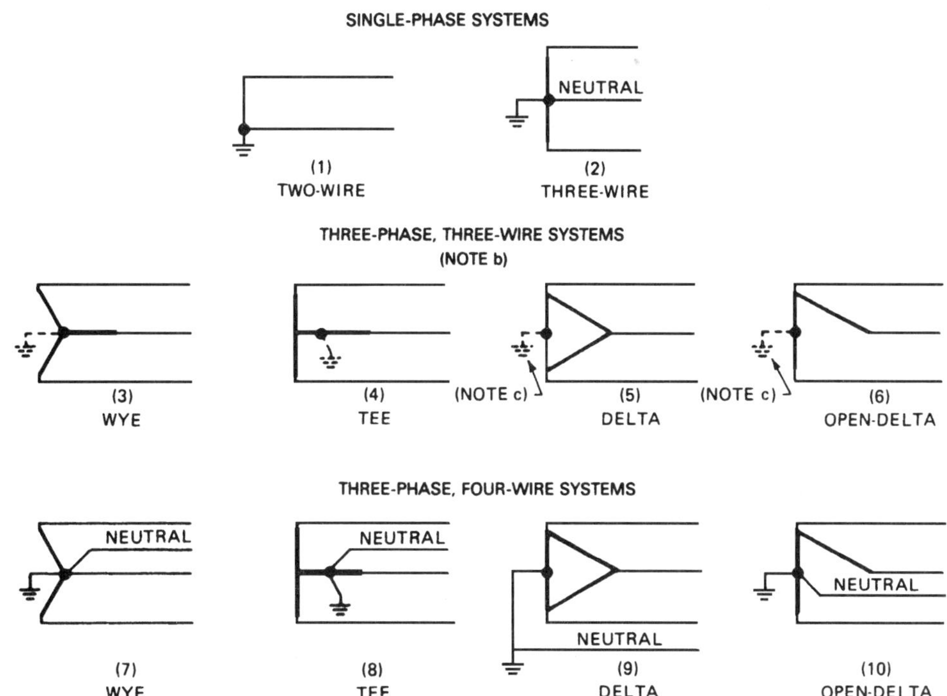

NOTES: (a) The above diagrams show connections of transformer secondary windings to supply the nominal system voltages of Table 11.1.3. Systems of more than 600 V are normally three-phase and supplied by connections (3), (5) ungrounded, or (7). Systems of 120-600 V may be either single-phase or three-phase, and all the connections shown are used to some extent for some systems in this voltage range.

(b) Three-phase, three-wire systems may be solidly grounded, impedance grounded, or ungrounded, but are not intended to supply loads connected phase-to-neutral (as are the four-wire systems).

(c) In connections (5) and (6), the ground may be connected to the midpoint of one winding as shown (if available), to one phase conductor (*corner* grounded), or omitted entirely (ungrounded).

(d) Single-phase services and loads may be supplied from single-phase or three-phase systems. They are connected phase-to-phase when supplied from three-phase, three-wire systems and either phase-to-phase or phase-to-neutral from three-phase, four-wire systems.

In general, equipment rated for use in the United States cannot be used outside the United States, and vice versa. If electrical equipment made for use in the United States must be used outside the United States, or vice versa, information on the voltage, frequency, and type of plug required should be obtained. If the difference is only in the voltage, transformers generally are available to convert the supply voltage to the equipment voltage.

11.1.7 System Voltage Tolerance Limits

Table 11.1.8 lists two voltage ranges in order to provide a practical application of voltage tolerance limits to distribution systems.

Electric supply systems are to be designed and operated so that most service voltages fall within the range A limits. User systems are to be designed and operated so that when the service voltages are within range A, the utilization voltages are within range A. Utilization equipment is to be designed and rated to give fully satisfactory performance within range A limits for utilization voltages.

Range B is provided to allow limited excursions of voltage outside the range A limits that necessarily result from practical design and operating conditions. The supplying utility is expected to take action within a reasonable time to restore service voltages to range A limits. The user is expected to take action within a reasonable time to restore utilization voltages to range A limits. Insofar as practical, utilization equipment may be expected to give acceptable performance outside range A but within range B. When voltages occur outside the limits of range B, prompt corrective action should be taken.

The voltage tolerance limits in ANSI C84.1-1989 are based on ANSI/NEMA MG1-1978, "Motors and Generators," which establishes the voltage tolerance limits of the standard low-voltage induction motor at ±10 percent of nameplate voltage ratings of 230 and 460 V. Since motors represent the major component of utilization equipment, they were given primary consideration in the establishment of this voltage standard.

The best way to show the voltages in a distribution system is by using a 120-V base. This cancels the transformation ratios between systems so that the actual voltages vary solely on the basis of voltage drops in the system. Any voltage may be converted to a 120-V base by dividing the actual voltage by the ratio of transformation to the 120-V base. For example, the ratio of transformation for a 480-V system is 480/120, or 4, so 460 V in a 480-V system would be 460/4, or 115 V.

The tolerance limits of the 460-V motor as they relate to the 120-V base become 115 V + 10 percent, or 126.5 V, and 115 V − 10 percent, or 103.5 V. The problem is to decide how this tolerance range of 23 V should be divided between the primary distribution system, the distribution transformer, and the secondary distribution system that make up the regulated distribution system. The solution adopted by American National Standards Committee C84 is shown in Table 11.1.8.

11.1.8 Standard Voltage Profile for a Regulated Power-Distribution System, 120-V Base

TABLE 11.1.8

	Range A	Range B
Maximum allowable voltage	126(125*)	127
Voltage-drop allowance for the primary distribution feeder	9	13
Minimum primary service voltage	117	114
Voltage-drop allowance for the distribution transformer	3	4
Minimum low-voltage service voltage	114	110
Voltage-drop allowance for the building wiring	6(4†)	6(4†)
Minimum utilization voltage	108(110†)	104(106†)

*For utilization voltages of 120-600 V.
†For building wiring circuits supplying lighting equipment.

11.1.9 Voltage Profile of the Limits of Range A, ANSI C84.1-1989 (*see page 11.7*)

11.1.10 Voltage Ratings of Standard Motors (*see page 11.7*)

TABLE 11.1.9

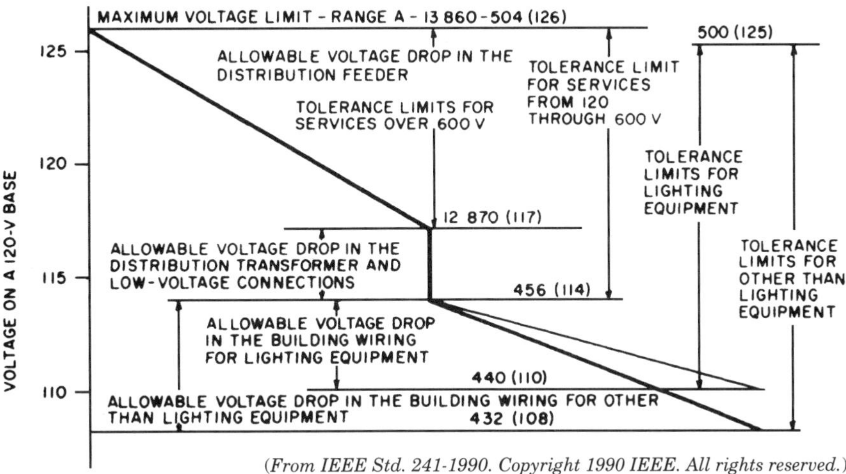

TABLE 11.1.10

Nominal System Voltage	Nameplate Voltage
Single-phase motors	
120	115
240	230
Three-phase motors	
208	200
240	230
480	460
600	575
2400	2300
4160	4000
4800	4600
6900	6600
13 800	13 200

11.1.11 General Effect of Voltage Variations on Induction Motor Characteristics (*see page 11.8*)

11.1.12 Voltage-Drop Calculations

Electrical design professionals designing building wiring systems should have a working knowledge of voltage-drop calculations not only to meet *NEC,* Articles 210.19(A) and 215.2 requirements (recommended, not mandatory) but also to ensure that the voltage applied to utilization equipment is maintained within proper limits. Due to the vector relationships of the circuit parameters, a working knowledge of trigonometry is needed, especially for making exact calculations. Fortunately, most voltage-drop calculations are based on assumed limiting conditions, and approximate formulas are adequate. Within the context of this book, voltage-drop tables and charts are sufficiently accurate to determine the approximate voltage drop for most problems; thus formulas will not be needed.

TABLE 11.1.11 General Effect of Voltage Variations on Induction Motor Characteristics

Characteristic	Function of Voltage	Voltage Variation 90% Voltage	110% Voltage
Starting and maximum running torque	$(\text{Voltage})^2$	Decrease 19%	Increase 21%
Synchronous speed	Constant	No change	No change
Percent slip	$1/(\text{Voltage})^2$	Increase 23%	Decrease 17%
Full-load speed	Synchronous speed-slip	Decrease 1.5%	Increase 1%
Efficiency			
Full load	—	Decrease 2%	Increase 0.5 to 1%
¾ load	—	Practically no change	Practically no change
½ load	—	Increase 1 to 2%	Decrease 1 to 2%
Power factor			
Full load	—	Increase 1%	Decrease 3%
¾ load	—	Increase 2 to 3%	Decrease 4%
½ load	—	Increase 4 to 5%	Decrease 5 to 6%
Full-load current	—	Increase 11%	Decrease 7%
Starting current	Voltage	Decrease 10 to 12%	Increase 10 to 12%
Temperature rise, full load	—	Increase 6 to 7 °C	Decrease 1 to 2 °C
Maximum overload capacity	$(\text{Voltage})^2$	Decrease 19%	Increase 21%
Magnetic noise - no load in particular	—	Decrease slightly	Increase slightly

11.1.13 Voltage-Drop Tables

These tables, reading directly in volts, give values for the voltage drop found in aluminum and copper cables under various circumstances.

1. In magnetic conduit, ac
   a. 70 percent power factor (PF)
   b. 80 percent power factor
   c. 90 percent power factor
   d. 95 percent power factor
   e. 100 percent power factor
2. In nonmagnetic conduit, ac
   a. 70 percent power factor (PF)
   b. 80 percent power factor
   c. 90 percent power factor
   d. 95 percent power factor
   e. 100 percent power factor
3. In direct current (dc) circuits

All voltage drops are calculated at 60 Hz and 60°C. This temperature represents a typical conductor temperature encountered in service. No error of practical significance is involved in using the table for any conductor temperature of 75°C or less.

Space limitations make it necessary to prepare the following pages with the "Ampere Feet" column in abbreviated form. For example, reference to the proper table will show that the voltage drop encountered in a 253,000 A·ft circuit using 1000 Kcmil aluminum cable would be (for 80 percent power factor, magnetic conduit)

17.6 + 4.4 + 0.3, or 22.3 V. These voltage drops are the individual drops given by the table for 200,000, 50,000, and 3000 A·ft, respectively, for a total of 253,000 A·ft. *Note that the length of run refers to the length of the physical circuit, i.e., circuit feet, not the footage of conductor.*

Factors are given at the bottom of each table to make the tables usable in any of the common ac circuits.

11.1.14 Voltage Drop for AL Conductor, Direct Current (*see page 11.10*)

11.1.15 Voltage Drop for AL Conductor in Magnetic Conduit, 70 Percent PF (*see page 11.11*)

11.1.16 Voltage Drop for AL Conductor in Magnetic Conduit, 80 Percent PF (*see page 11.12*)

11.1.17 Voltage Drop for AL Conductor in Magnetic Conduit, 90 Percent PF (*see page 11.13*)

11.1.18 Voltage Drop for AL Conductor in Magnetic Conduit, 95 Percent PF (*see page 11.14*)

11.1.19 Voltage Drop for AL Conductor in Magnetic Conduit, 100 Percent PF (*see page 11.15*)

11.1.20 Voltage Drop for AL Conductor in Nonmagnetic Conduit, 70 Percent PF (*see page 11.16*)

11.1.21 Voltage Drop for AL Conductor in Nonmagnetic Conduit, 80 Percent PF (*see page 11.17*)

11.1.22 Voltage Drop for AL Conductor in Nonmagnetic Conduit, 90 Percent PF (*see page 11.18*)

11.1.23 Voltage Drop for AL Conductor in Nonmagnetic Conduit, 95 Percent PF (*see page 11.19*)

11.1.24 Voltage Drop for AL Conductor in Nonmagnetic Conduit, 100 Percent PF (*see page 11.20*)

TABLE 11.1.14 Voltage Drop for Aluminum Conductor—Direct Current

WIRE SIZE AWG or MCM	1000	900	800	750	700	600	500	400	350	300	250	4/0	3/0	2/0	1/0	1	2	4	6	8*	10*	12*
Ampere Feet	Volts Drop																					
500,000	20.2	22.3	25.2	26.9	28.8	33.6	40.2	50.4	57.5	67.2	80.5	95.9	120.0	151.0	191.0	240.0	303.0	483.0	—	—	—	—
400,000	16.1	17.8	20.2	21.5	23.0	26.9	32.1	40.4	46.0	53.7	64.4	76.8	96.0	121.0	153.0	192.0	241.0	386.0	—	—	—	—
300,000	12.1	13.4	15.1	16.1	17.3	20.2	24.1	30.3	34.5	40.3	48.3	57.6	72.0	90.6	115.0	144.0	182.0	290.0	460.0	—	—	—
200,000	8.1	8.9	10.1	10.8	11.5	13.4	16.1	20.2	23.0	26.9	32.2	38.4	48.0	60.4	76.4	96.0	121.0	193.0	307.0	478.0	—	—
100,000	4.0	4.5	5.0	5.4	5.8	6.7	8.0	10.1	11.5	13.4	16.1	19.2	24.0	30.2	38.2	48.0	60.6	96.6	153.0	239.0	380.0	—
90,000	3.6	4.0	4.5	4.8	5.2	6.1	7.2	9.1	10.3	12.1	14.5	17.3	21.6	27.2	34.4	43.2	54.6	87.0	138.0	215.0	342.0	—
80,000	3.2	3.6	4.0	4.3	4.6	5.4	6.4	8.1	9.2	10.8	12.9	15.3	19.2	24.2	30.5	38.4	48.5	77.0	123.0	191.0	304.0	483.0
70,000	2.8	3.1	3.5	3.8	4.0	4.7	5.6	7.1	8.1	9.4	11.3	13.4	16.8	21.1	26.7	33.6	42.4	67.6	107.0	167.0	266.0	423.0
60,000	2.4	2.7	3.0	3.2	3.5	4.0	4.8	6.1	6.9	8.1	9.7	11.5	14.4	18.1	22.9	28.8	36.4	58.0	92.0	144.0	228.0	362.0
50,000	2.0	2.2	2.5	2.7	2.9	3.4	4.0	5.0	5.8	6.7	8.1	9.6	12.0	15.1	19.1	24.0	30.3	48.3	76.7	120.0	190.0	302.0
40,000	1.6	1.8	2.0	2.2	2.3	2.7	3.2	4.0	4.6	5.4	6.4	7.7	9.6	12.1	15.3	19.2	24.1	38.6	61.4	95.6	152.0	242.0
30,000	1.2	1.3	1.5	1.6	1.7	2.0	2.4	3.0	3.5	4.0	4.8	5.8	7.2	9.1	11.5	14.4	18.2	29.0	46.0	71.7	114.0	181.0
20,000	0.8	0.9	1.0	1.1	1.2	1.3	1.6	2.0	2.3	2.7	3.2	3.8	4.8	6.0	7.6	9.6	12.1	19.3	30.7	47.8	76.0	121.0
10,000	0.4	0.5	0.5	0.5	0.6	0.7	0.8	1.0	1.2	1.3	1.6	1.9	2.4	3.0	3.8	4.8	6.1	9.7	15.3	23.9	38.0	60.4
9,000	0.4	0.4	0.5	0.5	0.5	0.6	0.7	0.9	1.0	1.2	1.5	1.5	2.2	2.7	3.4	4.3	5.5	8.7	13.8	21.5	34.2	54.4
8,000	0.3	0.4	0.4	0.4	0.5	0.5	0.6	0.8	0.9	1.1	1.3	1.5	1.9	2.4	3.1	3.8	4.9	7.7	12.3	19.1	30.4	48.3
7,000	0.3	0.3	0.4	0.4	0.4	0.5	0.6	0.7	0.8	0.9	1.1	1.3	1.7	2.1	2.7	3.4	4.2	6.8	10.7	16.7	26.6	42.3
6,000	0.2	0.3	0.3	0.3	0.4	0.4	0.5	0.6	0.7	0.8	1.0	1.2	1.4	1.8	2.3	2.9	3.6	5.8	9.2	14.4	22.8	36.2
5,000	0.2	0.2	0.3	0.3	0.3	0.3	0.4	0.5	0.6	0.7	0.8	1.0	1.2	1.5	1.9	2.4	3.0	4.8	7.7	12.0	19.0	30.2
4,000	0.2	0.2	0.2	0.2	0.2	0.3	0.3	0.4	0.5	0.5	0.6	0.8	1.0	1.2	1.5	1.9	2.4	3.9	6.1	9.6	15.2	24.2
3,000	0.1	0.1	0.2	0.2	0.2	0.2	0.2	0.3	0.4	0.4	0.5	0.6	0.7	0.9	1.2	1.4	1.8	2.9	4.6	7.2	11.4	18.1
2,000	0.1	0.1	0.1	0.1	0.1	0.1	0.2	0.2	0.2	0.3	0.3	0.4	0.5	0.6	0.8	1.0	1.2	1.9	3.1	4.8	7.6	12.1
1,000	—	0.1	0.1	0.1	0.1	0.1	0.1	0.1	0.1	0.1	0.2	0.2	0.2	0.3	0.4	0.5	0.6	1.0	1.5	2.4	3.8	6.0
900	—	—	0.1	0.1	0.1	0.1	0.1	0.1	0.1	0.1	0.2	0.2	0.2	0.3	0.3	0.4	0.6	0.9	1.4	2.2	3.4	5.4
800	—	—	—	—	0.1	0.1	0.1	0.1	0.1	0.1	0.1	0.2	0.2	0.2	0.3	0.4	0.5	0.8	1.2	1.9	3.0	4.8
700	—	—	—	—	—	0.1	0.1	0.1	0.1	0.1	0.1	0.1	0.2	0.2	0.3	0.3	0.4	0.7	1.1	1.7	2.7	4.2
600	—	—	—	—	—	—	0.1	0.1	0.1	0.1	0.1	0.1	0.1	0.2	0.2	0.3	0.4	0.6	0.9	1.4	2.3	3.6
500	—	—	—	—	—	—	—	0.1	0.1	0.1	0.1	0.1	0.1	0.2	0.2	0.2	0.3	0.5	0.8	1.2	1.9	3.0
400	—	—	—	—	—	—	—	—	0.1	0.1	0.1	0.1	0.1	0.2	0.2	0.2	0.2	0.4	0.6	1.0	1.5	2.4
300	—	—	—	—	—	—	—	—	—	—	0.1	0.1	0.1	0.1	0.1	0.1	0.2	0.3	0.5	0.7	1.1	1.8
200	—	—	—	—	—	—	—	—	—	—	—	—	0.1	0.1	0.1	0.1	0.1	0.2	0.3	0.5	0.8	1.2
100	—	—	—	—	—	—	—	—	—	—	—	—	—	—	—	0.1	0.1	0.1	0.2	0.2	0.4	0.6

* Solid Conductors. Other conductors are stranded.

Note 1—The footage employed in the tabulated ampere feet refers to the length of run of the circuit rather than to the footage of individual conductor.

Note 2—The above table is figured at 60°C since this is an estimate of the average temperature which may be anticipated in service. The table may be used without significant error for conductor temperatures up to and including 75°C.

TABLE 11.1.15 Voltage Drop for Aluminum Conductor in Magnetic Conduit (Power Factor, 70%; Single Phase, 2 Wire; 60 Cycles)

WIRE SIZE AWG or MCM	1000	900	800	750	700	600	500	400	350	300	250	4/0	3/0	2/0	1/0	1	2	4	6	8*	10*	12*
Ampere Feet	Volts Drop																					
500,000	46.4	48.0	50.2	51.5	53.0	56.5	60.8	68.5	73.6	80.9	90.6	101.0	119.0	141.0	170.0	205.0	249.0	373.0	—	—	—	—
400,000	37.1	38.4	40.2	41.2	42.4	45.2	48.6	54.8	58.9	64.7	72.5	80.8	95.2	113.0	136.0	164.0	199.0	298.0	463.0	—	—	—
300,000	27.8	28.8	30.2	30.9	31.8	33.9	36.5	41.1	44.2	48.5	54.4	60.6	71.4	84.6	102.0	123.0	149.0	224.0	347.0	—	—	—
200,000	18.6	19.2	20.1	20.6	21.2	22.6	24.4	27.4	29.5	32.4	36.2	40.4	47.6	56.4	68.0	82.0	99.6	149.0	232.0	352.0	—	—
100,000	9.3	9.6	10.1	10.3	10.6	11.3	12.2	13.7	14.7	16.2	18.1	20.2	23.8	28.2	34.0	41.0	49.8	74.7	116.0	176.0	275.0	431.0
90,000	8.4	8.6	9.1	9.3	9.5	10.2	11.0	12.3	13.3	14.6	16.3	18.2	21.4	25.4	30.6	36.9	44.8	67.2	104.0	159.0	247.0	389.0
80,000	7.4	7.7	8.0	8.2	8.5	9.0	9.7	11.0	11.8	12.9	14.5	16.2	19.0	22.6	27.2	32.8	39.8	59.7	92.6	141.0	210.0	346.0
70,000	6.5	6.7	7.0	7.2	7.4	7.9	8.5	9.6	10.3	11.3	12.7	14.1	16.7	19.7	23.8	28.7	34.8	52.2	81.1	123.0	192.0	302.0
60,000	5.6	5.8	6.0	6.2	6.4	6.8	7.3	8.2	8.8	9.7	10.9	12.1	14.3	16.9	20.4	24.6	29.8	44.7	69.4	106.0	165.0	259.0
50,000	4.6	4.8	5.0	5.2	5.3	5.7	6.1	6.9	7.4	8.1	9.1	10.1	11.9	14.1	17.0	20.5	24.9	37.3	57.9	88.1	137.0	216.0
40,000	3.7	3.8	4.0	4.1	4.2	4.5	4.9	5.5	5.9	6.5	7.3	8.1	9.5	11.3	13.6	16.4	19.9	29.8	46.3	70.4	110.0	173.0
30,000	2.8	2.9	3.0	3.1	3.2	3.4	3.7	4.1	4.4	4.9	5.4	6.1	7.1	8.5	10.2	12.3	14.9	22.4	34.7	52.8	82.5	129.0
20,000	1.9	1.9	2.0	2.1	2.1	2.3	2.4	2.7	3.0	3.2	3.6	4.0	4.8	5.6	6.8	8.2	10.0	14.9	23.2	35.2	55.0	86.4
10,000	0.9	1.0	1.0	1.0	1.1	1.1	1.2	1.4	1.5	1.6	1.8	2.0	2.4	2.8	3.4	4.1	5.0	7.5	11.6	17.6	27.5	43.2
9,000	0.8	0.9	0.9	0.9	1.0	1.0	1.1	1.2	1.3	1.5	1.6	1.8	2.1	2.5	3.1	3.7	4.5	6.7	10.4	15.9	24.7	38.9
8,000	0.7	0.8	0.8	0.8	0.9	0.9	1.0	1.1	1.2	1.3	1.5	1.6	1.9	2.3	2.7	3.3	4.0	6.0	9.3	14.1	21.0	34.6
7,000	0.7	0.7	0.7	0.7	0.7	0.8	0.9	1.0	1.0	1.1	1.3	1.4	1.7	2.0	2.4	2.9	3.5	5.2	8.1	12.3	19.2	30.2
6,000	0.6	0.6	0.6	0.6	0.6	0.7	0.7	0.8	0.9	1.0	1.1	1.2	1.4	1.7	2.0	2.5	3.0	4.5	6.9	10.6	16.5	25.9
5,000	0.5	0.5	0.5	0.5	0.5	0.6	0.6	0.7	0.7	0.8	0.9	1.0	1.2	1.4	1.7	2.1	2.5	3.7	5.8	8.8	13.7	21.6
4,000	0.4	0.4	0.4	0.4	0.4	0.5	0.5	0.6	0.6	0.7	0.7	0.8	1.0	1.1	1.4	1.6	2.0	3.0	4.6	7.0	11.0	17.3
3,000	0.3	0.3	0.3	0.3	0.3	0.4	0.4	0.4	0.4	0.5	0.5	0.6	0.7	0.9	1.0	1.2	1.5	2.2	3.5	5.3	8.3	12.9
2,000	0.2	0.2	0.2	0.2	0.2	0.2	0.2	0.3	0.3	0.3	0.4	0.4	0.5	0.6	0.7	0.8	1.0	1.5	2.3	3.5	5.5	8.6
1,000	0.1	0.1	0.1	0.1	0.1	0.1	0.1	0.1	0.2	0.2	0.2	0.2	0.2	0.3	0.3	0.4	0.5	0.8	1.2	1.8	2.8	4.3
900	0.1	0.1	0.1	0.1	0.1	0.1	0.1	0.1	0.1	0.2	0.2	0.2	0.2	0.3	0.3	0.4	0.5	0.7	1.0	1.6	2.5	3.9
800	0.1	0.1	0.1	0.1	0.1	0.1	0.1	0.1	0.1	0.1	0.2	0.2	0.2	0.2	0.3	0.3	0.4	0.6	0.9	1.4	2.1	3.5
700	0.1	0.1	0.1	0.1	0.1	0.1	0.1	0.1	0.1	0.1	0.1	0.1	0.2	0.2	0.2	0.3	0.4	0.5	0.8	1.2	1.9	3.0
600	0.1	0.1	0.1	0.1	0.1	0.1	0.1	0.1	0.1	0.1	0.1	0.1	0.1	0.2	0.2	0.3	0.3	0.5	0.7	1.1	1.7	2.6
500	0.1	0.1	0.1	0.1	0.1	0.1	0.1	0.1	0.1	0.1	0.1	0.1	0.1	0.1	0.2	0.2	0.3	0.4	0.6	0.9	1.4	2.2
400	—	—	—	—	—	0.1	0.1	0.1	0.1	0.1	0.1	0.1	0.1	0.1	0.1	0.2	0.2	0.3	0.5	0.7	1.1	1.7
300	—	—	—	—	—	—	—	—	—	0.1	0.1	0.1	0.1	0.1	0.1	0.1	0.2	0.2	0.4	0.5	0.8	1.3
200	—	—	—	—	—	—	—	—	—	—	—	—	0.1	0.1	0.1	0.1	0.1	0.2	0.2	0.4	0.6	0.9
100	—	—	—	—	—	—	—	—	—	—	—	—	—	—	—	—	0.1	0.1	0.1	0.2	0.3	0.4

* Solid Conductors. Other conductors are stranded.

Note 1—The above table gives voltage drops encountered in a single phase two-wire system. The voltage drops in other systems may be obtained through multiplication by appropriate factors listed below:

System for Which Voltage Drop Is Desired	Multiplying Factors for Modification of Values in Table
Single Phase—3 Wire—Line to Line	1.00
Single Phase—3 Wire—Line to Neutral	0.50
Three Phase—3 Wire—Line to Line	0.866
Three Phase—4 Wire—Line to Line	0.866
Three Phase—4 Wire—Line to Neutral	0.50

Note 2—Allowable voltage drops for systems other than single phase, two wire cannot be used directly in the above table. Such drops should be modified through multiplication by the appropriate factor listed below. The voltage thus modified may then be used to obtain the proper wire size directly from the table.

System for Which Allowable Voltage Drop Is Known	Multiplying Factor for Modification of Known Value to Permit Direct Use of Table
Single Phase—3 Wire—Line to Line	1.00
Single Phase—3 Wire—Line to Neutral	2.00
Three Phase—3 Wire—Line to Line	1.155
Three Phase—4 Wire—Line to Line	1.155
Three Phase—4 Wire—Line to Neutral	2.00

Note 3—The footage employed in the tabulated ampere feet refers to the length of run of the circuit rather than to the footage of individual conductor.

Note 4—The above table is figured at 60°C since this is an estimate of the average temperature which may be anticipated in service. The table may be used without significant error for conductor temperatures up to and including 75°C.

TABLE 11.1.16 Voltage Drop for Aluminum Conductor in Magnetic Conduit (Power Factor, 80%; Single Phase, 2 Wire; 60 Cycles)

WIRE SIZE AWG or MCM	1000	900	800	750	700	600	500	400	350	300	250	4/0	3/0	2/0	1/0	1	2	4	6	8*	10*	12*
Ampere Feet	Volts Drop																					
500,000	44.0	46.0	48.0	49.0	51.0	55.0	60.0	69.0	74.0	82.0	94.0	105.0	125.0	150.0	183.0	223.0	273.0	419.0	—	—	—	—
400,000	35.2	36.8	38.4	39.2	40.8	44.0	48.0	55.2	59.2	65.6	75.2	84.0	100.0	120.0	146.0	178.0	218.0	335.0	—	—	—	—
300,000	26.4	27.6	28.8	29.4	30.6	33.0	36.0	41.4	44.4	49.2	61.4	63.0	75.0	90.0	110.0	134.0	164.0	251.0	389.0	—	—	—
200,000	17.6	18.4	19.2	19.6	20.4	22.0	24.0	27.6	29.6	32.8	37.6	42.0	50.0	60.0	73.2	89.2	109.0	168.0	259.0	398.0	—	—
100,000	8.8	9.2	9.6	9.8	10.2	11.0	12.0	13.8	14.8	16.4	18.8	21.0	25.0	30.0	36.6	44.6	54.6	83.8	130.0	199.0	311.0	492.0
90,000	7.9	8.3	8.6	8.8	9.2	9.9	10.8	12.4	13.3	14.8	16.9	18.9	22.5	27.0	32.9	41.4	49.1	75.4	117.0	179.0	280.0	443.0
80,000	7.0	7.4	7.7	7.8	8.2	8.8	9.6	11.0	11.8	13.1	15.0	16.8	20.0	24.0	29.3	35.7	43.7	67.0	104.0	159.0	249.0	394.0
70,000	6.2	6.4	6.7	6.9	7.1	7.7	8.4	9.7	10.4	11.8	13.2	14.7	17.5	21.0	25.6	31.2	38.2	58.7	90.9	139.0	218.0	345.0
60,000	5.3	5.5	5.8	5.8	6.1	6.6	7.2	8.3	8.9	9.8	11.3	12.6	15.0	18.0	21.9	26.8	32.8	50.3	77.9	119.0	187.0	295.0
50,000	4.4	4.6	4.8	4.9	5.1	5.5	6.0	6.9	7.4	8.2	9.4	10.5	12.5	15.0	18.3	22.3	27.3	41.9	64.9	99.4	156.0	246.0
40,000	3.5	3.7	3.8	3.9	4.1	4.4	4.8	5.5	5.9	6.6	7.5	8.4	10.0	12.0	14.6	17.8	21.8	33.5	51.9	79.6	124.0	197.0
30,000	2.6	2.8	2.9	2.9	3.1	3.3	3.6	4.1	4.4	4.9	6.1	6.3	7.5	9.0	11.0	13.4	16.4	25.1	38.9	59.7	93.3	148.0
20,000	1.8	1.8	1.9	2.0	2.0	2.2	2.4	2.8	2.9	3.3	3.8	4.2	5.0	6.0	7.3	8.9	10.9	16.8	25.9	39.8	62.2	98.4
10,000	0.9	0.9	1.0	1.0	1.0	1.1	1.2	1.4	1.5	1.6	1.9	2.1	2.5	3.0	3.7	4.5	5.5	8.4	13.0	19.9	31.1	49.2
9,000	0.8	0.8	0.9	0.9	0.9	1.0	1.1	1.2	1.3	1.5	1.7	1.9	2.3	2.7	3.3	4.1	4.9	7.5	11.7	17.9	28.0	44.3
8,000	0.7	0.7	0.8	0.8	0.8	0.9	1.0	1.1	1.2	1.3	1.5	1.7	2.0	2.4	2.9	3.6	4.4	6.7	10.4	15.9	24.9	39.4
7,000	0.6	0.6	0.7	0.7	0.7	0.8	0.8	1.0	1.0	1.2	1.3	1.5	1.8	2.1	2.6	3.1	3.8	5.9	9.1	13.9	21.8	34.5
6,000	0.5	0.6	0.6	0.6	0.6	0.7	0.7	0.8	0.9	1.0	1.1	1.3	1.5	1.8	2.2	2.7	3.3	5.0	7.8	11.9	18.7	29.5
5,000	0.4	0.5	0.5	0.5	0.5	0.6	0.6	0.7	0.7	0.8	0.9	1.1	1.3	1.5	1.8	2.2	2.7	4.2	6.5	9.9	15.6	24.6
4,000	0.4	0.4	0.4	0.4	0.4	0.4	0.5	0.6	0.6	0.6	0.8	0.8	1.0	1.2	1.5	1.8	2.2	3.4	5.2	8.0	12.4	19.7
3,000	0.3	0.3	0.3	0.3	0.3	0.3	0.4	0.4	0.4	0.5	0.6	0.6	0.8	0.9	1.1	1.3	1.6	2.5	3.9	5.9	9.3	14.8
2,000	0.2	0.2	0.2	0.2	0.2	0.2	0.2	0.3	0.3	0.3	0.4	0.4	0.5	0.6	0.7	0.9	1.1	1.7	2.6	4.0	6.2	9.8
1,000	0.1	0.1	0.1	0.1	0.1	0.1	0.1	0.1	0.2	0.2	0.2	0.2	0.3	0.3	0.4	0.5	0.6	0.8	1.3	2.0	3.1	4.9
900	0.1	0.1	0.1	0.1	0.1	0.1	0.1	0.1	0.1	0.2	0.2	0.2	0.2	0.3	0.3	0.4	0.5	0.8	1.2	1.8	2.8	4.4
800	0.1	0.1	0.1	0.1	0.1	0.1	0.1	0.1	0.1	0.1	0.2	0.2	0.2	0.2	0.3	0.4	0.4	0.7	1.0	1.6	2.5	3.9
700	0.1	0.1	0.1	0.1	0.1	0.1	0.1	0.1	0.1	0.1	0.1	0.2	0.2	0.2	0.3	0.3	0.4	0.6	0.9	1.4	2.2	3.5
600	0.1	0.1	0.1	0.1	0.1	0.1	0.1	0.1	0.1	0.1	0.1	0.1	0.2	0.2	0.2	0.3	0.3	0.5	0.8	1.2	1.9	3.0
500	—	0.1	0.1	0.1	0.1	0.1	0.1	0.1	0.1	0.1	0.1	0.1	0.1	0.2	0.2	0.2	0.3	0.4	0.7	1.0	1.6	2.5
400	—	—	—	—	—	—	0.1	0.1	0.1	0.1	0.1	0.1	0.1	0.1	0.2	0.2	0.2	0.3	0.5	0.8	1.2	2.0
300	—	—	—	—	—	—	—	—	—	0.1	0.1	0.1	0.1	0.1	0.1	0.1	0.2	0.3	0.4	0.6	0.9	1.5
200	—	—	—	—	—	—	—	—	—	—	—	—	0.1	0.1	0.1	0.1	0.1	0.2	0.3	0.4	0.6	1.0
100	—	—	—	—	—	—	—	—	—	—	—	—	—	—	—	0.1	0.1	0.1	0.1	0.2	0.3	0.5

* Solid Conductors. Other conductors are stranded.

Note 1—The above table gives voltage drops encountered in a single phase two-wire system. The voltage drops in other systems may be obtained through multiplication by appropriate factors listed below:

System for Which Voltage Drop is Desired	Multiplying Factors for Modification of Values in Table
Single Phase—3 Wire—Line to Line	1.00
Single Phase—3 Wire—Line to Neutral	0.50
Three Phase—3 Wire—Line to Line	0.866
Three Phase—4 Wire—Line to Line	0.866
Three Phase—4 Wire—Line to Neutral	0.50

Note 2—Allowable voltage drops for systems other than single phase, two wire cannot be used directly in the above table. Such drops should be modified through multiplication by the appropriate factor listed below. The voltage thus modified may then be used to obtain the proper wire size directly from the table.

System for Which Allowable Voltage Drop Is Known	Multiplying Factor for Modification of Known Value to Permit Direct Use of Table
Single Phase—3 Wire—Line to Line	1.00
Single Phase—3 Wire—Line to Neutral	2.00
Three Phase—3 Wire—Line to Line	1.155
Three Phase—4 Wire—Line to Line	1.155
Three Phase—4 Wire—Line to Neutral	2.00

Note 3—The footage employed in the tabulated ampere feet refers to the length of run of the circuit rather than to the footage of individual conductor.

Note 4—The above table is figured at 60°C since this is an estimate of the average temperature which may be anticipated in service. The table may be used without significant error for conductor temperatures up to and including *75°C*.

TABLE 11.1.17 Voltage Drop for Aluminum Conductor in Magnetic Conduit (Power Factor, 90%; Single Phase, 2 Wire; 60 Cycles)

WIRE SIZE AWG or MCM	1000	900	800	750	700	600	500	400	350	300	250	4/0	3/0	2/0	1/0	1	2	4	6	8*	10*	12*
Ampere Feet										Volts Drop												
500,000	39.6	41.5	44.0	45.6	47.4	51.0	57.3	66.5	73.0	81.7	93.9	107.0	130.0	157.0	194.0	239.0	295.0	459.0	—	—	—	—
400,000	31.7	33.2	35.2	36.4	37.9	40.8	45.8	53.2	58.4	65.4	75.1	85.6	104.0	125.0	155.0	191.0	236.0	367.0	—	—	—	—
300,000	23.8	24.9	26.4	27.4	28.4	30.6	34.4	39.9	43.8	49.0	56.3	64.2	78.0	94.2	116.0	143.0	177.0	275.0	430.0	—	—	—
200,000	15.9	16.6	17.6	18.2	18.9	20.4	22.9	26.6	29.2	32.7	37.5	42.8	52.0	62.8	77.6	95.6	118.0	184.0	287.0	440.0	—	—
100,000	7.9	8.3	8.8	9.1	9.5	10.2	11.5	13.3	14.6	16.3	18.8	21.4	26.0	31.4	38.8	47.8	59.0	91.8	143.0	220.0	247.0	—
90,000	7.1	7.5	7.9	8.2	8.5	9.2	10.3	12.0	13.1	14.7	16.9	19.3	23.4	28.2	35.0	43.0	53.1	82.7	129.0	198.0	313.0	495.0
80,000	6.3	6.6	7.0	7.3	7.6	8.2	9.2	10.6	11.7	13.1	15.0	17.1	20.8	25.1	31.0	38.2	47.2	73.5	115.0	177.0	278.0	440.0
70,000	5.5	5.8	6.2	6.4	6.6	7.1	8.0	9.3	10.2	11.4	13.1	15.0	18.2	22.0	27.2	33.5	41.3	64.3	100.0	155.0	243.0	385.0
60,000	4.8	5.0	5.3	5.5	5.7	6.1	6.9	8.0	8.8	9.8	11.3	12.8	15.6	18.8	23.3	28.7	35.4	55.1	86.0	133.0	208.0	330.0
50,000	4.0	4.2	4.4	4.6	4.7	5.1	5.7	6.7	7.3	8.2	9.4	10.7	13.0	15.7	19.4	23.9	29.5	45.9	71.6	110.0	174.0	275.0
40,000	3.2	3.3	3.5	3.6	4.0	4.1	4.6	5.3	5.8	6.5	7.5	8.6	10.4	12.5	15.5	19.1	23.6	36.7	57.3	88.0	139.0	220.0
30,000	2.4	2.5	2.6	2.7	2.8	3.1	3.4	4.0	4.4	4.9	5.6	6.4	7.8	9.4	11.6	14.3	17.7	27.5	43.0	66.0	104.0	165.0
20,000	1.6	1.7	1.8	1.8	1.9	2.0	2.3	2.7	2.9	3.3	3.8	4.3	5.2	6.3	7.8	9.6	11.8	18.4	28.7	44.0	69.4	110.0
10,000	0.8	0.8	0.9	0.9	1.0	1.0	1.2	1.3	1.5	1.6	1.9	2.1	2.6	3.1	3.9	4.8	5.9	9.2	14.3	22.0	34.7	55.0
9,000	0.7	0.8	0.8	0.8	0.9	0.9	1.0	1.2	1.3	1.5	1.7	1.9	2.3	2.8	3.5	4.3	5.3	8.3	12.8	19.9	31.3	49.5
8,000	0.6	0.7	0.7	0.7	0.8	0.8	0.9	1.1	1.2	1.3	1.5	1.7	2.1	2.5	3.1	3.8	4.7	7.4	11.5	17.7	27.8	44.0
7,000	0.6	0.6	0.6	0.6	0.7	0.7	0.8	0.9	1.0	1.1	1.3	1.5	1.8	2.2	2.7	3.4	4.2	6.4	10.0	15.5	24.3	38.5
6,000	0.5	0.5	0.5	0.6	0.6	0.6	0.7	0.8	0.9	1.0	1.1	1.3	1.6	1.9	2.3	2.9	3.5	5.5	8.6	13.3	20.8	33.0
5,000	0.4	0.4	0.4	0.5	0.5	0.5	0.6	0.7	0.7	0.8	0.9	1.1	1.3	1.6	1.9	2.4	3.0	4.6	7.2	11.0	17.4	27.5
4,000	0.3	0.3	0.4	0.4	0.4	0.4	0.5	0.5	0.6	0.7	0.8	0.9	1.0	1.3	1.6	1.9	2.4	3.7	5.7	8.8	13.9	22.0
3,000	0.2	0.2	0.3	0.3	0.3	0.3	0.3	0.4	0.4	0.5	0.6	0.6	0.8	0.9	1.2	1.4	1.8	2.8	4.3	6.6	10.4	16.5
2,000	0.2	0.2	0.2	0.2	0.2	0.2	0.2	0.3	0.3	0.3	0.4	0.4	0.5	0.6	0.8	1.0	1.2	1.8	2.9	4.4	6.9	11.0
1,000	0.1	0.1	0.1	0.1	0.1	0.1	0.1	0.1	0.2	0.2	0.2	0.2	0.3	0.3	0.4	0.5	0.6	0.9	1.4	2.2	3.5	5.5
900	0.1	0.1	0.1	0.1	0.1	0.1	0.1	0.1	0.1	0.2	0.2	0.2	0.2	0.3	0.4	0.4	0.5	0.8	1.3	2.0	3.1	5.0
800	0.1	0.1	0.1	0.1	0.1	0.1	0.1	0.1	0.1	0.1	0.2	0.2	0.2	0.3	0.3	0.4	0.5	0.7	1.2	1.8	2.8	4.4
700	0.1	0.1	0.1	0.1	0.1	0.1	0.1	0.1	0.1	0.1	0.1	0.2	0.2	0.2	0.3	0.3	0.4	0.6	1.0	1.6	2.4	3.9
600	0.1	0.1	0.1	0.1	0.1	0.1	0.1	0.1	0.1	0.1	0.1	0.1	0.2	0.2	0.2	0.3	0.4	0.6	0.9	1.3	2.1	3.3
500	—	—	—	0.1	0.1	0.1	0.1	0.1	0.1	0.1	0.1	0.1	0.1	0.2	0.2	0.2	0.3	0.5	0.7	1.1	1.7	2.8
400	—	—	—	—	—	—	0.1	0.1	0.1	0.1	0.1	0.1	0.1	0.1	0.2	0.2	0.2	0.4	0.6	0.9	1.4	2.2
300	—	—	—	—	—	—	—	—	—	0.1	0.1	0.1	0.1	0.1	0.1	0.1	0.2	0.3	0.4	0.7	1.0	1.7
200	—	—	—	—	—	—	—	—	—	—	—	—	0.1	0.1	0.1	0.1	0.1	0.2	0.3	0.4	0.7	1.1
100	—	—	—	—	—	—	—	—	—	—	—	—	—	—	—	0.1	0.1	0.1	0.1	0.2	0.4	0.6

* Solid Conductors. Other conductors are stranded.

Note 1—The above table gives voltage drops encountered in a single phase two-wire system. The voltage drops in other systems may be obtained through multiplication by appropriate factors listed below:

System for Which Voltage Drop is Desired	Multiplying Factors for Modification of Values in Table
Single Phase—3 Wire—Line to Line	1.00
Single Phase—3 Wire—Line to Neutral	0.50
Three Phase—3 Wire—Line to Line	0.866
Three Phase—4 Wire—Line to Line	0.866
Three Phase—4 Wire—Line to Neutral	0.50

Note 2—Allowable voltage drops for systems other than single phase, two wire cannot be used directly in the above table. Such drops should be modified through multiplication by the appropriate factor listed below. The voltage thus modified may then be used to obtain the proper wire size directly from the table.

System for Which Allowable Voltage Drop is Known	Multiplying Factor for Modification of Known Value to Permit Direct Use of Table
Single Phase—3 Wire—Line to Line	1.00
Single Phase—3 Wire—Line to Neutral	2.00
Three Phase—3 Wire—Line to Line	1.155
Three Phase—4 Wire—Line to Line	1.155
Three Phase—4 Wire—Line to Neutral	2.00

Note 3—The footage employed in the tabulated ampere feet refers to the length of run of the circuit rather than to the footage of individual conductor.

Note 4—The above table is figured at 60°C since this is an estimate of the average temperature which may be anticipated in service. The table may be used without significant error for conductor temperatures up to and including 75°C.

TABLE 11.1.18 Voltage Drop for Aluminum Conductor in Magnetic Conduit (Power Factor, 95%; Single Phase, 2 Wire; 60 Cycles)

WIRE SIZE AWG or MCM	1000	900	800	750	700	600	500	400	350	300	250	4/0	3/0	2/0	1/0	1	2	4	6	8*	10*	12*
Ampere Feet	Volts Drop																					
500,000	35.7	37.7	40.3	41.9	43.8	48.0	54.0	63.6	70.4	79.4	92.2	106.0	130.0	158.0	197.0	244.0	304.0	476.0	—	—	—	—
400,000	28.6	30.2	32.2	33.5	35.0	38.4	43.2	50.9	56.3	65.5	73.7	84.8	104.0	126.0	158.0	195.0	243.0	380.0	—	—	—	—
300,000	21.4	22.6	24.2	25.2	26.3	28.8	32.4	38.2	42.2	47.6	55.3	63.6	78.0	94.8	118.0	146.0	182.0	285.0	448.0	—	—	—
200,000	14.3	15.1	16.1	16.8	17.5	19.2	21.6	25.4	28.2	31.8	36.9	42.4	52.0	63.2	78.8	97.6	122.0	190.0	299.0	462.0	—	—
100,000	7.1	7.5	8.1	8.4	8.8	9.6	10.8	12.6	14.1	15.9	18.5	21.2	26.0	31.6	39.4	48.8	60.8	95.2	149.0	231.0	365.0	—
90,000	6.4	6.8	7.3	7.6	7.9	8.6	9.7	11.5	12.7	14.1	16.6	19.1	23.4	28.4	35.4	43.9	54.8	85.7	134.0	208.0	328.0	—
80,000	5.7	6.0	6.5	6.7	7.0	7.7	8.6	10.2	11.3	12.7	14.8	17.0	20.8	25.3	31.5	39.1	48.7	76.2	119.0	185.0	292.0	463.0
70,000	5.0	5.3	5.6	5.9	6.1	6.7	7.6	8.9	9.9	11.1	12.9	14.8	18.2	22.1	27.6	34.1	42.6	66.7	105.0	162.0	255.0	404.0
60,000	4.3	4.5	4.8	5.0	5.3	5.8	6.5	7.6	8.5	9.5	11.1	12.7	15.6	19.0	23.6	29.3	36.5	57.2	89.6	139.0	219.0	347.0
50,000	3.6	3.8	4.0	4.2	4.4	4.8	5.4	6.4	7.0	7.9	9.2	10.6	13.0	15.8	19.7	24.4	30.4	47.6	74.7	116.0	182.0	289.0
40,000	2.9	3.0	3.2	3.4	3.5	3.8	4.3	5.1	5.6	6.6	7.4	8.5	10.4	12.6	15.8	19.5	24.3	38.0	59.8	92.4	146.0	231.0
30,000	2.1	2.3	2.4	2.5	2.6	2.9	3.2	3.8	4.2	4.8	5.5	6.4	7.8	9.5	11.8	14.6	18.2	28.5	44.8	69.3	109.0	173.0
20,000	1.4	1.5	1.6	1.7	1.8	1.9	2.2	2.5	2.8	3.2	3.7	4.2	5.2	6.3	7.9	9.8	12.2	19.0	29.9	46.2	73.0	116.0
10,000	0.7	0.8	0.8	0.8	0.9	1.0	1.1	1.3	1.4	1.6	1.9	2.1	2.8	3.2	3.9	4.9	6.1	9.5	14.9	23.1	36.5	57.8
9,000	0.6	0.7	0.7	0.8	0.8	0.9	1.0	1.2	1.3	1.4	1.7	1.9	2.3	2.8	3.5	4.4	5.5	8.6	13.4	20.8	32.8	52.0
8,000	0.6	0.6	0.7	0.7	0.7	0.8	0.9	1.0	1.1	1.3	1.5	1.7	2.1	2.5	3.2	3.9	4.9	7.6	11.9	18.5	29.2	46.3
7,000	0.5	0.5	0.6	0.6	0.6	0.7	0.8	0.9	1.0	1.1	1.3	1.5	1.8	2.2	2.8	3.4	4.3	6.7	10.5	16.2	25.5	40.4
6,000	0.4	0.5	0.5	0.5	0.5	0.6	0.7	0.8	0.9	1.0	1.1	1.3	1.6	1.9	2.4	2.9	3.7	5.7	9.0	13.9	21.9	34.7
5,000	0.4	0.4	0.4	0.4	0.4	0.5	0.5	0.6	0.7	0.8	0.9	1.1	1.3	1.6	2.0	2.4	3.0	4.8	7.5	11.6	18.2	28.9
4,000	0.3	0.3	0.3	0.3	0.4	0.4	0.4	0.5	0.6	0.7	0.7	0.9	1.0	1.3	1.6	2.0	2.4	3.8	6.0	9.2	14.6	23.1
3,000	0.2	0.2	0.2	0.3	0.3	0.3	0.3	0.4	0.4	0.5	0.6	0.6	0.8	1.0	1.2	1.5	1.8	2.9	4.5	6.9	10.9	17.3
2,000	0.1	0.2	0.2	0.2	0.2	0.2	0.2	0.3	0.3	0.3	0.4	0.4	0.5	0.6	0.8	1.0	1.2	1.9	3.0	4.6	7.3	11.6
1,000	0.1	0.1	0.1	0.1	0.1	0.1	0.1	0.1	0.1	0.2	0.2	0.2	0.3	0.3	0.4	0.5	0.6	1.0	1.5	2.3	3.7	5.8
900	0.1	0.1	0.1	0.1	0.1	0.1	0.1	0.1	0.1	0.1	0.2	0.2	0.2	0.3	0.4	0.4	0.6	0.9	1.3	2.1	3.3	5.2
800	0.1	0.1	0.1	0.1	0.1	0.1	0.1	0.1	0.1	0.1	0.2	0.2	0.2	0.3	0.3	0.4	0.5	0.8	1.2	1.9	2.9	4.6
700	0.1	0.1	0.1	0.1	0.1	0.1	0.1	0.1	0.1	0.1	0.1	0.2	0.2	0.2	0.3	0.3	0.4	0.7	1.1	1.6	2.6	4.0
600	—	0.1	0.1	0.1	0.1	0.1	0.1	0.1	0.1	0.1	0.1	0.1	0.2	0.2	0.2	0.3	0.4	0.6	0.9	1.4	2.2	3.5
500	—	—	—	—	—	0.1	0.1	0.1	0.1	0.1	0.1	0.1	0.1	0.2	0.2	0.2	0.3	0.5	0.8	1.2	1.8	2.9
400	—	—	—	—	—	—	—	0.1	0.1	0.1	0.1	0.1	0.1	0.1	0.2	0.2	0.2	0.4	0.6	0.9	1.5	2.3
300	—	—	—	—	—	—	—	—	—	0.1	0.1	0.1	0.1	0.1	0.1	0.2	0.2	0.3	0.5	0.7	1.1	1.7
200	—	—	—	—	—	—	—	—	—	—	—	—	0.1	0.1	0.1	0.1	0.1	0.2	0.3	0.5	0.7	1.2
100	—	—	—	—	—	—	—	—	—	—	—	—	—	—	—	0.1	0.1	0.1	0.2	0.2	0.4	0.6

* Solid Conductors. Other conductors are stranded.

Note 1—The above table gives voltage drops encountered in a single phase two-wire system. The voltage drops in other systems may be obtained through multiplication by appropriate factors listed below:

System for Which Voltage Drop Is Desired	Multiplying Factors for Modification of Values in Table
Single Phase—3 Wire—Line to Line	1.00
Single Phase—3 Wire—Line to Neutral	0.50
Three Phase—3 Wire—Line to Line	0.866
Three Phase—4 Wire—Line to Line	0.866
Three Phase—4 Wire—Line to Neutral	0.50

Note 2—Allowable voltage drops for systems other than single phase, two wire cannot be used directly in the above table. Such drops should be modified through multiplication by the appropriate factor listed below. The voltage thus modified may then be used to obtain the proper wire size directly from the table.

System for Which Allowable Voltage Drop Is Known	Multiplying Factor for Modification of Known Value to Permit Direct Use of Table
Single Phase—3 Wire—Line to Line	1.00
Single Phase—3 Wire—Line to Neutral	2.00
Three Phase—3 Wire—Line to Line	1.155
Three Phase—4 Wire—Line to Line	1.155
Three Phase—4 Wire—Line to Neutral	2.00

Note 3—The footage employed in the tabulated ampere feet refers to the length of run of the circuit rather than to the footage of individual conductor.

Note 4—The above table is figured at 60°C since this is an estimate of the average temperature which may be anticipated in service. The table may be used without significant error for conductor temperatures up to and including 75°C.

TABLE 11.1.19 Voltage Drop for Aluminum Conductor in Magnetic Conduit (Power Factor, 100%; Single Phase, 2 Wire; 60 Cycles)

WIRE SIZE AWG or MCM	1000	900	800	750	700	600	500	400	350	300	250	4/0	3/0	2/0	1/0	1	2	4	6	8*	10*	12*
Ampere Feet	Volts Drop																					
500,000	24.0	25.9	28.5	30.1	32.0	36.2	42.6	52.4	59.2	68.5	81.8	96.8	121.0	151.0	191.0	240.0	303.0	483.0	—	—	—	—
400,000	19.2	20.7	22.8	24.1	25.6	29.0	34.1	41.9	47.4	54.8	65.4	77.4	96.8	121.0	153.0	192.0	241.0	386.0	—	—	—	—
300,000	14.4	15.5	17.1	18.1	19.2	21.7	25.5	31.4	35.5	41.1	49.1	58.1	72.6	90.6	115.0	144.0	182.0	290.0	460.0	—	—	—
200,000	9.6	10.3	11.4	12.0	12.8	14.5	17.0	21.0	27.7	27.4	32.7	38.7	48.4	60.4	76.4	96.0	121.0	193.0	307.0	478.0	—	—
100,000	4.8	5.2	5.7	6.0	6.4	7.2	8.5	10.5	11.8	13.7	16.4	19.4	24.2	30.2	38.2	48.0	60.6	96.6	153.0	239.0	380.0	—
90,000	4.3	4.7	5.1	5.4	5.8	6.5	7.7	9.4	10.7	12.3	14.7	17.4	21.8	27.2	34.4	43.2	54.6	87.0	138.0	215.0	342.0	—
80,000	3.8	4.1	4.6	4.8	5.1	5.8	6.8	8.4	9.5	11.0	13.1	15.5	19.4	24.2	30.5	38.4	48.5	77.3	123.0	191.0	304.0	483.0
70,000	3.4	3.6	4.0	4.2	4.5	5.1	6.0	7.3	8.3	9.6	11.5	13.5	16.9	21.1	26.7	33.6	42.4	67.6	107.0	167.0	266.0	423.0
60,000	2.9	3.1	3.4	3.6	3.8	4.3	5.1	6.3	7.1	8.2	9.8	11.6	14.5	18.1	22.9	28.8	36.4	58.0	92.0	144.0	228.0	382.0
50,000	2.4	2.6	2.9	3.0	3.2	3.6	4.3	5.2	5.9	6.9	8.2	9.7	12.1	15.1	19.1	24.0	30.3	48.3	76.7	120.0	190.0	302.0
40,000	1.9	2.1	2.3	2.4	2.6	2.9	3.4	4.2	4.7	5.5	6.5	7.7	9.7	12.1	15.3	19.2	24.1	38.6	61.4	95.6	152.0	242.0
30,000	1.4	1.6	1.7	1.8	1.9	2.2	2.6	3.1	3.6	4.1	4.9	5.8	7.3	9.1	11.5	14.4	18.2	29.0	46.0	71.7	114.0	181.0
20,000	1.0	1.0	1.1	1.2	1.3	1.5	1.7	2.1	2.8	2.7	3.3	3.9	4.8	6.0	7.6	9.6	12.1	19.3	30.7	47.8	76.0	121.0
10,000	0.5	0.5	0.6	0.6	0.6	0.7	0.9	1.1	1.2	1.4	1.6	1.9	2.4	3.0	3.8	4.8	6.1	9.7	15.3	23.9	38.0	60.4
9,000	0.4	0.5	0.5	0.5	0.6	0.7	0.8	0.9	1.1	1.2	1.5	1.7	2.2	2.7	3.4	4.3	5.5	8.7	13.8	21.5	34.2	54.4
8,000	0.4	0.4	0.5	0.5	0.5	0.6	0.7	0.8	1.0	1.1	1.3	1.6	1.9	2.4	3.1	3.8	4.9	7.7	12.3	19.1	30.4	48.3
7,000	0.3	0.4	0.4	0.4	0.5	0.5	0.6	0.7	0.8	1.0	1.2	1.4	1.7	2.1	2.7	3.4	4.2	6.8	10.7	16.7	26.6	42.3
6,000	0.3	0.3	0.3	0.4	0.4	0.4	0.5	0.6	0.7	0.8	1.0	1.2	1.5	1.8	2.3	2.9	3.6	5.8	9.2	14.4	22.8	36.2
5,000	0.2	0.3	0.3	0.3	0.3	0.4	0.4	0.5	0.6	0.7	0.8	1.0	1.2	1.5	1.9	2.4	3.0	4.8	7.7	12.0	19.0	30.2
4,000	0.2	0.2	0.2	0.2	0.3	0.3	0.3	0.4	0.5	0.6	0.7	0.8	1.0	1.2	1.5	1.9	2.4	3.9	6.1	9.6	15.2	24.2
3,000	0.1	0.2	0.2	0.2	0.2	0.2	0.3	0.3	0.4	0.4	0.5	0.6	0.7	0.9	1.2	1.4	1.8	2.9	4.6	7.2	11.4	18.1
2,000	0.1	0.1	0.1	0.1	0.1	0.2	0.2	0.2	0.3	0.3	0.3	0.4	0.5	0.6	0.8	1.0	1.2	1.9	3.1	4.8	7.6	12.1
1,000	0.1	0.1	0.1	0.1	0.1	0.1	0.1	0.1	0.1	0.1	0.2	0.2	0.2	0.3	0.4	0.5	0.6	1.0	1.5	2.4	3.8	6.0
900	—	0.1	0.1	0.1	0.1	0.1	0.1	0.1	0.1	0.1	0.2	0.2	0.2	0.3	0.3	0.4	0.6	0.9	1.4	2.2	3.4	5.4
800	—	—	0.1	0.1	0.1	0.1	0.1	0.1	0.1	0.1	0.1	0.2	0.2	0.2	0.3	0.4	0.5	0.8	1.2	1.9	3.0	4.8
700	—	—	—	—	0.1	0.1	0.1	0.1	0.1	0.1	0.1	0.1	0.2	0.2	0.3	0.3	0.4	0.7	1.1	1.7	2.7	4.2
600	—	—	—	—	—	—	0.1	0.1	0.1	0.1	0.1	0.1	0.1	0.2	0.2	0.3	0.4	0.6	0.9	1.4	2.3	3.6
500	—	—	—	—	—	—	—	0.1	0.1	0.1	0.1	0.1	0.1	0.2	0.2	0.2	0.3	0.5	0.8	1.2	1.9	3.0
400	—	—	—	—	—	—	—	—	0.1	0.1	0.1	0.1	0.1	0.1	0.2	0.2	0.2	0.4	0.6	1.0	1.5	2.4
300	—	—	—	—	—	—	—	—	—	—	0.1	0.1	0.1	0.1	0.1	0.1	0.2	0.3	0.5	0.7	1.1	1.8
200	—	—	—	—	—	—	—	—	—	—	—	—	0.1	0.1	0.1	0.1	0.1	0.2	0.3	0.5	0.8	1.2
100	—	—	—	—	—	—	—	—	—	—	—	—	—	—	—	0.1	0.1	0.1	0.2	0.2	0.4	0.6

* Solid Conductors. Other conductors are stranded.

Note 1—The above table gives voltage drops encountered in a single phase two-wire system. The voltage drops in other systems may be obtained through multiplication by appropriate factors listed below:

System for Which Voltage Drop Is Desired	Multiplying Factors for Modification of Values in Table
Single Phase—3 Wire—Line to Line	1.00
Single Phase—3 Wire—Line to Neutral	0.50
Three Phase—3 Wire—Line to Line	0.866
Three Phase—4 Wire—Line to Line	0.866
Three Phase—4 Wire—Line to Neutral	0.50

Note 2—Allowable voltage drops for systems other than single phase, two wire cannot be used directly in the above table. Such drops should be modified through multiplication by the appropriate factor listed below. The voltage thus modified may then be used to obtain the proper wire size directly from the table.

System for Which Allowable Voltage Drop Is Known	Multiplying Factor for Modification of Known Value to Permit Direct Use of Table
Single Phase—3 Wire—Line to Line	1.00
Single Phase—3 Wire—Line to Neutral	2.00
Three Phase—3 Wire—Line to Line	1.155
Three Phase—4 Wire—Line to Line	1.155
Three Phase—4 Wire—Line to Neutral	2.00

Note 3—The footage employed in the tabulated ampere feet refers to the length of run of the circuit rather than to the footage of individual conductor.

Note 4—The above table is figured at 60°C since this is an estimate of the average temperature which may be anticipated in service. The table may be used without significant error for conductor temperatures up to and including 75°C.

TABLE 11.1.20 Voltage Drop for Aluminum Conductor in Nonmagnetic Conduit (Power Factor, 70%; Single Phase, 2 Wire; 60 Cycles)

WIRE SIZE AWG or MCM	1000	900	800	750	700	600	500	400	350	300	250	4/0	3/0	2/0	1/0	1	2	4	6	8*	10*	12*
Ampere Feet	Volts Drop																					
500,000	38.1	39.9	42.1	43.4	44.8	47.9	53.0	60.8	66.2	73.3	83.0	93.3	101.0	134.0	163.0	198.0	242.0	370.0	—	—	—	—
400,000	30.5	31.9	33.6	34.7	35.9	38.3	42.4	48.6	52.9	58.7	66.4	74.6	80.8	107.0	130.0	158.0	194.0	296.0	456.0	—	—	—
300,000	22.8	23.9	25.2	26.1	26.9	28.7	31.8	36.5	39.7	44.0	49.8	55.9	60.6	80.4	97.8	119.0	145.0	222.0	342.0	—	—	—
200,000	15.2	15.9	16.8	17.4	17.9	19.2	21.2	24.4	26.4	29.3	33.2	37.3	40.4	53.6	65.2	79.2	96.8	148.0	228.0	348.0	—	—
100,000	7.6	8.0	8.4	8.7	9.0	9.6	10.6	12.2	13.2	14.7	16.6	18.6	20.2	26.8	32.6	39.6	48.4	74.0	114.0	174.0	273.0	430.0
90,000	6.9	7.2	7.6	7.8	8.1	8.6	9.5	11.0	11.9	13.2	14.9	16.8	18.2	24.1	29.3	35.6	43.6	66.6	103.0	157.0	246.0	387.0
80,000	6.1	6.4	6.7	7.0	7.2	7.7	8.5	9.7	10.6	11.7	13.3	14.9	16.2	21.4	26.1	31.7	38.7	59.2	91.2	139.0	219.0	344.0
70,000	5.3	5.6	5.9	6.1	6.3	6.7	7.4	8.5	9.3	10.3	11.6	13.1	14.1	18.8	22.8	27.7	33.9	51.8	79.8	122.0	191.0	301.0
60,000	4.6	4.8	5.1	5.2	5.4	5.8	6.4	7.3	8.0	8.8	10.0	11.2	12.1	16.1	19.6	23.8	29.1	44.4	68.4	105.0	164.0	258.0
50,000	3.8	4.0	4.2	4.3	4.5	4.8	5.3	6.1	6.6	7.3	8.3	9.3	10.1	13.4	16.3	19.8	24.2	37.0	57.0	87.2	137.0	215.0
40,000	3.1	3.2	3.4	3.5	3.6	3.8	4.2	4.9	5.3	5.9	6.6	7.5	8.1	10.7	13.0	15.8	19.4	29.6	45.6	69.6	109.0	172.0
30,000	2.3	2.4	2.6	2.6	2.7	2.9	3.2	3.7	4.0	4.4	5.0	5.6	6.1	8.0	9.8	11.9	14.5	22.2	34.2	52.5	81.9	129.0
20,000	1.5	1.6	1.7	1.7	1.8	1.9	2.1	2.4	2.6	2.9	3.3	3.7	4.0	5.4	6.5	7.9	9.7	14.8	22.8	34.8	54.6	86.0
10,000	0.8	0.8	0.8	0.9	0.9	1.0	1.1	1.2	1.3	1.5	1.7	1.9	2.0	2.7	3.3	4.0	4.8	7.4	11.4	17.4	27.3	43.0
9,000	0.7	0.7	0.8	0.8	0.8	0.9	1.0	1.1	1.2	1.3	1.5	1.7	1.8	2.4	2.9	3.6	4.4	6.7	10.3	15.7	24.6	38.7
8,000	0.6	0.6	0.7	0.7	0.7	0.8	0.9	1.0	1.1	1.2	1.3	1.5	1.6	2.1	2.6	3.2	3.9	5.9	9.1	13.9	21.9	34.4
7,000	0.5	0.6	0.6	0.6	0.6	0.7	0.7	0.9	0.9	1.0	1.2	1.3	1.4	1.9	2.3	2.8	3.4	5.2	8.0	12.2	19.1	30.1
6,000	0.5	0.5	0.5	0.5	0.5	0.6	0.6	0.7	0.8	0.9	1.0	1.1	1.2	1.6	2.0	2.4	2.9	4.4	6.8	10.5	16.4	25.8
5,000	0.4	0.4	0.4	0.4	0.5	0.5	0.5	0.6	0.7	0.7	0.8	0.9	1.0	1.3	1.6	2.0	2.4	3.7	5.7	8.7	13.7	21.5
4,000	0.3	0.3	0.3	0.4	0.4	0.4	0.4	0.5	0.5	0.6	0.7	0.8	0.8	1.1	1.3	1.6	1.9	3.0	4.6	7.0	10.9	17.2
3,000	0.2	0.2	0.3	0.3	0.3	0.3	0.3	0.4	0.4	0.4	0.5	0.6	0.6	0.8	1.0	1.2	1.5	2.2	3.4	5.3	8.2	12.9
2,000	0.2	0.2	0.2	0.2	0.2	0.2	0.2	0.2	0.3	0.3	0.3	0.4	0.4	0.5	0.7	0.8	1.0	1.5	2.3	3.5	5.5	8.6
1,000	0.1	0.1	0.1	0.1	0.1	0.1	0.1	0.1	0.1	0.2	0.2	0.2	0.2	0.3	0.3	0.4	0.5	0.7	1.1	1.7	2.7	4.3
900	0.1	0.1	0.1	0.1	0.1	0.1	0.1	0.1	0.1	0.1	0.2	0.2	0.2	0.2	0.3	0.4	0.4	0.7	1.0	1.6	2.5	3.9
800	0.1	0.1	0.1	0.1	0.1	0.1	0.1	0.1	0.1	0.1	0.1	0.2	0.2	0.2	0.3	0.3	0.4	0.6	0.9	1.4	2.2	3.4
700	0.1	0.1	0.1	0.1	0.1	0.1	0.1	0.1	0.1	0.1	0.1	0.1	0.1	0.2	0.2	0.3	0.3	0.5	0.8	1.2	1.9	3.0
600	0.1	0.1	0.1	0.1	0.1	0.1	0.1	0.1	0.1	0.1	0.1	0.1	0.1	0.2	0.2	0.2	0.3	0.4	0.7	1.1	1.6	2.6
500	—	—	—	—	0.1	0.1	0.1	0.1	0.1	0.1	0.1	0.1	0.1	0.1	0.2	0.2	0.2	0.4	0.6	0.9	1.4	2.2
400	—	—	—	—	—	—	—	0.1	0.1	0.1	0.1	0.1	0.1	0.1	0.1	0.2	0.2	0.3	0.5	0.7	1.1	1.7
300	—	—	—	—	—	—	—	—	—	—	0.1	0.1	0.1	0.1	0.1	0.1	0.2	0.2	0.3	0.5	0.8	1.3
200	—	—	—	—	—	—	—	—	—	—	—	—	—	0.1	0.1	0.1	0.1	0.2	0.2	0.4	0.6	0.9
100	—	—	—	—	—	—	—	—	—	—	—	—	—	—	—	—	0.1	0.1	0.1	0.2	0.3	0.4

* Solid Conductors. Other conductors are stranded.

Note 1—The above table gives voltage drops encountered in a single phase two-wire system. The voltage drops in other systems may be obtained through multiplication by appropriate factors listed below:

System for Which Voltage Drop Is Desired	Multiplying Factors for Modification of Values in Table
Single Phase—3 Wire—Line to Line	1.00
Single Phase—3 Wire—Line to Neutral	0.50
Three Phase—3 Wire—Line to Line	0.866
Three Phase—4 Wire—Line to Line	0.866
Three Phase—4 Wire—Line to Neutral	0.50

Note 2—Allowable voltage drops for systems other than single phase, two wire cannot be used directly in the above table. Such drops should be modified through multiplication by the appropriate factor listed below. The voltage thus modified may then be used to obtain the proper wire size directly from the table.

System for Which Allowable Voltage Drop Is Known	Multiplying Factor for Modification of Known Value to Permit Direct Use of Table
Single Phase—3 Wire—Line to Line	1.00
Single Phase—3 Wire—Line to Neutral	2.00
Three Phase—3 Wire—Line to Line	1.155
Three Phase—4 Wire—Line to Line	1.155
Three Phase—4 Wire—Line to Neutral	2.00

Note 3—The footage employed in the tabulated ampere feet refers to the length of run of the circuit rather than to the footage of individual conductor.

Note 4—The above table is figured at 60°C since this is an estimate of the average temperature which may be anticipated in service. The table may be used without significant error for conductor temperatures up to and including 75°C.

TABLE 11.1.21 Voltage Drop for Aluminum Conductor in Nonmagnetic Conduit (Power Factor, 80%; Single Phase, 2 Wire; 60 Cycles)

WIRE SIZE AWG or MCM	1000	900	800	750	700	600	500	400	350	300	250	4/0	3/0	2/0	1/0	1	2	4	6	8*	10*	12*
Ampere Feet	Volts Drop																					
500,000	36.5	38.3	40.8	42.2	43.9	48.0	53.1	61.9	67.7	75.8	86.8	98.7	119.0	145.0	177.0	217.0	267.0	413.0	—	—	—	—
400,000	29.2	30.6	32.6	33.8	35.1	38.4	42.4	49.5	54.1	60.7	69.4	78.9	95.2	116.0	142.0	174.0	214.0	330.0	—	—	—	—
300,000	21.9	23.0	24.5	25.3	26.3	28.8	31.8	37.1	40.6	45.5	52.1	69.2	71.4	87.0	106.0	130.0	160.0	248.0	385.0	—	—	—
200,000	14.6	15.3	16.3	16.9	17.6	19.2	21.2	24.8	27.1	30.4	34.7	39.4	47.6	58.0	70.8	86.8	107.0	165.0	257.0	394.0	—	—
100,000	7.3	7.7	8.2	8.4	8.8	9.6	10.6	12.4	13.5	15.2	17.4	19.7	23.8	29.0	35.4	43.4	53.4	82.6	128.0	197.0	310.0	490.0
90,000	6.6	6.9	7.3	7.6	7.9	8.6	9.6	11.1	12.2	13.6	15.6	17.7	21.4	26.1	31.9	39.0	48.1	74.3	116.0	177.0	279.0	441.0
80,000	5.8	6.1	6.5	6.8	7.0	7.7	8.5	9.9	10.8	12.1	13.9	15.8	19.0	23.2	28.3	34.7	42.7	66.1	103.0	158.0	248.0	392.0
70,000	5.1	5.4	5.7	5.9	6.1	6.7	7.4	8.7	9.5	10.6	12.1	13.8	16.7	20.3	24.8	30.4	37.4	57.8	88.9	138.0	217.0	343.0
60,000	4.4	4.6	4.9	5.1	5.3	5.8	6.3	7.4	8.1	9.1	10.4	11.8	14.3	17.4	21.2	26.0	32.0	49.6	77.1	118.0	186.0	294.0
50,000	3.7	3.8	4.1	4.2	4.4	4.8	5.3	6.2	6.8	7.6	8.7	9.9	11.9	14.5	17.7	21.7	26.7	41.3	64.2	98.6	155.0	245.0
40,000	2.9	3.1	3.3	3.4	3.5	3.8	4.2	5.0	5.4	6.1	6.9	7.9	9.5	11.6	14.2	17.4	21.4	33.0	51.4	78.8	124.0	196.0
30,000	2.2	2.3	2.5	2.5	2.6	2.9	3.2	3.7	4.1	4.6	5.2	5.9	7.1	8.7	10.6	13.0	16.0	24.8	38.5	59.1	93.0	147.0
20,000	1.5	1.5	1.6	1.7	1.8	1.9	2.1	2.5	2.7	3.0	3.5	3.9	4.8	5.8	7.1	8.7	10.7	16.5	25.7	39.4	62.0	98.0
10,000	0.7	0.8	0.8	0.8	0.9	1.0	1.1	1.2	1.4	1.5	1.7	2.0	2.4	2.9	3.5	4.3	5.3	8.3	12.8	19.7	31.0	49.0
9,000	0.7	0.7	0.7	0.8	0.8	0.9	1.0	1.1	1.2	1.4	1.6	1.8	2.1	2.6	3.2	3.9	4.8	7.4	11.6	17.7	27.9	44.1
8,000	0.6	0.6	0.7	0.7	0.7	0.8	0.9	1.0	1.1	1.2	1.4	1.6	1.9	2.3	2.8	3.5	4.3	6.6	10.3	15.8	24.8	39.2
7,000	0.5	0.5	0.6	0.6	0.6	0.7	0.7	0.9	1.0	1.1	1.2	1.4	1.7	2.0	2.5	3.0	3.7	5.8	8.9	13.8	21.7	34.3
6,000	0.4	0.5	0.5	0.5	0.5	0.6	0.6	0.7	0.9	0.9	1.0	1.2	1.4	1.7	2.1	2.6	3.2	5.0	7.7	11.8	18.6	29.4
5,000	0.4	0.4	0.4	0.4	0.4	0.5	0.5	0.6	0.7	0.8	0.9	1.0	1.2	1.5	1.8	2.2	2.7	4.1	6.4	9.9	15.5	24.5
4,000	0.3	0.3	0.3	0.3	0.4	0.4	0.4	0.5	0.5	0.6	0.7	0.8	1.0	1.2	1.4	1.7	2.1	3.3	5.1	7.9	12.4	19.6
3,000	0.2	0.2	0.3	0.3	0.3	0.3	0.3	0.4	0.4	0.5	0.5	0.6	0.7	0.9	1.1	1.3	1.6	2.5	3.9	5.9	9.3	14.7
2,000	0.2	0.2	0.2	0.2	0.2	0.2	0.2	0.3	0.3	0.3	0.4	0.4	0.5	0.6	0.7	0.9	1.1	1.7	2.6	3.9	6.2	9.8
1,000	0.1	0.1	0.1	0.1	0.1	0.1	0.1	0.1	0.1	0.2	0.2	0.2	0.2	0.3	0.4	0.4	0.5	0.8	1.3	2.0	3.1	4.9
900	0.1	0.1	0.1	0.1	0.1	0.1	0.1	0.1	0.1	0.1	0.2	0.2	0.2	0.3	0.3	0.4	0.5	0.7	1.2	1.8	2.8	4.4
800	0.1	0.1	0.1	0.1	0.1	0.1	0.1	0.1	0.1	0.1	0.1	0.2	0.2	0.2	0.3	0.4	0.4	0.7	1.0	1.6	2.5	3.9
700	0.1	0.1	0.1	0.1	0.1	0.1	0.1	0.1	0.1	0.1	0.1	0.1	0.2	0.2	0.3	0.3	0.4	0.6	0.9	1.4	2.2	3.4
600	—	0.1	0.1	0.1	0.1	0.1	0.1	0.1	0.1	0.1	0.1	0.1	0.1	0.2	0.2	0.3	0.3	0.5	0.8	1.2	1.9	2.9
500	—	—	—	—	—	0.1	0.1	0.1	0.1	0.1	0.1	0.1	0.1	0.2	0.2	0.2	0.3	0.4	0.6	1.0	1.6	2.5
400	—	—	—	—	—	—	—	0.1	0.1	0.1	0.1	0.1	0.1	0.1	0.1	0.2	0.2	0.3	0.5	0.8	1.2	2.0
300	—	—	—	—	—	—	—	—	—	0.1	0.1	0.1	0.1	0.1	0.1	0.1	0.2	0.3	0.4	0.6	0.9	1.5
200	—	—	—	—	—	—	—	—	—	—	—	—	0.1	0.1	0.1	0.1	0.1	0.2	0.3	0.4	0.6	1.0
100	—	—	—	—	—	—	—	—	—	—	—	—	—	—	—	—	0.1	0.1	0.1	0.2	0.3	0.5

* Solid Conductors. Other conductors are stranded.

Note 1—The above table gives voltage drops encountered in a single phase two-wire system. The voltage drops in other systems may be obtained through multiplication by appropriate factors listed below:

System for Which Voltage Drop is Desired	Multiplying Factors for Modification of Values in Table
Single Phase—3 Wire—Line to Line	1.00
Single Phase—3 Wire—Line to Neutral	0.50
Three Phase—3 Wire—Line to Line	0.866
Three Phase—4 Wire—Line to Line	0.866
Three Phase—4 Wire—Line to Neutral	0.50

Note 2—Allowable voltage drops for systems other than single phase, two wire cannot be used directly in the above table. Such drops should be modified through multiplication by the appropriate factor listed below. The voltage thus modified may then be used to obtain the proper wire size directly from the table.

System for Which Allowable Voltage Drop is Known	Multiplying Factor for Modification of Known Value to Permit Direct Use of Table
Single Phase—3 Wire—Line to Line	1.00
Single Phase—3 Wire—Line to Neutral	2.00
Three Phase—3 Wire—Line to Line	1.155
Three Phase—4 Wire—Line to Line	1.155
Three Phase—4 Wire—Line to Neutral	2.00

Note 3—The footage employed in the tabulated ampere feet refers to the length of run of the circuit rather than to the footage of individual conductor.

Note 4—The above table is figured at 60°C since this is an estimate of the average temperature which may be anticipated in service. The table may be used without significant error for conductor temperatures up to and including 75°C.

TABLE 11.1.22 Voltage Drop for Aluminum Conductor in Nonmagnetic Conduit (Power Factor, 90%; Single Phase, 2 Wire; 60 Cycles)

WIRE SIZE AWG or MCM	1000	900	800	750	700	600	500	400	350	300	250	4/0	3/0	2/0	1/0	1	2	4	6	8*	10*	12*
Ampere Feet	Volts Drop																					
500,000	33.0	35.1	37.8	39.4	41.2	45.7	51.5	61.1	67.5	76.5	88.7	102.0	125.0	153.0	189.0	234.0	290.0	454.0	—	—	—	—
400,000	26.4	28.1	30.2	31.5	32.9	36.5	41.2	48.8	54.0	61.2	71.0	81.6	100.0	122.0	151.0	187.0	232.0	363.0	—	—	—	—
300,000	19.8	21.1	22.7	23.6	24.7	27.4	30.9	36.6	40.5	45.9	53.3	61.2	75.0	91.8	113.0	141.0	174.0	272.0	427.0	—	—	—
200,000	13.2	14.1	15.1	15.8	16.5	18.5	20.6	24.4	27.0	30.6	35.5	40.8	50.0	61.2	75.6	93.6	116.0	182.0	284.0	438.0	—	—
100,000	6.6	7.0	7.6	7.9	8.2	9.1	10.3	12.2	13.5	15.3	17.7	20.4	25.0	30.6	37.8	46.8	58.0	90.8	142.0	219.0	346.0	—
90,000	5.9	6.3	6.8	7.1	7.4	8.2	9.3	11.0	12.2	13.8	16.0	18.4	22.5	27.5	34.0	42.1	52.2	81.7	128.0	197.0	312.0	494.0
80,000	5.3	5.6	6.1	6.3	6.6	7.3	8.2	9.8	10.8	12.2	14.2	16.3	20.0	24.5	30.2	37.4	46.4	72.6	114.0	175.0	277.0	438.0
70,000	4.6	4.9	5.3	5.5	5.8	6.4	7.2	8.6	9.5	10.7	12.4	14.3	17.5	21.4	26.4	32.8	40.6	63.5	99.6	153.0	242.0	384.0
60,000	3.9	4.2	4.5	4.7	4.9	5.5	6.2	7.3	8.1	9.2	10.6	12.2	15.0	18.4	22.5	28.1	34.8	54.4	85.4	132.0	208.0	329.0
50,000	3.3	3.5	3.8	3.9	4.1	4.6	5.2	6.1	6.8	7.7	8.9	10.2	12.5	15.3	18.9	23.4	29.0	45.4	71.1	109.0	173.0	274.0
40,000	2.6	2.8	3.0	3.2	3.3	3.7	4.1	4.9	5.4	6.1	7.1	8.2	10.0	12.2	15.1	18.7	23.2	36.3	56.8	87.6	138.0	219.0
30,000	2.0	2.1	2.3	2.4	2.5	2.7	3.1	3.7	4.1	4.6	5.3	6.1	7.5	9.2	11.3	14.1	17.4	27.2	42.7	65.7	104.0	165.0
20,000	1.3	1.4	1.5	1.6	1.7	1.9	2.1	2.4	2.7	3.1	3.6	4.1	5.0	6.1	7.6	9.4	11.6	18.2	28.4	43.8	69.2	110.0
10,000	0.7	0.7	0.8	0.8	0.8	0.9	1.0	1.2	1.4	1.5	1.8	2.0	2.5	3.1	3.8	4.7	5.8	9.1	14.2	21.9	34.6	54.9
9,000	0.6	0.6	0.7	0.7	0.7	0.8	0.9	1.1	1.2	1.4	1.6	1.8	2.3	2.8	3.4	4.2	5.2	8.2	12.8	19.7	31.2	49.4
8,000	0.5	0.6	0.6	0.6	0.7	0.7	0.8	1.0	1.1	1.2	1.4	1.6	2.0	2.5	3.0	3.7	4.6	7.3	11.4	17.5	27.7	43.8
7,000	0.5	0.5	0.5	0.6	0.6	0.6	0.7	0.9	1.0	1.1	1.2	1.4	1.8	2.1	2.6	3.3	4.1	6.4	10.0	15.3	24.2	38.4
6,000	0.4	0.4	0.5	0.5	0.5	0.6	0.6	0.7	0.8	0.9	1.1	1.2	1.5	1.8	2.3	2.8	3.5	5.4	8.5	13.2	20.8	32.9
5,000	0.3	0.4	0.4	0.4	0.4	0.5	0.5	0.6	0.7	0.8	0.9	1.0	1.3	1.5	1.9	2.3	2.9	4.5	7.1	10.9	17.3	27.4
4,000	0.3	0.3	0.3	0.3	0.3	0.4	0.4	0.5	0.5	0.6	0.7	0.8	1.0	1.2	1.5	1.9	2.3	3.6	5.7	8.8	13.8	21.9
3,000	0.2	0.2	0.2	0.2	0.3	0.3	0.3	0.4	0.4	0.5	0.5	0.6	0.8	0.9	1.1	1.4	1.7	2.7	4.3	6.6	10.4	16.5
2,000	0.1	0.1	0.2	0.2	0.2	0.2	0.2	0.2	0.3	0.3	0.4	0.4	0.5	0.6	0.7	0.9	1.2	1.8	2.8	4.4	6.9	11.0
1,000	0.1	0.1	0.1	0.1	0.1	0.1	0.1	0.1	0.1	0.2	0.2	0.2	0.3	0.3	0.4	0.5	0.6	0.9	1.4	2.2	3.5	5.5
900	0.1	0.1	0.1	0.1	0.1	0.1	0.1	0.1	0.1	0.1	0.2	0.2	0.2	0.3	0.3	0.4	0.5	0.8	1.3	2.0	3.1	4.9
800	0.1	0.1	0.1	0.1	0.1	0.1	0.1	0.1	0.1	0.1	0.1	0.2	0.2	0.3	0.3	0.4	0.5	0.7	1.1	1.8	2.8	4.4
700	0.1	0.1	0.1	0.1	0.1	0.1	0.1	0.1	0.1	0.1	0.1	0.1	0.2	0.2	0.3	0.3	0.4	0.6	1.0	1.5	2.4	3.8
600	—	—	0.1	0.1	0.1	0.1	0.1	0.1	0.1	0.1	0.1	0.1	0.2	0.2	0.2	0.3	0.4	0.5	0.9	1.3	2.1	3.3
500	—	—	—	—	—	0.1	0.1	0.1	0.1	0.1	0.1	0.1	0.1	0.2	0.2	0.2	0.3	0.5	0.7	1.1	1.7	2.7
400	—	—	—	—	—	—	—	0.1	0.1	0.1	0.1	0.1	0.1	0.1	0.2	0.2	0.2	0.4	0.6	0.9	1.4	2.2
300	—	—	—	—	—	—	—	—	—	0.1	0.1	0.1	0.1	0.1	0.1	0.1	0.2	0.3	0.4	0.7	1.0	1.7
200	—	—	—	—	—	—	—	—	—	—	—	—	0.1	0.1	0.1	0.1	0.1	0.2	0.3	0.4	0.7	1.1
100	—	—	—	—	—	—	—	—	—	—	—	—	—	—	—	0.1	0.1	0.1	0.1	0.2	0.4	0.6

* Solid Conductors. Other conductors are stranded.

Note 1—The above table gives voltage drops encountered in a single phase two-wire system. The voltage drops in other systems may be obtained through multiplication by appropriate factors listed below:

System for Which Voltage Drop is Desired	Multiplying Factors for Modification of Values in Table
Single Phase—3 Wire—Line to Line	1.00
Single Phase—3 Wire—Line to Neutral	0.50
Three Phase—3 Wire—Line to Line	0.866
Three Phase—4 Wire—Line to Line	0.866
Three Phase—4 Wire—Line to Neutral	0.50

Note 2—Allowable voltage drops for systems other than single phase, two wire cannot be used directly in the above table. Such drops should be modified through multiplication by the appropriate factor listed below. The voltage thus modified may then be used to obtain the proper wire size directly from the table.

System for Which Allowable Voltage Drop is Known	Multiplying Factor for Modification of Known Value to Permit Direct Use of Table
Single Phase—3 Wire—Line to Line	1.00
Single Phase—3 Wire—Line to Neutral	2.00
Three Phase—3 Wire—Line to Line	1.155
Three Phase—4 Wire—Line to Line	1.155
Three Phase—4 Wire—Line to Neutral	2.00

Note 3—The footage employed in the tabulated ampere feet refers to the length of run of the circuit rather than to the footage of in lvidual conductor.

Note 4—The above table is figured at 60°C since this is an estimate of the average temperature which may be anticipated in service. The table may be used without significant error for conductor temperatures up to and including 75°C.

TABLE 11.1.23 Voltage Drop for Aluminum Conductor in Nonmagnetic Conduit (Power Factor, 95%; Single Phase, 2 Wire; 60 Cycles)

WIRE SIZE AWG or MCM	1000	900	800	750	700	600	500	400	350	300	250	4/0	3/0	2/0	1/0	1	2	4	6	8*	10*	12*
Ampere Feet	Volts Drop																					
500,000	30.0	32.2	34.9	36.6	38.5	43.0	49.2	59.3	65.9	75.3	88.1	103.0	126.0	156.0	194.0	241.0	301.0	473.0	—	—	—	—
400,000	24.0	25.8	27.9	29.3	30.8	34.4	38.4	47.4	52.7	60.3	70.5	82.4	100.0	125.0	155.0	193.0	241.0	378.0	—	—	—	—
300,000	18.0	19.3	21.0	22.0	23.1	25.8	29.5	35.5	39.5	45.2	52.9	61.8	75.6	93.6	116.0	145.0	181.0	284.0	447.0	—	—	—
200,000	12.0	12.9	14.0	14.6	15.4	17.2	19.7	23.7	26.4	30.2	35.3	41.2	50.4	62.4	77.6	96.4	120.0	189.0	398.0	460.0	—	—
100,000	6.0	6.4	7.0	7.3	7.7	8.6	9.8	11.8	13.2	15.1	17.6	20.6	25.2	31.2	38.8	48.2	60.2	94.6	149.0	230.0	364.0	—
90,000	5.4	5.8	6.3	6.6	6.9	7.7	8.9	10.7	11.9	13.6	15.9	18.5	22.7	28.1	35.0	43.3	54.2	85.1	134.0	207.0	328.0	—
80,000	4.8	5.2	5.6	5.9	6.2	6.9	7.9	9.5	10.5	12.1	14.1	16.5	20.2	24.9	31.0	38.5	48.2	75.6	119.0	184.0	291.0	462.0
70,000	4.2	4.5	4.9	5.1	5.4	6.0	6.9	8.3	9.2	10.5	12.3	14.4	17.6	21.8	27.2	33.7	42.2	66.2	104.0	161.0	255.0	404.0
60,000	3.6	3.9	4.2	4.4	4.6	5.2	5.9	7.1	7.9	9.0	10.6	12.4	15.1	18.7	23.3	28.9	36.1	56.8	89.3	138.0	218.0	346.0
50,000	3.0	3.2	3.5	3.7	3.9	4.3	4.9	5.9	6.6	7.5	8.8	10.3	12.6	15.6	19.4	24.1	30.1	47.3	74.4	115.0	182.0	289.0
40,000	2.4	2.6	2.8	2.9	3.1	3.4	3.8	4.7	5.3	6.0	7.1	8.2	10.0	12.5	15.5	19.3	24.1	37.8	59.5	92.0	146.0	231.0
30,000	1.8	1.9	2.1	2.2	2.3	2.6	3.0	3.6	4.0	4.5	5.3	6.2	7.6	9.4	11.6	14.5	18.1	28.4	44.7	69.0	109.0	173.0
20,000	1.2	1.3	1.4	1.5	1.6	1.7	2.0	2.4	2.6	3.0	3.5	4.1	5.0	6.2	7.8	9.6	12.0	18.9	39.8	46.0	72.8	115.0
10,000	0.6	0.6	0.7	0.7	0.8	0.9	1.0	1.2	1.3	1.5	1.8	2.1	2.5	3.1	3.9	4.8	6.0	9.5	14.9	23.0	36.4	57.7
9,000	0.6	0.6	0.6	0.7	0.7	0.8	0.9	1.1	1.2	1.4	1.6	1.9	2.3	2.8	3.5	4.3	5.4	8.5	13.4	20.7	32.8	52.0
8,000	0.5	0.5	0.6	0.6	0.6	0.7	0.8	1.0	1.1	1.2	1.4	1.7	2.0	2.5	3.1	3.9	4.8	7.6	11.9	18.4	29.1	46.2
7,000	0.4	0.5	0.5	0.5	0.5	0.6	0.7	0.8	0.9	1.1	1.2	1.4	1.8	2.2	2.8	3.4	4.2	6.6	10.4	16.1	25.5	40.4
6,000	0.4	0.4	0.4	0.4	0.5	0.5	0.6	0.7	0.8	0.9	1.1	1.2	1.5	1.9	2.3	2.9	3.6	5.7	8.9	13.8	21.8	34.6
5,000	0.3	0.3	0.4	0.4	0.4	0.4	0.5	0.6	0.7	0.8	0.9	1.0	1.3	1.6	1.9	2.4	3.0	4.7	7.4	11.5	18.2	28.9
4,000	0.2	0.3	0.3	0.3	0.3	0.3	0.4	0.5	0.5	0.6	0.7	0.8	1.0	1.3	1.6	1.9	2.4	3.8	6.0	9.2	14.6	23.1
3,000	0.2	0.2	0.2	0.2	0.2	0.3	0.3	0.4	0.4	0.5	0.5	0.6	0.8	0.9	1.2	1.5	1.8	2.8	4.5	6.9	10.9	17.3
2,000	0.1	0.1	0.1	0.2	0.2	0.2	0.2	0.2	0.3	0.3	0.4	0.4	0.5	0.6	0.8	1.0	1.2	1.9	4.0	4.6	7.3	11.5
1,000	0.1	0.1	0.1	0.1	0.1	0.1	0.1	0.1	0.1	0.2	0.2	0.2	0.3	0.3	0.4	0.5	0.6	1.0	1.5	2.3	3.6	5.8
900	0.1	0.1	0.1	0.1	0.1	0.1	0.1	0.1	0.1	0.1	0.2	0.2	0.2	0.3	0.4	0.4	0.5	0.9	1.3	2.1	3.3	5.2
800	0.1	0.1	0.1	0.1	0.1	0.1	0.1	0.1	0.1	0.1	0.1	0.2	0.2	0.3	0.3	0.4	0.5	0.8	1.2	1.8	2.9	4.6
700	—	0.1	0.1	0.1	0.1	0.1	0.1	0.1	0.1	0.1	0.1	0.1	0.2	0.2	0.3	0.3	0.4	0.7	1.0	1.6	2.6	4.0
600	—	—	—	—	0.1	0.1	0.1	0.1	0.1	0.1	0.1	0.1	0.2	0.2	0.2	0.3	0.4	0.6	0.9	1.4	2.2	3.5
500	—	—	—	—	—	—	0.1	0.1	0.1	0.1	0.1	0.1	0.1	0.2	0.2	0.2	0.3	0.5	0.7	1.2	1.8	2.9
400	—	—	—	—	—	—	—	0.1	0.1	0.1	0.1	0.1	0.1	0.1	0.2	0.2	0.2	0.4	0.6	0.9	1.5	2.3
300	—	—	—	—	—	—	—	—	—	0.1	0.1	0.1	0.1	0.1	0.1	0.2	0.2	0.3	0.5	0.7	1.1	1.7
200	—	—	—	—	—	—	—	—	—				0.1	0.1	0.1	0.1	0.1	0.2	0.4	0.5	0.7	1.2
100	—	—	—	—	—	—	—	—	—	—	—	—	—	—	—	0.1	0.1	0.1	0.2	0.2	0.4	0.6

* Solid Conductors. Other conductors are stranded.

Note 1—The above table gives voltage drops encountered in a single phase two-wire system. The voltage drops in other systems may be obtained through multiplication by appropriate factors listed below:

System for Which Voltage Drop is Desired	Multiplying Factors for Modification of Values in Table
Single Phase—3 Wire—Line to Line	1.00
Single Phase—3 Wire—Line to Neutral	0.50
Three Phase—3 Wire—Line to Line	0.866
Three Phase—4 Wire—Line to Line	0.866
Three Phase—4 Wire—Line to Neutral	0.50

Note 2—Allowable voltage drops for systems other than single phase, two wire cannot be used directly in the above table. Such drops should be modified through multiplication by the appropriate factor listed below. The voltage thus modified may then be used to obtain the proper wire size directly from the table.

System for Which Allowable Voltage Drop is Known	Multiplying Factor for Modification of Known Value to Permit Direct Use of Table
Single Phase—3 Wire—Line to Line	1.00
Single Phase—3 Wire—Line to Neutral	2.00
Three Phase—3 Wire—Line to Line	1.155
Three Phase—4 Wire—Line to Line	1.155
Three Phase—4 Wire—Line to Neutral	2.00

Note 3—The footage employed in the tabulated ampere feet refers to the length of run of the circuit rather than to the footage of individual conductor.

Note 4—The above table is figured at 60°C since this is an estimate of the average temperature which may be anticipated in service. The table may be used without significant error for conductor temperatures up to and including 75°C.

TABLE 11.1.24 Voltage Drop for Aluminum Conductor in Nonmagnetic Conduit (Power Factor, 100%; Single Phase, 2 Wire; 60 Cycles)

WIRE SIZE AWG or MCM	1000	900	800	750	700	600	500	400	350	300	250	4/0	3/0	2/0	1/0	1	2	4	6	8*	10*	12*
Ampere Feet	Volts Drop																					
500,000	20.7	22.8	25.6	27.3	29.2	33.9	40.4	50.7	57.5	67.2	80.5	95.9	120.0	151.0	191.0	240.0	303.0	483.0	—	—	—	—
400,000	16.5	18.2	20.5	21.8	23.4	27.1	32.3	40.6	46.0	53.7	64.4	76.8	96.0	121.0	153.0	192.0	241.0	386.0	—	—	—	—
300,000	12.4	13.7	15.3	16.4	17.5	20.3	24.3	30.4	34.5	40.6	48.3	57.6	72.0	90.6	115.0	144.0	182.0	290.0	460.0	—	—	—
200,000	8.2	9.1	10.2	10.9	11.7	13.5	16.2	20.3	23.0	26.9	32.2	38.4	48.0	60.4	76.4	96.0	121.0	193.0	307.0	478.0	—	—
100,000	4.1	4.6	5.1	5.5	5.8	6.8	8.1	10.1	11.5	13.4	16.1	19.2	24.0	30.2	38.2	48.0	60.6	96.6	153.0	239.0	380.0	—
90,000	3.7	4.1	4.6	4.9	5.3	6.1	7.3	9.1	10.3	12.1	14.5	17.3	21.6	27.2	34.4	43.2	54.6	87.0	138.0	215.0	342.0	—
80,000	3.3	3.7	4.1	4.4	4.7	5.4	6.5	8.1	9.2	10.8	12.9	15.3	19.2	24.2	30.5	38.4	48.5	77.3	123.0	191.0	304.0	483.0
70,000	2.9	3.2	3.6	3.8	4.1	4.8	5.7	7.1	8.1	9.4	11.3	13.4	16.8	21.1	27.6	33.6	42.4	67.6	107.0	167.0	266.0	423.0
60,000	2.5	2.7	3.1	3.3	3.5	4.1	4.9	6.1	6.9	8.1	9.7	11.5	14.4	18.1	22.9	28.8	36.4	58.0	92.0	144.0	228.0	362.0
50,000	2.1	2.3	2.6	2.7	2.9	3.4	4.0	5.1	5.8	6.7	8.1	9.6	12.0	15.1	19.1	24.0	30.3	48.3	76.7	120.0	190.0	302.0
40,000	1.7	1.8	2.1	2.2	2.3	2.7	3.2	4.1	4.6	5.4	6.4	7.7	9.6	12.1	15.3	19.2	24.1	38.6	61.4	95.6	152.0	242.0
30,000	1.2	1.4	1.5	1.6	1.8	2.0	2.4	3.0	3.5	4.1	4.8	5.8	7.2	9.1	11.5	14.4	18.2	29.0	46.0	71.7	114.0	181.0
20,000	0.8	0.9	1.0	1.1	1.2	1.4	1.6	2.0	2.3	2.7	3.2	3.8	4.8	6.0	7.6	9.6	12.1	19.3	30.7	47.8	76.0	121.0
10,000	0.4	0.5	0.5	0.6	0.6	0.7	0.8	1.0	1.2	1.3	1.6	1.9	2.4	3.0	3.8	4.8	6.1	9.7	15.3	23.9	38.0	60.4
9,000	0.4	0.4	0.5	0.5	0.5	0.6	0.7	0.9	1.0	1.2	1.5	1.7	2.2	2.7	3.4	4.3	5.5	8.7	13.8	21.5	34.2	54.4
8,000	0.3	0.4	0.4	0.4	0.5	0.5	0.7	0.8	0.9	1.1	1.3	1.5	1.9	2.4	3.1	3.8	4.9	7.7	12.3	19.1	30.4	48.3
7,000	0.3	0.3	0.4	0.4	0.4	0.5	0.6	0.7	0.8	0.9	1.1	1.3	1.7	2.1	2.8	3.4	4.2	6.8	10.7	16.7	26.6	42.3
6,000	0.3	0.3	0.3	0.3	0.4	0.4	0.5	0.6	0.7	0.8	1.0	1.2	1.4	1.8	2.3	2.9	3.6	5.8	9.2	14.4	22.8	36.2
5,000	0.2	0.2	0.3	0.3	0.3	0.3	0.4	0.5	0.6	0.7	0.8	1.0	1.2	1.5	1.9	2.4	3.0	4.8	7.7	12.0	19.0	30.2
4,000	0.2	0.2	0.2	0.2	0.2	0.3	0.3	0.4	0.5	0.5	0.6	0.7	1.0	1.2	1.5	1.9	2.4	3.9	6.1	9.6	15.2	24.2
3,000	0.1	0.1	0.2	0.2	0.2	0.2	0.2	0.3	0.4	0.4	0.5	0.6	0.7	0.9	1.2	1.4	1.8	2.9	4.6	7.2	11.4	18.1
2,000	0.1	0.1	0.1	0.1	0.1	0.1	0.2	0.2	0.2	0.3	0.3	0.4	0.5	0.6	0.8	1.0	1.2	1.9	3.1	4.8	7.6	12.1
1,000	—	0.1	0.1	0.1	0.1	0.1	0.1	0.1	0.1	0.1	0.2	0.2	0.2	0.3	0.4	0.5	0.6	1.0	1.5	2.4	3.8	6.0
900	—	—	0.1	0.1	0.1	0.1	0.1	0.1	0.1	0.1	0.2	0.2	0.2	0.3	0.3	0.4	0.6	0.9	1.4	2.2	3.4	5.4
800	—	—	—	—	0.1	0.1	0.1	0.1	0.1	0.1	0.1	0.2	0.2	0.2	0.3	0.4	0.5	0.8	1.2	1.9	3.0	4.8
700	—	—	—	—	—	0.1	0.1	0.1	0.1	0.1	0.1	0.1	0.2	0.2	0.3	0.3	0.4	0.7	1.1	1.7	2.7	4.2
600	—	—	—	—	—	—	0.1	0.1	0.1	0.1	0.1	0.1	0.1	0.2	0.2	0.3	0.4	0.6	0.9	1.4	2.3	3.6
500	—	—	—	—	—	—	—	0.1	0.1	0.1	0.1	0.1	0.1	0.2	0.2	0.2	0.3	0.5	0.8	1.2	1.9	3.0
400	—	—	—	—	—	—	—	—	0.1	0.1	0.1	0.1	0.1	0.1	0.2	0.2	0.2	0.3	0.6	1.0	1.5	2.4
300	—	—	—	—	—	—	—	—	—	—	0.1	0.1	0.1	0.1	0.1	0.1	0.2	0.3	0.5	0.7	1.1	1.8
200	—	—	—	—	—	—	—	—	—	—	—	—	0.1	0.1	0.1	0.1	0.1	0.2	0.3	0.5	0.8	1.2
100	—	—	—	—	—	—	—	—	—	—	—	—	—	—	—	0.1	0.1	0.1	0.2	0.2	0.4	0.6

* Solid Conductors. Other conductors are stranded.

Note 1—The above table gives voltage drops encountered in a single phase two-wire system. The voltage drops in other systems may be obtained through multiplication by appropriate factors listed below:

System for Which Voltage Drop Is Desired	Multiplying Factors for Modification of Values in Table
Single Phase—3 Wire—Line to Line	1.00
Single Phase—3 Wire—Line to Neutral	0.50
Three Phase—3 Wire—Line to Line	0.866
Three Phase—4 Wire—Line to Line	0.866
Three Phase—4 Wire—Line to Neutral	0.50

Note 2—Allowable voltage drops for systems other than single phase, two wire cannot be used directly in the above table. Such drops should be modified through multiplication by the appropriate factor listed below. The voltage thus modified may then be used to obtain the proper wire size directly from the table.

System for Which Allowable Voltage Drop Is Known	Multiplying Factor for Modification of Known Value to Permit Direct Use of Table
Single Phase—3 Wire—Line to Line	1.00
Single Phase—3 Wire—Line to Neutral	2.00
Three Phase—3 Wire—Line to Line	1.155
Three Phase—4 Wire—Line to Line	1.155
Three Phase—4 Wire—Line to Neutral	2.00

Note 3—The footage employed in the tabulated ampere feet refers to the length of run of the circuit rather than to the footage of individual conductor.

Note 4—The above table is figured at 60°C since this is an estimate of the average temperature which may be anticipated in service. The table may be used without significant error for conductor temperatures up to and including 75°C.

11.1.25 Voltage Drop for CU Conductor, Direct Current (*see page 11.22*)

11.1.26 Voltage Drop for CU Conductor in Magnetic Conduit, 70 Percent PF (*see page 11.23*)

11.1.27 Voltage Drop for CU Conductor in Magnetic Conduit, 80 Percent PF (*see page 11.24*)

11.1.28 Voltage Drop for CU Conductor in Magnetic Conduit, 90 Percent PF (*see page 11.25*)

11.1.29 Voltage Drop for CU Conductor in Magnetic Conduit, 95 Percent PF (*see page 11.26*)

11.1.30 Voltage Drop for CU Conductor in Magnetic Conduit, 100 Percent PF (*see page 11.27*)

11.1.31 Voltage Drop for CU Conductor in Nonmagnetic Conduit, 70 Percent PF (*see page 11.28*)

11.1.32 Voltage Drop for CU Conductor in Nonmagnetic Conduit, 80 Percent PF (*see page 11.29*)

11.1.33 Voltage Drop for CU Conductor in Nonmagnetic Conduit, 90 Percent PF (*see page 11.30*)

11.1.34 Voltage Drop for CU Conductor in Nonmagnetic Conduit, 95 Percent PF (*see page 11.31*)

11.1.35 Voltage Drop for CU Conductor in Nonmagnetic Conduit, 100 Percent PF (*see page 11.32*)

11.1.36 Voltage-Drop Curves for Typical Interleaved Construction of Copper Busway at Rated Load, Assuming 70°C (158°F) as the Operating Temperature (*see page 11.33*)

TABLE 11.1.25 Voltage Drop for Copper Conductor—Direct Current

WIRE SIZE AWG or MCM	1000	900	800	750	700	600	500	400	350	300	250	4/0	3/0	2/0	1/0	1	2	4	6	8*	10*	12*	14*
Ampere Feet	Volts Drop																						
500,000	13.0	14.0	16.0	17.0	18.0	21.0	26.0	32.0	36.0	42.0	50.0	59.0	76.0	96.0	121.0	152.0	192.0	306.0	483.0	—	—	—	—
400,000	10.4	11.2	12.8	13.6	14.4	16.8	20.8	25.6	28.8	33.6	40.0	47.2	60.8	76.8	96.8	122.0	153.0	244.0	386.0	—	—	—	—
300,000	7.8	8.4	9.6	10.2	10.8	12.6	15.6	19.2	21.6	25.2	30.0	35.4	45.6	57.6	72.6	91.2	115.0	183.0	290.0	450.0	—	—	—
200,000	5.2	5.6	6.4	6.8	7.2	8.4	10.4	12.8	14.4	16.8	20.0	23.6	30.4	38.4	48.4	60.8	76.8	122.0	193.0	300.0	480.0	—	—
100 000	2.6	2.8	3.2	3.4	3.6	4.2	5.2	6.4	7.2	8.4	10.0	11.8	15.2	19.2	24.2	30.4	38.4	61.2	96.6	150.0	240.0	384.0	—
90,000	2.3	2.5	2.9	3.1	3.2	3.8	4.7	5.8	6.5	7.6	9.0	10.6	13.7	17.3	21.8	27.4	34.6	55.0	87.0	135.0	216.0	345.0	—
80,000	2.0	2.2	2.6	2.7	2.9	3.4	4.2	5.1	5.8	6.7	8.0	9.4	12.2	15.4	19.4	24.3	30.7	49.0	77.3	120.0	192.0	307.0	487.0
70,000	1.8	2.0	2.2	2.4	2.5	2.9	3.6	4.5	5.0	5.9	7.0	8.3	10.6	13.4	16.9	21.3	26.9	42.8	67.6	105.0	168.0	269.0	426.0
60,000	1.6	1.7	1.9	2.0	2.2	2.5	3.1	3.8	4.3	5.0	6.0	7.1	9.6	11.5	14.5	18.2	23.0	36.7	58.0	90.0	144.0	230.0	365.0
50,000	1.3	1.4	1.6	1.7	1.8	2.1	2.6	3.2	3.6	4.2	5.0	5.9	7.6	9.6	12.1	15.2	19.2	30.6	48.3	74.9	120.0	192.0	304.0
40,000	1.0	1.1	1.3	1.4	1.4	1.7	2.1	2.6	2.9	3.4	4.0	4.7	6.1	7.7	9.7	12.2	15.3	24.4	38.6	60.0	96.0	154.0	243.0
30,000	0.7	0.8	0.9	1.0	1.1	1.3	1.6	1.9	2.2	2.5	3.0	3.5	4.6	5.8	7.3	9.1	11.5	18.3	29.0	45.0	72.0	115.0	182.0
20,000	0.5	0.6	0.6	0.7	0.7	0.8	1.0	1.3	1.4	1.7	2.0	2.4	3.0	3.8	4.8	6.1	7.7	12.2	19.3	30.0	48.0	76.8	122.0
10,000	0.3	0.3	0.3	0.3	0.4	0.4	0.5	0.6	0.7	0.8	1.0	1.2	1.5	1.9	2.4	3.0	3.8	6.1	9.7	20.0	24.0	38.4	60.8
9,000	0.2	0.3	0.3	0.3	0.3	0.4	0.5	0.6	0.7	0.8	0.9	1.1	1.4	1.7	2.2	2.7	3.5	5.5	8.7	13.5	21.6	34.5	54.7
8,000	0.2	0.2	0.3	0.3	0.3	0.3	0.4	0.5	0.6	0.7	0.8	0.9	1.2	1.5	1.9	2.4	3.1	4.9	7.7	12.0	19.2	30.7	48.7
7,000	0.2	0.2	0.2	0.2	0.2	0.3	0.4	0.5	0.5	0.6	0.7	0.8	1.1	1.3	1.7	2.1	2.7	4.3	6.8	10.5	16.8	26.9	42.6
6,000	0.1	0.2	0.2	0.2	0.2	0.2	0.3	0.4	0.4	0.5	0.6	0.7	1.0	1.2	1.5	1.8	2.3	3.7	5.8	9.0	14.4	23.0	36.5
5,000	0.1	0.1	0.2	0.2	0.2	0.2	0.3	0.3	0.4	0.4	0.5	0.6	0.7	1.0	1.2	1.5	1.9	3.1	4.8	7.5	12.0	19.2	30.4
4,000	0.1	0.1	0.1	0.1	0.1	0.2	0.2	0.3	0.3	0.3	0.4	0.5	0.6	0.8	0.9	1.2	1.5	2.4	3.9	6.0	9.6	15.4	24.3
3,000	0.1	0.1	0.1	0.1	0.1	0.1	0.2	0.2	0.2	0.2	0.3	0.4	0.5	0.6	0.7	0.9	1.2	1.8	2.9	4.5	7.2	11.5	18.2
2,000	0.1	0.1	0.1	0.1	0.1	0.1	0.1	0.1	0.1	0.2	0.2	0.2	0.3	0.4	0.5	0.6	0.8	1.2	1.9	3.0	4.8	7.7	12.2
1,000	—	—	—	—	—	—	0.1	0.1	0.1	0.1	0.1	0.1	0.2	0.2	0.2	0.3	0.4	0.6	1.0	1.5	2.4	3.8	6.1
900	—	—	—	—	—	—	—	0.1	0.1	0.1	0.1	0.1	0.1	0.2	0.2	0.3	0.4	0.6	0.9	1.4	2.2	3.5	5.5
800	—	—	—	—	—	—	—	—	0.1	0.1	0.1	0.1	0.1	0.2	0.2	0.2	0.3	0.5	0.8	1.2	1.9	3.1	4.9
700	—	—	—	—	—	—	—	—	0.1	0.1	0.1	0.1	0.1	0.1	0.2	0.2	0.3	0.4	0.7	1.1	1.7	2.7	4.3
600	—	—	—	—	—	—	—	—	—	0.1	0.1	0.1	0.1	0.1	0.2	0.2	0.2	0.3	0.6	0.9	1.4	2.3	3.7
500	—	—	—	—	—	—	—	—	—	—	0.1	0.1	0.1	0.1	0.1	0.2	0.2	0.3	0.5	0.8	1.2	1.9	3.0
400	—	—	—	—	—	—	—	—	—	—	—	—	0.1	0.1	0.1	0.1	0.2	0.2	0.4	0.6	1.0	1.5	2.4
300	—	—	—	—	—	—	—	—	—	—	—	—	—	0.1	0.1	0.1	0.1	0.2	0.3	0.5	0.7	1.2	1.8
200	—	—	—	—	—	—	—	—	—	—	—	—	—	—	—	0.1	0.1	0.1	0.2	0.3	0.5	0.8	1.2
100	—	—	—	—	—	—	—	—	—	—	—	—	—	—	—	—	—	—	0.1	0.2	0.2	0.4	0.6

* Solid Conductors. Other conductors are stranded.

Note 1—The footage employed in the tabulated ampere feet refers to the length of run of the circuit rather than to the footage of individual conductor.

Note 2—The above table is figured at 60°C since this is an estimate of the average temperature which may be anticipated in service. The table may be used without significant error for conductor temperatures up to and including 75°C.

TABLE 11.1.26 Voltage Drop for Copper Conductor in Magnetic Conduit (Power Factor, 70%; Single Phase, 2 Wire; 60 Cycles)

WIRE SIZE AWG or MCM	1000	900	800	750	700	600	500	400	350	300	250	4/0	3/0	2/0	1/0	1	2	4	6	8*	10*	12*	14*
Ampere Feet	Volts Drop																						
500,000	41.0	42.0	44.0	45.0	46.0	48.0	51.0	56.0	60.0	64.0	71.0	77.0	89.0	104.0	122.0	144.0	173.0	254.0	380.0	—	—	—	—
400,000	32.8	33.6	35.2	36.0	36.8	38.4	40.8	44.8	48.0	51.2	56.8	61.6	71.2	84.2	97.6	115.0	138.0	203.0	304.0	456.0	—	—	—
300,000	24.6	25.2	26.4	27.0	27.6	28.8	30.6	33.6	36.0	38.4	42.6	46.2	53.4	62.4	73.2	86.4	103.0	152.0	228.0	342.0	—	—	—
200,000	16.4	16.8	17.6	18.0	18.4	19.2	20.4	22.4	24.0	25.6	28.4	30.8	35.6	41.6	48.8	57.6	69.2	101.0	152.0	228.0	354.0	—	—
100,000	8.2	8.4	8.8	9.0	9.2	9.6	10.2	11.2	12.0	12.8	14.2	15.4	17.8	20.8	24.4	28.8	34.6	50.8	76.0	114.0	177.0	278.0	436.0
90,000	7.4	7.5	7.9	8.1	8.3	8.6	9.2	10.1	10.8	11.5	12.7	13.8	16.0	18.7	21.9	25.9	31.1	45.7	68.4	103.0	159.0	250.0	392.0
80,000	6.6	6.7	7.0	7.2	7.4	7.7	8.2	8.9	9.6	10.2	11.4	12.3	14.2	16.6	19.5	23.0	27.7	40.6	60.8	91.2	142.0	222.0	349.0
70,000	5.7	5.9	6.2	6.3	6.4	6.7	7.1	7.8	8.4	8.9	9.9	10.8	12.5	14.5	17.1	20.2	24.2	35.5	53.2	79.7	124.0	194.0	306.0
60,000	4.9	5.0	5.3	5.4	5.5	5.8	6.1	6.7	7.2	7.7	8.5	9.2	10.7	12.5	14.6	17.3	20.8	30.5	45.6	68.3	106.0	167.0	262.0
50,000	4.1	4.2	4.4	4.5	4.6	4.8	5.1	5.6	6.0	6.4	7.1	7.7	8.9	10.4	12.2	14.4	17.3	25.4	38.0	56.9	88.5	139.0	218.0
40,000	3.3	3.4	3.5	3.6	3.7	3.8	4.1	4.5	4.8	5.1	5.7	6.2	7.1	8.4	9.8	11.5	13.8	20.3	30.4	45.6	70.8	111.0	174.0
30,000	2.5	2.5	2.6	2.7	2.8	2.9	3.1	3.4	3.6	3.8	4.3	4.6	5.3	6.2	7.3	8.6	10.3	15.2	22.8	34.2	53.1	83.4	131.0
20,000	1.6	1.7	1.8	1.8	1.8	1.9	2.0	2.2	2.4	2.6	2.8	3.1	3.6	4.2	4.9	5.8	6.9	10.1	15.2	22.8	35.4	55.6	87.2
10,000	0.8	0.8	0.9	0.9	0.9	1.0	1.0	1.1	1.2	1.3	1.4	1.5	1.8	2.1	2.4	2.9	3.5	5.1	7.6	11.4	17.7	27.8	43.6
9,000	0.7	0.8	0.8	0.8	0.8	0.9	0.9	1.0	1.1	1.2	1.3	1.4	1.6	1.9	2.2	2.6	3.1	4.6	6.8	10.3	15.9	25.0	39.2
8,000	0.7	0.7	0.7	0.7	0.7	0.8	0.8	0.9	1.0	1.0	1.1	1.2	1.4	1.7	2.0	2.3	2.8	4.1	6.1	9.1	14.2	22.2	34.9
7,000	0.6	0.6	0.6	0.6	0.6	0.7	0.7	0.8	0.8	0.9	1.0	1.1	1.3	1.5	1.7	2.0	2.4	3.6	5.3	7.8	12.4	19.4	30.5
6,000	0.5	0.5	0.5	0.5	0.6	0.6	0.6	0.7	0.7	0.8	0.9	0.9	1.1	1.3	1.5	1.7	2.1	3.1	4.6	6.8	10.6	16.7	26.2
5,000	0.4	0.4	0.4	0.5	0.5	0.5	0.5	0.6	0.6	0.6	0.7	0.8	0.9	1.0	1.2	1.4	1.7	2.5	3.8	5.7	8.9	13.9	21.8
4,000	0.3	0.3	0.4	0.4	0.4	0.4	0.4	0.5	0.5	0.5	0.6	0.6	0.7	0.8	1.0	1.2	1.4	2.0	3.0	4.6	7.1	11.1	17.4
3,000	0.3	0.3	0.3	0.3	0.3	0.3	0.3	0.3	0.4	0.4	0.4	0.5	0.5	0.6	0.7	0.9	1.0	1.5	2.3	3.4	5.3	8.3	13.1
2,000	0.2	0.2	0.2	0.2	0.2	0.2	0.2	0.2	0.2	0.3	0.3	0.3	0.4	0.4	0.5	0.6	0.7	1.0	1.5	2.3	3.5	5.6	8.7
1,000	0.1	0.1	0.1	0.1	0.1	0.1	0.1	0.1	0.1	0.1	0.1	0.2	0.2	0.2	0.2	0.3	0.4	0.5	0.8	1.1	1.8	2.8	4.4
900	0.1	0.1	0.1	0.1	0.1	0.1	0.1	0.1	0.1	0.1	0.1	0.1	0.2	0.2	0.2	0.3	0.3	0.5	0.7	1.0	1.6	2.5	3.9
800	0.1	0.1	0.1	0.1	0.1	0.1	0.1	0.1	0.1	0.1	0.1	0.1	0.1	0.2	0.2	0.2	0.3	0.4	0.6	0.9	1.4	2.2	3.5
700	0.1	0.1	0.1	0.1	0.1	0.1	0.1	0.1	0.1	0.1	0.1	0.1	0.1	0.2	0.2	0.2	0.2	0.4	0.5	0.8	1.2	1.9	3.1
600	0.1	0.1	0.1	0.1	0.1	0.1	0.1	0.1	0.1	0.1	0.1	0.1	0.1	0.1	0.2	0.2	0.2	0.3	0.5	0.7	1.1	1.7	2.6
500	—	—	—	0.1	0.1	0.1	0.1	0.1	0.1	0.1	0.1	0.1	0.1	0.1	0.1	0.1	0.2	0.3	0.4	0.6	0.9	1.4	2.2
400	—	—	—	—	—	—	—	0.1	0.1	0.1	0.1	0.1	0.1	0.1	0.1	0.1	0.1	0.2	0.3	0.5	0.7	1.1	1.7
300	—	—	—	—	—	—	—	—	—	—	—	0.1	0.1	0.1	0.1	0.1	0.1	0.2	0.2	0.3	0.5	0.8	1.3
200	—	—	—	—	—	—	—	—	—	—	—	—	—	—	0.1	0.1	0.1	0.1	0.2	0.2	0.4	0.6	0.9
100	—	—	—	—	—	—	—	—	—	—	—	—	—	—	—	—	—	0.1	0.1	0.1	0.2	0.3	0.4

* Solid Conductors. Other conductors are stranded.

Note 1—The above table gives voltage drops encountered in a single phase two-wire system. The voltage drops in other systems may be obtained through multiplication by appropriate factors listed below:

System for Which Voltage Drop Is Desired	Multiplying Factors for Modification of Values in Table
Single Phase—3 Wire—Line to Line	1.00
Single Phase—3 Wire—Line to Neutral	0.50
Three Phase—3 Wire—Line to Line	0.866
Three Phase—4 Wire—Line to Line	0.866
Three Phase—4 Wire—Line to Neutral	0.50

Note 2—Allowable voltage drops for systems other than single phase, two-wire cannot be used directly in the above table. Such drops should be modified through multiplication by the appropriate factor listed below. The voltage thus modified may then be used to obtain the proper wire size directly from the table.

System for Which Allowable Voltage Drop Is Known	Multiplying Factor for Modification of Known Value to Permit Direct Use of Table
Single Phase—3 Wire—Line to Line	1.00
Single Phase—3 Wire—Line to Neutral	2.00
Three Phase—3 Wire—Line to Line	1.155
Three Phase—4 Wire—Line to Line	1.155
Three Phase—4 Wire—Line to Neutral	2.00

Note 3—The footage employed in the tabulated ampere feet refers to the length of run of the circuit rather than to the footage of individual conductor.

Note 4—The above table is figured at 60°C since this is an estimate of the average temperature which may be anticipated in service. The table may be used without significant error for conductor temperatures up to and including 75°C.

TABLE 11.1.27 Voltage Drop for Copper Conductor in Magnetic Conduit (Power Factor, 80%; Single Phase, 2 Wire; 60 Cycles)

WIRE SIZE AWG or MCM	1000	900	800	750	700	600	500	400	350	300	250	4/0	3/0	2/0	1/0	1	2	4	6	8*	10*	12*	14*
Ampere Feet	Volts Drop																						
500,000	38.0	39.0	41.0	42.0	43.0	46.0	49.0	55.0	58.0	64.0	71.0	78.0	92.0	108.0	130.0	153.0	186.0	278.0	421.0	—	—	—	—
400,000	30.4	31.2	32.8	33.6	34.4	36.8	39.2	44.0	46.4	51.2	56.8	62.4	73.6	86.4	104.0	122.0	148.0	222.0	336.0	—	—	—	—
300,000	22.8	23.4	24.6	25.2	25.8	27.6	29.4	33.0	34.8	38.4	42.6	46.8	55.2	64.8	78.0	91.8	111.0	166.0	252.0	381.0	—	—	—
200,000	15.2	15.6	16.4	16.8	17.2	18.4	19.6	22.0	23.2	25.6	28.4	31.2	36.8	43.2	52.0	61.2	74.0	111.0	168.0	254.0	398.0	—	—
100,000	7.6	7.8	8.2	8.4	8.6	9.2	9.8	11.0	11.6	12.8	14.2	15.6	18.4	21.6	26.0	30.6	37.2	55.6	84.2	127.0	199.0	314.0	494.0
90,000	6.9	7.0	7.4	7.6	7.7	8.3	8.8	9.9	10.4	11.5	12.8	14.0	16.6	19.4	23.4	27.5	33.5	50.0	75.8	115.0	179.0	283.0	445.0
80,000	6.1	6.2	6.6	6.7	6.9	7.4	7.8	8.8	9.3	10.2	11.4	12.5	14.7	17.3	20.8	24.5	29.8	44.5	67.4	102.0	160.0	252.0	395.0
70,000	5.3	5.5	5.7	5.9	6.0	6.4	6.9	7.7	8.1	8.9	9.9	10.9	12.9	15.1	18.2	21.4	26.0	38.9	58.9	89.3	140.0	220.0	346.0
60,000	4.6	4.7	4.9	5.0	5.2	5.5	5.9	6.6	6.9	7.7	8.5	9.4	11.0	12.9	15.6	18.4	22.3	33.4	50.5	76.5	120.0	188.0	296.0
50,000	3.8	3.9	4.1	4.2	4.3	4.6	4.9	5.5	5.8	6.4	7.1	7.8	9.2	10.8	13.0	15.3	18.6	27.8	42.1	63.7	99.7	157.0	247.0
40,000	3.0	3.1	3.3	3.4	3.4	3.7	3.9	4.4	4.6	5.1	5.7	6.2	7.4	8.6	10.4	12.2	14.8	22.2	33.6	50.8	79.6	126.0	199.0
30,000	2.3	2.3	2.5	2.5	2.6	2.8	2.9	3.3	3.5	3.8	4.3	4.7	5.5	6.5	7.8	9.2	11.1	16.6	25.2	38.1	59.7	94.2	149.0
20,000	1.5	1.6	1.6	1.7	1.7	1.8	2.0	2.2	2.3	2.6	2.8	3.1	3.7	4.3	5.2	6.1	7.4	11.1	16.8	25.4	39.8	62.8	99.8
10,000	0.8	0.8	0.8	0.8	0.9	0.9	1.0	1.1	1.2	1.3	1.4	1.6	1.8	2.2	2.6	3.1	3.7	5.6	8.4	12.7	19.9	31.4	49.4
9,000	0.7	0.7	0.7	0.8	0.8	0.8	0.9	1.0	1.0	1.2	1.3	1.4	1.7	1.9	2.3	2.8	3.4	5.0	7.6	11.5	17.9	28.3	44.5
8,000	0.6	0.6	0.7	0.7	0.7	0.7	0.8	0.9	0.9	1.0	1.1	1.3	1.5	1.7	2.1	2.5	3.0	4.5	6.7	10.2	16.0	25.2	39.5
7,000	0.5	0.6	0.6	0.6	0.6	0.6	0.7	0.8	0.8	0.9	1.0	1.1	1.3	1.5	1.8	2.1	2.6	3.9	5.9	8.9	14.0	22.0	34.6
6,000	0.5	0.5	0.5	0.5	0.5	0.6	0.6	0.7	0.7	0.8	0.9	0.9	1.1	1.3	1.6	1.8	2.2	3.3	5.1	7.7	12.0	18.8	29.6
5,000	0.4	0.4	0.4	0.4	0.4	0.5	0.5	0.6	0.6	0.6	0.7	0.8	0.9	1.1	1.3	1.5	1.9	2.8	4.2	6.4	10.0	15.7	24.7
4,000	0.3	0.3	0.3	0.3	0.3	0.4	0.4	0.4	0.5	0.5	0.6	0.6	0.7	0.9	1.0	1.2	1.5	2.2	3.4	5.1	8.0	12.6	19.9
3,000	0.2	0.2	0.3	0.3	0.3	0.3	0.3	0.3	0.4	0.4	0.4	0.5	0.6	0.7	0.8	0.9	1.1	1.7	2.5	3.8	6.0	9.4	14.9
2,000	0.2	0.2	0.2	0.2	0.2	0.2	0.2	0.2	0.2	0.3	0.3	0.3	0.4	0.4	0.5	0.6	0.7	1.1	1.7	2.5	4.0	6.3	10.0
1,000	0.1	0.1	0.1	0.1	0.1	0.1	0.1	0.1	0.1	0.1	0.1	0.2	0.2	0.2	0.3	0.3	0.4	0.6	0.8	1.3	2.0	3.1	4.9
900	0.1	0.1	0.1	0.1	0.1	0.1	0.1	0.1	0.1	0.1	0.1	0.1	0.2	0.2	0.2	0.3	0.3	0.5	0.8	1.2	1.8	2.8	4.5
800	0.1	0.1	0.1	0.1	0.1	0.1	0.1	0.1	0.1	0.1	0.1	0.1	0.2	0.2	0.2	0.3	0.3	0.5	0.7	1.0	1.6	2.5	3.9
700	0.1	0.1	0.1	0.1	0.1	0.1	0.1	0.1	0.1	0.1	0.1	0.1	0.1	0.2	0.2	0.2	0.3	0.4	0.6	0.9	1.4	2.2	3.5
600	0.1	0.1	0.1	0.1	0.1	0.1	0.1	0.1	0.1	0.1	0.1	0.1	0.1	0.1	0.2	0.2	0.2	0.3	0.5	0.8	1.2	1.9	2.9
500	—	—	—	—	—	0.1	0.1	0.1	0.1	0.1	0.1	0.1	0.1	0.1	0.1	0.2	0.2	0.3	0.4	0.6	1.0	1.6	2.5
400	—	—	—	—	—	—	—	—	0.1	0.1	0.1	0.1	0.1	0.1	0.1	0.1	0.2	0.2	0.3	0.5	0.8	1.3	2.0
300	—	—	—	—	—	—	—	—	—	—	—	0.1	0.1	0.1	0.1	0.1	0.1	0.2	0.3	0.4	0.6	0.9	1.5
200	—	—	—	—	—	—	—	—	—	—	—	—	—	—	0.1	0.1	0.1	0.1	0.2	0.3	0.4	0.6	1.0
100	—	—	—	—	—	—	—	—	—	—	—	—	—	—	—	—	—	0.1	0.1	0.1	0.2	0.3	0.5

* Solid Conductors. Other conductors are stranded.

Note 1—The above table gives voltage drops encountered in a single phase two-wire system. The voltage drops in other systems may be obtained through multiplication by appropriate factors listed below:

System for Which Voltage Drop is Desired	Multiplying Factors for Modification of Values in Table
Single Phase—3 Wire—Line to Line	1.00
Single Phase—3 Wire—Line to Neutral	0.50
Three Phase—3 Wire—Line to Line	0.866
Three Phase—4 Wire—Line to Line	0.866
Three Phase—4 Wire—Line to Neutral	0.50

Note 2 Allowable voltage drops for systems other than single phase, two-wire cannot be used directly in the above table. Such drops should be modified through multiplication by the appropriate factor listed below. The voltage thus modified may then be used to obtain the proper wire size directly from the table.

System for Which Allowable Voltage Drop is Known	Multiplying Factor for Modification of Known Value to Permit Direct Use of Table
Single Phase—3 Wire—Line to Line	1.00
Single Phase—3 Wire—Line to Neutral	2.00
Three Phase—3 Wire—Line to Line	1.155
Three Phase—4 Wire—Line to Line	1.155
Three Phase—4 Wire—Line to Neutral	2.00

Note 3—The footage employed in the tabulated ampere feet refers to the length of run of the circuit rather than to the footage of individual conductor.

Note 4—The above table is figured at 60°C since this is an estimate of the average temperature which may be anticipated in service. The table may be used without significant error for conductor temperatures up to and including 75°C.

TABLE 11.1.28 Voltage Drop for Copper Conductor in Magnetic Conduit (Power Factor, 90%; Single Phase, 2 Wire; 60 Cycles)

WIRE SIZE AWG or MCM	1000	900	800	750	700	600	500	400	350	300	250	4/0	3/0	2/0	1/0	1	2	4	6	8*	10*	12*	14*
Ampere Feet	Volts Drop																						
500,000	33.0	34.0	36.0	37.0	38.0	41.0	45.0	51.0	55.0	61.0	68.0	76.0	91.0	110.0	133.0	161.0	198.0	300.0	461.0	—	—	—	—
400,000	26.4	26.2	28.8	29.6	30.4	32.8	36.0	40.8	44.0	48.8	54.4	60.8	72.8	88.0	106.0	129.0	158.0	240.0	369.0	—	—	—	—
300,000	19.8	20.4	21.6	22.2	22.8	24.6	27.0	30.6	33.0	36.6	40.8	45.6	54.6	66.0	79.8	96.6	119.0	180.0	277.0	420.0	—	—	—
200,000	13.2	13.6	14.4	14.8	15.2	16.4	18.0	20.4	22.0	24.4	27.2	30.4	36.4	44.0	53.2	64.4	79.2	120.0	184.0	280.0	442.0	—	—
100,000	6.6	6.8	7.2	7.4	7.6	8.2	9.0	10.2	11.0	12.2	13.6	15.2	18.2	22.0	26.6	32.2	39.6	60.0	92.2	140.0	221.0	351.0	—
90,000	5.9	6.2	6.4	6.8	6.5	7.4	8.1	9.2	9.9	10.9	12.3	13.6	16.3	19.8	23.8	28.9	35.8	54.0	82.9	126.0	199.0	316.0	498.0
80,000	5.2	5.5	5.7	6.0	6.1	6.6	7.2	8.2	8.8	9.7	10.9	12.1	14.5	17.6	21.2	25.7	31.8	48.0	73.7	112.0	177.0	281.0	443.0
70,000	4.6	4.8	5.0	5.2	5.3	5.7	6.3	7.1	7.7	8.5	9.5	10.6	12.7	15.4	18.6	22.5	27.8	42.0	64.5	98.4	155.0	246.0	388.0
60,000	4.0	4.1	4.3	4.4	4.6	4.9	5.4	6.1	6.6	7.3	8.1	9.1	10.9	13.2	16.0	19.3	23.8	36.0	55.3	84.3	133.0	210.0	332.0
50,000	3.3	3.4	3.6	3.7	3.8	4.1	4.5	5.1	5.5	6.1	6.8	7.6	9.1	11.0	13.3	16.1	19.8	30.0	46.1	70.2	111.0	176.0	277.0
40,000	2.6	2.6	2.8	2.9	3.0	3.3	3.6	4.1	4.4	4.9	5.4	6.1	7.3	8.8	10.6	12.9	15.8	24.0	36.9	56.0	88.4	140.0	222.0
30,000	1.9	2.0	2.1	2.2	2.3	2.5	2.7	3.1	3.3	3.7	4.1	4.6	5.5	6.6	7.9	9.7	11.9	18.0	27.7	42.0	66.3	105.0	166.0
20,000	1.3	1.4	1.4	1.5	1.5	1.6	1.8	2.0	2.2	2.4	2.7	3.0	3.6	4.4	5.3	6.4	7.9	12.0	18.4	28.0	44.2	70.2	111.0
10,000	0.7	0.7	0.7	0.7	0.7	0.8	1.0	1.0	1.1	1.2	1.4	1.5	1.8	2.2	2.7	3.2	3.9	6.0	9.2	14.0	22.1	35.1	55.4
9,000	0.6	0.6	0.6	0.7	0.7	0.7	0.8	0.9	1.0	1.1	1.2	1.4	1.6	1.9	2.4	2.9	3.6	5.4	8.3	12.6	19.9	31.6	49.8
8,000	0.5	0.6	0.6	0.6	0.6	0.7	0.7	0.8	0.9	1.0	1.1	1.2	1.5	1.8	2.1	2.6	3.2	4.8	7.4	11.2	17.7	28.1	44.3
7,000	0.4	0.5	0.5	0.5	0.5	0.6	0.6	0.7	0.8	0.9	0.9	1.1	1.3	1.5	1.9	2.3	2.8	4.2	6.5	9.8	15.5	24.6	38.8
6,000	0.4	0.4	0.4	0.4	0.5	0.5	0.5	0.6	0.7	0.7	0.8	0.9	1.1	1.3	1.6	1.9	2.4	3.6	5.5	8.4	13.3	21.0	33.2
5,000	0.3	0.3	0.4	0.4	0.4	0.4	0.5	0.5	0.6	0.6	0.7	0.8	0.9	1.1	1.3	1.6	1.9	3.0	4.6	7.0	11.1	17.6	27.7
4,000	0.3	0.3	0.3	0.3	0.3	0.3	0.4	0.4	0.4	0.5	0.5	0.6	0.7	0.9	1.1	1.3	1.6	2.4	3.7	5.6	8.8	14.0	22.2
3,000	0.2	0.2	0.2	0.2	0.2	0.2	0.3	0.3	0.3	0.4	0.4	0.5	0.5	0.7	0.8	1.0	1.2	1.8	2.8	4.2	6.6	10.5	16.6
2,000	0.1	0.1	0.1	0.2	0.2	0.2	0.2	0.2	0.2	0.2	0.3	0.3	0.4	0.4	0.5	0.6	0.8	1.2	1.8	2.8	4.4	7.0	11.1
1,000	0.1	0.1	0.1	0.1	0.1	0.1	0.1	0.1	0.1	0.1	0.1	0.2	0.2	0.2	0.3	0.3	0.4	0.6	0.9	1.4	2.2	3.5	5.5
900	—	0.1	0.1	0.1	0.1	0.1	0.1	0.1	0.1	0.1	0.1	0.1	0.1	0.2	0.2	0.3	0.4	0.5	0.8	1.3	2.0	3.2	5.0
800	—	0.1	0.1	0.1	0.1	0.1	0.1	0.1	0.1	0.1	0.1	0.1	0.1	0.2	0.2	0.2	0.3	0.5	0.7	1.1	1.8	2.8	4.4
700	—	—	0.1	0.1	0.1	0.1	0.1	0.1	0.1	0.1	0.1	0.1	0.1	0.1	0.2	0.2	0.3	0.4	0.7	1.0	1.6	2.5	3.9
600	—	—	—	—	—	—	0.1	0.1	0.1	0.1	0.1	0.1	0.1	0.1	0.2	0.2	0.2	0.4	0.6	0.8	1.3	2.1	3.3
500	—	—	—	—	—	—	—	0.1	0.1	0.1	0.1	0.1	0.1	0.1	0.1	0.2	0.2	0.3	0.5	0.7	1.1	1.8	2.8
400	—	—	—	—	—	—	—	—	—	—	0.1	0.1	0.1	0.1	0.1	0.1	0.2	0.2	0.4	0.6	0.9	1.4	2.2
300	—	—	—	—	—	—	—	—	—	—	—	—	—	0.1	0.1	0.1	0.1	0.2	0.3	0.4	0.7	1.1	1.7
200	—	—	—	—	—	—	—	—	—	—	—	—	—	—	0.1	0.1	0.1	0.1	0.2	0.3	0.4	0.7	1.1
100	—	—	—	—	—	—	—	—	—	—	—	—	—	—	—	—	—	0.1	0.1	0.1	0.2	0.4	0.6

* Solid Conductors. Other conductors are stranded.

Note 1—The above table gives voltage drops encountered in a single phase two-wire system. The voltage drops in other systems may be obtained through multiplication by appropriate factors listed below:

System for Which Voltage Drop Is Desired	Multiplying Factors for Modification of Values in Table
Single Phase—3 Wire—Line to Line	1.00
Single Phase—3 Wire—Line to Neutral	0.50
Three Phase—3 Wire—Line to Line	0.866
Three Phase—4 Wire—Line to Line	0.866
Three Phase—4 Wire—Line to Neutral	0.50

Note 2—Allowable voltage drops for systems other than single phase, two-wire cannot be used directly in the above table. Such drops should be modified through multiplication by the appropriate factor listed below. The voltage thus modified may then be used to obtain the proper wire size directly from the table.

System for Which Allowable Voltage Drop Is Known	Multiplying Factor for Modification of Known Value to Permit Direct Use of Table
Single Phase—3 Wire—Line to Line	1.00
Single Phase—3 Wire—Line to Neutral	2.00
Three Phase—3 Wire—Line to Line	1.155
Three Phase—4 Wire—Line to Line	1.155
Three Phase—4 Wire—Line to Neutral	2.00

Note 3—The footage employed in the tabulated ampere feet refers to the length of run of the circuit rather than to the footage of individual conductor.

Note 4—The above table is figured at 60°C since this is an estimate of the average temperature which may be anticipated in service. The table may be used without significant error for conductor temperatures up to and including 75°C.

TABLE 11.1.29 Voltage Drop for Copper Conductor in Magnetic Conduit (Power Factor, 95%; Single Phase, 2 Wire; 60 Cycles)

WIRE SIZE AWG or MCM	1000	900	800	750	700	600	500	400	350	300	250	4/0	3/0	2/0	1/0	1	2	4	6	8*	10*	12*	14*
Ampere Feet	Volts Drop																						
600,000	29.0	30.0	32.0	33.0	34.0	37.0	41.0	47.0	51.0	58.0	65.0	74.0	89.0	109.0	133.0	161.0	200.0	308.0	476.0	—	—	—	—
400,000	23.2	24.0	25.6	26.4	26.2	29.6	32.8	37.6	40.8	46.4	52.0	59.2	71.2	87.2	106.0	129.0	160.0	245.0	380.0	—	—	—	—
300,000	17.4	18.0	19.2	19.8	20.4	22.2	24.6	28.2	30.6	34.8	39.0	44.4	53.4	65.4	79.8	96.6	120.0	184.0	285.0	438.0	—	—	—
200,000	11.6	12.0	12.8	13.2	13.6	14.8	16.4	18.8	20.4	23.2	26.0	29.6	35.6	43.6	53.2	64.4	80.0	123.0	190.0	292.0	464.0	—	—
100,000	5.8	6.0	6.4	6.6	6.8	7.4	8.2	9.4	10.2	11.6	13.0	14.8	17.8	21.8	26.6	32.2	40.0	61.6	95.2	146.0	232.0	369.0	—
90,000	5.3	5.4	5.8	5.9	6.2	6.8	7.3	8.6	9.1	10.5	11.7	13.4	16.1	19.7	23.8	28.9	36.0	55.6	85.4	132.0	209.0	332.0	—
80,000	4.7	4.8	5.2	5.2	5.5	6.0	6.5	7.6	8.1	9.3	10.4	11.9	14.3	17.5	21.2	25.7	32.0	49.4	76.0	117.0	185.0	295.0	466.0
70,000	4.1	4.2	4.5	4.6	4.8	5.2	5.7	6.6	7.1	8.1	9.1	10.4	12.5	15.3	18.6	22.5	28.0	43.2	66.6	103.0	162.0	258.0	408.0
60,000	3.5	3.6	3.8	4.0	4.1	4.4	4.9	5.6	6.1	6.9	7.8	8.9	10.7	13.1	16.0	19.3	24.0	37.0	57.2	88.0	139.0	221.0	349.0
50,000	2.9	3.0	3.2	3.3	3.4	3.7	4.1	4.7	5.1	5.8	6.5	7.4	8.9	10.9	13.3	16.1	20.0	30.8	47.6	73.2	116.0	184.0	291.0
40,000	2.3	2.4	2.5	2.6	2.6	2.9	3.3	3.8	4.1	4.6	5.2	5.9	7.1	8.7	10.6	12.9	16.0	24.5	38.0	58.4	92.8	148.0	233.0
30,000	1.7	1.8	1.9	2.0	2.0	2.2	2.5	2.8	3.1	3.5	3.9	4.4	5.3	6.5	7.9	9.7	12.0	18.4	28.5	43.8	69.6	111.0	175.0
20,000	1.2	1.2	1.3	1.3	1.4	1.5	1.6	1.9	2.0	2.3	2.6	2.9	3.6	4.4	5.3	6.4	8.0	12.3	19.0	29.2	46.4	73.8	116.0
10,000	0.6	0.6	0.6	0.7	0.7	0.7	0.8	0.9	1.0	1.2	1.3	1.5	1.8	2.2	2.7	3.2	4.0	6.2	9.5	14.6	23.2	36.9	58.2
9,000	0.5	0.5	0.6	0.6	0.6	0.7	0.7	0.9	0.9	1.1	1.2	1.3	1.6	1.9	2.4	2.9	3.6	5.6	8.5	13.2	20.9	33.2	52.4
8,000	0.5	0.5	0.5	0.5	0.6	0.6	0.7	0.8	0.8	0.9	1.0	1.2	1.4	1.8	2.1	2.6	3.2	4.9	7.6	11.7	18.5	29.5	46.6
7,000	0.4	0.4	0.5	0.5	0.5	0.5	0.6	0.7	0.7	0.8	0.9	1.0	1.3	1.5	1.9	2.3	2.8	4.3	6.7	10.3	16.2	25.8	40.8
6,000	0.4	0.4	0.4	0.4	0.4	0.4	0.5	0.6	0.6	0.7	0.8	0.9	1.1	1.3	1.6	1.9	2.4	3.7	5.7	8.8	13.9	22.1	34.9
5,000	0.3	0.3	0.3	0.3	0.3	0.4	0.4	0.5	0.5	0.6	0.7	0.7	0.9	1.1	1.3	1.6	2.0	3.1	4.8	7.3	11.6	18.4	29.1
4,000	0.2	0.2	0.3	0.3	0.3	0.3	0.3	0.4	0.4	0.5	0.5	0.6	0.7	0.9	1.1	1.3	1.6	2.5	3.8	5.8	9.3	14.8	23.1
3,000	0.1	0.2	0.2	0.2	0.2	0.2	0.3	0.3	0.3	0.4	0.4	0.4	0.5	0.7	0.8	1.0	1.2	1.8	2.9	4.4	7.0	11.1	17.5
2,000	0.1	0.1	0.1	0.1	0.1	0.2	0.2	0.2	0.2	0.2	0.3	0.3	0.4	0.4	0.5	0.6	0.8	1.2	1.9	2.9	4.6	7.4	11.6
1,000	0.1	0.1	0.1	0.1	0.1	0.1	0.1	0.1	0.1	0.1	0.1	0.2	0.2	0.2	0.3	0.3	0.4	0.6	1.0	1.5	2.3	3.7	5.8
900	0.1	0.1	0.1	0.1	0.1	0.1	0.1	0.1	0.1	0.1	0.1	0.1	0.2	0.2	0.2	0.3	0.4	0.6	0.9	1.3	2.1	3.3	5.2
800	0.1	0.1	0.1	0.1	0.1	0.1	0.1	0.1	0.1	0.1	0.1	0.1	0.1	0.2	0.2	0.3	0.3	0.5	0.8	1.2	1.9	2.9	4.7
700	—	—	—	—	—	0.1	0.1	0.1	0.1	0.1	0.1	0.1	0.1	0.2	0.2	0.2	0.3	0.4	0.7	1.0	1.6	2.6	4.1
600	—	—	—	—	—	—	—	0.1	0.1	0.1	0.1	0.1	0.1	0.1	0.2	0.2	0.2	0.4	0.6	0.9	1.4	2.2	3.5
500	—	—	—	—	—	—	—	—	0.1	0.1	0.1	0.1	0.1	0.1	0.1	0.2	0.2	0.3	0.5	0.7	1.2	1.8	2.9
400	—	—	—	—	—	—	—	—	—	0.1	0.1	0.1	0.1	0.1	0.1	0.1	0.2	0.3	0.4	0.6	0.9	1.5	2.3
300	—	—	—	—	—	—	—	—	—	—	—	—	0.1	0.1	0.1	0.1	0.1	0.2	0.3	0.4	0.7	1.1	1.8
200	—	—	—	—	—	—	—	—	—	—	—	—	—	—	0.1	0.1	0.1	0.1	0.2	0.3	0.5	0.7	1.2
100	—	—	—	—	—	—	—	—	—	—	—	—	—	—	—	—	—	0.1	0.1	0.2	0.2	0.4	0.6

* Solid Conductors. Other conductors are stranded.

Note 1—The above table gives voltage drops encountered in a single phase two-wire system. The voltage drops in other systems may be obtained through multiplication by appropriate factors listed below:

System for Which Voltage Drop Is Desired	Multiplying Factors for Modification of Values in Table
Single Phase—3 Wire—Line to Line	1.00
Single Phase—3 Wire—Line to Neutral	0.50
Three Phase—3 Wire—Line to Line	0.866
Three Phase—4 Wire—Line to Line	0.866
Three Phase—4 Wire—Line to Neutral	0.50

Note 2—Allowable voltage drops for systems other than single phase, two-wire cannot be used directly in the above table. Such drops should be modified through multiplication by the appropriate factor listed below. The voltage thus modified may then be used to obtain the proper wire size directly from the table.

System for Which Allowable Voltage Drop Is Known	Multiplying Factor for Modification of Known Value to Permit Direct Use of Table
Single Phase—3 Wire—Line to Line	1.00
Single Phase—3 Wire—Line to Neutral	2.00
Three Phase—3 Wire—Line to Line	1.155
Three Phase—4 Wire—Line to Line	1.155
Three Phase—4 Wire—Line to Neutral	2.00

Note 3—The footage employed in the tabulated ampere feet refers to the length of run of the circuit rather than to the footage of individual conductor.

Note 4—The above table is figured at 60°C since this is an estimate of the average temperature which may be anticipated in service. The table may be used without significant error for conductor temperatures up to and including 75°C.

TABLE 11.1.30 Voltage Drop for Copper Conductor in Magnetic Conduit (Power Factor, 100%; Single Phase, 2 Wire; 60 Cycles)

WIRE SIZE AWG or MCM	1000	900	800	750	700	600	500	400	350	300	250	4/0	3/0	2/0	1/0	1	2	4	6	8*	10*	12*	14*
Ampere Feet	Volts Drop																						
500,000	16.4	17.9	19.6	20.3	21.5	24.4	28.8	34.8	39.0	45.0	53.4	62.4	78.6	98.5	123.0	153.0	194.0	306.0	483.0	—	—	—	—
400,000	13.1	14.3	15.7	16.2	17.2	19.5	23.0	27.8	31.2	36.0	42.8	50.0	62.9	78.8	98.4	122.0	155.0	244.0	386.0	—	—	—	—
300,000	9.9	10.7	11.8	12.2	12.9	14.6	17.3	20.9	23.4	27.0	32.0	37.4	47.2	59.1	73.9	91.8	116.0	184.0	290.0	450.0	—	—	—
200,000	6.6	7.2	7.8	8.1	8.6	9.8	11.6	14.0	15.6	18.0	21.4	25.0	31.4	39.4	49.2	61.2	77.6	122.0	193.0	300.0	480.0	—	—
100,000	3.3	3.6	3.9	4.1	4.3	4.9	5.8	7.0	7.8	9.0	10.7	12.5	15.7	19.7	24.6	30.6	38.8	61.2	96.6	150.0	240.0	384.0	—
90,000	2.9	3.2	3.5	3.7	3.9	4.4	5.2	6.3	7.0	8.1	9.6	11.2	14.2	17.7	22.2	27.5	34.9	55.1	87.0	135.0	216.0	345.0	—
80,000	2.6	2.9	3.1	3.2	3.4	3.9	4.6	5.6	6.2	7.2	8.5	10.0	12.6	15.8	19.7	24.5	31.0	49.0	77.3	120.0	192.0	307.0	487.0
70,000	2.3	2.5	2.7	2.8	3.0	3.4	4.0	4.9	5.4	6.3	7.5	8.7	11.0	13.8	17.2	21.4	27.2	42.8	67.6	105.0	168.0	269.0	426.0
60,000	2.0	2.1	2.3	2.4	2.6	2.9	3.4	4.2	4.7	5.4	6.4	7.5	9.4	11.8	14.8	18.4	23.3	36.7	58.0	90.0	144.0	230.0	365.0
50,000	1.6	1.8	1.9	2.0	2.2	2.4	2.9	3.5	3.9	4.5	5.3	6.2	7.9	9.9	12.3	15.3	19.4	30.6	48.3	74.9	120.0	192.0	304.0
40,000	1.3	1.4	1.6	1.6	1.7	1.9	2.3	2.8	3.1	3.6	4.3	5.0	6.3	7.9	9.8	12.2	15.5	24.4	38.6	60.0	96.0	154.0	243.0
30,000	1.0	1.1	1.2	1.2	1.3	1.5	1.7	2.1	2.3	2.7	3.2	3.7	4.7	5.9	7.4	9.2	11.6	18.4	29.0	45.0	72.0	115.0	182.0
20,000	0.7	0.7	0.8	0.8	0.9	1.0	1.2	1.4	1.6	1.8	2.1	2.5	3.1	3.9	4.9	6.1	7.8	12.2	19.3	30.0	48.0	76.8	122.0
10,000	0.3	0.4	0.4	0.4	0.4	0.5	0.6	0.7	0.8	0.9	1.1	1.3	1.6	1.9	2.5	3.1	3.9	6.1	9.7	15.0	24.0	38.4	60.8
9,000	0.3	0.3	0.4	0.4	0.4	0.4	0.5	0.6	0.7	0.8	1.0	1.1	1.4	1.8	2.2	2.8	3.5	5.5	8.7	13.5	21.6	34.5	54.7
8,000	0.3	0.3	0.3	0.3	0.3	0.4	0.5	0.6	0.6	0.7	0.9	1.0	1.3	1.6	1.9	2.5	3.1	4.9	7.7	12.0	19.2	30.7	48.7
7,000	0.2	0.3	0.3	0.3	0.3	0.3	0.4	0.5	0.5	0.6	0.8	0.9	1.1	1.4	1.7	2.1	2.7	4.3	6.8	10.5	16.8	26.9	42.6
6,000	0.2	0.2	0.2	0.2	0.3	0.3	0.3	0.4	0.5	0.5	0.6	0.8	0.9	1.2	1.5	1.8	2.3	3.7	5.8	9.0	14.4	23.0	36.5
5,000	0.2	0.2	0.2	0.2	0.2	0.2	0.3	0.4	0.4	0.5	0.5	0.6	0.8	1.0	1.2	1.5	1.9	3.1	4.8	7.5	12.0	19.2	30.4
4,000	0.1	0.1	0.2	0.2	0.2	0.2	0.2	0.3	0.3	0.4	0.4	0.5	0.6	0.8	1.0	1.2	1.5	2.4	3.8	6.0	9.6	15.4	24.3
3,000	0.1	0.1	0.1	0.1	0.1	0.2	0.2	0.2	0.2	0.3	0.3	0.4	0.5	0.6	0.7	0.9	1.2	1.8	2.9	4.5	7.2	11.5	18.2
2,000	0.1	0.1	0.1	0.1	0.1	0.1	0.1	0.1	0.2	0.2	0.2	0.3	0.3	0.4	0.5	0.6	0.8	1.2	1.9	3.0	4.8	7.7	12.2
1,000	—	—	—	—	—	0.1	0.1	0.1	0.1	0.1	0.1	0.1	0.2	0.2	0.3	0.3	0.4	0.6	1.0	1.5	2.4	3.9	6.1
900	—	—	—	—	—	—	0.1	0.1	0.1	0.1	0.1	0.1	0.1	0.2	0.2	0.3	0.4	0.6	0.9	1.4	2.2	3.5	5.5
800	—	—	—	—	—	—	0.1	0.1	0.1	0.1	0.1	0.1	0.1	0.2	0.2	0.3	0.3	0.5	0.8	1.2	1.9	3.1	4.9
700	—	—	—	—	—	—	—	0.1	0.1	0.1	0.1	0.1	0.1	0.1	0.2	0.2	0.3	0.4	0.7	1.1	1.7	2.7	4.3
600	—	—	—	—	—	—	—	—	0.1	0.1	0.1	0.1	0.1	0.1	0.2	0.2	0.2	0.4	0.6	0.9	1.4	2.3	3.7
500	—	—	—	—	—	—	—	—	—	0.1	0.1	0.1	0.1	0.1	0.1	0.2	0.2	0.3	0.5	0.8	1.2	1.9	3.0
400	—	—	—	—	—	—	—	—	—	—	—	0.1	0.1	0.1	0.1	0.1	0.2	0.2	0.4	0.6	1.0	1.5	2.4
300	—	—	—	—	—	—	—	—	—	—	—	—	0.1	0.1	0.1	0.1	0.1	0.2	0.3	0.5	0.7	1.2	1.8
200	—	—	—	—	—	—	—	—	—	—	—	—	—	—	0.1	0.1	0.1	0.1	0.2	0.3	0.5	0.8	1.2
100	—	—	—	—	—	—	—	—	—	—	—	—	—	—	—	—	—	0.1	0.1	0.2	0.2	0.4	0.6

* Solid Conductors. Other conductors are stranded.

Note 1—The above table gives voltage drops encountered in a single phase two-wire system. The voltage drops in other systems may be obtained through multiplication by appropriate factors listed below:

System for Which Voltage Drop is Desired	Multiplying Factors for Modification of Values in Table
Single Phase—3 Wire—Line to Line	1.00
Single Phase—3 Wire—Line to Neutral	0.50
Three Phase—3 Wire—Line to Line	0.866
Three Phase—4 Wire—Line to Line	0.866
Three Phase—4 Wire—Line to Neutral	0.50

Note 2—Allowable voltage drops for systems other than single phase, two-wire cannot be used directly in the above table. Such drops should be modified through multiplication by the appropriate factor listed below. The voltage thus modified may then be used to obtain the proper wire size directly from the table.

System for Which Allowable Voltage Drop is Known	Multiplying Factor for Modification of Known Value to Permit Direct Use of Table
Single Phase—3 Wire—Line to Line	1.00
Single Phase—3 Wire—Line to Neutral	2.00
Three Phase—3 Wire—Line to Line	1.155
Three Phase—4 Wire—Line to Line	1.155
Three Phase—4 Wire—Line to Neutral	2.00

Note 3—The footage employed in the tabulated ampere feet refers to the length of run of the circuit rather than to the footage of individual conductor.

Note 4—The above table is figured at 60°C since this is an estimate of the average temperature which may be anticipated in service. The table may be used without significant error for conductor temperatures up to and including 75°C.

TABLE 11.1.31 Voltage Drop for Copper Conductor in Nonmagnetic Conduit (Power Factor, 70%; Single Phase, 2 Wire; 60 Cycles)

WIRE SIZE AWG or MCM	1000	900	800	750	700	600	500	400	350	300	250	4/0	3/0	2/0	1/0	1	2	4	6	8*	10*	12*	14*
Ampere Feet	Volts Drop																						
500,000	33.0	34.0	36.0	37.0	38.0	40.0	43.0	48.0	51.0	6.0	62.0	68.0	81.0	95.0	114.0	136.0	164.0	246.0	372.0	—	—	—	—
400,000	26.4	27.2	28.8	29.6	30.4	32.0	34.4	38.4	40.8	44.8	49.6	54.4	64.8	76.0	91.2	109.0	131.0	196.0	297.0	448.0	—	—	—
300,000	19.8	20.4	21.6	22.2	22.8	24.0	25.8	28.8	30.6	33.6	37.2	40.8	48.6	57.0	68.4	81.6	98.4	147.0	223.0	336.0	—	—	—
200,000	13.2	13.6	14.4	14.8	15.2	16.0	17.2	19.2	20.4	22.4	24.8	27.2	32.4	38.0	45.6	54.4	65.6	98.0	148.0	224.0	350.0	—	—
100,000	6.6	6.8	7.2	7.4	7.6	8.0	8.6	9.6	10.2	11.2	12.4	13.6	16.2	19.0	22.8	27.2	32.8	49.2	74.4	112.0	175.0	276.0	434.0
90,000	5.9	6.2	6.4	6.7	6.8	7.2	7.7	8.6	9.2	10.1	11.3	12.3	14.6	17.1	20.5	24.5	29.5	44.2	67.1	101.0	158.0	248.0	390.0
80,000	5.2	5.5	5.7	6.0	6.1	6.4	6.9	7.7	8.2	8.9	10.0	10.9	13.0	15.2	18.3	21.8	26.2	39.3	59.6	89.6	140.0	221.0	347.0
70,000	4.6	4.8	5.0	5.2	5.3	5.6	6.0	6.7	7.1	7.8	8.7	9.5	11.3	13.3	16.0	19.0	23.0	34.4	52.1	78.4	123.0	193.0	304.0
60,000	4.0	4.1	4.3	4.4	4.6	4.8	5.2	5.8	6.1	6.7	7.4	8.1	9.7	11.4	13.7	16.3	19.7	29.5	44.6	67.3	105.0	166.0	260.0
50,000	3.3	3.4	3.6	3.7	3.8	4.0	4.3	4.8	5.1	5.6	6.2	6.8	8.1	9.5	11.4	13.6	16.4	24.6	37.2	56.0	87.6	138.0	217.0
40,000	2.6	2.6	2.9	2.9	3.0	3.2	3.4	3.8	4.1	4.5	4.9	5.4	6.5	7.6	9.1	10.9	13.1	19.6	29.7	44.8	70.0	110.0	174.0
30,000	1.9	2.0	2.1	2.2	2.3	2.4	2.6	2.9	3.1	3.4	3.7	4.1	4.8	5.7	6.8	8.2	9.8	14.7	22.3	33.6	52.5	82.8	130.0
20,000	1.3	1.3	1.4	1.5	1.5	1.6	1.7	1.9	2.0	2.2	2.5	2.7	3.2	3.8	4.6	5.4	6.6	9.8	14.8	22.4	35.0	55.2	86.8
10,000	0.7	0.7	0.7	0.7	0.8	0.8	0.9	1.0	1.0	1.1	1.2	1.4	1.6	1.9	2.3	2.7	3.3	4.9	7.4	11.2	17.5	27.6	43.4
9,000	0.6	0.6	0.6	0.7	0.7	0.7	0.8	0.9	0.9	1.0	1.1	1.2	1.5	1.7	2.1	2.5	2.9	4.4	6.7	10.1	15.8	24.5	39.0
8,000	0.5	0.6	0.6	0.6	0.6	0.6	0.7	0.8	0.8	0.9	1.0	1.1	1.3	1.5	1.8	2.2	2.6	3.9	5.9	8.9	14.0	22.1	34.7
7,000	0.5	0.5	0.5	0.5	0.5	0.6	0.6	0.7	0.7	0.8	0.9	1.0	1.1	1.3	1.6	1.9	2.3	3.4	5.2	7.8	12.3	19.3	30.4
6,000	0.4	0.4	0.4	0.4	0.5	0.5	0.5	0.6	0.6	0.7	0.7	0.8	1.0	1.1	1.4	1.6	1.9	2.9	4.5	6.7	10.5	16.6	26.0
5,000	0.3	0.3	0.4	0.4	0.4	0.4	0.4	0.5	0.5	0.6	0.6	0.7	0.8	1.0	1.1	1.4	1.6	2.5	3.7	5.6	8.8	13.8	21.7
4,000	0.3	0.3	0.3	0.3	0.3	0.3	0.3	0.4	0.4	0.5	0.5	0.5	0.7	0.8	0.9	1.1	1.3	1.9	2.9	4.5	7.0	11.0	17.4
3,000	0.2	0.2	0.2	0.2	0.2	0.2	0.3	0.3	0.3	0.3	0.4	0.4	0.5	0.6	0.7	0.8	1.0	1.5	2.2	3.4	5.3	8.3	13.0
2,000	0.1	0.1	0.1	0.2	0.2	0.2	0.2	0.2	0.2	0.2	0.3	0.3	0.3	0.4	0.5	0.5	0.7	1.0	1.5	2.2	3.5	5.5	8.7
1,000	0.1	0.1	0.1	0.1	0.1	0.1	0.1	0.1	0.1	0.1	0.1	0.1	0.2	0.2	0.2	0.3	0.3	0.5	0.7	1.1	1.8	2.8	4.3
900	0.1	0.1	0.1	0.1	0.1	0.1	0.1	0.1	0.1	0.1	0.1	0.1	0.2	0.2	0.2	0.3	0.3	0.4	0.7	1.0	1.6	2.5	3.9
800	0.1	0.1	0.1	0.1	0.1	0.1	0.1	0.1	0.1	0.1	0.1	0.1	0.1	0.2	0.2	0.2	0.3	0.4	0.6	0.9	1.4	2.2	3.6
700	0.1	0.1	0.1	0.1	0.1	0.1	0.1	0.1	0.1	0.1	0.1	0.1	0.1	0.1	0.2	0.2	0.2	0.3	0.5	0.8	1.2	1.9	3.0
600	—	—	—	—	0.1	0.1	0.1	0.1	0.1	0.1	0.1	0.1	0.1	0.1	0.1	0.2	0.2	0.3	0.5	0.7	1.1	1.7	2.6
500	—	—	—	—	—	—	—	0.1	0.1	0.1	0.1	0.1	0.1	0.1	0.1	0.1	0.2	0.3	0.4	0.6	0.9	1.4	2.2
400	—	—	—	—	—	—	—	—	—	0.1	0.1	0.1	0.1	0.1	0.1	0.1	0.1	0.2	0.3	0.5	0.7	1.1	1.7
300	—	—	—	—	—	—	—	—	—	—	-	—	0.1	0.1	0.1	0.1	0.1	0.1	0.2	0.3	0.5	0.8	1.3
200	—	—	—	—	—	—	—	—	—	—	—	—	—	—	0.1	0.1	0.1	0.1	0.2	0.2	0.4	0.6	0.9
100	—	—	—	—	—	—	—	—	—	—	—	—	—	—	—	—	—	0.1	0.1	0.1	0.2	0.3	0.4

* Solid Conductors. Other conductors are stranded.

Note 1—The above table gives voltage drops encountered in a single phase two-wire system. The voltage drops in other systems may be obtained through multiplication by appropriate factors listed below:

System for Which Voltage Drop is Desired	Multiplying Factors for Modification of Values in Table
Single Phase—3 Wire—Line to Line	1.00
Single Phase—3 Wire—Line to Neutral	0.50
Three Phase—3 Wire—Line to Line	0.866
Three Phase—4 Wire—Line to Line	0.866
Three Phase—4 Wire—Line to Neutral	0.50

Note 2—Allowable voltage drops for systems other than single phase, two wire cannot be used directly in the above table. Such drops should be modified through multiplication by the appropriate factor listed below. The voltage thus modified may then be used to obtain the proper wire size directly from the table.

System for Which Allowable Voltage Drop is Known	Multiplying Factor for Modification of Known Value to Permit Direct Use of Table
Single Phase—3 Wire—Line to Line	1.00
Single Phase—3 Wire—Line to Neutral	2.00
Three Phase—3 Wire—Line to Line	1.155
Three Phase—4 Wire—Line to Line	1.155
Three Phase—4 Wire—Line to Neutral	2.00

Note 3—The footage employed in the tabulated ampere feet refers to the length of run of the circuit rather than to the footage of individual conductor.

Note 4—The above table is figured at 60°C since this is an estimate of the average temperature which may be anticipated in service. The table may be used without significant error for conductor temperatures up to and including 75°C.

TABLE 11.1.32 Voltage Drop for Copper Conductor in Nonmagnetic Conduit (Power Factor, 80%; Single Phase, 2 Wire; 60 Cycles)

WIRE SIZE AWG or MCM	1000	900	800	750	700	600	500	400	350	300	250	4/0	3/0	2/0	1/0	1	2	4	6	8*	10*	12*	14*
Ampere Feet	Volts Drop																						
500,000	31.0	32.0	33.0	34.0	36.0	38.0	42.0	47.0	51.0	56.0	63.0	70.0	83.0	100.0	121.0	147.0	179.0	272.0	414.0	—	—	—	—
400,000	24.8	25.6	26.4	26.2	28.8	30.4	33.6	37.6	40.8	44.8	50.4	56.0	66.4	80.0	96.8	118.0	143.0	217.0	329.0	—	—	—	—
300,000	18.6	19.2	19.8	20.4	21.6	22.8	25.2	28.2	30.6	33.6	37.8	42.0	49.8	60.0	72.6	88.2	107.0	163.0	247.0	378.0	—	—	—
200,000	12.4	12.8	13.2	13.6	14.4	15.2	16.8	18.8	20.4	22.4	25.2	28.0	33.2	40.0	48.4	58.8	71.6	109.0	165.0	252.0	396.0	—	—
100,000	6.2	6.4	6.6	6.8	7.2	7.6	8.4	9.4	10.2	11.2	12.6	14.0	16.6	20.0	24.2	29.4	35.8	54.4	82.8	126.0	198.0	313.0	492.0
90,000	5.5	5.8	5.9	6.2	6.4	6.5	7.6	8.6	9.1	10.1	11.3	12.6	14.9	18.0	21.8	26.6	32.2	49.1	74.6	113.0	178.0	282.0	443.0
80,000	4.9	5.2	5.2	5.5	5.7	6.1	6.7	7.6	8.1	8.9	10.1	11.2	13.3	16.0	19.4	23.6	28.6	43.6	66.3	101.0	158.0	250.0	394.0
70,000	4.3	4.5	4.6	4.8	5.0	5.3	5.9	6.6	7.1	7.8	8.8	9.8	11.6	14.0	17.0	20.6	25.0	38.1	58.0	88.2	139.0	219.0	345.0
60,000	3.7	3.8	4.0	4.1	4.3	4.6	5.0	5.6	6.1	6.7	7.5	8.4	10.0	12.0	14.5	17.6	21.5	32.6	49.7	75.6	119.0	188.0	295.0
50,000	3.1	3.2	3.3	3.4	3.6	3.8	4.2	4.7	5.1	5.6	6.3	7.0	8.3	10.0	12.1	14.7	17.9	27.2	41.4	62.9	99.0	157.0	246.0
40,000	2.5	2.6	2.6	2.6	2.9	3.0	3.4	3.7	4.1	4.5	5.0	5.6	6.6	8.0	9.7	11.8	14.3	21.7	32.9	50.4	79.2	125.0	197.0
30,000	1.8	1.9	1.9	2.0	2.2	2.3	2.5	2.8	3.1	3.4	3.8	4.2	4.9	6.0	7.3	8.8	10.7	16.3	24.7	37.8	59.4	93.9	148.0
20,000	1.2	1.3	1.3	1.4	1.4	1.5	1.7	1.9	2.0	2.2	2.5	2.8	3.3	4.0	4.8	5.9	7.2	10.9	16.5	25.2	39.6	62.6	98.4
10,000	0.6	0.6	0.7	0.7	0.7	0.8	0.8	0.9	1.0	1.1	1.3	1.4	1.7	2.0	2.4	2.9	3.6	5.4	8.3	12.6	19.8	31.3	49.2
9,000	0.6	0.6	0.6	0.6	0.6	0.7	0.8	0.9	0.9	1.0	1.1	1.3	1.5	1.8	2.2	2.7	3.2	4.9	7.5	11.3	17.8	28.2	44.3
8,000	0.5	0.5	0.5	0.6	0.6	0.6	0.7	0.8	0.8	0.9	1.0	1.1	1.3	1.6	1.9	2.4	2.9	4.4	6.6	10.1	15.8	25.0	39.4
7,000	0.4	0.5	0.5	0.5	0.5	0.5	0.6	0.7	0.7	0.8	0.9	1.0	1.2	1.4	1.7	2.1	2.5	3.8	5.8	8.8	13.9	21.9	34.5
6,000	0.4	0.4	0.4	0.4	0.4	0.5	0.5	0.6	0.6	0.7	0.8	0.8	1.0	1.2	1.5	1.8	2.2	3.3	4.9	7.6	11.9	18.8	29.5
5,000	0.3	0.3	0.3	0.3	0.4	0.4	0.4	0.5	0.5	0.6	0.6	0.7	0.8	1.0	1.2	1.5	1.8	2.7	4.1	6.3	9.9	15.7	24.6
4,000	0.3	0.3	0.3	0.3	0.3	0.3	0.3	0.4	0.4	0.5	0.5	0.6	0.7	0.8	1.0	1.2	1.4	2.2	3.3	5.0	7.9	12.5	19.7
3,000	0.2	0.2	0.2	0.2	0.2	0.2	0.3	0.3	0.3	0.3	0.4	0.4	0.5	0.6	0.7	0.9	1.1	1.6	2.5	3.8	5.9	9.4	14.8
2,000	0.1	0.1	0.1	0.1	0.1	0.2	0.2	0.2	0.2	0.2	0.3	0.3	0.3	0.4	0.5	0.6	0.7	1.1	1.7	2.5	3.9	6.3	9.8
1,000	0.1	0.1	0.1	0.1	0.1	0.1	0.1	0.1	0.1	0.1	0.1	0.1	0.2	0.2	0.2	0.3	0.4	0.5	0.8	1.3	1.9	3.1	4.9
900	0.1	0.1	0.1	0.1	0.1	0.1	0.1	0.1	0.1	0.1	0.1	0.1	0.2	0.2	0.2	0.3	0.3	0.5	0.8	1.1	1.8	2.8	4.4
800	0.1	0.1	0.1	0.1	0.1	0.1	0.1	0.1	0.1	0.1	0.1	0.1	0.1	0.2	0.2	0.2	0.3	0.4	0.7	1.0	1.6	2.5	3.9
700	—	0.1	0.1	0.1	0.1	0.1	0.1	0.1	0.1	0.1	0.1	0.1	0.1	0.1	0.2	0.2	0.3	0.4	0.6	0.9	1.4	2.2	3.5
600	—	—	—	—	—	0.1	0.1	0.1	0.1	0.1	0.1	0.1	0.1	0.1	0.2	0.2	0.2	0.3	0.5	0.8	1.2	1.9	2.9
500	—	—	—	—	—	—	—	0.1	0.1	0.1	0.1	0.1	0.1	0.1	0.1	0.2	0.2	0.3	0.4	0.6	1.0	1.6	2.5
400	—	—	—	—	—	—	—	—	—	0.1	0.1	0.1	0.1	0.1	0.1	0.1	0.1	0.2	0.3	0.5	0.8	1.3	1.9
300	—	—	—	—	—	—	—	—	—	—	—	—	0.1	0.1	0.1	0.1	0.1	0.2	0.3	0.4	0.6	0.9	1.5
200	—	—	—	—	—	—	—	—	—	—	—	—	—	—	0.1	0.1	0.1	0.1	0.2	0.3	0.4	0.6	1.0
100	—	—	—	—	—	—	—	—	—	—	—	—	—	—	—	—	—	0.1	0.1	0.1	0.2	0.3	0.5

* Solid Conductors. Other conductors are stranded.

Note 1—The above table gives voltage drops encountered in a single phase two-wire system. The voltage drops in other systems may be obtained through multiplication by appropriate factors listed below:

System for Which Voltage Drop is Desired	Multiplying Factors for Modification of Values in Table
Single Phase—3 Wire—Line to Line	1.00
Single Phase—3 Wire—Line to Neutral	0.50
Three Phase—3 Wire—Line to Line	0.866
Three Phase—4 Wire—Line to Line	0.866
Three Phase—4 Wire—Line to Neutral	0.50

Note 2—Allowable voltage drops for systems other than single phase, two wire cannot be used directly in the above table. Such drops should be modified through multiplication by the appropriate factor listed below. The voltage thus modified may then be used to obtain the proper wire size directly from the table.

System for Which Allowable Voltage Drop is Known	Multiplying Factor for Modification of Known Value to Permit Direct Use of Table
Single Phase—3 Wire—Line to Line	1.00
Single Phase—3 Wire—Line to Neutral	2.00
Three Phase—3 Wire—Line to Line	1.155
Three Phase—4 Wire—Line to Line	1.155
Three Phase—4 Wire—Line to Neutral	2.00

Note 3—The footage employed in the tabulated ampere feet refers to the length of run of the circuit rather than to the footage of individual conductor.

Note 4—The above table is figured at 60°C since this is an estimate of the average temperature which may be anticipated in service. The table may be used without significant error for conductor temperatures up to and including 75°C.

TABLE 11.1.33 Voltage Drop for Copper Conductor in Nonmagnetic Conduit (Power Factor, 90%; Single Phase, 2 Wire; 60 Cycles)

WIRE SIZE AWG or MCM	1000	900	800	750	700	600	500	400	350	300	250	4/0	3/0	2/0	1/0	1	2	4	6	8*	10*	12*	14*
Ampere Feet	Volts Drop																						
500,000	27.0	28.0	30.0	31.0	32.0	34.0	39.0	44.0	49.0	55.0	62.0	69.0	85.0	103.0	127.0	155.0	191.0	295.0	456.0	—	—	—	—
400,000	21.6	22.4	24.0	24.8	25.6	26.2	31.2	35.2	39.2	44.0	49.6	55.2	68.0	82.4	102.0	124.0	153.0	236.0	364.0	—	—	—	—
300,000	16.2	16.8	18.0	18.6	19.2	20.4	23.4	26.4	29.4	33.0	37.2	41.4	51.0	61.8	76.2	93.0	115.0	177.0	273.0	417.0	—	—	—
200,000	10.8	11.2	12.0	12.4	12.8	13.6	15.6	17.6	19.6	22.0	24.8	27.6	34.0	41.2	50.8	62.0	76.4	118.0	182.0	278.0	440.0	—	—
100,000	5.4	5.6	6.0	6.2	6.4	6.8	7.8	8.8	9.8	11.0	12.4	13.8	17.0	20.6	25.4	31.0	38.2	59.0	91.2	139.0	220.0	350.0	—
90,000	4.8	4.9	5.4	5.5	5.8	6.2	7.0	7.9	8.8	9.9	11.3	12.4	15.3	18.5	22.9	27.9	34.4	53.2	81.7	125.0	198.0	315.0	497.0
80,000	4.3	4.4	4.8	4.9	5.2	5.5	6.2	7.0	7.8	8.8	10.0	11.0	13.6	16.5	20.3	24.8	30.6	47.4	72.7	111.0	176.0	280.0	442.0
70,000	3.8	3.9	4.2	4.3	4.5	4.8	5.5	6.2	6.9	7.7	8.7	9.6	11.9	14.4	17.8	21.7	26.8	41.4	63.7	97.3	154.0	245.0	386.0
60,000	3.2	3.4	3.6	3.7	3.8	4.1	4.7	5.3	5.9	6.6	7.4	8.3	10.2	12.4	15.2	18.6	22.9	35.4	54.7	83.5	132.0	210.0	331.0
50,000	2.7	2.8	3.0	3.1	3.2	3.4	3.9	4.4	4.9	5.5	6.2	6.9	8.5	10.3	12.7	15.5	19.1	29.5	45.6	69.6	110.0	175.0	276.0
40,000	2.2	2.2	2.4	2.5	2.6	2.6	3.1	3.5	3.9	4.4	4.9	5.5	6.8	8.2	10.2	12.4	15.3	23.6	36.4	55.6	88.0	140.0	221.0
30,000	1.6	1.7	1.8	1.8	1.9	2.0	2.3	2.6	2.9	3.3	3.7	4.1	5.1	6.2	7.6	9.3	11.5	17.7	27.3	41.7	66.0	105.0	166.0
20,000	1.1	1.1	1.2	1.2	1.3	1.4	1.6	1.8	1.9	2.2	2.5	2.8	3.4	4.1	5.1	6.2	7.6	11.8	18.2	27.8	44.0	70.0	110.0
10,000	0.5	0.6	0.6	0.6	0.6	0.7	0.8	0.9	1.0	1.1	1.2	1.4	1.7	2.1	2.6	3.1	3.8	5.9	9.1	13.9	22.0	35.0	55.2
9,000	0.5	0.5	0.5	0.6	0.6	0.6	0.7	0.8	0.9	1.0	1.1	1.2	1.5	1.9	2.3	2.8	3.4	5.3	8.2	12.5	19.8	31.5	49.7
8,000	0.4	0.4	0.5	0.5	0.5	0.6	0.6	0.7	0.8	0.9	1.0	1.1	1.4	1.7	2.0	2.5	3.1	4.7	7.3	11.1	17.6	28.0	44.2
7,000	0.4	0.4	0.4	0.4	0.5	0.5	0.6	0.6	0.7	0.8	0.9	1.0	1.2	1.4	1.8	2.2	2.7	4.1	6.4	9.7	15.4	24.5	38.6
6,000	0.3	0.3	0.4	0.4	0.4	0.4	0.5	0.5	0.6	0.7	0.7	0.8	1.0	1.2	1.5	1.9	2.3	3.5	5.5	8.4	13.2	21.0	33.1
5,000	0.3	0.3	0.3	0.3	0.3	0.3	0.4	0.4	0.5	0.6	0.6	0.7	0.9	1.0	1.3	1.6	1.9	2.9	4.6	6.9	11.0	17.5	27.6
4,000	0.2	0.2	0.2	0.3	0.3	0.3	0.3	0.4	0.4	0.4	0.5	0.6	0.7	0.8	1.0	1.2	1.5	2.4	3.6	5.6	8.8	14.0	22.1
3,000	0.2	0.2	0.2	0.2	0.2	0.2	0.2	0.3	0.3	0.3	0.4	0.4	0.5	0.6	0.8	0.9	1.2	1.8	2.7	4.2	6.6	10.5	16.6
2,000	0.1	0.1	0.1	0.1	0.1	0.1	0.2	0.2	0.2	0.2	0.3	0.3	0.3	0.4	0.5	0.6	0.8	1.2	1.8	2.8	4.4	7.0	11.0
1,000	0.1	0.1	0.1	0.1	0.1	0.1	0.1	0.1	0.1	0.1	0.1	0.1	0.2	0.2	0.3	0.3	0.4	0.6	0.9	1.4	2.2	3.5	5.5
900	0.1	0.1	0.1	0.1	0.1	0.1	0.1	0.1	0.1	0.1	0.1	0.1	0.2	0.2	0.2	0.3	0.3	0.5	0.8	1.3	1.9	3.2	4.9
800	—	—	0.1	0.1	0.1	0.1	0.1	0.1	0.1	0.1	0.1	0.1	0.1	0.2	0.2	0.3	0.3	0.5	0.7	1.1	1.8	2.8	4.4
700	—	—	—	—	0.1	0.1	0.1	0.1	0.1	0.1	0.1	0.1	0.1	0.1	0.2	0.2	0.3	0.4	0.6	1.0	1.5	2.5	3.9
600	—	—	—	—	—	—	0.1	0.1	0.1	0.1	0.1	0.1	0.1	0.1	0.2	0.2	0.2	0.4	0.6	0.8	1.3	2.1	3.3
500	—	—	—	—	—	—	—	—	0.1	0.1	0.1	0.1	0.1	0.1	0.1	0.2	0.2	0.3	0.5	0.7	1.1	1.8	2.8
400	—	—	—	—	—	—	—	—	—	—	0.1	0.1	0.1	0.1	0.1	0.1	0.2	0.2	0.4	0.6	0.9	1.4	2.2
300	—	—	—	—	—	—	—	—	—	—	—	—	0.1	0.1	0.1	0.1	0.1	0.2	0.3	0.4	0.7	1.1	1.7
200	—	—	—	—	—	—	—	—	—	—	—	—	—	—	0.1	0.1	0.1	0.1	0.2	0.3	0.4	0.7	1.1
100	—	—	—	—	—	—	—	—	—	—	—	—	—	—	—	—	—	0.1	0.1	0.1	0.2	0.4	0.6

* Solid Conductors. Other conductors are stranded.

Note 1—The above table gives voltage drops encountered in a single phase two-wire system. The voltage drops in other systems may be obtained through multiplication by appropriate factors listed below:

System for Which Voltage Drop is Desired	Multiplying Factors for Modification of Values in Table
Single Phase—3 Wire—Line to Line	1.00
Single Phase—3 Wire—Line to Neutral	0.50
Three Phase—3 Wire—Line to Line	0.866
Three Phase—4 Wire—Line to Line	0.866
Three Phase—4 Wire—Line to Neutral	0.50

Note 2—Allowable voltage drops for systems other than single phase, two wire cannot be used directly in the above table. Such drops should be modified through multiplication by the appropriate factor listed below. The voltage thus modified may then be used to obtain the proper wire size directly from the table.

System for Which Allowable Voltage Drop is Known	Multiplying Factor for Modification of Known Value to Permit Direct Use of Table
Single Phase—3 Wire—Line to Line	1.00
Single Phase—3 Wire—Line to Neutral	2.00
Three Phase—3 Wire—Line to Line	1.155
Three Phase—4 Wire—Line to Line	1.155
Three Phase—4 Wire—Line to Neutral	2.00

Note 3—The footage employed in the tabulated ampere feet refers to the length of run of the circuit rather than to the footage of individual conductor.

Note 4—The above table is figured at 60°C since this is an estimate of the average temperature which may be anticipated in service. The table may be used without significant error for conductor temperatures up to and including 75°C.

TABLE 11.1.34 Voltage Drop for Copper Conductor in Nonmagnetic Conduit (Power Factor, 95%; Single Phase, 2 Wire; 60 Cycles)

WIRE SIZE AWG or MCM	1000	900	800	750	700	600	500	400	350	300	250	4/0	3/0	2/0	1/0	1	2	4	6	8*	10*	12*	14*
Ampere Feet	Volts Drop																						
500,000	23.0	25.0	26.0	27.0	29.0	31.0	36.0	41.0	46.0	53.0	60.0	67.0	84.0	103.0	128.0	157.0	195.0	305.0	473.0	—	—	—	—
400,000	18.4	20.0	20.8	21.6	23.2	24.8	28.8	32.8	36.8	42.4	48.0	53.6	67.2	82.4	102.0	126.0	156.0	244.0	378.0	—	—	—	—
300,000	13.8	15.0	15.6	16.2	17.4	18.6	21.6	24.6	27.6	31.8	36.0	40.2	50.4	61.8	76.8	94.2	117.0	183.0	283.0	435.0	—	—	—
200,000	9.2	10.0	10.4	10.8	11.6	12.4	14.4	16.4	18.4	21.2	24.0	26.8	33.6	41.2	51.2	62.8	78.0	122.0	189.0	290.0	462.0	—	—
100,000	4.6	5.0	5.2	5.4	5.8	6.2	7.2	8.2	9.2	10.6	12.0	13.4	16.8	20.6	25.6	31.4	39.0	61.0	94.5	145.0	231.0	368.0	—
90,000	4.1	4.5	4.6	4.8	5.3	5.5	6.4	7.3	8.3	9.4	10.8	12.1	15.1	18.5	23.0	28.2	35.1	54.6	85.2	131.0	208.0	331.0	—
80,000	3.7	4.0	4.1	4.3	4.7	4.9	5.7	6.5	7.4	8.4	9.6	10.7	13.4	16.5	20.5	25.1	31.2	48.6	75.7	116.0	185.0	294.0	466.0
70,000	3.2	3.5	3.6	3.8	4.1	4.3	5.0	5.7	6.4	7.4	8.4	9.4	11.8	14.4	17.9	22.0	27.3	42.6	66.2	102.0	162.0	258.0	408.0
60,000	2.7	3.0	3.1	3.2	3.5	3.7	4.3	4.9	5.5	6.4	7.2	8.0	10.1	12.4	15.4	18.8	23.4	36.6	56.7	87.5	139.0	221.0	350.0
50,000	2.3	2.5	2.6	2.7	2.9	3.1	3.6	4.1	4.6	5.3	6.0	6.7	8.4	10.3	12.8	15.7	19.5	30.5	47.3	72.8	116.0	184.0	291.0
40,000	1.8	2.0	2.1	2.2	2.3	2.5	2.9	3.3	3.7	4.2	4.8	5.4	6.7	8.2	10.2	12.6	15.6	24.4	37.8	58.0	92.4	147.0	232.0
30,000	1.4	1.5	1.5	1.6	1.7	1.9	2.2	2.5	2.8	3.2	3.6	4.0	5.0	6.2	7.7	9.4	11.7	18.3	28.3	43.5	69.3	110.0	174.0
20,000	0.9	1.0	1.0	1.1	1.2	1.2	1.4	1.6	1.8	2.1	2.4	2.7	3.4	4.1	5.1	6.3	7.8	12.2	18.9	29.0	46.2	73.6	116.0
10,000	0.5	0.5	0.5	0.5	0.6	0.6	0.7	0.8	0.9	1.1	1.2	1.3	1.7	2.1	2.6	3.1	3.9	6.1	9.5	14.5	23.1	36.8	58.1
9,000	0.4	0.5	0.5	0.5	0.5	0.6	0.6	0.7	0.8	0.9	1.1	1.2	1.5	1.9	2.3	2.8	3.5	5.5	8.5	13.1	20.8	33.1	52.4
8,000	0.4	0.4	0.4	0.4	0.5	0.5	0.6	0.7	0.7	0.8	1.0	1.1	1.3	1.7	2.1	2.5	3.1	4.9	7.6	11.6	18.5	29.4	46.6
7,000	0.3	0.4	0.4	0.4	0.4	0.4	0.5	0.6	0.6	0.7	0.8	0.9	1.2	1.4	1.8	2.2	2.7	4.3	6.6	10.2	16.2	25.8	40.8
6,000	0.3	0.3	0.3	0.3	0.4	0.4	0.4	0.5	0.6	0.6	0.7	0.8	1.0	1.2	1.5	1.9	2.3	3.7	5.7	8.8	13.9	22.1	35.0
5,000	0.2	0.3	0.3	0.3	0.3	0.3	0.4	0.4	0.5	0.5	0.6	0.7	0.8	1.0	1.3	1.6	1.9	3.1	4.7	7.3	11.6	18.4	29.1
4,000	0.2	0.2	0.2	0.2	0.2	0.3	0.3	0.3	0.4	0.4	0.5	0.5	0.7	0.8	1.0	1.3	1.6	2.4	3.8	5.8	9.2	14.7	23.2
3,000	0.1	0.2	0.2	0.2	0.2	0.2	0.2	0.3	0.3	0.3	0.4	0.4	0.5	0.6	0.8	0.9	1.2	1.8	2.8	4.4	6.9	11.0	17.4
2,000	0.1	0.1	0.1	0.1	0.1	0.1	0.1	0.2	0.2	0.2	0.2	0.3	0.3	0.4	0.5	0.6	0.8	1.2	1.9	2.9	4.6	7.4	11.6
1,000	0.1	0.1	0.1	0.1	0.1	0.1	0.1	0.1	0.1	0.1	0.1	0.1	0.2	0.2	0.3	0.3	0.4	0.6	1.0	1.5	2.3	3.7	5.8
900	—	0.1	0.1	0.1	0.1	0.1	0.1	0.1	0.1	0.1	0.1	0.1	0.2	0.2	0.2	0.3	0.4	0.6	0.9	1.3	2.1	3.3	5.2
800	—	—	—	—	0.1	0.1	0.1	0.1	0.1	0.1	0.1	0.1	0.1	0.2	0.2	0.3	0.3	0.5	0.8	1.2	1.9	2.9	4.7
700	—	—	—	—	—	—	0.1	0.1	0.1	0.1	0.1	0.1	0.1	0.1	0.2	0.2	0.3	0.4	0.7	1.0	1.6	2.6	4.1
600	—	—	—	—	—	—	—	0.1	0.1	0.1	0.1	0.1	0.1	0.1	0.2	0.2	0.2	0.4	0.6	0.9	1.4	2.2	3.5
500	—	—	—	—	—	—	—	—	0.1	0.1	0.1	0.1	0.1	0.1	0.1	0.2	0.2	0.3	0.5	0.7	1.2	1.8	2.9
400	—	—	—	—	—	—	—	—	—	—	0.1	0.1	0.1	0.1	0.1	0.1	0.2	0.2	0.4	0.6	0.9	1.5	2.3
300	—	—	—	—	—	—	—	—	—	—	—	—	0.1	0.1	0.1	0.1	0.1	0.2	0.3	0.4	0.7	1.1	1.7
200	—	—	—	—	—	—	—	—	—	—	—	—	—	—	0.1	0.1	0.1	0.1	0.2	0.3	0.5	0.7	1.2
100	—	—	—	—	—	—	—	—	—	—	—	—	—	—	—	—	—	0.1	0.1	0.2	0.2	0.4	0.6

* Solid Conductors. Other conductors are stranded.

Note 1—The above table gives voltage drops encountered in a single phase two-wire system. The voltage drops in other systems may be obtained through multiplication by appropriate factors listed below:

System for Which Voltage Drop Is Desired	Multiplying Factors for Modification of Values in Table
Single Phase—3 Wire—Line to Line	1.00
Single Phase—3 Wire—Line to Neutral	0.50
Three Phase—3 Wire—Line to Line	0.866
Three Phase—4 Wire—Line to Line	0.866
Three Phase—4 Wire—Line to Neutral	0.50

Note 2—Allowable voltage drops for systems other than single phase, two wire cannot be used directly in the above table. Such drops should be modified through multiplication by the appropriate factor listed below. The voltage thus modified may then be used to obtain the proper wire size directly from the table.

System for Which Allowable Voltage Drop Is Known	Multiplying Factor for Modification of Known Value to Permit Direct Use of Table
Single Phase—3 Wire—Line to Line	1.00
Single Phase—3 Wire—Line to Neutral	2.00
Three Phase—3 Wire—Line to Line	1.155
Three Phase—4 Wire—Line to Line	1.155
Three Phase—4 Wire—Line to Neutral	2.00

Note 3—The footage employed in the tabulated ampere feet refers to the length of run of the circuit rather than to the footage of individual conductor.

Note 4—The above table is figured at 60°C since this is an estimate of the average temperature which may be anticipated in service. The table may be used without significant error for conductor temperatures up to and including 75°C.

TABLE 11.1.35 Voltage Drop for Copper Conductor in Nonmagnetic Conduit (Power Factor, 100%; Single Phase, 2 Wire; 60 Cycles)

WIRE SIZE AWG or MCM	1000	900	800	750	700	600	500	400	350	300	250	4/0	3/0	2/0	1/0	1	2	4	6	8*	10*	12*	14*
Ampere Feet	Volts Drop																						
500,000	13.0	15.0	16.0	17.0	19.0	22.0	26.0	32.0	36.0	42.0	51.0	59.0	75.0	95.0	121.0	152.0	192.0	306.0	483.0	—	—	—	—
400,000	10.4	12.0	12.8	13.6	15.2	17.6	20.8	25.6	28.8	33.6	40.8	47.2	60.0	76.0	96.8	122.0	154.0	244.0	386.0	—	—	—	—
300,000	7.8	9.0	9.6	10.2	11.4	13.2	15.6	19.2	21.6	25.2	30.6	35.4	45.0	57.0	72.6	91.2	115.0	184.0	290.0	450.0	—	—	—
200,000	5.2	6.0	6.4	6.8	7.6	8.8	10.4	12.8	14.4	16.8	20.4	23.6	30.0	38.0	48.4	60.8	76.8	122.0	193.0	300.0	480.0	—	—
100,000	2.6	3.0	3.2	3.4	3.8	4.4	5.2	6.4	7.2	8.4	10.2	11.8	15.0	19.0	24.2	30.4	38.4	61.2	96.6	150.0	240.0	384.0	—
90,000	2.3	2.7	2.9	3.0	3.4	4.0	4.6	5.8	6.4	7.6	9.2	10.6	13.5	17.1	21.8	27.4	34.5	55.1	87.0	135.0	216.0	345.0	—
80,000	2.1	2.4	2.5	2.7	3.0	3.5	4.1	5.2	5.7	6.7	8.2	9.4	12.0	15.2	19.4	24.3	30.7	49.0	77.3	120.0	192.0	307.0	487.0
70,000	1.8	2.1	2.2	2.4	2.7	3.1	3.6	4.5	5.0	5.9	7.1	8.3	10.5	13.3	17.0	21.2	26.8	42.8	67.6	105.0	168.0	269.0	426.0
60,000	1.6	1.8	1.9	2.0	2.3	2.6	3.1	3.8	4.3	5.0	6.1	7.1	9.0	11.4	14.5	18.2	23.0	36.7	58.0	90.0	144.0	230.0	365.0
50,000	1.3	1.5	1.6	1.7	1.9	2.2	2.6	3.2	3.6	4.2	5.1	5.9	7.5	9.5	12.1	15.2	19.2	30.6	48.3	74.9	120.0	192.0	304.0
40,000	1.0	1.2	1.3	1.3	1.5	1.7	2.1	2.6	2.9	3.4	4.1	4.7	6.0	7.6	9.7	12.2	15.4	24.4	38.6	60.0	96.0	154.0	243.0
30,000	0.8	0.9	1.0	1.0	1.1	1.3	1.6	1.9	2.2	2.5	3.1	3.5	4.5	5.7	7.3	9.1	11.5	18.4	29.0	45.0	72.0	115.0	182.0
20,000	0.5	0.6	0.6	0.7	0.8	0.9	1.0	1.3	1.4	1.7	2.0	2.4	3.0	3.8	4.8	6.1	7.7	12.2	19.3	30.0	48.0	76.8	122.0
10,000	0.3	0.3	0.3	0.3	0.4	0.4	0.5	0.6	0.7	0.8	1.0	1.2	1.5	1.9	2.4	3.0	3.8	6.1	9.7	15.0	24.0	38.4	60.8
9,000	0.2	0.3	0.3	0.3	0.3	0.4	0.5	0.6	0.6	0.8	0.9	1.1	1.4	1.7	2.2	2.7	3.5	5.5	8.7	13.5	21.6	34.5	54.7
8,000	0.2	0.2	0.3	0.3	0.3	0.4	0.4	0.5	0.6	0.7	0.8	0.9	1.2	1.5	1.9	2.4	3.1	4.9	7.7	12.0	19.2	30.7	48.7
7,000	0.2	0.2	0.2	0.2	0.3	0.3	0.4	0.5	0.5	0.6	0.7	0.8	1.1	1.3	1.7	2.1	2.7	4.3	6.8	10.5	16.8	26.9	42.6
6,000	0.2	0.2	0.2	0.2	0.2	0.3	0.3	0.4	0.4	0.5	0.6	0.7	0.9	1.1	1.5	1.8	2.3	3.7	5.8	9.0	14.4	23.0	36.5
5,000	0.1	0.2	0.2	0.2	0.2	0.2	0.3	0.3	0.4	0.4	0.5	0.6	0.8	1.0	1.2	1.5	1.9	3.1	4.8	7.5	12.0	19.2	30.4
4,000	0.1	0.1	0.1	0.1	0.2	0.2	0.2	0.3	0.3	0.3	0.4	0.5	0.6	0.8	1.0	1.2	1.5	2.4	3.9	6.0	9.6	15.4	24.3
3,000	0.1	0.1	0.1	0.1	0.1	0.1	0.2	0.2	0.2	0.3	0.3	0.4	0.5	0.6	0.7	0.9	1.2	1.8	2.9	4.5	7.2	11.5	18.2
2,000	0.1	0.1	0.1	0.1	0.1	0.1	0.1	0.1	0.1	0.2	0.2	0.2	0.3	0.4	0.5	0.6	0.8	1.2	1.9	3.0	4.8	7.7	12.2
1,000	—	—	—	—	—	—	0.1	0.1	0.1	0.1	0.1	0.1	0.2	0.2	0.2	0.3	0.4	0.6	1.0	1.5	2.4	3.8	6.1
900	—	—	—	—	—	—	0.1	0.1	0.1	0.1	0.1	0.1	0.1	0.2	0.2	0.3	0.4	0.6	0.9	1.4	2.2	3.5	5.5
800	—	—	—	—	—	—	—	0.1	0.1	0.1	0.1	0.1	0.1	0.2	0.2	0.2	0.3	0.5	0.8	1.2	1.9	3.1	4.9
700	—	—	—	—	—	—	—	0.1	0.1	0.1	0.1	0.1	0.1	0.1	0.2	0.2	0.3	0.4	0.7	1.1	1.7	2.7	4.3
600	—	—	—	—	—	—	—	—	—	0.1	0.1	0.1	0.1	0.1	0.2	0.2	0.2	0.4	0.6	0.9	1.4	2.3	3.7
500	—	—	—	—	—	—	—	—	—	—	0.1	0.1	0.1	0.1	0.1	0.2	0.2	0.3	0.5	0.8	1.2	1.9	3.0
400	—	—	—	—	—	—	—	—	—	—	—	0.1	0.1	0.1	0.1	0.1	0.2	0.2	0.4	0.6	1.0	1.5	2.4
300	—	—	—	—	—	—	—	—	—	—	—	—	0.1	0.1	0.1	0.1	0.1	0.2	0.3	0.5	0.7	1.2	1.8
200	—	—	—	—	—	—	—	—	—	—	—	—	—	—	0.1	0.1	0.1	0.1	0.2	0.3	0.5	0.8	1.2
100	—	—	—	—	—	—	—	—	—	—	—	—	—	—	—	—	—	0.1	0.1	0.2	0.2	0.4	0.6

* Solid Conductors. Other conductors are stranded.

Note 1—The above table gives voltage drops encountered in a single phase two-wire system. The voltage drops in other systems may be obtained through multiplication by appropriate factors listed below:

System for Which Voltage Drop is Desired	Multiplying Factors for Modification of Values in Table
Single Phase—3 Wire—Line to Line	1.00
Single Phase—3 Wire—Line to Neutral	0.50
Three Phase—3 Wire—Line to Line	0.866
Three Phase—4 Wire—Line to Line	0.866
Three Phase—4 Wire—Line to Neutral	0.50

Note 2—Allowable voltage drops for systems other than single phase, two wire cannot be used directly in the above table. Such drops should be modified through multiplication by the appropriate factor listed below. The voltage thus modified may then be used to obtain the proper wire size directly from the table.

System for Which Allowable Voltage Drop is Known	Multiplying Factor for Modification of Known Value to Permit Direct Use of Table
Single Phase—3 Wire—Line to Line	1.00
Single Phase—3 Wire—Line to Neutral	2.00
Three Phase—3 Wire—Line to Line	1.155
Three Phase—4 Wire—Line to Line	1.155
Three Phase—4 Wire—Line to Neutral	2.00

Note 3—The footage employed in the tabulated ampere feet refers to the length of run of the circuit rather than to the footage of individual conductor.

Note 4—The above table is figured at 60°C since this is an estimate of the average temperature which may be anticipated in service. The table may be used without significant error for conductor temperatures up to and including 75°C.

11.1.36 Voltage-Drop Curves for Typical Interleaved Construction of Copper Busway at Rated Load, Assuming 70°C (158°F) Operating Temperature

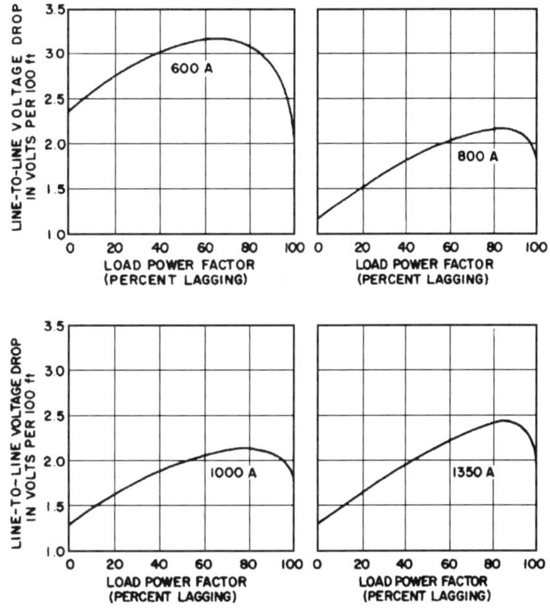

(From IEEE Std. 241-1990. Copyright 1990 IEEE. All rights reserved.)

11.1.37 Voltage-Drop Values for Three-Phase Busways with Copper Bus Bars, in Volts per 100 ft, Line-to-Line, at Rated Current with Balanced Entire Load at End

TABLE 11.1.37

Rating (amperes)	Power Factor 20	30	40	50	60	70	80	90	95	100
Low-voltage-drop ventilated feeder										
800	3.66	3.88	4.04	4.14	4.20	4.20	4.16	3.92	3.60	2.72
1000	1.84	2.06	2.22	2.40	2.54	2.64	2.72	2.70	2.62	2.30
1350	2.24	2.44	2.62	2.74	2.86	2.94	2.96	2.90	2.78	2.30
1600	1.88	2.10	2.30	2.46	2.62	2.74	2.82	2.84	2.76	2.42
2000	2.16	2.34	2.52	2.66	2.78	2.84	2.90	2.80	2.68	2.30
2500	2.04	2.18	2.38	2.48	2.62	2.68	2.72	2.62	2.50	2.14
3000	1.96	2.12	2.28	2.40	2.52	2.58	2.60	2.52	2.40	2.06
4000	2.18	2.36	2.54	2.68	2.80	2.80	2.90	2.80	2.68	2.28
5000	2.00	2.16	2.30	2.40	2.50	2.60	2.68	2.60	2.40	2.10
Low-voltage-drop ventilated plug-in										
800	6.80	6.86	6.92	6.86	6.72	6.52	6.04	5.26	4.64	2.76
1000	2.26	2.56	2.70	2.86	2.96	3.00	3.00	2.92	2.80	2.28
1350	2.98	3.16	3.32	3.38	3.44	3.46	3.40	3.22	3.00	2.32
1600	2.28	2.44	2.62	2.78	2.90	3.00	2.96	2.94	2.88	2.44
2000	2.58	2.78	2.92	3.02	3.10	3.16	3.08	3.00	2.82	2.28
2500	2.32	2.50	2.66	2.76	2.86	2.90	2.86	2.78	2.66	2.18
3000	2.18	2.34	2.48	2.60	2.70	2.74	2.72	2.66	2.58	2.10
4000	2.42	2.56	2.76	2.88	3.00	3.02	3.00	2.96	2.84	2.36
5000	2.22	2.30	2.48	2.60	2.70	2.76	2.74	2.68	2.60	2.16
Plug-in										
225	2.82	2.94	3.04	3.12	3.18	3.18	3.10	2.86	2.70	2.04
400	4.94	5.08	5.16	5.18	5.16	5.02	4.98	4.30	3.94	2.64
600	5.24	5.34	5.40	5.40	5.36	5.00	4.50	2.10	3.62	2.92
800	5.06	5.12	5.16	5.06	5.00	4.74	4.50	3.84	3.32	1.94
1000	5.80	5.88	5.84	5.76	5.56	5.30	4.82	4.12	3.52	1.94
Trolley busway										
100	1.2	1.38	1.58	1.74	1.80	2.06	2.20	2.30	2.30	2.18
Current-limiting ventilated										
1000	12.3	12.5	12.3	12.2	11.8	11.1	10.1	8.65	7.45	3.8
1350	15.5	15.6	15.4	15.3	14.7	13.9	12.6	10.7	9.2	4.7
1600	18.2	18.2	18.0	17.5	16.6	15.6	14.1	11.5	9.5	4.0
2000	20.4	20.3	20.0	19.4	18.4	17.0	13.9	12.1	10.1	3.8
2500	23.8	23.6	23.0	22.2	21.0	19.2	17.2	13.5	10.7	3.8
3000	26.0	26.2	25.8	24.8	23.4	21.5	19.1	15.1	12.0	4.0
4000	29.1	28.8	28.2	27.2	25.6	25.2	21.0	16.6	13.0	4.1

(From IEEE Std. 241-1990. Copyright 1990 IEEE. All rights reserved.)

11.1.38 Voltage-Drop Values for Three-Phase Busways with Aluminum Bus Bars, in Volts per 100 ft, Line-to-Line, at Rated Current with Balanced Entire Load at End

TABLE 11.1.38

Rating	Power Factor									
(amperes)	20	30	40	50	60	70	80	90	95	100
Low-voltage-drop ventilated feeder										
800	1.68	1.96	2.20	2.46	2.68	2.88	3.04	3.12	3.14	2.90
1000	1.90	2.16	2.38	2.60	2.80	2.96	3.06	3.14	3.12	2.82
1350	1.88	2.20	2.48	2.74	3.02	3.24	3.44	3.56	3.58	2.38
1600	1.66	1.92	2.18	2.42	2.64	2.84	3.02	3.12	3.16	2.94
2000	1.82	2.06	2.30	2.50	2.70	2.88	3.02	3.10	3.04	2.80
2500	1.86	2.10	2.34	2.56	2.74	2.90	3.04	3.10	3.08	2.78
3000	1.76	2.06	2.26	2.52	2.68	2.86	2.98	3.06	3.04	2.78
4000	1.74	1.98	2.24	2.48	2.70	2.88	3.04	3.08	3.12	2.88
5000	1.72	1.98	2.20	2.42	2.62	2.80	2.92	3.02	3.02	2.80
Low-voltage-drop ventilated plug-in										
800	2.12	2.38	2.58	2.80	3.00	3.16	3.26	3.30	3.24	2.90
1000	2.44	2.66	2.86	3.06	3.22	3.36	3.42	3.38	3.28	2.84
1350	2.22	2.48	2.78	3.00	3.24	3.46	3.60	3.68	3.64	3.30
1600	1.82	2.12	2.38	2.62	2.80	2.96	3.08	3.16	3.14	2.88
2000	2.00	2.30	2.50	2.76	2.92	3.06	3.12	3.18	3.12	2.80
2500	2.00	2.28	2.50	2.70	2.92	3.02	3.12	3.16	3.08	1.78
3000	1.98	2.26	2.44	2.66	2.86	3.00	3.10	3.18	3.14	2.82
4000	1.94	2.20	2.48	2.64	2.86	3.00	3.12	3.18	3.16	2.88
5000	1.90	2.16	2.38	2.58	2.76	2.92	3.06	3.10	3.08	2.52
Plug-in										
100	1.58	2.10	2.62	3.14	3.56	4.00	4.46	4.94	5.10	5.20
225	2.30	2.54	2.76	3.68	3.12	3.26	3.32	3.32	3.26	2.86
400	3.38	3.64	3.90	4.12	4.22	4.34	4.38	4.28	4.12	3.42
600	3.46	3.68	3.84	3.96	4.00	4.04	3.96	3.74	3.52	2.48
800	3.88	4.02	4.08	4.20	4.20	4.14	4.00	3.66	3.40	2.40
1000	3.30	3.48	3.62	3.72	3.78	3.80	3.72	3.50	3.30	2.50
Small plug-in										
50	2.2	2.6	3.0	3.5	3.8	4.1	4.5	4.7	4.8	4.6
Current-limiting ventilated										
1000	12.3	12.3	12.1	11.8	11.2	10.9	9.5	8.0	6.6	3.1
1350	16.3	16.3	16.1	15.6	14.7	13.7	12.1	8.1	8.0	3.1
1600	18.0	17.9	17.7	17.0	16.1	14.9	13.4	10.7	8.6	3.3
2000	22.5	22.4	21.8	21.2	19.9	18.2	16.0	12.7	9.9	3.1
2500	25.0	24.6	23.9	23.1	21.7	19.9	17.5	13.7	10.8	3.0
3000	26.2	25.8	25.1	24.1	22.7	20.8	18.2	14.2	10.9	2.9
4000	31.4	31.0	30.2	28.8	27.4	24.8	21.5	16.5	12.7	2.9

11.1.39 Voltage-Drop Curves for Typical Plug-in-Type CU Busway at Balanced Rated Load, Assuming 70°C (158°F) as the Operating Temperature

11.1.39

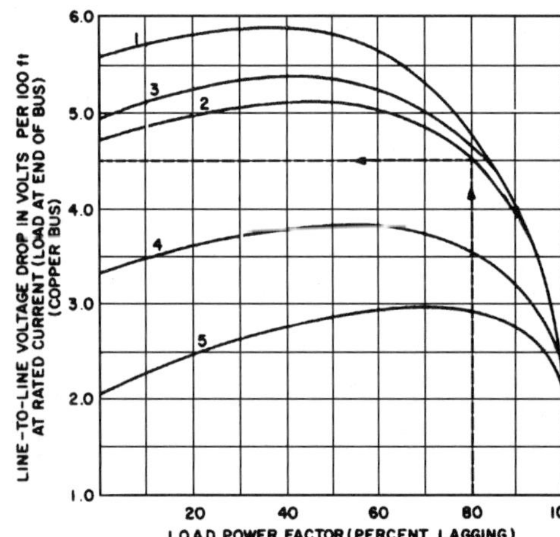

Curve	Bars per Phase	Size of Bar	Rated Current (amperes)
1	1	¼ × 4	1000
2	1	3/16 × 4	800
3	1	¼ × 2	600
4	1	3/16 × 2	400
5	1	3/16 × 1	225

11.1.40 Voltage-Drop Curves for Typical CU Feeder Busways at Balanced Rated Load Mounted Flat Horizontally, Assuming 70°C (158°F) as the Operating Temperature

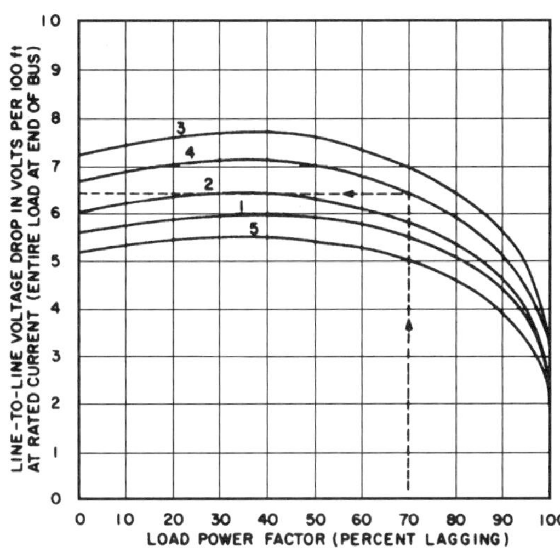

Curve	Bars per Phase	Size of Bar	Rated Current (amperes)
1	1	¼ x 2	600
2	1	¼ x 4	1000
3	2	¼ x 2½	1200
4	2	¼ x 4	1500
5	4	¼ x 3	2300

11.1.41 Voltage-Drop Curve versus Power Factor for Typical Light-Duty Trolley Busway Carrying Rated Load, Assuming 70°C (158°F) as the Operating Temperature

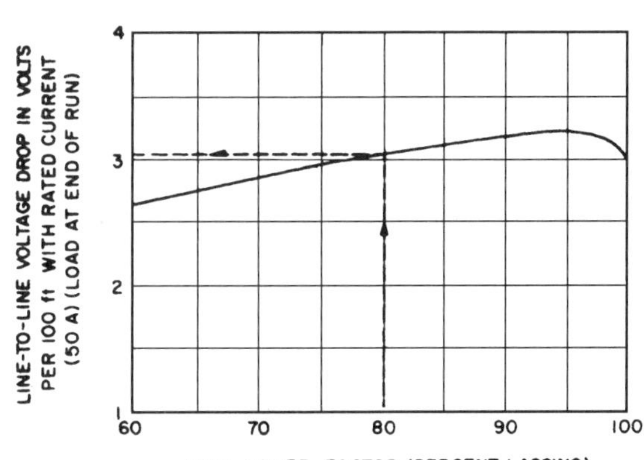

11.1.42 Voltage-Drop Curves for Three-Phase Transformers, 225 to 10,000 kVA, 5 to 25 kV (*see page 11.36*)

11.1.42 Voltage-Drop Curves for Three-Phase Transformers, 225 to 10,000 kVA, 5 to 25 kV

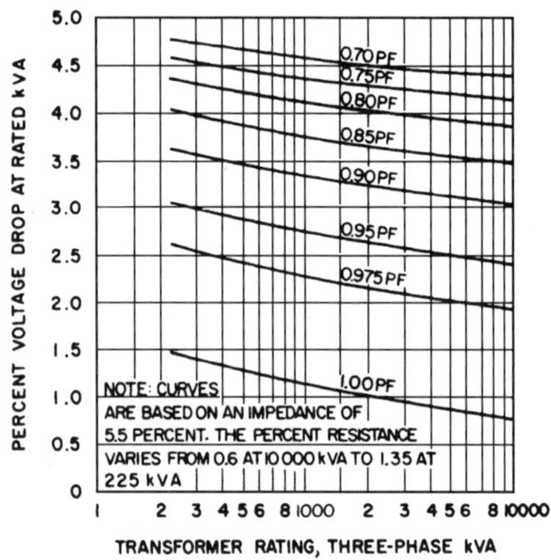

Note: This figure applies to 5.5 percent impedance transformers. For transformers of substantially different impedance, the information for the calculation should be obtained from the manufacturer.

11.1.43 Application Tips

1. Always locate the source of the low-voltage supply (service transformer and service equipment, distribution transformers, distribution panels, generators, and UPS systems) as close to the center of load as possible.
2. When you oversize a feeder or branch circuit for voltage drop compensation, note it as such on the design drawings. This prevents confusion for the electrical contractor(s) bidding on and/or installing the work.
3. *Rule of thumb.* When the distance in circuit feet equals the nominal system voltage (e.g., you are at 120 circuit feet and the nominal system voltage is 120 V), it serves as a "flag" that you should check the voltage drop. In practice, experience generally has shown that it is safe to go another 50 percent in circuit feet without a voltage-drop problem (180 circuit feet for the example given).
4. As is the case with short-circuit calculations, the only significant circuit impedance parameters generally needed for the voltage-drop calculations are those of transformers, busways, and conductors in conduit. Devices such as switches, circuit breakers, transfer switches, etc. contribute negligible impedance and generally can be ignored.
5. The *NEC* recommends (not mandatory) that the voltage drop from the point of service entrance to the furthest extremity of the electrical distribution system not exceed 5 percent. With this guideline, it is generally good practice to limit the voltage drop to distribution panels to a maximum of 2 to 3 percent, leaving the remaining 2 to 3 percent for the smaller branch circuits to the extremities of the system. For example, limiting the voltage drop to 2 percent to a distribution panel would allow up to 3 percent voltage drop for the branch circuits served by that panel.

11.1.44 Flicker of Incandescent Lamps Caused by Recurrent Voltage Dips

11.1.44

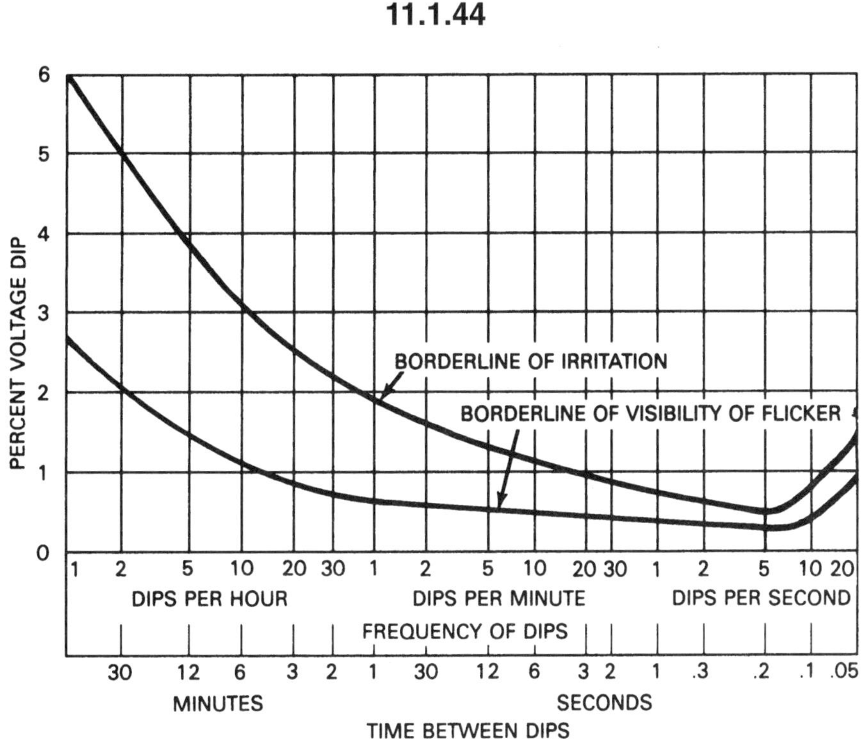

11.1.45 Effect of Voltage Variations on Incandescent Lamps

TABLE 11.1.45

Applied Voltage (volts)	Lamp Rating 120 V Percent Life	Percent Light	125 V Percent Life	Percent Light	130 V Percent Life	Percent Light
105	575	64	880	55	—	—
110	310	74	525	65	880	57
115	175	87	295	76	500	66
120	100	100	170	88	280	76
125	58	118	100	100	165	88
130	34	132	59	113	100	100

11.1.46 General Effect of Voltage Variations on Induction Motor Characteristics

TABLE 11.1.46

Characteristic	Function of Voltage	Voltage Variation 90% Voltage	110% Voltage
Starting and maximum running torque	$(\text{Voltage})^2$	Decrease 19%	Increase 21%
Synchronous speed	Constant	No change	No change
Percent slip	$1/(\text{Voltage})^2$	Increase 23%	Decrease 17%
Full-load speed	Synchronous speed-slip	Decrease 1.5%	Increase 1%
Efficiency			
Full load	—	Decrease 2%	Increase 0.5 to 1%
¾ load	—	Practically no change	Practically no change
½ load	—	Increase 1 to 2%	Decrease 1 to 2%
Power factor			
Full load	—	Increase 1%	Decrease 3%
¾ load	—	Increase 2 to 3%	Decrease 4%
½ load	—	Increase 4 to 5%	Decrease 5 to 6%
Full-load current	—	Increase 11%	Decrease 7%
Starting current	Voltage	Decrease 10 to 12%	Increase 10 to 12%
Temperature rise, full load	—	Increase 6 to 7 °C	Decrease 1 to 2 °C
Maximum overload capacity	$(\text{Voltage})^2$	Decrease 19%	Increase 21%
Magnetic noise - no load in particular	—	Decrease slightly	Increase slightly

(From IEEE Std. 241-1990.

11.1.47 Calculation of Voltage Dips (Momentary Voltage Variations)

One source of voltage dips in commercial buildings is the inrush current while starting large motors on a distribution transformer that also supplies incandescent lights. A quick way to estimate flicker problems from motor starting is to multiply the motor locked-rotor starting kVA by the supply transformer impedance. A typical motor may draw 5 kVA/hp, and a transformer impedance may be 6 percent. The equation below estimates flicker while starting a 15-hp motor on a 150-kVA transformer:

$$15 \text{ hp} \times 5 \text{ kVA/hp} \times 6\%/150 \text{ kVA} = 3\% \text{ flicker}$$

The estimated 3 percent dip associated with starting this motor reaches the borderline of irritation at 10 starts per hour. If the voltage dip combined with the starting frequency approaches the objectionable zone, more accurate calculations should be made using the actual locked-rotor current of the motor. Accurate locked-rotor kVA for motors is available from the motor manufacturer and from the starting code letter on the motor nameplate. The values for the code letters are listed in ANSI/NEMA MG1-1978 and in the *NEC,* Article 430. More accurate methods for calculating motor starting voltage dips are beyond the scope of this book.

One slightly more accurate method of calculating voltage dip for a quick calculation is to ratio the inrush current or kVA to the available short-circuit current or kVA (if known) $\times$ 100 percent, to that point in the system of concern. This takes into account all impedance to the point in the system.

When the amount of the voltage dip in combination with the frequency falls within the objectionable range, then consideration should be given to methods of reducing the dip to acceptable values, such as using two or more smaller motors, providing a separate transformer for motors, separating motor feeders from other feeders, or using reduced-voltage motor starting.

Section

12

Transformers

12.1.0 Introduction
12.1.1 Typical Transformer Weights (lb) by kVA
12.1.2 Transformer Full-Load Current, Three-Phase, Self-Cooled Ratings
12.1.3 Typical Impedances, Three-Phase, Liquid-Filled Transformers
12.1.4 Approximate Transformer Loss and Impedance Data
12.1.5 Transformer Primary (480-V, Three-Phase, Delta) and Secondary (208-Y/120-V, Three-Phase, Four-Wire) Overcurrent Protection, Conductors and Grounding
12.1.6 *NEC* Table 450.3 (A), Maximum Rating or Setting of Overcurrent Protection for Transformers Over 600 V (as a Percentage of Transformer Rated Current)
12.1.7 *NEC* Table 450.3 (B), Maximum Rating or Setting of Overcurrent Protection for Transformers 600 V and Less (as a Percentage of Transformer Rated Current)
12.2.1 Electrical Connection Diagrams
12.3.1 Auto Zigzag Grounding Transformers for Deriving a Neutral, Schematic and Wiring Diagram
12.3.2 Auto Zigzag Transformer Ratings
12.4.1 Buck-Boost Transformer Three-Phase Connection Summary
12.4.2 Wiring Diagrams for Low-Voltage Single-Phase Buck-Boost Transformers
12.4.3 Connection Diagrams for Buck-Boost Transformers in Autotransformer Arrangement for Single-Phase System
12.4.4 Connection Diagrams for Buck-Boost Transformers in Autotransformer Arrangement for Three-Phase System
12.5.1 Maximum Average Sound Levels for Transformers
12.5.2 Typical Building Ambient Sound Levels
12.6.1 Transformer Insulation System Temperature Ratings
12.7.1 *k*-Rated Transformers

12.1.0 Introduction

Transformers are a critical part of electrical distribution systems because they are used most often to change voltage levels. This affects voltage, current (both load and fault current levels), and system capacity. They also can be used to isolate, suppress harmonics, derive neutrals through a zigzag grounding arrangement, and reregulate voltage. The information that follows provides useful design and installation data.

12.1.1 Typical Transformer Weights (lb) by kVA

TABLE 12.1.1

Oil Filled 3 Phase 5/15 KV To 480/277			
kVA	Lbs.	kVA	Lbs.
150	1800	1000	6200
300	2900	1500	8400
500	4700	2000	9700
750	5300	3000	15000
Dry 240/480 To 120/240 Volt			
1 Phase		3 Phase	
kVA	Lbs.	kVA	Lbs.
1	23	3	90
2	36	6	135
3	59	9	170
5	73	15	220
7.5	131	30	310
10	149	45	400
15	205	75	600
25	255	112.5	950
37.5	295	150	1140
50	340	225	1575
75	550	300	1870
100	670	500	2850
167	900	750	4300

12.1.2 Transformer Full-Load Current, Three-Phase, Self-Cooled Ratings

TABLE 12.1.2

Voltage, Line-to-Line													
kVA	208	240	480	600	2,400	4,160	7,200	12,000	12,470	13,200	13,800	22,900	34,400
30	83.3	72.2	36.1	28.9	7.22	4.16	2.41	1.44	1.39	1.31	1.26	0.75	0.50
45	125	108	54.1	43.3	10.8	6.25	3.61	2.17	2.08	1.97	1.88	1.13	0.76
75	208	180	90.2	72.2	18.0	10.4	6.01	3.61	3.47	3.28	3.14	1.89	1.26
112½	312	271	135	108	27.1	15.6	9.02	5.41	5.21	4.92	4.71	2.84	1.89
150	416	361	180	144	36.1	20.8	12.0	7.22	6.94	6.56	6.28	3.78	2.52
225	625	541	271	217	54.1	31.2	18.0	10.8	10.4	9.84	9.41	5.67	3.78
300	833	722	361	289	72.2	41.6	24.1	14.4	13.9	13.1	12.6	7.56	5.04
500	1,388	1,203	601	481	120	69.4	40.1	24.1	23.1	21.9	20.9	12.6	8.39
750	2,082	1,804	902	722	180	104	60.1	36.1	34.7	32.8	31.4	18.9	12.6
1,000	2,776	2,406	1,203	962	241	139	80.2	48.1	46.3	43.7	41.8	25.2	16.8
1,500	4,164	3,608	1,804	1,443	361	208	120	72.2	69.4	65.6	62.8	37.8	25.2
2,000	....	4,811	2,406	1,925	481	278	160	96.2	92.6	87.5	83.7	50.4	33.6
2,500	....	....	3,007	2,406	601	347	200	120	116	109	105	63.0	42.0
3,000	....	....	3,609	2,887	722	416	241	144	139	131	126	75.6	50.4
3,750	....	....	4,511	3,608	902	520	301	180	174	164	157	94.5	62.9
5,000	....	....	....	4,811	1,203	694	401	241	231	219	209	126	83.9
7,500	....	....	....	....	1,804	1,041	601	361	347	328	314	189	126
10,000	....	....	....	....	2,406	1,388	802	481	463	437	418	252	168

12.1.3 Typical Impedances, Three-Phase, Liquid-Filled Transformers (*see page 12.3*)

12.1.4 Approximate Transformer Loss and Impedance Data (*see page 12.3*)

TABLE 12.1.3 Typical Impedances—Three-Phase Transformers

kVA	Liquid-Filled	
	Network	Padmount
37.5	....	....
45	....	....
50	....	....
75	....	3.4
112.5	....	3.2
150	....	2.4
225	....	3.3
300	5.00	3.4
500	5.00	4.6
750	5.00	5.75
1000	5.00	5.75
1500	7.00	5.75
2000	7.00	5.75
2500	7.00	5.75
3000	....	6.50
3750	....	6.50
5000	....	6.50

NOTE: Values are typical. Refer to transformer manufacturer for exact values.

Note: Values are typical. Refer to transformer manufacturer for exact values.

TABLE 12.1.4 Approximate Loss and Impedance Data

15 kV Class Oil Liquid-Filled Transformers

65°C Rise						
kVA	No Load Watts Loss	Full Load Watts Loss	%Z	%R	%X	X/R
112.5	550	2470	5.00	1.71	4.70	2.75
150	545	3360	5.00	1.88	4.63	2.47
225	650	4800	5.00	1.84	4.65	2.52
300	950	5000	5.00	1.35	4.81	3.57
500	1200	8700	5.00	1.50	4.77	3.18
750	1600	12160	5.75	1.41	5.57	3.96
1000	1800	15100	5.75	1.33	5.59	4.21
1500	3000	19800	5.75	1.12	5.64	5.04
2000	4000	22600	5.75	0.93	5.67	6.10
2500	4500	26000	5.75	0.86	5.69	6.61

15 kV Class Primary – Dry-Type Transformers Class H

150°C Rise						
kVA	No Load Watts Loss	Full Load Watts Loss	%Z	%R	%X	X/R
300	1600	10200	4.50	2.87	3.47	1.21
500	1900	15200	5.75	2.66	5.10	1.92
750	2700	21200	5.75	2.47	5.19	2.11
1000	3400	25000	5.75	2.16	5.33	2.47
1500	4500	32600	5.75	1.87	5.44	2.90
2000	5700	44200	5.75	1.93	5.42	2.81
2500	7300	50800	5.75	1.74	5.48	3.15
80°C Rise						
300	1800	7600	4.50	1.93	4.06	2.10
500	2300	9500	5.75	1.44	5.57	3.87
750	3400	13000	5.75	1.28	5.61	4.38
1000	4200	13500	5.75	0.93	5.67	6.10
1500	5900	19000	5.75	0.87	5.68	6.51
2000	6900	20000	5.75	0.66	5.71	8.72
2500	7200	21200	5.75	0.56	5.72	10.22

600-Volt Primary Class Dry-Type Transformers

150°C Rise						
kVA	No Load Watts Loss	Full Load Watts Loss	%Z	%R	%X	X/R
3	33	231	7.93	6.60	4.40	0.67
6	58	255	3.70	3.28	1.71	0.52
9	77	252	3.42	1.94	2.81	1.45
15	150	875	5.20	4.83	1.92	0.40
30	200	1600	5.60	4.67	3.10	0.66
45	300	1900	4.50	3.56	2.76	0.78
75	400	3000	4.90	3.47	3.46	1.00
112.5	500	4900	5.90	3.91	4.42	1.13
150	600	6700	6.20	4.07	4.68	1.15
225	700	8600	6.40	3.51	5.35	1.52
300	800	10200	7.10	3.13	6.37	2.03
500	1700	9000	5.50	1.46	5.30	3.63
750	2200	11700	6.30	1.27	6.17	4.87
1000	2800	13600	6.50	1.08	6.41	5.93

600-Volt Primary Class Dry-Type Transformers

115°C Rise						
kVA	No Load Watts Loss	Full Load Watts Loss	%Z	%R	%X	X/R
15	150	700	5.20	3.67	3.69	1.01
30	200	1500	4.60	4.33	1.54	0.36
45	300	1700	3.70	3.11	2.00	0.64
75	400	2300	4.60	2.53	3.84	1.52
112.5	500	3100	6.50	2.31	6.08	2.63
150	600	5900	6.20	3.53	5.09	1.44
225	700	6000	7.20	2.36	6.80	2.89
300	800	6600	6.30	1.93	6.00	3.10
500	1700	6800	5.50	1.02	5.40	5.30
750	1500	9000	4.10	1.00	3.98	3.98

600-Volt Primary Class Dry-Type Transformers

80°C Rise						
kVA	No Load Watts Loss	Full Load Watts Loss	%Z	%R	%X	X/R
15	200	500	2.30	2.00	1.14	0.57
30	300	975	2.90	2.25	1.83	0.81
45	300	1100	2.90	1.78	2.29	1.29
75	400	1950	3.70	2.07	3.07	1.49
112.5	600	3400	4.30	2.49	3.51	1.41
150	700	3250	4.10	1.70	3.73	2.19
225	800	4000	5.30	1.42	5.11	3.59
300	1300	4300	3.30	1.00	3.14	3.14
500	2200	5300	4.50	0.62	4.46	7.19

12.1.5 Transformer Primary (480-V, Three-Phase, Delta) and Secondary (208-Y/120-V, Three-Phase, Four-Wire) Overcurrent Protection, Conductors and Grounding

TABLE 12.1.5

Primary CB/Fuse	Recommended Primary Wire and Ground (CB/Fuse)	Primary FLA	Transformer KVA	Secondary FLA	Recommended Secondary Wire and Ground (CB/Fuse)	Secondary CB/Fuse
15/10	#12 and #12	9.0	7.5	20.8	#10 and #10/#16	25/25
25/20	#12 and #10 or #12	18.1	15	41.7	#8 and #10	60/50
40/35	#10 and #10	30.1	25	69.4	#4 and #8	90/80
50/45	#6 and #10	36.1	30	83.3	#3 and #6/#8	110/100
70/60	#4 and #8 or #10	54.2	45	125	#1 and #8	150/125
125/110	#1 and #6	90.3	75	208.3	#4/0 and #4	250/225
175/150	#2/0 and #6	135.4	112.5	312.5	350 and #3	400/350
225/200	#4/0 and #4 or #6	180.5	150	416.6	500 and #2	500/450
350/300	350 and #3 or #4	270.8	225	625	2 sets [350 and #1/0]	800/700
450/400	500 and #2 or #3	361.0	300	833.3	3 sets [500 and #2/0]	1000/1000
800/700	2 sets [350 and #1/0]	601.7	500	1388.8	5 sets [500 and #4/0]	1600/1600
1200/1000	3 sets [500 and #3/0 or #2/0]	902.5	750	2083.3	7 sets [500 and 350]	2500/2500
1600/1200	4 sets [350 and #4/0 or #3/0]	1203.4	1000	2777.8	8 sets [500 and 400]	3000/3000
2000/1600	5 sets [500 and 250/#4/0]	1504.2	1250	3472.2	11 sets [500 and 500]	4000/4000
2500/2000	6 sets [500 and 350/250]	1805.1	1500	4166.7	14 sets [500 and 700]	5000/5000

Notes:

1. C/B sizes based on maximum permissible per NEC Table 450.3(B) and are for 80% rated thermal magnetic C/Bs. Primary/Secondary coordination is taken into consideration. Fuse sizes are for dual element time delay RK-5 fuses.
2. Use fuse sizes for 100% rated C/Bs.
3. Lower C/B values are commonly used to align with other equipment ratings. In particular, 30 KVA: primary = 45A, secondary = 100A with 80% rated C/Bs; 75 KVA: primary = 100A, secondary = 225A with 80% rated C/Bs; 112.5 KVA: primary = 150A, secondary = 350A with 80% rated C/Bs; 150 KVA: primary = 200A, secondary = 400A with 80% rated C/Bs; 500 KVA: primary = 400A, secondary = 800Awith 80% rated C/Bs.
4. Equipment grounding conductor size is based on NEC Table 250.122 and assumes separately derived ground at the transformer secondary.
5. Recommended wire size is based on NEC Table 310.16 (75°C types THWN and XHHW), 240.3(B) and 240.3(C). Wire sizes may need to be increased due to voltage drop.
6. Maximum conduit fill depends on the wire insulation type and the type of conduit. These values have been calculated for conductors of the same size and are given in tabular form in Annex C of NFPA 70, which has been included in Section 15 of this book for convenient reference.

   EMT - Table C1
   IMC - Table C4
   RSC - Table C8
   Schedule 40 rigid PVC - Table C10
   Schedule 80 rigid PVC - Table C9

12.1.6 Maximum Rating or Setting of Overcurrent Protection for Transformers Over 600 V (as a Percentage of Transformer Rated Current) [*NEC* Table 450.3(A)] (*see page 12.5*)

TABLE 12.1.6

Location Limitations	Transformer Rated Impedance	Primary Protection Over 600 Volts: Circuit Breaker (See Note 4.)	Primary Protection Over 600 Volts: Fuse Rating	Secondary Protection (See Note 2.) Over 600 Volts: Circuit Breaker (See Note 4.)	Secondary Protection (See Note 2.) Over 600 Volts: Fuse Rating	Secondary Protection (See Note 2.) 600 Volts or Below: Circuit Breaker or Fuse Rating
Any location	Not more than 6%	600% (See Note 1.)	300% (See Note 1.)	300% (See Note 1.)	250% (See Note 1.)	125% (See Note 1.)
	More than 6% and not more than 10%	400% (See Note 1.)	300% (See Note 1.)	250% (See Note 1.)	225% (See Note 1.)	125% (See Note 1.)
Supervised locations only (See Note 3.)	Any	300% (See Note 1.)	250% (See Note 1.)	Not required	Not required	Not required
	Not more than 6%	600%	300%	300% (See Note 5.)	250% (See Note 5.)	250% (See Note 5.)
	More than 6% and not more than 10%	400%	300%	250% (See Note 5.)	225% (See Note 5.)	250% (See Note 5.)

Notes:

1. Where the required fuse rating or circuit breaker setting does not correspond to a standard rating or setting, a higher rating or setting that does not exceed the next higher standard rating or setting shall be permitted.
2. Where secondary overcurrent protection is required, the secondary overcurrent device shall be permitted to consist of not more than six circuit breakers or six sets of fuses grouped in one location. Where multiple overcurrent devices are utilized, the total of all the device ratings shall not exceed the allowed value of a single overcurrent device. If both circuit breakers and fuses are used as the overcurrent device, the total of the device ratings shall not exceed that allowed for fuses.
3. A supervised location is a location where conditions of maintenance and supervision ensure that only qualified persons monitor and service the transformer installation.
4. Electronically actuated fuses that may be set to open at a specific current shall be set in accordance with settings for circuit breakers.
5. A transformer equipped with a coordinated thermal overload protection by the manufacturer shall be permitted to have separate secondary protection omitted.

(©2001, NFPA)

12.1.7 Maximum Rating or Setting of Overcurrent Protection for Transformers 600 V and Less (as a Percentage of Transformer Rated Current)

TABLE 12.1.7

Protection Method	Primary Protection: Currents of 9 Amperes or More	Primary Protection: Currents Less Than 9 Amperes	Primary Protection: Currents Less Than 2 Amperes	Secondary Protection (See Note 2.): Currents of 9 Amperes or More	Secondary Protection (See Note 2.): Currents Less Than 9 Amperes
Primary only protection	125% (See Note 1.)	167%	300%	Not required	Not required
Primary and secondary protection	250% (See Note 3.)	250% (See Note 3.)	250% (See Note 3.)	125% (See Note 1.)	167%

Notes:

1. Where 125 percent of this current does not correspond to a standard rating of a fuse or nonadjustable circuit breaker, a higher rating that does not exceed the next higher standard rating shall be permitted.
2. Where secondary overcurrent protection is required, the secondary overcurrent device shall be permitted to consist of not more than six circuit breakers or six sets of fuses grouped in one location. Where multiple overcurrent devices are utilized, the total of all the device ratings shall not exceed the allowed value of a single overcurrent device. If both breakers and fuses are utilized as the overcurrent device, the total of the device ratings shall not exceed that allowed for fuses.
3. A transformer equipped with coordinated thermal overload protection by the manufacturer and arranged to interrupt the primary current shall be permitted to have primary overcurrent protection rated or set at a current value that is not more than six times the rated current of the transformer for transformers having not more than 6 percent impedance and not more than four times the rated current of the transformer for transformers having more than 6 percent but not more than 10 percent impedance.

(©2001, NFPA)

12.2.1 Electrical Connection Diagrams

12.2.1 Connection diagrams for transformers.

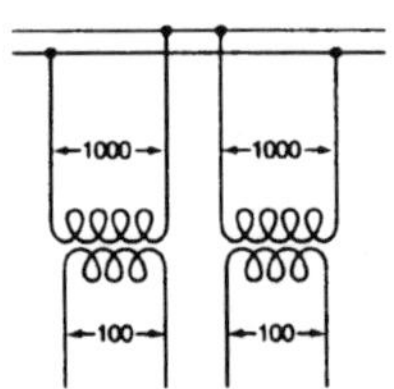

Single-phase transformers on a single phase system.

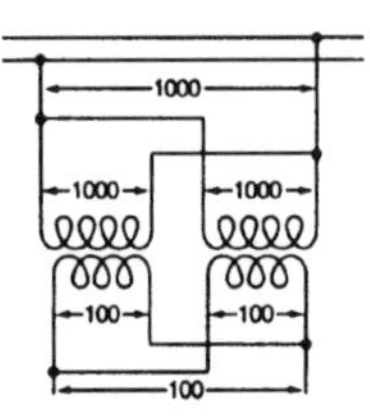

Single-phase transformers, secondaries in parallel.

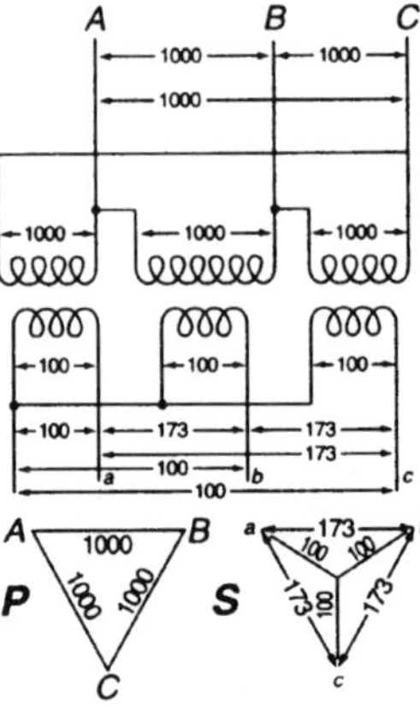

Three single-phase transformers connected delta-star to a three-phase system.

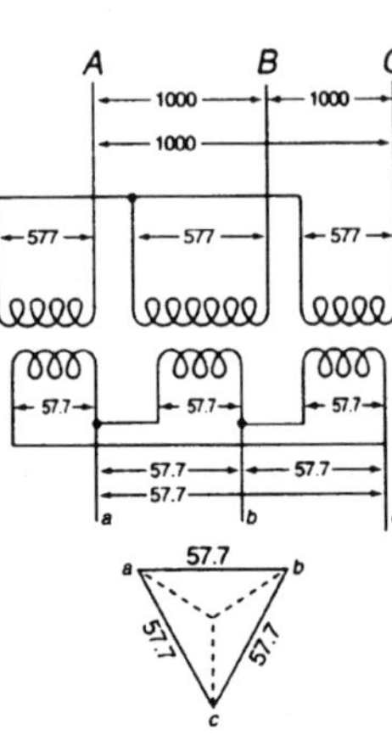

Three single-phase transformers connected star-delta to a three phase system.

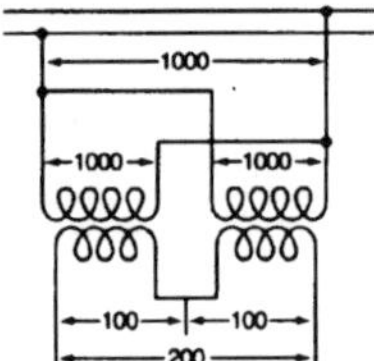

Single-phase transformers secondaries in series.

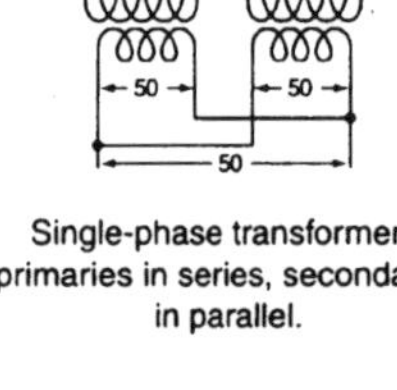

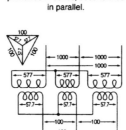

Single-phase transformers primaries in series, secondaries in parallel.

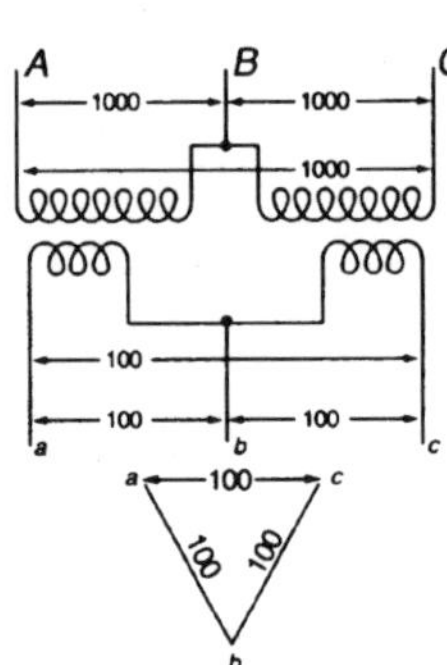

Two single-phase transformers connected open-delta to a three-phase system.

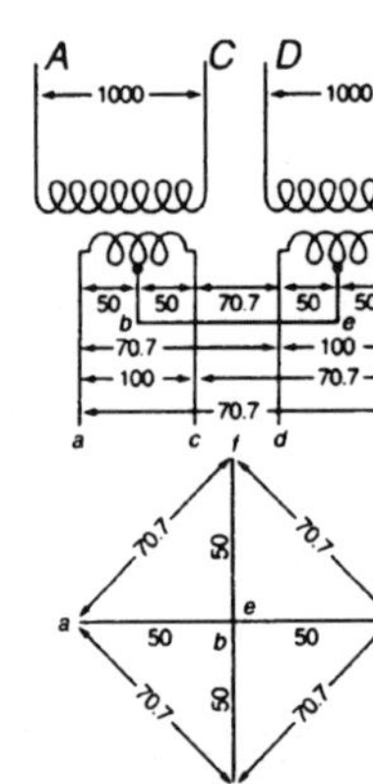

Two single-phase transformers connected star to a four-wire two-phase system.

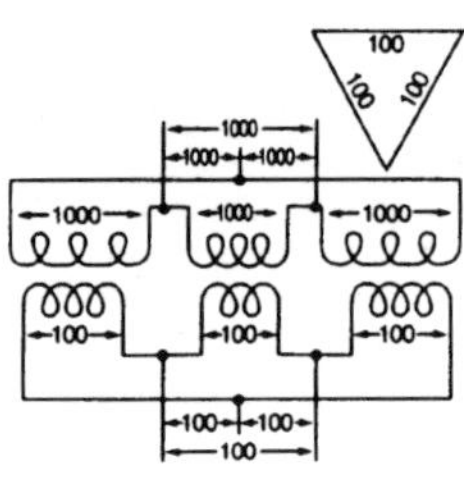

Three single-phase transformers connected delta-delta to a three-phase system.

Three single-phase transformers connected star-star to a three-phase system.

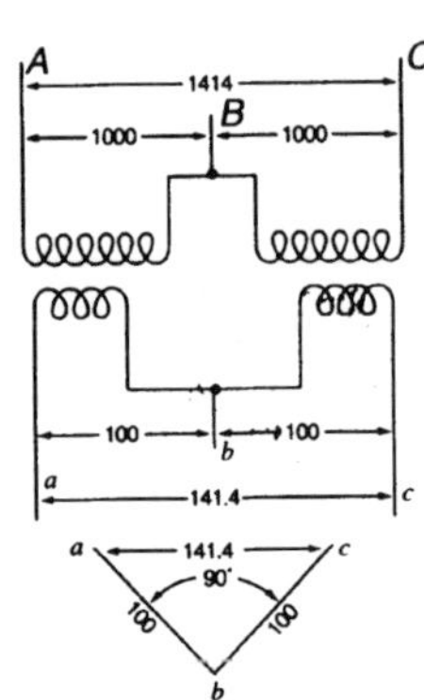

Two single-phase transformers connected to a three-wire two-phase system.

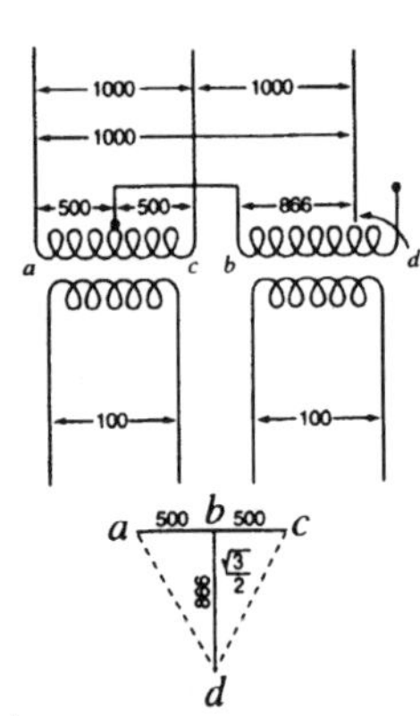

Two single-phase transformers connected T to a three-phase two-phase system. Scott Connection.

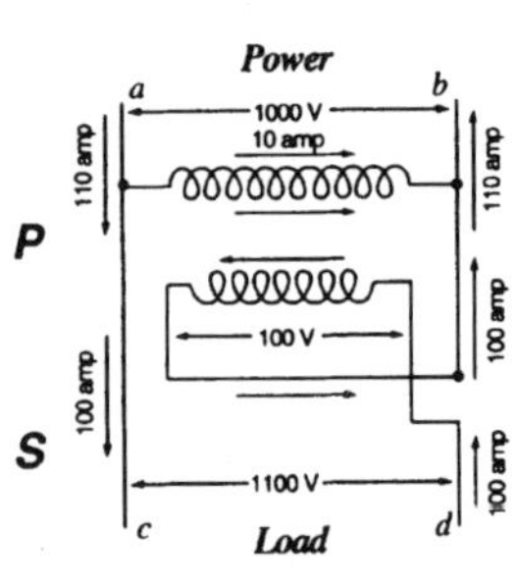

Single phase transformer used as a booster.

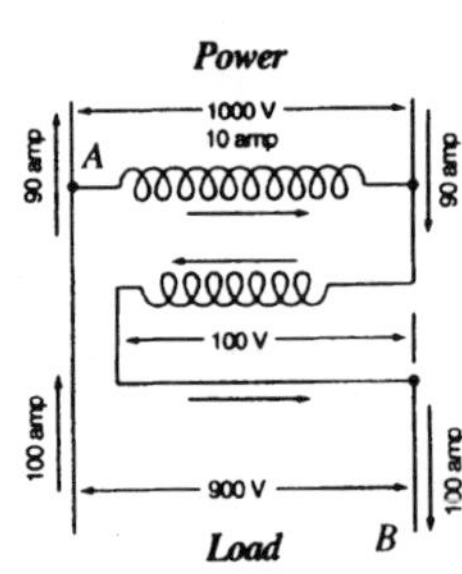

Single phase transformer connected to lower the E.M.F.

12.3.1 Auto Zigzag Grounding Transformers for Deriving a Neutral, Schematic and Wiring Diagram

12.3.1 Auto zig-zag transformers, for developing a neutral from a three-phase, 3-wire supply.

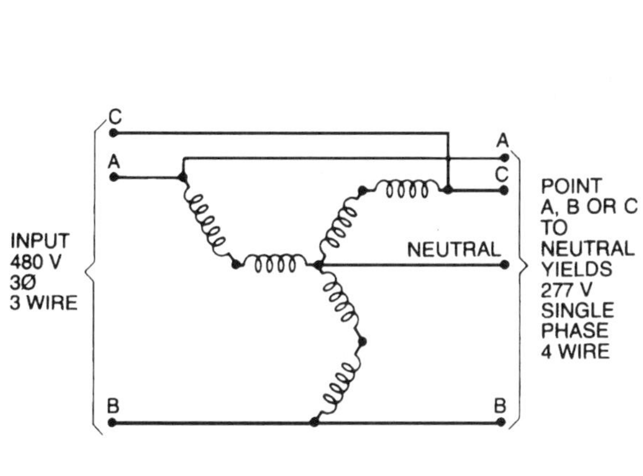

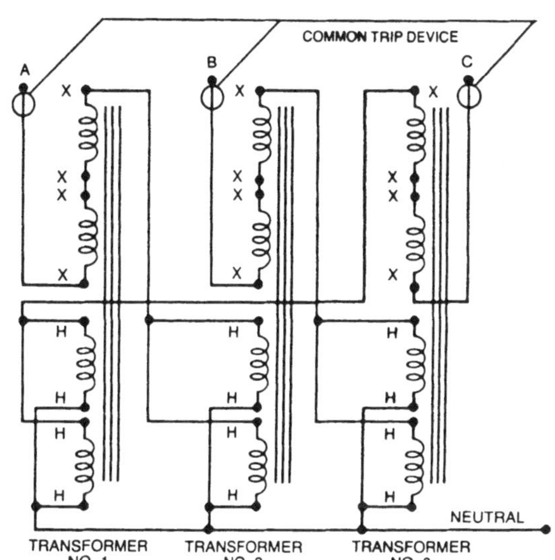

12.3.2 Auto Zigzag Transformer Ratings

TABLE 12.3.2

PRIMARY (INPUT): 480 VOLTS 3Ø, 3 WIRE ***50/60 Hz** **SECONDARY (OUTPUT): 480Y/277 VOLTS 3Ø, 4 WIRE**

Use 3 Pieces of Type No.	Available In	Nameplate KVA For Each Tfmr.	No. of Tfmr. Required	Three Phase KVA	Max. Continuous Amp. Load Per Phase (277 Volts)
T-2-53010-S	No Taps Only	1.0	3	10.8	12.50
T-2-53011-S	No Taps Only	1.5	3	15.6	18.75
T-2-53012-S	No Taps Only	2.0	3	20.7	25.00
T-2-53013-4S	Taps & No Taps	3.0	3	31.2	37.50
T-2-53014-4S	Taps & No Taps	5.0	3	51.9	62.50
T-2-53515-3S	With Taps Only	7.5	3	78.0	93.50
T-2-53516-3S	With Taps Only	10.0	3	103.8	125.00
T-2-53517-3S	With Taps Only	15.0	3	156.0	187.50
T-2-53518-3S	With Taps Only	25.0	3	259.5	312.00
T-1-53019-3S	With Taps Only	37.5	3	390.0	468.00
T-1-53020-3S	With Taps Only	50.0	3	519.0	625.00
T-1-53021-3S	With Taps Only	75.0	3	780.0	935.00
T-1-53022-3S	With Taps Only	100.0	3	1038.0	1250.00
T-1-53023-3S	With Taps Only	167.0	3	1734.0	2085.00

Connection diagram (using 3 pieces of 1 phase, 60 hertz transformers connected zig-zag auto) for developing a neutral (4th wire) from a 3 phase, 3 wire supply.
*Applicable for the above connection only.

12.4.1 Buck-Boost Transformer Three-Phase Connection Summary

TABLE 12.4.1 Three-Phase Connections

3 PHASE CONNECTIONS

INPUT (SUPPLY SYSTEM)	DESIRED OUTPUT CONNECTION	
DELTA 3 wire	WYE 3 or 4 wire	DO NOT USE
OPEN DELTA 3 wire	WYE 3 or 4 wire	DO NOT USE
WYE 3 or 4 wire	CLOSED DELTA 3 wire	DO NOT USE
WYE 4 wire	WYE 3 or 4 wire	OK
WYE 3 or 4 wire	OPEN DELTA 3 wire	OK
CLOSED DELTA 3 wire	OPEN DELTA 3 wire	OK

12.4.2 Wiring Diagrams for Low-Voltage Single-Phase Buck-Boost Transformers

12.4.2 Low-voltage buck-boost drawings.

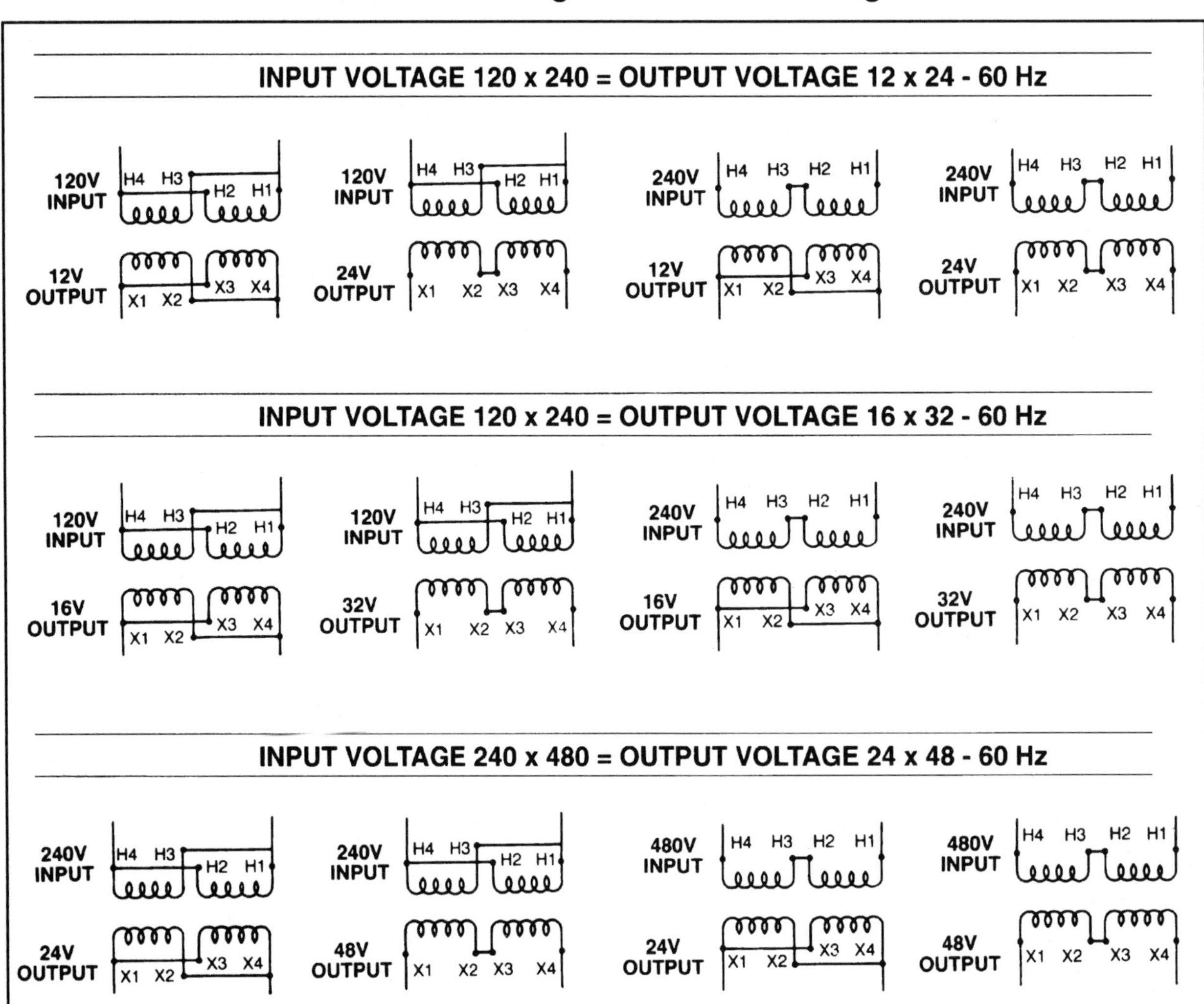

12.4.3 Connection Diagrams for Buck-Boost Transformers in Autotransformer Arrangement for Single-Phase System

12.4.3 Connection diagrams, single-phase.

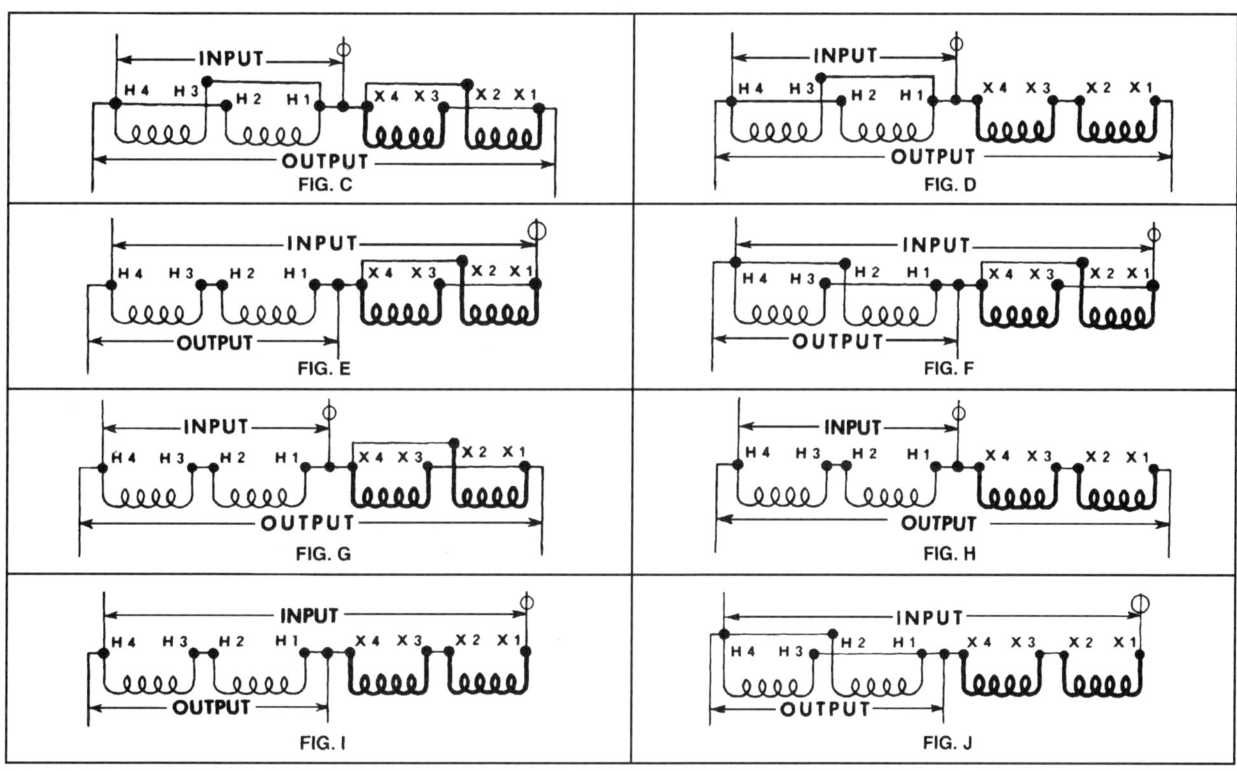

12.4.4 Connection Diagrams for Buck-Boost Transformers in Autotransformer Arrangement for Three-Phase System (*see page 12.10*)

12.4.4 Connection diagrams, three-phase.

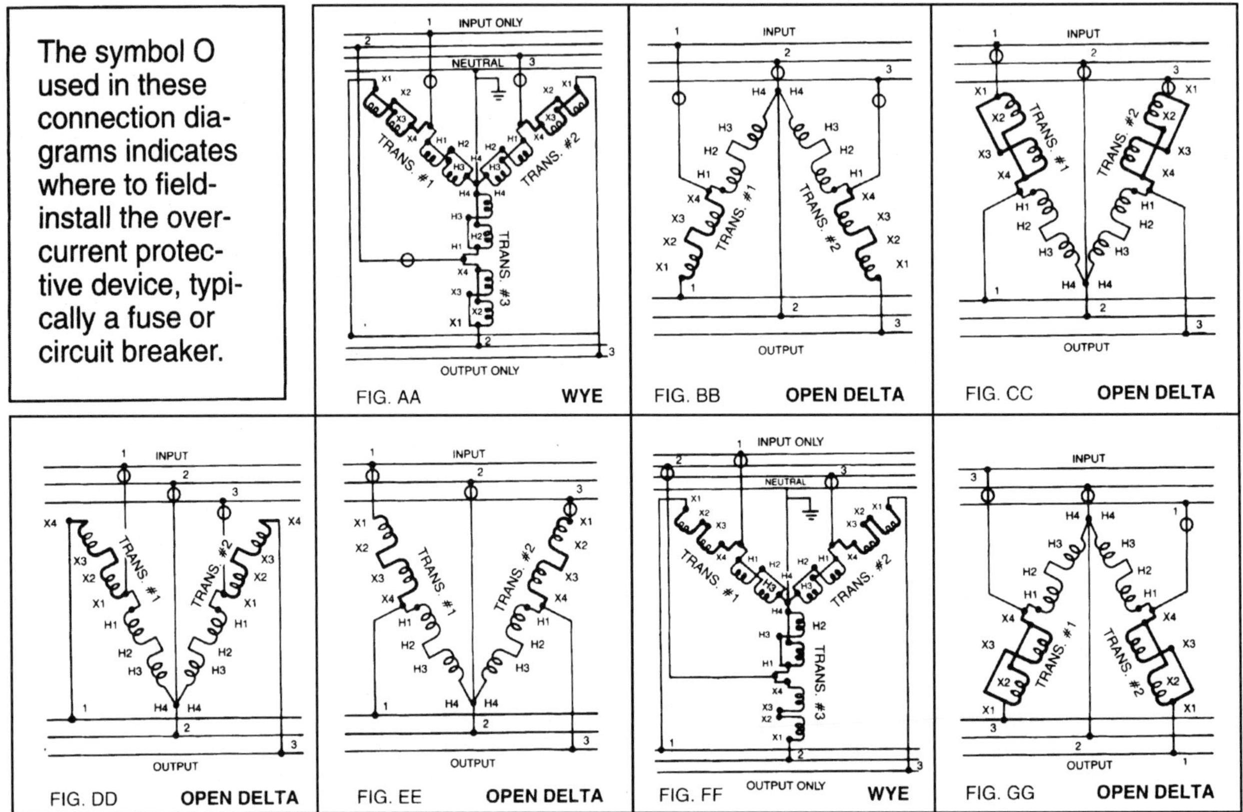

12.5.1 Maximum Average Sound Levels for Transformers

TABLE 12.5.1 Maximum Average Sound Levels (dB)

kVA	Dry-Type		Liquid-Filled	
	Self-Cooled Rating (AA)	Forced-Air Cooling (FA)	Self-Cooled Rating (OA)	Forced-Air Cooling (FA)
0-50	50	..	..	..
51-150	55	..	..	..
151-300	58	67	55	67
301-500	60	67	56	67
501-700	62	67	57	67
701-1000	64	67	58	67
1001-1500	65	68	60	67
1501-2000	66	69	61	67
2001-2500	68	71	62	67
2501-3000	70	71	63	67
3001-4000	71	73	64	67
4001-5000	72	74	65	67
5001-6000	73	75	66	68
6001-7500	..	76	67	69
7501-10000	..	76	68	70

12.5.2 Typical Building Ambient Sound Levels

TABLE 12.5.2 Typical Sound Levels (dB)

Radio, Recording and TV Studios	25-30 db
Theatres and Music Rooms	30-35
Hospitals, Auditoriums and Churches	35-40
Classrooms and Lecture Rooms	35-40
Apartments and Hotels	35-45
Private Offices and Conference Rooms	40-45
Stores	45-55
Residence (Radio, TV Off) and Small Offices	53
Medium Office (3 to 10 Desks)	58
Residence (Radio, TV On)	60
Large Store (5 or More Clerks)	61
Factory Office	61
Large Office	64
Average Factory	70
Average Street	80

12.6.1 Transformer Insulation System Temperature Ratings

12.6.1 Total winding temperature, °C.

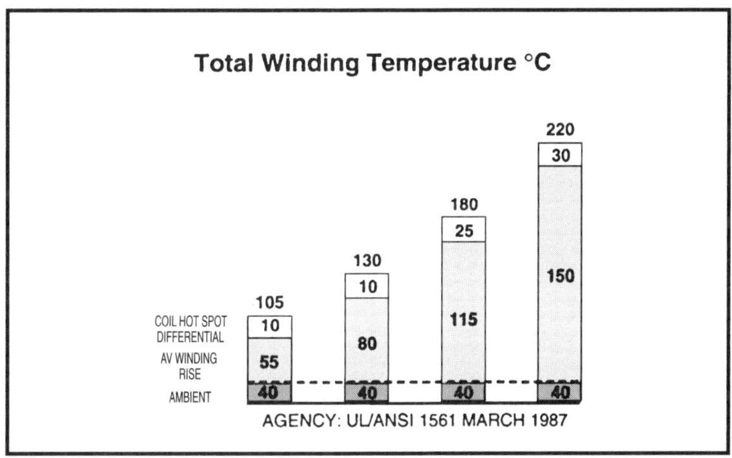

12.7.1 *k*-Rated Transformers

Transformers used for supplying the nonsinusoidal high-harmonic-content (>5 percent) loads that are increasingly prevalent must be designed and listed for these loads. ANSI C57.110-1986, "Recommended Practice for Establishing Transformer Capability When Supplying Non-Sinusoidal Load Currents," provides a method for calculating the heating effect in a transformer when high harmonic currents are present. This method generates a number called the *k-factor,* which is a multiplier that relates eddy-current losses in the transformer core due to harmonics to increased transformer heating. Transformer manufacturers use this information to design transformer core/coil and insulation systems that are more tolerant of the higher internal heating load than a standard design. Simply put, a *k*-rated transformer can tolerate approximately k times more internal heat than a similar standard design transformer (e.g., a k-4 transformer can handle approximately four times the internal heating load of a similar ANSI standard non-harmonic-rated transformer with no life expectancy reduction).

The k rating of a transformer addresses only increased internal heating. It does not address mitigation of the harmonic content of the transformer load.

Section

13

Grounding, Ground-Fault, and Lightning Protection

13.1.0 Introduction
13.1.1 *NEC* Table 250.122, Minimum Size Equipment Grounding Conductors for Grounding Raceway and Equipment
13.2.1 Solidly Grounded Systems
13.2.2 Ungrounded Systems
13.2.3 Resistance-Grounded Systems
13.2.4 Grounding-Electrode System (*NEC* Articles 250.50 and 250.52)
13.2.5 Grounding-Electrode Conductor for Alternating-Current Systems (*NEC* Table 250.66)
13.3.0 Ground-Fault Protection: Introduction
13.3.1 Ground-Return Sensing Method
13.3.2 Zero-Sequence Sensing Method
13.3.3 Residual Sensing Method
13.3.4 Dual-Source System: Single-Point Grounding
13.4.0 Lightning Protection
13.4.1 Annual Isokeraunic Map Showing the Average Number of Thunderstorm Days per Year: (a) USA and (b) Canada
13.4.2 Rolling-Ball Theory
13.4.3 Cone of Protection

13.1.0 Introduction

Grounding encompasses several different but interrelated aspects of electrical distribution system design and construction, all of which are essential to the safety and proper operation of the system and equipment supplied by it. Among these are equipment grounding, system grounding, static and lightning protection, and connection to earth as a reference (zero) potential.

13.1.1 *NEC* Table 250.122, Minimum Size Equipment Grounding Conductors for Grounding Raceway and Equipment (*see page 13.2*)

TABLE 13.1.1 Minimum Size Equipment Grounding Conductors for Grounding Raceway and Equipment

Rating or Setting of Automatic Overcurrent Device in Circuit Ahead of Equipment, Conduit, etc., Not Exceeding (Amperes)	Size (AWG or kcmil)	
	Copper	Aluminum or Copper-Clad Aluminum*
15	14	12
20	12	10
30	10	8
40	10	8
60	10	8
100	8	6
200	6	4
300	4	2
400	3	1
500	2	1/0
600	1	2/0
800	1/0	3/0
1000	2/0	4/0
1200	3/0	250
1600	4/0	350
2000	250	400
2500	350	600
3000	400	600
4000	500	800
5000	700	1200
6000	800	1200

Note: Where necessary to comply with Section 250.4(A)(5) or 250.4(B)(4), the equipment grounding conductor shall be sized larger than given in this table.
*See installation restrictions in Section 250.120.

13.2.1 Solidly Grounded Systems

13.2.1

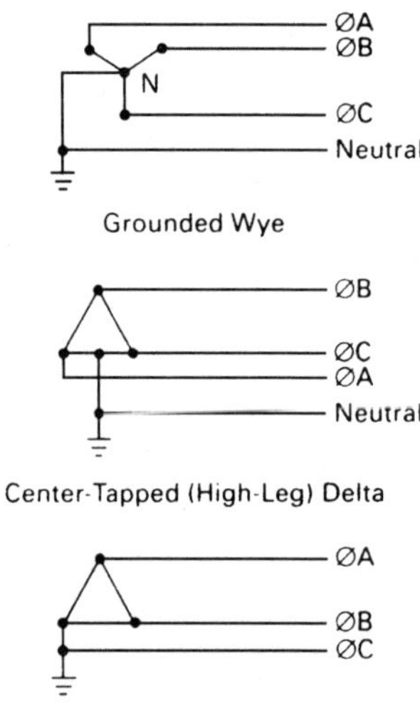

Solidly grounded three-phase systems usually are wye-connected, with the neutral point grounded. Less common is the *red-leg* or *high-leg* delta, a 240-V system supplied by some utilities with one winding center tapped to provide 120 V to ground for lighting and receptacles. This 240-V, three-phase, four-wire system is used where 120-V lighting load is small compared with 240-V power load because the installation is low in cost to the utility. A corner-grounded three-phase delta system is sometimes found, with one phase grounded to stabilize all voltages to ground. Better solutions are available for new installations.

13.2.2 Ungrounded Systems

Ungrounded systems can be either wye or delta, although the ungrounded delta is far more common.

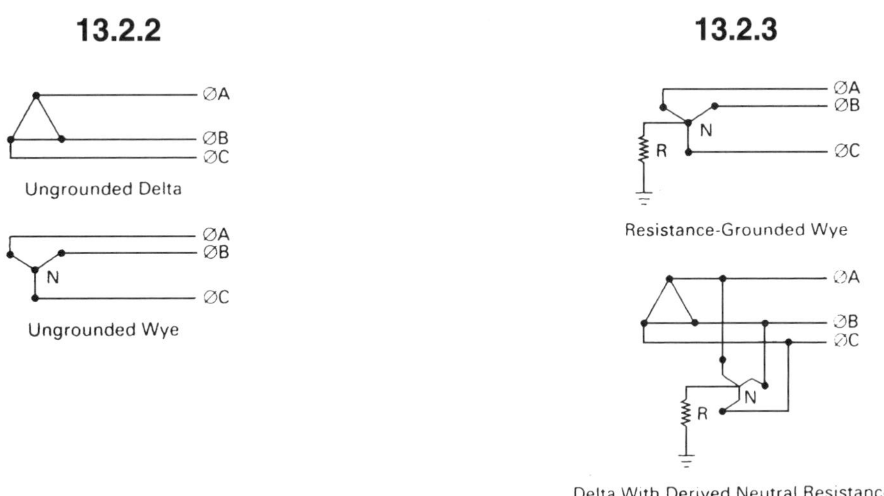

13.2.3 Resistance-Grounded Systems

Resistance-grounded systems are simplest with a wye connection, grounding the neutral point directly through the resistor. Delta systems can be grounded by means of a zigzag or other grounding transformer. Open-delta transformer banks also may be used.

This drives a neutral point, which can be either solidly or impedance grounded. If the grounding transformer has sufficient capacity, the neutral created can be solidly grounded and used as a part of a three-phase, four-wire system. Most transformer-supplied systems are either solidly grounded or resistance grounded. Generator neutrals often are grounded through a reactor to limit ground-fault (zero sequence) currents to values the generator can withstand. Generators that operate in parallel are sometimes resistance grounded to suppress circulating harmonics.

13.2.4 Grounding-Electrode System (*NEC* Articles 250.50 and 250.52) (*see page 13.4*)

13.2.4 Grounding electrode system (*NEC* Articles 250.50 and 250.52).

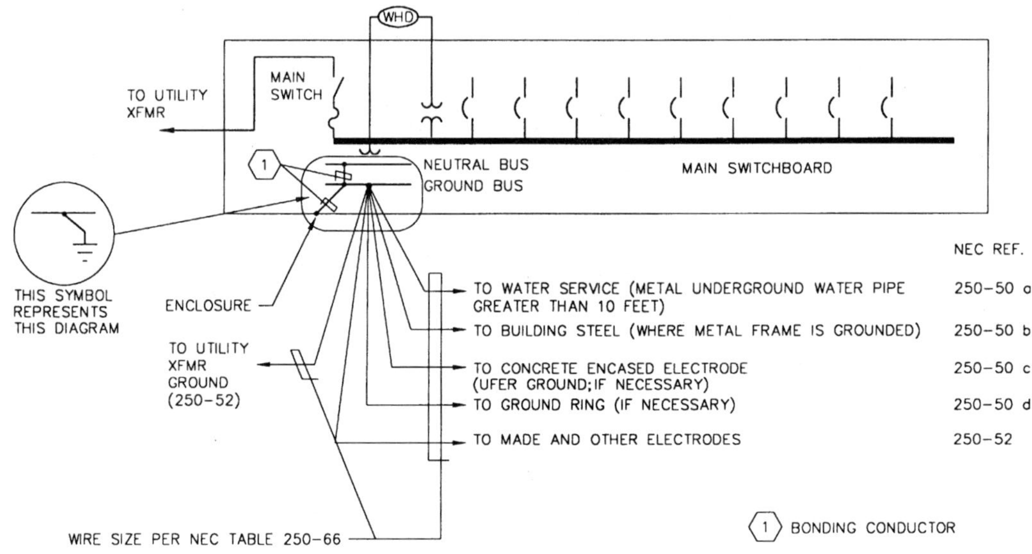

DESIGN NOTES:

1. ARTICLE 250-50 IS MANDATORY. IT IS ALSO A PRIORITY LIST WITH (a) BEING HIGHEST AND (d) BEING LOWEST. ITEMS (c) & (d) WOULD TYPICALLY BE INSTALLED IF (a) OR (b) ARE NOT AVAILABLE.

2. ARTICLE 250-52 CANNOT REPLACE 250-50 UNLESS ITEMS a, b, c, & d ARE NOT PRESENT

3. WATER SERVICE AND BUILDING STEEL CONNECTION LOCATIONS SHOULD BE SHOWN ON THE FLOOR PLANS.

13.2.5 Grounding-Electrode Conductor for Alternating-Current Systems (*NEC* Table 250.66)

TABLE 13.2.5

Size of Largest Service-Entrance Conductor or Equivalent Area for Parallel Conductors[1]		Size of Grounding Electrode Conductor	
Copper	Aluminum or Copper-Clad Aluminum	Copper	Aluminum or Copper-Clad Aluminum[2]
2 or smaller	1/0 or smaller	8	6
1 or 1/0	2/0 or 3/0	6	4
2/0 or 3/0	4/0 or 250 kcmil	4	2
Over 3/0 through 350 kcmil	Over 250 kcmil through 500 kcmil	2	1/0
Over 350 kcmil through 600 kcmil	Over 500 kcmil through 900 kcmil	1/0	3/0
Over 600 kcmil through 1100 kcmil	Over 900 kcmil through 1750 kcmil	2/0	4/0
Over 1100 kcmil	Over 1750 kcmil	3/0	250 kcmil

Notes:

1. Where multiple sets of service-entrance conductors are used as permitted in Section 230.40, Exception No. 2, the equivalent size of the largest service-entrance conductor shall be determined by the largest sum of the areas of the corresponding conductors of each set.

2. Where there are no service-entrance conductors, the grounding electrode conductor size shall be determined by the equivalent size of the largest service-entrance conductor required for the load to be served.

[1]This table also applies to the derived conductors of separately derived ac systems.

[2]See installation restrictions in Section 250.64(A).

13.3.0 Ground-Fault Protection: Introduction

Overcurrent protection is designed to protect conductors and equipment against currents that exceed their ampacity or rating under prescribed time values. An overcurrent can result from an overload, short-circuit, or high-level ground-fault condition. When currents flow outside the normal current path to ground, supplementary ground-fault protection equipment will be required to sense low-level ground-fault currents and initiate the protection required. Normal phase overcurrent protection devices provide no protection against low-level ground faults.

There are three basic means of sensing ground faults, namely, the ground-return sensing method, the zero-sequence sensing method, and the residual sensing method. The following subsections briefly describe these methods.

13.3.1 Ground-Return Sensing Method

13.3.1

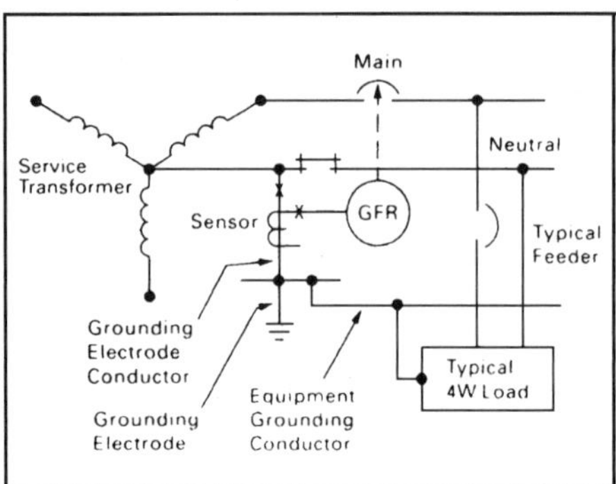

This is the most simple and direct method. This sensing method is based on the fact that all currents supplied by a transformer must return to that transformer.

When an energized conductor faults to grounded metal, the fault current returns along the ground-return path to the neutral of the source transformer. This path includes the grounding electrode conductor, sometimes called the *ground strap,* as shown. A current sensor on this conductor (which can be a conventional bar-type or window-type CT) will respond to ground-fault currents only. Normal neutral currents resulting from unbalanced loads will return along the neutral conductor and will not be detected by the ground-return sensor.

This is an inexpensive method of sensing ground faults where only minimum protection per *NEC* Article 230.95 is desired. For it to operate properly, the neutral must be grounded in only one location, as indicated. In many installations, the servicing utility grounds the neutral at the transformer, and additional grounding is required in the service equipment per *NEC* Article 250.24(A). In such cases and others, including multiple sources with multiple interconnected neutral ground points, residual or zero-sequence sensing methods should be employed.

13.3.2 Zero-Sequence Sensing Method

13.3.2

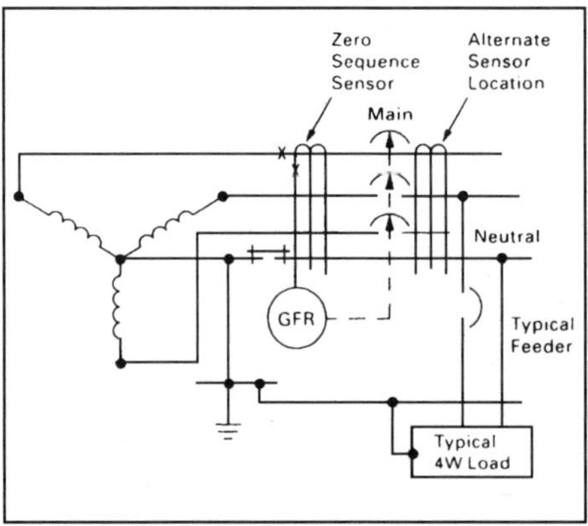

This sensing method requires a single, specially designed sensor either of a torriodal or rectangular configuration. This core balance current transformer surrounds all the phase and neutral conductors in a typical three-phase, four-wire distribution system.

The sensing method is based on the fact that the vectorial sum of the phase and neutral currents in any distribution circuit will equal zero unless a ground-fault condition exists downstream from the sensor. All currents that flow only in the circuit conductors, including balanced or unbalanced phase-to-phase and phase-to-neutral normal or fault currents and harmonic currents, will result in zero sensor output. However, should any conductor become grounded, the fault current will return along the ground path—not the normal circuit conductors—and the sensor will have an unbalanced magnetic flux condition, generating a sensor output to actuate the ground-fault relay.

This method of sensing ground faults can be employed on the main disconnect, where minimum protection per *NEC* Article 230-95 is desired. It also can be employed in multitier systems, where additional ground-fault protection is desired for added service continuity. Additional grounding points may be employed upstream of the sensor, but not on the load side.

Ground-fault protection employing ground-return or zero-sequence sensing methods can be accomplished by the use of separate ground-fault relays (GFRs) and disconnects equipped with standard shunt trip devices or by circuit breakers with integral ground-fault protection with external connections arranged for these modes of sensing.

13.3.3 Residual Sensing Method

13.3.3

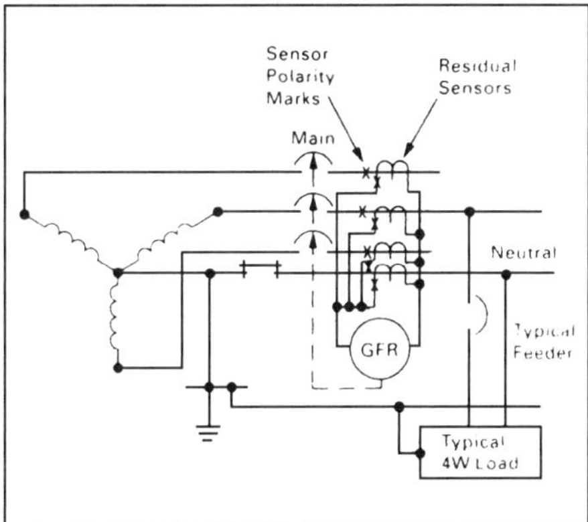

This is a very common sensing method used with circuit breakers equipped with electronic trip units and integral ground-fault protection. The three-phase sensors are required for normal phase overcurrent protection. Ground-fault sensing is obtained with the addition of an identically rated sensor mounted on the neutral. In a residual sensing scheme, the relationship of the polarity markings—as noted by the X on each sensor—is critical. Since the vectorial sum of the currents in all the conductors will total zero under normal, non-ground-faulted conditions, it is imperative that proper polarity connections are employed to reflect this condition.

As with the zero-sequence sensing method, the resulting residual sensor output to the ground-fault relay or integral ground-fault tripping circuit will be zero if all currents flow only in the circuit conductors. Should a ground fault occur, the current from the faulted conductor will return along the ground path rather than on the other circuit conductors, and the residual sum of the sensor outputs will not be zero. When the level of ground-fault current exceeds the preset current and time-delay settings, a ground-fault tripping action will be initiated.

This method of sensing ground faults can be economically applied on main service disconnects where circuit breakers with integral ground-fault protection are provided. It can be used in minimum-protection schemes per *NEC* Article 230.95 or in multitier schemes where additional levels of ground-fault protection are desired for added service continuity. Additional grounding points may be employed upstream of the residual sensors, but not on the load side.

13.3.4 Dual-Source System: Single-Point Grounding

13.3.4

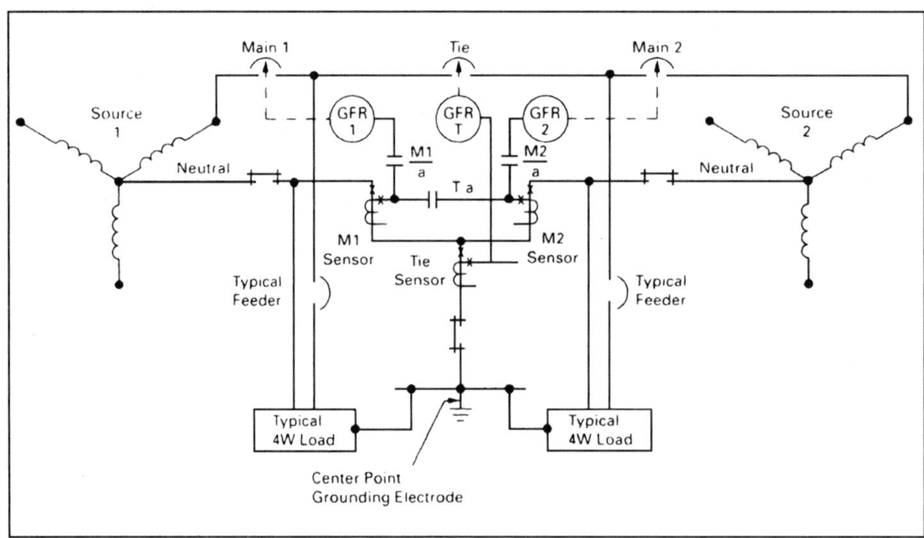

This system uses individual sensors connected in ground-return fashion. Under tie-breaker closed operating conditions, either the M1 sensor or the M2 sensor could see neutral unbalance current and possibly initiate an improper tripping operation. However, with the polarity arrangements of these two sensors along with the tie-breaker auxiliary switch (T/a) and the interconnections as shown, this possibility is eliminated. Selective ground-fault tripping coordination between the tie breaker and the two main circuit breakers is achieved by preset current pickup and time-delay settings between devices GFR/1, GFR/2, and GFR/T.

The advantages of increased service continuity offered by this system can only be used effectively if additional levels of ground-fault protection are added on each downstream feeder. Some users prefer individual grounding of the transformer neutrals. In such cases, a partial differential ground-fault scheme should be used for the mains and tie breaker.

An infinite number of ground-fault protection schemes can be developed depending on the number of alternate sources and the number of grounding points and system interconnections involved. Depending on the individual system configuration, either mode of sensing or a combination of all types may be employed to accomplish the desired end results.

13.4.0 Lightning Protection

Lighting protection deals with the protection of buildings and other structures due to direct damage from lightning. Requirements will vary with geographic location, building type and environment, and many other factors. Any lightning-protection system must be grounded, and the lightning-protection ground must be bonded to the electrical equipment grounding system. Installations must be installed in conformance with NFPA 780.

Nature of Lightning. Lightning is an electric discharge between clouds or between clouds and earth. Charges of one polarity are accumulated in the clouds and of the opposite polarity in the earth. When the charge increases to the point that the insulation between can no longer contain it, a discharge takes place. This discharge is evidenced by a flow of current, usually great in magnitude but extremely short in time.

Damage to buildings and structures is the result of heat and mechanical forces produced by the passage of current through resistance in the path of discharge. Although the discharge takes place at the point where the potential difference exceeds the dielectric strength of the insulation, which implies low resistance relative to other paths, it is not uncommon for the current to follow the path of high resistance. This may be a tree, a masonry structure, or a porcelain insulator. Obviously, damage due to direct stroke can be minimized by providing a direct path of low resistance to earth.

Lightning can cause damage to structures by direct stroke and to equipment by surges coming in over exposed power lines. Surges may be the result of direct strokes to the line at some distance away, or they may be electrostatically induced voltages.

Need for Protection. Damage to structures and equipment due to surge effect is a subject in itself, and protection against this type of damage is not within the scope of this text except as grounding is involved.

It is not possible to positively protect a structure against damage from a direct stroke except by completely enclosing it with metal. The extent to which lightning protection should be provided is governed by weighing the cost of protection against the possible consequences of being struck. The following factors are to be considered:

1. Frequency and severity of thunderstorms
2. Value and nature of structure or contents
3. Personnel hazards
4. Consequential loss such as a loss of production, salaries of workers, damage suits, and other indirect losses
5. Effect on insurance premiums

These factors are listed primarily to call attention to their importance. No general conclusions can be drawn as to the relative importance of each or to the necessity for or extent of lightning protection for any given combinations of conditions. As a matter of interest, a map showing the frequency of thunderstorm days for various areas of the United States and Canada is given in Figure 13.4.1. It should be noted, however, that the severity of storms is much greater in some local areas than in others, and therefore, the need for protection is not necessarily in direct proportion to the frequency.

Equipment and Structures That Should Be Considered for Protection. The nature of buildings and their content are important in deciding whether lightning protection is desirable. Some of the structures that should be considered are as follows:

- All-metal structures
- Metal-roofed and metal-clad buildings
- Metal-frame buildings with nonmetallic facings
- Buildings of wood, stone, brick, tile, and other nonconducting materials
- Spires, steeples, and flagpoles
- Buildings of historical value
- Buildings containing readily combustible or explosive materials
- Tanks and tank farms
- Transmission lines
- Power plants, substations, and water-pumping stations
- High stacks and chimneys
- Water towers, silos, and similar structures
- Buildings containing a significant amount of sensitive electronic equipment such as data centers
- Hospitals and health care facilities
- High-rise buildings

Metal buildings and structures offer a very satisfactory path to earth and require little in the way of additional protection. Metal-frame buildings with nonmetallic facings require more extensive measures. Buildings made entirely of nonconducting materials require complete lightning-protection systems.

In special cases, buildings may have historical value out of proportion to their intrinsic value and may justify extensive protection systems. Power stations, substations, and water-pumping stations providing extremely important functions to outside facilities may demand protective measures far more extensive than normally would be warranted by the value of the structure. By the same token, structures containing combustible or explosive materials and liquids and gases of a toxic nature or otherwise harmful to personnel or property if allowed to escape from their confining enclosures may justify extensive protection systems.

Requirements for Good Protection. The fundamental theory of lightning protection of structures is to provide means by which a discharge may enter or leave the earth without passing through paths of high resistance. Such a condition is usually met by grounded steel-frame structures. Suitable protection is nearly always provided by the installation of lightning conductors.

A lightning-conductor system consists of terminals projecting into the air above the uppermost parts of the structure with interconnecting and ground conductors. Terminals should be placed so as to project above all points that are likely to be struck. Conductors should present the least possible impedance to earth. There should be no sharp bends or loops. Each projecting terminal above the structure should have at least two connecting paths to earth and more if practicable.

Each conductor running down from the terminals on top of the structure should have an earth connection. Properly made connections to earth are an essential feature of a lightning-rod system for protection of buildings. It is more important to provide ample distribution of metallic contacts in the earth than to provide low-resistance connections. Low-resistance connections are desirable, however, and should be provided where practicable. Earth connections should be made at uniform intervals about the structure, avoiding as much as possible the grouping of connections on one side. Electrodes should be at least 2 ft (0.6 m) away from and should extend below building foundations (except when using reinforcing bars for grounds). They should make contact with the earth from the surface downward to avoid flashing at the surface.

Interior metal parts of buildings or structures should be grounded independently, and if they are within 6 ft (1.8 m) of metallic roofs, walls, or conductors running down from the terminals on top of the structure, they should be securely connected thereto.

Terminals projecting above the structure should be of ample length to bring the top point at least 10 in (0.25 m) above the object to be protected. In many cases, a greater height is desirable. Experiments have indicated that a vertical conductor, or point, will divert to itself direct hits that otherwise might fall within a cone-shaped space, of which the apex is the point and the base is a circle whose radius is approximately equal to the height of the point (only for single aerial terminals).

The foregoing outlines requirements for good protection of buildings. Good protection of electrical substations, power stations, tanks and tank farms, and other special applications is beyond the scope of this book. For further information, refer to IEEE Standard 142.

Application Tips

- As a practical matter, once it is decided that a lightning-protection system is needed, consulting electrical engineers generally write a performance specification calling for an Underwriters Laboratories (UL) master label system. The system is actually designed and installed by a qualified lightning-protection contractor.
- When considering a lighting-protection system for a building, it is important to verify the history of frequency and severity of thunderstorms in the immediate area of the building being considered. This could be checked through the weather service and building owners in the local area.
- Experience has shown that adding a lightning-protection system to a building increases its susceptibility to lightning strokes.
- If a lightning-protection system is to be provided for a building addition, it also must be added to all existing contiguous buildings to obtain a UL master label. Even if the contiguous buildings already have a lightning-protection system, their lightning-protection system may have to be upgraded to obtain a UL master label.

13.4.1 Annual Isokeraunic Map Showing the Average Number of Thunderstorm Days per Year: (a) USA and (b) Canada (*see page 13.12*)

13.4.2 Rolling-Ball Theory (*see page 13.13*)

The rolling-ball theory of protection is a frequently used concept to determine the area of protection around a building or structure from lightning strikes. Basically, the zone of protection is thought to include the space not intruded on by the rolling ball, which has a radius of 150 ft (45.75 m). In other words, if the rolling ball were to touch two air terminals, there must be a gap between the bottom of the rolling ball and the structure to be in the zone of protection (ref.: NFPA 780, Section 3-10.3.1).

13.4.1A Annual isokeraunic map showing number of thunderstorm days per year (United States).

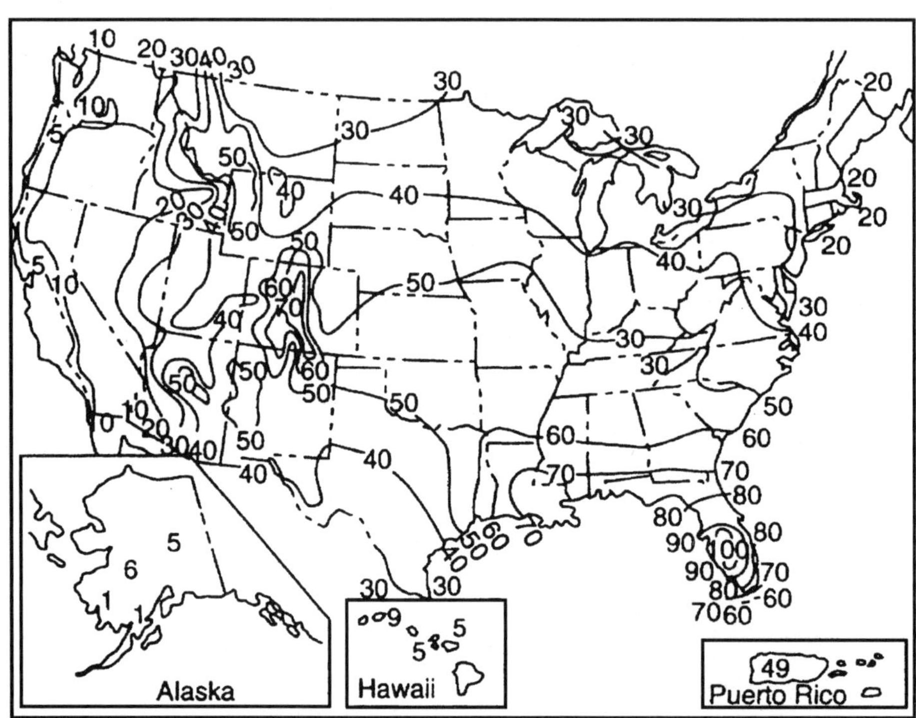

13.4.1B Annual isokeraunic map showing number of thunderstorm days per year (Canada).

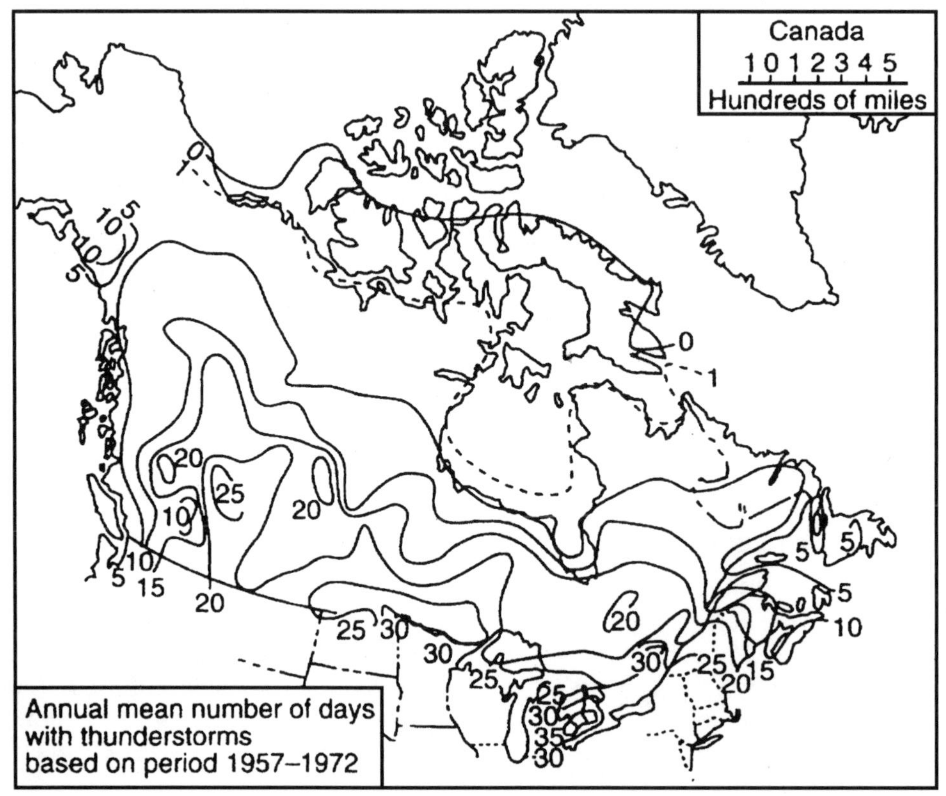

13.4.2 Rolling-Ball Theory

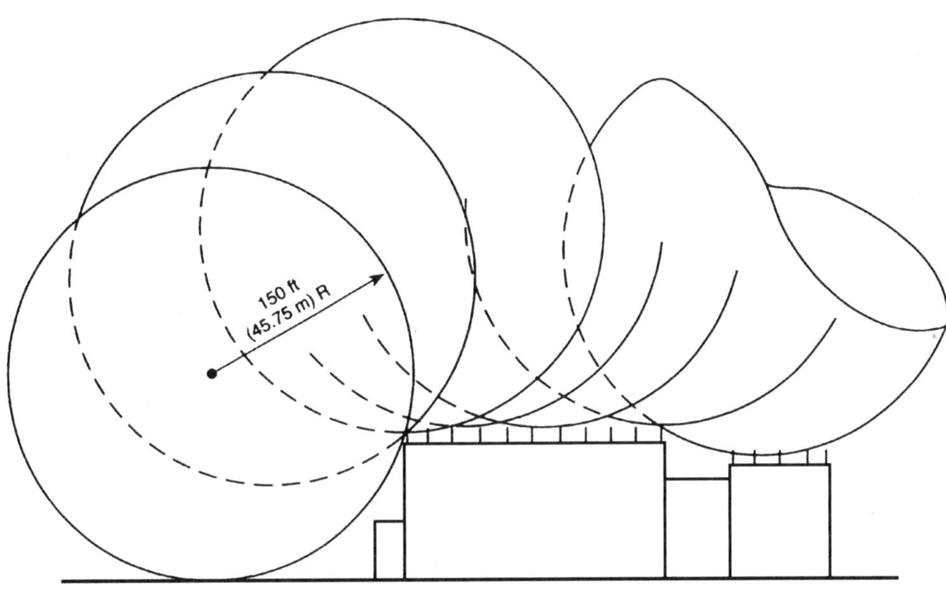

13.4.3 Cone of Protection

The area of protection for a well-grounded object is considered to be a conical zone (cone of protection) below and around such object that is based on a 45-degree angle or 30-degrees from vertical (where appropriate), respectively.

In other words, the grounded object throws a protective "shadow" over and below things located within such shadow, and lightning strikes normally will not enter this shadow zone.

Section

14

Emergency and Standby Power Systems

14.1.0 Introduction
14.1.1 Summary of Codes for Emergency Power in the United States by States and Major Cities (Completed September 1984)
14.1.2 Condensed General Criteria for Preliminary Consideration
14.1.3 Typical Emergency/Standby Lighting Recommendations
14.2.0 Emergency/Standby Power Source Options and Arrangements
14.2.1 Two-Utility-Source System Using One Automatic Transfer Switch
14.2.2 Two-Utility-Source System Where Any Two Circuit Breakers Can Be Closed
14.2.3 Diagram Illustrating Multiple Automatic Double-Throw Transfer Switches Providing Varying Degrees of Emergency and Standby Power
14.2.4 Typical Transfer Switching Methods (a) Total Transfer and (b) Critical-Load Transfer
14.2.5 Typical Multiengine Automatic Paralleling System
14.2.6 Elevator Emergency Power Transfer System
14.2.7 Typical Hospital Installation with a Nonautomatic Transfer Switch and Several Automatic Transfer Switches
14.3.0 Engine and Generator-Set Sizing: Introduction
14.3.1 Engine Rating Considerations
14.3.2 Engine-Generator Set Load Factor
14.3.3 Load Management
14.3.4 Standards
14.3.5 Generator Set Sizing Example
14.3.6 Critical Installation Considerations
14.3.7 Illustration Showing a Typical Emergency Standby Generator Installation
14.4.0 Uninterruptible Power Supply (UPS) Systems: Introduction
14.4.1 Nonredundant UPS System Configuration
14.4.2 "Cold" Standby Redundant UPS System
14.4.3 Parallel Redundant UPS System
14.4.4 Isolated Redundant UPS System
14.4.5 Application of UPS
14.5.0 Power-System Configuration for 60-Hz Distribution
14.5.1 Single-Module UPS System
14.5.2 Parallel-Capacity UPS System
14.5.3 Parallel Redundant UPS System
14.5.4 Dual Redundant UPS System
14.5.5 Isolated Redundant UPS System
14.5.6 Parallel Tandem UPS System
14.5.7 Hot Tied-Bus UPS System

14.5.8 Superredundant Parallel System: Hot Tied-Bus UPS System
14.5.9 Uninterruptible Power with Dual Utility Sources and Static Transfer Switches
14.5.10 Power-System Configuration with 60-Hz UPS
14.5.11 UPS Distribution Systems
14.6.1 Power-System Configuration for 400-Hz Distribution

14.1.0 Introduction

Emergency electric services are required for protection of life, property, or business where loss might be the result of an interruption of the electric service. The extent of the emergency services required depends on the type of occupancy, the consequences of a power interruption, and the frequency and duration of expected power interruptions.

Municipal, state, and federal codes define minimum requirements for emergency systems for some types of public buildings and institutions. These shall be adhered to, but economics or other advantages may result in making provisions beyond these minimums (see the *NEC,* Articles 517, 700, 701, and 702). This chapter presents some of the basic information on emergency and standby power systems. For additional information, design details, and maintenance requirements, see ANSI/IEEE Standard 446-1987, "IEEE Recommended Practice for Emergency and Standby Power Systems for Industrial and Commercial Applications," ANSI/NFPA 110, "Emergency and Standby Power Systems," and ANSI/NFPA 110A, "Stored Energy Systems."

Emergency power systems should be separated from the normal power systems by using separate raceways and panelboards. The *NEC* requires that each item of emergency equipment be clearly marked as to its purpose. In large public buildings, physical separation of the emergency system from the normal system elements would enhance the reliability of the emergency system in the event of fire or other contingencies. Also, more and more states are requiring that the emergency systems not only be separated from the normal systems but that they be enclosed in 2-hour fire-rated construction.

14.1.1 Summary of Codes for Emergency Power in the United States by States and Major Cities (Completed September 1984) (*see pages 14.3–14.5*)

Table 14.1.1 is a guide to state codes and regulations for emergency power systems in the United States. All the latest codes and regulations for the area in which the industrial or commercial facility is located must be consulted and followed.

14.1.2 Condensed General Criteria for Preliminary Consideration (*see pages 14.6–14.10*)

Table 14.1.2 lists the needs in 13 general categories with some breakdown under each to indicate major requirements. Ranges under the columns "Maximum Tolerance Duration of Power Failure" and "Recommended Minimum Auxiliary Supply Time" are assigned based on experience. Written standards have been referenced where applicable.

In some cases, under the column "Type of Auxiliary Power System," both emergency and standby have been indicated as required. An emergency supply of limited time capacity may be used at a low cost for immediate or interruptible power until a

TABLE 14.1.1 Codes for Emergency Power by States and Major Cities (Completed September 1984)

State/City	Does State/City Have Legislation?	Legislation Code Type?	Hospitals	Nursing Homes	Schools	Theaters (Public Gathering Places)	Office Buildings	Hotels	Apartment Buildings	Airports	Fire and Police Stations	Water Treatment Plants	Sewage Treatment Plants	All Public Buildings, State	All Public Buildings, Commercial	Applicable Government Agency	Smokeproof Enclosures in High-Rise Buildings
Alabama	Yes	4,6	A,C,D	A,C,D	C,D	C,D	C,D	C,D	C,D	C,D				C,D	C,D	M,N,O	B
Birmingham	Yes	4,6	A,C,D	A,C,D	C,D	C,D	C,D	C,D	C,D	C,D	C	C	C	C,D	C,D	Q,S	A,B
Mobile **	Yes	1,4	A,C,D	A,C,D	C,D	A,C,D	A,C,D	A,C,D	A,C,D	C,D	C				C	S	
Alaska	Yes	1,3	A,C,D	A,C,D	A,C,D	A,C,D	A,C,D	A,C,D	A,C,D	A,C,D	C,D	C,D	C,D	C,D	A,C,D	M,P	B
Arizona	No																
Phoenix **	Yes	1	A,C,D	A,C,D	C,D	C,D	C,D		A,C,D					C,D	C,D	O	
Arkansas	Yes	1,6	A,C,D	A,C,D	A,C,D	A,C,D	A,C,D	A,C,D	A,C,D	A,C,D	C,D	C,D	C,D	C,D	A,C,D	M	B
California	Yes	2,3,4	A,C,D	A,C,D	C,D	C,D	C,D	C,D	C,D	C,D				C,D	C,D	O,T	A
Anaheim	Yes	2,3,4,7	A,C,D	A,C,D												M,,N,O,Q	
Berkeley **	Yes	1,3	A,C,D	A,C,D	C,D	C,D	C,D	C,D	C,D	C,D	C,D	A,C,D	C,D	C,D	C,D	M,Q	
Fresno	Yes	3,4	A,C,D	A,C,D	A,C,D	A,C,D	A,C,D	A,C,D	A,C,D	C,D	C,D			C,D	C,D	O,S	B
Glendale	Yes	3,4	A,C,D	A,C,D	A,C,D	A,C,D	A,C,D	A,C,D	A,C,D	C,D	C,D			C,D	C,D	M,Q	B
Long Beach **	Yes	3	A,C,D	A,C,D	A,C,D	A,C,D	C,D	C,D	C,D	A,C,D	A,C,D	C,D	C,D	C,D	C,D	M,O	
Los Angeles	Yes	3,4,8	A,C,D	A,C,D	A,C,D	A,C,D	A,C,D	A,C,D	A,C,D	B,C,D	C,D	C,D	C,D	C,D	C,D	M,Q,S	B
Oakland	Yes	1,3,4,8	A,C,D	A,C,D	A,C,D	A,C,D	A,C,D	A,C,D	A,C,D	C,D	C,D	C,D	C,D	C,D	C,D	M,O	B
Pasadena	Yes	1,2,3,4	A,C,D	C,D	C,D	A,C,D	C,D	C,D	C,D	C,D	C,D	C,D	C,D		C,D	O,Q	
San Diego	Yes	3,4	A,C,D	A,C,D	A,C,D	A,C,D	A,C,D	A,C,D	A,C,D	C,D	C,D	C,D	C,D	C,D	C,D	S	B
San Francisco	Yes	3,4	A,C,D	A,C,D	C,D	A,C,D	A,C,D	A,C,D	A,C,D	A,C,D	C,D	C,D	C,D	C,D	C,D	O	B
San Jose **	Yes	4														M,S	
Santa Ana **	Yes	3,4														M,Q	
Colorado	Yes	3,4	A,C,D	A,C,D	C,D	C,D	A,C,D	A,C,D	A,C,D	C,D			C,D	C,D	C,D	Q	B
Denver	Yes	4,8	A,C,D	A,C,D	C,D	A,C,D	A,C,D	A,C,D	A,C,D	A,C,D					A,C,D	Q	B
Connecticut	Yes	2,4	A,C,D	A,C,D	A,C,D	A,C,D	A,C,D	A,C,D	C,D	C,D	C,D	C,D	C,D	C,D	C,D	P	A
Hartford **	Yes	2	A,C,D	A,C,D	A,C,D	A,C,D	A,C,D	A,C,D	A,C,D	A,C,D	A,C,D	A,C,D	A,C,D	A,C,D	A,C,D	M,R	
New Haven **	Yes	1,2,4	A,C,D	A,C,D	A,C,D	A,C,D	A,C,D	A,C,D	A,C,D	A,C,D	A,C,D	A,C,D	A,C,D	A,C,D	A,C,D	M,Q	
Delaware	Yes	1,2,4	A,C,D	A,C,D	C,D	C,D	C,D	C,D	C,D	C,D					C,D	M	A

NOTE: An explanation of the numbers and letters used is given at the end of the table (see p 14.5).
Table 14.1.1 courtesy of the Electrical Generating Systems Marketing Association (April 1975).

State/City	Does State/City Have Legislation?	Legislation Code Type?	Hospitals	Nursing Homes	Schools	Theaters (Public Gathering Places)	Office Buildings	Hotels	Apartment Buildings	Airports	Fire and Police Stations	Water Treatment Plants	Sewage Treatment Plants	All Public Buildings, State	All Public Buildings, Commercial	Applicable Government Agency	Smokeproof Enclosures in High-Rise Buildings
District of Columbia	Yes	2,4	A,C,D	A,C,D	C,D	C,D	C,D	C,D	C,D	C,D	C,D	C,D	C,D	C,D	C,D	M,Q	A
Florida	Yes	1,2,4,6	A,C,D	A,C,D		C,D	C,D	C,D	C,D	C,D					C,D	M,O	A
Jacksonville	Yes	8	A,C,D	A,C,D	C,D	C,D	C,D	C,D	C,D	C,D	C,D	C,D	C,D	C,D	C,D	Q,S	
St Petersburg **	Yes	1,4,7	A,C,D	A,C,D	A,C,D	A,C,D	A,C,D	A,C,D	A,C,D	A,C,D					A,C,D	U	
Tampa	Yes	1,4,8	A,C,D	A,C,D	C,D	C,D	C,D	C,D	C,D		C,D			C,D	C,D	M	A
Georgia	Yes	1,4,7	A,C,D	A,C,D	C,D	C,D	C,D	C,D	C,D	A,C,D		B	B	C,D	C,D	M	
Atlanta	Yes	1,4,8	A,C,D	A,C,D	C,D	C,D	C,D	C,D	C,D	C,D	C,D			C,D	C,D	O	B
Columbus **	Yes	4,6	A,C,D	B,C,D	B,C,D	B,C,D	C,D	B,C,D	C,D	C,D	C,D	C,D	C,D	C,D	C,D	S	
Savannah	Yes	4	A,C,D	A,C,D	C,D	C,D	C,D	C,D	C,D	C,D					C,D	S	
Hawaii	Yes	1,3	A,C,D	A,C,D	A,C,D	C,D	C,D	C,D	C,D	C,D	A,C,D	C,D	C,D	C,D	C,D	M,Q	B
Honolulu	Yes	1,3,8	A,C,D	A,C,D	C,D	C,D	C,D	C,D	C,D	C,D	C,D			C,D	C,D	M,Q	B
Idaho	Yes	1,3,4	A,C,D	A,C,D	C,D	A,C,D	A,C,D	C,D	C,D	C,D	C,D	C,D	C,D	C,D	C,D	M,R	B
Illinois	No	2	A,C,D	A,C,D	C,D	A,C,D	A,C,D	A,C,D	A,C,D	A,C,D		B	B	C,D	A,C,D	M,N	
Chicago **	Yes	8	A,C,D	A,C,D	A,C,D	A,C,D	A,C,D	A,C,D	A,C,D	A,C,D	A,C,D	A,C,D	A,C,D	A,C,D	A,C,D	S	
Rockford **	Yes	1,4	A,C,D	A,C,D	A,C,D	A,C,D	A,C,D	A,C,D	A,C,D					A,C,D	A,C,D	S	
Indiana	Yes	2,3,4	A,C,D	A,C,D	A,C,D	A,C,D	A,C,D	A,C,D	A,C,D	A,C,D	A,C,D	C,D	C,D	A,C,D	A,C,D	M,Q,R	B
Evansville	Yes	3,4	A,C,D	A,C,D	C,D	A,C,D	C,D	A,C,D	C,D	C,D	A,C,D			C,D	C,D	Q	B
Fort Wayne	Yes	1,3,4	A,C,D	A,C,D	A,C,D	A,C,D	A,C,D	A,C,D	A,C,D	A,C,D	A,C,D	A,C,D	B,C,D	A,C,D	A,C,D	M,Q	B
Gary	Yes	1,4	A,C,D	A,C,D	A,C,D	C,D	C,D	C,D	C,D	C,D	C,D	C,D	C,D	C,D	C,D	S	B
Indianapolis **	Yes	2	A,C,D	A,C,D	C,D	C,D	A,C,D	C,D	C,D	A,C,D	A,C,D			A,C,D	A,C,D	S	
South Bend	Yes	1,3,4	A,C,D	A,C,D	A,C,D	A,C,D	A,C,D	A,C,D	C,D	A,C,D	A,C,D	C,D	C,D	A,C,D	A,C,D	M	B
Iowa	Yes	1,4	A,C,D	A,C,D	C,D	C,D	C,D	C,D	C,D	C,D	C,D			C,D	C,D	M,Q	B
Des Moines	Yes	3,4	A,C,D	A,C,D	C,D	C,D	C,D	C,D	C,D	C,D	C,D			C,D	C,D	M,Q	B
Kansas	Yes	1,3,4	A,C,D	A,C,D	C,D											M,O	B
Kansas City **	Yes	3,4	A,C,D	A,C,D	C,D											S	B
Wichita **	Yes	4	A,C,D	A,C,D	C,D											S	B
Kentucky	Yes	1,2,4,5	A,C,D	A,C,D	C,D	C,D	C,D	C,D	C,D	C,D				C,D	C,D	O,Q	B

NOTE: An explanation of the numbers and letters used is given at the end of the table (see p 14.5).

(continued)

TABLE 14.1.1 Codes for Emergency Power by States and Major Cities (Completed September 1984) *(Continued)*

State/City	Does State/City Have Legislation?	Legislation Code Type?	Hospitals	Nursing Homes	Schools	Theaters (Public Gathering Places)	Office Buildings	Hotels	Apartment Buildings	Airports	Fire and Police Stations	Water Treatment Plants	Sewage Treatment Plants	All Public Buildings, State	All Public Buildings, Commercial	Applicable Government Agency	Smokeproof Enclosures in High-Rise Buildings
Louisiana	Yes	1,4	A,C,D	A,C,D	C,D	C,D	C,D	C,D	C,D	C,D				C,D	C,D	M,Q	
Baton Rouge **	Yes	4	A,C,D	A,C,D	B,C,D	A,C,D	B,C,D	A,C	C	A,C,D	A,C,D	B	B		C,D	S	
New Orleans **	Yes	4	A,C,D	A,C,D	C,D	C,D	A,C,D*	A,C,D*	C,D	C,D	C,D		C,D	C,D	C,D	S	
Shreveport	Yes	1,4	A,C,D	A,C,D	B,C,D	C,D	A,C,D	C,D	C,D	A,C,D	A,C,D	C,D	C,D	C,D	A,C,D	S	
Maine	Yes	1,2,4	A,C,D	C,D	C,D	C,D	C,D	C,D	C,D	C,D					C,D	M,S	A
Maryland	Yes	1,2,4,5	A,C,D	A,C,D	C,D	C,D	C,D	C,D	C,D	A,C,D	C,D			C,D	C,D	M,N	A
Baltimore **	Yes	4,8	A,C,D	C,D	C,D	A,C,D	C,D	A,C,D	C,D		A,C,D	C,D	C,D	C,D	C,D	S	
Massachusetts	Yes	2,5	A,C,D	A,C,D	A,C,D	C,D	C,D	C,D	C,D	C,D	A,C,D	C,D	C,D	C,D	C,D	P,Q	B
Bedford **	Yes	2	A,C,D	C,D	C,D	C,D	C,D	C,D	C,D	C,D	C,D	C,D	C,D	C,D	C,D	S	
Boston **	Yes	2		A,C,D	A,C,D	A,C,D	A,C,D	A,C,D	A,C,D							S	
Cambridge **	Yes	2	A,B	A,C,D	A,C,D	A,C,D	A,C,D	A,C,D	A,C,D		A,C,D	A,C,D	A,C,D	A,C,D	A,C,D	S	
Springfield **	Yes	2	A,C,D	A,C,D	A,C,D	A,C,D	A,C,D	A,C,D	A,C,D		A,C,D			A,C,D	A,C,D	Q	
Worcester **	Yes	2	A,C	A,C	A,C	A,C		A,C	A,C	A,C	A,C					Q	
Michigan	Yes	2,4,5	A,C,D	A,C,D	A,C,D	A,C,D	C,D	A,C,D	C,D	A,C,D	C,D	B,C,D	B,C,D	C,D	C,D	R	B
Detroit	Yes	1,4,5	A,C,D	A,C,D	A,C,D	A,C,D	A,C,D	A,C,D	A,C,D	A,C,D	B,C,D	B,C,D	B,C,D		A,C,D	M,S	B
Flint **	Yes	4,5	A,C,D	A,C,D	A,C,D	A,C,D					A,C,D					M,S	
Grand Rapids	Yes	1,4,5	A,C,D	A,C,D	A,C,D	A,C,D	C,D	A,C,D	C,D	C,D	C,D			C,D	C,D	M,S	B
Lansing **	Yes	2	A,C,D	A,C,D	A,C,D	A,C,D	A,C,D	A,C,D	A,C,D			B	B		D	S	
Minnesota	Yes	2,3,4,7	A,C,D	A,C,D	A,C,D	A,C,D	A,C,D	A,C,D	A,C,D	A,C,D	C,D	C,D	C,D	C,D	C,D	M,N,O	B
Minneapolis	Yes	1,3,4	A,C,D	A,C,D	A,C,D	A,C,D	A,C,D	A,C,D	A,C,D	A,C,D	C,D	C,D	C,D	C,D	C,D	M,S	B
Saint Paul	Yes	1,3,4	A,C,D	A,C,D	C,D	C,D	A,C,D	A,C,D	C,D	C,D	A	A,C,D	A,C,D	A,C,D	A,C,D	M,S	B
Mississippi	Yes	1,4,6	A,C,D	C,D	A,C,D	A,C,D	A,C,D	A,C,D	C,D	C,D	C,D	C,D	C,D	C,D	C,D	M	
Jackson	Yes	1,4	A,C,D	C,D												S	
Misssouri	No																
Kansas City	Yes	4	A,C,D														
Montana	Yes	2,3,4	A,C,D	A,C,D	C,D	C,D	C,D	C,D	C,D	C,D	C,D			C,D	C,D	M,Q	A
Nebraska	Yes	1,4	A,C,D	A,C,D	C,D	C,D	C,D	C,D	C,D	C,D					C,D	M	

NOTE: An explanation of the numbers and letters used is given at the end of the table (see p 14.5).

State/City	Does State/City Have Legislation?	Legislation Code Type?	Hospitals	Nursing Homes	Schools	Theaters (Public Gathering Places)	Office Buildings	Hotels	Apartment Buildings	Airports	Fire and Police Stations	Water Treatment Plants	Sewage Treatment Plants	All Public Buildings, State	All Public Buildings, Commercial	Applicable Government Agency	Smokeproof Enclosures in High-Rise Buildings
Lincoln	Yes	4,8	A,C,D	A,C,D	C,D	C,D	C,D	C,D	C,D	C,D					C,D	M,S	
Omaha	Yes	4,8	A,C,D	A,C,D		C,D	C,D	C,D	C,D	C,D					C,D	S	
Nevada	Yes	1,2,3,4	A,C,D	A,C,D	C,D	C,D	C,D	C,D	C,D	C,D				A,C,D	C,D	M,O,Q	A*
New Hampshire	Yes	1,2,4	A,C,D	A,C,D	C,D	C,D	C,D	C,D	C,D	C,D					C,D	M	
New Jersey	Yes	2,3,4,5	A,C,D	A,C,D	C,D	C,D	C,D	C,D	C,D	C,D				C,D	C,D	Q	A
New Mexico	Yes	1,2,3,4	A,C,D	A,C,D	A,C,D	A,C,D	A,C,D	A,C,D	C,D	A,C,D	C,D	C,D	C,D	C,D	A,C,D	Q	
Albuquerque	Yes	3,4	A,C,D	A,C,D	C,D	C,D	C,D	C,D	C,D	C,D	C,D	C,D	C,D	C,D	C,D	S	
New York	Yes	2,4	A,C,D	A,C,D	C,D	C,D	C,D	C,D	C,D	C,D		B,C,D	B,C,D	C,D	C,D	O	A
Albany	Yes	2,4	A,C,D	A,C,D	C,D	C,D	C,D	C,D	C,D	C,D		B,C,D	B,C,D	C,D	C,D	Q	
Buffalo	Yes	2,8	A,C,D	A,C,D	C,D	C,D	C,D	C,D	C,D	C,D	C,D	B,C,D	B,C,D	C,D	C,D	N,O,R	
New York	Yes	2,8	A,C,D	A,C,D	C,D	C,D	C,D	C,D								Q	
Syracuse	Yes	3,8	A,C,D	A,C,D	C,D	C,D	C,D	C,D	C,D	C,D					C,D	M,Q	B
North Carolina	Yes	2,4,6	A,C,D	A,C,D	C,D	C,D	C,D	C,D	C,D	C,D				C,D	C,D	O,T	A
North Dakota	Yes	1,3,4	A,C,D	A,C,D	C,D	C,D	C,D	C,D	C,D	C,D	C,D	C,D	C,D	C,D	C,D	M,O	
Ohio	Yes	2,4,5	A,C,D	A,C,D	C,D	C,D	C,D	C,D	C,D	C,D	C,D	C,D	C,D	C,D	C,D	Q,S	A
Akron	Yes	2,4,5	A,C,D	A,C,D	C,D	A,C,D	A,C,D	A,C,D	A,C,D	C,D	C,D	C,D	C,D	C,D	C,D	O	A
Cincinnati	Yes	2,4,5	A,C,D	A,C,D	C,D	A,C,D	A,C,D	A,C,D	A,C,D	C,D	C,D	C,D	C,D	C,D	C,D	Q,S	A
Cleveland	Yes	2,4,5	A,C,D	A,C,D	C,D	C,D	C,D	C,D	C,D	C,D	C,D	C,D	C,D	C,D	C,D	M,Q	A
Dayton	Yes	2,4	A,C,D	A,C,D	C,D	C,D	C,D	C,D	C,D	C,D	C,D	C,D	C,D	C,D	C,D	Q	A
Youngstown	Yes	2,4,5	A,C,D	A,C,D	C,D	C,D	C,D	C,D	C,D	C,D	C,D	C,D	C,D	C,D	C,D	Q	A
Oklahoma	Yes†	1,2	A,C,D	C,D	C,D	C,D								C,D		M	
Oregon	Yes	1,2,3,4	A,C,D	A,C,D	C,D	C,D	C,D	C,D	C,D	A,C,D	A,C,D	C,D	C,D	C,D	C,D	M,Q	B
Portland **	Yes	1,2,4	A,C,D	C,D	C,D	C,D	C,D	A,C,D	A,C,D	C,D					C,D	M,Q	
Pennsylvania	Yes	1,2	A,C,D	A,C,D	A,C,D	A,C,D	A,C,D	A,C,D	A,C,D	A,C,D	A,C,D			A,C,D	A,C,D	R	A
Philadelphia	Yes	4,8	A,C,D	A,C,D	A,C,D	A,C,D	A,C,D	A,C,D	A,C,D	A,C,D	C,D	C,D	C,D		A,C,D	M,Q	A
Rhode Island	Yes	2,4,5	A,C,D	A,C,D	C,D	C,D	C,D	C,D	C,D	C,D	C,D	C,D	C,D	C,D	C,D	M,Q	A
South Carolina	Yes	1,4,6	A,C,D	A,C,D	C,D	C,D	C,D	C,D	C,D	C,D				C,D	C,D	M,O	

NOTE: An explanation of the numbers and letters used is given at the end of the table (see p 14.5).
†State buildings only.

(continued)

TABLE 14.1.1 Codes for Emergency Power by States and Major Cities (Completed September 1984) *(Continued)*

State/City		Does State/City Have Legislation?	Legislation Code Type?	Hospitals	Nursing Homes	Schools	Theaters (Public Gathering Places)	Office Buildings	Hotels	Apartment Buildings	Airports	Fire and Police Stations	Water Treatment Plants	Sewage Treatment Plants	All Public Buildings, State	All Public Buildings, Commercial	Applicable Government Agency	Smokeproof Enclosures in High-Rise Buildings
South Dakota		Yes	1,2,4	A,C,D	A,C,D	C,D	C,D		C,D						C,D	C,D	M	
Tenessee		Yes	1,4,6	A,C,D	A,C,D	C,D	C,D	C,D	C,D	C,D	C,D				C,D	C,D	M,N,O	A
Texas		Yes	2	A,C,D	B,C,D												M,O,U	
Amarillo	**	Yes	8	A,C	A,C	C,D	C,D	C,D	C,D	C	A,C	A,C	A	A	A	C	O	
Austin		Yes	3	A,C,D	A,C,D	C,D	C,D	C,D	C,D	C,D	C,D	C,D				C,D	Q	
Corpus Christi	**	Yes	3	A,C,D	A,C,D	C,D	A,C,D	C,D	C,D	C,D	A,C,D				C,D	C,D	Q	
Dallas	**	Yes	3,4	A,C,D	A,C,D	C,D	A,C,D	A,C,D	A,C,D	A,C,D	C,D	C,D	C,D	C,D	C,D	C,D	S	
El Paso		Yes	4,6	A,C,D	A,C,D	C,D	C,D	C,D	C,D	C,D					A,C,D	A,C,D	S	A
Fort Worth	**	Yes	3,4	A,C,D	C,D	C,D	C,D	C,D	C,D	C,D	C,D	A,C,D	C,D	C,D	C,D	C,D	S	
Houston		Yes	3,4,8	A,C,D	A,C,D	C,D	C,D	B,C,D*	B,C,D*	B,C,D*	B,C,D*	B,C,D*	C,D	C,D	B,C,D*	B,C,D*	S	
Lubbock		Yes	1,3,4	A,C,D	A,C,D	C,D	C,D	C,D	C,D	C,D	C,D	C,D	C,D	C,D	C,D	C,D	M,Q	B
San Antonio	**	Yes	4,8	A,C,D	A,C,D	C,D	A,C,D	C,D	C,D	C,D	A,C,D	C,D	C,D	C,D	C,D	A,C,D	M	
Wichita Falls	**	Yes	7	A,C,D	A,C,D	C,D	C,D	C,D	C,D	C,D	C,D	C,D	C,D	C,D	C,D	C,D	S	
Utah		Yes	1,2,3,4	A,C,D	A,C,D	C,D	C,D	C,D	C,D	C,D	C,D	C,D			C,D	C,D	M,Q	B
Salt Lake City	**	Yes	3,8	C,D	C,D	C,D	C,D	C,D	C,D	C,D	C,D	C,D	C,D	C,D		C,D	M,Q	
Vermont		Yes	1,2,4,5	A,C,D	A,C,D	C,D	C,D	C,D	C,D	C,D	C,D	C,D			C,D	C,D	R	A
Virginia		Yes	2,4,5	A,C,D	C,D	C,D	C,D	C,D	C,D	C,D	C,D	C,D	C,D	C,D	C,D	C,D	O	B
Richmond		Yes	1,4,5	A,C,D	A,C,D	C,D	C,D	C,D	C,D	C,D	C,D	C,D	C,D	C,D	C,D	C,D	O,Q	
Virginia Beach		Yes	4,5	A,C,D	A,C,D	A,C,D	A,C,D	C,D	C,D	C,D	A,C,D	B,C,D			C,D	C,D	Q	
Washington		Yes	2,3,4	A,C,D	A,C,D	C,D	C,D	C,D	C,D	C,D	C,D	C,D			C,D	C,D	R	B
Seattle	**	Yes	3,4	A,C,D	A,C,D	C,D	C,D	C,D	A,C,D	A,C,D		C,D	A,B	A,B	A,B	C,D	Q	
West Virginia		Yes	1,2,4	A,C,D	A,C,D	C,D	C,D	C,D	C,D	C,D	C,D	C,D	C,D	C,D	C,D	C,D	M,O	
Wisconsin		Yes	2,4	A,C,D	A,C,D	C,D	C,D	C,D	C,D	C,D	C,D	C,D	C,D	C,D	C,D	C,D	R,S	
Madison		Yes	2,4	A,C,D	A,C,D	C,D	C,D	C,D	C,D	C,D	C,D	C,D	C,D	C,D	C,D	C,D	O	
Milwaukee		Yes	2,4,8	A,C,D	A,C,D	C,D	C,D	C,D	C,D	C,D	C,D	C,D	C,D	C,D	C,D	C,D	Q	
Wyoming		Yes	1,2,3	A,C,D	A,C,D	A,C,D	C,D	C,D	C,D	C,D	C,D	C,D	C,D	C,D	C,D	C,D	M,N,P	

Explanation of Numbers and Letters Used in Table 14.1.1

Legislation Code Type

1. Life Safety Code, ANSI/NFPA 101-1985 [11]
2. State
3. Uniform Building Code [24]
4. National Electrical Code, ANSI/NFPA 70-1987 [9]
5. Building Officials and Code Administration (BOCA)
6. Standard Building Code [23]
7. Health Care Facilities Code, ANSI/NFPA 99-1984 [10]
8. City

Power Source

A. Emergency Power
B. Standby Power
C. Exit Lighting
D. Egress Lighting

Governing Agency

M. Fire Marshal or Division of Fire
N. Department of Public Health
O. Local Government Units
P. Public Safety
Q. Building Commission or Department
R. Department of Labor
S. Inspection Department
T. Department of Insurance
U. Various, but usually depends on occupancy

* High-rise building.
** No changes made since previous report.

(From IEEE Std. 446-1995. Copyright 1995 IEEE. All rights reserved.)

standby supply can be brought on line. An example would be the case where battery lighting units come on until a standby generator can be started and transferred to critical loads.

Readers using this text may find that various combinations of general needs will require an indepth system and cost analysis that will modify the recommended equipment and systems to best meet all requirements.

Small commercial establishments and manufacturing plants usually will find their requirements under two or three of the general need guidelines given in this chapter. Large manufacturers and commercial facilities will find that portions or all of the need guidelines given here apply to their operations and justify or require emergency and backup standby electric power.

TABLE 14.1.2 Condensed General Criteria for Preliminary Consideration

General Need	Specific Need	Maximum Tolerance Duration of Power Failure	Recommended Minimum Auxiliary Supply Time	Type of Auxiliary Power System: Emergency	Type of Auxiliary Power System: Standby	System Justification
Lighting	Evacuation of personnel	Up to 10 s, preferably not more than 3 s	2 h	×		Prevention of panic, injury, loss of life Compliance with building codes and local, state, and federal laws Lower insurance rates Prevention of property damage Lessening of losses due to legal suits
	Perimeter and security	10 s	10–12 h during all dark hours	×	×	Lower losses from theft and property damage Lower insurance rates Prevention of injury
	Warning	From 10 s up to 2 or 3 min	To return to prime power source	×		Prevention or reduction of property loss Compliance with building codes and local, state, and federal laws Prevention of injury and loss of life
	Restoration of normal power system	1 s to indefinite depending on available light	Until repairs completed and power restored	×	×	Risk of extended power and light outage due to a longer repair time
	General lighting	Indefinite; depends on analysis and evaluation	Indefinite; depends on analysis and evaluation		×	Prevention of loss of sales Reduction of production losses Lower risk of theft Lower insurance rates
	Hospitals and medical areas	Uninterruptible to 10 s ANSI/NFPA 99-1984 [10], 101-1985 (11] allow 10 s for alternate power source to start and transfer	To return of prime power	×	×	Facilitate continuous patient care by surgeons, medical doctors, nurses, and aids Compliance with all codes, standards, and laws Prevention of injury or loss of life Lessening of losses due to legal suits
	Orderly shutdown time	0.1 s to 1 h	10 min to several hours	×		Prevention of injury or loss of life Prevention of property loss by a more orderly and rapid shutdown of critical systems Lower risk of theft Lower insurance rates
Startup power	Boilers	3 s	To return of prime power	×	×	Return to production Prevention of property damage due to freezing Provision of required electric power
	Air compressors	1 min	To return of prime power		×	Return to production Provision for instrument control
Transportation	Elevators	15 s to 1 min	1 h to return of prime power		×	Personnel safety Building evacuation Continuation of normal activity
	Material handling	15 s to 1 min	1 h to return of prime power		×	Completion of production run Orderly shutdown Continuation of normal activity
	Escalators	15 s to no requirement for power	Zero to return of prime power		×	Orderly evacuation Continuation of normal activity

(continued)

TABLE 14.1.2 Condensed General Criteria for Preliminary Consideration *(Continued)*

General Need	Specific Need	Maximum Tolerance Duration of Power Failure	Recommended Minimum Auxiliary Supply Time	Type of Auxiliary Power System: Emergency	Type of Auxiliary Power System: Standby	System Justification
	Conveyors	15 s to 1 min	As analyzed and economically justified		×	Completion of production run Completion of customer order Orderly shutdown Continuation of normal activity
Mechanical utility systems	Water (cooling and general use)	15 s	½ h to return of prime power		×	Continuation of production Prevention of damage to equipment Supply of fire protection
	Water (drinking and sanitary)	1 min to no requirement	Indefinite until evaluated		×	Providing of customer service Maintaining personnel performance
	Boiler power	0.1 s	1 h to return of prime power	×	×	Prevention of loss of electric generation and steam Maintaining production Prevention of damage to equipment
	Pumps for water, sanitation, and production fluids	10 s to no requirement	Indefinite until evaluated		×	Prevention of flooding Maintaining cooling facilities Providing sanitary needs Continuation of production Maintaining boiler operation
	Fans and blowers for ventilation and heating	0.1 s to return of normal power	Indefinite until evaluated	×	×	Maintaining boiler operation Providing for gas-fired unit venting and purging Maintaining cooling and heating functions for buildings and production
Heating	Food preparation	5 min	To return of prime power		×	Prevention of loss of sales and profit Prevention of spoilage of in-process preparation
	Process	5 min	Indefinite until evaluated; normally for time for orderly shutdown, or to return of prime power		×	Prevention of in-process product damage Prevention of property damage Continued production Prevention of payment to workers during no production Lower insurance rates
Refrigeration	Special equipment or devices which have critical warmup (cryogenics)	5 min	To return of prime power		×	Prevention of equipment or product damage
	Depositories of critical nature (blood banks, etc)	5 min (10 s per ANSI/NFPA 99-1984 [10]	To return of prime power		×	Prevention of loss of material stored
	Depositories of noncritical nature (meat, produce, etc)	2 h	Indefinite until evaluated		×	Prevention of loss of material stored Lower insurance rates
Production	Critical process power (sugar factory, steel mills, chemical processes, glass products, etc)	1 min	To return of prime power or until orderly shutdown		×	Prevention of product and equipment damage Continued normal production Reduction of payment to workers on guaranteed wages during nonproductive period Lower insurance rates Prevention of prolonged shutdown due to nonorderly shutdown

(continued)

TABLE 14.1.2 Condensed General Criteria for Preliminary Consideration *(Continued)*

General Need	Specific Need	Maximum Tolerance Duration of Power Failure	Recommended Minimum Auxiliary Supply Time	Type of Auxiliary Power System: Emergency	Type of Auxiliary Power System: Standby	System Justification
	Process control power	Uninterruptible (UPS) to 1 min	To return of prime power	×	×	Prevention of loss of machine and process computer control program Maintaining production Prevention of safety hazards from developing Prevention of out-of-tolerance products
Space conditioning	Temperature (critical application)	10 s	1 min to return of prime power	×	×	Prevention of personnel hazards Prevention of product or property damage Lower insurance rates Continuation of normal activities Prevention of loss of computer function
	Pressure (critical) pos/neg atmosphere	1 min	1 min to return of prime power	×	×	Prevention of personnel hazards Continuation of normal activities Prevention of product or property damage Lower insurance rates Compliance with local, state, and federal codes, standards, and laws
	Humidity (critical)	1 min	To return of prime power		×	Prevention of loss of computer functions Maintenance of normal operations and tests Prevention of explosions or other hazards
	Static charge	10 s or less	To return of prime power	×	×	Prevention of static electric charge and associated hazards Continuation of normal production (printing press operation, painting spray operations)
	Building heating and cooling	30 min	To return of prime power		×	Prevention of loss due to freezing Maintenance of personnel efficiency Continuation of normal activities
	Ventilation (toxic fumes)	15 s	To return of prime power or orderly shutdown	×	×	Reduction of health hazards Compliance with local, state, and federal codes, standards, and laws Reduction of pollution
	Ventilation (explosive atmosphere)	10 s	To return of prime power or orderly shutdown	×	×	Reduction of explosion hazard Prevention of property damage Lower insurance rates Compliance with local, state, and federal codes, standards, and laws Lower hazard of fire Reduce hazards to personnel
	Ventilation (building general)	1 min	To return of prime power		×	Maintaining of personnel efficiency Providing make-up air in building

(continued)

TABLE 14.1.2 Condensed General Criteria for Preliminary Consideration (*Continued*)

General Need	Specified Need	Maximum Tolerance Duration of Power Failure	Recommended Minimum Auxiliary Supply Time	Type of Auxiliary Power System: Emergency	Type of Auxiliary Power System: Standby	System Justification
	Ventilation (special equipment)	15 s	To return of prime power or orderly shutdown	×	×	Purging operation to provide safe shutdown or startup Lowering of hazards to personnel and property Meeting requirements of insurance company Compliance with local, state, and federal codes, standards, and laws Continuation of normal operation
	Ventilation (all categories noncritical)	1 min	Optional		×	Maintaining comfort Preventing loss of tests
	Air pollution control	1 min	Indefinite until evaluated; compliance or shutdowns are options	×	×	Continuation of normal operation Compliance with local, state, and federal codes, standards, and laws
Fire protection	Annunciator alarms	1 s	To return of prime power	×		Compliance with local, state, and federal codes, standards, and laws Lower insurance rates Minimizing life and property damage
	Fire pumps	10 s	To return of prime power		×	Compliance with local, state, and federal codes, standards, and laws Lower insurance rates Minimizing life and property damage
	Auxiliary lighting	10 s	5 min to return of prime power	×	×	Servicing of fire pump engine should it fail to start Providing visual guidance for fire-fighting personnel
Data processing	CPU memory tape/disk storage, peripherals	½ cycle	To return of prime power or orderly shutdown	×	×	Prevention of program loss Maintaining normal operations for payroll, process control, machine control, warehousing, reservations, etc
	Humidity and temperature control	5 to 15 min (1 min for water-cooled equipment)	To return of prime power or orderly shutdown		×	Maintenance of conditions to prevent malfunctions in data processing system Prevention of damage to equipment Continuation of normal activity
Life support and life safety systems (medical field, hospitals, clinics, etc)	X-ray	Milliseconds to several hours	From no requirement to return of prime power, as evaluated	×	×	Maintenance of exposure quality Availability for emergencies
	Light	Milliseconds to several hours	To return of prime power	×	×	Compliance with local, state, and federal codes, standards, and laws Preventing interruption to operation and operating needs
	Critical to life, machines, and services	½ cycle to 10 s	To return of prime power	×	×	Maintenance of life Prevention of interruption of treatment or surgery Continuation of normal activity Compliance with local, state, and federal codes, standards, and laws

(*continued*)

TABLE 14.1.2 Condensed General Criteria for Preliminary Consideration *(Continued)*

General Need	Specified Need	Maximum Tolerance Duration of Power Failure	Recommended Minimum Auxiliary Supply Time	Type of Auxiliary Power System: Emergency	Type of Auxiliary Power System: Standby	System Justification
	Refrigeration	5 min	To return of prime power		×	Maintaining blood, plasma, and related stored material at recommended temperature and in prime condition
Communication systems	Teletypewriter	5 min	To return of prime power		×	Maintenance of customer services Maintenance of production control and warehousing Continuation of normal communication to prevent economic loss
	Inner building telephone	10 s	To return of prime power	×		Continuation of normal activity and control
	Television (closed circuit and commercial)	10 s	To return of prime power		×	Continuation of sales Meeting of contracts Maintenance of security Continuation of production
	Radio systems	10 s	To return of prime power	×	×	Maintenance of security and fire alarms Providing evacuation instructions Continuation of service to customers Prevention of economic loss Directing vehicles normally
	Intercommunication systems	10 s	To return of prime power	×	×	Providing evacuation instructions Directing activities during emergency Providing for continuation of normal activities Maintaining security
	Paging systems	10 s	½ h	×	×	Locating of responsible persons concerned with power outage Providing evacuation instructions Prevention of panic
Signal circuits	Alarms and annunciation	1 to 10 s	To return of prime power	×	×	Prevention of loss from theft, arson, or riot Maintaining security systems Compliance with codes, standards, and laws Lower insurance rates Alarm for critical out-of-tolerance temperature, pressure, water level, and other hazardous or dangerous conditions Prevention of economic loss
	Land-based aircraft, railroad, and ship warning systems	1 s to 1 min	To return of prime power	×	×	Compliance with local, state, and federal codes, standards, and laws Prevention of personnel injury Prevention of property and economic loss

14.1.3 Typical Emergency/Standby Lighting Recommendations

For short-time-duration primarily lighting for personnel safety and evacuation purposes, battery units are satisfactory. Where longer service and heavier loads are required, an engine- or turbine-driven generator is usually used that starts automatically on failure of the prime power source with the load applied by an automatic transfer switch. It is generally considered that an average level of 0.4 fc is adequate where passage is required and no precise operations are expected.

Table 14.1.3 summarizes the user's needs for emergency and standby electric power for lighting by application and area.

TABLE 14.1.3

Standby*	Immediate, Short-Term†	Immediate, Long-Term‡
Security lighting	Evacuation lighting	Hazardous areas
Outdoor perimeters	Exit signs	Laboratories
Closed circuit TV	Exit lights	Warning lights
Night lights	Stairwells	Storage areas
Guard stations	Open areas	Process areas
Entrance gates	Tunnels	
	Halls	Warning lights
Production lighting		Beacons
Machine areas	Miscellaneous	Hazardous areas
Raw materials storage	Standby generator areas	Traffic signals
Packaging	Hazardous machines	
Inspection		Health care facilities
Warehousing		Operating rooms
Offices		Delivery rooms
		Intensive care areas
Commercial lighting		Emergency treatment areas
Displays		
Product shelves		Miscellaneous
Sales counters		Switchgear rooms
Offices		Elevators
		Boiler rooms
Miscellaneous		Control rooms
Switchgear rooms		
Landscape lighting		
Boiler rooms		
Computer rooms		

* An example of a standby lighting system is an engine-driven generator.
† An example of an immediate short-term lighting system is the common unit battery equipment.
‡ An example of an immediate long-term lighting system is a central battery bank rated to handle the required lighting load only until a standby engine-driven generator is placed on-line.

(From IEEE Std. 446-1995. Copyright 1995 IEEE. All rights reserved.)

14.2.0 Emergency/Standby Power Source Options and Arrangements

Sources of emergency power may include batteries, local generation, a separate source over separate lines from the electric utility, or various combinations of these. The quality of service required, the amount of load to be served, and the characteristics of the load will determine which type of emergency supply is required.

Some arrangements commonly found for multiple services and/or engine-driven local generation are as follows.

…ource System Using One …witch (*see page 14.12*)

…le utility services may be used as an emergency or standby source of power. …red is an additional utility service from a separate source and the required …witching equipment. The figure shows automatic transfer between two low-voltage utility supplies. Utility source 1 is the normal power line, and utility source 2 is a separate utility supply providing emergency power. Both circuit breakers are normally closed. The load must be able to tolerate the few cycles of interruption while the automatic transfer device operates.

14.2.1

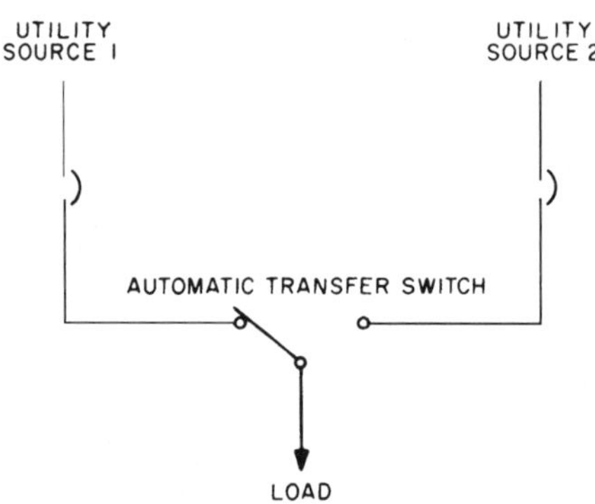

14.2.2 Two-Utility-Source System Where Any Two Circuit Breakers Can Be Closed

14.2.2

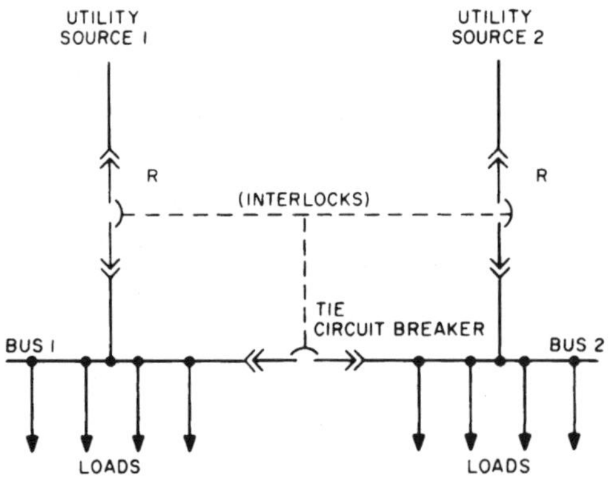

Automatic switching equipment may consist of three circuit breakers with suitable control and interlocks, as shown in the figure. Circuit breakers generally are used for primary switching where the voltage exceeds 600 V. They are more expensive but safer to operate, and the use of fuses for overcurrent protection is avoided.

Relaying is provided to transfer the load automatically to either source if the other one fails, provided that the circuit is energized. The supplying utility normally will designate which source is for normal use and which is for emergency use. If either supply is not able to carry the entire load, provisions must be made to drop noncritical loads before the transfer takes place. If the load can be taken from both services, the two R circuit breakers are closed, and the tie circuit breaker is open. This mode of operation generally is preferred by the supplying utility and the customer. The three circuit breakers are interlocked to permit any two to be closed but prevent all three from being closed. The advantages of this arrangement are that the momentary transfer outage will occur only on the load supplied from the circuit that is lost, the loads can be balanced between the two buses, and the supplying utility does not have to keep track of reserve capacity for the emergency feeder. However, the supplying utility may not allow the load to be taken from both sources, especially since a more expensive totalizing meter may be required. A manual override of the interlock system should be provided so that a closed-transition transfer can be made if the supplying utility wants to take either line out of service for maintenance or repair and a momentary tie is permitted.

If the supplying utility will not permit power to be taken from both sources, the control system must be arranged so that the circuit breaker on the normal source is closed, the tie circuit breaker is closed, and the emergency-source circuit breaker is open. If the utility will not permit dual or totalized metering, the two sources must be connected together to provide a common metering point and then connected to the distribution switchboard. In this case, the tie circuit breaker can be eliminated, and the two circuit breakers act as a transfer device (sometimes called a *transfer pair*). Under these conditions, the cost of an extra circuit breaker rarely can be justified.

The arrangement shown only provides protection against failure of the normal utility service. Continuity of power to critical loads also can be disrupted by

1. An open circuit within the building (load side of the incoming service)
2. An overload or fault tripping out a circuit
3. Electrical or mechanical failure of the electric power distribution system within the building

It may be desirable to locate transfer devices close to the load and have the operation of the transfer devices independent of overcurrent protection. Multiple transfer devices of lower current rating, each supplying a part of the load, may be used rather than one transfer device for the entire load.

The arrangement shown can represent the secondary of a double-ended substation configuration or a primary service. It is sometimes referred to as a *main-tie-main* configuration.

14.2.3 Diagram Illustrating Multiple Automatic Double-Throw Transfer Switches Providing Varying Degrees of Emergency and Standby Power (*see page 14.14*)

Availability of multiple utility service systems can be improved by adding a standby engine-generator set capable of supplying the more critical load. Such an arrangement, using multiple automatic transfer switches, is shown in the figure.

14.2.4 Typical Transfer Switching Methods (a) Total Transfer and (b) Critical-Load Transfer

Figure 14.2.4(a) shows a typical switching arrangement in which a local emergency generator is used to supply the entire load on loss of the normal power supply. All emergency loads normally are supplied through device A. Device B is open, and the generator is at rest. When the normal supply fails, the transfer switch undervoltage relay is deenergized and, after a predetermined time delay, closes its engine-starting contacts. The time delay is introduced so that the generator will not be

14.2.3

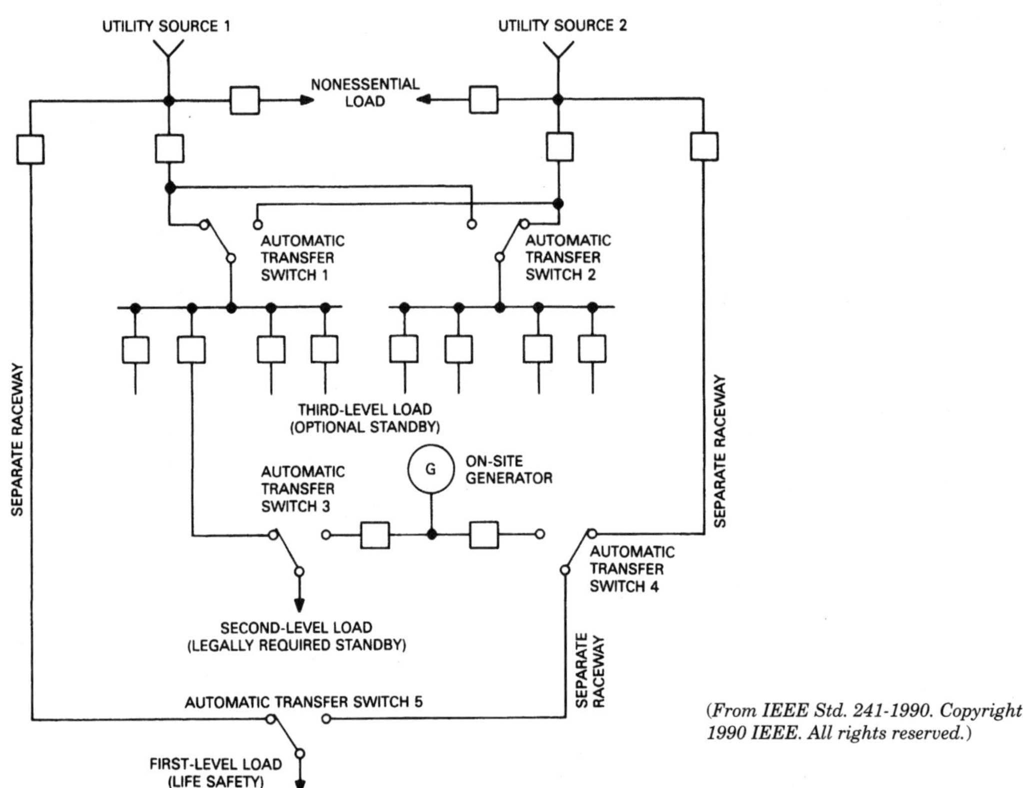

(From IEEE Std. 241-1990. Copyright 1990 IEEE. All rights reserved.)

14.2.4

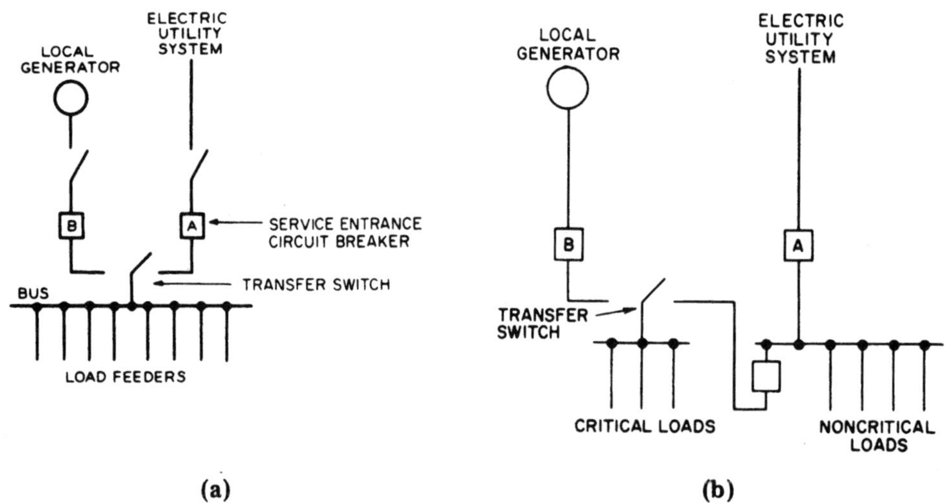

(a) **(b)**

(From IEEE Std. 241-1990. Copyright 1990 IEEE. All rights reserved.)

started unnecessarily during transient voltage dips and momentary outages. When the alternate source is a generator, sufficient time or speed monitoring should be allowed to permit the generator to reach acceptable speed (thus frequency and voltage) before transfer and application of load. It should be noted that the arrangement shown in Figure 14.2.4(a) does not provide complete protection against power disruption within the building.

Figure 14.2.4(b) shows a typical switching arrangement in which only the critical loads are transferred to the emergency source, in this case an emergency generator. For maximum protection, the transfer switch is located close to the critical loads.

Other transfer methods are illustrated in the preceding discussion of multiple utility services.

14.2.5 Typical Multiengine Automatic Paralleling System

14.2.5 Typical multiengine automatic paralleling system.

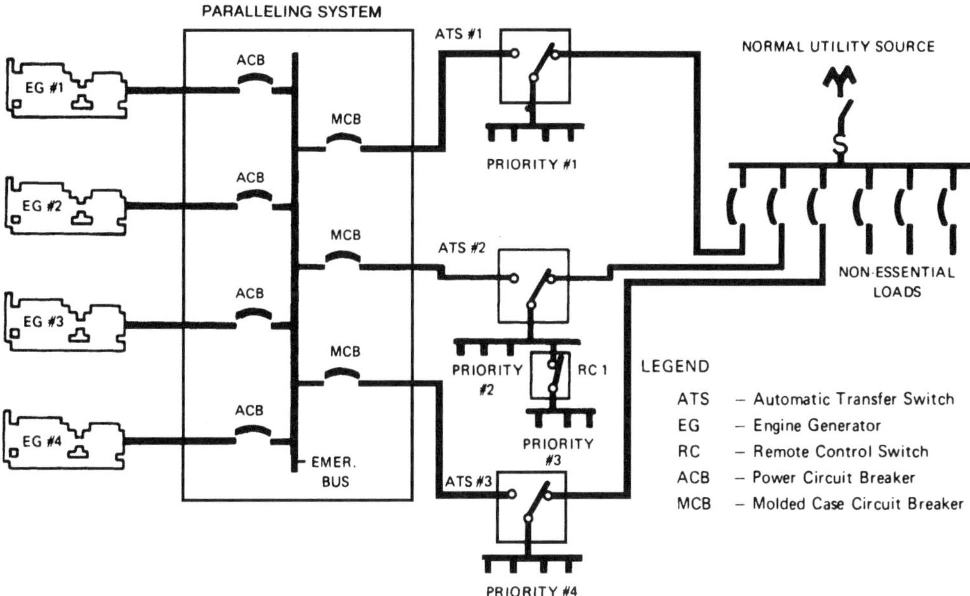

The operation is for a random-access paralleling system, and the loads are connected to the bus in random order as they become available.

The loads, however, are always connected to the emergency bus in ascending order of priority beginning with priority one. For load shedding, the loads are disconnected in descending order of priority beginning with the last priority of load to be connected.

On a loss of normal-source voltage, as determined by any one or more of the automatic transfer switches shown, a signal initiates starting of all engine-generator sets. The first set to come up to 90 percent of nominal voltage and frequency is connected to the alternate-source bus. Critical and life-safety loads are then transferred via ATS 1 and 2 to the bus on sensing availability of power on the bus. As the remaining engine-generator sets achieve 90 percent of the nominal voltage and frequency, their respective synchronizing monitors will control the voltage and frequency of these oncoming units to produce synchronism with the bus. Once the oncoming unit is matched in voltage, frequency, and phase angle with the bus, its synchronizer will initiate paralleling. On connection to the bus, the governor will cause the engine generator set to share the connected load with the other on-line sets.

Each time an additional set is added to the emergency bus, the next load is transferred in a numbered sequence via additional transfer switches, such as ATS 3, until all sets and essential loads are connected to the bus. Control circuitry should prevent the automatic transfer or connection of loads to the bus until there is sufficient capacity to carry these loads. Provision is made for manual override of the load addition circuits for supervised operation.

On restoration of the normal source of supply, as determined by the automatic transfer switches, the engines are run for a period of up to 15 minutes for cooling down and then are shut down. All controls automatically reset in readiness for the next automatic operation.

The system is designed so that reduced operation is initiated automatically on failure of any plant through load dumping. This mode overrides any previous manual controls to prevent overloading the emergency bus. On sensing a failure mode on an engine, the controls automatically initiate disconnect, shutdown, and lockout of the failed engine and reduction of the connected load to within the capacity of the remaining plants. Controls should require manual reset under these conditions.

Protection of the engine and generator against motorization is provided. A reverse power monitor, on sensing a motorizing condition on any plant, will initiate load shedding, disconnect the failing plant, and shut it down.

Sometimes a higher level of reliability is economically justifiable in a parallel-generation arrangement for critical loads such as hospitals and data centers. This is known as providing an $(N + 1)$ level of reliability (redundancy), i.e., providing one more generator than is needed to serve the emergency load. Thus, if one of the emergency generators fails to start or is out of service for any reason, the remaining plants can serve the entire emergency load. This precludes the need for automatic load shedding, which can be expensive in itself. This thus provides for two levels of contingency operation, the first being loss of the normal source of power and the second being loss of one of the emergency/standby generators. Providing an even higher level of reliability is rarely justifiable.

14.2.6 Elevator Emergency Power Transfer System

Elevators present a unique emergency power situation. Where elevator service is critical for personnel and patients, it is desirable to have automatic power transfer with manual supervision. Operators and maintenance personnel may not be available in time if the power failure occurs on a weekend or at night.

1. *Typical elevator system.* The figure shows an elevator emergency power transfer system whereby one preferred elevator is fed from a vital load bus through an emergency riser, whereas the rest of the elevators are fed from the normal service. By providing an automatic transfer switch for each elevator and a remote selector station, it is possible to select individual elevators, thus permitting complete evacuation in the event of power failure. The engine-generator set and emergency riser need only be sized for one elevator, thus minimizing the installation cost. The controls for the remote selector, automatic transfer switches, and engine starting are independent of the elevator controls, thereby simplifying installation.
2. *Regenerated power.* Regenerated power is a concern for motor-generator-type elevator applications. In some elevator applications, the motor is used as a brake when the elevator is descending and generates electricity. Electric power is then pumped back into the power source. If the source is commercial utility power, it can be absorbed easily. If the power source is an engine-driven generator, the regenerated power can cause the generating set and the elevator to overspeed. To prevent overspeeding of the elevator, the maximum amount of power that can be pumped back into the generating set must be known. The permissible amount of absorption is approximately 20 percent of the generating set's rating in kilowatts. If the amount pumped back is greater than 20 percent, other loads must be connected to

14.2.6

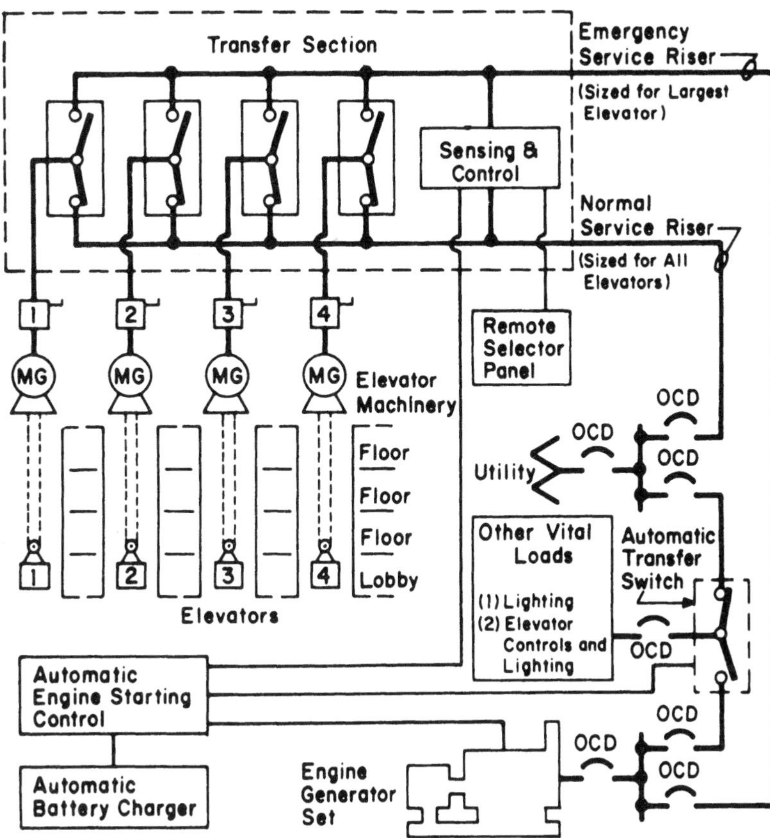

(From IEEE Std. 602-1996. Copyright 1996 IEEE. All rights reserved.)

the generating set, such as emergency lights or "dummy (parasitic)" load resistances. Emergency lighting should be connected permanently to the generating set for maximum safety. A dummy (parasitic) load also can be automatically switched on the line whenever the elevator is operating from an engine-driven generator.

14.2.7 Typical Hospital Installation with a Nonautomatic Transfer Switch and Several Automatic Transfer Switches (*see page 14.18*)

Hospital/health care facilities present a unique situation. ANSI/NFPA 99 mandates that emergency loads be broken into three distinct branches, namely, critical, life safety, and equipment, as illustrated in the figure.

This arrangement provides a very high level of reliability and integrity. Critical, life-safety, and essential-equipment loads are transferred automatically and immediately (i.e., with no intentional delay) to the emergency source on loss of commercial power. Lower-priority nonessential loads are transferred manually via nonautomatic transfer switches when the system has stabilized in the emergency mode and available capacity has been verified.

14.3.1 Engine and Generator-Set Sizing: Introduction

Proper sizing of a generator is an important task. The following guidelines represent the general and specific considerations that must be taken into account in properly sizing a generator for a specific application. These guidelines are based on

14.2.7

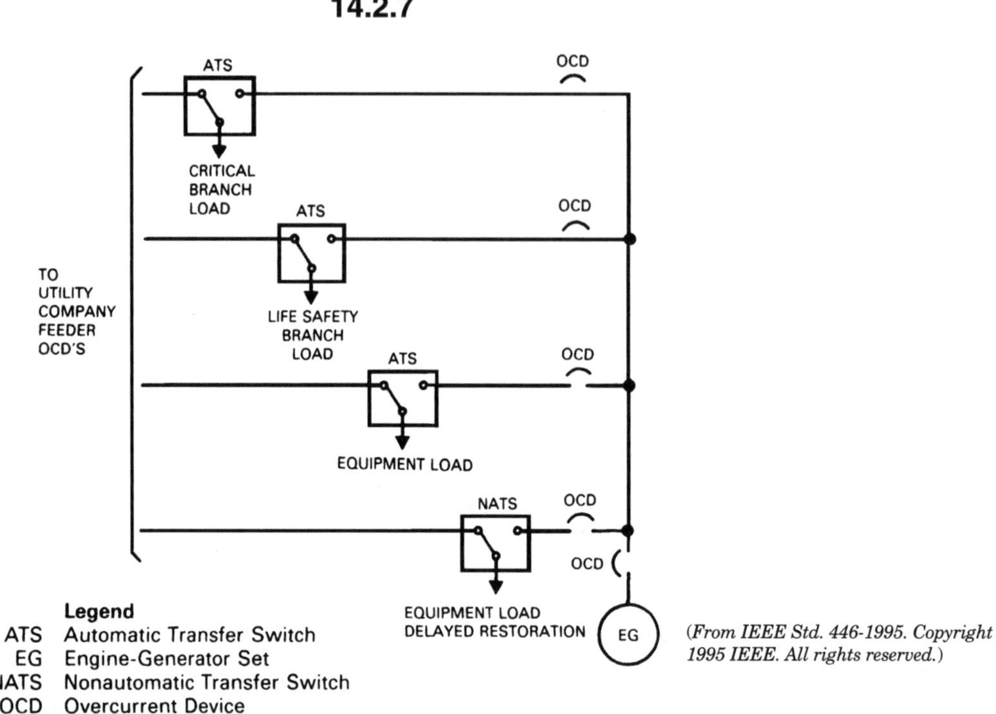

(From IEEE Std. 446-1995. Copyright 1995 IEEE. All rights reserved.)

Caterpillar generator sets as an industry leader. A common practice in the industry is to base a given design around a specific manufacturer of a major piece of equipment, such as a generator, and to make allowances for idiosyncratic differences that allow competitive bids and supply to the purchaser. Most generator manufacturers now use computer software programs for proper sizing of generators in specific applications. The following is provided to give a basic understanding of the methodology and can be used for preliminary calculations. It is in this context that the Caterpillar guidelines are offered.

14.3.1 Engine Rating Considerations

A generator set (genset) consists of an engine and a generator. However, it is best to consider the engine and generator as a system. Individually, each has unique characteristics; but together these qualities have a significant impact on the performance and sizing of the genset system.

Capabilities of both engine and generator are considered individually and collectively when selecting generator sets. Enginges produce brake horsepower (or kilowatts) while controlling speed or frequency. Generators influence engine behavior, but are primarily responsible for changing engine power into kilovolt-amperes (kVA) and electrical kilowatts (kW). They also must satisfy high "magnetizing current" draws (kVAR), or transient conditions from electrical equipment.

Normally, a generator set is furnished with a generator which matches the engine output capability. Engines are sized according to the actual power in kW required to meet the needs of the facility. The generator, on the other hand, must be capable of handling the maximum apparent power which is measured in kVA. There are several ways in which the actual power can be identified. It can be calculated by adding the nameplate ratings of the equipment to be powered by the generator. If this is done, the efficiencies of the equipment must also be considered. The actual power can be determined by performing a load analysis on the facility. This involves making a survey of the power requirements over a period of time.

$$\text{ekW} = \text{pf} \times \text{kVA}$$

$$\text{bkW} = \frac{\text{ekW}}{\text{eff}} + \text{Fan demand}$$

kVA = kVA output of generator

pf = power factor of connected load

ekW = electrical power (electrical kW)

bkW = engine power (brake kW)

eff = generator efficiency

When kW is neither qualified as electrical (ekW) or brake (bkW), it is important to clarify between the two when performing calculations or product comparisons.

14.3.2 Engine—Generator Set Load Factor

Load factor of a generator set is used as one criterion for rating a genset. It is calculated by finding the product of various loads:

$$\text{Percent of time} \times \text{percent of load}$$

$$\text{Percent of time} = \frac{\text{time at specific load}}{\text{total operating time}}$$

$$\text{Percent of load} = \frac{\text{specific load}}{\text{rated load}}$$

Extended idling time and the time when the generator set is not operating does not enter into the calculation for load factor.

For example, assume a facility has a genset rated at 550 kW and runs it two hours a week. During those two hours it runs at 400 kW for 1½ hours. Find the load factor.

Using the formulas we find:

$$\text{Percent of load} = \frac{400\text{ kW}}{550\text{ kW}} = 0.73$$

$$\text{Percent of time} = \frac{90\text{ min}}{120\text{ min}} = 0.75$$

$$\text{Load factor} = 0.73 \times 0.75 = 54.75\%$$

This load factor would indicate that the genset could be used as a standby rated genset because it meets the load factor and other criteria of standby.

Rating definitions for Caterpillar Generator sets are based on typical load factor, hours of use per year, peak demand and application use. Caterpillar Genset Ratings are as follows:

Standby rating: Output available with varying load for the duration of the interruption for the normal power source

Typical load factor = 60 percent or less

Typical hours/year = 500 hours

Typical peak demand = 80 percent of standby rated ekW with 100 percent of rating available for the duration of an emergency outage

Typical application = Building Services standby and enclosure/sheltered environment

Prime rating: Output available with varying load for an unlimited time

Typical load factor = 60 to 70 percent

Typical hours/year = no limit

Typical peak demand = 100 percent of prime rating used occasionally

Typical application = industrial, pumping, construction, peak shaving, or cogeneration

Continuous rating: Output available without varying load for an unlimited time

Typical load factor = 70 to 100 percent

Typical hours/year = no limit

Typical peak demand = 100 percent of continuous rating used 100 percent of the time

Typical application = base load, utility, cogeneration, or parallel operation

Operating above these rating definitions will result in shorter life and higher generating costs per year.

The International Standards Organization (ISO) 8528-1 defines three types of duty:

Continuous operating power (COP)

Prime running power (PRP)

Limited-time running power (LTP)

Continuous Operating Power (COP) is the power a generator set can operate at a *continuous* load for an unlimited number of hours under stated ambient conditions. Maintenance according to the manufacturer must be followed to reach these standards.

Prime Running Power (PRP) is the maximum power a generator set has during a *variable* power sequence for an unlimited number of hours under stated ambient conditions. Maintenance according to the manufacturer must be followed to reach these standards.

Limited-Time Running Power (LTP) is the *maximum* power that a generator set delivers for up to 500 hours per year under stated ambient conditions. Only 300 hours can be continuous running. Maintenance according to the manufacturer must be followed to reach these standards.

Many times specifications will be written with ISO standards in mind. The following chart matches Caterpillar genset ratings to ISO genset ratings.

ISO Ratings	*Caterpillar Ratings*
LTP	Standby
COP	Continuous
PRP	Prime

14.3.3 Load Management

Load management is the deliberate control of loads on a genset and/or utility so as to have the lowest possible electrical costs. In addition, Caterpillar has two broad rating categories in terms of load management: isolated from a utility and paralleled with a utility.

Load management when isolated from the utility under 750 hours per year: Output available with varying load for less than 6 hours per day.

Typical load factor = 60 percent or less

Typical hours/year = less than 750 hours

Typical peak demand = 80% of rated ekW with emergency outage

Typical application = interruptible utility rates, peak sharing

Load management when isolated from the utility over 750 hours per year: Output available without varying load for over 750 hours per year and less than 6 hours per day.

Typical load factor = 60% to 70%

Typical hours per year = more than 750 hours

Typical peak demand = 100% of prime plus 10% rating used occasionally

Typical application = peak sharing or cogeneration

Load management when paralleled with the utility under 750 hours per year: Output available without varying load for under 750 hours

Typical load factor = 60% to 70%

Typical hours/year = less than 750 hours

Typical peak demand = 100% of prime rating used occasionally

Typical application = peak sharing

Load management when paralleled with the utility over 750 hours per year: Output available without varying load for unlimited time.

Typical load factor = 70% to 100%

Typical hours/year = no limit

Typical peak demand = 100% of continuous rating used 100% of the time

Typical application = base load, utility, peak sharing, cogeneration, parallel operation

Depending on how the genset will be applied will determine the size needed. For example, if for a given load, the genset will only be used as standby for a critical load only, then a smaller kW genset can be used than one used for prime power.

14.3.4 Standards

Caterpillar generator sets are in accordance with the International Standardization Organization (ISO) and the Society for Automotive Engineers (SAE) standards. Each organization uses different techniques and tolerances for power ratings and fuel consumption.

ISO standard 3046-1 is specific to engines and ISO 8528 is followed for the generator set.

14.3.5 Generator Set Sizing Example

Figures 14.3.5.1 and 14.3.5.2 will be used to detail how to size a Caterpillar generator set, using load information. Figure 14.3.5.2 is used when motors with NEMA code letters are considered.

1. Fill out information in Part I and II from nameplate information. The motor efficiency can be estimated using a chart of approximate efficiencies (see Table 14.3.5.3 at end of this example).
2. Total engine load is determined by calculating motor efficiencies and adding to resistive load: Add the lighting loads, other non-motor loads, and motor loads together.
3. Select engine (Part IV) which matches frequency (Hz), configuration (gas, diesel, turbocharged, aftercooled, naturally aspirated), speed, and load. Hz and configuration are found in Section I. Speed is by customer preference and the load was determined in Section II.
4. Generator sizing: In Section V fill in Part A for all motors from nameplate. Minimize starting requirements by starting the largest motors first. The motors rating and NEMA code can be found in Section II. The skVA/hp is found using the NEMA code letter chart (see Table 14.3.5.4).

   Find the starting skVA using the following formula:

$$\text{SkVA} = \frac{\text{LRA} \times \text{V} \times 1732}{1000}$$

14.3.5.1 Generator Sizing Chart

Customer ____________ Project ____________ Analyst ____________ Date ____________

I. APPLICATION DATA

Prime/Standby Power Gas/Diesel Fuel ______ Volts ______ Phase ______ Hz

II. LOADS **III. ENGINE SIZING**

A. Lighting Loads .... ______kW

B. Other Non-Motor Loads .... ______kW

C. Motors

Starting Sequence	Motor kW	Full Load Amps	Locked Rotor Amps	Nameplate Data Reduced Voltage Starting Type	Acceptable Voltage Dip Percent	Motor Eff. (Chart 5)	kW (Engine) = kW (Motor) / Motor Efficiency (Chart 5)
1	____	____	____	____	____	____	______kW
2	____	____	____	____	____	____	______kW
3	____	____	____	____	____	____	______kW
4	____	____	____	____	____	____	______kW
5	____	____	____	____	____	____	______kW

Total Motor Load ______kW

Total Engine Load (A + B + C) ______kW

IV. ENGINE SELECTION Model: ________ Frame: ________ Rating (With Fan): ______kW ______Hz ______rpm

V. GENERATOR SIZING Start Sequence	Motor(s) 1	Motor(s) 2	Motor(s) 3	Motor(s) 4	Motor (s) 5
A. Starting kV•A (SKVA)					
1. Locked Rotor Amps (Use 6.0 x Full Load Amperes if LRA Unknown), Refer to Chart 4 if Full Load Amps Unknown	____LRA	____LRA	____LRA	____LRA	____LRA
2. SKVA = $\frac{\text{LRA} \times \text{V} \times 1.732}{1{,}000}$	____SKVA	____SKVA	____SKVA	____SKVA	____SKVA
B. Effective SKVA					
1. All Motors Running	0 kW	____kW	____kW	____kW	____kW
2. All Motors Running & Motor Being Started	____kW	____kW	____kW	____kW	____kW
3. $\frac{\text{B.1}}{\text{B.2}} \times 100$	0 %	____%	____%	____%	____%
4. Compensation for Motors Already Started (Chart 2)	1.0	____	____	____	____
5. Step A.2. x Step B.4.	____SKVA	____SKVA	____SKVA	____SKVA	____SKVA
6. Reduced Voltage Factor (Chart 3) (use 1.0 if no starting aid used)	____	____	____	____	____
7. Effective SKVA = Step B.5. x B.6.	____SKVA	____SKVA	____SKVA	____SKVA	____SKVA
8. Acceptable Voltage Dip (10, 20, 30%)	____%	____%	____%	____%	____%
C. Generator Selection (Chart 1)					
1. Frame	____	____	____	____	____
2. Rating	____kW	____kW	____kW	____kW	____kW
3. SKVA at Selected Voltage Dip	____	____	____	____	____

VI. GENERATOR SET SIZING

Select Largest Generator Set Model of Step IV and Step V.C.1.

Model: ____________ Frame: ________ Rating: ______ kW Prime/Standby ______ Hz ______ rpm

14.3.5.2 Generator Sizing Chart (When Using NEMA Code Letters)

Customer ______________ Project ______________ Analyst ______________ Date ______________

I. APPLICATION DATA
Prime/Standby Power Gas/Diesel Fuel ________ Volts ________ Phase ________ Hz

II. LOADS **III. ENGINE SIZING**
A. Lighting Loads .... ________kW
B. Other Non-Motor Loads .... ________kW
C. Motors

$$\text{kW (Engine)} = \frac{\text{hp (Motor)} \times 0.746}{\text{Motor Efficiency (Chart 5)}}$$

Starting Sequence	hp	Nema Code	Nameplate Data: Reduced Voltage Starting Type	Acceptable Voltage Dip Percent	Motor Eff. (Chart 5)	kW (Engine)
1	____	________	________	________	________	________kW
2	____	________	________	________	________	________kW
3	____	________	________	________	________	________kW
4	____	________	________	________	________	________kW
5	____	________	________	________	________	________kW

Total Motor Load ________kW
Total Engine Load (A + B + C) ________kW

IV. ENGINE SELECTION Model: ________ Frame: ________ Rating (With Fan): ________kW ________Hz ________rpm

V. GENERATOR SIZING Start Sequence	Motor(s) 1	Motor(s) 2	Motor(s) 3	Motor(s) 4	Motor (s) 5
A. Starting kV•A (SKVA)					
1. Motor Ratings	____hp	____hp	____hp	____hp	____hp
2. NEMA Code	____	____	____	____	____
3. SKVA/hp (Use 6.0 if Code Letter Unknown)	____	____	____	____	____
4. SKVA/hp x Motor hp (A.1 x A.3)	____SKVA	____SKVA	____SKVA	____SKVA	____SKVA
B. Effective SKVA					
1. All Motors Running	0 kW	____kW	____kW	____kW	____kW
2. All Motors Running & Motor Being Started	____kW	____kW	____kW	____kW	____kW
3. $\frac{B.1}{B.2} \times 100$	0 %	____%	____%	____%	____%
4. Compensation for Motors Already Started (Chart 2)	1.0	____	____	____	____
5. Step A.4. x Step B.4.	____SKVA	____SKVA	____SKVA	____SKVA	____SKVA
6. Reduced Voltage Factor (Chart 3) (use 1.0 if no starting aid used)	____	____	____	____	____
7. Effective SKVA = Step B.5. x B.6.	____SKVA	____SKVA	____SKVA	____SKVA	____SKVA
8. Acceptable Voltage Dip (10, 20, 30%)	____%	____%	____%	____%	____%
C. Generator Selection (Chart 1)					
1. Frame	____	____	____	____	____
2. Rating	____kW	____kW	____kW	____kW	____kW
3. SKVA at Selected Voltage Dip	____	____	____	____	____

VI. GENERATOR SET SIZING
Select Largest Generator Set Model of Step IV and Step V.C.1.
Model: ________ Frame: ________ Rating: ________ kW Prime/Standby ________ Hz ________ rpm

NEMA

TABLE 14.3.5.3

Approximate Efficiencies Squirrel Cage Induction Motors		
hp	**kW**	**Full-Load Efficiency**
5-7½	4-6	0.83
10	7.5	0.85
15	11	0.86
20-25	15-19	0.89
30-50	22-37	0.90
60-75	45-56	0.91
100-300	74.6-224	0.92
350-600	261-448	0.93

TABLE 14.3.5.4

Identifying Code Letters on AC Motors	
NEMA Code Letter	**Starting kVA/A/hp**
A	0.00-3.14
B	3.15-3.54
C	3.55-3.99
D	4.00-4.49
E	4.50-4.99
F	5.00-5.59
G	5.60-6.29
H	6.30-7.09
J	7.10-7.99
K	8.00-8.99
L	9.00-9.99
M	10.00-11.19
N	11.20-12.49
P	12.50-13.99
R	14.00-15.99
S	16.00-17.99
T	18.00-19.99
U	20.00-22.39
V	22.40 -

5. Motors on-line diminish skVA to start additional motors. For column B1 the first motor is always 0 if there is more than one motor. Motor #2 is the kW in Section II from first sequence motor. Additional motor's kW are added to the total for each motor starting.

   B2 columns are filled in using Section II information as written. B3 uses the Motor Preload Multiplier formula to determine the percentage of motor load: B1/B3 or

$$\text{Percent motor load} = \frac{\text{All motors running}}{\text{All motors running \& being started}} \times 100$$

   Under 40%, the multiplier is 1.0. Over 40% use the chart in Figure 14.3.5.5.

$$\text{Percent motor load} = \frac{\text{All motors running}}{\text{All motors running \& being started}} \times 100$$

$$\text{Percent motor load} < 40\%, \text{multiplier} = 1.0$$

   If reduced voltage starting is used multiply the skVA by the factor found on the Table 14.3.5.6 and fill in space B5 and B7.

   If no starting aid is used, use 1.0.

   Columns B8 can be filled by using information in Section II.

6. Use the effective skVA and acceptable voltage dip numbers to find on TMI the appropriate generator. Fill in C1, C2, and C3 using the numbers from TMI.
7. Generator set sizing (VI) can be found by selecting the largest generator set model of Step IV and Step V. (C1)

14.3.6 Critical Installation Considerations

The following summary contains important points to remember for a successful generator installation:

1. The generator set must be sized properly for the installation. Determine the duty cycle—continuous, prime, standby, or peak shaving or sharing (paralleled or not paralleled with the utility).

14.3.5.5

Motor Preload Multiplier

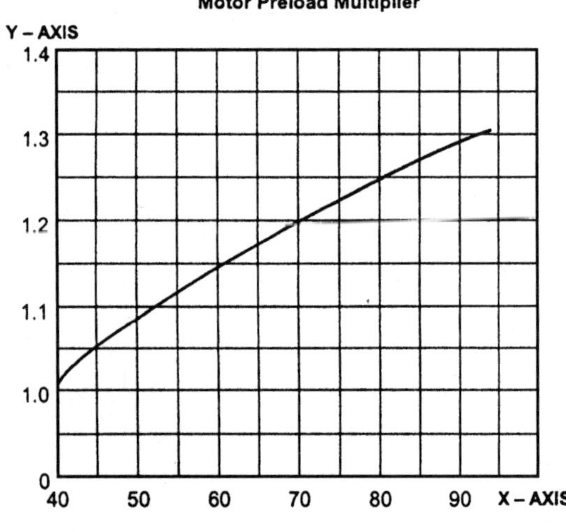

TABLE 14.3.5.6

Reduced Voltage Starting Factors	
Type	**Multiply skVA By**
Resistor, Reactor, Impedance	
80% Tap	0.80
65% Tap	0.65
50% Tap	0.50
45% Tap	0.45
Autotransformer	
80% Tap	0.68
65% Tap	0.46
50% Tap	0.29
Y Start, Run	0.33
Solid State: Adjustable, consult manufacturer or estimate 300% of full load kVA (Use 1 if no reduced voltage starting aids used)	

- *Continuous.* Output available without varying load for an unlimited time.
- *Prime.* Output available with varying load for an unlimited time.
- *Standby.* Output available with varying load for the duration of the interruption of the normal source of power. Usually sized initially for 60 percent of actual load, since loads tend to increase during the 30-year life of the unit. Normal hours of operation are less than 100 hours per year.
- *Peak Shaving / Sharing.* Prime if paralleled with the utility. Standby if not paralleled with the utility and if the load meets the definition of prime or standby. Normally, peak shaving/sharing is less than 200 hours per year of operation.

Loads that are too light cause engine slobber. Overloading causes excessive piston loading and high exhaust temperatures. Standby engines that must be exercised regularly but cannot be loaded should only be run long enough to achieve normal oil pressure and then shut off—less than 5 minutes of running time. Good practice dictates that this be done weekly and that once a month the generators be run under load for a half hour or so and then unloaded briefly for cool-down. The load should be at least two-thirds of capacity, either using a "dummy" resistive load bank or preferably under actual building load. The latter requirement is mandatory for hospitals under NFPA 99.

2. The generator set must be installed properly in an atmosphere that allows it to achieve the required life.

- *Air flow.* Provide adequate clean, cool air for cooling and combustion. High engine room temperatures may require ducting cooler outside air to the engine intake to avoid power derating. Restriction of radiator air reduces its cooling capability.
- *Exhaust.* Isolate exhaust piping from the engine with flexible connections. Wrap the piping with a thermal blanket to keep exhaust heat out of the engine room. The exhaust stack and muffler need to be sized so that the exhaust backpressure at the turbocharger outlet does not exceed 6.7 kPa (27 inH_2O). Excessive backpressure raises exhaust temperatures and reduces engine life.
- *Fuel.* Use clean fuel. Fuel day tanks should be below the level of the injectors.
- *Mounting.* The generator sets must have a flat and secure mounting surface. The generator-set mounting must allow adequate space around the generator set for maintenance and repairs.
- *Starting.* Batteries should be close to the starter and protected from very cold temperatures. Do not disconnect batteries from a running engine or a plugged-in battery charger.

3. SCR loads can affect generator output waveform. Make sure the SCR supplier is aware of the possible problems.

Every generator-set installation is unique and requires careful consideration of the particular application and site-specific conditions. It is therefore best to determine the foundation, ventilation, exhaust, fuel, vibration isolation, and other requirements in conjunction with the generator-set manufacturer(s) for the specific application and site conditions.

14.3.7 Illustration Showing a Typical Emergency Standby Generator Installation (*see page 14.26*)

14.3.7 Illustration showing a typical emergency standby generator installation.

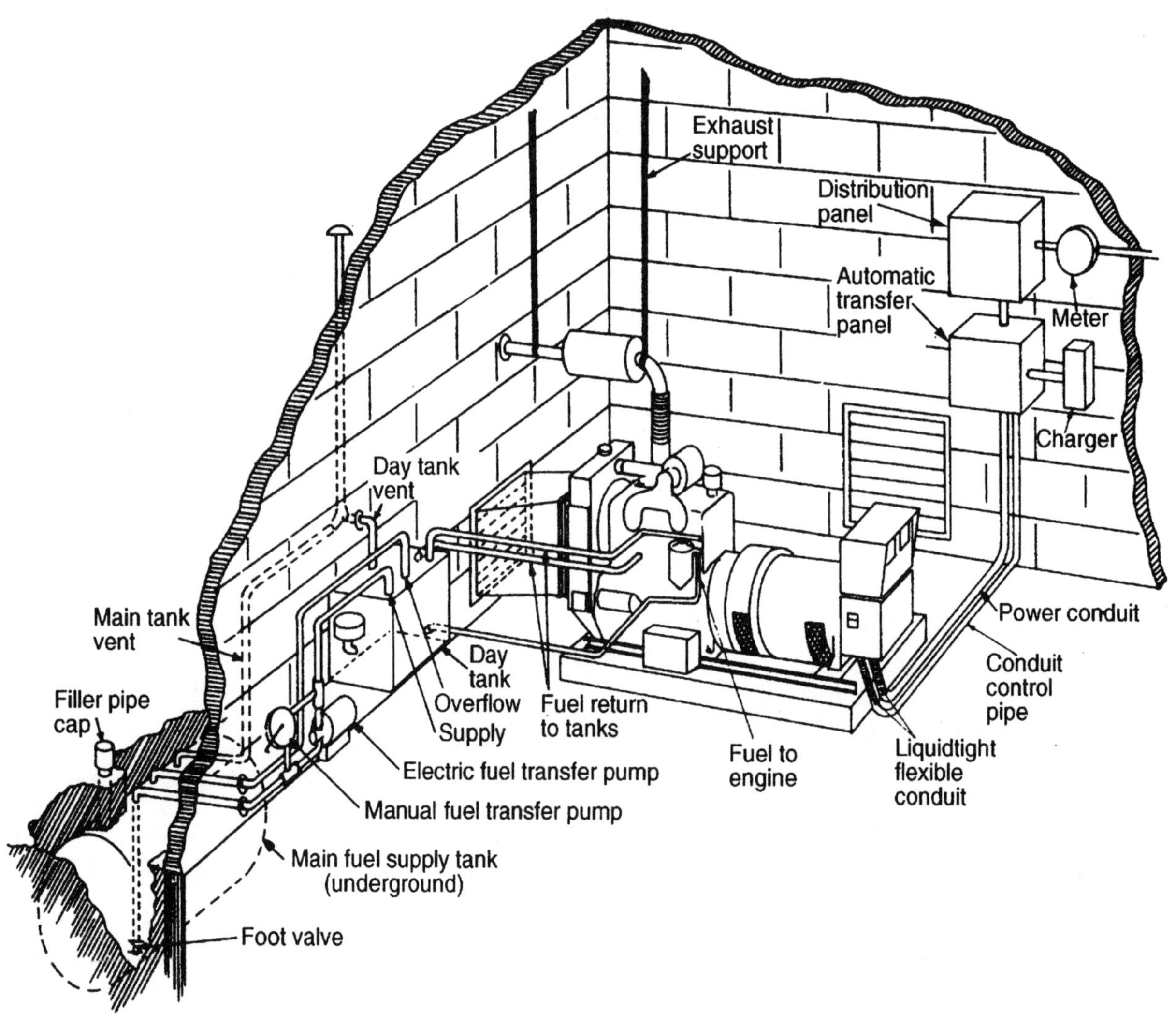

14.4.0 Uninterruptible Power Supply (UPS) Systems: Introduction

A UPS is a device or system that provides quality and continuity of an ac power source. Every UPS should maintain some specified degree of continuity of load for a specified stored energy time on ac input failure (see NEMA PE1-1990, "Uninterruptible Power Systems"). The term *UPS* commonly includes equipment, backup power source(s), environmental equipment (enclosure, heating and ventilating equipment), switchgear, and controls, which together provide a reliable continuous-quality electric power system.

The following definitions are given for clarification:

1. *Critical load.* That part of the load which requires continuous-quality electric power for its successful operation.

2. *Uninterruptible power supply (UPS) system.* Consists of one or more UPS modules and energy-storage battery (per module or common battery) and accessories (as required) to provide a reliable and high-quality power supply. The UPS isolates the load from the primary and emergency sources and, in the event of a power interruption, provides regulated power to the critical load for a specified period depending on the battery capacity. (The battery is normally sized to provide a capacity of 15 minutes when operating at full load.)
3. *UPS module.* The power-conversion portion of the UPS system. A UPS module may be made entirely of solid-state electronic construction or a hybrid combining rotary equipment (motor-generator) and solid-state electronic equipment. A solid-state electronic UPS consists of a rectifier, an inverter, and associated controls along with synchronizing, protective, and auxiliary devices. UPS modules may be designed to operate either individually or in parallel. A rotary UPS consists of a pony motor, a motor-generator, or alternatively, a synchronous machine in which the synchronous motor and generator have been combined into a single unit. This comprises a stator whose slots carry alternate motor and generator windings and a rotor with dc excitation, a rectifier, an inverter, a solid-state transfer switch, and associated controls along with synchronizing, protective, and auxiliary devices.

14.4.1 Nonredundant UPS System Configuration (*see page 14.28*)

This system consists of one or more UPS modules operating in parallel with a bypass circuit transfer switch and a battery. The rating and number of UPS modules are chosen to supply the critical load with no intentional excess capacity. On the failure of any UPS module, the bypass circuit automatically transfers the critical load to the bypass source without an interruption. The solid-state electronic UPS configuration relies on a static transfer switch for transfer within 4.17 ms. The rotary UPS configuration relies on the stored energy of a flywheel to propel the generator and maintain normal voltage and frequency for the time that the electromechanical circuit breakers are transferring the critical load to the alternate source. All operational transfers are "make before break."

14.4.2 "Cold" Standby Redundant UPS System (*see page 14.29*)

This system consists of two independent, nonredundant modules with either individual module batteries or a common battery. One UPS module operates on the line, and the other UPS module is turned off. Should the operating UPS module fail, its static bypass circuit automatically will transfer the critical load to the bypass source without an interruption to the critical load. The second UPS module is then energized manually and placed on the bypass mode of operation. To transfer the critical load, external "make before break" nonautomatic circuit breakers are operated to place the load on the second UPS bypass circuit. Finally, the critical load is returned from the bypass to the second UPS module via the bypass transfer switch. The two UPS modules cannot operate in parallel; therefore, a safety interlock circuit should be provided to prevent this condition. This configuration is used rarely.

14.4.1 Nonredundant UPS system configuration.

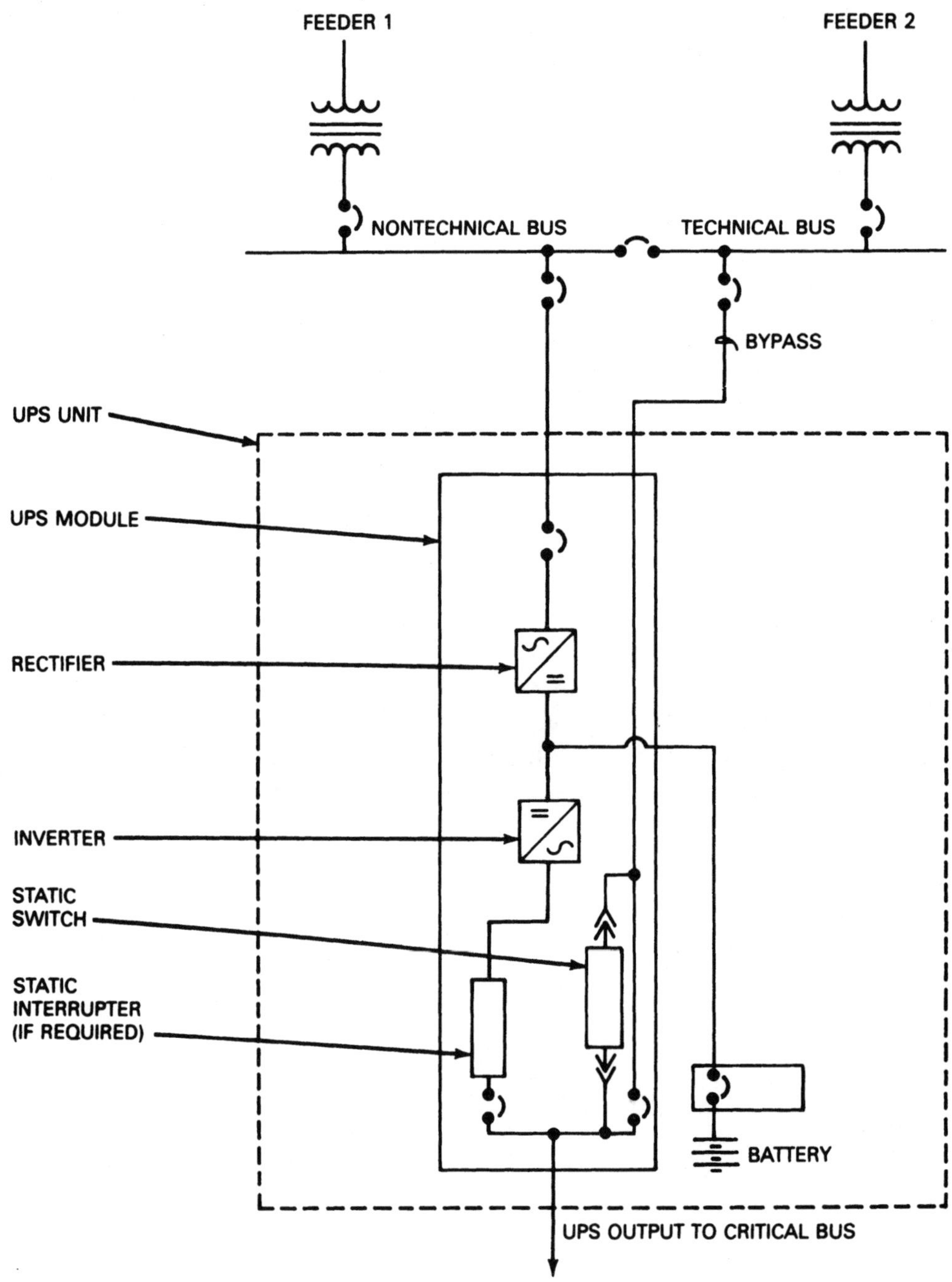

14.4.2 "Cold" Standby Redundant UPS System

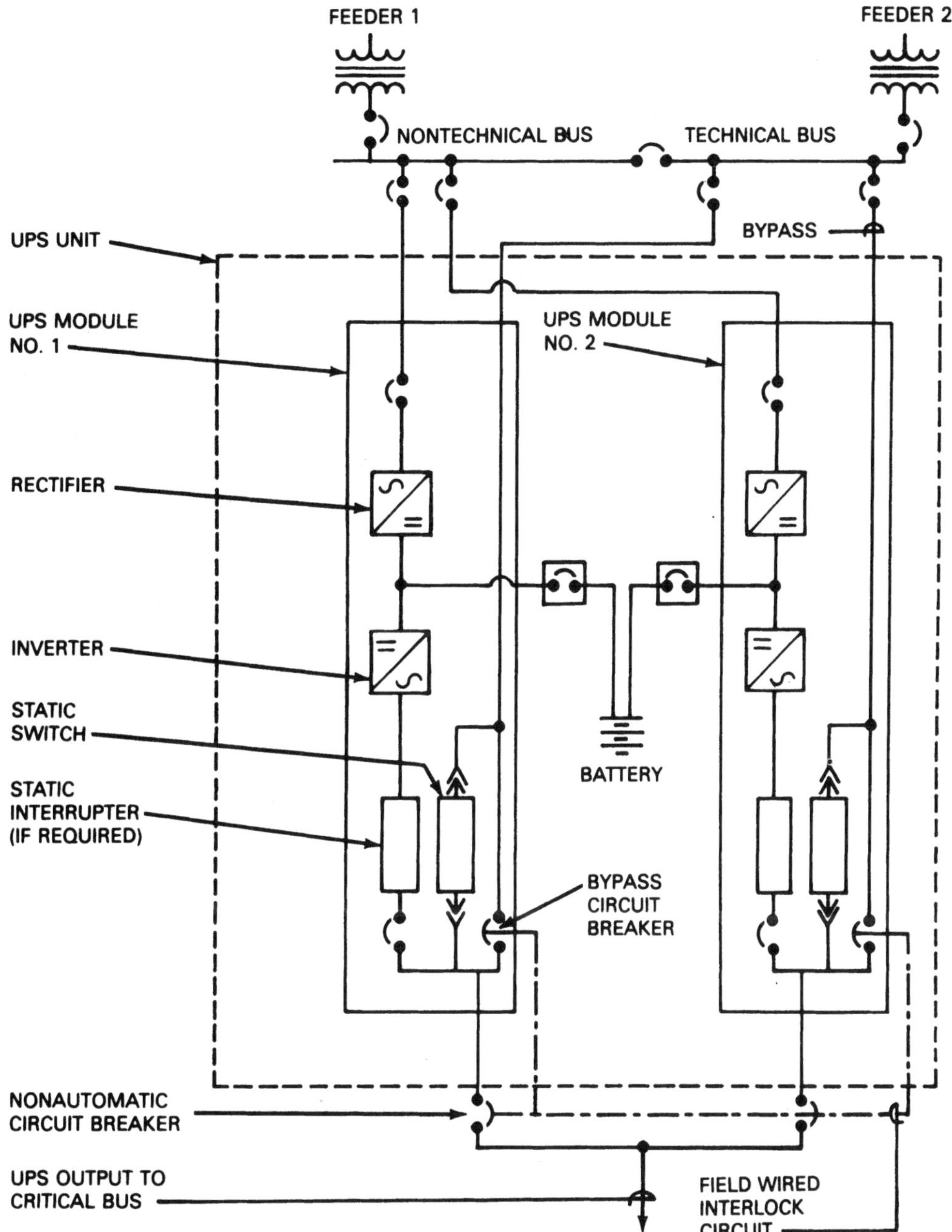

l Redundant UPS System

14.4.3

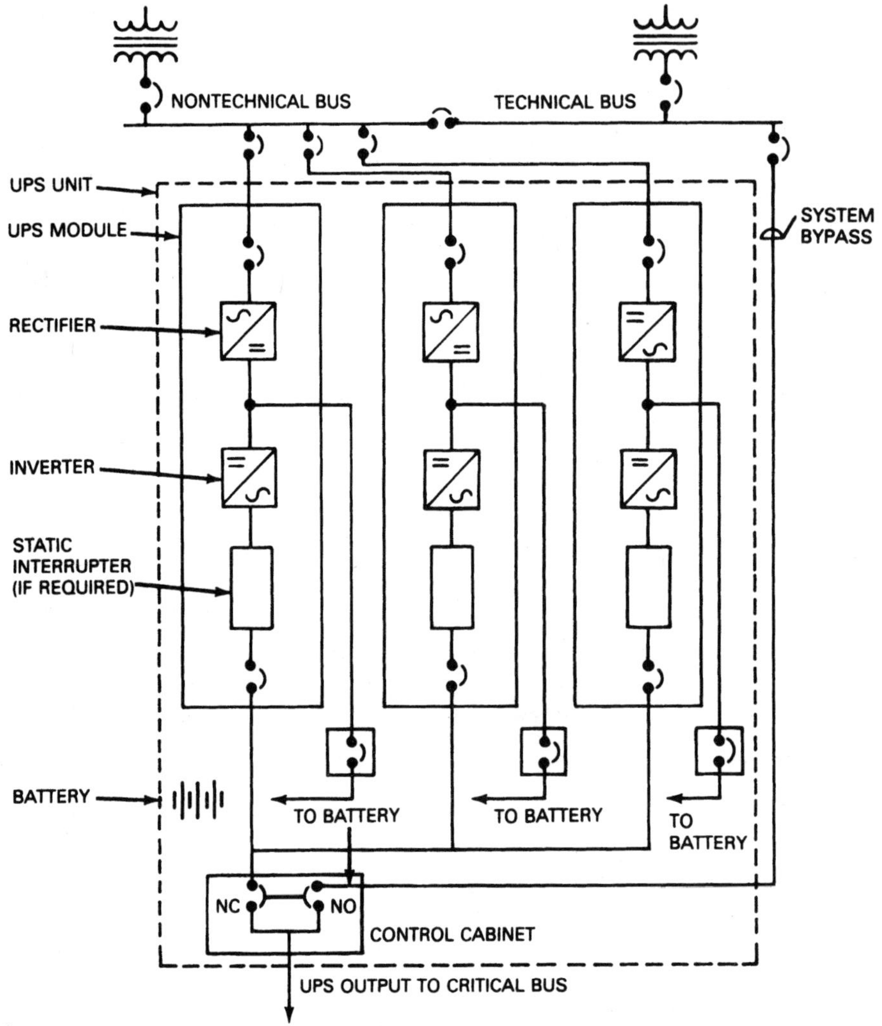

(From IEEE Std. 241-1990. Copyright 1990 IEEE. All rights reserved.)

This system consists of two or more UPS modules with static inverter turnoff(s), a system control cabinet, and either individual module batteries or a common battery. The UPS modules operate in parallel and normally share the load, and the system is capable of supplying the rated critical load on failure of any one UPS module. A static interrupter will disconnect the failed UPS module from the other UPS modules without an interruption to the critical load. A system bypass usually is included to permit system maintenance.

14.4.4 Isolated Redundant UPS System

This system uses a combination of automatic transfer switches and a reserve system to serve as the bypass source for any of the active systems (in this case, a system consists of a single module with its own system switchgear). The use of

14.4.4

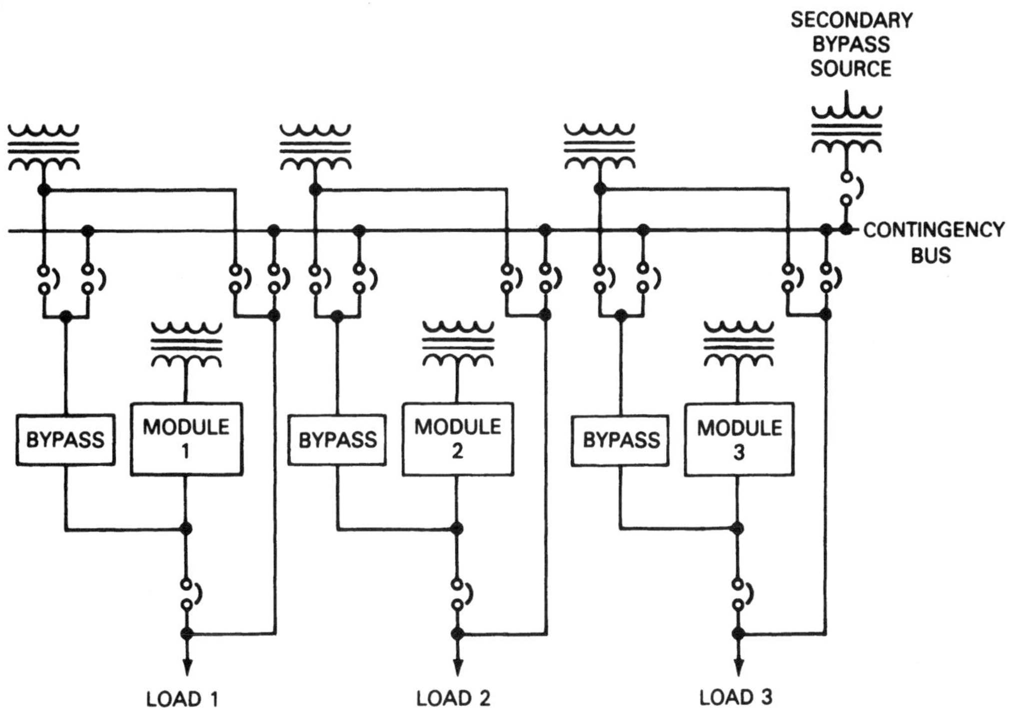

(From IEEE Std. 241-1990. Copyright 1990 IEEE. All rights reserved.)

this configuration requires each active system to serve an isolated/independent load. The advantage of this type of configuration is minimization of single-point failure modes; i.e., systems do not communicate via logic connections with each other (the systems operate independently of one another). The disadvantage of this type of system is that each system requires its own separate feeder to its dedicated load.

14.4.5 Application of UPS

1. The nonredundant UPS may be satisfactory for many critical-load applications.
2. Parallel redundant UPS system. When the criticality of the load demands the greatest protection and the load cannot be divided into suitable blocks, then installation of a parallel redundant UPS system is justified.

14.5.0 Power-System Configuration for 60-Hz Distribution

In 60-Hz power-distribution systems, the following basic concepts are used.

14.5.1 Single-Module UPS System (*see page 14.32*)

This is a single unit that is capable of supplying power to the total load. In the event of an overload, or if the unit fails, the critical-load bus is transferred to the bypass source via the bypass transfer switch. Transfer is uninterrupted.

14.5.1

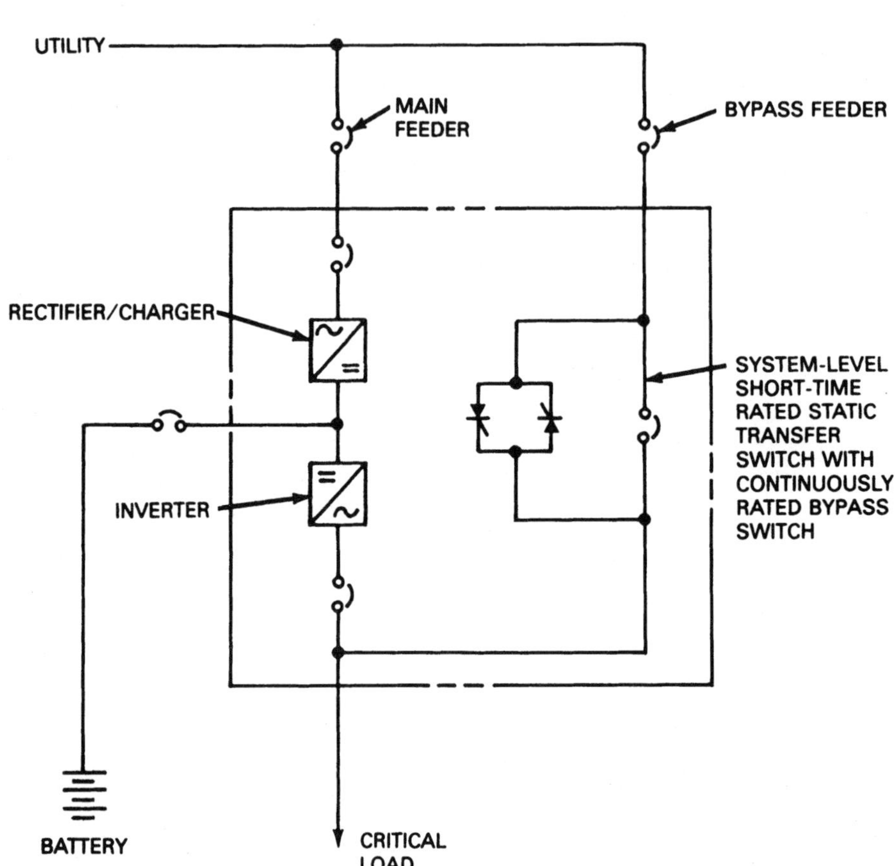

14.5.2 Parallel-Capacity UPS System (*see page 14.33)*

This system consists of two or more units capable of supplying power to the total load. In the event of overload, or if either unit fails, the critical-load bus is transferred to the bypass source via the bypass transfer switch. Transfer is uninterrupted. The battery may be common or separate.

14.5.3 Parallel Redundant UPS System (*see page 14.33)*

This system consists of two or more units with more capacity than is required by the total load. If any unit fails, the remaining units should be capable of carrying the total load. If more than one unit fails, the critical-load bus will be transferred to the bypass source via the bypass transfer switch. The battery may be common or separate per module.

14.5.4 Dual Redundant UPS System (*see page 14.34)*

In this system, one UPS module is standing by, running unloaded. If the loaded module fails, the load is transferred to the standby module. Each rating is limited to the size of the largest available module.

14.5.2 Parallel capacity UPS system.

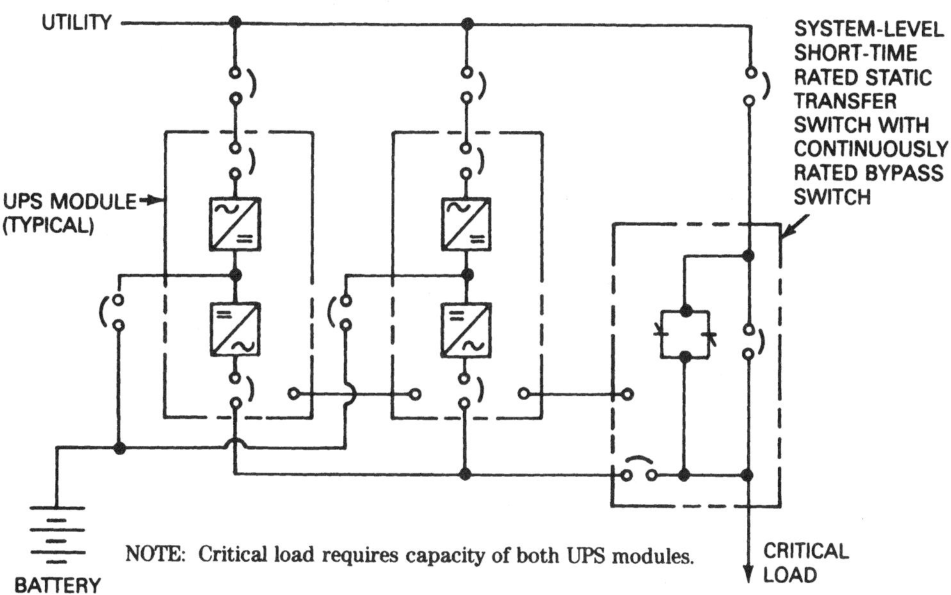

14.5.3 Parallel redundant UPS system.

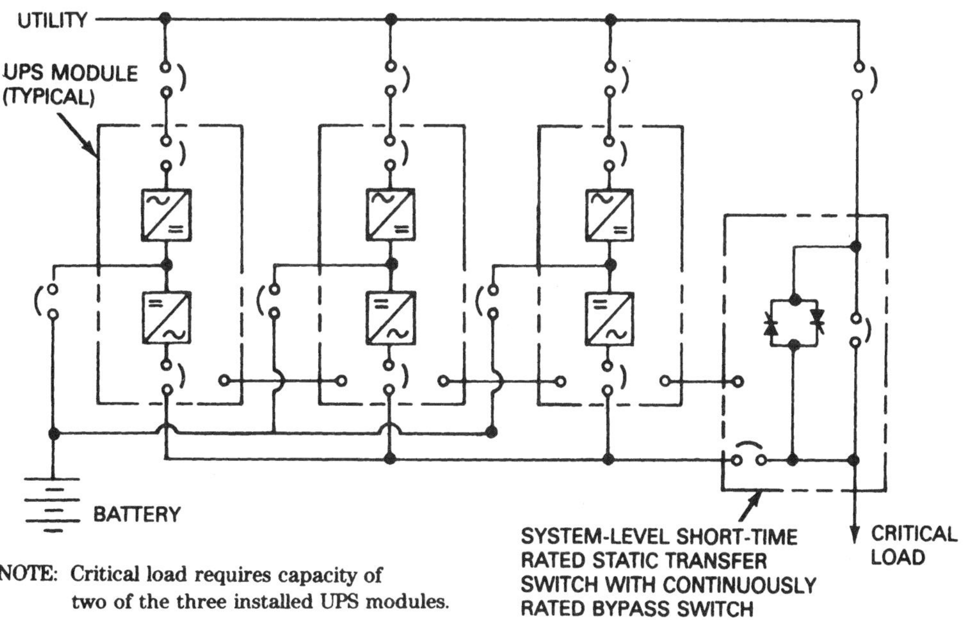

14.5.4 Dual redundant UPS system.

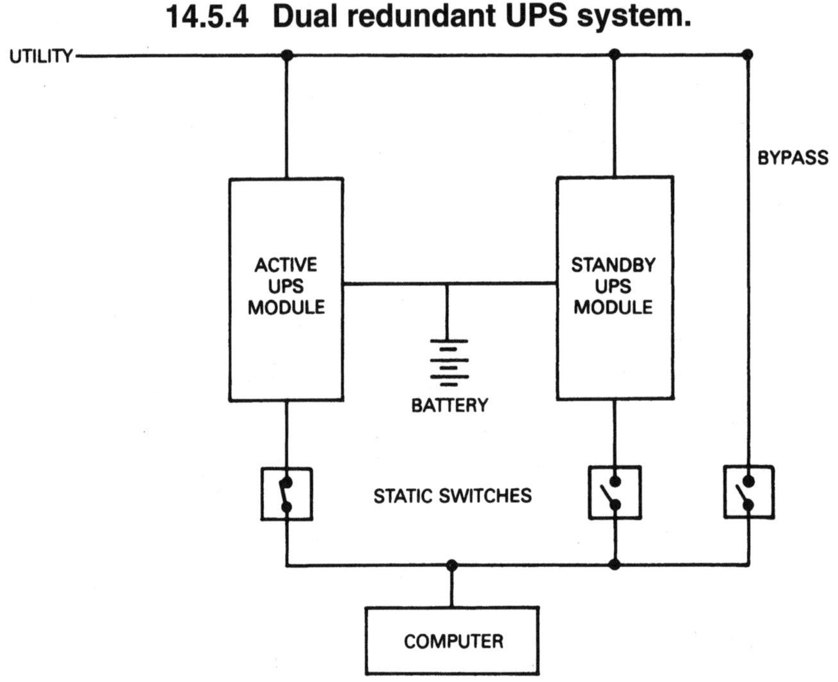

NOTE: Each module is capable of supplying the load.

14.5.5 Isolated Redundant UPS System

14.5.5

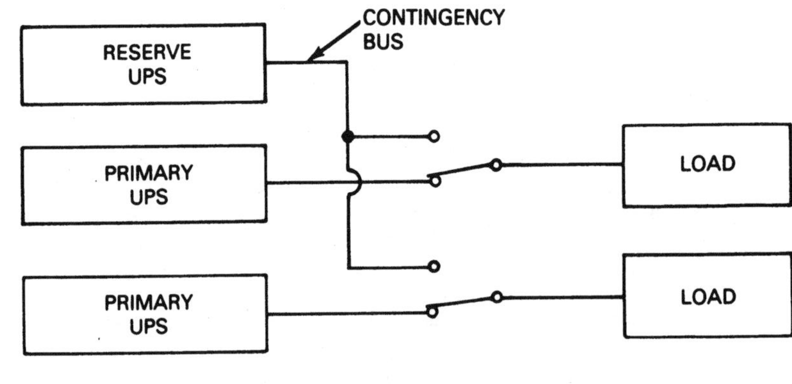

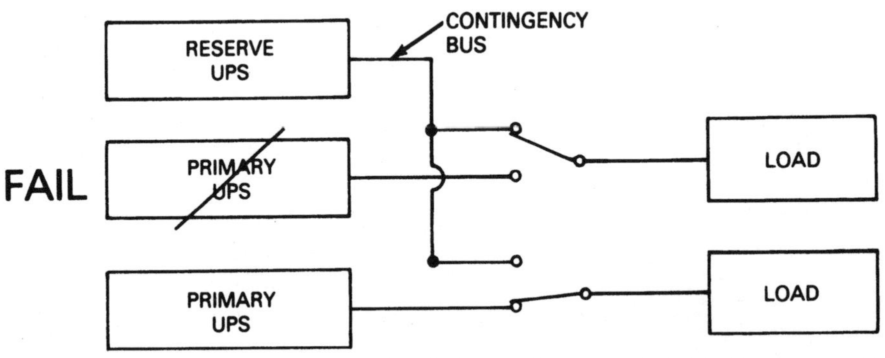

In this system, multiple UPS modules, usually three, are individually supplied from transformer sources. Each UPS module supplies a critical load and is available to supply a common contingency bus. The common contingency bus supplies the bypass circuit for each UPS module. In addition to being supplied from the common contingency bus, the bypass switch of each module is supplied from an individual transformer source. Furthermore, the common contingency bus is also supplied from a separate standby transformer called a *secondary bypass source.* The arrangement includes one UPS module in reserve as a "hot" standby. When a primary UPS module fails, the reserve UPS module is transferred to the load.

14.5.6 Parallel Tandem UPS System

14.5.6

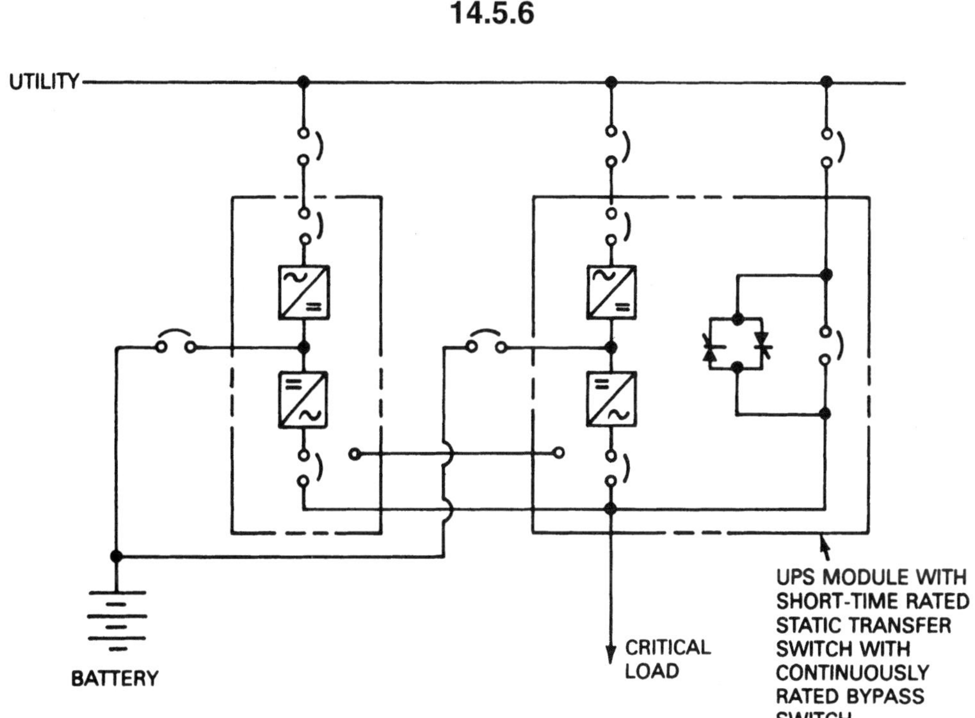

(From IEEE Std. 241-1990. Copyright 1990 IEEE. All rights reserved.)

The tandem configuration is a special case of two modules in parallel redundancy. In this arrangement, both modules have rectifier/chargers, dc links, and inverters; also, one of the modules houses the system-level static transfer switch. Either module can support full system load while the other has scheduled or corrective maintenance performed.

14.5.7 Hot Tied-Bus UPS System (*see page 14.36)*

The UPS tied-bus arrangement consists of two individual UPS systems (single module, parallel capacity, or redundant), with each one supplying a critical-load bus. The two critical-load buses can be paralleled via a tie breaker (normally open) while remaining on inverter power, which allows greater user flexibility for scheduled maintenance or damage control due to various failures.

14.5.7 Hot tied-bus UPS system.

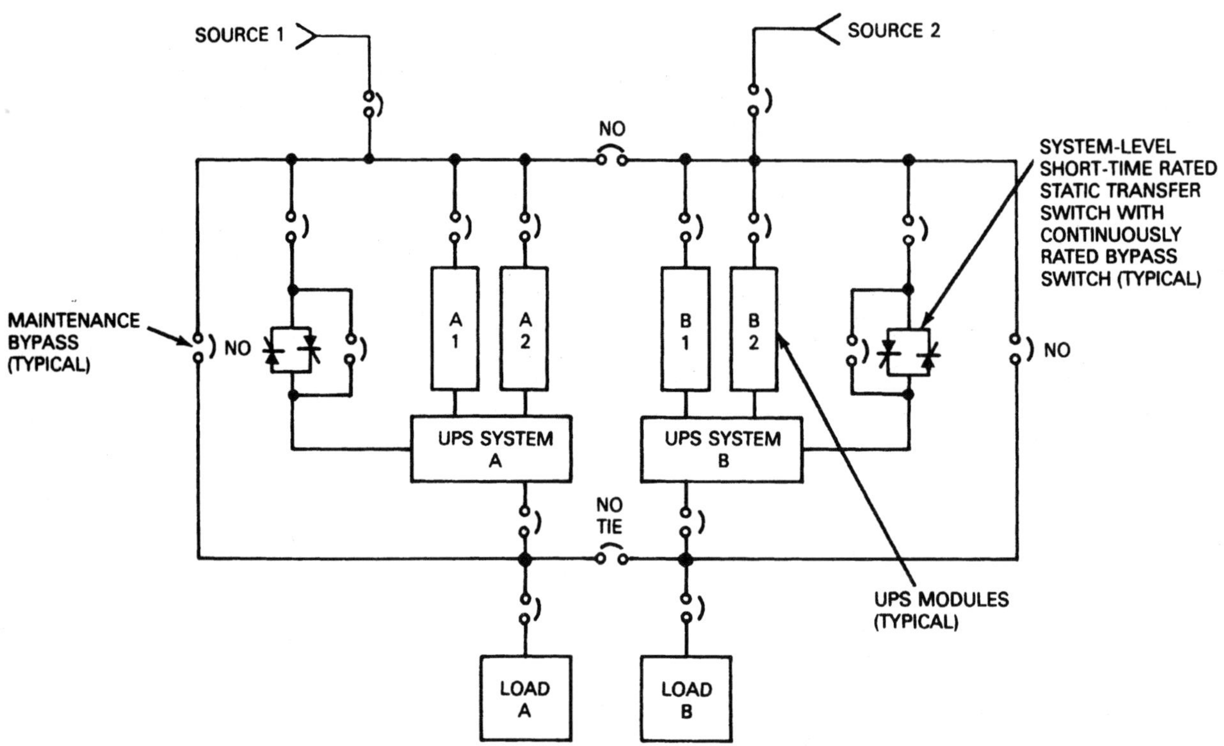

14.5.8 Superredundant Parallel System: Hot Tied-Bus UPS System

14.5.8

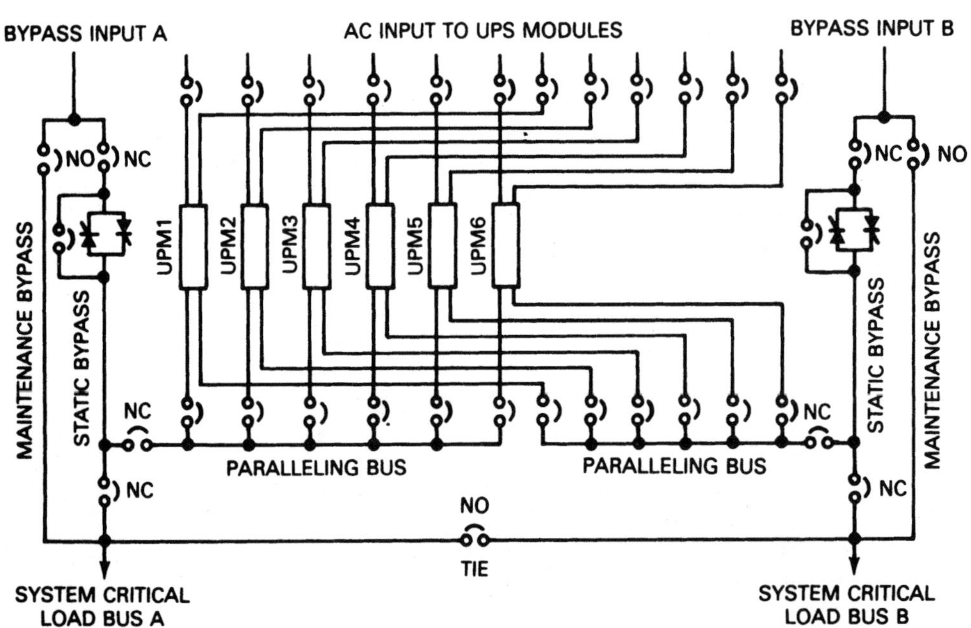

The superredundant UPS arrangement consists of n UPS modules (limited by a 4000-A bus). Each UPS module is supplied from dual sources (either/or) to supply two critical "paralleling buses." Each paralleling bus is connected via a circuit breaker to a "common bus" in parallel with the output feeder of one of the system static bypass switches. This junction is connected via a breaker to a "system critical-load bus." A tie enables the two system critical-load buses to be paralleled. Bypass sources for each system supply their own respective static bypass switches and maintenance bypasses. The superredundant UPS arrangement normally operates with the tie breaker open between the two system critical-load buses. When all UPS modules are supplying one paralleling bus, then the tie breaker is closed. All operations are preselected, automatic, and allow the user to do module- and system-level reconfiguration without submitting either critical load to utility power.

14.5.9 Uninterruptible Power with Dual Utility Sources and Static Transfer Switches *(see page 14.38)*

Essentially, uninterruptible electric power to the critical load may be achieved by the installation of dual utility sources, preferably from two separate substations, supplying secondary buses via step-down transformers as required. Feeders from each of two source buses are connected to static transfer switches as source 1 and source 2. A feeder from the load connection of the static transfer switch supplies a power-line conditioner, if needed. The power-line conditioner filters transients and provides voltage regulation. Filtered and regulated power is then supplied from the power-line conditioner to the critical-load distribution switchgear. This system eliminates the need for energy-storage batteries, emergency generators, and other equipment. The reliability of this system depends on the two utility sources and power conditioners.

14.5.10 Power-System Configuration with 60-Hz UPS

1. *Electric service and bypass connectors.* Two separate electric sources, one to the UPS rectifier circuit and the other to the UPS bypass circuit, should be provided. When possible, they should emanate from two separate buses, with the UPS bypass connected to the noncyclic load bus (also called the *technical bus*). This connection provides for the isolation of sensitive technical loads from the effects of UPS rectifier harmonic distortion and motor startup current inrush.
2. *Maintenance bypass provisions.* To provide for the maintenance of equipment, bypass provisions are necessary to isolate each UPS module or system.

14.5.11 UPS Distribution Systems

The UPS serves critical loads only. Noncritical loads are served by separate distribution systems that are supplied from either the noncyclic load bus (technical bus) or the cyclic load bus (nontechnical bus) as appropriate.

1. *Critical-load protection.* Critical-load overcurrent devices equipped with fast-acting fuses to shorten the transient effects of undervoltage caused by short circuits will result in a reliable system. Solid-state transient suppression (metal-oxide type) also should be supplied to lessen the overvoltage transients caused by reactive load switching.

14.5.9 Uninterruptible power with dual utility sources and static transfer switches.

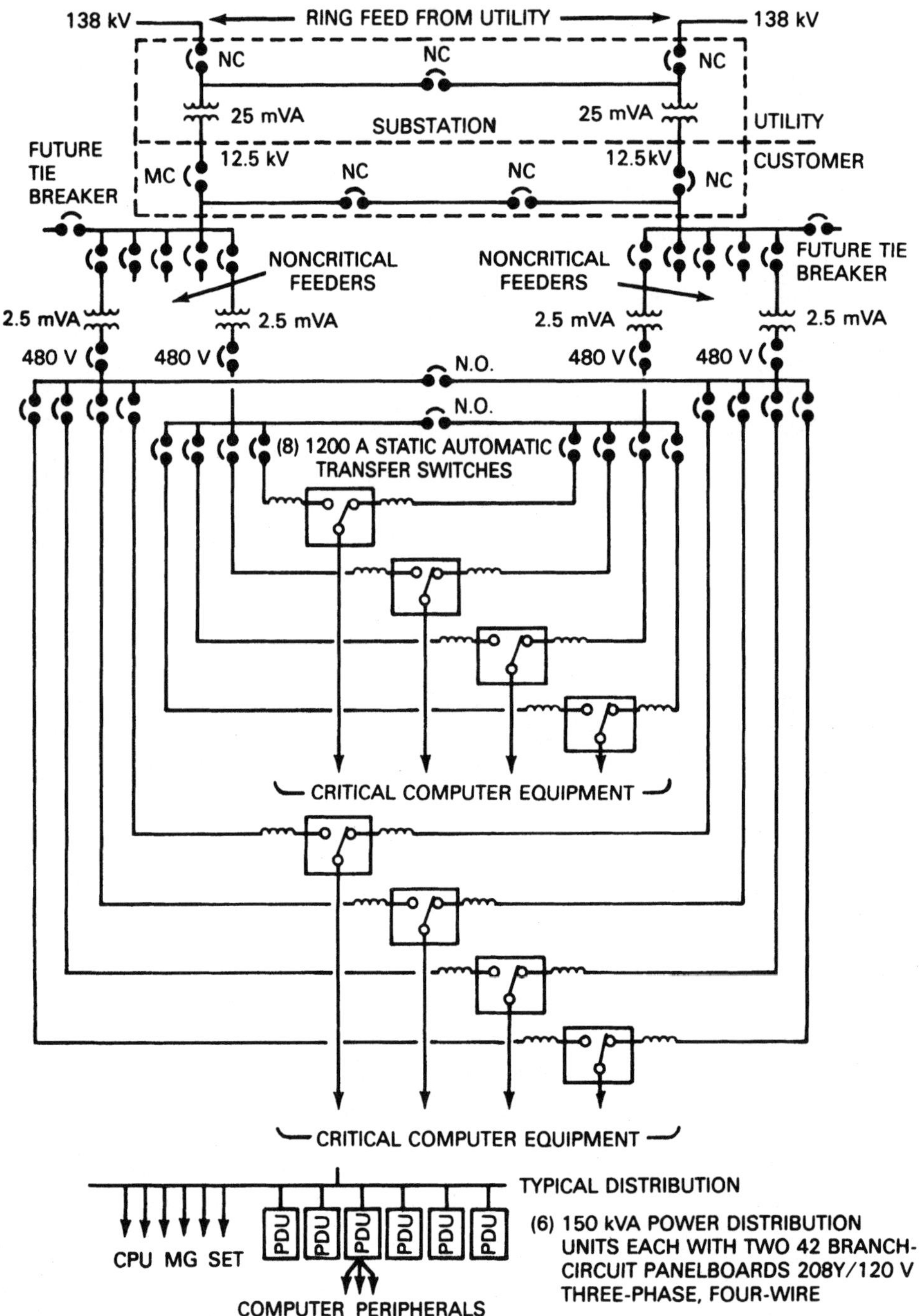

2. *Critical motor loads.* Due to the energy losses and starting current inrush inherent in motors, the connection of motors to the UPS bus should be limited to frequency-conversion applications, i.e., motor-generator sets. Generally, due to the current inrush, motor-generator sets are started on the UPS bypass circuit. Motor-generator sets may be started on the rectifier/inverter mode of operation under the following conditions:

   a. When the rating of the motor-generator set is less than 10 percent of the UPS rating.

b. When reduced-voltage and peak-current starters, such as the wye-delta closed transition type, are used for each motor load.
c. When more than one motor-generator set is connected to the critical bus, each set should be energized sequentially rather than simultaneously.

Refer all applications requiring connection of induction and synchronous motor loads to the UPS manufacturer. Application rules differ depending on the design and rating of the UPS.

14.6.1 Power-System Configuration for 400-Hz Distribution

In 400-Hz power distribution systems, the following basic concepts are used:

1. *Direct-utility supply to dual rotary-frequency converters parallel at the output critical-load bus.* Each frequency converter is sized for 100 percent load, or the arrangement has redundant capacity. The frequency converters may be equipped with an inverter/charger and battery on utility failure. Transfer from the utility line to the inverter occurs by synchronizing the inverter to the residual voltage of the motor.
2. *Dual-utility supply.* Dual-utility feeders supply an automatic transfer switch. The automatic transfer switch supplies multiple rotary-frequency converters (flywheel-equipped). The frequency converters are parallel at the critical-load bus. Transfer from one utility line to another occurs within the ride-through capability of the rotary-frequency converters.
3. *UPS.* A static or rotary UPS supplies multiple-frequency converters and other 60-Hz loads.
4. *UPS with local generation backup.* Both the utility feeder (connected to the normal terminals) and the feeder from the backup generator (connected to the emergency terminals) supply the automatic transfer switch. The automatic transfer switch in turn supplies the UPS. Critical-load distribution is as described above.
5. *Parallel 400-Hz single-CPU configuration.* Two or more 60- to 400-Hz frequency converters normally are connected in a redundant configuration to supply the critical load. There is no static switch or bypass breaker. Note that, on static converters, it is possible to use a 400-Hz motor-generator as a bypass source.
6. *Common UPS for single mainframe computer site.* Two 60- to 400-Hz frequency converters normally are connected in a redundant configuration supplying the mainframe computer, whereas a 60-Hz UPS supplies the peripherals.
7. *Alternative combination UPS for single mainframe computer site.* A 60-Hz UPS supplies a critical-load bus that, in turn, supplies the peripherals plus the input to a motor-generator set frequency converter (60 to 400 Hz).
8. *Combination UPS for multiple mainframe computer site.* A utility source supplies a redundant 400-Hz UPS system. This paralleled system supplies a 400-Hz critical-load distribution bus. Feeders from the 400-Hz distribution bus, equipped with line-drop compensators (LDCs) to reactive voltage drop, supply computer mainframes. A utility source also supplies a parallel redundant 60-Hz UPS system. This system supplies the critical peripheral load.
9. *Remote redundant 400-Hz UPS.* A 60-Hz UPS and a downstream parallel redundant 400-Hz motor-generator frequency conversion system with paralleling and distribution switchgear and line-drop compensators, which are all installed in the facility power equipment room with 60- and 400-Hz feeders distributed into the computer room.

10. *Point-of-use redundant 400-Hz UPS.* A 60-Hz UPS and a parallel redundant frequency conversion system as in item 9, except that the motor-generators are equipped with silencing enclosures and are installed in the computer room near the mainframes.
11. *Point-of-use 400-Hz UPS.* A 60-Hz UPS and a nonparalleled point-of-use static or rotary 400-Hz frequency converter installed in the computer room adjacent to each mainframe.
12. *Remote 400-Hz UPS.* A 60-Hz UPS and a separate parallel redundant 400-Hz UPS installed in the power equipment room, which is similar to item 8.
13. *Wiring.* For 400-Hz circuits, the reactance of circuit conductors may produce unacceptable voltage drops. Multiple-conductor cables and use of conductors in parallel, if necessary, should be installed in accordance with the *NEC,* Article 310-4. Also, use of nonmagnetic conduit will help in reducing voltage drop.

It should be noted that 400-Hz (actually 415-Hz) mainframe computers are rarely used today. Most mainframe computers are now 60 Hz.

Section

15

NEC Chapter 9 Tables and Annexes B and C

15.1.1 *NEC* Chapter 9, Table 1, Percent of Cross Section of Conduit and Tubing for Conductors

15.1.2 *NEC* Chapter 9, Table 4, Dimensions and Percent Area of Conduit and Tubing (Areas of Conduit or Tubing for the Combinations of Wires Permitted in Table 1, Chapter 9)

15.1.3 *NEC* Chapter 9, Table 5, Dimensions of Insulated Conductors and Fixture Wires

15.1.4 *NEC* Chapter 9, Table 5A, Compact Aluminum Building Wire Nominal Dimensions and Areas

15.1.5 *NEC* Chapter 9, Table 8, Conductor Properties

15.1.6 *NEC* Chapter 9, Table 9, Alternating-Current Resistance and Reactance for 600-V Cables, Three-Phase, 60-Hz, 75°C (167°F): Three Single Conductors in Conduit

15.1.7 *NEC* Chapter 9, Tables 11(A) and 11(B), Class 2 and Class 3, Alternating-Current and Direct-Current Power-Source Limitations, Respectively

15.1.8 *NEC* Chapter 9, Tables 12(A) and 12(B), PLFA Alternating-Current and Direct-Current Power-Source Limitations, Respectively

15.2.1 *NEC* (Annex B), Table B.310.1, Ampacities of Two or Three Insulated Conductors, Rated 0 through 2000 V, Within an Overall Covering (Multiconductor Cable), in Raceway in Free Air

15.2.2 *NEC* (Annex B), Table B.310.3, Ampacities of Multiconductor Cables with Not More Than Three Insulated Conductors, Rated 0 through 2000 V, in Free Air (for Type TC, MC, MI, UF, and USE Cables)

15.2.3 *NEC* (Annex B), Table B.310.5, Ampacities of Single-Insulated Conductors, Rated 0 through 2000 V, in Nonmagnetic Underground Electrical Ducts (One Conductor per Electrical Duct)

15.2.4 *NEC* (Annex B), Table B.310.6, Ampacities of Three Insulated Conductors, Rated 0 through 2000 V, Within an Overall Covering (Three-Conductor Cable) in Underground Electrical Ducts (One Cable per Electrical Duct)

15.2.5 *NEC* (Annex B), Table B.310.7, Ampacities of Three Single-Insulated Conductors, Rated 0 through 2000 V, in Underground Electrical Ducts (Three Conductors per Electrical Duct)

15.2.6 *NEC* (Annex B), Table B.310.8, Ampacities of Two or Three Insulated Conductors, Rated 0 through 2000 V, Cabled Within an Overall (Two- or Three-Conductor) Covering, Directly Buried in Earth

15.2.7 *NEC* (Annex B), Table B.310.9, Ampacities of Three Triplexed Single-Insulated Conductors, Rated 0 through 2000 V, Directly Buried in Earth

15.2.8 *NEC* (Annex B), Table B.310.10, Ampacities of Three Single-Insulated Conductors, Rated 0 through 2000 V, Directly Buried in Earth

15.3.1 *NEC* (Annex B), Figure B.310.1, Interpolation Chart for Cables in a Duct Bank I_1 = Ampacity for Rho = 60, 50 LF (Load Factor); I_2 = Ampacity for Rho = 120, 100 LF; Desired Ampacity = $F \times I_1$
15.3.2 *NEC* (Annex B), Figure B.310.2, Cable Installation Dimensions for Use with *NEC* Tables B.310.5 through B.310.10
15.3.3 *NEC* (Annex B), Figure B.310.3, Ampacities of Single-Insulated Conductors, Rated 0 through 5000 V, in Underground Electrical Ducts (Three Conductors per Electrical Duct), Nine Single-Conductor Cables per Phase
15.3.4 *NEC* (Annex B), Figure B.310.4, Ampacities of Single-Insulated Conductors, Rated 0 through 5000 V, in Nonmagnetic Underground Electrical Ducts (One Conductor per Electrical Duct), Four Single-Conductor Cables per Phase
15.3.5 *NEC* (Annex B), Figure B.310.5, Ampacities of Single-Insulated Conductors, Rated 0 through 5000 V, in Nonmagnetic Underground Electrical Ducts (One Conductor per Electrical Duct), Five Single-Conductor Cables per Phase
15.3.6 *NEC* (Annex B), Table B.310.11, Adjustment Factors for More Than Three Current-Carrying Conductors in a Raceway or Cable with Load Diversity
15.4.0 *NEC* Annex C, Conduit and Tube Fill Tables for Conductors and Fixture Wires of the Same Size
15.4.1 Table C1, Maximum Number of Conductors or Fixture Wires in Electrical Metallic Tubing
15.4.2 Table C1(A), Maximum Number of Compact Conductors in Electrical Metallic Tubing
15.4.3 Table C2, Maximum Number of Conductors or Fixture Wires in Electrical Nonmetallic Tubing
15.4.4 Table C2(A), Maximum Number of Compact Conductors in Electrical Nonmetallic Tubing
15.4.5 Table C3, Maximum Number of Conductors or Fixture Wires in Flexible Metal Conduit
15.4.6 Table C3(A), Maximum Number of Compact Conductors in Flexible Metal Conduit
15.4.7 Table C4, Maximum Number of Conductors or Fixture Wires in Intermediate Metal Conduit
15.4.8 Table C4(A), Maximum Number of Compact Conductors in Intermediate Metal Conduit
15.4.9 Table C5, Maximum Number of Conductors or Fixture Wires in Liquidtight Flexible Nonmetallic Conduit (Type LFNC-B)
15.4.10 Table C5(A), Maximum Number of Compact Conductors in Liquidtight Flexible Nonmetallic Conduit (Type LFNC-B)
15.4.11 Table C6, Maximum Number of Conductors or Fixture Wires in Liquidtight Flexible Nonmetallic Conduit (Type LFNC-A)
15.4.12 Table C6(A), Maximum Number of Compact Conductors in Liquidtight Flexible Nonmetallic Conduit (Type LFNC-A)
15.4.13 Table C7, Maximum Number of Conductors or Fixture Wires in Liquidtight Flexible Metal Conduit (LFMC)
15.4.14 Table C7(A), Maximum Number of Compact Conductors in Liquidtight Flexible Metal Conduit
15.4.15 Table C8, Maximum Number of Conductors or Fixture Wires in Rigid Metal Conduit
15.4.16 Table C8(A), Maximum Number of Compact Conductors in Rigid Metal Conduit
15.4.17 Table C9, Maximum Number of Conductors or Fixture Wires in Rigid PVC Conduit, Schedule 80
15.4.18 Table C9(A), Maximum Number of Compact Conductors in Rigid PVC Conduit, Schedule 80

15.4.19 Table C10, Maximum Number of Conductors or Fixture Wires in Rigid PVC Conduit, Schedule 40 and HDPE Conduit
15.4.20 Table C10(A), Maximum Number of Compact Conductors in Rigid PVC Conduit, Schedule 40 and HDPE Conduit
15.4.21 Table C11, Maximum Number of Conductors or Fixture Wires in Type A Rigid PVC Conduit
15.4.22 Table C11(A), Maximum Number of Compact Conductors in Type A Rigid PVC Conduit
15.4.23 Table C12, Maximum Number of Conductors in Type EB PVC Conduit
15.4.24 Table C12(A), Maximum Number of Compact Conductors in Type EB PVC Conduit

15.1.1 *NEC* Chapter 9, Table 1, Percent of Cross Section of Conduit and Tubing for Conductors

TABLE 15.1.1

Number of Conductors	All Conductor Types
1	53
2	31
Over 2	40

FPN No. 1: Table 1 is based on common conditions of proper cabling and alignment of conductors where the length of the pull and the number of bends are within reasonable limits. It should be recognized that, for certain conditions, a larger size conduit or a lesser conduit fill should be considered.

FPN No. 2: When pulling three conductors or cables into a raceway, if the ratio of the raceway (inside diameter) to the conductor or cable (outside diameter) is between 2.8 and 3.2, jamming can occur. While jamming can occur when pulling four or more conductors or cables into a raceway, the probability is very low.

15.1.2 *NEC* Chapter 9, Table 4, Dimensions and Percent Area of Conduit and Tubing (Areas of Conduit or Tubing for the Combinations of Wires Permitted in Table 1, Chapter 9)

1. See Annex C for the maximum number of conductors and fixture wires, all of the same size (total cross-sectional area including insulation) permitted in trade sizes of the applicable conduit or tubing.
2. Table 1 applies only to complete conduit or tubing systems and is not intended to apply to sections of conduit or tubing used to protect exposed wiring from physical damage.
3. Equipment grounding or bonding conductors, where installed, shall be included when calculating conduit or tubing fill. The actual dimensions of the equipment grounding or bonding conductor (insulated or bare) shall be used in the calculation.
4. Where conduit or tubing nipples having a maximum length not to exceed 600 mm (24 in.) are installed between boxes, cabinets, and similar enclosures, the nipples shall be permitted to be filled to 60 percent of their total cross-sectional area, and 310.15(B)(2)(a) adjustment factors need not apply to this condition.
5. For conductors not included in Chapter 9, such as multiconductor cables, the actual dimensions shall be used.
6. For combinations of conductors of different sizes, use Table 5 and Table 5A for dimensions of conductors and Table 4 for the applicable conduit or tubing dimensions.
7. When calculating the maximum number of conductors permitted in a conduit or tubing, all of the same size (total cross-sectional area including insulation), the next higher whole number shall be used to determine the maximum number of conductors permitted when the calculation results in a decimal of 0.8 or larger.
8. Where bare conductors are permitted by other sections of this *Code*, the dimensions for bare conductors in Table 8 shall be permitted.
9. A multiconductor cable of two or more conductors shall be treated as a single conductor for calculating percentage conduit fill area. For cables that have elliptical cross sections, the cross-sectional area calculation shall be based on using the major diameter of the ellipse as a circle diameter.

TABLE 15.1.2

Article 358 — Electrical Metallic Tubing (EMT)

Metric Designator	Trade Size	Nominal Internal Diameter		Total Area 100%		2 Wires 31%		Over 2 Wires 40%		1 Wire 53%		60%	
		mm	in.	mm²	in.²	mm²	in.²	mm²	in.²	mm²	in.²	mm²	in.²
16	½	15.8	0.622	196	0.304	61	0.094	78	0.122	104	0.161	118	0.182
21	¾	20.9	0.824	343	0.533	106	0.165	137	0.213	182	0.283	206	0.320
27	1	26.6	1.049	556	0.864	172	0.268	222	0.346	295	0.458	333	0.519
35	1¼	35.1	1.380	968	1.496	300	0.464	387	0.598	513	0.793	581	0.897
41	1½	40.9	1.610	1314	2.036	407	0.631	526	0.814	696	1.079	788	1.221
53	2	52.5	2.067	2165	3.356	671	1.040	866	1.342	1147	1.778	1299	2.013
63	2½	69.4	2.731	3783	5.858	1173	1.816	1513	2.343	2005	3.105	2270	3.515
78	3	85.2	3.356	5701	8.846	1767	2.742	2280	3.538	3022	4.688	3421	5.307
91	3½	97.4	3.834	7451	11.545	2310	3.579	2980	4.618	3949	6.119	4471	6.927
103	4	110.1	4.334	9521	14.753	2951	4.573	3808	5.901	5046	7.819	5712	8.852

(continued)

TABLE 15.1.2 Dimensions and Percent Area of Conduit and Tubing (Areas of Conduit or Tubing for the Combinations of Wires Permitted in Table 15.1.1) *(Continued)*

Article 362 — Electrical Nonmetallic Tubing (ENT)

Metric Designator	Trade Size	Nominal Internal Diameter mm	Nominal Internal Diameter in.	Total Area 100% mm²	Total Area 100% in.²	2 Wires 31% mm²	2 Wires 31% in.²	Over 2 Wires 40% mm²	Over 2 Wires 40% in.²	1 Wire 53% mm²	1 Wire 53% in.²	60% mm²	60% in.²
16	½	14.2	0.560	158	0.246	49	0.076	63	0.099	84	0.131	95	0.148
21	¾	19.3	0.760	293	0.454	91	0.141	117	0.181	155	0.240	176	0.272
27	1	25.4	1.000	507	0.785	157	0.243	203	0.314	269	0.416	304	0.471
35	1¼	34.0	1.340	908	1.410	281	0.437	363	0.564	481	0.747	545	0.846
41	1½	39.9	1.570	1250	1.936	388	0.600	500	0.774	663	1.026	750	1.162
53	2	51.3	2.020	2067	3.205	641	0.993	827	1.282	1095	1.699	1240	1.923
63	2½	—	—	—	—	—	—	—	—	—	—	—	—
78	3	—	—	—	—	—	—	—	—	—	—	—	—
91	3½	—	—	—	—	—	—	—	—	—	—	—	—

Article 348 — Flexible Metal Conduit (FMT)

Metric Designator	Trade Size	Nominal Internal Diameter mm	Nominal Internal Diameter in.	Total Area 100% mm²	Total Area 100% in.²	2 Wires 31% mm²	2 Wires 31% in.²	Over 2 Wires 40% mm²	Over 2 Wires 40% in.²	1 Wire 53% mm²	1 Wire 53% in.²	60% mm²	60% in.²
12	⅜	9.7	0.384	74	0.116	23	0.036	30	0.046	39	0.061	44	0.069
16	½	16.1	0.635	204	0.317	63	0.098	81	0.127	108	0.168	122	0.190
21	¾	20.9	0.824	343	0.533	106	0.165	137	0.213	182	0.283	206	0.320
27	1	25.9	1.020	527	0.817	163	0.253	211	0.327	279	0.433	316	0.490
35	1¼	32.4	1.275	824	1.277	256	0.396	330	0.511	437	0.677	495	0.766
41	1½	39.1	1.538	1201	1.858	372	0.576	480	0.743	636	0.985	720	1.115
53	2	51.8	2.040	2107	3.269	653	1.013	843	1.307	1117	1.732	1264	1.961
63	2½	63.5	2.500	3167	4.909	982	1.522	1267	1.963	1678	2.602	1900	2.945
78	3	76.2	3.000	4560	7.069	1414	2.191	1824	2.827	2417	3.746	2736	4.241
91	3½	88.9	3.500	6207	9.621	1924	2.983	2483	3.848	3290	5.099	3724	5.773
103	4	101.6	4.000	8107	12.566	2513	3.896	3243	5.027	4297	6.660	4864	7.540

Article 342 — Intermediate Metal Conduit (IMC)

Metric Designator	Trade Size	Nominal Internal Diameter mm	Nominal Internal Diameter in.	Total Area 100% mm²	Total Area 100% in.²	2 Wires 31% mm²	2 Wires 31% in.²	Over 2 Wires 40% mm²	Over 2 Wires 40% in.²	1 Wire 53% mm²	1 Wire 53% in.²	60% mm²	60% in.²
12	⅜	—	—	—	—	—	—	—	—	—	—	—	—
16	½	16.8	0.660	222	0.342	69	0.106	89	0.137	117	0.181	133	0.205
21	¾	21.9	0.864	377	0.586	117	0.182	151	0.235	200	0.311	226	0.352
27	1	28.1	1.105	620	0.959	192	0.297	248	0.384	329	0.508	372	0.575
35	1¼	36.8	1.448	1064	1.647	330	0.510	425	0.659	564	0.873	638	0.988
41	1½	42.7	1.683	1432	2.225	444	0.690	573	0.890	759	1.179	859	1.335
53	2	54.6	2.150	2341	3.630	726	1.125	937	1.452	1241	1.924	1405	2.178
63	2½	64.9	2.557	3308	5.135	1026	1.592	1323	2.054	1753	2.722	1985	3.081
78	3	80.7	3.176	5115	7.922	1586	2.456	2046	3.169	2711	4.199	3069	4.753
91	3½	93.2	3.671	6822	10.584	2115	3.281	2729	4.234	3616	5.610	4093	6.351
103	4	105.4	4.166	8725	13.631	2705	4.226	3490	5.452	4624	7.224	5235	8.179

(continued)

TABLE 15.1.2 Dimensions and Percent Area of Conduit and Tubing (Areas of Conduit or Tubing for the Combinations of Wires Permitted in Table 15.1.1) *(Continued)*

Article 356— Liquidtight Flexible Nonmetallic Conduit (LFNC-B*)

Metric Designator	Trade Size	Nominal Internal Diameter		Total Area 100%		2 Wires 31%		Over 2 Wires 40%		1 Wire 53%		60%	
		mm	in.	mm^2	in.^{2}	mm^2	in.^{2}	mm^2	in.^{2}	mm^2	in.^{2}	mm^2	in.^{2}
12	3/8	12.5	0.494	123	0.192	38	0.059	49	0.077	65	0.102	74	0.115
16	1/2	16.1	0.632	204	0.314	63	0.097	81	0.125	108	0.166	122	0.188
21	3/4	21.1	0.830	350	0.541	108	0.168	140	0.216	185	0.287	210	0.325
27	1	26.8	1.054	564	0.873	175	0.270	226	0.349	299	0.462	338	0.524
35	1 1/4	35.4	1.395	984	1.528	305	0.474	394	0.611	522	0.810	591	0.917
41	1 1/2	40.3	1.588	1276	1.981	395	0.614	510	0.792	676	1.050	765	1.188
53	2	51.6	2.033	2091	3.246	648	1.006	836	1.298	1108	1.720	1255	1.948

*Corresponds to 356.2(2)

Article 356 — Liquidtight Flexible Nonmetallic Conduit (LFNC-A*)

Metric Designator	Trade Size	Nominal Internal Diameter		Total Area 100%		2 Wires 31%		Over 2 Wires 40%		1 Wire 53%		60%	
		mm	in.	mm^2	in.^{2}	mm^2	in.^{2}	mm^2	in.^{2}	mm^2	in.^{2}	mm^2	in.^{2}
12	3/8	12.6	0.495	125	0.192	39	0.060	50	0.077	66	0.102	75	0.115
16	1/2	16.0	0.630	201	0.312	62	0.097	80	0.125	107	0.165	121	0.187
21	3/4	21.0	0.825	346	0.535	107	0.166	139	0.214	184	0.283	208	0.321
27	1	26.5	1.043	552	0.854	171	0.265	221	0.342	292	0.453	331	0.513
35	1 1/4	35.1	1.383	968	1.502	300	0.466	387	0.601	513	0.796	581	0.901
41	1 1/2	40.7	1.603	1301	2.018	403	0.626	520	0.807	690	1.070	781	1.211
53	2	52.4	2.063	2157	3.343	669	1.036	863	1.337	1143	1.772	1294	2.006

*Corresponds to 356.2(1)

Article 350 — Liquidtight Flexible Metal Conduit (LFMC)

Metric Designator	Trade Size	Nominal Internal Diameter		Total Area 100%		2 Wires 31%		Over 2 Wires 40%		1 Wire 53%		60%	
		mm	in.	mm^2	in.^{2}	mm^2	in.^{2}	mm^2	in.^{2}	mm^2	in.^{2}	mm^2	in.^{2}
12	3/8	12.5	0.494	123	0.192	38	0.059	49	0.077	65	0.102	74	0.115
16	1/2	16.1	0.632	204	0.314	63	0.097	81	0.125	108	0.166	122	0.188
21	3/4	21.1	0.830	350	0.541	108	0.168	140	0.216	185	0.287	210	0.325
27	1	26.8	1.054	564	0.873	175	0.270	226	0.349	299	0.462	338	0.524
35	1 1/4	35.4	1.395	984	1.528	305	0.474	394	0.611	522	0.810	591	0.917
41	1 1/2	40.3	1.588	1276	1.981	395	0.614	510	0.792	676	1.050	765	1.188
53	2	51.6	2.033	2091	3.246	648	1.006	836	1.298	1108	1.720	1255	1.948
63	2 1/2	63.3	2.493	3147	4.881	976	1.513	1259	1.953	1668	2.587	1888	2.929
78	3	78.4	3.085	4827	7.475	1497	2.317	1931	2.990	2559	3.962	2896	4.485
91	3 1/2	89.4	3.520	6277	9.731	1946	3.017	2511	3.893	3327	5.158	3766	5.839
103	4	102.1	4.020	8187	12.692	2538	3.935	3275	5.077	4339	6.727	4912	7.615
129	5	—	—	—	—	—	—	—	—	—	—	—	—
155	6	—	—	—	—	—	—	—	—	—	—	—	—

(continued)

TABLE 15.1.2 Dimensions and Percent Area of Conduit and Tubing (Areas of Conduit or Tubing for the Combinations of Wires Permitted in Table 15.1.1) *(Continued)*

Article 344 — Rigid Metal Conduit (RMC)

Metric Designator	Trade Size	Nominal Internal Diameter		Total Area 100%		2 Wires 31%		Over 2 Wires 40%		1 Wire 53%		60%	
		mm	in.	mm²	in.²	mm²	in.²	mm²	in.²	mm²	in.²	mm²	in.²
12	3/8	—	—	—	—	—	—	—	—	—	—	—	—
16	1/2	16.1	0.632	204	0.314	63	0.097	81	0.125	108	0.166	122	0.188
21	3/4	21.2	0.836	353	0.549	109	0.170	141	0.220	187	0.291	212	0.329
27	1	27.0	1.063	573	0.887	177	0.275	229	0.355	303	0.470	344	0.532
35	1¼	35.4	1.394	984	1.526	305	0.473	394	0.610	522	0.809	591	0.916
41	1½	41.2	1.624	1333	2.071	413	0.642	533	0.829	707	1.098	800	1.243
53	2	52.9	2.083	2198	3.408	681	1.056	879	1.363	1165	1.806	1319	2.045
63	2½	63.2	2.489	3137	4.866	972	1.508	1255	1.946	1663	2.579	1882	2.919
78	3	78.5	3.090	4840	7.499	1500	2.325	1936	3.000	2565	3.974	2904	4.499
91	3½	90.7	3.570	6461	10.010	2003	3.103	2584	4.004	3424	5.305	3877	6.006
103	4	102.9	4.050	8316	12.882	2578	3.994	3326	5.153	4408	6.828	4990	7.729
129	5	128.9	5.073	13050	20.212	4045	6.266	5220	8.085	6916	10.713	7830	12.127
155	6	154.8	6.093	18821	29.158	5834	9.039	7528	11.663	9975	15.454	11292	17.495

Article 352 — Rigid PVC Conduit (RNC), Schedule 80

Metric Designator	Trade Size	Nominal Internal Diameter		Total Area 100%		2 Wires 31%		Over 2 Wires 40%		1 Wire 53%		60%	
		mm	in.	mm²	in.²	mm²	in.²	mm²	in.²	mm²	in.²	mm²	in.²
12	3/8	—	—	—	—	—	—	—	—	—	—	—	—
16	1/2	13.4	0.526	141	0.217	44	0.067	56	0.087	75	0.115	85	0.130
21	3/4	18.3	0.722	263	0.409	82	0.127	105	0.164	139	0.217	158	0.246
27	1	23.8	0.936	445	0.688	138	0.213	178	0.275	236	0.365	267	0.413
35	1¼	31.9	1.255	799	1.237	248	0.383	320	0.495	424	0.656	480	0.742
41	1½	37.5	1.476	1104	1.711	342	0.530	442	0.684	585	0.907	663	1.027
53	2	48.6	1.913	1855	2.874	575	0.891	742	1.150	983	1.523	1113	1.725
63	2½	58.2	2.290	2660	4.119	825	1.277	1064	1.647	1410	2.183	1596	2.471
78	3	72.7	2.864	4151	6.442	1287	1.997	1660	2.577	2200	3.414	2491	3.865
91	3½	84.5	3.326	5608	8.688	1738	2.693	2243	3.475	2972	4.605	3365	5.213
103	4	96.2	3.786	7268	11.258	2253	3.490	2907	4.503	3852	5.967	4361	6.755
129	5	121.1	4.768	11518	17.855	3571	5.535	4607	7.142	6105	9.463	6911	10.713
155	6	145.0	5.709	16513	25.598	5119	7.935	6605	10.239	8752	13.567	9908	15.359

Article 352 — Rigid PVC Conduit (RNC), Schedule 40, and HDPE Conduit

Metric Designator	Trade Size	Nominal Internal Diameter		Total Area 100%		2 Wires 31%		Over 2 Wires 40%		1 Wire 53%		60%	
		mm	in.	mm²	in.²	mm²	in.²	mm²	in.²	mm²	in.²	mm²	in.²
12	3/8	—	—	—	—	—	—	—	—	—	—	—	—
16	1/2	15.3	0.602	184	0.285	57	0.088	74	0.114	97	0.151	110	0.171
21	3/4	20.4	0.804	327	0.508	101	0.157	131	0.203	173	0.269	196	0.305
27	1	26.1	1.029	535	0.832	166	0.258	214	0.333	284	0.441	321	0.499
35	1¼	34.5	1.360	935	1.453	290	0.450	374	0.581	495	0.770	561	0.872
41	1½	40.4	1.590	1282	1.986	397	0.616	513	0.794	679	1.052	769	1.191
53	2	52.0	2.047	2124	3.291	658	1.020	849	1.316	1126	1.744	1274	1.975

(continued)

TABLE 15.1.2 Dimensions and Percent Area of Conduit and Tubing (Areas of Conduit or Tubing for the Combinations of Wires Permitted in Table 15.1.1) *(Continued)*

Article 352 — Rigid PVC Conduit (RNC), Schedule 40, and HDPE Conduit

Metric Designator	Trade Size	Nominal Internal Diameter		Total Area 100%		2 Wires 31%		Over 2 Wires 40%		1 Wire 53%		60%	
		mm	in.	mm²	in.²	mm²	in.²	mm²	in.²	mm²	in.²	mm²	in.²
63	2½	62.1	2.445	3029	4.695	939	1.455	1212	1.878	1605	2.488	1817	2.817
78	3	77.3	3.042	4693	7.268	1455	2.253	1877	2.907	2487	3.852	2816	4.361
91	3½	89.4	3.521	6277	9.737	1946	3.018	2511	3.895	3327	5.161	3766	5.842
103	4	101.5	3.998	8091	12.554	2508	3.892	3237	5.022	4288	6.654	4855	7.532
129	5	127.4	5.016	12748	19.761	3952	6.126	5099	7.904	6756	10.473	7649	11.856
155	6	153.2	6.031	18433	28.567	5714	8.856	7373	11.427	9770	15.141	11060	17.140

Article 352 — Type A, Rigid PVC Conduit (RNC)

Metric Designator	Trade Size	Nominal Internal Diameter		Total Area 100%		2 Wires 31%		Over 2 Wires 40%		1 Wire 53%		60%	
		mm	in.	mm²	in.²	mm²	in.²	mm²	in.²	mm²	in.²	mm²	in.²
16	½	17.8	0.700	249	0.385	77	0.119	100	0.154	132	0.204	149	0.231
21	¾	23.1	0.910	419	0.650	130	0.202	168	0.260	222	0.345	251	0.390
27	1	29.8	1.175	697	1.084	216	0.336	279	0.434	370	0.575	418	0.651
35	1¼	38.1	1.500	1140	1.767	353	0.548	456	0.707	604	0.937	684	1.060
41	1½	43.7	1.720	1500	2.324	465	0.720	600	0.929	795	1.231	900	1.394
53	2	54.7	2.155	2350	3.647	728	1.131	940	1.459	1245	1.933	1410	2.188
63	2½	66.9	2.635	3515	5.453	1090	1.690	1406	2.181	1863	2.890	2109	3.272
78	3	82.0	3.230	5281	8.194	1637	2.540	2112	3.278	2799	4.343	3169	4.916
91	3½	93.7	3.690	6896	10.694	2138	3.315	2758	4.278	3655	5.668	4137	6.416
103	4	106.2	4.180	8858	13.723	2746	4.254	3543	5.489	4695	7.273	5315	8.234
129	5	—	—	—	—	—	—	—	—	—	—	—	—
155	6	—	—	—	—	—	—	—	—	—	—	—	—

Article 352 — Type EB, PVC Conduit (RNC)

Metric Designator	Trade Size	Nominal Internal Diameter		Total Area 100%		2 Wires 31%		Over 2 Wires 40%		1 Wire 53%		60%	
		mm	in.	mm²	in.²	mm²	in.²	mm²	in.²	mm²	in.²	mm²	in.²
16	½	—	—	—	—	—	—	—	—	—	—	—	—
21	¾	—	—	—	—	—	—	—	—	—	—	—	—
27	1	—	—	—	—	—	—	—	—	—	—	—	—
35	1¼	—	—	—	—	—	—	—	—	—	—	—	—
41	1½	—	—	—	—	—	—	—	—	—	—	—	—
53	2	56.4	2.221	2498	3.874	774	1.201	999	1.550	1324	2.053	1499	2.325
63	2½	—	—	—	—	—	—	—	—	—	—	—	—
78	3	84.6	3.330	5621	8.709	1743	2.700	2248	3.484	2979	4.616	3373	5.226
91	3½	96.6	3.804	7329	11.365	2272	3.523	2932	4.546	3884	6.023	4397	6.819
103	4	108.9	4.289	9314	14.448	2887	4.479	3726	5.779	4937	7.657	5589	8.669
129	5	135.0	5.316	14314	22.195	4437	6.881	5726	8.878	7586	11.763	8588	13.317
155	6	160.9	6.336	20333	31.530	6303	9.774	8133	12.612	10776	16.711	12200	18.918

15.1.3 *NEC* Chapter 9, Table 5, Dimensions of Insulated Conductors and Fixture Wires

TABLE 15.1.3

Type	Size (AWG or kcmil)	Approximate Diameter mm	Approximate Diameter in.	Approximate Area mm^2	Approximate Area $in.^2$
Type: FFH-2, RFH-1, RFH-2, RHH*, RHW*, RHW-2*, RHH, RHW, RHW-2, SF-1, SF-2, SFF-1, SFF-2, TF, TFF, THHW, THW, THW-2, TW, XF, XFF					
RFH-2, FFH-2	18	3.454	0.136	9.355	0.0145
	16	3.759	0.148	11.10	0.0172
RHW-2, RHH, RHW	14	4.902	0.193	18.90	0.0293
	12	5.385	0.212	22.77	0.0353
	10	5.994	0.236	28.19	0.0437
	8	8.280	0.326	53.87	0.0835
	6	9.246	0.364	67.16	0.1041
	4	10.46	0.412	86.00	0.1333
	3	11.18	0.440	98.13	0.1521
	2	11.99	0.472	112.9	0.1750
	1	14.78	0.582	171.6	0.2660
	1/0	15.80	0.622	196.1	0.3039
	2/0	16.97	0.668	226.1	0.3505
	3/0	18.29	0.720	262.7	0.4072
	4/0	19.76	0.778	306.7	0.4754
	250	22.73	0.895	405.9	0.6291
	300	24.13	0.950	457.3	0.7088
	350	25.43	1.001	507.7	0.7870
	400	26.62	1.048	556.5	0.8626
	500	28.78	1.133	650.5	1.0082
	600	31.57	1.243	782.9	1.2135
	700	33.38	1.314	874.9	1.3561
	750	34.24	1.348	920.8	1.4272
	800	35.05	1.380	965.0	1.4957
	900	36.68	1.444	1057	1.6377
	1000	38.15	1.502	1143	1.7719
	1250	43.92	1.729	1515	2.3479
	1500	47.04	1.852	1738	2.6938
	1750	49.94	1.966	1959	3.0357
	2000	52.63	2.072	2175	3.3719
SF-2, SFF-2	18	3.073	0.121	7.419	0.0115
	16	3.378	0.133	8.968	0.0139
	14	3.759	0.148	11.10	0.0172
SF-1, SFF-1	18	2.311	0.091	4.194	0.0065
RFH-1, XF, XFF	18	2.692	0.106	5.161	0.0080
TF, TFF, XF, XFF	16	2.997	0.118	7.032	0.0109
TW, XF, XFF, THHW, THW, THW-2	14	3.378	0.133	8.968	0.0139
TW, THHW, THW, THW-2	12	3.861	0.152	11.68	0.0181
	10	4.470	0.176	15.68	0.0243
	8	5.994	0.236	28.19	0.0437
RHH*, RHW*, RHW-2*	14	4.140	0.163	13.48	0.0209
	12	4.623	0.182	16.77	0.0260

Type	Size (AWG or kcmil)	Approximate Diameter mm	Approximate Diameter in.	Approximate Area mm^2	Approximate Area $in.^2$
Type: RHH*, RHW*, RHW-2*, THHN, THHW, THW, THW-2, TFN, TFFN, THWN, THWN-2, XF, XFF					
THHW, THW, AF, XF, XFF	10	5.232	0.206	21.48	0.0333
RHH*, RHW*, RHW-2*	8	6.756	0.266	35.87	0.0556
TW, THW, THHW, THW-2, RHH*, RHW*, RHW-2*	6	7.722	0.304	46.84	0.0726
	4	8.941	0.352	62.77	0.0973
	3	9.652	0.380	73.16	0.1134
	2	10.46	0.412	86.00	0.1333
	1	12.50	0.492	122.6	0.1901
	1/0	13.51	0.532	143.4	0.2223
	2/0	14.68	0.578	169.3	0.2624
	3/0	16.00	0.630	201.1	0.3117
	4/0	17.48	0.688	239.9	0.3718
	250	19.43	0.765	296.5	0.4596
	300	20.83	0.820	340.7	0.5281
	350	22.12	0.871	384.4	0.5958
	400	23.32	0.918	427.0	0.6619
	500	25.48	1.003	509.7	0.7901
	600	28.27	1.113	627.7	0.9729
	700	30.07	1.184	710.3	1.1010
	750	30.94	1.218	751.7	1.1652
	800	31.75	1.250	791.7	1.2272
	900	33.38	1.314	874.9	1.3561
	1000	34.85	1.372	953.8	1.4784
	1250	39.09	1.539	1200	1.8602
	1500	42.21	1.662	1400	2.1695
	1750	45.11	1.776	1598	2.4773
	2000	47.80	1.882	1795	2.7818
TFN, TFFN	18	2.134	0.084	3.548	0.0055
	16	2.438	0.096	4.645	0.0072
THHN, THWN, THWN-2	14	2.819	0.111	6.258	0.0097
	12	3.302	0.130	8.581	0.0133
	10	4.166	0.164	13.61	0.0211
	8	5.486	0.216	23.61	0.0366
	6	6.452	0.254	32.71	0.0507
	4	8.230	0.324	53.16	0.0824
	3	8.941	0.352	62.77	0.0973
	2	9.754	0.384	74.71	0.1158
	1	11.33	0.446	100.8	0.1562
	1/0	12.34	0.486	119.7	0.1855
	2/0	13.51	0.532	143.4	0.2223
	3/0	14.83	0.584	172.8	0.2679
	4/0	16.31	0.642	208.8	0.3237
	250	18.06	0.711	256.1	0.3970
	300	19.46	0.766	297.3	0.4608

(continued)

TABLE 15.1.3 Dimensions of Insulated Conductors and Fixture Wires *(Continued)*

Type	Size (AWG or kcmil)	Approximate Diameter		Approximate Area	
		mm	in.	mm²	in.²
Type: FEP, FEPB, PAF, PAFF, PF, PFA, PFAH, PFF, PGF, PGFF, PTF, PTFF, TFE, THHN, THWN, THWN-2, Z, ZF, ZFF					
THHN, THWN, THWN-2	350	20.75	0.817	338.2	0.5242
	400	21.95	0.864	378.3	0.5863
	500	24.10	0.949	456.3	0.7073
	600	26.70	1.051	559.7	0.8676
	700	28.50	1.12 2	637.9	0.9887
	750	29.36	1.156	677.2	1.0496
	800	30.18	1.188	715.2	1.1085
	900	31.80	1.252	794.3	1.2311
	1000	33.27	1.310	869.5	1.3478
PF, PGFF, PGF, PFF, PTF, PAF, PTFF, PAFF	18	2.184	0.086	3.742	0.0058
	16	2.489	0.098	4.839	0.0075
PF, PGFF, PGF, PFF, PTF, PAF, PTFF, PAFF, TFE, FEP, PFA, FEPB, PFAH	14	2.870	0.113	6.452	0.0100
TFE, FEP, PFA, FEPB, PFAH	12	3.353	0.132	8.839	0.0137
	10	3.962	0.156	12.32	0.0191
	8	5.232	0.206	21.48	0.0333
	6	6.198	0.244	30.19	0.0468
	4	7.417	0.292	43.23	0.0670
	3	8.128	0.320	51.87	0.0804
	2	8.941	0.352	62.77	0.0973
TFE, PFAH	1	10.72	0.422	90.26	0.1399
TFE, PFA PFAH, Z	1/0	11.73	0.462	108.1	0.1676
	2/0	12.90	0.508	130.8	0.2027
	3/0	14.22	0.560	158.9	0.2463
	4/0	15.70	0.618	193.5	0.3000
ZF, ZFF	18	1.930	0.076	2.903	0.0045
	16	2.235	0.088	3.935	0.0061
Z, ZF, ZFF	14	2.616	0.103	5.355	0.0083
Z	12	3.099	0.122	7.548	0.0117
	10	3.962	0.156	12.32	0.0191
	8	4.978	0.196	19.48	0.0302
	6	5.944	0.234	27.74	0.0430
	4	7.163	0.282	40.32	0.0625
	3	8.382	0.330	55.16	0.0855
	2	9.195	0.362	66.39	0.1029
	1	10.21	0.402	81.87	0.1269

Type	Size (AWG or kcmil)	Approximate Diameter		Approximate Area	
		mm	in.	mm²	in.²
Type: KF-1, KF-2, KFF-1, KFF-2, XHH, XHHW, XHHW-2, ZW					
XHHW, ZW, XHHW-2, XHH	14	3.378	0.133	8.968	0.0139
	12	3.861	0.152	11.68	0.0181
	10	4.470	0.176	15.68	0.0243
	8	5.994	0.236	28.19	0.0437
	6	6.960	0.274	38.06	0.0590
	4	8.179	0.322	52.52	0.0814
	3	8.890	0.350	62.06	0.0962
	2	9.703	0.382	73.94	0.1146
XHHW, XHHW-2, XHH	1	11.23	0.442	98.97	0.1534
	1/0	12.24	0.482	117.7	0.1825
	2/0	13.41	0.528	141.3	0.2190
	3/0	14.73	0.58	170.5	0.2642
	4/0	16.21	0.638	206.3	0.3197
	250	17.91	0.705	251.9	0.3904
	300	19.30	0.76	292.6	0.4536
	350	20.60	0.811	333.3	0.5166
	400	21.79	0.858	373.0	0.5782
	500	23.95	0.943	450.6	0.6984
	600	26.75	1.053	561.9	0.8709
	700	28.55	1.124	640.2	0.9923
	750	29.41	1.158	679.5	1.0532
	800	30.23	1.190	717.5	1.1122
	900	31.85	1.254	796.8	1.2351
	1000	33.32	1.312	872.2	1.3519
	1250	37.57	1.479	1108	1.7180
	1500	40.69	1.602	1300	2.0157
	1750	43.59	1.716	1492	2.3127
	2000	46.28	1.822	1682	2.6073
KF-2, KFF-2	18	1.600	0.063	2.000	0.0031
	16	1.905	0.075	2.839	0.0044
	14	2.286	0.090	4.129	0.0064
	12	2.769	0.109	6.000	0.0093
	10	3.378	0.133	8.968	0.0139
KF-1, KFF-1	18	1.448	0.057	1.677	0.0026
	16	1.753	0.069	2.387	0.0037
	14	2.134	0.084	3.548	0.0055
	12	2.616	0.103	5.355	0.0083
	10	3.226	0.127	8.194	0.0127

*Types RHH, RHW, and RHW-2 without outer covering.

15.1.4 *NEC* Chapter 9, Table 5A, Compact Aluminum Building Wire Nominal Dimensions* and Areas

TABLE 15.1.4

Size (AWG or kcmil)	Bare Conductor			Types THW and THHW				Type THHN				Type XHHW				Size (AWG or kcmil)
	Number of Strands	Diameter		Approximate Diameter		Approximate Area		Approximate Diameter		Approximate Area		Approximate Diameter		Approximate Area		
		mm	in.	mm	in.	mm²	in.²	mm	in.	mm²	in.²	mm	in.	mm²	in.²	
8	7	3.404	0.134	6.477	0.255	32.90	0.0510	—	—	—	—	5.690	0.224	25.42	0.0394	8
6	7	4.293	0.169	7.366	0.290	42.58	0.0660	6.096	0.240	29.16	0.0452	6.604	0.260	34.19	0.0530	6
4	7	5.410	0.213	8.509	0.335	56.84	0.0881	7.747	0.305	47.10	0.0730	7.747	0.305	47.10	0.0730	4
2	7	6.807	0.268	9.906	0.390	77.03	0.1194	9.144	0.360	65.61	0.1017	9.144	0.360	65.61	0.1017	2
1	19	7.595	0.299	11.81	0.465	109.5	0.1698	10.54	0.415	87.23	0.1352	10.54	0.415	87.23	0.1352	1
1/0	19	8.534	0.336	12.70	0.500	126.6	0.1963	11.43	0.450	102.6	0.1590	11.43	0.450	102.6	0.1590	1/0
2/0	19	9.550	0.376	13.84	0.545	150.5	0.2332	12.57	0.495	124.1	0.1924	12.45	0.490	121.6	0.1885	2/0
3/0	19	10.74	0.423	14.99	0.590	176.3	0.2733	13.72	0.540	147.7	0.2290	13.72	0.540	147.7	0.2290	3/0
4/0	19	12.07	0.475	16.38	0.645	210.8	0.3267	15.11	0.595	179.4	0.2780	14.99	0.590	176.3	0.2733	4/0
250	37	13.21	0.520	18.42	0.725	266.3	0.4128	17.02	0.670	227.4	0.3525	16.76	0.660	220.7	0.3421	250
300	37	14.48	0.570	19.69	0.775	304.3	0.4717	18.29	0.720	262.6	0.4071	18.16	0.715	259.0	0.4015	300
350	37	15.65	0.616	20.83	0.820	340.7	0.5281	19.56	0.770	300.4	0.4656	19.30	0.760	292.6	0.4536	350
400	37	16.74	0.659	21.97	0.865	379.1	0.5876	20.70	0.815	336.5	0.5216	20.32	0.800	324.3	0.5026	400
500	37	18.69	0.736	23.88	0.940	447.7	0.6939	22.48	0.885	396.8	0.6151	22.35	0.880	392.4	0.6082	500
600	61	20.65	0.813	26.67	1.050	558.6	0.8659	25.02	0.985	491.6	0.7620	24.89	0.980	486.6	0.7542	600
700	61	22.28	0.877	28.19	1.110	624.3	0.9676	26.67	1.050	558.6	0.8659	26.67	1.050	558.6	0.8659	700
750	61	23.06	0.908	29.21	1.150	670.1	1.0386	27.31	1.075	585.5	0.9076	27.69	1.090	602.0	0.9331	750
1000	61	26.92	1.060	32.64	1.285	836.6	1.2968	31.88	1.255	798.1	1.2370	31.24	1.230	766.6	1.1882	1000

*Dimensions are from industry sources.

15.1.5 *NEC* Chapter 9, Table 8, Conductor Properties

TABLE 15.1.5

	Conductors									Direct-Current Resistance at 75°C (167°F)					
			Stranding			Overall				Copper				Aluminum	
	Area			Diameter		Diameter		Area		Uncoated		Coated			
Size (AWG or kcmil)	mm^2	Circular mils	Quantity	mm	in.	mm	in.	mm^2	$in.^2$	ohm/km	ohm/kFT	ohm/km	ohm/kFT	ohm/km	ohm/kFT
18	0.823	1620	1	—	—	1.02	0.040	0.823	0.001	25.5	7.77	26.5	8.08	42.0	12.8
18	0.823	1620	7	0.39	0.015	1.16	0.046	1.06	0.002	26.1	7.95	27.7	8.45	42.8	13.1
16	1.31	2580	1	—	—	1.29	0.051	1.31	0.002	16.0	4.89	16.7	5.08	26.4	8.05
16	1.31	2580	7	0.49	0.019	1.46	0.058	1.68	0.003	16.4	4.99	17.3	5.29	26.9	8.21
14	2.08	4110	1	—	—	1.63	0.064	2.08	0.003	10.1	3.07	10.4	3.19	16.6	5.06
14	2.08	4110	7	0.62	0.024	1.85	0.073	2.68	0.004	10.3	3.14	10.7	3.26	16.9	5.17
12	3.31	6530	1	—	—	2.05	0.081	3.31	0.005	6.34	1.93	6.57	2.01	10.45	3.18
12	3.31	6530	7	0.78	0.030	2.32	0.092	4.25	0.006	6.50	1.98	6.73	2.05	10.69	3.25
10	5.261	10380	1	—	—	2.588	0.102	5.26	0.008	3.984	1.21	4.148	1.26	6.561	2.00
10	5.261	10380	7	0.98	0.038	2.95	0.116	6.76	0.011	4.070	1.24	4.226	1.29	6.679	2.04
8	8.367	16510	1	—	—	3.264	0.128	8.37	0.013	2.506	0.764	2.579	0.786	4.125	1.26
8	8.367	16510	7	1.23	0.049	3.71	0.146	10.76	0.017	2.551	0.778	2.653	0.809	4.204	1.28
6	13.30	26240	7	1.56	0.061	4.67	0.184	17.09	0.027	1.608	0.491	1.671	0.510	2.652	0.808
4	21.15	41740	7	1.96	0.077	5.89	0.232	27.19	0.042	1.010	0.308	1.053	0.321	1.666	0.508
3	26.67	52620	7	2.20	0.087	6.60	0.260	34.28	0.053	0.802	0.245	0.833	0.254	1.320	0.403
2	33.62	66360	7	2.47	0.097	7.42	0.292	43.23	0.067	0.634	0.194	0.661	0.201	1.045	0.319
1	42.41	83690	19	1.69	0.066	8.43	0.332	55.80	0.087	0.505	0.154	0.524	0.160	0.829	0.253
1/0	53.49	105600	19	1.89	0.074	9.45	0.372	70.41	0.109	0.399	0.122	0.415	0.127	0.660	0.201
2/0	67.43	133100	19	2.13	0.084	10.62	0.418	88.74	0.137	0.3170	0.0967	0.329	0.101	0.523	0.159
3/0	85.01	167800	19	2.39	0.094	11.94	0.470	111.9	0.173	0.2512	0.0766	0.2610	0.0797	0.413	0.126
4/0	107.2	211600	19	2.68	0.106	13.41	0.528	141.1	0.219	0.1996	0.0608	0.2050	0.0626	0.328	0.100
250		—	37	2.09	0.082	14.61	0.575	168	0.260	0.1687	0.0515	0.1753	0.0535	0.2778	0.0847
300		—	37	2.29	0.090	16.00	0.630	201	0.312	0.1409	0.0429	0.1463	0.0446	0.2318	0.0707
350		—	37	2.47	0.097	17.30	0.681	235	0.364	0.1205	0.0367	0.1252	0.0382	0.1984	0.0605
400		—	37	2.64	0.104	18.49	0.728	268	0.416	0.1053	0.0321	0.1084	0.0331	0.1737	0.0529
500		—	37	2.95	0.116	20.65	0.813	336	0.519	0.0845	0.0258	0.0869	0.0265	0.1391	0.0424
600		—	61	2.52	0.099	22.68	0.893	404	0.626	0.0704	0.0214	0.0732	0.0223	0.1159	0.0353
700		—	61	2.72	0.107	24.49	0.964	471	0.730	0.0603	0.0184	0.0622	0.0189	0.0994	0.0303
750		—	61	2.82	0.111	25.35	0.998	505	0.782	0.0563	0.0171	0.0579	0.0176	0.0927	0.0282
800		—	61	2.91	0.114	26.16	1.030	538	0.834	0.0528	0.0161	0.0544	0.0166	0.0868	0.0265
900		—	61	3.09	0.122	27.79	1.094	606	0.940	0.0470	0.0143	0.0481	0.0147	0.0770	0.0235
1000		—	61	3.25	0.128	29.26	1.152	673	1.042	0.0423	0.0129	0.0434	0.0132	0.0695	0.0212
1250		—	91	2.98	0.117	32.74	1.289	842	1.305	0.0338	0.0103	0.0347	0.0106	0.0554	0.0169
1500		—	91	3.26	0.128	35.86	1.412	1011	1.566	0.02814	0.00858	0.02814	0.00883	0.0464	0.0141
1750		—	127	2.98	0.117	38.76	1.526	1180	1.829	0.02410	0.00735	0.02410	0.00756	0.0397	0.0121
2000		—	127	3.19	0.126	41.45	1.632	1349	2.092	0.02109	0.00643	0.02109	0.00662	0.0348	0.0106

Notes:

1. These resistance values are valid **only** for the parameters as given. Using conductors having coated strands, different stranding type, and, especially, other temperatures changes the resistance.
2. Formula for temperature change: $R_2 = R_1 [1 + \alpha (T_2 - 75)]$ where $\alpha_{cu} = 0.00323$, $\alpha_{AL} = 0.00330$ at 75°C.
3. Conductors with compact and compressed stranding have about 9 percent and 3 percent, respectively, smaller bare conductor diameters than those shown. See Table 5A for actual compact cable dimensions.
4. The IACS conductivities used: bare copper = 100%, aluminum = 61%.
5. Class B stranding is listed as well as solid for some sizes. Its overall diameter and area is that of its circumscribing circle.

FPN: The construction information is per NEMA WC8-1992 or ANSI/UL 1581-1998. The resistance is calculated per National Bureau of Standards Handbook 100, dated 1966, and Handbook 109, dated 1972.

15.1.6 *NEC* Chapter 9, Table 9, Alternating-Current Resistance and Reactance for 600-V Cables, Three-Phase, 60 Hz, 75°C (167°F): Three Single Conductors in Conduit

TABLE 15.1.6

Size (AWG or kcmil)	Ohms to Neutral per Kilometer / Ohms to Neutral per 1000 Feet														Size (AWG or kcmil)
	X_L (Reactance) for All Wires		Alternating-Current Resistance for Uncoated Copper Wires			Alternating-Current Resistance for Aluminum Wires			Effective *Z* at 0.85 *PF* for Uncoated Copper Wires			Effective *Z* at 0.85 *PF* for Aluminum Wires			
	PVC, Aluminum Conduits	Steel Conduit	PVC Conduit	Aluminum Conduit	Steel Conduit	PVC Conduit	Aluminum Conduit	Steel Conduit	PVC Conduit	Aluminum Conduit	Steel Conduit	PVC Conduit	Aluminum Conduit	Steel Conduit	
14	0.190 0.058	0.240 0.073	10.2 3.1	10.2 3.1	10.2 3.1	— —	— —	— —	8.9 2.7	8.9 2.7	8.9 2.7	— —	— —	— —	14
12	0.177 0.054	0.223 0.068	6.6 2.0	6.6 2.0	6.6 2.0	10.5 3.2	10.5 3.2	10.5 3.2	5.6 1.7	5.6 1.7	5.6 1.7	9.2 2.8	9.2 2.8	9.2 2.8	12
10	0.164 0.050	0.207 0.063	3.9 1.2	3.9 1.2	3.9 1.2	6.6 2.0	6.6 2.0	6.6 2.0	3.6 1.1	3.6 1.1	3.6 1.1	5.9 1.8	5.9 1.8	5.9 1.8	10
8	0.171 0.052	0.213 0.065	2.56 0.78	2.56 0.78	2.56 0.78	4.3 1.3	4.3 1.3	4.3 1.3	2.26 0.69	2.26 0.69	2.30 0.70	3.6 1.1	3.6 1.1	3.6 1.1	8
6	0.167 0.051	0.210 0.064	1.61 0.49	1.61 0.49	1.61 0.49	2.66 0.81	2.66 0.81	2.66 0.81	1.44 0.44	1.48 0.45	1.48 0.45	2.33 0.71	2.36 0.72	2.36 0.72	6
4	0.157 0.048	0.197 0.060	1.02 0.31	1.02 0.31	1.02 0.31	1.67 0.51	1.67 0.51	1.67 0.51	0.95 0.29	0.95 0.29	0.98 0.30	1.51 0.46	1.51 0.46	1.51 0.46	4
3	0.154 0.047	0.194 0.059	0.82 0.25	0.82 0.25	0.82 0.25	1.31 0.40	1.35 0.41	1.31 0.40	0.75 0.23	0.79 0.24	0.79 0.24	1.21 0.37	1.21 0.37	1.21 0.37	3
2	0.148 0.045	0.187 0.057	0.62 0.19	0.66 0.20	0.66 0.20	1.05 0.32	1.05 0.32	1.05 0.32	0.62 0.19	0.62 0.19	0.66 0.20	0.98 0.30	0.98 0.30	0.98 0.30	2
1	0.151 0.046	0.187 0.057	0.49 0.15	0.52 0.16	0.52 0.16	0.82 0.25	0.85 0.26	0.82 0.25	0.52 0.16	0.52 0.16	0.52 0.16	0.79 0.24	0.79 0.24	0.82 0.25	1
1/0	0.144 0.044	0.180 0.055	0.39 0.12	0.43 0.13	0.39 0.12	0.66 0.20	0.69 0.21	0.66 0.20	0.43 0.13	0.43 0.13	0.43 0.13	0.62 0.19	0.66 0.20	0.66 0.20	1/0
2/0	0.141 0.043	0.177 0.054	0.33 0.10	0.33 0.10	0.33 0.10	0.52 0.16	0.52 0.16	0.52 0.16	0.36 0.11	0.36 0.11	0.36 0.11	0.52 0.16	0.52 0.16	0.52 0.16	2/0
3/0	0.138 0.042	0.171 0.052	0.253 0.077	0.269 0.082	0.259 0.079	0.43 0.13	0.43 0.13	0.43 0.13	0.289 0.088	0.302 0.092	0.308 0.094	0.43 0.13	0.43 0.13	0.46 0.14	3/0
4/0	0.135 0.041	0.167 0.051	0.203 0.062	0.220 0.067	0.207 0.063	0.33 0.10	0.36 0.11	0.33 0.10	0.243 0.074	0.256 0.078	0.262 0.080	0.36 0.11	0.36 0.11	0.36 0.11	4/0
250	0.135 0.041	0.171 0.052	0.171 0.052	0.187 0.057	0.177 0.054	0.279 0.085	0.295 0.090	0.282 0.086	0.217 0.066	0.230 0.070	0.240 0.073	0.308 0.094	0.322 0.098	0.33 0.10	250
300	0.135 0.041	0.167 0.051	0.144 0.044	0.161 0.049	0.148 0.045	0.233 0.071	0.249 0.076	0.236 0.072	0.194 0.059	0.207 0.063	0.213 0.065	0.269 0.082	0.282 0.086	0.289 0.088	300
350	0.131 0.040	0.164 0.050	0.125 0.038	0.141 0.043	0.128 0.039	0.200 0.061	0.217 0.066	0.207 0.063	0.174 0.053	0.190 0.058	0.197 0.060	0.240 0.073	0.253 0.077	0.262 0.080	350
400	0.131 0.040	0.161 0.049	0.108 0.033	0.125 0.038	0.115 0.035	0.177 0.054	0.194 0.059	0.180 0.055	0.161 0.049	0.174 0.053	0.184 0.056	0.217 0.066	0.233 0.071	0.240 0.073	400
500	0.128 0.039	0.157 0.048	0.089 0.027	0.105 0.032	0.095 0.029	0.141 0.043	0.157 0.048	0.148 0.045	0.141 0.043	0.157 0.048	0.164 0.050	0.187 0.057	0.200 0.061	0.210 0.064	500
600	0.128 0.039	0.157 0.048	0.075 0.023	0.092 0.028	0.082 0.025	0.118 0.036	0.135 0.041	0.125 0.038	0.131 0.040	0.144 0.044	0.154 0.047	0.167 0.051	0.180 0.055	0.190 0.058	600

(continued)

TABLE 15.1.6 Alternating-Current Resistance and Reactance for 600-V Cables, Three-Phase, 60 Hz, 75°C (167°F): Three Single Conductors in Conduit *(Continued)*

Size (AWG or kcmil)	Ohms to Neutral per Kilometer / Ohms to Neutral per 1000 Feet														Size (AWG or kcmil)
	X_L (Reactance) for All Wires		Alternating-Current Resistance for Uncoated Copper Wires			Alternating-Current Resistance for Aluminum Wires			Effective Z at 0.85 *PF* for Uncoated Copper Wires			Effective Z at 0.85 *PF* for Aluminum Wires			
	PVC, Aluminum Conduits	Steel Conduit	PVC Conduit	Aluminum Conduit	Steel Conduit	PVC Conduit	Aluminum Conduit	Steel Conduit	PVC Conduit	Aluminum Conduit	Steel Conduit	PVC Conduit	Aluminum Conduit	Steel Conduit	
750	0.125 0.038	0.157 0.048	0.062 0.019	0.079 0.024	0.069 0.021	0.095 0.029	0.112 0.034	0.102 0.031	0.118 0.036	0.131 0.040	0.141 0.043	0.148 0.045	0.161 0.049	0.171 0.052	750
1000	0.121 0.037	0.151 0.046	0.049 0.015	0.062 0.019	0.059 0.018	0.075 0.023	0.089 0.027	0.082 0.025	0.105 0.032	0.118 0.036	0.131 0.040	0.128 0.039	0.138 0.042	0.151 0.046	1000

Notes:

1. These values are based on the following constants: UL-Type RHH wires with Class B stranding, in cradled configuration. Wire conductivities are 100 percent IACS copper and 61 percent IACS aluminum, and aluminum conduit is 45 percent IACS. Capacitive reactance is ignored, since it is negligible at these voltages. These resistance values are valid only at 75°C (167°F) and for the parameters as given, but are representative for 600-volt wire types operating at 60 Hz.
2. *Effective Z* is defined as $R \cos(\theta) + X \sin(\theta)$, where θ is the power factor angle of the circuit. Multiplying current by effective impedance gives a good approximation for line-to-neutral voltage drop. Effective impedance values shown in this table are valid only at 0.85 power factor. For another circuit power factor (*PF*), effective impedance (*Ze*) can be calculated from *R* and X_L values given in this table as follows: $Ze = R \times PF + X_L \sin[\arccos(PF)]$.

15.1.7 *NEC* Chapter 9, Tables 11(A) and 11(B), Class 2 and Class 3, Alternating-Current and Direct-Current Power-Source Limitations, Respectively

TABLE 15.1.7A Class 2 and Class 3 Alternating-Current Power-Source Limitations

Power Source		Inherently Limited Power Source (Overcurrent Protection Not Required)				Not Inherently Limited Power Source (Overcurrent Protection Required)			
		Class 2			Class 3	Class 2		Class 3	
Source voltage V_{max} (volts) (see Note 1)		0 through 20*	Over 20 and through 30*	Over 30 and through 150	Over 30 and through 100	0 through 20*	Over 20 and through 30*	Over 30 and through 100	Over 100 and through 150
Power limitations VA_{max} (volt-amperes) (see Note 1)		—	—	—	—	250 (see Note 3)	250	250	N.A.
Current limitations I_{max} (amperes) (see Note 1)		8.0	8.0	0.005	$150/V_{max}$	$1000/V_{max}$	$1000/V_{max}$	$1000/V_{max}$	1.0
Maximum overcurrent protection (amperes)		—	—	—	—	5.0	$100/V_{max}$	$100/V_{max}$	1.0
Power source maximum nameplate rating	VA (volt-amperes)	$5.0 \times V_{max}$	100	$0.005 \times V_{max}$	100	$5.0 \times V_{max}$	100	100	100
	Current (amperes)	5.0	$100/V_{max}$	0.005	$100/V_{max}$	5.0	$100/V_{max}$	$100/V_{max}$	$100/V_{max}$

*Voltage ranges shown are for sinusoidal ac in indoor locations or where wet contact is not likely to occur. For nonsinusoidal or wet contact conditions, see Note 2.

TABLE 15.1.7B Class 2 and Class 3 Direct-Current Power-Source Limitations

	Inherently Limited Power Source (Overcurrent Protection Not Required)					Not Inherently Limited Power Source (Overcurrent Protection Required)			
Power Source	Class 2				Class 3	Class 2		Class 3	
Source voltage V_{max} (volts) (see Note 1)	0 through 20*	Over 20 and through 30*	Over 30 and through 60*	Over 60 and through 150	Over 60 and through 100	0 through 20*	Over 20 and through 60*	Over 60 and through 100	Over 100 and through 150
Power limitations VA_{max} (volt-amperes) (see Note 1)	—	—	—	—	—	250 (see Note 3)	250	250	N.A.
Current limitations I_{max} (amperes) (see Note 1)	8.0	8.0	$150/V_{max}$	0.005	$150/V_{max}$	$1000/V_{max}$	$1000/V_{max}$	$1000/V_{max}$	1.0
Maximum overcurrent protection (amperes)	—	—	—	—	—	5.0	$100/V_{max}$	$100/V_{max}$	1.0
Power source maximum nameplate rating: VA (volt-amperes)	$5.0 \times V_{max}$	100	100	$0.005 \times V_{max}$	100	$5.0 \times V_{max}$	100	100	100
Power source maximum nameplate rating: Current (amperes)	5.0	$100/V_{max}$	$100/V_{max}$	0.005	$100/V_{max}$	5.0	$100/V_{max}$	$100/V_{max}$	$100/V_{max}$

*Voltage ranges shown are for continuous dc in indoor locations or where wet contact is not likely to occur. For interrupted dc or wet contact conditions, see Note 4.

Notes for Tables 15.1.7(A) and 15.1.7(B)

1. V_{max}, I_{max}, and VA_{max} are determined with the current-limiting impedance in the circuit (not bypassed) as follows:

V_{max}: Maximum output voltage regardless of load with rated input applied.

I_{max}: Maximum output current under any noncapacitive load, including short circuit, and with overcurrent protection bypassed if used. Where a transformer limits the output current, I_{max} limits apply after 1 minute of operation. Where a current-limiting impedance, listed for the purpose, or as part of a listed product, is used in combination with a nonpower-limited transformer or a stored energy source, e.g., storage battery, to limit the output current, I_{max} limits apply after 5 seconds.

VA_{max}: Maximum volt-ampere output after 1 minute of operation regardless of load and overcurrent protection bypassed if used.

2. For nonsinusoidal ac, V_{max} shall not be greater than 42.4 volts peak. Where wet contact (immersion not included) is likely to occur, Class 3 wiring methods shall be used or V_{max} shall not be greater than 15 volts for sinusoidal ac and 21.2 volts peak for nonsinusoidal ac.

3. If the power source is a transformer, VA_{max} is 350 or less when V_{max} is 15 or less.

4. For dc interrupted at a rate of 10 to 200 Hz, V_{max} shall not be greater than 24.8 volts peak. Where wet contact (immersion not included) is likely to occur, Class 3 wiring methods shall be used, or V_{max} shall not be greater than 30 volts for continuous dc; 12.4 volts peak for dc that is interrupted at a rate of 10 to 200 Hz.

15.1.8 *NEC* Chapter 9, Tables 12(A) and 12(B), PLFA Alternating-Current and Direct-Current Power-Source Limitations, Respectively

TABLE 15.1.8A PLFA Alternating-Current Power-Source Limitations

Power Source		Inherently Limited Power Source (Overcurrent Protection Not Required)			Not Inherently Limited Power Source (Overcurrent Protection Required)		
Circuit voltage V_{max}(volts) (see Note 1)		0 through 20	Over 20 and through 30	Over 30 and through 100	0 through 20	Over 20 and through 100	Over 100 and through 150
Power limitations VA_{max} (volt-amperes) (see Note 1)		—	—	—	250 (see Note 2)	250	N.A.
Current limitations I_{max}(amperes) (see Note 1)		8.0	8.0	$150/V_{max}$	$1000/V_{max}$	$1000/V_{max}$	1.0
Maximum overcurrent protection (amperes)		—	—	—	5.0	$100/V_{max}$	1.0
Power source maximum nameplate ratings	VA (volt- amperes)	$5.0 \times V_{max}$	100	100	$5.0 \times V_{max}$	100	100
	Current (amperes)	5.0	$100/V_{max}$	$100/V_{max}$	5.0	$100/V_{max}$	$100/V_{max}$

TABLE 15.1.8B PLFA Direct-Current Power-Source Limitations

Power Source		Inherently Limited Power Source (Overcurrent Protection Not Required)				Not Inherently Limited Power Source (Overcurrent Protection Required)		
Circuit voltage V_{max} (volts) (see Note 1)		0 through 20	Over 20 and through 30	Over 30 and through 100	Over 100 and through 250	0 through 20	Over 20 and through 100	Over 100 and through 150
Power limitations VA_{max} (volt-amperes) (see Note 1)		—	—	—	—	250 (see Note 2)	250	N.A.
Current limitations I_{max} (amperes) (see Note 1)		8.0	8.0	$150/V_{max}$	0.030	$1000/V_{max}$	$1000/V_{max}$	1.0
Maximum overcurrent protection (amperes)		—	—	—	—	5.0	$100/V_{max}$	1.0
Power source maximum nameplate ratings	VA (volt-amperes)	$5.0 \times V_{max}$	100	100	$0.030 \times V_{max}$	$5.0 \times V_{max}$	100	100
	Current (amperes)	5.0	$100/V_{max}$	$100/V_{max}$	0.030	5.0	$100/V_{max}$	$100/V_{max}$

Notes for Tables 15.1.8(A) and 15.1.8(B)

1. V_{max}, I_{max}, and VA_{max} are determined as follows:

V_{max}: Maximum output voltage regardless of load with rated input applied.

I_{max}: Maximum output current under any noncapacitive load, including short circuit, and with overcurrent protection bypassed if used. Where a transformer limits the output current, I_{max} limits apply after 1 minute of operation. Where a current-limiting impedance, listed for the purpose, is used in combination with a nonpower-limited transformer or a stored energy source, e.g., storage battery, to limit the output current, I_{max} limits apply after 5 seconds.

VA_{max}: Maximum volt-ampere output after 1 minute of operation regardless of load and overcurrent protection bypassed if used. Current limiting impedance shall not be bypassed when determining I_{max} and VA_{max}.

2. If the power source is a transformer, VA_{max} is 350 or less when V_{max} is 15 or less.

15.2.1 *NEC* (Annex B), Table B.310.1, Ampacities of Two or Three Insulated Conductors, Rated 0 through 2000 V, Within an Overall Covering (Multiconductor Cable), in Raceway in Free Air, Based on Ambient Air Temperature of 30°C (86°F)

TABLE 15.2.1

Size (AWG or kcmil)	Temperature Rating of Conductor. (See Table 310.13.)						Size (AWG or kcmil)
	60°C (140°F)	75°C (167°F)	90°C (194°F)	60°C (140°F)	75°C (167°F)	90°C (194°F)	
	Types TW, UF	Types RHW, THHW, THW, THWN, XHHW, ZW	Types THHN, THHW, THW-2, THWN-2, RHH, RWH-2, USE-2, XHHW, XHHW-2, ZW-2	Type TW	Types RHW, THHW, THW, THWN, XHHW	Types THHN, THHW, THW-2, THWN-2, RHH, RWH-2, USE-2, XHHW, XHHW-2, ZW-2	
	COPPER			ALUMINUM OR COPPER-CLAD ALUMINUM			
14	16*	18*	21*				14
12	20*	24*	27*	16*	18*	21*	12
10	27*	33*	36*	21*	25*	28*	10
8	36	43	48	28	33	37	8
6	48	58	65	38	45	51	6
4	66	79	89	51	61	69	4
3	76	90	102	59	70	79	3
2	88	105	119	69	83	93	2
1	102	121	137	80	95	106	1
1/0	121	145	163	94	113	127	1/0
2/0	138	166	186	108	129	146	2/0
3/0	158	189	214	124	147	167	3/0
4/0	187	223	253	147	176	197	4/0
250	205	245	276	160	192	217	250
300	234	281	317	185	221	250	300
350	255	305	345	202	242	273	350
400	274	328	371	218	261	295	400
500	315	378	427	254	303	342	500
600	343	413	468	279	335	378	600
700	376	452	514	310	371	420	700
750	387	466	529	321	384	435	750
800	397	479	543	331	397	450	800
900	415	500	570	350	421	477	900
1000	448	542	617	382	460	521	1000

Correction Factors

Ambient Temp. (°C)	For ambient temperatures other than 30°C (86°F), multiply the ampacities shown above by the appropriate factor shown below.						Ambient Temp. (°F)
21–25	1.08	1.05	1.04	1.08	1.05	1.04	70–77
26–30	1.00	1.00	1.00	1.00	1.00	1.00	79–86
31–35	0.91	0.94	0.96	0.91	0.94	0.96	88–95
36–40	0.82	0.88	0.91	0.82	0.88	0.91	97–104
41–45	0.71	0.82	0.87	0.71	0.82	0.87	106–113
46–50	0.58	0.75	0.82	0.58	0.75	0.82	115–122
51–55	0.41	0.67	0.76	0.41	0.67	0.76	124–131
56–60	—	0.58	0.71	—	0.58	0.71	133–140
61–70	—	0.33	0.58	—	0.33	0.58	142–158
71–80	—	—	0.41	—	—	0.41	160–176

*Unless otherwise specifically permitted elsewhere in this *Code*, the overcurrent protection for these conductor types shall not exceed 15 amperes for 14 AWG, 20 amperes for 12 AWG, and 30 amperes for 10 AWG copper; or 15 amperes for 12 AWG and 25 amperes for 10 AWG aluminum and copper-clad aluminum.

15.2.2 *NEC* (Annex B), Table B.310.3, Ampacities of Multiconductor Cables with Not More Than Three Insulated Conductors, Rated 0 through 2000 V, in Free Air, Based on Ambient Air Temperature of 40°C (104°F) (for Type TC,MC, MI, UF, and USE Cables)

TABLE 15.2.2

Size (AWG or kcmil)	Temperature Rating of Conductor. (See Table 310.13.)								Size (AWG or kcmil)
	60°C (140°F)	75°C (167°F)	85°C (185°F)	90°C (194°F)	60°C (140°F)	75°C (167°F)	85°C (185°F)	90°C (194°F)	
	COPPER				ALUMINUM OR COPPER-CLAD ALUMINUM				
18	—	—	—	11*	—	—	—	—	18
16	—	—	—	16*	—	—	—	—	16
14	18*	21*	24*	25*	—	—	—	—	14
12	21*	28*	30*	32*	18*	21*	24*	25*	12
10	28*	36*	41*	43*	21*	28*	30*	32*	10
8	39	50	56	59	30	39	44	46	8
6	52	68	75	79	41	53	59	61	6
4	69	89	100	104	54	70	78	81	4
3	81	104	116	121	63	81	91	95	3
2	92	118	132	138	72	92	103	108	2
1	107	138	154	161	84	108	120	126	1
1/0	124	160	178	186	97	125	139	145	1/0
2/0	143	184	206	215	111	144	160	168	2/0
3/0	165	213	238	249	129	166	185	194	3/0
4/0	190	245	274	287	149	192	214	224	4/0
250	212	274	305	320	166	214	239	250	250
300	237	306	341	357	186	240	268	280	300
350	261	337	377	394	205	265	296	309	350
400	281	363	406	425	222	287	317	334	400
500	321	416	465	487	255	330	368	385	500
600	354	459	513	538	284	368	410	429	600
700	387	502	562	589	306	405	462	473	700
750	404	523	586	615	328	424	473	495	750
800	415	539	604	633	339	439	490	513	800
900	438	570	639	670	362	469	514	548	900
1000	461	601	674	707	385	499	558	584	1000

Correction Factors

Ambient Temp. (°C)	For ambient temperatures other than 40°C (104°F), multiply the ampacities shown above by the appropriate factor shown below.								Ambient Temp. (°F)
21–25	1.32	1.20	1.15	1.14	1.32	1.20	1.15	1.14	70–77
26–30	1.22	1.13	1.11	1.10	1.22	1.13	1.11	1.10	79–86
31–35	1.12	1.07	1.05	1.05	1.12	1.07	1.05	1.05	88–95
36–40	1.00	1.00	1.00	1.00	1.00	1.00	1.00	1.00	97–104
41–45	0.87	0.93	0.94	0.95	0.87	0.93	0.94	0.95	106–113
46–50	0.71	0.85	0.88	0.89	0.71	0.85	0.88	0.89	115–122
51–55	0.50	0.76	0.82	0.84	0.50	0.76	0.82	0.84	124–131
56–60	—	0.65	0.75	0.77	—	0.65	0.75	0.77	133–140
61–70	—	0.38	0.58	0.63	—	0.38	0.58	0.63	142–158
71–80	—	—	0.33	0.44	—	—	0.33	0.44	160–176

*Unless otherwise specifically permitted elsewhere in this *Code*, the overcurrent protection for these conductor types shall not exceed 15 amperes for 14 AWG, 20 amperes for 12 AWG, and 30 amperes for 10 AWG copper; or 15 amperes for 12 AWG and 25 amperes for 10 AWG aluminum and copper-clad aluminum.

15.2.3 *NEC* (Annex B), Table B.310.5, Ampacities of Single-Insulated Conductors, Rated 0 through 2000 V, in Nonmagnetic Underground Electrical Ducts (One Conductor per Electrical Duct), Based on Ambient Earth Temperature of 20°C (68°F), Electrical Duct Arrangement per *NEC* Figure B.310.2, Conductor Temperature 75°C (167°F)

TABLE 15.2.3

	3 Electrical Ducts (Fig. B.310.2, Detail 2)			6 Electrical Ducts (Fig. B.310.2, Detail 3)			9 Electrical Ducts (Fig. B.310.2, Detail 4)			3 Electrical Ducts (Fig. B.310.2, Detail 2)			6 Electrical Ducts (Fig. B.310.2, Detail 3)			9 Electrical Ducts (Fig. B.310.2, Detail 4)			
	Types RHW, THHW, THW, THWN, XHHW, USE			Types RHW, THHW, THW, THWN, XHHW, USE			Types RHW, THHW, THW, THWN, XHHW, USE			Types RHW, THHW, THW, THWN, XHHW, USE			Types RHW, THHW, THW, THWN, XHHW, USE			Types RHW, THHW, THW, THWN, XHHW, USE			
	COPPER									ALUMINUM OR COPPER-CLAD ALUMINUM									
Size (kcmil)	RHO 60 LF 50	RHO 90 LF 100	RHO 120 LF 100	RHO 60 LF 50	RHO 90 LF 100	RHO 120 LF 100	RHO 60 LF 50	RHO 90 LF 100	RHO 120 LF 100	RHO 60 LF 50	RHO 90 LF 100	RHO 120 LF 100	RHO 60 LF 50	RHO 90 LF 100	RHO 120 LF 100	RHO 60 LF 50	RHO 90 LF 100	RHO 120 LF 100	Size (kcmil)
250	410	344	327	386	295	275	369	270	252	320	269	256	302	230	214	288	211	197	250
350	503	418	396	472	355	330	446	322	299	393	327	310	369	277	258	350	252	235	350
500	624	511	484	583	431	400	545	387	360	489	401	379	457	337	313	430	305	284	500
750	794	640	603	736	534	494	674	469	434	626	505	475	581	421	389	538	375	347	750
1000	936	745	700	864	617	570	776	533	493	744	593	557	687	491	453	629	432	399	1000
1250	1055	832	781	970	686	632	854	581	536	848	668	627	779	551	508	703	478	441	1250
1500	1160	907	849	1063	744	685	918	619	571	941	736	689	863	604	556	767	517	477	1500
1750	1250	970	907	1142	793	729	975	651	599	1026	796	745	937	651	598	823	550	507	1750
2000	1332	1027	959	1213	836	768	1030	683	628	1103	850	794	1005	693	636	877	581	535	2000
Ambient Temp. (°C)	Correction Factors																		Ambient Temp. (°F)
6–10		1.09			1.09			1.09			1.09			1.09			1.09		43–50
11–15		1.04			1.04			1.04			1.04			1.04			1.04		52–59
16–20		1.00			1.00			1.00			1.00			1.00			1.00		61–68
21–25		0.95			0.95			0.95			0.95			0.95			0.95		70–77
26–30		0.90			0.90			0.90			0.90			0.90			0.90		79–86

15.2.4 *NEC* (Annex B), Table B.310.6, Ampacities of Three Insulated Conductors, Rated 0 through 2000 V, Within an Overall Covering (Three-Conductor Cable) in Underground Electrical Ducts (One Cable per Electrical Duct), Based on Ambient Earth Temperature of 20°C (68°F), Electrical Duct Arrangement per *NEC* Figure B.310.2, Conductor Temperature 75°C (167°F)

TABLE 15.2.4

	1 Electrical Duct (Fig. B.310.2, Detail 1)			3 Electrical Ducts (Fig. B.310.2, Detail 2)			6 Electrical Ducts (Fig. B.310.2, Detail 3)			1 Electrical Duct (Fig. B.310.2, Detail 1)			3 Electrical Ducts (Fig. B.310.2, Detail 2)			6 Electrical Ducts (Fig. B.310.2, Detail 3)			
	Types RHW, THHW, THW, THWN, XHHW, USE			Types RHW, THHW, THW, THWN, XHHW, USE			Types RHW, THHW, THW, THWN, XHHW, USE			Types RHW, THHW, THW, THWN, XHHW, USE			Types RHW, THHW, THW, THWN, XHHW, USE			Types RHW, THHW, THW, THWN, XHHW, USE			
	COPPER									ALUMINUM OR COPPER-CLAD ALUMINUM									
Size (AWG or kcmil)	RHO 60 LF 50	RHO 90 LF 100	RHO 120 LF 100	RHO 60 LF 50	RHO 90 LF 100	RHO 120 LF 100	RHO 60 LF 50	RHO 90 LF 100	RHO 120 LF 100	RHO 60 LF 50	RHO 90 LF 100	RHO 120 LF 100	RHO 60 LF 50	RHO 90 LF 100	RHO 120 LF 100	RHO 60 LF 50	RHO 90 LF 100	RHO 120 LF 100	Size (AWG or kcmil)
8	58	54	53	56	48	46	53	42	39	45	42	41	43	37	36	41	32	30	8
6	77	71	69	74	63	60	70	54	51	60	55	54	57	49	47	54	42	39	6
4	101	93	91	96	81	77	91	69	65	78	72	71	75	63	60	71	54	51	4
2	132	121	118	126	105	100	119	89	83	103	94	92	98	82	78	92	70	65	2
1	154	140	136	146	121	114	137	102	95	120	109	106	114	94	89	107	79	74	1
1/0	177	160	156	168	137	130	157	116	107	138	125	122	131	107	101	122	90	84	1/0
2/0	203	183	178	192	156	147	179	131	121	158	143	139	150	122	115	140	102	95	2/0
3/0	233	210	204	221	178	158	205	148	137	182	164	159	172	139	131	160	116	107	3/0
4/0	268	240	232	253	202	190	234	168	155	209	187	182	198	158	149	183	131	121	4/0
250	297	265	256	280	222	209	258	184	169	233	207	201	219	174	163	202	144	132	250
350	363	321	310	340	267	250	312	219	202	285	252	244	267	209	196	245	172	158	350
500	444	389	375	414	320	299	377	261	240	352	308	297	328	254	237	299	207	190	500
750	552	478	459	511	388	362	462	314	288	446	386	372	413	314	293	374	254	233	750
1000	628	539	518	579	435	405	522	351	321	521	447	430	480	361	336	433	291	266	1000
Ambient Temp. (°C)	Correction Factors																		Ambient Temp (°F)
6–10		1.09			1.09			1.09			1.09			1.09			1.09		43–50
11–15		1.04			1.04			1.04			1.04			1.04			1.04		52–59
16–20		1.00			1.00			1.00			1.00			1.00			1.00		61–68
21–25		0.95			0.95			0.95			0.95			0.95			0.95		70–77
26–30		0.90			0.90			0.90			0.90			0.90			0.90		79–86

15.2.5 *NEC* (Annex B), Table B.310.7, Ampacities of Three Single-Insulated Conductors, Rated 0 through 2000 V, in Underground Electrical Ducts (Three Conductors per Electrical Duct), Based on Ambient Earth Temperature of 20°C (68°F), Electrical Duct Arrangement per *NEC* Figure B.310.2, Conductor Temperature 75°C (167°F)

TABLE 15.2.5

	1 Electrical Duct (Fig. B.310.2, Detail 1)			3 Electrical Ducts (Fig. B.310.2, Detail 2)			6 Electrical Ducts (Fig. B.310.2, Detail 3)			1 Electrical Duct (Fig. B.310.2, Detail 1)			3 Electrical Ducts (Fig. B.310.2, Detail 2)			6 Electrical Ducts (Fig. B.310.2, Detail 3)			
	Types RHW, THHW, THW, THWN, XHHW, USE			Types RHW, THHW, THW, THWN, XHHW, USE			Types RHW, THHW, THW, THWN, XHHW, USE			Types RHW, THHW, THW, THWN, XHHW, USE			Types RHW, THHW, THW, THWN, XHHW, USE			Types RHW, THHW, THW, THWN, XHHW, USE			
	COPPER									ALUMINUM OR COPPER-CLAD ALUMINUM									
Size (AWG or kcmil)	RHO 60 LF 50	RHO 90 LF 100	RHO 120 LF 100	RHO 60 LF 50	RHO 90 LF 100	RHO 120 LF 100	RHO 60 LF 50	RHO 90 LF 100	RHO 120 LF 100	RHO 60 LF 50	RHO 90 LF 100	RHO 120 LF 100	RHO 60 LF 50	RHO 90 LF 100	RHO 120 LF 100	RHO 60 LF 50	RHO 90 LF 100	RHO 120 LF 100	Size (AWG or kcmil)
8	63	58	57	61	51	49	57	44	41	49	45	44	47	40	38	45	34	32	8
6	84	77	75	80	67	63	75	56	53	66	60	58	63	52	49	59	44	41	6
4	111	100	98	105	86	81	98	73	67	86	78	76	79	67	63	77	57	52	4
3	129	116	113	122	99	94	113	83	77	101	91	89	83	77	73	84	65	60	3
2	147	132	128	139	112	106	129	93	86	115	103	100	108	87	82	101	73	67	2
1	171	153	148	161	128	121	149	106	98	133	119	115	126	100	94	116	83	77	1
1/0	197	175	169	185	146	137	170	121	111	153	136	132	144	114	107	133	94	87	1/0
2/0	226	200	193	212	166	156	194	136	126	176	156	151	165	130	121	151	106	98	2/0
3/0	260	228	220	243	189	177	222	154	142	203	178	172	189	147	138	173	121	111	3/0
4/0	301	263	253	280	215	201	255	175	161	235	205	198	219	168	157	199	137	126	4/0
250	334	290	279	310	236	220	281	192	176	261	227	218	242	185	172	220	150	137	250
300	373	321	308	344	260	242	310	210	192	293	252	242	272	204	190	245	165	151	300
350	409	351	337	377	283	264	340	228	209	321	276	265	296	222	207	266	179	164	350
400	442	376	361	394	302	280	368	243	223	349	297	284	321	238	220	288	191	174	400
500	503	427	409	460	341	316	412	273	249	397	338	323	364	270	250	326	216	197	500
600	552	468	447	511	371	343	457	296	270	446	373	356	408	296	274	365	236	215	600
700	602	509	486	553	402	371	492	319	291	488	408	389	443	321	297	394	255	232	700
750	632	529	505	574	417	385	509	330	301	508	425	405	461	334	309	409	265	241	750
800	654	544	520	597	428	395	527	338	308	530	439	418	481	344	318	427	273	247	800
900	692	575	549	628	450	415	554	355	323	563	466	444	510	365	337	450	288	261	900
1000	730	605	576	659	472	435	581	372	338	597	494	471	538	385	355	475	304	276	1000
Ambient Temp. (°C)	Correction Factors																		Ambient Temp. (°F)
6–10	1.09			1.09			1.09			1.09			1.09			1.09			43–50
11–15	1.04			1.04			1.04			1.04			1.04			1.04			52–59
16–20	1.00			1.00			1.00			1.00			1.00			1.00			61–68
21–25	0.95			0.95			0.95			0.95			0.95			0.95			70–77
26–30	0.90			0.90			0.90			0.90			0.90			0.90			79–86

15.2.6 *NEC* (Annex B), Table B.310.8, Ampacities of Two or Three Insulated Conductors, Rated 0 through 2000 V, Cabled within an Overall (Two- or Three-Conductor) Covering, Directly Buried in Earth, Based on Ambient Earth Temperature of 20°C (68°F), Arrangement per *NEC* Figure B.310.2, 100 Percent Load Factor, Thermal Resistance (Rho) of 90

TABLE 15.2.6

	1 Cable (Fig. B.310.2, Detail 5)		2 Cables (Fig. B.310.2, Detail 6)		1 Cable (Fig. B.310.2, Detail 5)		2 Cables (Fig. B.310.2, Detail 6)		
	60°C (140°F)	75°C (167°F)	60°C (140°F)	75°C (167°F)	60°C (140°F)	75°C (167°F)	60°C (140°F)	75°C (167°F)	
	TYPES				TYPES				
	UF	RHW, THHW, THW, THWN, XHHW, USE	UF	RHW, THHW, THW, THWN, XHHW, USE	UF	RHW, THHW, THW, THWN, XHHW, USE	UF	RHW, THHW, THW, THWN, XHHW, USE	
Size (AWG or kcmil)	COPPER				ALUMINUM OR COPPER-CLAD ALUMINUM				Size (AWG or kcmil)
8	64	75	60	70	51	59	47	55	8
6	85	100	81	95	68	75	60	70	6
4	107	125	100	117	83	97	78	91	4
2	137	161	128	150	107	126	110	117	2
1	155	182	145	170	121	142	113	132	1
1/0	177	208	165	193	138	162	129	151	1/0
2/0	201	236	188	220	157	184	146	171	2/0
3/0	229	269	213	250	179	210	166	195	3/0
4/0	259	304	241	282	203	238	188	220	4/0
250	—	333	—	308	—	261	—	241	250
350	—	401	—	370	—	315	—	290	350
500	—	481	—	442	—	381	—	350	500
750	—	585	—	535	—	473	—	433	750
1000	—	657	—	600	—	545	—	497	1000
Ambient Temp. (°C)	Correction Factors								Ambient Temp. (°F)
6–10	1.12	1.09	1.12	1.09	1.12	1.09	1.12	1.09	43–50
11–15	1.06	1.04	1.06	1.04	1.06	1.04	1.06	1.04	52–59
16–20	1.00	1.00	1.00	1.00	1.00	1.00	1.00	1.00	61–68
21–25	0.94	0.95	0.94	0.95	0.94	0.95	0.94	0.95	70–77
26–30	0.87	0.90	0.87	0.90	0.87	0.90	0.87	0.90	79–86

Note: For ampacities of Type UF cable in underground electrical ducts, multiply the ampacities shown in the table by 0.74.

15.2.7 *NEC* (Annex B), Table B.310.9, Ampacities of Three Triplexed Single-Insulated Conductors, Rated 0 through 2000 V, Directly Buried in Earth, Based on Ambient Earth Temperature of 20°C (68°F), Arrangement per *NEC* Figure B.310.2, 100 Percent Load Factor, Thermal Resistance (Rho) of 90

TABLE 15.2.7

Size (AWG or kcmil)	See Fig. B.310.2, Detail 7		See Fig. B.310.2, Detail 8		See Fig. B.310.2, Detail 7		See Fig. B.310.2, Detail 8		Size (AWG or kcmil)
	60°C (140°F)	75°C (167°F)	60°C (140°F)	75°C (167°F)	60°C (140°F)	75°C (167°F)	60°C (140°F)	75°C (167°F)	
	TYPES				TYPES				
	UF	USE	UF	USE	UF	USE	UF	USE	
	COPPER				ALUMINUM OR COPPER-CLAD ALUMINUM				
8	72	84	66	77	55	65	51	60	8
6	91	107	84	99	72	84	66	77	6
4	119	139	109	128	92	108	85	100	4
2	153	179	140	164	119	139	109	128	2
1	173	203	159	186	135	158	124	145	1
1/0	197	231	181	212	154	180	141	165	1/0
2/0	223	262	205	240	175	205	159	187	2/0
3/0	254	298	232	272	199	233	181	212	3/0
4/0	289	339	263	308	226	265	206	241	4/0
250	—	370	—	336	—	289	—	263	250
350	—	445	—	403	—	349	—	316	350
500	—	536	—	483	—	424	—	382	500
750	—	654	—	587	—	525	—	471	750
1000	—	744	—	665	—	608	—	544	1000
Ambient Temp. (°C)	Correction Factors								Ambient Temp. (°F)
6–10	1.12	1.09	1.12	1.09	1.12	1.09	1.12	1.09	43–50
11–15	1.06	1.04	1.06	1.04	1.06	1.04	1.06	1.04	52–59
16–20	1.00	1.00	1.00	1.00	1.00	1.00	1.00	1.00	61–68
21–25	0.94	0.95	0.94	0.95	0.94	0.95	0.94	0.95	70–77
26–30	0.87	0.90	0.87	0.90	0.87	0.90	0.87	0.90	79–86

15.2.8 *NEC* (Annex B), Table B.310.10, Ampacities of Three Single-Insulated Conductors, Rated 0 through 2000 V, Directly Buried in Earth, Based on Ambient Earth Temperature of 20°C (68°F), Arrangement per *NEC* Figure B.310.2, 100 Percent Load Factor, Thermal Resistance (Rho) of 90

TABLE 15.2.8

Size (AWG or kcmil)	See Fig. B.310.2, Detail 9		See Fig. B.310.2, Detail 10		See Fig. B.310.2, Detail 9		See Fig. B.310.2, Detail 10		Size (AWG or kcmil)
	60°C (140°F)	75°C (167°F)	60°C (140°F)	75°C (167°F)	60°C (140°F)	75°C (167°F)	60°C (140°F)	75°C (167°F)	
	TYPES				TYPES				
	UF	USE	UF	USE	UF	USE	UF	USE	
	COPPER				ALUMINUM OR COPPER-CLAD ALUMINUM				
8	84	98	78	92	66	77	61	72	8
6	107	126	101	118	84	98	78	92	6
4	139	163	130	152	108	127	101	118	4
2	178	209	165	194	139	163	129	151	2
1	201	236	187	219	157	184	146	171	1
1/0	230	270	212	249	179	210	165	194	1/0
2/0	261	306	241	283	204	239	188	220	2/0
3/0	297	348	274	321	232	272	213	250	3/0
4/0	336	394	309	362	262	307	241	283	4/0
250	—	429	—	394	—	335	—	308	250
350	—	516	—	474	—	403	—	370	350
500	—	626	—	572	—	490	—	448	500
750	—	767	—	700	—	605	—	552	750
1000	—	887	—	808	—	706	—	642	1000
1250	—	979	—	891	—	787	—	716	1250
1500	—	1063	—	965	—	862	—	783	1500
1750	—	1133	—	1027	—	930	—	843	1750
2000	—	1195	—	1082	—	990	—	897	2000
Ambient Temp. (°C)	Correction Factors								Ambient Temp. (°F)
6–10	1.12	1.09	1.12	1.09	1.12	1.09	1.12	1.09	43–50
11–15	1.06	1.04	1.06	1.04	1.06	1.04	1.06	1.04	52–59
16–20	1.00	1.00	1.00	1.00	1.00	1.00	1.00	1.00	61–68
21–25	0.94	0.95	0.94	0.95	0.94	0.95	0.94	0.95	70–77
26–30	0.87	0.90	0.87	0.90	0.87	0.90	0.87	0.90	79–86

15.3.1 *NEC* (Annex B), Figure B.310.1, Interpolation Chart for Cables in a Duct Bank I_1 = Ampacity for Rho = 60, 50 LF (Load Factor); I_2 = Ampacity for Rho = 120, 100 LF; Desired Ampacity = F × I_1

15.3.1

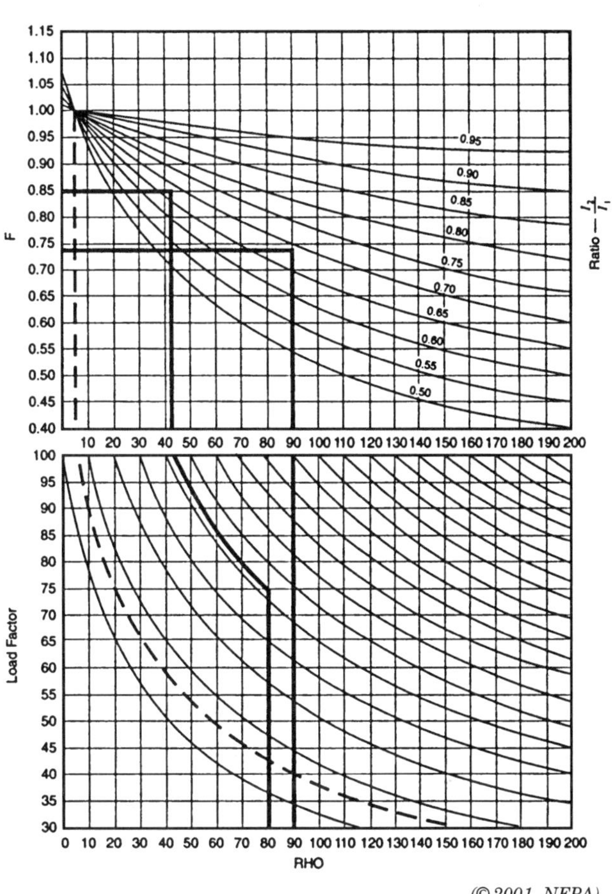

15.3.2 *NEC* (Annex B), Figure B.310.2, Cable Installation Dimensions for Use with *NEC* Tables B.310.5 through B.310.10

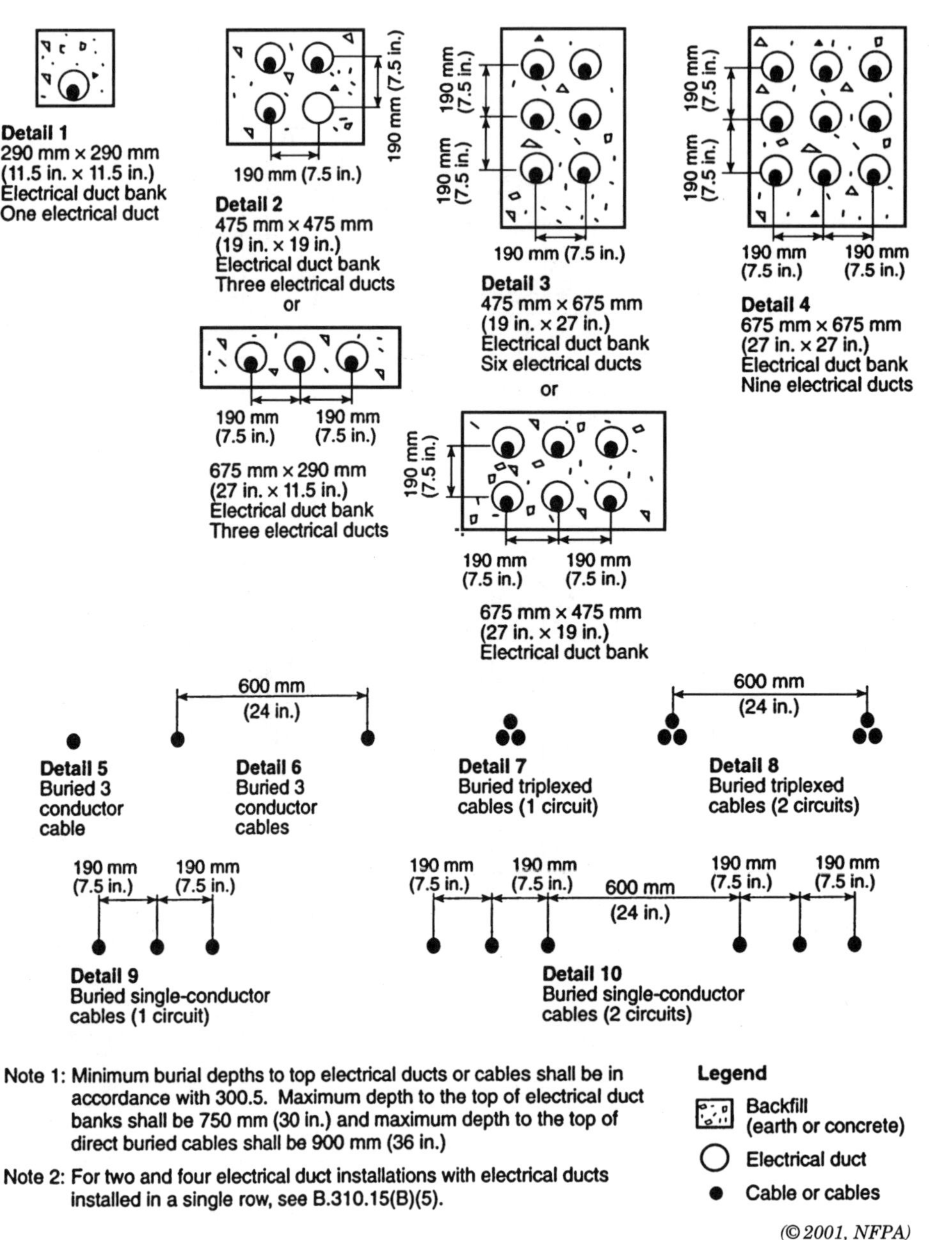

Note 1: Minimum burial depths to top electrical ducts or cables shall be in accordance with 300.5. Maximum depth to the top of electrical duct banks shall be 750 mm (30 in.) and maximum depth to the top of direct buried cables shall be 900 mm (36 in.)

Note 2: For two and four electrical duct installations with electrical ducts installed in a single row, see B.310.15(B)(5).

15.3.3 *NEC* (Annex B), Figure B.310.3, Ampacities of Single-Insulated Conductors, Rated 0 through 5000 V, in Underground Electrical Ducts (Three Conductors per Electrical Duct), Nine Single-Conductor Cables per Phase, Based on Ambient Earth Temperature of 20°C (68°F), Conductor Temperature 75°C (167°F)

15.3.3

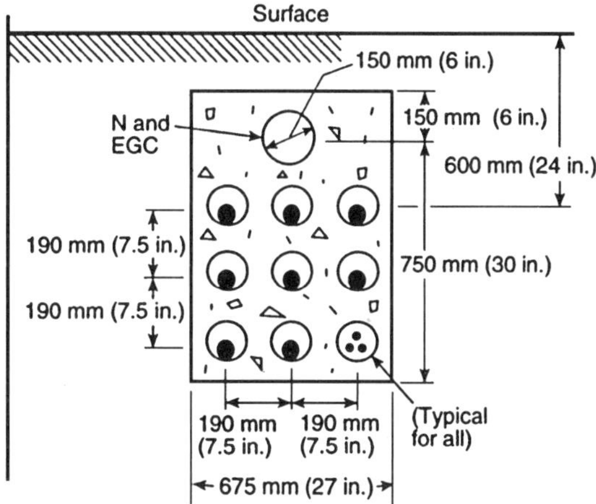

Design Criteria
Neutral and Equipment
Grounding conductor (EGC)
Duct = 150 mm (6 in.)
Phase Ducts = 75 to 125 mm (3 to 5 in.)
Conductor Material = Copper
Number of Cables per Duct = 3

Number of Cables per Phase = 9
Rho concrete = Rho Earth – 5

Rho PVC Duct = 650
Rho Cable Insulation = 500
Rho Cable Jacket = 650

Notes:

1. Neutral configuration per 300.5(I), Exception No. 2, for isolated phase installations in nonmagnetic ducts.
2. Phasing is A, B, C in rows or columns. Where magnetic electrical ducts are used, conductors are installed A, B, C per electrical duct with the neutral and all equipment grounding conductors in the same electrical duct. In this case, the 6-in. trade size neutral duct is eliminated.
3. Maximum harmonic loading on the neutral conductor cannot exceed 50 percent of the phase current for the ampacities shown in the table.
4. Metallic shields of Type MV-90 cable shall be grounded at one point only where using A, B, C phasing in rows or columns.

Size kcmil	TYPES RHW, THHW, THW, THWN, XHHW, USE, OR MV-90*			Size kcmil
	Total per Phase Ampere Rating			
	RHO EARTH 60 LF 50	RHO EARTH 90 LF 100	RHO EARTH 120 LF 100	
250	2340 (260A/Cable)	1530 (170A/Cable)	1395 (155A/Cable)	250
350	2790 (310A/Cable)	1800 (200A/Cable)	1665 (185A/Cable)	350
500	3375 (375A/Cable)	2160 (240A/Cable)	1980 (220A/Cable)	350

Ambient Temp. (°C)	For ambient temperatures other than 20°C (68°F), multiply the ampacities shown above by the appropriate factor shown below.					Ambient Temp. (°F)
6–10	1.09	1.09	1.09	1.09	1.09	43–50
11–15	1.04	1.04	1.04	1.04	1.04	52–59
16–20	1.00	1.00	1.00	1.00	1.00	61–68
21–25	0.95	0.95	0.95	0.95	0.95	70–77
26–30	0.90	0.90	0.90	0.90	0.90	79–86

*Limited to 75°C conductor temperature.

15.3.4 *NEC* (Annex B), Figure B.310.4, Ampacities of Single-Insulated Conductors, Rated 0 through 5000 V, in Nonmagnetic Underground Electrical Ducts (One Conductor per Electrical Duct), Four Single-Conductor Cables per Phase, Based on Ambient Earth Temperature of 20°C (68°F), Conductor Temperature 75°C (167°F)

15.3.4

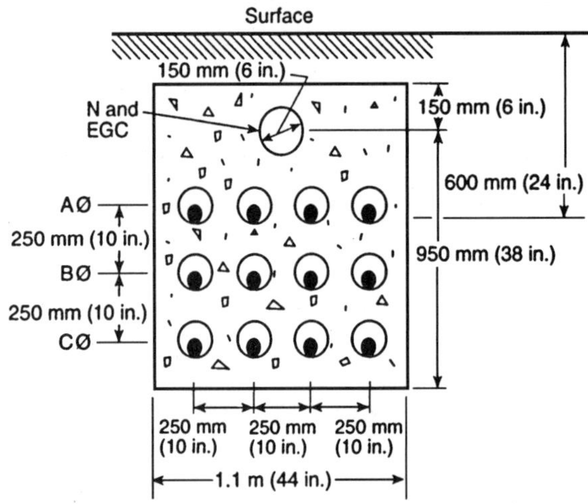

Design Criteria
Neutral and Equipment
Grounding Conductor (EGC)
Duct = 150 mm (6 in.)
Phase Ducts = 75 mm (3 in.)
Conductor Material = Copper
Number of Cables per Duct = 1

Number of Cables per Phase = 4
Rho Concrete = Rho Earth – 5
Rho PVC Duct = 650

Rho Cable Insulation = 500
Rho Cable Jacket = 650

Notes:
1. Neutral configuration per 300.5(I), Exception No. 2.
2. Maximum harmonic loading on the neutral conductor cannot exceed 50 percent of the phase current for the ampacities shown in the table.
3. Metallic shields of Type MV-90 cable shall be grounded at one point only.

Size kcmil	TYPES RHW, THHW, THW, THWN, XHHW, USE, OR MV-90* Total per Phase Ampere Rating			Size kcmil
	RHO EARTH 60 LF 50	RHO EARTH 90 LF 100	RHO EARTH 120 LF 100	
750	2520 (705A/Cable)	1860 (465A/Cable)	1680 (420A/Cable)	750
1000	3300 (825A/Cable)	2140 (535A/Cable)	1920 (480A/Cable)	1000
1250	3700 (925A/Cable)	2380 (595A/Cable)	2120 (530A/Cable)	1250
1500	4060 (1015A/Cable)	2580 (645A/Cable)	2300 (575A/Cable)	1500
1750	4360 (1090A/Cable)	2740 (685A/Cable)	2460 (615A/Cable)	1750

Ambient Temp. (°C)	For ambient temperatures other than 20°C (68°F), multiply the ampacities shown above by the appropriate factor shown below.					Ambient Temp. (°F)
6–10	1.09	1.09	1.09	1.09	1.09	43–50
11–15	1.04	1.04	1.04	1.04	1.04	52–59
16–20	1.00	1.00	1.00	1.00	1.00	61–68
21–25	0.95	0.95	0.95	0.95	0.95	70–77
26–30	0.90	0.90	0.90	0.90	0.90	79–86

*Limited to 75°C conductor temperature.

15.3.5 *NEC* (Annex B), Figure B.310.5, Ampacities of Single-Insulated Conductors, Rated 0 through 5000 V, in Nonmagnetic Underground Electrical Ducts (One Conductor per Electrical Duct), Five Single-Conductor Cables per Phase, Based on Ambient Earth Temperature of 20°C (68°F), Conductor Temperature 75°C (167°F)

15.3.5

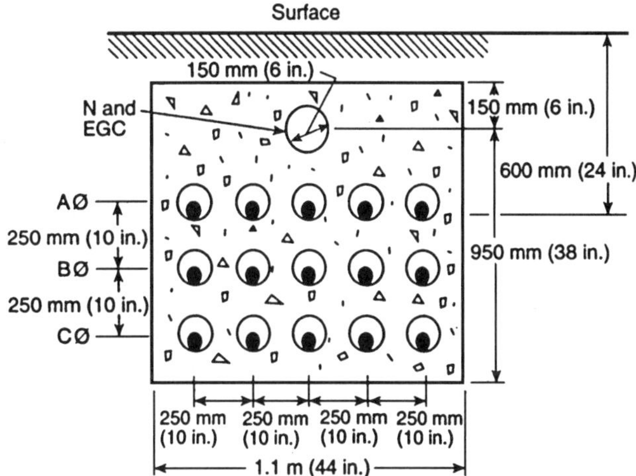

Design Criteria
Neutral and Equipment
Grounding Conductor (EGC)
Duct = 150 mm (6 in.)
Phase Ducts = 75 mm (3 in.)
Conductor Material = Copper
Number of Cables per Duct = 1

Number of Cables per Phase = 5
Rho Concrete = Rho Earth – 5
Rho PVC Duct = 650

Rho Cable Insulation = 500
Rho Cable Jacket = 650

Notes:
1. Neutral configuration per 300.5(I), Exception No. 2.
2. Maximum harmonic loading on the neutral conductor cannot exceed 50 percent of the phase current for the ampacities shown in the table.
3. Metallic shields of Type MV-90 cable shall be grounded at one point only.

Size kcmil	TYPES RHW, THHW, THW, THWN, XHHW, USE, OR MV-90* Total per Phase Ampere Rating			Size kcmil
	RHO EARTH 60 LF 50	RHO EARTH 90 LF 100	RHO EARTH 120 LF 100	
2000	5575 (1115A/Cable)	3375 (675A/Cable)	3000 (600A/Cable)	2000

Ambient Temp. (°C)	For ambient temperatures other than 20°C (68°F), multiply the ampacities shown above by the appropriate factor shown below.					Ambient Temp. (°F)
6–10	1.09	1.09	1.09	1.09	1.09	43–50
11–15	1.04	1.04	1.04	1.04	1.04	52–59
16–20	1.00	1.00	1.00	1.00	1.00	61–68
21–25	0.95	0.95	0.95	0.95	0.95	70–77
26–30	0.90	0.90	0.90	0.90	0.90	79–86

*Limited to 75°C conductor temperature.

15.3.6 *NEC* (Annex B), Table B.310.11, Adjustment Factors for More Than Three Current-Carrying Conductors in a Raceway or Cable with Load Diversity

TABLE 15.3.6

Number of Current-Carrying Conductors	Percent of Values in Tables as Adjusted for Ambient Temperature if Necessary
4 – 6	80
7 – 9	70
10 – 24	70*
25 – 42	60*
43 – 85	50*

*These factors include the effects of a load diversity of 50 percent.

FPN: The ampacity limit for the number of current-carrying conductors in 10 through 85 is based on the following formula. For greater than 85 conductors, special calculations are required that are beyond the scope of this table.

$$A_2 = \sqrt{\frac{0.5N}{E}} \times (A_1) \text{ or } A_1, \text{whichever is less}$$

where:

A_1 = ampacity from Tables 310.16; 310.18; B.310.1; B.310.6; and B.310.7 multiplied by the appropriate factor from Table B.310.11.

N = total number of conductors used to obtain multiplying factor from Table B.310.11

E = desired number of current-carrying conductors in the raceway or cable

A_2 = ampacity limit for the current-carrying conductors in the raceway or cable

Example 1

Calculate the ampacity limit for twelve 14 AWG THWN current-carrying conductors (75°C) in a raceway that contains 24 conductors.

$$A_2 = \sqrt{\frac{(0.5)(24)}{12}} \times 20(0.7)$$

$$= 14 \text{ amperes (i.e., 50 percent diversity)}$$

Example 2

Calculate the ampacity limit for eighteen 14 AWG THWN current-carrying conductors (75°C) in a raceway that contains 24 conductors.

$$A_2 = \sqrt{\frac{(0.5)(24)}{18}} \times 20(0.7) = 11.5 \text{ amperes}$$

15.4.0 *NEC* Annex C, Conduit and Tube Fill Tables for Conductors and Fixture Wires of the Same Size

In all the tables that follow, Tables C1 and C1(A) through C12 and C12(A), conduit and tube fills are based on *NEC* Chapter 9, Table 1.

15.4.1 Table C1, Maximum Number of Conductors or Fixture Wires in Electrical Metallic Tubing

TABLE 15.4.1

CONDUCTORS											
	Conductor	Metric Designator (Trade Size)									
Type	Size (AWG/kcmil)	16 (½)	21 (¾)	27 (1)	35 (1¼)	41 (1½)	53 (2)	63 (2½)	78 (3)	91 (3½)	103 (4)
RHH, RHW, RHW-2	14	4	7	11	20	27	46	80	120	157	201
	12	3	6	9	17	23	38	66	100	131	167
	10	2	5	8	13	18	30	53	81	105	135
	8	1	2	4	7	9	16	28	42	55	70
	6	1	1	3	5	8	13	22	34	44	56
	4	1	1	2	4	6	10	17	26	34	44
	3	1	1	1	4	5	9	15	23	30	38
	2	1	1	1	3	4	7	13	20	26	33
	1	0	1	1	1	3	5	9	13	17	22
	1/0	0	1	1	1	2	4	7	11	15	19
	2/0	0	1	1	1	2	4	6	10	13	17
	3/0	0	0	1	1	1	3	5	8	11	14
	4/0	0	0	1	1	1	3	5	7	9	12
	250	0	0	0	1	1	1	3	5	7	9
	300	0	0	0	1	1	1	3	5	6	8
	350	0	0	0	1	1	1	3	4	6	7
	400	0	0	0	1	1	1	2	4	5	7
	500	0	0	0	0	1	1	2	3	4	6
	600	0	0	0	0	1	1	1	3	4	5
	700	0	0	0	0	0	1	1	2	3	4
	750	0	0	0	0	0	1	1	2	3	4
	800	0	0	0	0	0	1	1	2	3	4
	900	0	0	0	0	0	1	1	1	3	3
	1000	0	0	0	0	0	1	1	1	2	3
	1250	0	0	0	0	0	0	1	1	1	2
	1500	0	0	0	0	0	0	1	1	1	1
	1750	0	0	0	0	0	0	1	1	1	1
	2000	0	0	0	0	0	0	1	1	1	1
TW, THHW, THW, THW-2	14	8	15	25	43	58	96	168	254	332	424
	12	6	11	19	33	45	74	129	195	255	326
	10	5	8	14	24	33	55	96	145	190	243
	8	2	5	8	13	18	30	53	81	105	135

(continued)

TABLE 15.4.1 Maximum Number of Conductors or Fixture Wires in Electrical Metallic Tubing *(Continued)*

CONDUCTORS											
	Conductor	Metric Designator (Trade Size)									
Type	Size (AWG/kcmil)	16 (½)	21 (¾)	27 (1)	35 (1¼)	41 (1½)	53 (2)	63 (2½)	78 (3)	91 (3½)	103 (4)
RHH*,	14	6	10	16	28	39	64	112	169	221	282
RHW*,	12	4	8	13	23	31	51	90	136	177	227
RHW-2*	10	3	6	10	18	24	40	70	106	138	177
	8	1	4	6	10	14	24	42	63	83	106
RHH*,	6	1	3	4	8	11	18	32	48	63	81
RHW*,	4	1	1	3	6	8	13	24	36	47	60
RHW-2*,	3	1	1	3	5	7	12	20	31	40	52
TW, THW,	2	1	1	2	4	6	10	17	26	34	44
THHW,	1	1	1	1	3	4	7	12	18	24	31
THW-2	1/0	0	1	1	2	3	6	10	16	20	26
	2/0	0	1	1	1	3	5	9	13	17	22
	3/0	0	1	1	1	2	4	7	11	15	19
	4/0	0	0	1	1	1	3	6	9	12	16
	250	0	0	1	1	1	3	5	7	10	13
	300	0	0	1	1	1	2	4	6	8	11
	350	0	0	0	1	1	1	4	6	7	10
	400	0	0	0	1	1	1	3	5	7	9
	500	0	0	0	1	1	1	3	4	6	7
	600	0	0	0	1	1	1	2	3	4	6
	700	0	0	0	0	1	1	1	3	4	5
	750	0	0	0	0	1	1	1	3	4	5
	800	0	0	0	0	1	1	1	3	3	5
	900	0	0	0	0	0	1	1	2	3	4
	1000	0	0	0	0	0	1	1	2	3	4
	1250	0	0	0	0	0	1	1	1	2	3
	1500	0	0	0	0	0	1	1	1	1	2
	1750	0	0	0	0	0	0	1	1	1	2
	2000	0	0	0	0	0	0	1	1	1	1
THHN,	14	12	22	35	61	84	138	241	364	476	608
THWN,	12	9	16	26	45	61	101	176	266	347	443
THWN-2	10	5	10	16	28	38	63	111	167	219	279
	8	3	6	9	16	22	36	64	96	126	161
	6	2	4	7	12	16	26	46	69	91	116
	4	1	2	4	7	10	16	28	43	56	71
	3	1	1	3	6	8	13	24	36	47	60
	2	1	1	3	5	7	11	20	30	40	51
	1	1	1	1	4	5	8	15	22	29	37
	1/0	1	1	1	3	4	7	12	19	25	32
	2/0	0	1	1	2	3	6	10	16	20	26
	3/0	0	1	1	1	3	5	8	13	17	22
	4/0	0	1	1	1	2	4	7	11	14	18
	250	0	0	1	1	1	3	6	9	11	15
	300	0	0	1	1	1	3	5	7	10	13
	350	0	0	1	1	1	2	4	6	9	11
	400	0	0	0	1	1	1	4	6	8	10
	500	0	0	0	1	1	1	3	5	6	8
	600	0	0	0	1	1	1	2	4	5	7
	700	0	0	0	1	1	1	2	3	4	6
	750	0	0	0	0	1	1	1	3	4	5
	800	0	0	0	0	1	1	1	3	4	5
	900	0	0	0	0	1	1	1	3	3	4
	1000	0	0	0	0	1	1	1	2	3	4
FEP,	14	12	21	34	60	81	134	234	354	462	590
FEPB,	12	9	15	25	43	59	98	171	258	337	430
PFA,	10	6	11	18	31	42	70	122	185	241	309
PFAH,	8	3	6	10	18	24	40	70	106	138	177
TFE	6	2	4	7	12	17	28	50	75	98	126
	4	1	3	5	9	12	20	35	53	69	88
	3	1	2	4	7	10	16	29	44	57	73
	2	1	1	3	6	8	13	24	36	47	60
PFA, PFAH, TFE	1	1	1	2	4	6	9	16	25	33	42
PFAH,	1/0	1	1	1	3	5	8	14	21	27	35
TFE PFA,	2/0	0	1	1	3	4	6	11	17	22	29
PFAH,	3/0	0	1	1	2	3	5	9	14	18	24
TFE, Z	4/0	0	1	1	1	2	4	8	11	15	19

(continued)

TABLE 15.4.1 Maximum Number of Conductors or Fixture Wires in Electrical Metallic Tubing *(Continued)*

CONDUCTORS

Type	Conductor Size (AWG/kcmil)	Metric Designator (Trade Size) 16 (½)	21 (¾)	27 (1)	35 (1¼)	41 (1½)	53 (2)	63 (2½)	78 (3)	91 (3½)	103 (4)
Z	14	14	25	41	72	98	161	282	426	556	711
	12	10	18	29	51	69	114	200	302	394	504
	10	6	11	18	31	42	70	122	185	241	309
	8	4	7	11	20	27	44	77	117	153	195
	6	3	5	8	14	19	31	54	82	107	137
	4	1	3	5	9	13	21	37	56	74	94
	3	1	2	4	7	9	15	27	41	54	69
	2	1	1	3	6	8	13	22	34	45	57
	1	1	1	2	4	6	10	18	28	36	46
XHH, XHHW, XHHW-2, ZW	14	8	15	25	43	58	96	168	254	332	424
	12	6	11	19	33	45	74	129	195	255	326
	10	5	8	14	24	33	55	96	145	190	243
	8	2	5	8	13	18	30	53	81	105	135
	6	1	3	6	10	14	22	39	60	78	100
	4	1	2	4	7	10	16	28	43	56	72
	3	1	1	3	6	8	14	24	36	48	61
	2	1	1	3	5	7	11	20	31	40	51
XHH, XHHW, XHHW-2	1	1	1	1	4	5	8	15	23	30	38
	1/0	1	1	1	3	4	7	13	19	25	32
	2/0	0	1	1	2	3	6	10	16	21	27
	3/0	0	1	1	1	3	5	9	13	17	22
	4/0	0	1	1	1	2	4	7	11	14	18
	250	0	0	1	1	1	3	6	9	12	15
	300	0	0	1	1	1	3	5	8	10	13
	350	0	0	1	1	1	2	4	7	9	11
	400	0	0	0	1	1	1	4	6	8	10
	500	0	0	0	1	1	1	3	5	6	8
	600	0	0	0	1	1	1	2	4	5	6
	700	0	0	0	0	1	1	2	3	4	6
	750	0	0	0	0	1	1	1	3	4	5
	800	0	0	0	0	1	1	1	3	4	5
	900	0	0	0	0	1	1	1	3	3	4
	1000	0	0	0	0	0	1	1	2	3	4
	1250	0	0	0	0	0	1	1	1	2	3
	1500	0	0	0	0	0	1	1	1	1	3
	1750	0	0	0	0	0	0	1	1	1	2
	2000	0	0	0	0	0	0	1	1	1	1

FIXTURE WIRES

Type	Conductor Size (AWG/kcmil)	Metric Designator (Trade Size) 16 (½)	21 (¾)	27 (1)	35 (1¼)	41 (1½)	53 (2)
FFH-2, RFH-2, RFHH-3	18	8	14	24	41	56	92
	16	7	12	20	34	47	78
SF-2, SFF-2	18	10	18	30	52	71	116
	16	8	15	25	43	58	96
	14	7	12	20	34	47	78
SF-1, SFF-1	18	18	33	53	92	125	206
RFH-1, RFHH-2, TF, TFF, XF, XFF	18	14	24	39	68	92	152
RFHH-2, TF, TFF, XF, XFF	16	11	19	31	55	74	123
XF, XFF	14	8	15	25	43	58	96
TFN, TFFN	18	22	38	63	108	148	244
	16	17	29	48	83	113	186
PF, PFF, PGF, PGFF, PAF, PTF, PTFF, PAFF	18	21	36	59	103	140	231
	16	16	28	46	79	108	179
	14	12	21	34	60	81	134
ZF, ZFF, ZHF, HF, HFF	18	27	47	77	133	181	298
	16	20	35	56	98	133	220
	14	14	25	41	72	98	161
KF-2, KFF-2	18	39	69	111	193	262	433
	16	27	48	78	136	185	305
	14	19	33	54	93	127	209
	12	13	23	37	64	87	144
	10	8	15	25	43	58	96
KF-1, KFF-1	18	46	82	133	230	313	516
	16	33	57	93	161	220	362
	14	22	38	63	108	148	244
	12	14	25	41	72	98	161
	10	9	16	27	47	64	105
XF, XFF	12	4	8	13	23	31	51
	10	3	6	10	18	24	40

Note; This table is for concentric stranded conductors only. For compact stranded conductors, Table C1(A) should be used.

*Types RHH, RHW, and RHW-2 without outer covering.

15.4.2 Table C1(A), Maximum Number of Compact Conductors in Electrical Metallic Tubing

TABLE 15.4.2

COMPACT CONDUCTORS

Type	Conductor Size (AWG/kcmil)	Metric Designator (Trade Size) 16 (½)	21 (¾)	27 (1)	35 (1¼)	41 (1½)	53 (2)	63 (2½)	78 (3)	91 (3½)	103 (4)
THW, THW-2, THHW	8	2	4	6	11	16	26	46	69	90	115
	6	1	3	5	9	12	20	35	53	70	89
	4	1	2	4	6	9	15	26	40	52	67
	2	1	1	3	5	7	11	19	29	38	49
	1	1	1	1	3	4	8	13	21	27	34
	1/0	1	1	1	3	4	7	12	18	23	30
	2/0	0	1	1	2	3	5	10	15	20	25
	3/0	0	1	1	1	3	5	8	13	17	21
	4/0	0	1	1	1	2	4	7	11	14	18
	250	0	0	1	1	1	3	5	8	11	14
	300	0	0	1	1	1	3	5	7	9	12
	350	0	0	1	1	1	2	4	6	8	11
	400	0	0	0	1	1	1	4	6	8	10
	500	0	0	0	1	1	1	3	5	6	8
	600	0	0	0	1	1	1	2	4	5	7
	700	0	0	0	1	1	1	2	3	4	6
	750	0	0	0	0	1	1	1	3	4	5
	1000	0	0	0	0	1	1	1	2	3	4
THHN, THWN, THWN-2	8	—	—	—	—	—	—	—	—	—	—
	6	2	4	7	13	18	29	52	78	102	130
	4	1	3	4	8	11	18	32	48	63	81
	2	1	1	3	6	8	13	23	34	45	58
	1	1	1	2	4	6	10	17	26	34	43
	1/0	1	1	1	3	5	8	14	22	29	37
	2/0	1	1	1	3	4	7	12	18	24	30
	3/0	0	1	1	2	3	6	10	15	20	25
	4/0	0	1	1	1	3	5	8	12	16	21
	250	0	1	1	1	1	4	6	10	13	16
	300	0	0	1	1	1	3	5	8	11	14
	350	0	0	1	1	1	3	5	7	10	12
	400	0	0	1	1	1	2	4	6	9	11
	500	0	0	0	1	1	1	4	5	7	9
	600	0	0	0	1	1	1	3	4	6	7
	700	0	0	0	1	1	1	2	4	5	7
	750	0	0	0	1	1	1	2	4	5	6
	1000	0	0	0	0	1	1	1	3	3	4
XHHW, XHHW-2	8	3	5	8	15	20	34	59	90	117	149
	6	1	4	6	11	15	25	44	66	87	111
	4	1	3	4	8	11	18	32	48	63	81
	2	1	1	3	6	8	13	23	34	45	58
	1	1	1	2	4	6	10	17	26	34	43
	1/0	1	1	1	3	5	8	14	22	29	37
	2/0	1	1	1	3	4	7	12	18	24	31
	3/0	0	1	1	2	3	6	10	15	20	25
	4/0	0	1	1	1	3	5	8	13	17	21
	250	0	1	1	1	2	4	7	10	13	17
	300	0	0	1	1	1	3	6	9	11	14
	350	0	0	1	1	1	3	5	8	10	13
	400	0	0	1	1	1	2	4	7	9	11
	500	0	0	0	1	1	1	4	6	7	9
	600	0	0	0	1	1	1	3	4	6	8
	700	0	0	0	1	1	1	2	4	5	7
	750	0	0	0	1	1	1	2	3	5	6
	1000	0	0	0	0	1	1	1	3	4	5

Definition: *Compact stranding* is the result of a manufacturing process where the standard conductor is compressed to the extent that the interstices (voids between strand wires) are virtually eliminated.

15.4.3 Table C2, Maximum Number of Conductors or Fixture Wires in Electrical Nonmetallic Tubing

TABLE 15.4.3

CONDUCTORS							
		Metric Designator (Trade Size)					
Type	Conductor Size (AWG/kcmil)	16 (½)	21 (¾)	27 (1)	35 (1¼)	41 (1½)	53 (2)
RHH, RHW, RHW-2	14	3	6	10	19	26	43
	12	2	5	9	16	22	36
RHH, RHW, RHW-2	10	1	4	7	13	17	29
	8	1	1	3	6	9	15
	6	1	1	3	5	7	12
	4	1	1	2	4	6	9
	3	1	1	1	3	5	8
	2	0	1	1	3	4	7
	1	0	1	1	1	3	5
	1/0	0	0	1	1	2	4
	2/0	0	0	1	1	1	3
	3/0	0	0	1	1	1	3
	4/0	0	0	1	1	1	2
	250	0	0	0	1	1	1
	300	0	0	0	1	1	1
	350	0	0	0	1	1	1
	400	0	0	0	1	1	1
	500	0	0	0	0	1	1
	600	0	0	0	0	1	1
	700	0	0	0	0	0	1
	750	0	0	0	0	0	1
	800	0	0	0	0	0	1
	900	0	0	0	0	0	1
	1000	0	0	0	0	0	1
	1250	0	0	0	0	0	0
	1500	0	0	0	0	0	0
	1750	0	0	0	0	0	0
	2000	0	0	0	0	0	0
TW, THHW, THW, THW-2	14	7	13	22	40	55	92
	12	5	10	17	31	42	71
	10	4	7	13	23	32	52
	8	1	4	7	13	17	29
RHH*, RHW*, RHW-2*	14	4	8	15	27	37	61
RHH*, RHW*, RHW-2*	12	3	7	12	21	29	49
	10	3	5	9	17	23	38
RHH*, RHW*, RHW-2*	8	1	3	5	10	14	23
RHH*, RHW*, RHW-2*, TW, THW, THHW, THW-2	6	1	2	4	7	10	17
	4	1	1	3	5	8	13
	3	1	1	2	5	7	11
	2	1	1	2	4	6	9
	1	0	1	1	3	4	6
	1/0	0	1	1	2	3	5
	2/0	0	1	1	1	3	5
	3/0	0	0	1	1	2	4
	4/0	0	0	1	1	1	3
	250	0	0	1	1	1	2
	300	0	0	0	1	1	2
	350	0	0	0	1	1	1
	400	0	0	0	1	1	1
	500	0	0	0	1	1	1
	600	0	0	0	0	1	1
	700	0	0	0	0	1	1
	750	0	0	0	0	1	1
	800	0	0	0	0	1	1
	900	0	0	0	0	0	1
	1000	0	0	0	0	0	1
	1250	0	0	0	0	0	1
	1500	0	0	0	0	0	0
	1750	0	0	0	0	0	0
	2000	0	0	0	0	0	0

(continued)

TABLE 15.4.3 Maximum Number of Conductors or Fixture Wires in Electrical Nonmetallic Tubing (*Continued*)

CONDUCTORS							
	Conductor Size	Metric Designator (Trade Size)					
Type	(AWG/kcmil)	16 (½)	21 (¾)	27 (1)	35 (1¼)	41 (1½)	53 (2)
THHN,	14	10	18	32	58	80	132
THWN,	12	7	13	23	42	58	96
THWN-2	10	4	8	15	26	36	60
	8	2	5	8	15	21	35
	6	1	3	6	11	15	25
	4	1	1	4	7	9	15
	3	1	1	3	5	8	13
	2	1	1	2	5	6	11
	1	1	1	1	3	5	8
	1/0	0	1	1	3	4	7
	2/0	0	1	1	2	3	5
	3/0	0	1	1	1	3	4
	4/0	0	0	1	1	2	4
	250	0	0	1	1	1	3
	300	0	0	1	1	1	2
	350	0	0	0	1	1	2
	400	0	0	0	1	1	1
	500	0	0	0	1	1	1
	600	0	0	0	1	1	1
	700	0	0	0	0	1	1
	750	0	0	0	0	1	1
	800	0	0	0	0	1	1
	900	0	0	0	0	1	1
	1000	0	0	0	0	0	1
FEP, FEPB,	14	10	18	31	56	77	128
PFA, PFAH,	12	7	13	23	41	56	93
TFE	10	5	9	16	29	40	67
	8	3	5	9	17	23	38
	6	1	4	6	12	16	27
	4	1	2	4	8	11	19
	3	1	1	4	7	9	16
	2	1	1	3	5	8	13
PFA, PFAH, TFE	1	1	1	1	4	5	9
PFA, PFAH,	1/0	0	1	1	3	4	7
TFE, Z	2/0	0	1	1	2	4	6
	3/0	0	1	1	1	3	5
	4/0	0	1	1	1	2	4
Z	14	12	22	38	68	93	154
	12	8	15	27	48	66	109
	10	5	9	16	29	40	67
	8	3	6	10	18	25	42
	6	1	4	7	13	18	30
	4	1	3	5	9	12	20
	3	1	1	3	6	9	15
	2	1	1	3	5	7	12
	1	1	1	2	4	6	10
XHH,	14	7	13	22	40	55	92
XHHW,	12	5	10	17	31	42	71
XHHW-2,	10	4	7	13	23	32	52
ZW	8	1	4	7	13	17	29
	6	1	3	5	9	13	21
	4	1	1	4	7	9	15
	3	1	1	3	6	8	13
	2	1	1	2	5	6	11
XHH,	1	1	1	1	3	5	8
XHHW,	1/0	0	1	1	3	4	7
XHHW-2	2/0	0	1	1	2	3	6
	3/0	0	1	1	1	3	5
	4/0	0	0	1	1	2	4
	250	0	0	1	1	1	3
	300	0	0	1	1	1	3
	350	0	0	1	1	1	2
	400	0	0	0	1	1	1
	500	0	0	0	1	1	1
	600	0	0	0	1	1	1
	700	0	0	0	0	1	1
	750	0	0	0	0	1	1
	800	0	0	0	0	1	1
	900	0	0	0	0	1	1
	1000	0	0	0	0	0	1
	1250	0	0	0	0	0	1
	1500	0	0	0	0	0	1
	1750	0	0	0	0	0	0
	2000	0	0	0	0	0	0

(continued)

TABLE 15.4.3 Maximum Number of Conductors or Fixture Wires in Electrical Nonmetallic Tubing (*Continued*)

FIXTURE WIRES

Type	Conductor Size (AWG/kcmil)	16 (½)	21 (¾)	27 (1)	35 (1¼)	41 (1½)	53 (2)
		Metric Designator (Trade Size)					
FFH-2, RFH-2, RFHH-3	18	6	12	21	39	53	88
	16	5	10	18	32	45	74
SF-2, SFF-2	18	8	15	27	49	67	111
	16	7	13	22	40	55	92
	14	5	10	18	32	45	74
SF-1, SFF-1	18	15	28	48	86	119	197
RFH-1, RFHH-2, TF, TFF, XF, XFF	18	11	20	35	64	88	145
RFHH-2, TF, TFF, XF, XFF	16	9	16	29	51	71	117
XF, XFF	14	7	13	22	40	55	92
TFN, TFFN	18	18	33	57	102	141	233
	16	13	25	43	78	107	178
PF, PFF, PGF, PGFF, PAF, PTF, PTFF, PAFF	18	17	31	54	97	133	221
	16	13	24	42	75	103	171
	14	10	18	31	56	77	128
ZF, ZFF, ZHF, HF, HFF	18	22	40	70	125	172	285
	16	16	29	51	92	127	210
	14	12	22	38	68	93	154
KF-2, KFF-2	18	31	58	101	182	250	413
	16	22	41	71	128	176	291
	14	15	28	49	88	121	200
	12	10	19	33	60	83	138
	10	7	13	22	40	55	92
KF-1, KFF-1	18	38	69	121	217	298	493
	16	26	49	85	152	209	346
	14	18	33	57	102	141	233
	12	12	22	38	68	93	154
	10	7	14	24	44	61	101
XF, XFF	12	3	7	12	21	29	49
	10	3	5	9	17	23	38

Note: This table is for concentric stranded conductors only. For compact stranded conductors, Table C2(A) should be used.

*Types RHH, RHW, and RHW-2 without outer covering.

15.4.4 Table C2(A), Maximum Number of Compact Conductors in Electrical Nonmetallic Tubing

TABLE 15.4.4

COMPACT CONDUCTORS

Type	Conductor Size (AWG/kcmil)	Metric Designator (Trade Size) 16 (½)	21 (¾)	27 (1)	35 (1¼)	41 (1½)	53 (2)
THW, THW-2, THHW	8	1	3	6	11	15	25
	6	1	2	4	8	11	19
	4	1	1	3	6	8	14
	2	1	1	2	4	6	10
	1	0	1	1	3	4	7
	1/0	0	1	1	3	4	6
	2/0	0	1	1	2	3	5
	3/0	0	1	1	1	3	4
	4/0	0	0	1	1	2	4
	250	0	0	1	1	1	3
	300	0	0	1	1	1	2
	350	0	0	0	1	1	2
	400	0	0	0	1	1	1
	500	0	0	0	1	1	1
	600	0	0	0	1	1	1
	700	0	0	0	0	1	1
	750	0	0	0	0	1	1
	1000	0	0	0	0	0	1
THHN, THWN, THWN-2	8	—	—	—	—	—	—
	6	1	4	7	12	17	28
	4	1	2	4	7	10	17
	2	1	1	3	5	7	12
	1	1	1	2	4	5	9
	1/0	1	1	1	3	5	8
	2/0	0	1	1	3	4	6
	3/0	0	1	1	2	3	5
	4/0	0	1	1	1	2	4
	250	0	0	1	1	1	3
	300	0	0	1	1	1	3
	350	0	0	1	1	1	2
	400	0	0	0	1	1	2
	500	0	0	0	1	1	1
	600	0	0	0	1	1	1
	700	0	0	0	1	1	1

COMPACT CONDUCTORS

Type	Conductor Size (AWG/kcmil)	Metric Designator (Trade Size) 16 (½)	21 (¾)	27 (1)	35 (1¼)	41 (1½)	53 (2)
THHN, THWN, THWN-2	750	0	0	0	1	1	1
	1000	0	0	0	0	1	1
XHHW, XHHW-2	8	2	4	8	14	19	32
	6	1	3	6	10	14	24
	4	1	2	4	7	10	17
	2	1	1	3	5	7	12
	1	1	1	2	4	5	9
	1/0	1	1	1	3	5	8
	2/0	0	1	1	3	4	7
	3/0	0	1	1	2	3	5
	4/0	0	1	1	1	3	4
	250	0	0	1	1	1	3
	300	0	0	1	1	1	3
	350	0	0	1	1	1	3
	400	0	0	1	1	1	2
	500	0	0	0	1	1	1
	600	0	0	0	1	1	1
	700	0	0	0	1	1	1
	750	0	0	0	1	1	1
	1000	0	0	0	0	1	1

Definition: *Compact stranding* is the result of a manufacturing process where the standard conductor is compressed to the extent that the interstices (voids between strand wires) are virtually eliminated.

15.4.5 Table C3, Maximum Number of Conductors or Fixture Wires in Flexible Metal Conduit

TABLE 15.4.5

CONDUCTORS											
	Conductor	Metric Designator (Trade Size)									
Type	Size (AWG/kcmil)	16 (½)	21 (¾)	27 (1)	35 (1¼)	41 (1½)	53 (2)	63 (2½)	78 (3)	91 (3½)	103 (4)
RHH, RHW, RHW-2	14	4	7	11	17	25	44	67	96	131	171
	12	3	6	9	14	21	37	55	80	109	142
RHH, RHW, RHW-2	10	3	5	7	11	17	30	45	64	88	115
	8	1	2	4	6	9	15	23	34	46	60
	6	1	1	3	5	7	12	19	27	37	48
	4	1	1	2	4	5	10	14	21	29	37
	3	1	1	1	3	5	8	13	18	25	33
	2	1	1	1	3	4	7	11	16	22	28
	1	0	1	1	1	2	5	7	10	14	19
	1/0	0	1	1	1	2	4	6	9	12	16
	2/0	0	1	1	1	1	3	5	8	11	14
	3/0	0	0	1	1	1	3	5	7	9	12
	4/0	0	0	1	1	1	2	4	6	8	10
	250	0	0	0	1	1	1	3	4	6	8
	300	0	0	0	1	1	1	2	4	5	7
	350	0	0	0	1	1	1	2	3	5	6
	400	0	0	0	0	1	1	1	3	4	6
	500	0	0	0	0	1	1	1	3	4	5
	600	0	0	0	0	1	1	1	2	3	4
	700	0	0	0	0	0	1	1	1	3	3
	750	0	0	0	0	0	1	1	1	2	3
	800	0	0	0	0	0	1	1	1	2	3
	900	0	0	0	0	0	1	1	1	2	3
	1000	0	0	0	0	0	1	1	1	1	3
	1250	0	0	0	0	0	0	1	1	1	1
	1500	0	0	0	0	0	0	1	1	1	1
	1750	0	0	0	0	0	0	1	1	1	1
	2000	0	0	0	0	0	0	0	1	1	1
TW, THHW, THW, THW-2	14	9	15	23	36	53	94	141	203	277	361
	12	7	11	18	28	41	72	108	156	212	277
	10	5	8	13	21	30	54	81	116	158	207
	8	3	5	7	11	17	30	45	64	88	115
RHH*, RHW*, RHW-2*	14	6	10	15	24	35	62	94	135	184	240
RHH*, RHW*, RHW-2*	12	5	8	12	19	28	50	75	108	148	193
	10	4	6	10	15	22	39	59	85	115	151
RHH*, RHW*, RHW-2*	8	1	4	6	9	13	23	35	51	69	90

(continued)

TABLE 15.4.5 Maximum Number of Conductors or Fixture Wires in Flexible Metal Conduit *(Continued)*

CONDUCTORS											
	Conductor	Metric Designator (Trade Size)									
	Size	16	21	27	35	41	53	63	78	91	103
Type	(AWG/kcmil)	(½)	(¾)	(1)	(1¼)	(1½)	(2)	(2½)	(3)	(3½)	(4)
RHH*, RHW*, RHW-2*, TW, THW, THHW, THW-2	6	1	3	4	7	10	18	27	39	53	69
	4	1	1	3	5	7	13	20	29	39	51
	3	1	1	3	4	6	11	17	25	34	44
	2	1	1	2	4	5	10	14	21	29	37
	1	1	1	1	2	4	7	10	15	20	26
	1/0	0	1	1	1	3	6	9	12	17	22
	2/0	0	1	1	1	3	5	7	10	14	19
	3/0	0	1	1	1	2	4	6	9	12	16
	4/0	0	0	1	1	1	3	5	7	10	13
	250	0	0	1	1	1	3	4	6	8	11
	300	0	0	1	1	1	2	3	5	7	9
	350	0	0	0	1	1	1	3	4	6	8
	400	0	0	0	1	1	1	3	4	6	7
	500	0	0	0	1	1	1	2	3	5	6
	600	0	0	0	0	1	1	1	3	4	5
	700	0	0	0	0	1	1	1	2	3	4
	750	0	0	0	0	1	1	1	2	3	4
	800	0	0	0	0	1	1	1	1	3	4
	900	0	0	0	0	0	1	1	1	3	3
	1000	0	0	0	0	0	1	1	1	2	3
	1250	0	0	0	0	0	1	1	1	1	2
	1500	0	0	0	0	0	0	1	1	1	1
	1750	0	0	0	0	0	0	1	1	1	1
	2000	0	0	0	0	0	0	1	1	1	1
THHN, THWN, THWN-2	14	13	22	33	52	76	134	202	291	396	518
	12	9	16	24	38	56	98	147	212	289	378
	10	6	10	15	24	35	62	93	134	182	238
	8	3	6	9	14	20	35	53	77	105	137
	6	2	4	6	10	14	25	38	55	76	99
	4	1	2	4	6	9	16	24	34	46	61
	3	1	1	3	5	7	13	20	29	39	51
	2	1	1	3	4	6	11	17	24	33	43
	1	1	1	1	3	4	8	12	18	24	32
	1/0	1	1	1	2	4	7	10	15	20	27
	2/0	0	1	1	1	3	6	9	12	17	22
	3/0	0	1	1	1	2	5	7	10	14	18
	4/0	0	1	1	1	1	4	6	8	12	15
	250	0	0	1	1	1	3	5	7	9	12
	300	0	0	1	1	1	3	4	6	8	11
	350	0	0	1	1	1	2	3	5	7	9
	400	0	0	0	1	1	1	3	5	6	8
	500	0	0	0	1	1	1	2	4	5	7
	600	0	0	0	0	1	1	1	3	4	5
	700	0	0	0	0	1	1	1	3	4	5
	750	0	0	0	0	1	1	1	2	3	4
	800	0	0	0	0	1	1	1	2	3	4
	900	0	0	0	0	0	1	1	1	3	4
	1000	0	0	0	0	0	1	1	1	3	3
FEP, FEPB, PFA, PFAH, TFE	14	12	21	32	51	74	130	196	282	385	502
	12	9	15	24	37	54	95	143	206	281	367
	10	6	11	17	26	39	68	103	148	201	263
	8	4	6	10	15	22	39	59	85	115	151
	6	2	4	7	11	16	28	42	60	82	107
	4	1	3	5	7	11	19	29	42	57	75
	3	1	2	4	6	9	16	24	35	48	62
	2	1	1	3	5	7	13	20	29	39	51
PFA, PFAH, TFE	1	1	1	2	3	5	9	14	20	27	36
PFA, PFAH, TFE, Z	1/0	1	1	1	3	4	8	11	17	23	30
	2/0	1	1	1	2	3	6	9	14	19	24
	3/0	0	1	1	1	3	5	8	11	15	20
	4/0	0	1	1	1	2	4	6	9	13	16
Z	14	15	25	39	61	89	157	236	340	463	605
	12	11	18	28	43	63	111	168	241	329	429
	10	6	11	17	26	39	68	103	148	201	263
	8	4	7	11	17	24	43	65	93	127	166
	6	3	5	7	12	17	30	45	65	89	117
	4	1	3	5	8	12	21	31	45	61	80
	3	1	2	4	6	8	15	23	33	45	58
	2	1	1	3	5	7	12	19	27	37	49
	1	1	1	2	4	6	10	15	22	30	39
XHH, XHHW, XHHW-2, ZW	14	9	15	23	36	53	94	141	203	277	361
	12	7	11	18	28	41	72	108	156	212	277
	10	5	8	13	21	30	54	81	116	158	207
	8	3	5	7	11	17	30	45	64	88	115
	6	1	3	5	8	12	22	33	48	65	85
	4	1	2	4	6	9	16	24	34	47	61
	3	1	1	3	5	7	13	20	29	40	52
	2	1	1	3	4	6	11	17	24	33	44

(continued)

TABLE 15.4.5 Maximum Number of Conductors or Fixture Wires in Flexible Metal Conduit (*Continued*)

CONDUCTORS											
	Conductor	Metric Designator (Trade Size)									
Type	Size (AWG/kcmil)	16 (½)	21 (¾)	27 (1)	35 (1¼)	41 (1½)	53 (2)	63 (2½)	78 (3)	91 (3½)	103 (4)
XHH, XHHW, XHHW-2	1	1	1	1	3	5	8	13	18	25	32
	1/0	1	1	1	2	4	7	10	15	21	27
	2/0	0	1	1	2	3	6	9	13	17	23
	3/0	0	1	1	1	3	5	7	10	14	19
	4/0	0	1	1	1	2	4	6	9	12	15
	250	0	0	1	1	1	3	5	7	10	13
	300	0	0	1	1	1	3	4	6	8	11
	350	0	0	1	1	1	2	4	5	7	9
	400	0	0	0	1	1	1	3	5	6	8
	500	0	0	0	1	1	1	3	4	5	7
	600	0	0	0	0	1	1	1	3	4	5
	700	0	0	0	0	1	1	1	3	4	5
	750	0	0	0	0	1	1	1	2	3	4
	800	0	0	0	0	1	1	1	2	3	4
	900	0	0	0	0	0	1	1	1	3	4
	1000	0	0	0	0	0	1	1	1	3	3
	1250	0	0	0	0	0	1	1	1	1	3
	1500	0	0	0	0	0	1	1	1	1	2
	1750	0	0	0	0	0	0	1	1	1	1
	2000	0	0	0	0	0	0	1	1	1	1

*Types RHH, RHW, and RHW-2 without outer covering.

FIXTURE WIRES							
		Metric Designator (Trade Size)					
Type	Conductor Size (AWG/kcmil)	16 (½)	21 (¾)	27 (1)	35 (1¼)	41 (1½)	53 (2)
FFH-2, RFH-2, RFHH-3	18	8	14	22	35	51	90
	16	7	12	19	29	43	76
SF-2, SFF-2	18	11	18	28	44	64	113
	16	9	15	23	36	53	94
	14	7	12	19	29	43	76
SF-1, SFF-1	18	19	32	50	78	114	201
RFH-1, RFHH-2, TF, TFF, XF, XFF	18	14	24	37	58	84	148
RFHH-2, TF, TFF, XF, XFF	16	11	19	30	47	68	120
XF, XFF	14	9	15	23	36	53	94
TFN, TFFN	18	23	38	59	93	135	237
	16	17	29	45	71	103	181
PF, PFF, PGF, PGFF, PAF, PTF, PTFF, PAFF	18	22	36	56	88	128	225
	16	17	28	43	68	99	174
	14	12	21	32	51	74	130
ZF, ZFF, ZHF, HF, HFF	18	28	47	72	113	165	290
	16	20	35	53	83	121	214
	14	15	25	39	61	89	157
KF-2, KFF-2	18	41	68	105	164	239	421
	16	28	48	74	116	168	297
	14	19	33	51	80	116	204
	12	13	23	35	55	80	140
	10	9	15	23	36	53	94
KF-1, KFF-1	18	48	82	125	196	285	503
	16	34	57	88	138	200	353
	14	23	38	59	93	135	237
	12	15	25	39	61	89	157
	10	10	16	25	40	58	103
XF, XFF	12	5	8	12	19	28	50
	10	4	6	10	15	22	39

Note: This table is for concentric stranded conductors only. For compact stranded conductors, Table C3(A) should be used.

15.4.6 Table C3(A), Maximum Number of Compact Conductors in Flexible Metal Conduit

TABLE 15.4.6

COMPACT CONDUCTORS											
		Metric Designator (Trade Size)									
Type	Conductor Size (AWG/kcmil)	16 (½)	21 (¾)	27 (1)	35 (1¼)	41 (1½)	53 (2)	63 (2½)	78 (3)	91 (3½)	103 (4)
THW, THHW, THW-2	8	2	4	6	10	14	25	38	55	75	98
	6	1	3	5	7	11	20	29	43	58	76
	4	1	2	3	5	8	15	22	32	43	57
	2	1	1	2	4	6	11	16	23	32	42
	1	1	1	1	3	4	7	11	16	22	29
	1/0	1	1	1	2	3	6	10	14	19	25
	2/0	0	1	1	1	3	5	8	12	16	21
	3/0	0	1	1	1	2	4	7	10	14	18
	4/0	0	1	1	1	1	4	6	8	11	15
	250	0	0	1	1	1	3	4	7	9	12
	300	0	0	1	1	1	2	4	6	8	10
	350	0	0	1	1	1	2	3	5	7	9
	400	0	0	0	1	1	1	3	5	6	8
	500	0	0	0	1	1	1	3	4	5	7
	600	0	0	0	0	1	1	1	3	4	6
	700	0	0	0	0	1	1	1	3	4	5
	750	0	0	0	0	1	1	1	2	3	5
	1000	0	0	0	0	0	1	1	1	3	4
THHN, THWN, THWN-2	8	—	—	—	—	—	—	—	—	—	—
	6	3	4	7	11	16	29	43	62	85	111
	4	1	3	4	7	10	18	27	38	52	69
	2	1	1	3	5	7	13	19	28	38	49
	1	1	1	2	3	5	9	14	21	28	37
	1/0	1	1	1	3	4	8	12	17	24	31
	2/0	1	1	1	2	4	6	10	14	20	26
	3/0	0	1	1	1	3	5	8	12	17	22
	4/0	0	1	1	1	2	4	7	10	14	18
	250	0	1	1	1	1	3	5	8	11	14
	300	0	0	1	1	1	3	5	7	9	12
	350	0	0	1	1	1	3	4	6	8	10
	400	0	0	1	1	1	2	3	5	7	9
	500	0	0	0	1	1	1	3	4	6	8
	600	0	0	0	1	1	1	2	3	5	6
	700	0	0	0	0	1	1	1	3	4	6
	750	0	0	0	0	1	1	1	3	4	5
	1000	0	0	0	0	0	1	1	1	3	4
XHHW, XHHW-2	8	3	5	8	13	19	33	50	71	97	127
	6	2	4	6	9	14	24	37	53	72	95
	4	1	3	4	7	10	18	27	38	52	69
	2	1	1	3	5	7	13	19	28	38	49
	1	1	1	2	3	5	9	14	21	28	37
	1/0	1	1	1	3	4	8	12	17	24	31
	2/0	1	1	1	2	4	7	10	15	20	26
	3/0	0	1	1	1	3	5	8	12	17	22
	4/0	0	1	1	1	2	4	7	10	14	18
	250	0	1	1	1	1	4	5	8	11	14
	300	0	0	1	1	1	3	5	7	9	12
	350	0	0	1	1	1	3	4	6	8	11
	400	0	0	1	1	1	2	4	5	7	10
	500	0	0	0	1	1	1	3	4	6	8
	600	0	0	0	1	1	1	2	3	5	6
	700	0	0	0	0	1	1	1	3	4	6
	750	0	0	0	0	1	1	1	3	4	5
	1000	0	0	0	0	1	1	1	2	3	4

Definition: *Compact stranding* is the result of a manufacturing process where the standard conductor is compressed to the extent that the interstices (voids between strand wires) are virtually eliminated.

15.4.7 Table C4, Maximum Number of Conductors or Fixture Wires in Intermediate Metal Conduit

TABLE 15.4.7

CONDUCTORS											
	Conductor	Metric Designator (Trade Size)									
Type	Size (AWG/kcmil)	16 (½)	21 (¾)	27 (1)	35 (1¼)	41 (1½)	53 (2)	63 (2½)	78 (3)	91 (3½)	103 (4)
RHH, RHW, RHW-2	14	4	8	13	22	30	49	70	108	144	186
	12	4	6	11	18	25	41	58	89	120	154
RHH, RHW, RHW-2	10	3	5	8	15	20	33	47	72	97	124
	8	1	3	4	8	10	17	24	38	50	65
	6	1	1	3	6	8	14	19	30	40	52
	4	1	1	3	5	6	11	15	23	31	41
	3	1	1	2	4	6	9	13	21	28	36
	2	1	1	1	3	5	8	11	18	24	31
	1	0	1	1	2	3	5	7	12	16	20
	1/0	0	1	1	1	3	4	6	10	14	18
	2/0	0	1	1	1	2	4	6	9	12	15
	3/0	0	0	1	1	1	3	5	7	10	13
	4/0	0	0	1	1	1	3	4	6	9	11
	250	0	0	1	1	1	1	3	5	6	8
	300	0	0	0	1	1	1	3	4	6	7
	350	0	0	0	1	1	1	2	4	5	7
	400	0	0	0	1	1	1	2	3	5	6
	500	0	0	0	1	1	1	1	3	4	5
	600	0	0	0	0	1	1	1	2	3	4
	700	0	0	0	0	1	1	1	2	3	4
	750	0	0	0	0	1	1	1	1	3	4
	800	0	0	0	0	0	1	1	1	3	3
	900	0	0	0	0	0	1	1	1	2	3
	1000	0	0	0	0	0	1	1	1	2	3
	1250	0	0	0	0	0	1	1	1	1	2
	1500	0	0	0	0	0	0	1	1	1	1
	1750	0	0	0	0	0	0	1	1	1	1
	2000	0	0	0	0	0	0	1	1	1	1
TW, THHW, THW, THW-2	14	10	17	27	47	64	104	147	228	304	392
	12	7	13	21	36	49	80	113	175	234	301
	10	5	9	15	27	36	59	84	130	174	224
	8	3	5	8	15	20	33	47	72	97	124
RHH*, RHW*, RHW-2	14	6	11	18	31	42	69	98	151	202	261
RHH*, RHW*, RHW-2*	12	5	9	14	25	34	56	79	122	163	209
	10	4	7	11	19	26	43	61	95	127	163
RHH*, RHW*, RHW-2*	8	2	4	7	12	16	26	37	57	76	98
RHH*, RHW*, RHW-2*,	6	1	3	5	9	12	20	28	43	58	75
	4	1	2	4	6	9	15	21	32	43	56
TW, THW, THHW, THW-2	3	1	1	3	6	8	13	18	28	37	48
	2	1	1	3	5	6	11	15	23	31	41
	1	1	1	1	3	4	7	11	16	22	28
	1/0	1	1	1	3	4	6	9	14	19	24
	2/0	0	1	1	2	3	5	8	12	16	20
	3/0	0	1	1	1	3	4	6	10	13	17
	4/0	0	1	1	1	2	4	5	8	11	14
	250	0	0	1	1	1	3	4	7	9	12
	300	0	0	1	1	1	2	4	6	8	10
	350	0	0	1	1	1	2	3	5	7	9
	400	0	0	0	1	1	1	3	4	6	8
	500	0	0	0	1	1	1	2	4	5	7
	600	0	0	0	1	1	1	1	3	4	5
	700	0	0	0	0	1	1	1	3	4	5
	750	0	0	0	0	1	1	1	2	3	4
	800	0	0	0	0	1	1	1	2	3	4
	900	0	0	0	0	1	1	1	2	3	4
	1000	0	0	0	0	0	1	1	1	3	3
	1250	0	0	0	0	0	1	1	1	1	3
	1500	0	0	0	0	0	1	1	1	1	2
	1750	0	0	0	0	0	0	1	1	1	1
	2000	0	0	0	0	0	0	1	1	1	1

(continued)

TABLE 15.4.7 Maximum Number of Conductors or Fixture Wires in Intermediate Metal Conduit (*Continued*)

CONDUCTORS											
	Conductor	Metric Designator (Trade Size)									
Type	Size (AWG/kcmil)	16 (½)	21 (¾)	27 (1)	35 (1¼)	41 (1½)	53 (2)	63 (2½)	78 (3)	91 (3½)	103 (4)
THHN, THWN, THWN-2	14	14	24	39	68	91	149	211	326	436	562
	12	10	17	29	49	67	109	154	238	318	410
	10	6	11	18	31	42	68	97	150	200	258
	8	3	6	10	18	24	39	56	86	115	149
	6	2	4	7	13	17	28	40	62	83	107
	4	1	3	4	8	10	17	25	38	51	66
	3	1	2	4	6	9	15	21	32	43	56
	2	1	1	3	5	7	12	17	27	36	47
	1	1	1	2	4	5	9	13	20	27	35
	1/0	1	1	1	3	4	8	11	17	23	29
	2/0	1	1	1	3	4	6	9	14	19	24
	3/0	0	1	1	2	3	5	7	12	16	20
	4/0	0	1	1	1	2	4	6	9	13	17
	250	0	0	1	1	1	3	5	8	10	13
	300	0	0	1	1	1	3	4	7	9	12
	350	0	0	1	1	1	2	4	6	8	10
	400	0	0	1	1	1	2	3	5	7	9
	500	0	0	0	1	1	1	3	4	6	7
	600	0	0	0	1	1	1	2	3	5	6
	700	0	0	0	1	1	1	1	3	4	5
	750	0	0	0	1	1	1	1	3	4	5
	800	0	0	0	0	1	1	1	3	4	5
	900	0	0	0	0	1	1	1	2	3	4
	1000	0	0	0	0	1	1	1	2	3	4
FEP, FEPB, PFA, PFAH, TFE	14	13	23	38	66	89	145	205	317	423	545
	12	10	17	28	48	65	106	150	231	309	398
	10	7	12	20	34	46	76	107	166	221	285
	8	4	7	11	19	26	43	61	95	127	163
	6	3	5	8	14	19	31	44	67	90	116
	4	1	3	5	10	13	21	30	47	63	81
	3	1	3	4	8	11	18	25	39	52	68
	2	1	2	4	6	9	15	21	32	43	56
PFA, PFAH, TFE	1	1	1	2	4	6	10	14	22	30	39
PFA, PFAH, TFE, Z	1/0	1	1	1	4	5	8	12	19	25	32
	2/0	1	1	1	3	4	7	10	15	21	27
	3/0	0	1	1	2	3	6	8	13	17	22
	4/0	0	1	1	1	3	5	7	10	14	18
Z	14	16	28	46	79	107	175	247	381	510	657
	12	11	20	32	56	76	124	175	271	362	466
	10	7	12	20	34	46	76	107	166	221	285
	8	4	7	12	21	29	48	68	105	140	180
	6	3	5	9	15	20	33	47	73	98	127
	4	1	3	6	10	14	23	33	50	67	87
	3	1	2	4	7	10	17	24	37	49	63
	2	1	1	3	6	8	14	20	30	41	53
	1	1	1	3	5	7	11	16	25	33	43
XHH, XHHW, XHHW-2, ZW	14	10	17	27	47	64	104	147	228	304	392
	12	7	13	21	36	49	80	113	175	234	301
	10	5	9	15	27	36	59	84	130	174	224
	8	3	5	8	15	20	33	47	72	97	124
	6	1	4	6	11	15	24	35	53	71	92
	4	1	3	4	8	11	18	25	39	52	67
	3	1	2	4	7	9	15	21	33	44	56
	2	1	1	3	5	7	12	18	27	37	47
XHH, XHHW, XHHW-2	1	1	1	2	4	5	9	13	20	27	35
	1/0	1	1	1	3	5	8	11	17	23	30
	2/0	1	1	1	3	4	6	9	14	19	25
	3/0	0	1	1	2	3	5	7	12	16	20
	4/0	0	1	1	1	2	4	6	10	13	17
	250	0	0	1	1	1	3	5	8	11	14
	300	0	0	1	1	1	3	4	7	9	12
	350	0	0	1	1	1	3	4	6	8	10
	400	0	0	1	1	1	2	3	5	7	9
	500	0	0	0	1	1	1	3	4	6	8
	600	0	0	0	1	1	1	2	3	5	6
	700	0	0	0	1	1	1	1	3	4	5
	750	0	0	0	1	1	1	1	3	4	5
	800	0	0	0	0	1	1	1	3	4	5
	900	0	0	0	0	1	1	1	2	3	4
	1000	0	0	0	0	1	1	1	2	3	4
	1250	0	0	0	0	0	1	1	1	2	3
	1500	0	0	0	0	0	1	1	1	1	2
	1750	0	0	0	0	0	1	1	1	1	2
	2000	0	0	0	0	0	0	1	1	1	1

(continued)

TABLE 15.4.7 Maximum Number of Conductors or Fixture Wires in Intermediate Metal Conduit (*Continued*)

FIXTURE WIRES

Type	Conductor Size (AWG/kcmil)	Metric Designator (Trade Size) 16 (½)	21 (¾)	27 (1)	35 (1¼)	41 (1½)	53 (2)
FHH-2, RFH-2, RFHH-3	18	9	16	26	45	61	100
	16	8	13	22	38	51	84
SF-2, SFF-2	18	12	20	33	57	77	126
	16	10	17	27	47	64	104
	14	8	13	22	38	51	84
SF-1, SFF-1	18	21	36	59	101	137	223
RFH-1, RFHH-2, TF, TFF, XF, XFF	18	15	26	43	75	101	165
RFH-2, TF, TFF, XF, XFF	16	12	21	35	60	81	133
XF, XFF	14	10	17	27	47	64	104
TFN, TFFN	18	25	42	69	119	161	264
	16	19	32	53	91	123	201
PF, PFF, PGF, PGFF, PAF, PTF, PTFF, PAFF	18	23	40	66	113	153	250
	16	18	31	51	87	118	193
	14	13	23	38	66	89	145
ZF, ZFF, ZHF, HF, HFF	18	30	52	85	146	197	322
	16	22	38	63	108	145	238
	14	16	28	46	79	107	175
KF-2, KFF-2	18	44	75	123	212	287	468
	16	31	53	87	149	202	330
	14	21	36	60	103	139	227
	12	14	25	41	70	95	156
	10	10	17	27	47	64	104
KF-1, KFF-1	18	52	90	147	253	342	558
	16	37	63	103	178	240	392
	14	25	42	69	119	161	264
	12	16	28	46	79	107	175
	10	10	18	30	52	70	114
XF, XFF	12	5	9	14	25	34	56
	10	4	7	11	19	26	43

Note: This table is for concentric stranded conductors only. For compact stranded conductors, Table C4(A) should be used.

*Types RHH, RHW, and RHW-2 without outer covering.

15.4.8 Table C4(A), Maximum Number of Compact Conductors in Intermediate Metal Conduit

TABLE 15.4.8

COMPACT CONDUCTORS

Type	Conductor Size (AWG/kcmil)	Metric Designator (Trade Size) 16 (½)	21 (¾)	27 (1)	35 (1¼)	41 (1½)	53 (2)	63 (2½)	78 (3)	91 (3½)	103 (4)
THW, THW-2, THHW	8	2	4	7	13	17	28	40	62	83	107
	6	1	3	6	10	13	22	31	48	64	82
	4	1	2	4	7	10	16	23	36	48	62
	2	1	1	3	5	7	12	17	26	35	45
	1	1	1	1	4	5	8	12	18	25	32
	1/0	1	1	1	3	4	7	10	16	21	27
	2/0	0	1	1	3	4	6	9	13	18	23
	3/0	0	1	1	2	3	5	7	11	15	20
	4/0	0	1	1	1	2	4	6	9	13	16
	250	0	0	1	1	1	3	5	7	10	13
	300	0	0	1	1	1	3	4	6	9	11
	350	0	0	1	1	1	2	4	6	8	10
	400	0	0	1	1	1	2	3	5	7	9
	500	0	0	0	1	1	1	3	4	6	8
	600	0	0	0	1	1	1	2	3	5	6
	700	0	0	0	1	1	1	1	3	4	5
	750	0	0	0	1	1	1	1	3	4	5
	1000	0	0	0	0	1	1	1	2	3	4

COMPACT CONDUCTORS

Type	Conductor Size (AWG/kcmil)	Metric Designator (Trade Size) 16 (½)	21 (¾)	27 (1)	35 (1¼)	41 (1½)	53 (2)	63 (2½)	78 (3)	91 (3½)	103 (4)
THHN, THWN, THWN-2	8	—	—	—	—	—	—	—	—	—	—
	6	3	5	8	14	19	32	45	70	93	120
	4	1	3	5	9	12	20	28	43	58	74
	2	1	1	3	6	8	14	20	31	41	53
	1	1	1	3	5	6	10	15	23	31	40
	1/0	1	1	2	4	5	9	13	20	26	34
	2/0	1	1	1	3	4	7	10	16	22	28
	3/0	0	1	1	3	4	6	9	14	18	24
	4/0	0	1	1	2	3	5	7	11	15	19
	250	0	1	1	1	2	4	6	9	12	15
	300	0	0	1	1	1	3	5	7	10	13
	350	0	0	1	1	1	3	4	7	9	11
	400	0	0	1	1	1	2	4	6	8	10
	500	0	0	1	1	1	2	3	5	7	9
	600	0	0	0	1	1	1	2	4	5	7
	700	0	0	0	1	1	1	2	3	5	6
	750	0	0	0	1	1	1	1	3	4	6
	1000	0	0	0	0	1	1	1	2	3	4
XHHW, XHHW-2	8	3	6	9	16	22	37	52	80	107	138
	6	2	4	7	12	16	27	38	59	80	103
	4	1	3	5	9	12	20	28	43	58	74
	2	1	1	3	6	8	14	20	31	41	53
	1	1	1	3	5	6	10	15	23	31	40
	1/0	1	1	2	4	5	9	13	20	26	34
	2/0	1	1	1	3	4	7	11	17	22	29
	3/0	0	1	1	3	4	6	9	14	18	24
	4/0	0	1	1	2	3	5	7	11	15	20
	250	0	1	1	1	2	4	6	9	12	16
	300	0	0	1	1	1	3	5	8	10	13
	350	0	0	1	1	1	3	4	7	9	12
	400	0	0	1	1	1	3	4	6	8	11
	500	0	0	1	1	1	2	3	5	7	9
	600	0	0	0	1	1	1	2	4	5	7
	700	0	0	0	1	1	1	2	3	5	6
	750	0	0	0	1	1	1	1	3	4	6
	1000	0	0	0	0	1	1	1	2	3	4

Definition: *Compact stranding* is the result of a manufacturing process where the standard conductor is compressed to the extent that interstices (voids between strand wires) are virtually eliminated.

15.4.9 Table C5, Maximum Number of Conductors or Fixture Wires in Liquidtight Flexible Nonmetallic Conduit (Type LFNC-B*)

TABLE 15.4.9

CONDUCTORS								
		Metric Designator (Trade Size)						
Type	Conductor Size (AWG/kcmil)	12 (⅜)	16 (½)	21 (¾)	27 (1)	35 (1¼)	41 (1½)	53 (2)
RHH, RHW, RHW-2	14	2	4	7	12	21	27	44
	12	1	3	6	10	17	22	36
RHH, RHW, RHW-2	10	1	3	5	8	14	18	29
	8	1	1	2	4	7	9	15
	6	1	1	1	3	6	7	12
	4	0	1	1	2	4	6	9
	3	0	1	1	1	4	5	8
	2	0	1	1	1	3	4	7
	1	0	0	1	1	1	3	5
	1/0	0	0	1	1	1	2	4
	2/0	0	0	1	1	1	1	3
	3/0	0	0	0	1	1	1	3
	4/0	0	0	0	1	1	1	2
	250	0	0	0	0	1	1	1
	300	0	0	0	0	1	1	1
	350	0	0	0	0	1	1	1
	400	0	0	0	0	1	1	1
	500	0	0	0	0	1	1	1
	600	0	0	0	0	0	1	1
	700	0	0	0	0	0	0	1
	750	0	0	0	0	0	0	1
	800	0	0	0	0	0	0	1
	900	0	0	0	0	0	0	1
	1000	0	0	0	0	0	0	1
	1250	0	0	0	0	0	0	0
	1500	0	0	0	0	0	0	0
	1750	0	0	0	0	0	0	0
	2000	0	0	0	0	0	0	0
TW, THHW, THW, THW-2	14	5	9	15	25	44	57	93
	12	4	7	12	19	33	43	71
	10	3	5	9	14	25	32	53
	8	1	3	5	8	14	18	29

(continued)

TABLE 15.4.9 Maximum Number of Conductors or Fixture Wires in Liquidtight Flexible Nonmetallic Conduit (Type LFNC-B*) (*Continued*)

CONDUCTORS								
		Metric Designator (Trade Size)						
Type	Conductor Size (AWG/kcmil)	12 (3/8)	16 (1/2)	21 (3/4)	27 (1)	35 (1¼)	41 (1½)	53 (2)
RHH†, RHW†, RHW-2†	14	3	6	10	16	29	38	62
RHH†, RHW†, RHW-2†	12	3	5	8	13	23	30	50
	10	1	3	6	10	18	23	39
RHH†, RHW†, RHW-2†	8	1	1	4	6	11	14	23
RHH†, RHW†, RHW-2†, TW, THW,	6	1	1	3	5	8	11	18
	4	1	1	1	3	6	8	13
	3	1	1	1	3	5	7	11
THHW, THW-2	2	0	1	1	2	4	6	9
	1	0	1	1	1	3	4	7
	1/0	0	0	1	1	2	3	6
	2/0	0	0	1	1	2	3	5
	3/0	0	0	1	1	1	2	4
	4/0	0	0	0	1	1	1	3
	250	0	0	0	1	1	1	3
	300	0	0	0	1	1	1	2
	350	0	0	0	0	1	1	1
	400	0	0	0	0	1	1	1
	500	0	0	0	0	1	1	1
	600	0	0	0	0	1	1	1
	700	0	0	0	0	0	1	1
	750	0	0	0	0	0	1	1
	800	0	0	0	0	0	1	1
	900	0	0	0	0	0	0	1
	1000	0	0	0	0	0	0	1
	1250	0	0	0	0	0	0	1
	1500	0	0	0	0	0	0	0
	1750	0	0	0	0	0	0	0
	2000	0	0	0	0	0	0	0
THHN, THWN, THWN-2	14	8	13	22	36	63	81	133
	12	5	9	16	26	46	59	97
	10	3	6	10	16	29	37	61
	8	1	3	6	9	16	21	35
	6	1	2	4	7	12	15	25
	4	1	1	2	4	7	9	15
	3	1	1	1	3	6	8	13
	2	1	1	1	3	5	7	11
	1	0	1	1	1	4	5	8
	1/0	0	1	1	1	3	4	7
	2/0	0	0	1	1	2	3	6
	3/0	0	0	1	1	1	3	5
	4/0	0	0	1	1	1	2	4
	250	0	0	0	1	1	1	3
	300	0	0	0	1	1	1	3
	350	0	0	0	1	1	1	2
	400	0	0	0	0	1	1	1
	500	0	0	0	0	1	1	1
	600	0	0	0	0	1	1	1
	700	0	0	0	0	1	1	1
	750	0	0	0	0	0	1	1
	800	0	0	0	0	0	1	1
	900	0	0	0	0	0	1	1
	1000	0	0	0	0	0	0	1
FEP, FEPB, PFA, PFAH, TFE	14	7	12	21	35	61	79	129
	12	5	9	15	25	44	57	94
	10	4	6	11	18	32	41	68
	8	1	3	6	10	18	23	39
	6	1	2	4	7	13	17	27
	4	1	1	3	5	9	12	19
	3	1	1	2	4	7	10	16
	2	1	1	1	3	6	8	13
PFA, PFAH, TFE	1	0	1	1	2	4	5	9
PFA, PFAH	1/0	0	1	1	1	3	4	7
TFE, Z	2/0	0	1	1	1	3	4	6
	3/0	0	0	1	1	2	3	5
	4/0	0	0	1	1	1	2	4
Z	14	9	15	26	42	73	95	156
	12	6	10	18	30	52	67	111
	10	4	6	11	18	32	41	68
	8	2	4	7	11	20	26	43
	6	1	3	5	8	14	18	30
	4	1	1	3	5	9	12	20
	3	1	1	2	4	7	9	15
	2	0	1	1	3	6	7	12
	1	0	1	1	2	5	6	10

(continued)

TABLE 15.4.9 Maximum Number of Conductors or Fixture Wires in Liquidtight Flexible Nonmetallic Conduit (Type LFNC-B*) (*Continued*)

		CONDUCTORS						
		Metric Designator (Trade Size)						
Type	Conductor Size (AWG/kcmil)	12 (3/8)	16 (1/2)	21 (3/4)	27 (1)	35 (1 1/4)	41 (1 1/2)	53 (2)
XHH, XHHW, XHHW-2, ZW	14	5	9	15	25	44	57	93
	12	4	7	12	19	33	43	71
	10	3	5	9	14	25	32	53
	8	1	3	5	8	14	18	29
	6	1	1	3	6	10	13	22
XHH, XHHW, XHHW-2, ZW	4	1	1	2	4	7	9	16
	3	1	1	1	3	6	8	13
	2	1	1	1	3	5	7	11
XHH, XHHW, XHHW-2	1	0	1	1	1	4	5	8
	1/0	0	1	1	1	3	4	7
	2/0	0	0	1	1	2	3	6
	3/0	0	0	1	1	1	3	5
	4/0	0	0	1	1	1	2	4
	250	0	0	0	1	1	1	3
	300	0	0	0	1	1	1	3
	350	0	0	0	1	1	1	2
	400	0	0	0	0	1	1	1
	500	0	0	0	0	1	1	1
	600	0	0	0	0	1	1	1
	700	0	0	0	0	1	1	1
	750	0	0	0	0	0	1	1
	800	0	0	0	0	0	1	1
	900	0	0	0	0	0	1	1
	1000	0	0	0	0	0	0	1
	1250	0	0	0	0	0	0	1
	1500	0	0	0	0	0	0	1
	1750	0	0	0	0	0	0	0
	2000	0	0	0	0	0	0	0
		FIXTURE WIRES						
FFH-2, RFH-2	18	5	8	15	24	42	54	89
	16	4	7	12	20	35	46	75
SF-2, SFF-2	18	6	11	19	30	53	69	113
	16	5	9	15	25	44	57	93
	14	4	7	12	20	35	46	75
SF-1, SFF-1	18	11	19	33	53	94	122	199
RFH-1, RFHH-2, TF, TFF, XF, XFF	18	8	14	24	39	69	90	147
RFHH-2, TF, TFF, XF, XFF	16	7	11	20	32	56	72	119
XF, XFF	14	5	9	15	25	44	57	93
TFN, TFFN	18	14	23	39	63	111	144	236
	16	10	17	30	48	85	110	180
PF, PFF, PGF, PGFF, PAF, PTF, PTFF, PAFF	18	13	21	37	60	105	136	223
	16	10	16	29	46	81	105	173
	14	7	12	21	35	61	79	129
HF, HFF, ZF, ZFF, ZHF	18	17	28	48	77	136	176	288
	16	12	20	35	57	100	129	212
	14	9	15	26	42	73	95	156
KF-2, KFF-2	18	24	40	70	112	197	255	418
	16	17	28	49	79	139	180	295
	14	12	19	34	54	95	123	202
	12	8	13	23	37	65	85	139
	10	5	9	15	25	44	57	93
KF-1, KFF-1	18	29	48	83	134	235	304	499
	16	20	34	58	94	165	214	350
	14	14	23	39	63	111	144	236
	12	9	15	26	42	73	95	156
	10	6	10	17	27	48	62	102
XF, XFF	12	3	5	8	13	23	30	50
	10	1	3	6	10	18	23	39

Note: This table is for concentric stranded conductors only. For compact stranded conductors, Table C5(A). should be used.

*Corresponds to 356.2(2).

†Types RHH, RHW, and RHW-2 without outer covering.

15.4.10 Table C5(A), Maximum Number of Compact Conductors in Liquidtight Flexible Nonmetallic Conduit (Type LFNC-B*)

TABLE 15.4.10

COMPACT CONDUCTORS

Type	Conductor Size (AWG/kcmil)	Metric Designator (Trade Size)						
		12 (⅜)	16 (½)	21 (¾)	27 (1)	35 (1¼)	41 (1½)	53 (2)
THW, THW-2, THHW	8	1	2	4	7	12	15	25
	6	1	1	3	5	9	12	19
	4	1	1	2	4	7	9	14
	2	1	1	1	3	5	6	11
	1	0	1	1	1	3	4	7
	1/0	0	1	1	1	3	4	6
	2/0	0	0	1	1	2	3	5
	3/0	0	0	1	1	1	3	4
	4/0	0	0	1	1	1	2	4
	250	0	0	0	1	1	1	3
	300	0	0	0	1	1	1	2
	350	0	0	0	1	1	1	2
	400	0	0	0	0	1	1	1
	500	0	0	0	0	1	1	1
	600	0	0	0	0	1	1	1
	700	0	0	0	0	1	1	1
	750	0	0	0	0	0	1	1
	1000	0	0	0	0	0	1	1
THHN, THWN, THWN-2	8	—	—	—	—	—	—	—
	6	1	2	4	7	13	17	28
	4	1	1	3	4	8	11	17
	2	1	1	1	3	6	7	12
	1	0	1	1	2	4	6	9
	1/0	0	1	1	1	4	5	8
	2/0	0	1	1	1	3	4	6
	3/0	0	0	1	1	2	3	5
	4/0	0	0	1	1	1	3	4
	250	0	0	1	1	1	1	3
	300	0	0	0	1	1	1	3
	350	0	0	0	1	1	1	2
	400	0	0	0	1	1	1	2
	500	0	0	0	0	1	1	1
	600	0	0	0	0	1	1	1
	700	0	0	0	0	1	1	1
	750	0	0	0	0	1	1	1
	1000	0	0	0	0	0	1	1
XHHW, XHHW-2	8	1	3	5	9	15	20	33
	6	1	2	4	6	11	15	24
	4	1	1	3	4	8	11	17
	2	1	1	1	3	6	7	12
	1	0	1	1	2	4	6	9
	1/0	0	1	1	1	4	5	8
	2/0	0	1	1	1	3	4	7
	3/0	0	0	1	1	2	3	5
	4/0	0	0	1	1	1	3	4
	250	0	0	1	1	1	1	3
	300	0	0	0	1	1	1	3
	350	0	0	0	1	1	1	3
	400	0	0	0	1	1	1	2
	500	0	0	0	0	1	1	1
	600	0	0	0	0	1	1	1
	700	0	0	0	0	1	1	1
	750	0	0	0	0	1	1	1
	1000	0	0	0	0	0	1	1

*Corresponds to 356.2(2).

Definition: *Compact stranding* is the result of a manufacturing process where the standard conductor is compressed to the extent that the interstices (voids between strand wires) are virtually eliminated.

15.4.11 Table C6, Maximum Number of Conductors or Fixture Wires in Liquidtight Flexible Nonmetallic Conduit (Type LFNC-A*)

TABLE 15.4.11

CONDUCTORS								
		Metric Designator (Trade Size)						
Type	Conductor Size (AWG/kcmil)	12 (3/8)	16 (1/2)	21 (3/4)	27 (1)	35 (1¼)	41 (1½)	53 (2)
RHH, RHW, RHW-2	14	2	4	7	11	20	27	45
	12	1	3	6	9	17	23	38
	10	1	3	5	8	13	18	30
	8	1	1	2	4	7	9	16
	6	1	1	1	3	5	7	13
	4	0	1	1	2	4	6	10
	3	0	1	1	1	4	5	8
	2	0	1	1	1	3	4	7
	1	0	0	1	1	1	3	5
	1/0	0	0	1	1	1	2	4
	2/0	0	0	1	1	1	1	4
	3/0	0	0	0	1	1	1	3
	4/0	0	0	0	1	1	1	3
	250	0	0	0	0	1	1	1
	300	0	0	0	0	1	1	1
	350	0	0	0	0	1	1	1
	400	0	0	0	0	1	1	1
	500	0	0	0	0	0	1	1
	600	0	0	0	0	0	1	1
	700	0	0	0	0	0	0	1
	750	0	0	0	0	0	0	1
	800	0	0	0	0	0	0	1
	900	0	0	0	0	0	0	1
	1000	0	0	0	0	0	0	1
	1250	0	0	0	0	0	0	0
	1500	0	0	0	0	0	0	0
	1750	0	0	0	0	0	0	0
	2000	0	0	0	0	0	0	0
TW, THHW, THW, THW-2	14	5	9	15	24	43	58	96
	12	4	7	12	19	33	44	74
	10	3	5	9	14	24	33	55
	8	1	3	5	8	13	18	30
RHH†, RHW†, RHW-2†	14	3	6	10	16	28	38	64
	12	3	4	8	13	23	31	51
	10	1	3	6	10	18	24	40
	8	1	1	4	6	10	14	24
RHH†, RHW†, RHW-2†, TW, THW, THHW, THW-2	6	1	1	3	4	8	11	18
	4	1	1	1	3	6	8	13
	3	1	1	1	3	5	7	11
	2	0	1	1	2	4	6	10
	1	0	1	1	1	3	4	7
	1/0	0	0	1	1	2	3	6
	2/0	0	0	1	1	1	3	5
	3/0	0	0	1	1	1	2	4
	4/0	0	0	0	1	1	1	3
	250	0	0	0	1	1	1	3
	300	0	0	0	1	1	1	2
	350	0	0	0	0	1	1	1
	400	0	0	0	0	1	1	1
	500	0	0	0	0	1	1	1
	600	0	0	0	0	1	1	1
	700	0	0	0	0	0	1	1
	750	0	0	0	0	0	1	1
	800	0	0	0	0	0	1	1
	900	0	0	0	0	0	0	1
	1000	0	0	0	0	0	0	1
	1250	0	0	0	0	0	0	1
	1500	0	0	0	0	0	0	1
	1750	0	0	0	0	0	0	0
	2000	0	0	0	0	0	0	0

(continued)

TABLE 15.4.11 Maximum Number of Conductors or Fixture Wires in Liquidtight Flexible Nonmetallic Conduit (Type LFNC-A*) (*Continued*)

		CONDUCTORS						
		Metric Designator (Trade Size)						
Type	Conductor Size (AWG/kcmil)	12 (⅜)	16 (½)	21 (¾)	27 (1)	35 (1¼)	41 (1½)	53 (2)
THHN, THWN, THWN-2	14	8	13	22	35	62	83	137
	12	5	9	16	25	45	60	100
	10	3	6	10	16	28	38	63
	8	1	3	6	9	16	22	36
	6	1	2	4	6	12	16	26
	4	1	1	2	4	7	9	16
	3	1	1	1	3	6	8	13
	2	1	1	1	3	5	7	11
	1	0	1	1	1	4	5	8
	1/0	0	1	1	1	3	4	7
	2/0	0	0	1	1	2	3	6
	3/0	0	0	1	1	1	3	5
	4/0	0	0	1	1	1	2	4
	250	0	0	0	1	1	1	3
	300	0	0	0	1	1	1	3
	350	0	0	0	1	1	1	2
	400	0	0	0	0	1	1	1
	500	0	0	0	0	1	1	1
	600	0	0	0	0	1	1	1
	700	0	0	0	0	1	1	1
	750	0	0	0	0	0	1	1
	800	0	0	0	0	0	1	1
	900	0	0	0	0	0	1	1
	1000	0	0	0	0	0	0	1
FEP, FEPB, PFA, PFAH, TFE	14	7	12	21	34	60	80	133
	12	5	9	15	25	44	59	97
	10	4	6	11	18	31	42	70
	8	1	3	6	10	18	24	40
	6	1	2	4	7	13	17	28
	4	1	1	3	5	9	12	20
	3	1	1	2	4	7	10	16
	2	1	1	1	3	6	8	13
PFA, PFAH, TFE	1	0	1	1	2	4	5	9
PFA, PFAH, TFE, Z	1/0	0	1	1	1	3	5	8
	2/0	0	1	1	1	3	4	6
	3/0	0	0	1	1	2	3	5
	4/0	0	0	1	1	1	2	4
Z	14	9	15	25	41	72	97	161
	12	6	10	18	29	51	69	114
	10	4	6	11	18	31	42	70
	8	2	4	7	11	20	26	44
	6	1	3	5	8	14	18	31
	4	1	1	3	5	9	13	21
	3	1	1	2	4	7	9	15
	2	1	1	1	3	6	8	13
	1	1	1	1	2	4	6	10
XHH, XHHW, XHHW-2, ZW	14	5	9	15	24	43	58	96
	12	4	7	12	19	33	44	74
	10	3	5	9	14	24	33	55
	8	1	3	5	8	13	18	30
	6	1	1	3	5	10	13	22
	4	1	1	2	4	7	10	16
	3	1	1	1	3	6	8	14
	2	1	1	1	3	5	7	11
XHH, XHHW, XHHW-2	1	0	1	1	1	4	5	8
	1/0	0	1	1	1	3	4	7
	2/0	0	0	1	1	2	3	6
	3/0	0	0	1	1	1	3	5
	4/0	0	0	1	1	1	2	4
	250	0	0	0	1	1	1	3
	300	0	0	0	1	1	1	3
	350	0	0	0	1	1	1	2
	400	0	0	0	0	1	1	1
	500	0	0	0	0	1	1	1
	600	0	0	0	0	1	1	1
	700	0	0	0	0	1	1	1
	750	0	0	0	0	0	1	1
	800	0	0	0	0	0	1	1
	900	0	0	0	0	0	1	1
	1000	0	0	0	0	0	0	1
	1250	0	0	0	0	0	0	1
	1500	0	0	0	0	0	0	1
	1750	0	0	0	0	0	0	0
	2000	0	0	0	0	0	0	0
		FIXTURE WIRES						
FFH-2, RFH-2, RFHH-3	18	5	8	14	23	41	55	92
	16	4	7	12	20	35	47	77
	16	4	7	12	20	35	47	77
SF-2, SFF-2	18	6	11	18	29	52	70	116
	16	5	9	15	24	43	58	96
	14	4	7	12	20	35	47	77

(*continued*)

TABLE 15.4.11 Maximum Number of Conductors or Fixture Wires in Liquidtight Flexible Nonmetallic Conduit (Type LFNC-A*) (*Continued*)

CONDUCTORS								
		Metric Designator (Trade Size)						
Type	Conductor Size (AWG/kcmil)	12 (3/8)	16 (1/2)	21 (3/4)	27 (1)	35 (1¼)	41 (1½)	53 (2)
SF-1, SFF-1	18	12	19	33	52	92	124	205
RFH-1, RFHH-2, TF, TFF, XF, XFF	18	8	14	24	39	68	91	152
RFHH-2, TF, TFF, XF, XFF	16	7	11	19	31	55	74	122
XF, XFF	14	5	9	15	24	43	58	96
TFN, TFFN	18	14	22	39	62	109	146	243
	16	10	17	29	47	83	112	185
PF, PFF, PGF, PGFF, PAF, PTF, PTFF, PAFF	18	13	21	37	59	103	139	230
	16	10	16	28	45	80	107	178
	14	7	12	21	34	60	80	133
HF, HFF, ZF, ZFF, ZHF	18	17	27	47	76	133	179	297
	16	12	20	35	56	98	132	219
	14	9	15	25	41	72	97	161
KF-2, KFF-2	18	25	40	69	110	193	260	431
	16	17	28	48	77	136	183	303
	14	12	19	33	53	94	126	209
	12	8	13	23	36	64	86	143
	10	5	9	15	24	43	58	96
KF-1, KFF-1	18	29	48	82	131	231	310	514
	16	21	33	57	92	162	218	361
	14	14	22	39	62	109	146	243
	12	9	15	25	41	72	97	161
	10	6	10	17	27	47	63	105
XF, XFF	12	3	4	8	13	23	31	51
	10	1	3	6	10	18	24	40

Note: This table is for concentric stranded conductors only. For compact stranded conductors, Table C6(A) should be used.

*Corresponds to 356.2(1).

†Types RHH, RHW, and RHW-2 without outer covering.

15.4.12 Table C6(A), Maximum Number of Compact Conductors in Liquidtight Flexible Nonmetallic Conduit (Type LFNC-A*)

TABLE 15.4.12

COMPACT CONDUCTORS								
		Metric Designator (Trade Size)						
Type	Conductor Size (AWG/kcmil)	12 (3/8)	16 (1/2)	21 (3/4)	27 (1)	35 (1¼)	41 (1½)	53 (2)
THW, THW-2, THHW	8	1	2	4	6	11	16	26
	6	1	1	3	5	9	12	20
	4	1	1	2	4	7	9	15
	2	1	1	1	3	5	6	11
	1	0	1	1	1	3	4	8
	1/0	0	1	1	1	3	4	7
	2/0	0	0	1	1	2	3	5
	3/0	0	0	1	1	1	3	5
	4/0	0	0	1	1	1	2	4
	250	0	0	0	1	1	1	3
	300	0	0	0	1	1	1	3
	350	0	0	0	1	1	1	2
	400	0	0	0	0	1	1	1
	500	0	0	0	0	1	1	1
	600	0	0	0	0	1	1	1
	700	0	0	0	0	1	1	1
	750	0	0	0	0	0	1	1
	1000	0	0	0	0	0	1	1
THHN, THWN, THWN-2	8	—	—	—	—	—	—	—
	6	1	2	4	7	13	18	29
	4	1	1	3	4	8	11	18
	2	1	1	1	3	6	8	13
	1	0	1	1	2	4	6	10
	1/0	0	1	1	1	3	5	8
	2/0	0	1	1	1	3	4	7
	3/0	0	0	1	1	2	3	6
	4/0	0	0	1	1	1	3	5
	250	0	0	1	1	1	1	3
	300	0	0	0	1	1	1	3
	350	0	0	0	1	1	1	3
	400	0	0	0	1	1	1	2
	500	0	0	0	0	1	1	1
	600	0	0	0	0	1	1	1
	700	0	0	0	0	1	1	1
	750	0	0	0	0	1	1	1
	1000	0	0	0	0	0	1	1

COMPACT CONDUCTORS								
		Metric Designator (Trade Size)						
Type	Conductor Size (AWG/kcmil)	12 (3/8)	16 (1/2)	21 (3/4)	27 (1)	35 (1¼)	41 (1½)	53 (2)
XHHW, XHHW-2	8	1	3	5	8	15	20	34
	6	1	2	4	6	11	15	25
	4	1	1	3	4	8	11	18
	2	1	1	1	3	6	8	13
	1	0	1	1	2	4	6	10
	1/0	0	1	1	1	3	5	8
	2/0	0	1	1	1	3	4	7
	3/0	0	0	1	1	2	3	6
	4/0	0	0	1	1	1	3	5
	250	0	0	1	1	1	2	4
	300	0	0	0	1	1	1	3
	350	0	0	0	1	1	1	3
	400	0	0	0	1	1	1	2
	500	0	0	0	0	1	1	1
	600	0	0	0	0	1	1	1
	700	0	0	0	0	1	1	1
	750	0	0	0	0	1	1	1
	1000	0	0	0	0	0	1	1

*Corresponds to 356.2(1).

Definition: *Compact stranding* is the result of a manufacturing process where the standard conductor is compressed to the extent that the interstices (voids between strand wires) are virtually eliminated.

15.4.13 Table C7, Maximum Number of Conductors or Fixture Wires in Liquidtight Flexible Metal Conduit (LFMC)

TABLE 15.4.13

CONDUCTORS											
	Conductor	Metric Designator (Trade Size)									
Type	Size (AWG/kcmil)	16 (½)	21(¾)	27 (1)	35 (1¼)	41 (1½)	53 (2)	63 (2½)	78 (3)	91 (3½)	103 (4)
RHH, RHW, RHW-2	14	4	7	12	21	27	44	66	102	133	173
	12	3	6	10	17	22	36	55	84	110	144
	10	3	5	8	14	18	29	44	68	89	116
	8	1	2	4	7	9	15	23	36	46	61
	6	1	1	3	6	7	12	18	28	37	48
	4	1	1	2	4	6	9	14	22	29	38
	3	1	1	1	4	5	8	13	19	25	33
	2	1	1	1	3	4	7	11	17	22	29
	1	0	1	1	1	3	5	7	11	14	19
	1/0	0	1	1	1	2	4	6	10	13	16
	2/0	0	1	1	1	1	3	5	8	11	14
	3/0	0	0	1	1	1	3	4	7	9	12
	4/0	0	0	1	1	1	2	4	6	8	10
	250	0	0	0	1	1	1	3	4	6	8
	300	0	0	0	1	1	1	2	4	5	7
	350	0	0	0	1	1	1	2	3	5	6
	400	0	0	0	1	1	1	1	3	4	6
	500	0	0	0	1	1	1	1	3	4	5
	600	0	0	0	0	1	1	1	2	3	4
	700	0	0	0	0	0	1	1	1	3	3
	750	0	0	0	0	0	1	1	1	2	3
	800	0	0	0	0	0	1	1	1	2	3
	900	0	0	0	0	0	1	1	1	2	3
	1000	0	0	0	0	0	1	1	1	1	3
	1250	0	0	0	0	0	0	1	1	1	1
	1500	0	0	0	0	0	0	1	1	1	1
	1750	0	0	0	0	0	0	1	1	1	1
	2000	0	0	0	0	0	0	0	1	1	1
TW, THHW, THW, THW-2	14	9	15	25	44	57	93	140	215	280	365
	12	7	12	19	33	43	71	108	165	215	280
	10	5	9	14	25	32	53	80	123	160	209
	8	3	5	8	14	18	29	44	68	89	116
RHH*, RHW*, RHW-2*	14	6	10	16	29	38	62	93	143	186	243

(continued)

TABLE 15.4.13 Maximum Number of Conductors or Fixture Wires in Liquidtight Flexible Metal Conduit (LFMC) *(Continued)*

CONDUCTORS											
		Metric Designator (Trade Size)									
Type	Conductor Size (AWG/kcmil)	16 (½)	21(¾)	27 (1)	35 (1¼)	41 (1½)	53 (2)	63 (2½)	78 (3)	91 (3½)	103 (4)
RHH*, RHW*, RHW-2*	12	5	8	13	23	30	50	75	115	149	195
	10	3	6	10	18	23	39	58	89	117	152
	8	1	4	6	11	14	23	35	53	70	91
RHH*, RHW*, RHW-2*, TW, THW, THHW, THW-2	6	1	3	5	8	11	18	27	41	53	70
	4	1	1	3	6	8	13	20	30	40	52
	3	1	1	3	5	7	11	17	26	34	44
	2	1	1	2	4	6	9	14	22	29	38
	1	1	1	1	3	4	7	10	15	20	26
	1/0	0	1	1	2	3	6	8	13	17	23
	2/0	0	1	1	2	3	5	7	11	15	19
	3/0	0	1	1	1	2	4	6	9	12	16
	4/0	0	0	1	1	1	3	5	8	10	13
	250	0	0	1	1	1	3	4	6	8	11
	300	0	0	1	1	1	2	3	5	7	9
	350	0	0	0	1	1	1	3	5	6	8
	400	0	0	0	1	1	1	3	4	6	7
	500	0	0	0	1	1	1	2	3	5	6
	600	0	0	0	1	1	1	1	3	4	5
	700	0	0	0	0	1	1	1	2	3	4
	750	0	0	0	0	1	1	1	2	3	4
	800	0	0	0	0	1	1	1	2	3	4
	900	0	0	0	0	0	1	1	1	3	3
	1000	0	0	0	0	0	1	1	1	2	3
	1250	0	0	0	0	0	1	1	1	1	2
	1500	0	0	0	0	0	0	1	1	1	2
	1750	0	0	0	0	0	0	1	1	1	1
	2000	0	0	0	0	0	0	1	1	1	1
THHN, THWN, THWN-2	14	13	22	36	63	81	133	201	308	401	523
	12	9	16	26	46	59	97	146	225	292	381
	10	6	10	16	29	37	61	92	141	184	240
	8	3	6	9	16	21	35	53	81	106	138
	6	2	4	7	12	15	25	38	59	76	100
	4	1	2	4	7	9	15	23	36	47	61
	3	1	1	3	6	8	13	20	30	40	52
	2	1	1	3	5	7	11	17	26	33	44
	1	1	1	1	4	5	8	12	19	25	32
	1/0	1	1	1	3	4	7	10	16	21	27
	2/0	0	1	1	2	3	6	8	13	17	23
	3/0	0	1	1	1	3	5	7	11	14	19
	4/0	0	1	1	1	2	4	6	9	12	15
	250	0	0	1	1	1	3	5	7	10	12
	300	0	0	1	1	1	3	4	6	8	11
	350	0	0	1	1	1	2	3	5	7	9
	400	0	0	0	1	1	1	3	5	6	8
	500	0	0	0	1	1	1	2	4	5	7
	600	0	0	0	1	1	1	1	3	4	6
	700	0	0	0	1	1	1	1	3	4	5
	750	0	0	0	0	1	1	1	3	3	5
	800	0	0	0	0	1	1	1	2	3	4
	900	0	0	0	0	1	1	1	2	3	4
	1000	0	0	0	0	0	1	1	1	3	3
FEP, FEPB, PFA, PFAH, TFE	14	12	21	35	61	79	129	195	299	389	507
	12	9	15	25	44	57	94	142	218	284	370
	10	6	11	18	32	41	68	102	156	203	266
	8	3	6	10	18	23	39	58	89	117	152
	6	2	4	7	13	17	27	41	64	83	108
	4	1	3	5	9	12	19	29	44	58	75
	3	1	2	4	7	10	16	24	37	48	63
	2	1	1	3	6	8	13	20	30	40	52
PFA, PFAH, TFE	1	1	1	2	4	5	9	14	21	28	36
PFA, PFAH, TFE, Z	1/0	1	1	1	3	4	7	11	18	23	30
	2/0	1	1	1	3	4	6	9	14	19	25
	3/0	0	1	1	2	3	5	8	12	16	20
	4/0	0	1	1	1	2	4	6	10	13	17
Z	14	20	26	42	73	95	156	235	360	469	611
	12	14	18	30	52	67	111	167	255	332	434
	10	8	11	18	32	41	68	102	156	203	266
	8	5	7	11	20	26	43	64	99	129	168
	6	4	5	8	14	18	30	45	69	90	118
	4	2	3	5	9	12	20	31	48	62	81
	3	2	2	4	7	9	15	23	35	45	59
	2	1	1	3	6	7	12	19	29	38	49
	1	1	1	2	5	6	10	15	23	30	40

(continued)

TABLE 15.4.13 Maximum Number of Conductors or Fixture Wires in Liquidtight Flexible Metal Conduit (LFMC) *(Continued)*

CONDUCTORS

Type	Conductor Size (AWG/kcmil)	Metric Designator (Trade Size)									
		16 (½)	21(¾)	27 (1)	35 (1¼)	41 (1½)	53 (2)	63 (2½)	78 (3)	91 (3½)	103 (4)
XHH, XHHW, XHHW-2, ZW	14	9	15	25	44	57	93	140	215	280	365
	12	7	12	19	33	43	71	108	165	215	280
	10	5	9	14	25	32	53	80	123	160	209
	8	3	5	8	14	18	29	44	68	89	116
	6	1	3	6	10	13	22	33	50	66	86
	4	1	2	4	7	9	16	24	36	48	62
	3	1	1	3	6	8	13	20	31	40	52
	2	1	1	3	5	7	11	17	26	34	44
XHH, XHHW, XHHW-2	1	1	1	1	4	5	8	12	19	25	33
	1/0	1	1	1	3	4	7	10	16	21	28
	2/0	0	1	1	2	3	6	9	13	17	23
	3/0	0	1	1	1	3	5	7	11	14	19
	4/0	0	1	1	1	2	4	6	9	12	16
	250	0	0	1	1	1	3	5	7	10	13
	300	0	0	1	1	1	3	4	6	8	11
	350	0	0	1	1	1	2	3	5	7	10
	400	0	0	0	1	1	1	3	5	6	8
	500	0	0	0	1	1	1	2	4	5	7
	600	0	0	0	1	1	1	1	3	4	6
	700	0	0	0	1	1	1	1	3	4	5
	750	0	0	0	0	1	1	1	3	3	5
	800	0	0	0	0	1	1	1	2	3	4
	900	0	0	0	0	1	1	1	2	3	4
	1000	0	0	0	0	0	1	1	1	3	3
	1250	0	0	0	0	0	1	1	1	1	3
	1500	0	0	0	0	0	1	1	1	1	2
	1750	0	0	0	0	0	0	1	1	1	2
	2000	0	0	0	0	0	0	1	1	1	2

*Types RHH, RHW, and RHW-2 without outer covering.

FIXTURE WIRES

Type	Conductor Size (AWG/kcmil)	Metric Designator (Trade Size)					
		16 (½)	21 (¾)	27 (1)	35 (1¼)	41 (1½)	53 (2)
FFH-2, RFH-2, RFHH-3	18	8	15	24	42	54	89
	16	7	12	20	35	46	75
SF-2, SFF-2	18	11	19	30	53	69	113
	16	9	15	25	44	57	93
	14	7	12	20	35	46	75
SF-1, SFF-1	18	19	33	53	94	122	199
RFH-1, RFHH-2, TF, TFF, XF, XFF	18	14	24	39	69	90	147
RFHH-2, TF, TFF, XF, XFF	16	11	20	32	56	72	119
XF, XFF	14	9	15	25	44	57	93
TFN, TFFN	18	23	39	63	111	144	236
	16	17	30	48	85	110	180
PF, PFF, PGF, PGFF, PAF, PTF, PTFF, PAFF	18	21	37	60	105	136	223
	16	16	29	46	81	105	173
	14	12	21	35	61	79	129
HF, HFF, ZF, ZFF, ZHF	18	28	48	77	136	176	288
	16	20	35	57	100	129	212
	14	15	26	42	73	95	156
KF-2, KFF-2	18	40	70	112	197	255	418
	16	28	49	79	139	180	295
	14	19	34	54	95	123	202
	12	13	23	37	65	85	139
	10	9	15	25	44	57	93
KF-1, KFF-1	18	48	83	134	235	304	499
	16	34	58	94	165	214	350
	14	23	39	63	111	144	236
	12	15	26	42	73	95	156
	10	10	17	27	48	62	102
XF, XFF	12	5	8	13	23	30	50
	10	3	6	10	18	23	39

Note: This table is for concentric stranded conductors only. For compact stranded conductors, Table C7(A) should be used.

15.4.14 Table C7(A), Maximum Number of Compact Conductors in Liquidtight Flexible Metal Conduit (LFMC)

TABLE 15.4.14

COMPACT CONDUCTORS

Type	Conductor Size (AWG/kcmil)	Metric Designator (Trade Size)										
		12 (3/8)	16 (1/2)	21 (3/4)	27 (1)	35 (1¼)	41 (1½)	53 (2)	63 (2½)	78 (3)	91 (3½)	103 (4)
THW,	8	1	2	4	7	12	15	25	38	58	76	99
THW-2,	6	1	1	3	5	9	12	19	29	45	59	77
THHW	4	1	1	2	4	7	9	14	22	34	44	57
	2	1	1	1	3	5	6	11	16	25	32	42
	1	0	1	1	1	3	4	7	11	17	23	30
	1/0	0	1	1	1	3	4	6	10	15	20	26
	2/0	0	0	1	1	2	3	5	8	13	16	21
	3/0	0	0	1	1	1	3	4	7	11	14	18
	4/0	0	0	1	1	1	2	4	6	9	12	15
	250	0	0	0	1	1	1	3	4	7	9	12
	300	0	0	0	1	1	1	2	4	6	8	10
	350	0	0	0	1	1	1	2	3	5	7	9
	400	0	0	0	0	1	1	1	3	5	6	8
	500	0	0	0	0	1	1	1	3	4	5	7
	600	0	0	0	0	1	1	1	1	3	4	6
	700	0	0	0	0	1	1	1	1	3	4	5
	750	0	0	0	0	0	1	1	1	3	3	5
	1000	0	0	0	0	0	1	1	1	1	3	4
THHN,	8	—	—	—	—	—	—	—	—	—	—	—
THWN,	6	1	2	4	7	13	17	28	43	66	86	112
THWN-2	4	1	1	3	4	8	11	17	26	41	53	69
	2	1	1	1	3	6	7	12	19	29	38	50
	1	0	1	1	2	4	6	9	14	22	28	37
	1/0	0	1	1	1	4	5	8	12	19	24	32
	2/0	0	1	1	1	3	4	6	10	15	20	26
	3/0	0	0	1	1	2	3	5	8	13	17	22
	4/0	0	0	1	1	1	3	4	7	10	14	18
	250	0	0	1	1	1	1	3	5	8	11	14
	300	0	0	0	1	1	1	3	4	7	9	12
	350	0	0	0	1	1	1	2	4	6	8	11
	400	0	0	0	1	1	1	2	3	5	7	9
	500	0	0	0	0	1	1	1	3	5	6	8
	600	0	0	0	0	1	1	1	2	4	5	6
	700	0	0	0	0	1	1	1	1	3	4	6
	750	0	0	0	0	1	1	1	1	3	4	5
	1000	0	0	0	0	0	1	1	1	2	3	4
XHHW,	8	1	3	5	9	15	20	33	49	76	98	129
XHHW-2	6	1	2	4	6	11	15	24	37	56	73	95
	4	1	1	3	4	8	11	17	26	41	53	69
	2	1	1	1	3	6	7	12	19	29	38	50
	1	0	1	1	2	4	6	9	14	22	28	37
	1/0	0	1	1	1	4	5	8	12	19	24	32
	2/0	0	1	1	1	3	4	7	10	16	20	27
	3/0	0	0	1	1	2	3	5	8	13	17	22
	4/0	0	0	1	1	1	3	4	7	11	14	18
	250	0	0	1	1	1	1	3	5	8	11	15
	300	0	0	0	1	1	1	3	5	7	9	12
	350	0	0	0	1	1	1	3	4	6	8	11
	400	0	0	0	1	1	1	2	4	6	7	10
	500	0	0	0	0	1	1	1	3	5	6	8
	600	0	0	0	0	1	1	1	2	4	5	6
	700	0	0	0	0	1	1	1	1	3	4	6
	750	0	0	0	0	1	1	1	1	3	4	5
	1000	0	0	0	0	0	1	1	1	2	3	4

Definition: *Compact stranding* is the result of a manufacturing process where the standard conductor is compressed to the extent that the interstices (voids between strand wires) are virtually eliminated.

15.4.15 Table C8, Maximum Number of Conductors or Fixture Wires in Rigid Metal Conduit (RMC)

TABLE 15.4.15

CONDUCTORS

Type	Conductor Size (AWG/kcmil)	Metric Designator (Trade Size) 16 (½)	21 (¾)	27 (1)	35 (1¼)	41 (1½)	53 (2)	63 (2½)	78 (3)	91 (3½)	103 (4)	129 (5)	155 (6)
RHH,	14	4	7	12	21	28	46	66	102	136	176	276	398
RHW,	12	3	6	10	17	23	38	55	85	113	146	229	330
RHW-2	10	3	5	8	14	19	31	44	68	91	118	185	267
	8	1	2	4	7	10	16	23	36	48	61	97	139
	6	1	1	3	6	8	13	18	29	38	49	77	112
	4	1	1	2	4	6	10	14	22	30	38	60	87
	3	1	1	2	4	5	9	12	19	26	34	53	76
	2	1	1	1	3	4	7	11	17	23	29	46	66
	1	0	1	1	1	3	5	7	11	15	19	30	44
	1/0	0	1	1	1	2	4	6	10	13	17	26	38
	2/0	0	1	1	1	2	4	5	8	11	14	23	33
	3/0	0	0	1	1	1	3	4	7	10	12	20	28
	4/0	0	0	1	1	1	3	4	6	8	11	17	24
	250	0	0	0	1	1	1	3	4	6	8	13	18
	300	0	0	0	1	1	1	2	4	5	7	11	16
	350	0	0	0	1	1	1	2	4	5	6	10	15
	400	0	0	0	1	1	1	1	3	4	6	9	13
	500	0	0	0	1	1	1	1	3	4	5	8	11
	600	0	0	0	0	1	1	1	2	3	4	6	9
	700	0	0	0	0	1	1	1	1	3	4	6	8
	750	0	0	0	0	0	1	1	1	3	3	5	8
	800	0	0	0	0	0	1	1	1	2	3	5	7
	900	0	0	0	0	0	1	1	1	2	3	5	7
	1000	0	0	0	0	0	1	1	1	1	3	4	6
	1250	0	0	0	0	0	0	1	1	1	1	3	5
	1500	0	0	0	0	0	0	1	1	1	1	3	4
	1750	0	0	0	0	0	0	1	1	1	1	2	4
	2000	0	0	0	0	0	0	0	1	1	1	2	3
TW,	14	9	15	25	44	59	98	140	216	288	370	581	839
THHW, THW,	12	7	12	19	33	45	75	107	165	221	284	446	644
THW-2	10	5	9	14	25	34	56	80	123	164	212	332	480
	8	3	5	8	14	19	31	44	68	91	118	185	267
RHH*, RHW*, RHW-2*	14	6	10	17	29	39	65	93	143	191	246	387	558
RHH*,	12	5	8	13	23	32	52	75	115	154	198	311	448
RHW*, RHW-2*	10	3	6	10	18	25	41	58	90	120	154	242	350
RHH*, RHW*, RHW-2*	8	1	4	6	11	15	24	35	54	72	92	145	209
RHH*,	6	1	3	5	8	11	18	27	41	55	71	111	160
RHW*,	4	1	1	3	6	8	14	20	31	41	53	83	120
RHW-2*,	3	1	1	3	5	7	12	17	26	35	45	71	103
TW,	2	1	1	2	4	6	10	14	22	30	38	60	87
THW,	1	1	1	1	3	4	7	10	15	21	27	42	61
THHW,	1/0	0	1	1	2	3	6	8	13	18	23	36	52
THW-2	2/0	0	1	1	2	3	5	7	11	15	19	31	44
	3/0	0	1	1	1	2	4	6	9	13	16	26	37
	4/0	0	0	1	1	1	3	5	8	10	14	21	31
	250	0	0	1	1	1	3	4	6	8	11	17	25
	300	0	0	1	1	1	2	3	5	7	9	15	22
	350	0	0	0	1	1	1	3	5	6	8	13	19
	400	0	0	0	1	1	1	3	4	6	7	12	17
	500	0	0	0	1	1	1	2	3	5	6	10	14
	600	0	0	0	1	1	1	1	3	4	5	8	12
	700	0	0	0	0	1	1	1	2	3	4	7	10
	750	0	0	0	0	1	1	1	2	3	4	7	10
	800	0	0	0	0	1	1	1	2	3	4	6	9
	900	0	0	0	0	1	1	1	1	3	4	6	8
	1000	0	0	0	0	0	1	1	1	2	3	5	8
	1250	0	0	0	0	0	1	1	1	1	2	4	6
	1500	0	0	0	0	0	1	1	1	1	2	3	5
	1750	0	0	0	0	0	0	1	1	1	1	3	4
	2000	0	0	0	0	0	0	1	1	1	1	3	4

(continued)

TABLE 15.4.15 Maximum Number of Conductors or Fixture Wires in Rigid Metal Conduit (RMC) (*Continued*)

		CONDUCTORS											
	Conductor	Metric Designator (Trade Size)											
Type	Size (AWG/kcmil)	16 (½)	21 (¾)	27 (1)	35 (1¼)	41 (1½)	53 (2)	63 (2½)	78 (3)	91 (3½)	103 (4)	129 (5)	155 (6)
THHN, THWN, THWN-2	14	13	22	36	63	85	140	200	309	412	531	833	1202
	12	9	16	26	46	62	102	146	225	301	387	608	877
	10	6	10	17	29	39	64	92	142	189	244	383	552
	8	3	6	9	16	22	37	53	82	109	140	221	318
	6	2	4	7	12	16	27	38	59	79	101	159	230
	4	1	2	4	7	10	16	23	36	48	62	98	141
	3	1	1	3	6	8	14	20	31	41	53	83	120
	2	1	1	3	5	7	11	17	26	34	44	70	100
	1	1	1	1	4	5	8	12	19	25	33	51	74
	1/0	1	1	1	3	4	7	10	16	21	27	43	63
	2/0	0	1	1	2	3	6	8	13	18	23	36	52
	3/0	0	1	1	1	3	5	7	11	15	19	30	43
	4/0	0	1	1	1	2	4	6	9	12	16	25	36
	250	0	0	1	1	1	3	5	7	10	13	20	29
	300	0	0	1	1	1	3	4	6	8	11	17	25
	350	0	0	1	1	1	2	3	5	7	10	15	22
	400	0	0	1	1	1	2	3	5	7	8	13	20
	500	0	0	0	1	1	1	2	4	5	7	11	16
	600	0	0	0	1	1	1	1	3	4	6	9	13
	700	0	0	0	1	1	1	1	3	4	5	8	11
	750	0	0	0	0	1	1	1	3	4	5	7	11
	800	0	0	0	0	1	1	1	2	3	4	7	10
	900	0	0	0	0	1	1	1	2	3	4	6	9
	1000	0	0	0	0	1	1	1	1	3	4	6	8
FEP, FEPB, PFA, PFAH, TFE	14	12	22	35	61	83	136	194	300	400	515	808	1166
	12	9	16	26	44	60	99	142	219	292	376	590	851
	10	6	11	18	32	43	71	102	157	209	269	423	610
	8	3	6	10	18	25	41	58	90	120	154	242	350
	6	2	4	7	13	17	29	41	64	85	110	172	249
	4	1	3	5	9	12	20	29	44	59	77	120	174
	3	1	2	4	7	10	17	24	37	50	64	100	145
	2	1	1	3	6	8	14	20	31	41	53	83	120
PFA, PFAH, TFE	1	1	1	2	4	6	9	14	21	28	37	57	83
PFA, PFAH, TFE, Z	1/0	1	1	1	3	5	8	11	18	24	30	48	69
	2/0	1	1	1	3	4	6	9	14	19	25	40	57
	3/0	0	1	1	2	3	5	8	12	16	21	33	47
	4/0	0	1	1	1	2	4	6	10	13	17	27	39
Z	14	15	26	42	73	100	164	234	361	482	621	974	1405
	12	10	18	30	52	71	116	166	256	342	440	691	997
	10	6	11	18	32	43	71	102	157	209	269	423	610
	8	4	7	11	20	27	45	64	99	132	170	267	386
	6	3	5	8	14	19	31	45	69	93	120	188	271
	4	1	3	5	9	13	22	31	48	64	82	129	186
	3	1	2	4	7	9	16	22	35	47	60	94	136
	2	1	1	3	6	8	13	19	29	39	50	78	113
	1	1	1	2	5	6	10	15	23	31	40	63	92
XHH, XHHW, XHHW-2, ZW	14	9	15	25	44	59	98	140	216	288	370	581	839
	12	7	12	19	33	45	75	107	165	221	284	446	644
	10	5	9	14	25	34	56	80	123	164	212	332	480
	8	3	5	8	14	19	31	44	68	91	118	185	267
	6	1	3	6	10	14	23	33	51	68	87	137	197
	4	1	2	4	7	10	16	24	37	49	63	99	143
	3	1	1	3	6	8	14	20	31	41	53	84	121
	2	1	1	3	5	7	12	17	26	35	45	70	101

(continued)

TABLE 15.4.15 Maximum Number of Conductors or Fixture Wires in Rigid Metal Conduit (RMC) *(Continued)*

CONDUCTORS													
		Metric Designator (Trade Size)											
Type	Conductor Size (AWG/kcmil)	16 (½)	21 (¾)	27 (1)	35 (1¼)	41 (1½)	53 (2)	63 (2½)	78 (3)	91 (3½)	103 (4)	129 (5)	155 (6)
XHH, XHHW, XHHW-2	1	1	1	1	4	5	9	12	19	26	33	52	76
	1/0	1	1	1	3	4	7	10	16	22	28	44	64
	2/0	0	1	1	2	3	6	9	13	18	23	37	53
	3/0	0	1	1	1	3	5	7	11	15	19	30	44
	4/0	0	1	1	1	2	4	6	9	12	16	25	36
	250	0	0	1	1	1	3	5	7	10	13	20	30
	300	0	0	1	1	1	3	4	6	9	11	18	25
	350	0	0	1	1	1	2	3	6	7	10	15	22
	400	0	0	1	1	1	2	3	5	7	9	14	20
	500	0	0	0	1	1	1	2	4	5	7	11	16
	600	0	0	0	1	1	1	1	3	4	6	9	13
	700	0	0	0	1	1	1	1	3	4	5	8	11
	750	0	0	0	0	1	1	1	3	4	5	7	11
	800	0	0	0	0	1	1	1	2	3	4	7	10
	900	0	0	0	0	1	1	1	2	3	4	6	9
	1000	0	0	0	0	1	1	1	1	3	4	6	8
	1250	0	0	0	0	0	1	1	1	2	3	4	6
	1500	0	0	0	0	0	1	1	1	1	2	4	5
	1750	0	0	0	0	0	0	1	1	1	1	3	5
	2000	0	0	0	0	0	0	1	1	1	1	3	4

FIXTURE WIRES							
		Metric Designator (Trade Size)					
Type	Conductor Size (AWG/kcmil)	16 (½)	21 (¾)	27 (1)	35 (1¼)	41 (1½)	53 (2)
FFH-2, RFH-2, RFHH-3	18	8	15	24	42	57	94
	16	7	12	20	35	48	79
SF-2, SFF-2	18	11	19	31	53	72	118
	16	9	15	25	44	59	98
	14	7	12	20	35	48	79
SF-1, SFF-1	18	19	33	54	94	127	209
RFH-1, RFHH-2, TF, TFF, XF, XFF	18	14	25	40	69	94	155
RFHH-2, TF, TFF, XF, XFF	16	11	20	32	56	76	125
XF, XFF	14	9	15	25	44	59	98
TFN, TFFN	18	23	40	64	111	150	248
	16	17	30	49	84	115	189
PF, PFF, PGF, PGFF, PAF, PTF, PTFF, PAFF	18	21	38	61	105	143	235
	16	16	29	47	81	110	181
	14	12	22	35	61	83	136
HF, HFF, ZF, ZFF, ZHF	18	28	48	79	135	184	303
	16	20	36	58	100	136	223
	14	15	26	42	73	100	164
KF-2, KFF-2	18	40	71	114	197	267	439
	16	28	50	80	138	188	310
	14	19	34	55	95	129	213
	12	13	23	38	65	89	146
	10	9	15	25	44	59	98
KF-1, KFF-1	18	48	84	136	235	318	524
	16	34	59	96	165	224	368
	14	23	40	64	111	150	248
	12	15	26	42	73	100	164
	10	10	17	28	48	65	107
XF, XFF	12	5	8	13	23	32	52
	10	3	6	10	18	25	41

Note: This table is for concentric stranded conductors only. For compact stranded conductors, Table C8(A) should be used.

*Types RHH, RHW, and RHW-2 without outer covering.

15.4.16 Table C8(A), Maximum Number of Compact Conductors in Rigid Metal Conduit (RMC)

TABLE 15.4.16

COMPACT CONDUCTORS

Type	Conductor Size (AWG/kcmil)	Metric Designator (Trade Size) 16 (½)	21 (¾)	27 (1)	35 (1¼)	41 (1½)	53 (2)	63 (2½)	78 (3)	91 (3½)	103 (4)	129 (5)	155 (6)
THW, THW-2, THHW	8	2	4	7	12	16	26	38	59	78	101	158	228
	6	1	3	5	9	12	20	29	45	60	78	122	176
	4	1	2	4	7	9	15	22	34	45	58	91	132
	2	1	1	3	5	7	11	16	25	33	43	67	97
	1	1	1	1	3	5	8	11	17	23	30	47	68
	1/0	1	1	1	3	4	7	10	15	20	26	41	59
	2/0	0	1	1	2	3	6	8	13	17	22	34	50
	3/0	0	1	1	1	3	5	7	11	14	19	29	42
	4/0	0	1	1	1	2	4	6	9	12	15	24	35
	250	0	0	1	1	1	3	4	7	9	12	19	28
	300	0	0	1	1	1	3	4	6	8	11	17	24
	350	0	0	1	1	1	2	3	5	7	9	15	22
	400	0	0	1	1	1	1	3	5	7	8	13	20
	500	0	0	0	1	1	1	3	4	5	7	11	17
	600	0	0	0	1	1	1	1	3	4	6	9	13
	700	0	0	0	1	1	1	1	3	4	5	8	12
	750	0	0	0	0	1	1	1	3	4	5	7	11
	1000	0	0	0	0	1	1	1	1	3	4	6	9
THHN, THWN, THWN-2	8	—	—	—	—	—	—	—	—	—	—	—	—
	6	2	5	8	13	18	30	43	66	88	114	179	258
	4	1	3	5	8	11	18	26	41	55	70	110	159
	2	1	1	3	6	8	13	19	29	39	50	79	114
	1	1	1	2	4	6	10	14	22	29	38	60	86
	1/0	1	1	1	4	5	8	12	19	25	32	51	73
	2/0	1	1	1	3	4	7	10	15	21	26	42	60
	3/0	0	1	1	2	3	6	8	13	17	22	35	51
	4/0	0	1	1	1	3	5	7	10	14	18	29	42
	250	0	1	1	1	2	4	5	8	11	14	23	33
	300	0	0	1	1	1	3	4	7	10	12	20	28
	350	0	0	1	1	1	3	4	6	8	11	17	25
	400	0	0	1	1	1	2	3	5	7	10	15	22
	500	0	0	0	1	1	1	3	5	6	8	13	19
	600	0	0	0	1	1	1	2	4	5	6	10	15
	700	0	0	0	1	1	1	1	3	4	6	9	13
	750	0	0	0	1	1	1	1	3	4	5	9	13
	1000	0	0	0	0	1	1	1	2	3	4	6	9
XHHW, XHHW-2	8	3	5	9	15	21	34	49	76	101	130	205	296
	6	2	4	6	11	15	25	36	56	75	97	152	220
	4	1	3	5	8	11	18	26	41	55	70	110	159
	2	1	1	3	6	8	13	19	29	39	50	79	114
	1	1	1	2	4	6	10	14	22	29	38	60	86
	1/0	1	1	1	4	5	8	12	19	25	32	51	73
	2/0	1	1	1	3	4	7	10	16	21	27	43	62
	3/0	0	1	1	2	3	6	8	13	17	22	35	51
	4/0	0	1	1	1	3	5	7	11	14	19	29	42
	250	0	1	1	1	2	4	5	8	11	15	23	34
	300	0	0	1	1	1	3	5	7	10	13	20	29
	350	0	0	1	1	1	3	4	6	9	11	18	25
	400	0	0	1	1	1	2	4	6	8	10	16	23
	500	0	0	0	1	1	1	3	5	6	8	13	19
	600	0	0	0	1	1	1	2	4	5	7	10	15
	700	0	0	0	1	1	1	1	3	4	6	9	13
	750	0	0	0	1	1	1	1	3	4	5	8	12
	1000	0	0	0	0	1	1	1	2	3	4	7	10

Definition: *Compact stranding* is the result of a manufacturing process where the standard conductor is compressed to the extent that the interstices (voids between strand wires) are virtually eliminated.

15.4.17 Table C9, Maximum Number of Conductors or Fixture Wires in Rigid PVC Conduit, Schedule 80

TABLE 15.4.17

		CONDUCTORS											
		Metric Designator (Trade Size)											
Type	Conductor Size (AWG/kcmil)	16 (½)	21 (¾)	27 (1)	35 (1¼)	41 (1½)	53 (2)	63 (2½)	78 (3)	91 (3½)	103 (4)	129 (5)	155 (6)
RHH,	14	3	5	9	17	23	39	56	88	118	153	243	349
RHW,	12	2	4	7	14	19	32	46	73	98	127	202	290
RHW-2													
	10	1	3	6	11	15	26	37	59	79	103	163	234
	8	1	1	3	6	8	13	19	31	41	54	85	122
	6	1	1	2	4	6	11	16	24	33	43	68	98
	4	1	1	1	3	5	8	12	19	26	33	53	77
	3	0	1	1	3	4	7	11	17	23	29	47	67
	2	0	1	1	3	4	6	9	14	20	25	41	58
	1	0	1	1	1	2	4	6	9	13	17	27	38
	1/0	0	0	1	1	1	3	5	8	11	15	23	33
	2/0	0	0	1	1	1	3	4	7	10	13	20	29
	3/0	0	0	1	1	1	3	4	6	8	11	17	25
	4/0	0	0	0	1	1	2	3	5	7	9	15	21
	250	0	0	0	1	1	1	2	4	5	7	11	16
	300	0	0	0	1	1	1	2	3	5	6	10	14
	350	0	0	0	1	1	1	1	3	4	5	9	13
	400	0	0	0	0	1	1	1	3	4	5	8	12
	500	0	0	0	0	1	1	1	2	3	4	7	10
	600	0	0	0	0	0	1	1	1	3	3	6	8
	700	0	0	0	0	0	1	1	1	2	3	5	7
	750	0	0	0	0	0	1	1	1	2	3	5	7
	800	0	0	0	0	0	1	1	1	2	3	4	7
	1000	0	0	0	0	0	1	1	1	1	2	4	5
	1250	0	0	0	0	0	0	1	1	1	1	3	4
	1500	0	0	0	0	0	0	1	1	1	1	2	4
	1750	0	0	0	0	0	0	0	1	1	1	2	3
	2000	0	0	0	0	0	0	0	1	1	1	1	3
TW,	14	6	11	20	35	49	82	118	185	250	324	514	736
THHW,	12	5	9	15	27	38	63	91	142	192	248	394	565
THW,	10	3	6	11	20	28	47	67	106	143	185	294	421
THW-2	8	1	3	6	11	15	26	37	59	79	103	163	234
RHH*,	14	4	8	13	23	32	55	79	123	166	215	341	490
RHW*,	12	3	6	10	19	26	44	63	99	133	173	274	394
RHW-2*	10	2	5	8	15	20	34	49	77	104	135	214	307
	8	1	3	5	9	12	20	29	46	62	81	128	184
RHH*,	6	1	1	3	7	9	16	22	35	48	62	98	141
RHW*,	4	1	1	3	5	7	12	17	26	35	46	73	105
RHW-2*,	3	1	1	2	4	6	10	14	22	30	39	63	90
TW,	2	1	1	1	3	5	8	12	19	26	33	53	77
THW.	1	0	1	1	2	3	6	8	13	18	23	37	54
THHW,	1/0	0	1	1	1	3	5	7	11	15	20	32	46
THW-2	2/0	0	1	1	1	2	4	6	10	13	17	27	39
	3/0	0	0	1	1	1	3	5	8	11	14	23	33
	4/0	0	0	1	1	1	3	4	7	9	12	19	27
	250	0	0	0	1	1	2	3	5	7	9	15	22
	300	0	0	0	1	1	1	3	5	6	8	13	19
	350	0	0	0	1	1	1	2	4	6	7	12	17
	400	0	0	0	1	1	1	2	4	5	7	10	15
	500	0	0	0	1	1	1	1	3	4	5	9	13
	600	0	0	0	0	1	1	1	2	3	4	7	10
	700	0	0	0	0	1	1	1	2	3	4	6	9
	750	0	0	0	0	0	1	1	1	3	4	6	8
	800	0	0	0	0	0	1	1	1	3	3	6	8
	900	0	0	0	0	0	1	1	1	2	3	5	7
	1000	0	0	0	0	0	1	1	1	2	3	5	7
	1250	0	0	0	0	0	1	1	1	1	2	4	5
	1500	0	0	0	0	0	0	1	1	1	1	3	4
	1750	0	0	0	0	0	0	1	1	1	1	3	4
	2000	0	0	0	0	0	0	0	1	1	1	2	3

(continued)

TABLE 15.4.17 Maximum Number of Conductors or Fixture Wires in Rigid PVC Conduit, Schedule 80 *(Continued)*

CONDUCTORS													
		Metric Designator (Trade Size)											
Type	Conductor Size (AWG/kcmil)	16 (½)	21 (¾)	27 (1)	35 (1¼)	41 (1½)	53 (2)	63 (2½)	78 (3)	91 (3½)	103 (4)	129 (5)	155 (6)
THHN, THWN, THWN-2	14	9	17	28	51	70	118	170	265	358	464	736	1055
	12	6	12	20	37	51	86	124	193	261	338	537	770
	10	4	7	13	23	32	54	78	122	164	213	338	485
	8	2	4	7	13	18	31	45	70	95	123	195	279
	6	1	3	5	9	13	22	32	51	68	89	141	202
	4	1	1	3	6	8	14	20	31	42	54	86	124
	3	1	1	3	5	7	12	17	26	35	46	73	105
	2	1	1	2	4	6	10	14	22	30	39	61	88
	1	0	1	1	3	4	7	10	16	22	29	45	65
	1/0	0	1	1	2	3	6	9	14	18	24	38	55
	2/0	0	1	1	1	3	5	7	11	15	20	32	46
	3/0	0	1	1	1	2	4	6	9	13	17	26	38
	4/0	0	0	1	1	1	3	5	8	10	14	22	31
	250	0	0	1	1	1	3	4	6	8	11	18	25
	300	0	0	0	1	1	2	3	5	7	9	15	22
	350	0	0	0	1	1	1	3	5	6	8	13	19
	400	0	0	0	1	1	1	3	4	6	7	12	17
	500	0	0	0	1	1	1	2	3	5	6	10	14
	600	0	0	0	0	1	1	1	3	4	5	8	12
	700	0	0	0	0	1	1	1	2	3	4	7	10
	750	0	0	0	0	1	1	1	2	3	4	7	9
	800	0	0	0	0	1	1	1	2	3	4	6	9
	900	0	0	0	0	0	1	1	1	3	3	6	8
	1000	0	0	0	0	0	1	1	1	2	3	5	7
FEP, FEPB, PFA, PFAH, TFE	14	8	16	27	49	68	115	164	257	347	450	714	1024
	12	6	12	20	36	50	84	120	188	253	328	521	747
	10	4	8	14	26	36	60	86	135	182	235	374	536
	8	2	5	8	15	20	34	49	77	104	135	214	307
	6	1	3	6	10	14	24	35	55	74	96	152	218
	4	1	2	4	7	10	17	24	38	52	67	106	153
	3	1	1	3	6	8	14	20	32	43	56	89	127
	2	1	1	3	5	7	12	17	26	35	46	73	105
PFA, PFAH, TFE	1	1	1	1	3	5	8	11	18	25	32	51	73
PFA, PFAH, TFE, Z	1/0	0	1	1	3	4	7	10	15	20	27	42	61
	2/0	0	1	1	2	3	5	8	12	17	22	35	50
	3/0	0	1	1	1	2	4	6	10	14	18	29	41
	4/0	0	0	1	1	1	4	5	8	11	15	24	34
Z	14	10	19	33	59	82	138	198	310	418	542	860	1233
	12	7	14	23	42	58	98	141	220	297	385	610	875
	10	4	8	14	26	36	60	86	135	182	235	374	536
	8	3	5	9	16	22	38	54	85	115	149	236	339
	6	2	4	6	11	16	26	38	60	81	104	166	238
	4	1	2	4	8	11	18	26	41	55	72	114	164
	3	1	2	3	5	8	13	19	30	40	52	83	119
	2	1	1	2	5	6	11	16	25	33	43	69	99
	1	0	1	2	4	5	9	13	20	27	35	56	80
XHH, XHHW, XHHW-2, ZW	14	6	11	20	35	49	82	118	185	250	324	514	736
	12	5	9	15	27	38	63	91	142	192	248	394	565
	10	3	6	11	20	28	47	67	106	143	185	294	421
	8	1	3	6	11	15	26	37	59	79	103	163	234
	6	1	2	4	8	11	19	28	43	59	76	121	173
	4	1	1	3	6	8	14	20	31	42	55	87	125
	3	1	1	3	5	7	12	17	26	36	47	74	106
	2	1	1	2	4	6	10	14	22	30	39	62	89

(continued)

TABLE 15.4.17 Maximum Number of Conductors or Fixture Wires in Rigid PVC Conduit, Schedule 80 *(Continued)*

CONDUCTORS													
		Metric Designator (Trade Size)											
Type	Conductor Size (AWG/kcmil)	16 (½)	21 (¾)	27 (1)	35 (1¼)	41 (1½)	53 (2)	63 (2½)	78 (3)	91 (3½)	103 (4)	129 (5)	155 (6)
XHH, XHHW, XHHW-2	1	0	1	1	3	4	7	10	16	22	29	46	66
	1/0	0	1	1	2	3	6	9	14	19	24	39	56
	2/0	0	1	1	1	3	5	7	11	16	20	32	46
	3/0	0	1	1	1	2	4	6	9	13	17	27	38
	4/0	0	0	1	1	1	3	5	8	11	14	22	32
	250	0	0	1	1	1	3	4	6	9	11	18	26
	300	0	0	1	1	1	2	3	5	7	10	15	22
	350	0	0	0	1	1	1	3	5	6	8	14	20
	400	0	0	0	1	1	1	3	4	6	7	12	17
	500	0	0	0	1	1	1	2	3	5	6	10	14
	600	0	0	0	0	1	1	1	3	4	5	8	11
	700	0	0	0	0	1	1	1	2	3	4	7	10
	750	0	0	0	0	1	1	1	2	3	4	6	9
	800	0	0	0	0	1	1	1	1	3	4	6	9
	900	0	0	0	0	0	1	1	—	3	3	5	8
	1000	0	0	0	0	0	1	1	1	2	3	5	7
	1250	0	0	0	0	0	1	1	1	1	2	4	6
	1500	0	0	0	0	0	0	1	1	1	1	3	5
	1750	0	0	0	0	0	0	1	1	1	1	3	4
	2000	0	0	0	0	0	0	1	1	1	1	2	4

FIXTURE WIRES							
		Metric Designator (Trade Size)					
Type	Conductor Size (AWG/kcmil)	16 (½)	21 (¾)	27 (1)	35 (1¼)	41 (1½)	53 (2)
FFH-2, RFH-2, RFHH-3	18	6	11	19	34	47	79
	16	5	9	16	28	39	67
SF-2, SFF-2	18	7	14	24	43	59	100
	16	6	11	20	35	49	82
	14	5	9	16	28	39	67
SF-1, SFF-1	18	13	25	42	76	105	177
RFH-1, RFHH-2, TF, TFF, XF, XFF	18	10	18	31	56	77	130
RFHH-2, TF, TFF, XF, XFF	16	8	15	25	45	62	105
XF, XFF	14	6	11	20	35	49	82
TFN, TFFN	18	16	29	50	90	124	209
	16	12	22	38	68	95	159
PF, PFF, PGF, PGFF, PAF, PTF, PTFF, PAFF	18	15	28	47	85	118	198
	16	11	22	36	66	91	153
	14	8	.16	27	49	68	115
HF, HFF, ZF, ZFF, ZHF	18	19	36	61	110	152	255
	16	14	27	45	81	112	188
	14	10	19	33	59	82	138
KF-2, KFF-2	18	28	53	88	159	220	371
	16	19	37	62	112	155	261
	14	13	25	43	77	107	179
	12	9	17	29	53	73	123
	10	6	11	20	35	49	82
KF-1, KFF-1	18	33	63	106	190	263	442
	16	23	44	74	133	185	310
	14	16	29	50	90	124	209
	12	10	19	33	59	82	138
	10	7	13	21	39	54	90
XF, XFF	12	3	6	10	19	26	44
	10	2	5	8	15	20	34

Note: This table is for concentric stranded conductors only. For compact stranded conductors, Table C9(A) should be used.

*Types RHH, RHW, and RHW-2 without outer covering.

15.4.18 Table C9(A), Maximum Number of Compact Conductors in Rigid PVC Conduit, Schedule 80

TABLE 15.4.18

COMPACT CONDUCTORS

Type	Conductor Size (AWG/kcmil)	Metric Designator (Trade Size) 16 (½)	21 (¾)	27 (1)	35 (1¼)	41 (1½)	53 (2)	63 (2½)	78 (3)	91 (3½)	103 (4)	129 (5)	155 (6)
THW, THW-2, THHW	8	1	3	5	9	13	22	32	50	68	88	140	200
	6	1	2	4	7	10	17	25	39	52	68	108	155
	4	1	1	3	5	7	13	18	29	39	51	81	116
	2	1	1	1	4	5	9	13	21	29	37	60	85
	1	0	1	1	3	4	6	9	15	20	26	42	60
	1/0	0	1	1	2	3	6	8	13	17	23	36	52
	2/0	0	1	1	1	3	5	7	11	15	19	30	44
	3/0	0	0	1	1	2	4	6	9	12	16	26	37
	4/0	0	0	1	1	1	3	5	8	10	13	22	31
	250	0	0	1	1	1	2	4	6	8	11	17	25
	300	0	0	0	1	1	2	3	5	7	9	15	21
	350	0	0	0	1	1	1	3	5	6	8	13	19
	400	0	0	0	1	1	1	3	4	6	7	12	17
	500	0	0	0	1	1	1	2	3	5	6	10	14
	600	0	0	0	0	1	1	1	3	4	5	8	12
	700	0	0	0	0	1	1	1	2	3	4	7	10
	750	0	0	0	0	1	1	1	2	3	4	7	10
	1000	0	0	0	0	0	1	1	1	2	3	5	8
THHN, THWN, THWN-2	8	—	—	—	—	—	—	—	—	—	—	—	—
	6	1	3	6	11	15	25	36	57	77	99	158	226
	4	1	1	3	6	9	15	22	35	47	61	98	140
	2	1	1	2	5	6	11	16	25	34	44	70	100
	1	1	1	1	3	5	8	12	19	25	33	53	75
	1/0	0	1	1	3	4	7	10	16	22	28	45	64
	2/0	0	1	1	2	3	6	8	13	18	23	37	53
	3/0	0	1	1	1	3	5	7	11	15	19	31	44
	4/0	0	0	1	1	2	4	6	9	12	16	25	37
	250	0	0	1	1	1	3	4	7	10	12	20	29
	300	0	0	1	1	1	3	4	6	8	11	17	25
	350	0	0	0	1	1	2	3	5	7	9	15	22
	400	0	0	0	1	1	1	3	5	6	8	13	19
	500	0	0	0	1	1	1	2	4	5	7	11	16
	600	0	0	0	1	1	1	1	3	4	6	9	13
	700	0	0	0	0	1	1	1	3	4	5	8	12
	750	0	0	0	0	1	1	1	3	4	5	8	11
	1000	0	0	0	0	0	1	1	1	3	3	5	8
XHHW, XHHW-2	8	1	4	7	12	17	29	42	65	88	114	181	260
	6	1	3	5	9	13	21	31	48	65	85	134	193
	4	1	1	3	6	9	15	22	35	47	61	98	140
	2	1	1	2	5	6	11	16	25	34	44	70	100
	1	1	1	1	3	5	8	12	19	25	33	53	75
	1/0	0	1	1	3	4	7	10	16	22	28	45	64
	2/0	0	1	1	2	3	6	8	13	18	24	38	54
	3/0	0	1	1	1	3	5	7	11	15	19	31	44
	4/0	0	0	1	1	2	4	6	9	12	16	26	37
	250	0	0	1	1	1	3	5	7	10	13	21	30
	300	0	0	1	1	1	3	4	6	8	11	17	25
	350	0	0	1	1	1	2	3	5	7	10	15	22
	400	0	0	0	1	1	1	3	5	7	9	14	20
	500	0	0	0	1	1	1	2	4	5	7	11	17
	600	0	0	0	1	1	1	1	3	4	6	9	13
	700	0	0	0	0	1	1	1	3	4	5	8	12
	750	0	0	0	0	1	1	1	2	3	5	7	11
	1000	0	0	0	0	0	1	1	1	3	3	6	8

Definition: *Compact stranding* is the result of a manufacturing process where the standard conductor is compressed to the extent that the interstices (voids between strand wires) are virtually eliminated.

15.4.19 Table C10, Maximum Number of Conductors or Fixture Wires in Rigid PVC Conduit, Schedule 40 and HDPE Conduit

TABLE 15.4.19

CONDUCTORS													
	Conductor	Metric Designator (Trade Size)											
Type	Size (AWG/kcmil)	16 (½)	21 (¾)	27 (1)	35 (1¼)	41 (1½)	53 (2)	63 (2½)	78 (3)	91 (3½)	103 (4)	129 (5)	155 (6)
RHH, RHW, RHW-2	14	4	7	11	20	27	45	64	99	133	171	269	390
	12	3	5	9	16	22	37	53	82	110	142	224	323
	10	2	4	7	13	18	30	43	66	89	115	181	261
	8	1	2	4	7	9	15	22	35	46	60	94	137
	6	1	1	3	5	7	12	18	28	37	48	76	109
	4	1	1	2	4	6	10	14	22	29	37	59	85
	3	1	1	1	4	5	8	12	19	25	33	52	75
	2	1	1	1	3	4	7	10	16	22	28	45	65
	1	0	1	1	1	3	5	7	11	14	19	29	43
	1/0	0	1	1	1	2	4	6	9	13	16	26	37
	2/0	0	0	1	1	1	3	5	8	11	14	22	32
	3/0	0	0	1	1	1	3	4	7	9	12	19	28
	4/0	0	0	1	1	1	2	4	6	8	10	16	24
	250	0	0	0	1	1	1	3	4	6	8	12	18
	300	0	0	0	1	1	1	2	4	5	7	11	16
	350	0	0	0	1	1	1	2	3	5	6	10	14
	400	0	0	0	1	1	1	1	3	4	6	9	13
	500	0	0	0	0	1	1	1	3	4	5	8	11
	600	0	0	0	0	1	1	1	2	3	4	6	9
	700	0	0	0	0	0	1	1	1	3	3	6	8
	750	0	0	0	0	0	1	1	1	2	3	5	8
	800	0	0	0	0	0	1	1	1	2	3	5	7
	900	0	0	0	0	0	1	1	1	2	3	5	7
	1000	0	0	0	0	0	1	1	1	1	3	4	6
	1250	0	0	0	0	0	0	1	1	1	1	3	5
	1500	0	0	0	0	0	0	1	1	1	1	3	4
	1750	0	0	0	0	0	0	1	1	1	1	2	3
	2000	0	0	0	0	0	0	0	1	1	1	2	3
TW, THHW, THW, THW-2	14	8	14	24	42	57	94	135	209	280	361	568	822
	12	6	11	18	32	44	72	103	160	215	277	436	631
	10	4	8	13	24	32	54	77	119	160	206	325	470
	8	2	4	7	13	18	30	43	66	89	115	181	261
RHH*, RHW*, RHW-2*	14	5	9	16	28	38	63	90	139	186	240	378	546
	12	4	8	12	22	30	50	72	112	150	193	304	439
	10	3	6	10	17	24	39	56	87	117	150	237	343
	8	1	3	6	10	14	23	33	52	70	90	142	205
TW, THW, THHW, THW-2	6	1	2	4	8	11	18	26	40	53	69	109	157
	4	1	1	3	6	8	13	19	30	40	51	81	117
	3	1	1	3	5	7	11	16	25	34	44	69	100
	2	1	1	2	4	6	10	14	22	29	37	59	85
	1	0	1	1	3	4	7	10	15	20	26	41	60
	1/0	0	1	1	2	3	6	8	13	17	22	35	51
	2/0	0	1	1	1	3	5	7	11	15	19	30	43
	3/0	0	1	1	1	2	4	6	9	12	16	25	36
	4/0	0	0	1	1	1	3	5	8	10	13	21	30
	250	0	0	1	1	1	3	4	6	8	11	17	25
	300	0	0	1	1	1	2	3	5	7	9	15	21
	350	0	0	0	1	1	1	3	5	6	8	13	19
	400	0	0	0	1	1	1	3	4	6	7	12	17
	500	0	0	0	1	1	1	2	3	5	6	10	14
	600	0	0	0	0	1	1	1	3	4	5	8	11
	700	0	0	0	0	1	1	1	2	3	4	7	10
	750	0	0	0	0	1	1	1	2	3	4	6	10
	800	0	0	0	0	1	1	1	2	3	4	6	9
	900	0	0	0	0	0	1	1	1	3	3	6	8
	1000	0	0	0	0	0	1	1	1	2	3	5	7
	1250	0	0	0	0	0	1	1	1	1	2	4	6
	1500	0	0	0	0	0	1	1	1	1	1	3	5
	1750	0	0	0	0	0	0	1	1	1	1	3	4
	2000	0	0	0	0	0	0	1	1	1	1	3	4

(continued)

TABLE 15.4.19 Maximum Number of Conductors or Fixture Wires in Rigid PVC Conduit, Schedule 40 and HDPE Conduit *(Continued)*

CONDUCTORS													
	Conductor	Metric Designator (Trade Size)											
Type	Size (AWG/kcmil)	16 (½)	21 (¾)	27 (1)	35 (1¼)	41 (1½)	53 (2)	63 (2½)	78 (3)	91 (3½)	103 (4)	129 (5)	155 (6)
THHN, THWN, THWN-2	14	11	21	34	60	82	135	193	299	401	517	815	1178
	12	8	15	25	43	59	99	141	218	293	377	594	859
	10	5	9	15	27	37	62	89	137	184	238	374	541
	8	3	5	9	16	21	36	51	79	106	137	216	312
	6	1	4	6	11	15	26	37	57	77	99	156	225
	4	1	2	4	7	9	16	22	35	47	61	96	138
	3	1	1	3	6	8	13	19	30	40	51	81	117
	2	1	1	3	5	7	11	16	25	33	43	68	98
	1	1	1	1	3	5	8	12	18	25	32	50	73
	1/0	1	1	1	3	4	7	10	15	21	27	42	61
	2/0	0	1	1	2	3	6	8	13	17	22	35	51
	3/0	0	1	1	1	3	5	7	11	14	18	29	42
	4/0	0	1	1	1	2	4	6	9	12	15	24	35
	250	0	0	1	1	1	3	4	7	10	12	20	28
	300	0	0	1	1	1	3	4	6	8	11	17	24
	350	0	0	1	1	1	2	3	5	7	9	15	21
	400	0	0	0	1	1	1	3	5	6	8	13	19
	500	0	0	0	1	1	1	2	4	5	7	11	16
	600	0	0	0	1	1	1	1	3	4	5	9	13
	700	0	0	0	0	1	1	1	3	4	5	8	11
	750	0	0	0	0	1	1	1	2	3	4	7	11
	800	0	0	0	0	1	1	1	2	3	4	7	10
	900	0	0	0	0	1	1	1	2	3	4	6	9
	1000	0	0	0	0	0	1	1	1	3	3	6	8
FEP, FEPB, PFA, PFAH, TFE	14	11	20	33	58	79	131	188	290	389	502	790	1142
	12	8	15	24	42	58	96	137	212	284	366	577	834
	10	6	10	17	30	41	69	98	152	204	263	414	598
	8	3	6	10	17	24	39	56	87	117	150	237	343
	6	2	4	7	12	17	28	40	62	83	107	169	244
	4	1	3	5	8	12	19	28	43	58	75	118	170
	3	1	2	4	7	10	16	23	36	48	62	98	142
	2	1	1	3	6	8	13	19	30	40	51	81	117
PFA, PFAH, TFE	1	1	1	2	4	5	9	13	20	28	36	56	81
PFA, PFAH, TFE, Z	1/0	1	1	1	3	4	8	11	17	23	30	47	68
	2/0	0	1	1	3	4	6	9	14	19	24	39	56
	3/0	0	1	1	2	3	5	7	12	16	20	32	46
	4/0	0	1	1	1	2	4	6	9	13	16	26	38
Z	14	13	24	40	70	95	158	226	350	469	605	952	1376
	12	9	17	28	49	68	112	160	248	333	429	675	976
	10	6	10	17	30	41	69	98	152	204	263	414	598
	8	3	6	11	19	26	43	62	96	129	166	261	378
	6	2	4	7	13	18	30	43	67	90	116	184	265
	4	1	3	5	9	12	21	30	46	62	80	126	183
	3	1	2	4	6	9	15	22	34	45	58	92	133
	2	1	1	3	5	7	12	18	28	38	49	77	111
	1	1	1	2	4	6	10	14	23	30	39	62	90
XHH, XHHW, XHHW-2, ZW	14	8	14	24	42	57	94	135	209	280	361	568	822
	12	6	11	18	32	44	72	103	160	215	277	436	631
	10	4	8	13	24	32	54	77	119	160	206	325	470
	8	2	4	7	13	18	30	43	66	89	115	181	261
	6	1	3	5	10	13	22	32	49	66	85	134	193
	4	1	2	4	7	9	16	23	35	48	61	97	140
	3	1	1	3	6	8	13	19	30	40	52	82	118
	2	1	1	3	5	7	11	16	25	34	44	69	99
XHH, XHHW, XHHW-2	1	1	1	1	3	5	8	12	19	25	32	51	74
	1/0	1	1	1	3	4	7	10	16	21	27	43	62
	2/0	0	1	1	2	3	6	8	13	17	23	36	52
	3/0	0	1	1	1	3	5	7	11	14	19	30	43
	4/0	0	1	1	1	2	4	6	9	12	15	24	35
	250	0	0	1	1	1	3	5	7	10	13	20	29
	300	0	0	1	1	1	3	4	6	8	11	17	25
	350	0	0	1	1	1	2	3	5	7	9	15	22
	400	0	0	0	1	1	1	3	5	6	8	13	19
	500	0	0	0	1	1	1	2	4	5	7	11	16
	600	0	0	0	1	1	1	1	3	4	5	9	13
	700	0	0	0	0	1	1	1	3	4	5	8	11
	750	0	0	0	0	1	1	1	2	3	4	7	11
	800	0	0	0	0	1	1	1	2	3	4	7	10
	900	0	0	0	0	1	1	1	2	3	4	6	9
	1000	0	0	0	0	0	1	1	1	3	3	6	8
	1250	0	0	0	0	0	1	1	1	1	3	4	6
	1500	0	0	0	0	0	1	1	1	1	2	4	5
	1750	0	0	0	0	0	0	1	1	1	1	3	5
	2000	0	0	0	0	0	0	1	1	1	1	3	4

(continued)

TABLE 15.4.19 Maximum Number of Conductors or Fixture Wires in Rigid PVC Conduit, Schedule 40 and HDPE Conduit *(Continued)*

FIXTURE WIRES							
		Metric Designator (Trade Size)					
Type	Conductor Size (AWG/kcmil)	16 (½)	21 (¾)	27 (1)	35 (1¼)	41 (1½)	53 (2)
FFH-2, RFH-2, RFHH-3	18	8	14	23	40	54	90
	16	6	12	19	33	46	76
SF-2, SFF-2	18	10	17	29	50	69	114
	16	8	14	24	42	57	94
	14	6	12	19	33	46	76
SF-1. SFF-1	18	17	31	51	89	122	202
RFHH-2, TF, TFF, XF, XFF RFH-1,	18	13	23	38	66	90	149
RFHH-2, TF, TFF, XF, XFF	16	10	18	30	53	73	120
XF. XFF	14	8	14	24	42	57	94
TFN, TFFN	18	20	37	60	105	144	239
	16	16	28	46	80	110	183
PF, PFF, PGF, PGFF, PAF, PTF, PTFF. PAFF	18	19	35	57	100	137	227
	16	15	27	44	77	106	175
	14	11	20	33	58	79	131
HF, HFF, ZF, ZFF, ZHF	18	25	45	74	129	176	292
	16	18	33	54	95	130	216
	14	13	24	40	70	95	158
KF-2, KFF-2	18	36	65	107	187	256	424
	16	26	46	75	132	180	299
	14	17	31	52	90	124	205
	12	12	22	35	62	85	141
	10	8	14	24	42	57	94
KF-1, KFF-1	18	43	78	128	223	305	506
	16	30	55	90	157	214	355
	14	20	37	60	105	144	239
	12	13	24	40	70	95	158
	10	9	16	26	45	62	103
XF, XFF	12	4	8	12	22	30	50
	10	3	6	10	17	24	39

Note: This table is for concentric stranded conductors only. For compact stranded conductors, Table C10(A) should be used.

*Types RHH, RHW, and RHW-2 without outer covering.

15.4.20 Table C10(A), Maximum Number of Compact Conductors in Rigid PVC Conduit, Schedule 40 and HDPE Conduit

TABLE 15.4.20

COMPACT CONDUCTORS

Type	Conductor Size (AWG/kcmil)	Metric Designator (Trade Size) 16 (½)	21 (¾)	27 (1)	35 (1¼)	41 (1½)	53 (2)	63 (2½)	78 (3)	91 (3½)	103 (4)	129 (5)	155 (6)
THW, THW-2, THHW	8	1	4	6	11	15	26	37	57	76	98	155	224
	6	1	3	5	9	12	20	28	44	59	76	119	173
	4	1	1	3	6	9	15	21	33	44	57	89	129
	2	1	1	2	5	6	11	15	24	32	42	66	95
	1	1	1	1	3	4	7	11	17	23	29	46	67
	1/0	0	1	1	3	4	6	9	15	20	25	40	58
	2/0	0	1	1	2	3	5	8	12	16	21	34	49
	3/0	0	1	1	1	3	5	7	10	14	18	29	42
	4/0	0	1	1	1	2	4	5	9	12	15	24	35
	250	0	0	1	1	1	3	4	7	9	12	19	27
	300	0	0	1	1	1	2	4	6	8	10	16	24
	350	0	0	1	1	1	2	3	5	7	9	15	21
	400	0	0	0	1	1	1	3	5	6	8	13	19
	500	0	0	0	1	1	1	2	4	5	7	11	16
	600	0	0	0	1	1	1	1	3	4	5	9	13
	700	0	0	0	0	1	1	1	3	4	5	8	12
	750	0	0	0	0	1	1	1	2	3	5	7	11
	1000	0	0	0	0	1	1	1	1	3	4	6	9

COMPACT CONDUCTORS

Type	Conductor Size (AWG/kcmil)	Metric Designator (Trade Size) 16 (½)	21 (¾)	27 (1)	35 (1¼)	41 (1½)	53 (2)	63 (2½)	78 (3)	91 (3½)	103 (4)	129 (5)	155 (6)
THHN, THWN, THWN-2	8	—	—	—	—	—	—	—	—	—	—	—	—
	6	2	4	7	13	17	29	41	64	86	111	175	253
	4	1	2	4	8	11	18	25	40	53	68	108	156
	2	1	1	3	5	8	13	18	28	38	49	77	112
	1	1	1	2	4	6	9	14	21	29	37	58	84
	1/0	1	1	1	3	5	8	12	18	24	31	49	72
	2/0	0	1	1	3	4	7	9	15	20	26	41	59
	3/0	0	1	1	2	3	5	8	12	17	22	34	50
	4/0	0	1	1	1	3	4	6	10	14	18	28	41
	250	0	0	1	1	1	3	5	8	11	14	22	32
	300	0	0	1	1	1	3	4	7	9	12	19	28
	350	0	0	1	1	1	3	4	6	8	10	17	24
	400	0	0	1	1	1	2	3	5	7	9	15	22
	500	0	0	0	1	1	1	3	4	6	8	13	18
	600	0	0	0	1	1	1	2	4	5	6	10	15
	700	0	0	0	1	1	1	1	3	4	5	9	13
	750	0	0	0	1	1	1	1	3	4	5	8	12
	1000	0	0	0	0	1	1	1	2	3	4	6	9
XHHW, XHHW-2	8	3	5	8	14	20	33	47	73	99	127	200	290
	6	1	4	6	11	15	25	35	55	73	94	149	215
	4	1	2	4	8	11	18	25	40	53	68	108	156
	2	1	1	3	5	8	13	18	28	38	49	77	112
	1	1	1	2	4	6	9	14	21	29	37	58	84
	1/0	1	1	1	3	5	8	12	18	24	31	49	72
	2/0	1	1	1	3	4	7	10	15	20	26	42	60
	3/0	0	1	1	2	3	5	8	12	17	22	34	50
	4/0	0	1	1	1	3	5	7	10	14	18	29	42
	250	0	0	1	1	1	4	5	8	11	14	23	33
	300	0	0	1	1	1	3	4	7	9	12	19	28
	350	0	0	1	1	1	3	4	6	8	11	17	25
	400	0	0	1	1	1	2	3	5	7	10	15	22
	500	0	0	0	1	1	1	3	4	6	8	13	18
	600	0	0	0	1	1	1	2	4	5	6	10	15
	700	0	0	0	1	1	1	1	3	4	5	9	13
	750	0	0	0	1	1	1	1	3	4	5	8	12
	1000	0	0	0	0	1	1	1	2	3	4	6	9

Definition: *Compact stranding* is the result of a manufacturing process where the standard conductor is compressed to the extent that the interstices (voids between strand wires) are virtually eliminated.

15.4.21 Table C11, Maximum Number of Conductors or Fixture Wires in Type A Rigid PVC Conduit

TABLE 15.4.21

CONDUCTORS											
	Conductor	Metric Designator (Trade Size)									
Type	Size (AWG/kcmil)	16 (½)	21 (¾)	27 (1)	35 (1¼)	41 (1½)	53 (2)	63 (2½)	78 (3)	91 (3½)	103 (4)
RHH, RHW RHW-2	14	5	9	15	24	31	49	74	112	146	187
	12	4	7	12	20	26	41	61	93	121	155
	10	3	6	10	16	21	33	50	75	98	125
	8	1	3	5	8	11	17	26	39	51	65
	6	1	2	4	6	9	14	21	31	41	52
	4	1	1	3	5	7	11	16	24	32	41
	3	1	1	3	4	6	9	14	21	28	36
	2	1	1	2	4	5	8	12	18	24	31
	1	0	1	1	2	3	5	8	12	16	20
	1/0	0	1	1	2	3	5	7	10	14	18
	2/0	0	1	1	1	2	4	6	9	12	15
	3/0	0	1	1	1	1	3	5	8	10	13
	4/0	0	0	1	1	1	3	4	7	9	11
	250	0	0	1	1	1	1	3	5	7	8
	300	0	0	1	1	1	1	3	4	6	7
	350	0	0	0	1	1	1	2	4	5	7
	400	0	0	0	1	1	1	2	4	5	6
	500	0	0	0	1	1	1	1	3	4	5
	600	0	0	0	0	1	1	1	2	3	4
	700	0	0	0	0	1	1	1	2	3	4
	750	0	0	0	0	1	1	1	1	3	4
	800	0	0	0	0	1	1	1	1	3	3
	900	0	0	0	0	0	1	1	1	2	3
	1000	0	0	0	0	0	1	1	1	2	3
	1250	0	0	0	0	0	1	1	1	1	2
	1500	0	0	0	0	0	0	1	1	1	1
	1750	0	0	0	0	0	0	1	1	1	1
	2000	0	0	0	0	0	0	1	1	1	1
TW,	14	11	18	31	51	67	105	157	235	307	395
THHW,	12	8	14	24	39	51	80	120	181	236	303
THW,	10	6	10	18	29	38	60	89	135	176	226
THW-2	8	3	6	10	16	21	33	50	75	98	125

(continued)

TABLE 15.4.21 Maximum Number of Conductors or Fixture Wires in Type A Rigid PVC Conduit (*Continued*)

CONDUCTORS											
	Conductor	Metric Designator (Trade Size)									
Type	Size (AWG/kcmil)	16 (½)	21 (¾)	27 (1)	35 (1¼)	41 (1½)	53 (2)	63 (2½)	78 (3)	91 (3½)	103 (4)
RHH*, RHW*, RHW-2*	14	7	12	20	34	44	70	104	157	204	262
	12	6	10	16	27	35	56	84	126	164	211
	10	4	8	13	21	28	44	65	98	128	165
	8	2	4	8	12	16	26	39	59	77	98
RHH, RHW*	6	1	3	6	9	13	20	30	45	59	75
TW, THW, THHW, THW-2	4	1	2	4	7	9	15	22	33	44	56
	3	1	1	4	6	8	13	19	29	37	48
	2	1	1	3	5	7	11	16	24	32	41
	1	1	1	1	3	5	7	11	17	22	29
	1/0	1	1	1	3	4	6	10	14	19	24
	2/0	0	1	1	2	3	5	8	12	16	21
	3/0	0	1	1	1	3	4	7	10	13	17
	4/0	0	1	1	1	2	4	6	9	11	14
	250	0	0	1	1	1	3	4	7	9	12
	300	0	0	1	1	1	2	4	6	8	10
	350	0	0	1	1	1	2	3	5	7	9
	400	0	0	1	1	1	1	3	5	6	8
	500	0	0	0	1	1	1	2	4	5	7
	600	0	0	0	1	1	1	1	3	4	5
	700	0	0	0	1	1	1	1	3	4	5
	750	0	0	0	1	1	1	1	3	3	4
	800	0	0	0	0	1	1	1	2	3	4
	900	0	0	0	0	1	1	1	2	3	4
	1000	0	0	0	0	1	1	1	1	3	3
	1250	0	0	0	0	0	1	1	1	1	3
	1500	0	0	0	0	0	1	1	1	1	2
	1750	0	0	0	0	0	0	1	1	1	1
	2000	0	0	0	0	0	0	1	1	1	1
THHN, THWN, THWN-2	14	16	27	44	73	96	150	225	338	441	566
	12	11	19	32	53	70	109	164	246	321	412
	10	7	12	20	33	44	69	103	155	202	260
	8	4	7	12	19	25	40	59	89	117	150
	6	3	5	8	14	18	28	43	64	84	108
	4	1	3	5	8	11	17	26	39	52	66
	3	1	2	4	7	9	15	22	33	44	56
	2	1	1	3	6	8	12	19	28	37	47
	1	1	1	2	4	6	9	14	21	27	35
	1/0	1	1	2	4	5	8	11	17	23	29
	2/0	1	1	1	3	4	6	10	14	19	24
	3/0	0	1	1	2	3	5	8	12	16	20
	4/0	0	1	1	1	3	4	6	10	13	17
	250	0	1	1	1	2	3	5	8	10	14
	300	0	0	1	1	1	3	4	7	9	12
	350	0	0	1	1	1	2	4	6	8	10
	400	0	0	1	1	1	2	3	5	7	9
	500	0	0	1	1	1	1	3	4	6	7
	600	0	0	0	1	1	1	2	3	5	6
	700	0	0	0	1	1	1	1	3	4	5
	750	0	0	0	1	1	1	1	3	4	5
	800	0	0	0	1	1	1	1	3	4	5
	900	0	0	0	0	1	1	1	2	3	4
	1000	0	0	0	0	1	1	1	2	3	4
FEP, FEPB, PFA, PFAH, TFE	14	15	26	43	70	93	146	218	327	427	549
	12	11	19	31	51	68	106	159	239	312	400
	10	8	13	22	37	48	76	114	171	224	287
	8	4	8	13	21	28	44	65	98	128	165
	6	3	5	9	15	20	31	46	70	91	117
	4	1	4	6	10	14	21	32	49	64	82
	3	1	3	5	8	11	18	27	40	53	68
	2	1	2	4	7	9	15	22	33	44	56
PFA, PFAH, TFE	1	1	1	3	5	6	10	15	23	30	39
PFA, PFAH, TFE, Z	1/0	1	1	2	4	5	8	13	19	25	32
	2/0	1	1	1	3	4	7	10	16	21	27
	3/0	1	1	1	3	3	6	9	13	17	22
	4/0	0	1	1	2	3	5	7	11	14	18
Z	14	18	31	52	85	112	175	263	395	515	661
	12	13	22	37	60	79	124	186	280	365	469
	10	8	13	22	37	48	76	114	171	224	287
	8	5	8	14	23	30	48	72	108	141	181
	6	3	6	10	16	21	34	50	76	99	127
	4	2	4	7	11	15	23	35	52	68	88
	3	1	3	5	8	11	17	25	38	50	64
	2	1	2	4	7	9	14	21	32	41	53
	1	1	1	3	5	7	11	17	26	33	43
XHH, XHHW, XHHW-2, ZW	14	11	18	31	51	67	105	157	235	307	395
	12	8	14	24	39	51	80	120	181	236	303
	10	6	10	18	29	38	60	89	135	176	226
	8	3	6	10	16	21	33	50	75	98	125
	6	2	4	7	12	15	24	37	55	72	93
	4	1	3	5	8	11	18	26	40	52	67
	3	1	2	4	7	9	15	22	34	44	57
	2	1	1	3	6	8	12	19	28	37	48

(continued)

TABLE 15.4.21 Maximum Number of Conductors or Fixture Wires in Type A Rigid PVC Conduit *(Continued)*

CONDUCTORS

Type	Conductor Size (AWG/kcmil)	Metric Designator (Trade Size)									
		16 (½)	21 (¾)	27 (1)	35 (1¼)	41 (1½)	53 (2)	63 (2½)	78 (3)	91 (3½)	103 (4)
XHH, XHHW, XHHW-2	1	1	1	3	4	6	9	14	21	28	35
	1/0	1	1	2	4	5	8	12	18	23	30
	2/0	1	1	1	3	4	6	10	15	19	25
	3/0	0	1	1	2	3	5	8	12	16	20
	4/0	0	1	1	1	3	4	7	10	13	17
	250	0	1	1	1	2	3	5	8	11	14
	300	0	0	1	1	1	3	5	7	9	12
	350	0	0	1	1	1	3	4	6	8	10
	400	0	0	1	1	1	2	3	5	7	9
	500	0	0	1	1	1	1	3	4	6	8
	600	0	0	0	1	1	1	2	3	5	6
	700	0	0	0	1	1	1	1	3	4	5
	750	0	0	0	1	1	1	1	3	4	5
	800	0	0	0	1	1	1	1	3	4	5
	900	0	0	0	0	1	1	1	2	3	4
	1000	0	0	0	0	1	1	1	2	3	4
	1250	0	0	0	0	0	1	1	1	2	3
	1500	0	0	0	0	0	1	1	1	1	2
	1750	0	0	0	0	0	1	1	1	1	2
	2000	0	0	0	0	0	0	1	1	1	1

FIXTURE WIRES

Type	Conductor Size (AWG/kcmil)	Metric Designator (Trade Size)					
		16 (½)	21 (¾)	27 (1)	35 (1¼)	41 (1½)	53 (2)
FFH-2, RFH-2, RFHH-3	18	10	18	30	48	64	100
	16	9	15	25	41	54	85
SF-2, SFF-2	18	13	22	37	61	81	127
	16	11	18	31	51	67	105
	14	9	15	25	41	54	85
SF-1, SFF-1	18	23	40	66	108	143	224
RFH-1, RFHH-2, TF, TFF, XF, XFF	18	17	29	49	80	105	165
RFHH-2, TF, TFF, XF, XFF	16	14	24	39	65	85	134
XF, XFF	14	11	18	31	51	67	105
TFN, TFFN	18	28	47	79	128	169	265
	16	21	36	60	98	129	202
PF, PFF, PGF, PGFF, PAF, PTF, PTFF, PAFF	18	26	45	74	122	160	251
	16	20	34	58	94	124	194
	14	15	26	43	70	93	146
HF, HFF, ZF, ZFF, ZHF	18	34	58	96	157	206	324
	16	25	42	71	116	152	239
	14	18	31	52	85	112	175
KF-2, KFF-2	18	49	84	140	228	300	470
	16	35	59	98	160	211	331
	14	24	40	67	110	145	228
	12	16	28	46	76	100	157
	10	11	18	31	51	67	105
KF-1, KFF-1	18	59	100	167	272	357	561
	16	41	70	117	191	251	394
	14	28	47	79	128	169	265
	12	18	31	52	85	112	175
	10	12	20	34	55	73	115
XF, XFF	12	6	10	16	27	35	56
	10	4	8	13	21	28	44

Note: This table is for concentric stranded conductors only. For compact stranded conductors, Table C11(A) should be used.

*Types RHH, RHW, and RWH-2 without outer covering.

15.4.22 Table C11(A), Maximum Number of Compact Conductors in Type A Rigid PVC Conduit

TABLE 15.4.22

COMPACT CONDUCTORS											
	Conductor	Metric Designator (Trade Size)									
Type	Size (AWG/kcmil)	16 (½)	21 (¾)	27 (1)	35 (1¼)	41 (1½)	53 (2)	63 (2½)	78 (3)	91 (3½)	103 (4)
THW, THW-2, THHW	8	3	5	8	14	18	28	42	64	84	107
	6	2	4	6	10	14	22	33	49	65	83
	4	1	3	5	8	10	16	24	37	48	62
	2	1	1	3	6	7	12	18	27	36	46
	1	1	1	2	4	5	8	13	19	25	32
	1/0	1	1	1	3	4	7	11	16	21	28
	2/0	1	1	1	3	4	6	9	14	18	23
	3/0	0	1	1	2	3	5	8	12	15	20
	4/0	0	1	1	1	3	4	6	10	13	17
	250	0	1	1	1	1	3	5	8	10	13
	300	0	0	1	1	1	3	4	7	9	11
	350	0	0	1	1	1	2	4	6	8	10
	400	0	0	1	1	1	2	3	5	7	9
	500	0	0	1	1	1	1	3	4	6	8
	600	0	0	0	1	1	1	2	3	5	6
	700	0	0	0	1	1	1	1	3	4	5
	750	0	0	0	1	1	1	1	3	4	5
	1000	0	0	0	0	1	1	1	2	3	4
THHN, THWN, THWN-2	8	—	—	—	—	—	—	—	—	—	—
	6	3	5	9	15	20	32	48	72	94	121
	4	1	3	6	9	12	20	30	45	58	75
	2	1	2	4	7	9	14	21	32	42	54
	1	1	1	3	5	7	10	16	24	31	40
	1/0	1	1	2	4	6	9	13	20	27	34
	2/0	1	1	1	3	5	7	11	17	22	28
	3/0	1	1	1	3	4	6	9	14	18	24
	4/0	0	1	1	2	3	5	8	11	15	19
	250	0	1	1	1	2	4	6	9	12	15
	300	0	1	1	1	1	3	5	8	10	13
	350	0	0	1	1	1	3	4	7	9	11
	400	0	0	1	1	1	2	4	6	8	10
	500	0	0	1	1	1	2	3	5	7	9
	600	0	0	0	1	1	1	3	4	5	7
	700	0	0	0	1	1	1	2	3	5	6
	750	0	0	0	1	1	1	2	3	4	6
	1000	0	0	0	0	1	1	1	2	3	4
XHHW, XHHW-2	8	4	6	11	18	23	37	55	83	108	139
	6	3	5	8	13	17	27	41	62	80	103
	4	1	3	6	9	12	20	30	45	58	75
	2	1	2	4	7	9	14	21	32	42	54
	1	1	1	3	5	7	10	16	24	31	40
	1/0	1	1	2	4	6	9	13	20	27	34
	2/0	1	1	1	3	5	7	11	17	22	29
	3/0	1	1	1	3	4	6	9	14	18	24
	4/0	0	1	1	2	3	5	8	12	15	20
	250	0	1	1	1	2	4	6	9	12	16
	300	0	1	1	1	1	3	5	8	10	13
	350	0	0	1	1	1	3	5	7	9	12
	400	0	0	1	1	1	3	4	6	8	11
	500	0	0	1	1	1	2	3	5	7	9
	600	0	0	0	1	1	1	3	4	5	7
	700	0	0	0	1	1	1	2	3	5	6
	750	0	0	0	1	1	1	2	3	4	6
	1000	0	0	0	0	1	1	1	2	3	4

Definition: *Compact stranding* is the result of a manufacturing process where the standard conductor is compressed to the extent that the interstices (voids between strand wires) are virtually eliminated.

15.4.23 Table C12, Maximum Number of Conductors in Type EB PVC Conduit

TABLE 15.4.23

		CONDUCTORS					
		Metric Designator (Trade Size)					
Type	Conductor Size (AWG/kcmil)	53 (2)	78 (3)	91 (3½)	103 (4)	129 (5)	155 (6)
RHH, RHW, RHW-2	14	53	119	155	197	303	430
	12	44	98	128	163	251	357
RHH, RHW, RHW-2	10	35	79	104	132	203	288
	8	18	41	54	69	106	151
	6	15	33	43	55	85	121
	4	11	26	34	43	66	94
	3	10	23	30	38	58	83
	2	9	20	26	33	50	72
	1	6	13	17	21	33	47
	1/0	5	11	15	19	29	41
	2/0	4	10	13	16	25	36
	3/0	4	8	11	14	22	31
	4/0	3	7	9	12	18	26
	250	2	5	7	9	14	20
	300	1	5	6	8	12	17
	350	1	4	5	7	11	16
	400	1	4	5	6	10	14
	500	1	3	4	5	9	12
	600	1	3	3	4	7	10
	700	1	2	3	4	6	9
	750	1	2	3	4	6	9
	800	1	2	3	4	6	8
	900	1	1	2	3	5	7
	1000	1	1	2	3	5	7
	1250	1	1	1	2	3	5
	1500	0	1	1	1	3	4
	1750	0	1	1	1	3	4
	2000	0	1	1	1	2	3
TW, THHW, THW, THW-2	14	111	250	327	415	638	907
	12	85	192	251	319	490	696
	10	63	143	187	238	365	519
	8	35	79	104	132	203	288
RHH*,RHW*, WH-2*	14	74	166	217	276	424	603
RHH*, HW*, RHW-2*	12	59	134	175	222	341	485
	10	46	104	136	173	266	378
RHH*, HW*,RHW-2*	8	28	62	81	104	159	227
RHH*, RHW*, RHW-2*, TW, THW, THHW, THW-2	6	21	48	62	79	122	173
	4	16	36	46	59	91	129
	3	13	30	40	51	78	111
	2	11	26	34	43	66	94
	1	8	18	24	30	46	66
	1/0	7	15	20	26	40	56
	2/0	6	13	17	22	34	48
	3/0	5	11	14	18	28	40
	4/0	4	9	12	15	24	34
	250	3	7	10	12	19	27
	300	3	6	8	11	17	24
	350	2	6	7	9	15	21
	400	2	5	7	8	13	19
	500	1	4	5	7	11	16
	600	1	3	4	6	9	13
	700	1	3	4	5	8	11
	750	1	3	4	5	7	11
	800	1	3	3	4	7	10
	900	1	2	3	4	6	9
	1000	1	2	3	4	6	8
	1250	1	1	2	3	4	6
	1500	1	1	1	2	4	6
	1750	1	1	1	2	3	5
	2000	0	1	1	1	3	4

(continued)

TABLE 15.4.23 Maximum Number of Conductors in Type EB PVC Conduit *(Continued)*

CONDUCTORS							
		Metric Designator (Trade Size)					
Type	Conductor Size (AWG/kcmil)	53 (2)	78 (3)	91 (3½)	103 (4)	129 (5)	155 (6)
THHN, THWN, THWN-2	14	159	359	468	595	915	1300
	12	116	262	342	434	667	948
	10	73	165	215	274	420	597
	8	42	95	124	158	242	344
	6	30	68	89	114	175	248
	4	19	42	55	70	107	153
	3	16	36	46	59	91	129
	2	13	30	39	50	76	109
	1	10	22	29	37	57	80
	1/0	8	18	24	31	48	68
	2/0	7	15	20	26	40	56
	3/0	5	13	17	21	33	47
	4/0	4	10	14	18	27	39
	250	4	8	11	14	22	31
	300	3	7	10	12	19	27
	350	3	6	8	11	17	24
	400	2	6	7	10	15	21
	500	1	5	6	8	12	18
	600	1	4	5	6	10	14
	700	1	3	4	6	9	12
	750	1	3	4	5	8	12
	800	1	3	4	5	8	11
	900	1	3	3	4	7	10
	1000	1	2	3	4	6	9
FEP, FEPB, PFA, PFAH, TFE	14	155	348	454	578	888	1261
	12	113	254	332	422	648	920
	10	81	182	238	302	465	660
	8	46	104	136	173	266	378
	6	33	74	97	123	189	269
	4	23	52	68	86	132	188
	3	19	43	56	72	110	157
	2	16	36	46	59	91	129
PFA, PFAH, TFE	1	11	25	32	41	63	90
PFA, PFAH, TFE, Z	1/0	9	20	27	34	53	75
	2/0	7	17	22	28	43	62
	3/0	6	14	18	23	36	51
	4/0	5	11	15	19	29	42
Z	14	186	419	547	696	1069	1519
	12	132	297	388	494	759	1078
	10	81	182	238	302	465	660
	8	51	115	150	191	294	417
	6	36	81	105	134	206	293
	4	24	55	72	92	142	201
	3	18	40	53	67	104	147
	2	15	34	44	56	86	122
	1	12	27	36	45	70	99
XHH, XHHW, XHHW-2, ZW	14	111	250	327	415	638	907
	12	85	192	251	319	490	696
	10	63	143	187	238	365	519
	8	35	79	104	132	203	288
	6	26	59	77	98	150	213
	4	19	42	56	71	109	155
	3	16	36	47	60	92	131
	2	13	30	39	50	77	110
XHH, XHHW, XHHW-2	1	10	22	29	37	58	82
	1/0	8	19	25	31	48	69
	2/0	7	16	20	26	40	57
	3/0	6	13	17	22	33	47
	4/0	5	11	14	18	27	39
	250	4	9	11	15	22	32
	300	3	7	10	12	19	28
	350	3	6	9	11	17	24
	400	2	6	8	10	15	22
	500	1	5	6	8	12	18
	600	1	4	5	6	10	14
	700	1	3	4	6	9	12
	750	1	3	4	5	8	12
	800	1	3	4	5	8	11
	900	1	3	3	4	7	10
	1000	1	2	3	4	6	9
	1250	1	1	2	3	5	7
	1500	1	1	1	3	4	6
	1750	1	1	1	2	4	5
	2000	0	1	1	1	3	5

Note: This table is for concentric stranded conductors only. For compact stranded conductors, Table C12(A) should be used.

*Types RHH, RHW, and RHW-2 without outer covering.

15.4.24 Table C12(A), Maximum Number of Compact Conductors in Type EB PVC Conduit

TABLE 15.4.24

COMPACT CONDUCTORS

Type	Conductor Size (AWG/kcmil)	Metric Designator (Trade Size) 53 (2)	78 (3)	91 (3½)	103 (4)	129 (5)	155 (6)
THW, THW-2, THHW	8	30	68	89	113	174	247
	6	23	52	69	87	134	191
	4	17	39	51	65	100	143
	2	13	29	38	48	74	105
	1	9	20	26	34	52	74
	1/0	8	17	23	29	45	64
	2/0	6	15	19	24	38	54
	3/0	5	12	16	21	32	46
	4/0	4	10	14	17	27	38
	250	3	8	11	14	21	30
	300	3	7	9	12	19	26
	350	3	6	8	11	17	24
	400	2	6	7	10	15	21
	500	1	5	6	8	12	18
	600	1	4	5	6	10	14
	700	1	3	4	6	9	13
	750	1	3	4	5	8	12
	1000	1	2	3	4	7	9
THHN, THWN, THWN-2	8	—	—	—	—	—	—
	6	34	77	100	128	196	279
	4	21	47	62	79	121	172
	2	15	34	44	57	87	124
	1	11	25	33	42	65	93
	1/0	9	22	28	36	56	79
	2/0	8	18	23	30	46	65
	3/0	6	15	20	25	38	55
	4/0	5	12	16	20	32	45
	250	4	10	13	16	25	35
	300	4	8	11	14	22	31
	350	3	7	9	12	19	27
	400	3	6	8	11	17	24
	500	2	5	7	9	14	20
	600	1	4	6	7	11	16
	700	1	4	5	6	10	14
	750	1	4	5	6	9	14
	1000	1	3	3	4	7	10

COMPACT CONDUCTORS

Type	Conductor Size (AWG/kcmil)	Metric Designator (Trade Size) 53 (2)	78 (3)	91 (3½)	103 (4)	129 (5)	155 (6)
XHHW, XHHW-2	8	39	88	115	146	225	320
	6	29	65	85	109	167	238
	4	21	47	62	79	121	172
	2	15	34	44	57	87	124
	1	11	25	33	42	65	93
	1/0	9	22	28	36	56	79
	2/0	8	18	24	30	47	67
	3/0	6	15	20	25	38	55
	4/0	5	12	16	21	32	46
	250	4	10	13	17	26	37
	300	4	8	11	14	22	31
	350	3	7	10	12	19	28
	400	3	7	9	11	17	25
	500	2	5	7	9	14	20
	600	1	4	6	7	11	16
	700	1	4	5	6	10	14
	750	1	3	5	6	9	13
	1000	1	3	4	5	7	10

Definition: *Compact stranding* is the result of a manufacturing process where the standard conductor is compressed to the extent that the interstices (voids between strand wires) are virtually eliminated.

Section

16

Lighting

16.1.1 Conversion Factors of Units of Illumination
16.1.2 U.S. and Canadian Standards for Ballast Efficacy Factor
16.1.3 Starting and Restrike Times among Different HID Lamps
16.1.4 How Light Affects Color
16.1.5 Summary of Light-Source Characteristics and Effects on Color
16.2.1 Determination of Illuminance Categories
16.3.1 Zonal Cavity Method of Calculating Illumination
16.3.2 Coefficients of Utilization for Typical Luminaires
16.3.3 Light-Loss Factor (*LLF*)
16.3.4 Light-Loss Factors by Groups
16.3.5 Light Output Change Due to Voltage Change
16.3.6 Lumen Output for HID Lamps as a Function of Operating Position
16.3.7 Lamp Lumen Depreciation
16.3.8 Procedure for Determining Luminaire Maintenance Categories
16.3.9 Evaluation of Operating Atmosphere
16.3.10 Five Degrees of Dirt Conditions
16.3.11 Luminaire Dirt Depreciation (*LDD*) Factors for Six Luminaire Categories (I through VI) and for the Five Degrees of Dirtiness as Determined from Figure 16.3.8 or Table 16.3.9
16.3.12 Room Surface Dirt Depreciation (*RSDD*) Factors
16.3.13 Step-by-Step Calculations for the Number of Luminaires Required for a Particular Room
16.3.14 Reflectance Values of Various Materials and Colors
16.3.15 Room Cavity Ratios
16 3.16 Percent Effective Ceiling or Floor Cavity Reflectances for Various Reflectance Combinations
16.3.17 Multiplying Factors for Effective Floor Cavity Reflectances Other than 20 Percent (0.2)
16.3.18 Characteristics of Typical Lamps
16.3.19 Guide to Lamp Selection
16.3.20 Recommended Reflectances of Interior Surfaces
16.3.21 Recommended Luminance Ratios
16.3.22 Average Illuminance Calculation Sheet

16.1.1 Conversion Factors of Units of Illumination

TABLE 16.1.1

Given	Multiply by	to obtain
Illuminance (E) in lux	0.0929	footcandles
Illuminance (E) in footcandles	10.764	lux
Luminance (L) in cd/sq. m	0.2919	footlamberts
Luminance (L) in footlamberts	3.4263	cd/sq. m
Intensity (I) candelas	1.0	candlepower

16.1.2 U.S. and Canadian Standards for Ballast Efficacy Factor

TABLE 16.1.2

Lamp Type	Ballast Input Voltage (V)	Total Nominal Lamp Active Power (W)	BEF (minimum)	
			U.S. Standard	Canadian Standard
One F40T12	120	40	1.805	1.805
	277	40	1.805	1.805
	347	40	N/A	1.750
Two F40T12	120	80	1.060	1.060
	277	80	1.050	1.050
	347	80	N/A	1.020
Two F96T12	120	150	0.570	0.570
	277	150	0.570	0.570
	347	150	N/A	0.560
Two F96T12/HO	120	220	0.390	0.390
	277	220	0.390	0.390
	347	220	N/A	0.380
Two F32T8	120	64	N/A	1.250
	277	64	N/A	1.230
	347	64	N/A	1.200

(Reprinted from the IESNA Lighting Handbook, 9th Edition, 2000, Courtesy of the Illuminating Engineering Society of North America.)

16.1.3 Starting and Restrike Times among Different HID Lamps

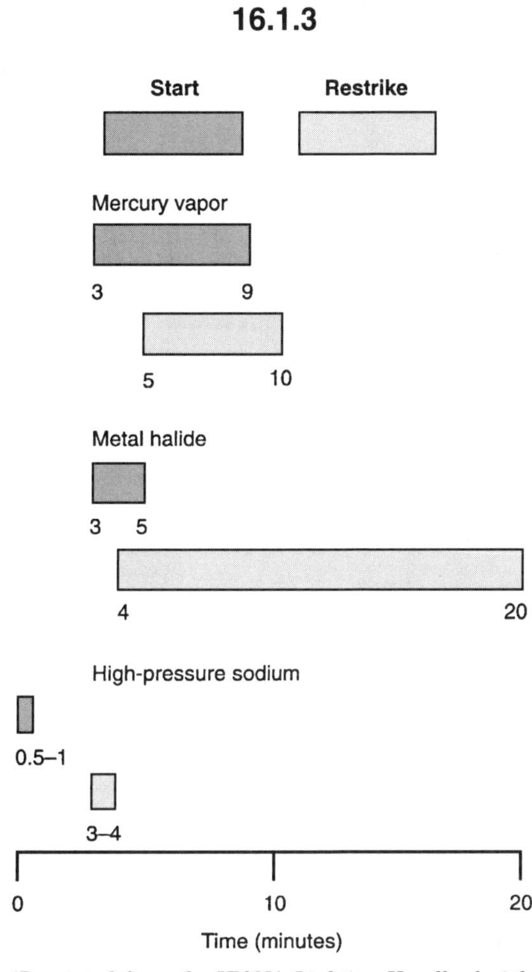

(Reprinted from the IESNA Lighting Handbook, 9th Edition, 2000, Courtesy of the Illuminating Engineering Society of North America.)

16.1.4 How Light Affects Color

Relationship of Light and Color. *Light* is the radiant energy produced by a light source. It may come to your eye directly from the source or be reflected or transmitted by some object.

Color is the interaction of the light source, the reflector or transmitter, and our own ability to detect the color of light. Remember, you cannot perceive color without light. Different light sources radiate different wavelengths of light, influencing the appearance of colored objects or surfaces.

Color Temperature. *Color temperature* describes how the lamp itself appears when lit. Color temperature is measured by Kelvin degrees, ranging from 9000 K (which appears blue) down to 1500 K (which appears orange-red). Light sources lie somewhere between the two, with those of higher color temperature—4000 K or more—being “cool” and those of lower color temperature—3100 K or less—being "warm." Certain fluorescent lamps are intermediate types, lying somewhere between cool and warm.

Color Rendition. *Color rendition* describes the effect a light source has on the appearance of colored objects. The color rendering capability of a lamp is measured as the *color rendering index* (CRI). In general, the higher the CRI, the less distortion of the object's color there is by the lamp's light output. The scale used ranges from 0 to 100. A CRI of 100 indicates that there is no color shift as compared with a reference source, and the lower the CRI, the more pronounced the shift may be.

It is important to recognize that the reference source (and thus the CRI scale) is different at different color temperatures. As a result, CRI values should only be compared between lamps of similar color temperatures.

Additional Factors Affect Color Appearance. The color rendering properties of a lamp are an important influence on the color appearance of an object. However, many other factors also affect color appearance, such as the finishes used on walls, floors, and furnishings; the intensity level of the lighting; and the presence of daylight in the room. All these factors should be considered in selecting the appropriate light source. Additionally, the room decor is a critical consideration in selecting a light source. If colors such as reds and oranges are the main element, a warm light source (color temperature below 3200 K) would be the best choice. Conversely, if blues and violets are being used, cool lamps (color temperature above 4000 K) should be used. For areas with mixed cool and warm elements, or where neutral colors such as gray predominate, an intermediate color temperature source (3400 to 3600 K) should be considered.

16.1.5 Summary of Light-Source Characteristics and Effects on Color

TABLE 16.1.5

Light Source	Characteristics	Effect on Color
Incandescent Color Temperatures from 2750K to 3400K. CRI: 95+	• Warm, inviting light • Standard light source • Relatively inefficient	• Brightens reds, oranges, yellows • Darkens blues and greens
Tungsten Halogen Color temperatures from 2850K to 3000K. CRI: 95+	• Brighter, whiter light than standard incandescent • More efficient than regular incandescent	• Brightens reds, oranges, yellows • Darkens blues and greens
Fluorescent Color temperatures from 2700K to 6300K. CRIs from 48 to 90	• Wide selection of phosphor colors—select warm to cool lighting atmosphere • Generally high efficiency • Much longer life	• Wide range of color temperatures and CRIs to light effectively any (basically indoor) area with a "warm" to "cool" environment as decor or task dictates
High Intensity Discharge Metal Halide (Metalarc®) High Pressure Sodium (Lumalux® and Unalux®) Mercury	• Different gases and phosphor colors create a variety of atmospheres • High efficiency • Very long life	• Sylvania Metalarc® (metal halide) lamps provide excellent color rendering. Mercury and High Pressure Sodium provide poor color rendering. Mercury gives a blue-green coloration and High Pressure Sodium imparts an orange-yellow color

16.2.1 Determination of Illuminance Categories

TABLE 16.2.1

Orientation and simple visual tasks. Visual performance is largely unimportant. These tasks are found in public spaces where reading and visual inspection are only occasionally performed. Higher levels are recommended for tasks where visual performance is occasionally important.		
A	Public spaces	30 lx (3 fc)
B	Simple orientation for short visits	50 lx (5 fc)
C	Working spaces where simple visual tasks are performed	100 lx (10 fc)
Common visual tasks. Visual performance is important. These tasks are found in commercial, industrial and residential applications. Recommended illuminance levels differ because of the characteristics of the visual task being illuminated. Higher levels are recommended for visual tasks with critical elements of low contrast or small size.		
D	Performance of visual tasks of high contrast and large size	300 lx (30 fc)
E	Performance of visual tasks of high contrast and small size, or visual tasks of low contrast and large size	500 lx (50 fc)
F	Performance of visual tasks of low contrast and small size	1000 lx (100 fc)
Special visual tasks. Visual performance is of critical importance. These tasks are very specialized, including those with very small or very low contrast critical elements. Recommended illuminance levels should be achieved with supplementary task lighting. Higher recommended levels are often achieved by moving the light source closer to the task.		
G	Performance of visual tasks near threshold	3000 to 10,000 lx (300 to 1000 fc)

* Expected accuracy in illuminance calculations are given in Chapter 9, Lighting Calculations. To account for both uncertainty in photometric measurements and uncertainty in space reflections, measured illuminances should be with ± 10% of the recommended value. It should be noted, however, that the final illuminance may deviate from these recommended values due to other lighting design criteria.

(Reprinted from the IESNA Lighting Handbook, 9th Edition, 2000, Courtesy of the Illuminating Engineering Society of North America.)

16.3.1 Zonal Cavity Method of Calculating Illumination

The number of luminaires required to light a space to a desired illumination level (footcandles) can be calculated knowing certain characteristics of the room and light source. The following method is the zonal cavity method of calculating illumination.

$$\frac{\text{Area}}{\text{Luminaire}} = \frac{N \times \text{lumens per lamp} \times CU \times LLF}{\text{Footcandles required } (E)}$$

where N = number of lamps
CU = coefficient of utilization
LLF = light-loss factor
E = recommended illumination (maintained)

This formula can be rewritten to find the number of luminaires or to determine the maintained footcandle level:

$$\text{Number of luminaires} = \frac{\text{Footcandles required} \times \text{area of room}}{N \times \text{lumens per lamp} \times CU \times LLF}$$

$$\text{Footcandles} = \frac{N \times \text{lumens per lamp} \times CU \times LLF}{\text{Area per luminaire}}$$

The coefficient of utilization (*CU*) is a factor that reflects the fact that not all the lumens produced by a luminaire reach the work surface. It depends on the particular light fixture used as well as the characteristics of the room in which it is placed, including the room size and the surface reflectances of the room. If you know the specific luminaire you want to use, obtain *CU* factors from the manufacturer. They are usually included in product catalogs.

If you do not know specifically what fixture you will be selecting, you can use general *CU* tables based on luminaire types (see Table 16.3.2).

16.3.2 Coefficients of Utilization for Typical Luminaires (*pages 16.7–16.17*)

16.3.3 Light-Loss Factor (*LLF*)

The *light-loss factor* (*LLF*) is a fraction that represents the amount of light that will be lost due to things such as dirt on lamps, reduction of light output of a lamp over time, and similar factors. The following items are the individual components of the *LLF*. The total *LLF* is calculated by multiplying all the individual factors together. No factor should be ignored (set equal to 1) until investigations justify doing so. Lighting calculations should not be attempted until all light-loss factors are considered.

16.3.4 Light-Loss Factors by Groups (*page 16.18*)

Light-loss factors are divided into two groups: recoverable and nonrecoverable. *Recoverable* factors are those which can be changed by regular maintenance, such as cleaning and relamping luminaires and cleaning or painting room surfaces. *Nonrecoverable* factors are those attributed to equipment and site conditions and cannot be changed with normal maintenance.

Nonrecoverable Factors

1. *Luminaire ambient temperature factor.* For normal indoor temperatures, use a factor of 1.
2. *Heat-extraction thermal factor.* For air-handling luminaires, use a factor of 1.10.
3. *Voltage-to-luminaire factor.* Assuming operation at rated nominal voltage, a factor of 1 can be used. For other conditions, refer to Figure 16.3.5.
4. *Ballast factor.* In general, refer to the manufacturer's data. In the absence of that, the ballast factor depends on the lamp as well as on the ballast, so a ballast factor developed for a standard lamp does not apply when, for example, an energy-conserving lamp is used, even though the ballast is the same. Magnetic ballasts bearing the label of certified ballast manufacturers (CBM) have a ballast factor that is not less than 0.925 for standard 30- and 40-W rapid-start lamps; the ballast factor for such ballasts is frequently estimated at between 0.94 and

TABLE 16.3.2

Instructions and Notes:

1. The luminaires in this table are organized by source type and luminaire form rather than by application, for convenience in locating luminaires. In some cases, the data are based on an actual luminaire; in other cases, they represent a composite of generic luminaire types. Therefore, whenever possible, specific luminaire data should be used in preference to those in this table.
2. The polar intensity sketch (intensity distribution curve) and the corresponding luminaire spacing criterion are representative of many luminaires of each type shown. A specific luminaire may differ in perpendicular-plane (crosswise) and parallel-plane (lengthwise) intensity distributions and in spacing criterion from the values shown. However, the various coefficients depend only on the average intensity at each polar angle from nadir. The average intensity values used to generate the coefficients are given at the end of the table, normalized to 1000 lamp lumens for reference.
3. The various coefficients depend only on the average intensity distribution curve and are linearly related to the total luminaire efficiency. Consequently, the tabulated coefficients can be applied to luminaires with similarly shaped average intensity distributions along with a correcting multiplier equal to the new luminaire total efficiency divided by the tabulated luminaire total efficiency. The use of polarizing lenses on fluorescent luminaires has no effect on the coefficients given in this table except as they affect the total luminaire efficiency.
4. Satisfactory installations depend on many factors, including the environment, space utilization, and luminous criteria, as well as the luminaire itself. Consequently, a definitive spacing recommendation cannot be assigned to the luminaire as such. The spacing criterion (SC) values given are only general guides. SC values are not assigned to semi-indirect and indirect luminaires, since the basis of this technique does not apply to such situations. Also, SC values are not given for those bat-wing luminaires that must be located by criteria other than that of horizontal illuminance.
5. Key:

   ρ_{CC} = ceiling cavity reflectance (percent),
   ρ_W = wall reflectance (percent),
   ρ_{FC} = floor cavity reflectance (percent),
   RCR = room cavity ratio,
   WDRC = wall direct radiation coefficient,
   SC = luminaire spacing criterion,
   NA = not applicable.
6. Many of the luminaires in this figure appeared in earlier editions of the IESNA *Lighting Handbook*. The identifying number may be different due to a reordering of the luminaires. In some cases, the data have been modified in view of more recent or more extensive information. The user should specifically refer to this edition of the *Lighting Handbook* when referencing luminaires.
7. Fluorescent lamp luminaire efficiencies, and consequently the coefficients, are a function of the number of lamps in relation to the size of the luminaire. This is due to temperature changes and to changes in the blocking of light. In this figure, fluorescent lamp luminaires have been chosen with typical luminaire sizes and numbers of lamps; these are identified under the typical luminaire drawings. Variations of the coefficients with size and number of lamps depend on the many details of luminaire construction. The following correction factors are average values to apply to a four-lamp luminaire 610 mm (2 ft) wide:

No. of lamps	Width, mm	Width, ft	Multiply by
8	1220	4	1.05
3	610	2	1.05
2	610	2	1.1
2	300	1	0.9

   Multiply the entries for two-lamp wraparound luminaires by 0.95 for four lamps.
8. Photometric data for fluorescent lamp luminaires in this table are based on tests using standard-wattage fluorescent lamps. Reduced-wattage fluorescent lamps cause lower lamp operating temperatures with some luminaires. Consequently, the efficiency and coefficients may be slightly increased. It is desirable to obtain specific correction factors from the manufacturers. Typical factors for reduced-wattage fluorescent lamps (approximately 10% below standard lamp wattages) are as follows:

Luminaire	Multiply by
2-lamp strip, surface mounted	1.03
4-lamp troffer, enclosed, non-air-handling	1.07
4-lamp wraparound, surface mounted	1.07
2-lamp industrial, vented	1.00

   Electronic ballasts can be designed for any arbitrary operating condition. The manufacturer must be consulted for specific data.

(continued)

TABLE 16.3.2 (*Continued*)

Typical Luminaire	Typical Intensity Distribution	ρcc →	80			70			50			30			10			0
		ρw →	70	50	30	70	50	30	50	30	10	50	30	10	50	30	10	0
1 Pendant diffusing sphere with incandescent lamp		RCR ↓	EFF = 80.5%			% DN = 55.9%			% UP = 44.1%			Lamp = 150A21IF SC (along, across, 45°) = 1.5, 1.5, 1.5						
		0	0.87	0.87	0.87	0.81	0.81	0.81	0.70	0.70	0.70	0.59	0.59	0.59	0.49	0.49	0.49	0.45
		1	0.76	0.71	0.66	0.70	0.66	0.61	0.56	0.52	0.50	0.46	0.44	0.42	0.38	0.36	0.34	0.30
		2	0.68	0.60	0.53	0.62	0.55	0.50	0.47	0.42	0.38	0.39	0.35	0.32	0.31	0.29	0.26	0.23
		3	0.61	0.52	0.44	0.56	0.48	0.41	0.40	0.35	0.31	0.33	0.29	0.26	0.27	0.24	0.21	0.18
		4	0.55	0.45	0.37	0.51	0.42	0.35	0.35	0.30	0.25	0.29	0.25	0.21	0.23	0.20	0.17	0.14
		5	0.51	0.40	0.32	0.46	0.37	0.30	0.31	0.25	0.21	0.26	0.21	0.18	0.21	0.17	0.14	0.12
		6	0.46	0.35	0.28	0.42	0.33	0.26	0.28	0.22	0.18	0.23	0.19	0.15	0.19	0.15	0.12	0.10
		7	0.43	0.32	0.25	0.39	0.29	0.23	0.25	0.20	0.16	0.21	0.16	0.13	0.17	0.13	0.11	0.09
		8	0.40	0.29	0.22	0.36	0.27	0.20	0.23	0.17	0.14	0.19	0.15	0.12	0.15	0.12	0.09	0.07
		9	0.37	0.26	0.19	0.34	0.24	0.18	0.21	0.16	0.12	0.17	0.13	0.10	0.14	0.11	0.08	0.07
		10	0.34	0.24	0.17	0.32	0.22	0.16	0.19	0.14	0.11	0.16	0.12	0.09	0.13	0.10	0.08	0.06
2 Porcelain-enameled ventilated standard dome with inc. lamp			EFF = 86.5%			% DN = 4.0%			% UP = 96.0%			Lamp = 100A21/5B SC (along, across, 45°) = N/A, N/A, N/A						
		0	0.99	0.99	0.99	0.97	0.97	0.97	0.93	0.93	0.93	0.89	0.89	0.89	0.85	0.85	0.85	0.83
		1	0.91	0.87	0.84	0.89	0.85	0.82	0.82	0.79	0.77	0.79	0.76	0.74	0.76	0.74	0.72	0.71
		2	0.83	0.76	0.70	0.80	0.74	0.69	0.71	0.67	0.63	0.69	0.65	0.62	0.66	0.63	0.60	0.59
		3	0.75	0.66	0.59	0.73	0.65	0.59	0.62	0.57	0.53	0.60	0.55	0.52	0.58	0.54	0.51	0.49
		4	0.69	0.58	0.51	0.67	0.57	0.50	0.55	0.49	0.44	0.53	0.48	0.44	0.51	0.47	0.43	0.41
		5	0.63	0.52	0.44	0.61	0.51	0.44	0.49	0.43	0.38	0.47	0.42	0.37	0.46	0.41	0.37	0.35
		6	0.58	0.46	0.39	0.56	0.46	0.38	0.44	0.38	0.33	0.43	0.37	0.33	0.41	0.36	0.32	0.31
		7	0.53	0.42	0.34	0.52	0.41	0.34	0.40	0.33	0.29	0.39	0.33	0.29	0.38	0.32	0.28	0.27
		8	0.50	0.38	0.31	0.48	0.37	0.31	0.36	0.30	0.26	0.35	0.30	0.25	0.34	0.29	0.25	0.24
		9	0.46	0.35	0.28	0.45	0.34	0.28	0.33	0.27	0.23	0.32	0.27	0.23	0.32	0.26	0.23	0.21
		10	0.43	0.32	0.25	0.42	0.32	0.25	0.31	0.25	0.21	0.30	0.24	0.21	0.29	0.24	0.20	0.19
3 Bare lamp PAR-38 flood			EFF = 100%			% DN = 100%			% UP = 0%			Lamp = 150PAR38FL SC (along, across, 45°) = 0.6, 0.6, 0.6						
		0	1.20	1.20	1.20	1.17	1.17	1.17	1.12	1.12	1.12	1.07	1.07	1.07	1.03	1.03	1.03	1.00
		1	1.14	1.11	1.08	1.11	1.08	1.06	1.04	1.02	1.00	1.01	0.99	0.97	0.97	0.96	0.95	0.93
		2	1.08	1.02	0.98	1.05	1.01	0.97	0.97	0.94	0.91	0.94	0.92	0.89	0.92	0.89	0.87	0.86
		3	1.02	0.95	0 90	1.00	0.94	0.89	0.91	0.87	0.84	0.89	0.85	0.82	0.86	0.84	0.81	0.79
		4	0.97	0.89	0.83	0.95	0.88	0.83	0.86	0.81	0.78	0.84	0.80	0.77	0.82	0.79	0.76	0.74
		5	0.92	0.84	0.78	0.91	0.83	0.77	0.81	0.76	0.72	0.79	0.75	0.72	0.78	0.74	0.71	0.70
		6	0.88	0.79	0.73	0.87	0.78	0.73	0.77	0.72	0.68	0.75	0.71	0.68	0.74	0.70	0.67	0.66
		7	0.84	0.75	0.69	0.83	0.74	0.69	0.73	0.68	0.64	0.72	0.67	0.64	0.71	0.67	0.64	0.62
		8	0.81	0.71	0.66	0.80	0.71	0.65	0.70	0.65	0.61	0.69	0.64	0.61	0.68	0.64	0.61	0.59
		9	0.78	0.68	0.62	0.77	0.68	0.62	0.67	0.62	0.58	0.66	0.61	0.58	0.65	0.61	0.58	0.57
		10	0.75	0.65	0.60	0.74	0.65	0.59	0.64	0.59	0.56	0.63	0.59	0.56	0.63	0.58	0.55	0.54
4 PAR-38 Flood with spec. anodized reflector (45 deg. cutoff)			EFF = 66.2%			% DN = 100			% UP = 0			Lamp = 150PAR38FL* SC (along, across, 45°) = 0.6, 0.6, 0.6						
		0	1.10	1.10	1.10	1.07	1.07	1.07	1.02	1.02	1.02	0.98	0.98	0.98	0.94	0.94	0.94	0.92
		1	1.06	1.03	1.02	1.03	1.01	1.00	0.98	0.96	0.95	0.94	0.93	0.92	0.91	0.90	0.90	0.88
		2	1.02	0.08	0.95	1.00	0.96	0.94	0.94	0.91	0.89	0.91	0.89	0.88	0.88	0.87	0.86	0.84
		3	0.98	0.93	0.90	0.96	0.92	0.89	0.90	0.87	0.85	0.88	0.85	0.83	0.86	0.84	0.82	0.81
		4	0.94	0.89	0.85	0.93	0.88	0.84	0.86	0.83	0.80	0.84	0.82	0.79	0.83	0.81	0.79	0.77
		5	0.91	0.85	0.81	0.90	0.84	0.80	0.83	0.79	0.77	0.81	0.78	0.76	0.80	0.77	0.75	0.74
		6	0.88	0.81	0.77	0.87	0.81	0.77	0.80	0.76	0.73	0.78	0.75	0.73	0.77	0.75	0.72	0.71
		7	0.85	0.78	0.74	0.84	0.78	0.74	0.77	0.73	0.70	0.76	0.72	0.70	0.75	0.72	0.70	0.69
		8	0.82	0.75	0.71	0.81	0.75	0.71	0.74	0.70	0.68	0.73	0.70	0.67	0.72	0.69	0.67	0.66
		9	0.79	0.72	0.68	0.78	0.72	0.68	0.71	0.68	0.65	0.70	0.67	0.65	0.70	0.67	0.65	0.64
		10	0.77	0.70	0.66	0.76	0.69	0.65	0.69	0.65	0.63	0.68	0.65	0.62	0.68	0.65	0.62	0.61
5 PAR-38 Flood with black baffle			EFF = 66.2%			% DN = 100			% UP = 0			Lamp = 150PAR38FL* SC (along, across, 45°) = 0.6, 0.6, 0.6						
		0	0.79	0.79	0.79	0.77	0.77	0.77	0.74	0.74	0.74	0.70	0.70	0.70	0.68	0.68	0.68	0.66
		1	0.76	0.75	0.73	0.75	0.73	0.72	0.71	0.70	0.69	0.68	0.68	0.67	0.66	0.66	0.65	0.64
		2	0.74	0.71	0.69	0.72	0.70	0.68	0.68	0.67	0.65	0.66	0.65	0.64	0.64	0.63	0.63	0.62
		3	0.71	0.68	0.66	0.70	0.67	0.65	0.66	0.64	0.62	0.64	0.63	0.61	0.63	0.61	0.60	0.60
		4	0.69	0.65	0.63	0.68	0.65	0.62	0.63	0.61	0.60	0.62	0.60	0.59	0.61	0.60	0.58	0.58
		5	0.67	0.63	0.60	0.66	0.62	0.60	0.61	0.59	0.57	0.60	0.58	0.57	0.59	0.58	0.56	0.56
		6	0.65	0.61	0.58	0.64	0.60	0.58	0.59	0.57	0.55	0.59	0.57	0.55	0.58	0.56	0.55	0.54
		7	0.63	0.59	0.56	0.62	0.58	0.56	0.57	0.55	0.53	0.57	0.55	0.53	0.56	0.54	0.53	0.52
		8	0.61	0.57	0.54	0.60	0.56	0.54	0.56	0.53	0.52	0.55	0.53	0.52	0.55	0.53	0.51	0.51
		9	0.59	0.55	0.52	0.59	0.55	0.52	0.54	0.52	0.50	0.54	0.51	0.50	0.53	0.51	0.50	0.49
		10	0.58	0.53	0.51	0.57	0.53	0.50	0.53	0.50	0.49	0.52	0.50	0.48	0.52	0.50	0.48	0.48

(continued)

TABLE 16.3.2 (*Continued*)

Typical Luminaire	Typical Intensity Distribution	ρcc →	80			70			50			30			10			0
		ρw →	70	50	30	70	50	30	50	30	10	50	30	10	50	30	10	0
		RCR ↓	EFF = 96.2%			% DN = 100%			% UP = 0%			Lamp = 150A21IF SC (along, across, 45°) = 1.2, 1.2, 1.1						
6		0	0.82	0.82	0.82	0.81	0.81	0.81	0.77	0.77	0.77	0.74	0.74	0.74	0.71	0.71	0.71	0.69
		1	0.78	0.76	0.75	0.77	0.75	0.73	0.72	0.71	0.70	0.70	0.68	0.68	0.67	0.66	0.66	0.64
		2	0.74	0.71	0.68	0.73	0.70	0.67	0.67	0.65	0.63	0.65	0.63	0.62	0.63	0.62	0.61	0.59
		3	0.70	0.66	0.62	0.69	0.65	0.61	0.63	0.60	0.58	0.61	0.59	0.57	0.59	0.57	0.56	0.55
		4	0.66	0.61	0.57	0.65	0.60	0.56	0.58	0.55	0.53	0.57	0.54	0.52	0.56	0.53	0.51	0.50
		5	0.63	0.56	0.52	0.61	0.56	0.52	0.55	0.51	0.48	0.53	0.50	0.48	0.52	0.50	0.48	0.46
		6	0.59	0.53	0.48	0.58	0.52	0.48	0.51	0.47	0.45	0.50	0.47	0.44	0.49	0.46	0.44	0.43
		7	0.56	0.49	0.45	0.55	0.49	0.44	0.48	0.44	0.41	0.47	0.43	0.41	0.46	0.43	0.41	0.40
		8	0.53	0.46	0.41	0.52	0.45	0.41	0.45	0.41	0.38	0.44	0.40	0.38	0.43	0.40	0.38	0.37
A-lamp downlight with spec. anodized reflector		9	0.50	0.43	0.39	0.49	0.43	0.38	0.42	0.38	0.35	0.41	0.38	0.35	0.41	0.37	0.35	0.34
		10	0.47	0.40	0.36	0.47	0.40	0.36	0.39	0.36	0.33	0.39	0.35	0.33	0.38	0.35	0.33	0.32
7			EFF = 64.9%			% DN = 100%			% UP = 0%			Lamp = CF6/32 SC (along, across, 45°) = 1.1, 1.1, 1.1						
		0	0.77	0.77	0.77	0.75	0.75	0.75	0.72	0.72	0.72	0.69	0.69	0.69	0.66	0.66	0.66	0.65
		1	0.73	0.72	0.70	0.72	0.70	0.69	0.67	0.66	0.65	0.65	0.64	0.63	0.63	0.62	0.61	0.60
		2	0.69	0.66	0.63	0.68	0.65	0.62	0.63	0.61	0.59	0.61	0.59	0.58	0.59	0.58	0.57	0.55
		3	0.66	0.61	0.58	0.64	0.60	0.57	0.59	0.56	0.54	0.57	0.55	0.53	0.55	0.54	0.52	0.51
		4	0.62	0.57	0.53	0.61	0.56	0.52	0.55	0.51	0.49	0.53	0.51	0.48	0.52	0.50	0.48	0.47
		5	0.58	0.53	0.49	0.57	0.52	0.48	0.51	0.48	0.45	0.50	0.47	0.45	0.49	0.46	0.44	0.43
		6	0.55	0.49	0.45	0.54	0.48	0.45	0.47	0.44	0.41	0.46	0.43	0.41	0.46	0.43	0.41	0.40
		7	0.52	0.46	0.41	0.51	0.45	0.41	0.44	0.41	0.38	0.44	0.40	0.38	0.43	0.40	0.38	0.37
		8	0.49	0.43	0.38	0.48	0.42	0.38	0.42	0.38	0.35	0.41	0.38	0.35	0.40	0.37	0.35	0.34
8″ Open reflector downlight (32W CFL)		9	0.47	0.40	0.36	0.46	0.40	0.36	0.39	0.35	0.33	0.38	0.35	0.33	0.38	0.35	0.33	0.32
		10	0.44	0.37	0.33	0.44	0.37	0.33	0.37	0.33	0.31	0.36	0.33	0.30	0.36	0.33	0.30	0.29
8			EFF = 61.6%			% DN = 100%			% UP = 0%			Lamp = (2) CFQ26 SC (along, across, 45°) = 1.5, 1.6, 1.5						
		0	0.73	0.73	0.73	0.72	0.72	0.72	0.68	0.68	0.68	0.66	0.66	0.66	0.63	0.63	0.63	0.62
		1	0.69	0.67	0.66	0.68	0.66	0.64	0.63	0.62	0.61	0.61	0.60	0.59	0.59	0.58	0.57	0.56
		2	0.65	0.61	0.59	0.64	0.60	0.58	0.58	0.56	0.54	0.56	0.55	0.53	0.55	0.53	0.52	0.51
		3	0.61	0.56	0.52	0.59	0.55	0.52	0.53	0.51	0.48	0.52	0.49	0.47	0.50	0.48	0.47	0.46
		4	0.57	0.51	0.47	0.56	0.50	0.47	0.49	0.46	0.43	0.48	0.45	0.43	0.47	0.44	0.42	0.41
		5	0.53	0.47	0.42	0.52	0.46	0.42	0.45	0.41	0.39	0.44	0.41	0.38	0.43	0.40	0.38	0.37
		6	0.50	0.43	0.38	0.48	0.42	0.38	0.41	0.38	0.35	0.40	0.37	0.35	0.40	0.37	0.34	0.33
		7	0.46	0.39	0.35	0.45	0.39	0.35	0.38	0.34	0.31	0.37	0.34	0.31	0.36	0.33	0.31	0.30
		8	0.43	0.36	0.32	0.42	0.36	0.32	0.35	0.31	0.29	0.34	0.31	0.28	0.34	0.31	0.28	0.27
8″ Open reflector downlight (2-26W CFL)		9	0.41	0.33	0.29	0.40	0.33	0.29	0.33	0.29	0.26	0.32	0.28	0.26	0.31	0.28	0.26	0.25
		10	0.38	0.31	0.27	0.37	0.31	0.27	0.30	0.26	0.24	0.30	0.26	0.24	0.29	0.26	0.24	0.23
9			EFF = 40.2%			% DN = 100			% UP = 0			Lamp = (2) CFQ26 SC (along, across, 45°) = 1.2, 1.2, 1.1						
		0	0.47	0.47	0.47	0.46	0.46	0.46	0.44	0.44	0.44	0.42	0.42	0.42	0.41	0.41	0.41	0.40
		1	0.45	0.44	0.43	0.44	0.43	0.42	0.41	0.41	0.40	0.40	0.39	0.39	0.39	0.38	0.38	0.37
		2	0.43	0.41	0.39	0.42	0.40	0.38	0.39	0.37	0.36	0.37	0.36	0.35	0.36	0.35	0.35	0.34
		3	0.40	0.37	0.35	0.39	0.37	0.35	0.36	0.34	0.33	0.35	0.33	0.32	0.34	0.33	0.32	0.31
		4	0.38	0.35	0.32	0.37	0.34	0.32	0.33	0.31	0.30	0.32	0.31	0.29	0.32	0.30	0.29	0.28
		5	0.36	0.32	0.29	0.35	0.32	0.29	0.31	0.29	0.27	0.30	0.28	0.27	0.30	0.28	0.27	0.26
		6	0.34	0.30	0.27	0.33	0.29	0.27	0.29	0.27	0.25	0.28	0.26	0.25	0.28	0.26	0.25	0.24
		7	0.32	0.28	0.25	0.31	0.27	0.25	0.27	0.25	0.23	0.26	0.24	0.23	0.26	0.24	0.23	0.22
		8	0.30	0.26	0.23	0.29	0.25	0.23	0.25	0.23	0.21	0.25	0.23	0.21	0.24	0.22	0.21	0.20
		9	0.28	0.24	0.21	0.28	0.24	0.21	0.23	0.21	0.20	0.23	0.21	0.19	0.23	0.21	0.19	0.19
8″ Round with cross baffles		10	0.27	0.22	0.20	0.26	0.22	0.20	0.22	0.20	0.18	0.22	0.20	0.18	0.21	0.19	0.18	0.17
10			EFF = 63.8%			% DN = 78.4			% UP = 21.6			Lamp = M100/C/U SC (along, across, 45°) = 1.2, 1.2, 1.1						
		0	0.76	0.76	0.76	0.74	0.74	0.74	0.71	0.71	0.71	0.68	0.68	0.68	0.65	0.65	0.65	0.64
		1	0.72	0.70	0.68	0.70	0.69	0.67	0.66	0.65	0.64	0.64	0.63	0.62	0.61	0.61	0.60	0.59
		2	0.68	0.64	0.62	0.66	0.63	0.61	0.61	0.59	0.57	0.59	0.57	0.56	0.57	0.56	0.55	0.54
		3	0.64	0.59	0.56	0.63	0.58	0.55	0.57	0.54	0.52	0.55	0.53	0.51	0.54	0.52	0.50	0.49
		4	0.60	0.55	0.51	0.59	0.54	0.50	0.52	0.49	0.47	0.51	0.48	0.46	0.50	0.48	0.46	0.45
		5	0.57	0.51	0.46	0.55	0.50	0.46	0.49	0.45	0.43	0.48	0.45	0.42	0.47	0.44	0.42	0.41
		6	0.53	0.47	0.43	0.52	0.46	0.42	0.45	0.42	0.39	0.44	0.41	0.39	0.43	0.41	0.38	0.37
		7	0.50	0.43	0.39	0.49	0.43	0.39	0.42	0.39	0.36	0.41	0.38	0.36	0.41	0.38	0.35	0.34
		8	0.47	0.41	0.36	0.46	0.40	0.36	0.39	0.36	0.33	0.39	0.35	0 33	0.38	0.35	0 33	0.32
		9	0.45	0.38	0.34	0.44	0.37	0.33	0.37	0.33	0.31	0.36	0.33	0.31	0.36	0.33	0.30	0.29
Metal halide downlight		10	0.42	0.35	0.31	0.42	0.35	0.31	0.35	0.31	0.28	0.34	0.31	0.28	0.34	0.30	0.28	0.27

(continued)

TABLE 16.3.2 (*Continued*)

Typical Luminaire	Typical Intensity Distribution	ρcc →	80			70			50			30			10			0
		ρw →	70	50	30	70	50	30	50	30	10	50	30	10	50	30	10	0
		RCR ↓	EFF = 55.9%			% DN = 76.2%			% UP = 23.8%			Lamp = 22 & 32W circ.* SC (along, across, 45°) = 1.3, 1.3, 1.5						
11		0	0.63	0.63	0.63	0.60	0.60	0.60	0.55	0.55	0.55	0.50	0.50	0.50	0.45	0.45	0.45	0.43
		1	0.57	0.54	0.51	0.54	0.51	0.48	0.46	0.44	0.42	0.42	0.40	0.39	0.38	0.36	0.35	0.33
		2	0.51	0.46	0.42	0.48	0.44	0.40	0.40	0.37	0.34	0.36	0.33	0.31	0.32	0.30	0.29	0.27
		3	0.46	0.40	0.35	0.44	0.38	0.34	0.35	0.31	0.28	0.31	0.28	0.26	0.28	0.26	0.24	0.22
		4	0.42	0.35	0.30	0.40	0.34	0.29	0.31	0.27	0.24	0.28	0.24	0.22	0.25	0.22	0.20	0.19
		5	0.39	0.31	0.26	0.37	0.30	0.25	0.27	0.23	0.20	0.25	0.21	0.19	0.22	0.20	0.17	0.16
		6	0.36	0.28	0.23	0.34	0.27	0.22	0.24	0.20	0.18	0.22	0.19	0.16	0.20	0.17	0.15	0.14
		7	0.33	0.25	0.20	0.31	0.24	0.20	0.22	0.18	0.15	0.20	0.17	0.14	0.18	0.16	0.13	0.12
		8	0.31	0.23	0.18	0.29	0.22	0.18	0.20	0.16	0.14	0.18	0.15	0.13	0.17	0.14	0.12	0.11
		9	0.28	0.21	0.16	0.27	0.20	0.16	0.18	0.15	0.12	0.17	0.14	0.11	0.16	0.13	0.11	0.10
CFL surface-mounted disk		10	0.27	0.19	0.15	0.25	0.19	0.14	0.17	0.13	0.11	0.16	0.13	0.10	0.14	0.12	0.10	0.09
12			EFF = 87.5%			% DN = 85.9%			% UP = 1.6%			Lamp = M400/C/U SC (along, across, 45°) = 1.1, 1.1, 1						
		0	1.04	1.04	1.04	1.01	1.01	1.01	0.96	0.96	0.96	0.92	0.92	0.92	0.88	0.88	0.88	0.86
		1	0.98	0.95	0.93	0.96	0.93	0.91	0.89	0.87	0.86	0.86	0.84	0.83	0.82	0.81	0.80	0.78
		2	0.92	0.87	0.83	0.90	0.85	0.82	0.82	0.79	0.76	0.79	0.77	0.74	0.77	0.75	0.73	0.71
		3	0.86	0.80	0.75	0.84	0.78	0.74	0.76	0.72	0.69	0.73	0.70	0.67	0.71	0.68	0.66	0.64
		4	0.81	0.73	0.68	0.79	0.72	0.67	0.70	0.66	0.62	0.68	0.64	0.61	0.66	0.63	0.60	0.59
		5	0.76	0.68	0.62	0.74	0.67	0.61	0.65	0.60	0.56	0.63	0.59	0.56	0.62	0.58	0.55	0.54
		6	0.72	0.63	0.57	0.70	0.62	0.56	0.60	0.55	0.52	0.59	0.54	0.51	0.57	0.54	0.51	0.49
		7	0.67	0.58	0.52	0.66	0.58	0.52	0.56	0.51	0.47	0.55	0.50	0.47	0.54	0.50	0.47	0.45
		8	0.64	0.54	0.48	0.62	0.54	0.48	0.53	0.47	0.44	0.51	0.47	0.43	0.50	0.46	0.43	0.42
High bay, open metal reflector,		9	0.60	0.51	0.45	0.59	0.50	0.45	0.49	0.44	0.41	0.48	0.44	0.40	0.47	0.43	0.40	0.39
narrow		10	0.57	0.48	0.42	0.56	0.47	0.42	0.46	0.41	0.38	0.45	0.41	0.38	0.45	0.40	0.37	0.36
13			EFF = 83.9%			% DN = 95.2%			% UP = 4.8%			Lamp = M400/C/U SC (along, across, 45°) = 1.6, 1.6, 1.4						
		0	0.99	0.99	0.99	0.96	0.96	0.96	0.91	0.91	0.91	0.86	0.86	0.86	0.82	0.82	0.82	0.80
		1	0.93	0.90	0.87	0.90	0.88	0.85	0.83	0.81	0.80	0.80	0.78	0.77	0.76	0.75	0.74	0.72
		2	0.86	0.81	0.77	0.84	0.79	0.75	0.76	0.73	0.70	0.73	0.70	0.68	0.70	0.68	0.66	0.64
		3	0.80	0.73	0.68	0.78	0.72	0.67	0.65	0.65	0.61	0.66	0.63	0.60	0.64	0.61	0.58	0.57
		4	0.75	0.67	0.61	0.73	0.65	0.60	0.63	0.58	0.54	0.61	0.57	0.53	0.58	0.55	0.52	0.51
		5	0.70	0.61	0.54	0.68	0.59	0.54	0.57	0.52	0.48	0.55	0.51	0.48	0.54	0.50	0.47	0.45
		6	0.65	0.55	0.49	0.63	0.54	0.48	0.53	0.47	0.43	0.51	0.46	0.43	0.49	0.45	0.42	0.41
		7	0.60	0.51	0.44	0.59	0.50	0.44	0.48	0.43	0.39	0.47	0.42	0.39	0.45	0.41	0.38	0.37
		8	0.56	0.47	0.40	0.55	0.46	0.40	0.44	0.39	0.35	0.43	0.38	0.35	0.42	0.38	0.34	0.33
High bay, open metal reflector,		9	0.53	0.43	0.37	0.52	0.42	0.36	0.41	0.36	0.32	0.40	0.35	0.32	0.39	0.35	0.31	0.30
medium		10	0.50	0.40	0.34	0.48	0.39	0.33	0.38	0.33	0.29	0.37	0.32	0.29	0.36	0.32	0.29	0.27
14			EFF = 83.8%			% DN = 97			% UP = 3			Lamp = M400/C/U SC (along, across, 45°) = 1.9, 1.9, 1.7						
		0	0.99	0.99	0.99	0.97	0.97	0.97	0.92	0.92	0.92	0.87	0.87	0.87	0.83	0.83	0.83	0.81
		1	0.92	0.89	0.86	0.90	0.87	0.84	0.83	0.81	0.79	0.80	0.78	0.76	0.76	0.75	0.74	0.72
		2	0.85	0.80	0.75	0.83	0.78	0.73	0.75	0.71	0.68	0.72	0.69	0.66	0.69	0.66	0.64	0.62
		3	0.79	0.71	0.65	0.76	0.70	0.64	0.67	0.62	0.58	0.64	0.60	0.57	0.62	0.59	0.56	0.54
		4	0.72	0.63	0.57	0.70	0.62	0.56	0.60	0.55	0.51	0.58	0.53	0.50	0.56	0.52	0.49	0.47
		5	0.67	0.57	0.50	0.65	0.56	0.50	0.54	0.48	0.44	0.52	0.47	0.43	0.50	0.46	0.43	0.41
		6	0.62	0.51	0.44	0.60	0.50	0.44	0.49	0.43	0.39	0.47	0.42	0.38	0.46	0.41	0.38	0.36
		7	0.57	0.46	0.40	0.56	0.46	0.39	0.44	0.38	0.34	0.43	0.38	0.34	0.42	0.37	0.33	0.32
		8	0.53	0.42	0.35	0.52	0.42	0.35	0.40	0.34	0.30	0.39	0.34	0.30	0.38	0.33	0.30	0.28
High bay, open metal reflector,		9	0.49	0.39	0.32	0.48	0.38	0.32	0.37	0.31	0.27	0.36	0.31	0.27	0.35	0.30	0.27	0.25
wide		10	0.46	0.35	0.29	0.45	0.35	0.29	0.34	0.28	0.24	0.33	0.28	0.24	0.32	0.27	0.24	0.22
15			EFF = 61.4%			% DN = 80.6			% UP = 19.4			Lamp = M400/C/U SC (along, across, 45°) = 1.1, 1.1, 1.1						
		0	0.70	0.70	0.70	0.67	0.67	0.67	0.62	0.62	0.62	0.56	0.56	0.56	0.52	0.52	0.52	0.49
		1	0.65	0.62	0.60	0.62	0.59	0.57	0.55	0.53	0.52	0.50	0.49	0.48	0.46	0.45	0.44	0.42
		2	0.60	0.55	0.52	0.57	0.53	0.50	0.49	0.47	0.45	0.46	0.44	0.42	0.42	0.41	0.39	0.37
		3	0.56	0.50	0.46	0.53	0.48	0.44	0.45	0.42	0.39	0.42	0.39	0.37	0.39	0.37	0.35	0.33
		4	0.52	0.45	0.41	0.49	0.44	0.40	0.41	0.38	0.35	0.38	0.35	0.33	0.36	0.33	0.32	0.30
		5	0.48	0.41	0.37	0.46	0.40	0.36	0.38	0.34	0.31	0.35	0.32	0.30	0.33	0.31	0.29	0.27
		6	0.45	0.38	0.33	0.43	0.37	0.33	0.35	0.31	0.28	0.33	0.30	0.27	0.31	0.28	0.26	0.25
		7	0.42	0.35	0.30	0.40	0.34	0.30	0.32	0.28	0.26	0.30	0.27	0.25	0.29	0.26	0.24	0.23
		8	0.40	0.32	0.28	0.38	0.31	0.27	0.30	0.26	0.24	0.28	0.25	0.23	0.27	0.24	0.22	0.21
High bay, open prismatic reflec-		9	0.37	0.30	0.26	0.36	0.29	0.25	0.28	0.24	0.22	0.26	0.23	0.21	0.25	0.22	0.20	0.19
tor, narrow		10	0.35	0.28	0.24	0.34	0.27	0.23	0.26	0.22	0.20	0.25	0.22	0.19	0.23	0.21	0.19	0.18

(*continued*)

TABLE 16.3.2 (*Continued*)

Typical Luminaire / Typical Intensity Distribution	ρcc →	80			70			50			30			10			0
	ρw →	70	50	30	70	50	30	50	30	10	50	30	10	50	30	10	0
	RCR ↓	EFF = 59%			% DN = 77.9%			% UP = 22.1%			Lamp = M400/C/U SC (along, across, 45°) = 1.3, 1.3, 1.2						
16	0	0.67	0.67	0.67	0.64	0.64	0.64	0.58	0.58	0.58	0.53	0.53	0.53	0.48	0.48	0.48	0.46
	1	0.62	0.59	0.57	0.59	0.57	0.55	0.52	0.50	0.49	0.48	0.46	0.45	0.43	0.43	0.42	0.40
	2	0.57	0.53	0.50	0.55	0.51	0.48	0.47	0.45	0.43	0.43	0.41	0.40	0.40	0.38	0.37	0.35
	3	0.53	0.48	0.44	0.51	0.46	0.43	0.43	0.40	0.37	0.40	0.37	0.35	0.36	0.35	0.33	0.31
	4	0.50	0.44	0.39	0.47	0.42	0.38	0.39	0.36	0.33	0.36	0.34	0.32	0.34	0.32	0.30	0.28
	5	0.46	0.40	0.35	0.44	0.38	0.34	0.36	0.33	0.30	0.33	0.31	0.28	0.31	0.29	0.27	0.26
	6	0.43	0.37	0.32	0.41	0.35	0.31	0.33	0.30	0.27	0.31	0.28	0.26	0.29	0.27	0.25	0.23
	7	0.40	0.34	0.29	0.39	0.33	0.28	0.31	0.27	0.25	0.29	0.26	0.24	0.27	0.24	0.23	0.21
	8	0.38	0.31	0.27	0.36	0.30	0.26	0.28	0.25	0.22	0.27	0.24	0.22	0.25	0.23	0.21	0.20
	9	0.36	0.29	0.24	0.34	0.28	0.24	0.26	0.23	0.21	0.25	0.22	0.20	0.23	0.21	0.19	0.18
High bay, open prismatic reflector, medium	10	0.33	0.27	0.23	0.32	0.26	0.22	0.24	0.21	0.19	0.23	0.20	0.18	0.22	0.19	0.18	0.17
17		EFF = 61.5%			% DN = 83.7%			% UP = 16.3%			Lamp = M400/C/U SC (along, across, 45°) = 2.2, 2.2, 1.8						
	0	0.71	0.71	0.71	0.68	0.68	0.68	0.63	0.63	0.63	0.58	0.58	0.58	0.54	0.54	0.54	0.51
	1	0.65	0.62	0.59	0.62	0.59	0.57	0.55	0.53	0.52	0.51	0.50	0.48	0.47	0.46	0.45	0.43
	2	0.59	0.55	0.51	0.57	0.53	0.49	0.49	0.46	0.43	0.45	0.43	0.41	0.42	0.40	0.39	0.37
	3	0.54	0.48	0.44	0.52	0.47	0.42	0.43	0.40	0.37	0.40	0.38	0.35	0.38	0.35	0.33	0.32
	4	0.50	0.43	0.38	0.48	0.42	0.37	0.39	0.35	0.32	0.36	0.33	0.30	0.34	0.31	0.29	0.27
	5	0.46	0.38	0.33	0.44	0.37	0.32	0.35	0.31	0.27	0.32	0.29	0.26	0.30	0.27	0.25	0.24
	6	0.42	0.34	0.29	0.40	0.33	0.28	0.31	0.27	0.24	0.29	0.26	0.23	0.27	0.24	0.22	0.21
	7	0.39	0.31	0.26	0.37	0.30	0.25	0.28	0.24	0.21	0.27	0.23	0.20	0.25	0.22	0.19	0.18
	8	0.36	0.28	0.23	0.35	0.27	0.22	0.26	0.21	0.18	0.24	0.20	0.18	0.23	0.19	0.17	0.16
	9	0.34	0.25	0.21	0.32	0.25	0.20	0.23	0.19	0.16	0.22	0.18	0.16	0.21	0.18	0.15	0.14
High bay, open prismatic reflector, wide	10	0.31	0.23	0.19	0.30	0.23	0.18	0.21	0.17	0.15	0.20	0.17	0.14	0.19	0.16	0.14	0.12
18		EFF = 72.5%			% DN = 97.8%			% UP = 2.2%			Lamp = M400/C/U SC (along, across, 45°) = 1.7, 1.7, 1.7						
	0	0.86	0.86	0.86	0.84	0.84	0.84	0.80	0.80	0.80	0.76	0.76	0.76	0.73	0.73	0.73	0.71
	1	0.78	0.75	0.71	0.76	0.73	0.70	0.69	0.67	0.65	0.66	0.64	0.63	0.63	0.62	0.60	0.59
	2	0.71	0.65	0.60	0.69	0.63	0.59	0.60	0.56	0.53	0.58	0.54	0.52	0.55	0.53	0.50	0.49
	3	0.64	0.56	0.50	0.62	0.55	0.50	0.53	0.48	0.44	0.51	0.46	0.43	0.48	0.45	0.42	0.40
	4	0.59	0.50	0.43	0.57	0.49	0.42	0.47	0.41	0.37	0.45	0.40	0.36	0.43	0.39	0.36	0.34
	5	0.54	0.44	0.37	0.52	0.43	0.37	0.41	0.36	0.32	0.40	0.35	0.31	0.38	0.34	0.31	0.29
	6	0.49	0.39	0.33	0.48	0.39	0.32	0.37	0.31	0.27	0.36	0.31	0.27	0.34	0.30	0.26	0.25
	7	0.46	0.35	0.29	0.44	0.35	0.28	0.33	0.28	0.24	0.32	0.27	0.23	0.31	0.27	0.23	0.22
	8	0.42	0.32	0.26	0.41	0.31	0.25	0.30	0.25	0.21	0.29	0.24	0.21	0.28	0.24	0.21	0.19
	9	0.39	0.29	0.23	0.38	0.29	0.23	0.28	0.22	0.19	0.27	0.22	0.18	0.26	0.22	0.18	0.17
Low bay with drop lens, narrow	10	0.37	0.27	0.21	0.36	0.26	0.21	0.26	0.20	0.17	0.25	0.20	0.17	0.24	0.20	0.16	0.15
19		EFF = 75.3%			% DN = 8.1			% UP = 91.9			Lamp = M175/C* SC (along, across, 45°) = 1.3, 1.3, 1.5						
	0	0.73	0.73	0.73	0.63	0.63	0.63	0.45	0.45	0.45	0.29	0.29	0.29	0.13	0.13	0.13	0.06
	1	0.66	0.63	0.60	0.57	0.55	0.52	0.39	0.38	0.36	0.25	0.24	0.23	0.11	0.11	0.11	0.05
	2	0.60	0.55	0.51	0.52	0.48	0.44	0.34	0.32	0.30	0.21	0.20	0.19	0.10	0.09	0.09	0.04
	3	0.55	0.48	0.43	0.47	0.42	0.37	0.30	0.27	0.25	0.19	0.17	0.16	0.09	0.08	0.07	0.03
	4	0.50	0.42	0.37	0.43	0.37	0.32	0.26	0.23	0.21	0.17	0.15	0.13	0.08	0.07	0.06	0.02
	5	0.46	0.37	0.32	0.39	0.33	0.28	0.23	0.20	0.18	0.15	0.13	0.11	0.07	0.06	0.05	0.02
	6	0.42	0.33	0.28	0.36	0.29	0.24	0.21	0.18	0.15	0.13	0.11	0.10	0.06	0.05	0.05	0.02
	7	0.39	0.30	0.24	0.33	0.26	0.21	0.19	0.16	0.13	0.12	0.10	0.09	0.05	0.05	0.04	0.02
	8	0.36	0.27	0.21	0.31	0.23	0.19	0.17	0.14	0.12	0.11	0.09	0.07	0.05	0.04	0.03	0.01
	9	0.33	0.24	0.19	0.28	0.21	0.17	0.15	0.12	0.10	0.10	0.08	0.07	0.05	0.04	0.03	0.01
Glowing suspended bowl, MH	10	0.31	0.22	0.17	0.26	0.19	0.15	0.14	0.11	0.09	0.09	0.07	0.06	0.04	0.03	0.03	0.01
20		EFF = 81.5%			% DN = 15.3			% UP = 84.7			Lamp = (4) FT39* SC (along, across, 45°) = 1.3, 1.3, 1.5						
	0	0.81	0.81	0.81	0.71	0.71	0.71	0.52	0.52	0.52	0.35	0.35	0.35	0.20	0.20	0.20	0.13
	1	0.73	0.69	0.66	0.64	0.61	0.58	0.45	0.43	0.42	0.30	0.29	0.28	0.17	0.16	0.16	0.10
	2	0.66	0.60	0.55	0.58	0.53	0.49	0.39	0.36	0.34	0.26	0.25	0.23	0.15	0.14	0.13	0.08
	3	0.60	0.53	0.47	0.53	0.46	0.42	0.34	0.31	0.29	0.23	0.21	0.20	0.13	0.12	0.11	0.06
	4	0.55	0.47	0.40	0.48	0.41	0.36	0.30	0.27	0.24	0.20	0.18	0.17	0.11	0.10	0.09	0.05
	5	0.50	0.41	0.35	0.44	0.36	0.31	0.27	0.23	0.20	0.18	0.16	0.14	0.10	0.09	0.08	0.05
	6	0.46	0.37	0.30	0.40	0.32	0.27	0.24	0.20	0.18	0.16	0.14	0.12	0.09	0.08	0.07	0.04
	7	0.42	0.33	0.27	0.37	0.29	0.24	0.22	0.18	0.15	0.15	0.12	0.11	0.08	0.07	0.06	0.04
	8	0.39	0.30	0.24	0.34	0.26	0.21	0.20	0.16	0.13	0.13	0.11	0.09	0.08	0.06	0.05	0.03
	9	0.36	0.27	0.21	0.32	0.24	0.19	0.18	0.14	0.12	0.12	0.10	0.08	0.07	0.06	0.05	0.03
Glowing suspended bowl, CFL	10	0.34	0.25	0.19	0.30	0.22	0.17	0.16	0.13	0.11	0.11	0.09	0.07	0.06	0.05	0.04	0.03

(*continued*)

TABLE 16.3.2 (*Continued*)

Typical Luminaire	Typical Intensity Distribution	ρcc →	80			70			50			30			10			0
		ρw →	70	50	30	70	50	30	50	30	10	50	30	10	50	30	10	0
21		RCR ↓	EFF = 90.5%			% DN = 78.2%			% UP = 21.8%			Lamp = (2) F40T12 SC (along, across, 45°) = 1.3, 1.5, 1.5						
		0	1.03	1.03	1.03	0.98	0.98	0.98	0.90	0.90	0.90	0.82	0.82	0.82	0.74	0.74	0.74	0.71
		1	0.93	0.89	0.85	0.89	0.85	0.81	0.77	0.74	0.72	0.70	0.68	0.66	0.64	0.62	0.61	0.58
		2	0.84	0.77	0.71	0.80	0.74	0.68	0.67	0.63	0.59	0.61	0.58	0.54	0.56	0.53	0.50	0.47
		3	0.77	0.67	0.60	0.73	0.64	0.58	0.59	0.53	0.49	0.54	0.49	0.45	0.49	0.45	0.42	0.40
		4	0.70	0.59	0.51	0.66	0.57	0.50	0.52	0.46	0.41	0.48	0.43	0.39	0.44	0.39	0.36	0.33
		5	0.64	0.53	0.45	0.61	0.51	0.43	0.46	0.40	0.35	0.43	0.37	0.33	0.39	0.35	0.31	0.29
		6	0.59	0.47	0.39	0.56	0.45	0.38	0.42	0.35	0.31	0.38	0.33	0.29	0.35	0.31	0.27	0.25
		7	0.55	0.43	0.35	0.52	0.41	0.34	0.38	0.32	0.27	0.35	0.30	0.26	0.32	0.28	0.24	0.22
		8	0.51	0.39	0.31	0.48	0.37	0.30	0.34	0.28	0.24	0.32	0.27	0.23	0.29	0.25	0.21	0.19
Industrial, white enamel reflector, 20% up		9	0.47	0.35	0.28	0.45	0.34	0.27	0.32	0.26	0.21	0.29	0.24	0.20	0.27	0.23	0.19	0.17
		10	0.44	0.33	0.26	0.42	0.31	0.25	0.29	0.23	0.19	0.27	0.22	0.18	0.25	0.21	0.17	0.16
22			EFF = 86.9%			% DN = 100%			% UP = 0%			Lamp = (2) F40T12 SC (along, across, 45°) = 1.3, 1.5, 1.5						
		0	1.03	1.03	1.03	1.01	1.01	1.01	0.97	0.97	0.97	0.92	0.92	0.92	0.89	0.89	0.89	0.87
		1	0.94	0.90	0.86	0.92	0.88	0.84	0.84	0.81	0.79	0.81	0.79	0.76	0.78	0.76	0.74	0.72
		2	0.85	0.78	0.72	0.83	0.76	0.70	0.73	0.68	0.64	0.70	0.66	0.63	0.67	0.64	0.61	0.59
		3	0.77	0.68	0.60	0.75	0.66	0.59	0.64	0.58	0.53	0.61	0.56	0.52	0.59	0.55	0.51	0.49
		4	0.70	0.60	0.52	0.68	0.58	0.51	0.56	0.50	0.45	0.54	0.48	0.44	0.52	0.47	0.43	0.41
		5	0.65	0.53	0.45	0.63	0.52	0.44	0.50	0.43	0.38	0.48	0.42	0.38	0.47	0.41	0.37	0.35
		6	0.59	0.47	0.39	0.58	0.47	0.39	0.45	0.38	0.33	0.43	0.37	0.33	0.42	0.37	0.32	0.31
		7	0.55	0.43	0.35	0.53	0.42	0.35	0.41	0.34	0.29	0.39	0.33	0.29	0.38	0.33	0.29	0.27
		8	0.51	0.39	0.31	0.50	0.38	0.31	0.37	0.30	0.26	0.36	0.30	0.26	0.35	0.29	0.25	0.24
Industrial, white enamel reflector, down only		9	0.48	0.36	0.28	0.46	0.35	0.28	0.34	0.28	0.23	0.33	0.27	0.23	0.32	0.27	0.23	0.21
		10	0.45	0.33	0.26	0.43	0.32	0.25	0.31	0.25	0.21	0.31	0.25	0.21	0.30	0.24	0.21	0.19
23			EFF = 89.3%			% DN = 86.4%			% UP = 13.6%			Lamp = (2) F40T12 SC (along, across, 45°) = 1.3, 1.5, 1.6						
		0	1.03	1.03	1.03	1.00	1.00	1.00	0.92	0.92	0.92	0.86	0.86	0.86	0.80	0.80	0.80	0.77
		1	0.93	0.88	0.83	0.89	0.84	0.80	0.78	0.75	0.72	0.73	0.70	0.68	0.67	0.65	0.63	0.61
		2	0.83	0.75	0.68	0.80	0.72	0.66	0.67	0.62	0.58	0.62	0.58	0.55	0.58	0.55	0.52	0.49
		3	0.75	0.65	0.57	0.72	0.63	0.56	0.58	0.52	0.47	0.54	0.49	0.45	0.50	0.46	0.43	0.40
		4	0.69	0.57	0.49	0.65	0.55	0.47	0.51	0.45	0.40	0.48	0.42	0.38	0.44	0.40	0.36	0.34
		5	0.63	0.51	0.42	0.60	0.49	0.41	0.46	0.39	0.34	0.43	0.37	0.32	0.40	0.35	0.31	0.29
		6	0.58	0.45	0.37	0.55	0.44	0.36	0.41	0.34	0.29	0.38	0.32	0.28	0.36	0.31	0.27	0.25
		7	0.53	0.41	0.33	0.51	0.40	0.32	0.37	0.30	0.26	0.35	0.29	0.25	0.33	0.27	0.24	0.22
		8	0.50	0.37	0.29	0.47	0.36	0.29	0.34	0.27	0.23	0.32	0.26	0.22	0.30	0.25	0.21	0.19
		9	0.46	0.34	0.26	0.44	0.33	0.26	0.31	0.25	0.20	0.29	0.24	0.19	0.27	0.22	0.19	0.17
2-Lamp bare strip		10	0.43	0.31	0.24	0.41	0.30	0.23	0.29	0.22	0.18	0.27	0.21	0.18	0.25	0.20	0.17	0.15
24			EFF = 72.7%			% DN = 100			% UP = 0			Lamp = (3) F32T8 SC (along, across, 45°) = 1.3, 1.6, 1.6						
		0	0.87	0.87	0.87	0.85	0.85	0.85	0.81	0.81	0.81	0.77	0.77	0.77	0.74	0.74	0.74	0.73
		1	0.81	0.78	0.76	0.79	0.77	0.74	0.74	0.72	0.70	0.71	0.69	0.68	0.68	0.67	0.66	0.65
		2	0.75	0.70	0.66	0.73	0.69	0.65	0.66	0.63	0.61	0.64	0.61	0.59	0.62	0.60	0.58	0.57
		3	0.69	0.63	0.58	0.68	0.62	0.57	0.60	0.56	0.52	0.58	0.54	0.52	0.56	0.53	0.51	0.49
		4	0.64	0.56	0.51	0.62	0.55	0.50	0.54	0.49	0.46	0.52	0.48	0.45	0.51	0.47	0.44	0.43
		5	0.59	0.51	0.45	0.58	0.50	0.44	0.48	0.44	0.40	0.47	0.43	0.40	0.46	0.42	0.39	0.38
		6	0.55	0.46	0.40	0.53	0.45	0.40	0.44	0.39	0.35	0.43	0.38	0.35	0.42	0.38	0.35	0.33
		7	0.51	0.42	0.36	0.50	0.41	0.36	0.40	0.35	0.31	0.39	0.35	0.31	0.38	0.34	0.31	0.30
2 × 4, 3-Lamp parabolic troffer with 3″ semi-spec. louvers, 18 cells		8	0.47	0.38	0.32	0.46	0.38	0.32	0.37	0.32	0.28	0.36	0.31	0.28	0.35	0.31	0.28	0.27
		9	0.44	0.35	0.29	0.43	0.35	0.29	0.34	0.29	0.25	0.33	0.29	0.25	0.32	0.28	0.25	0.24
		10	0.41	0.32	0.27	0.40	0.32	0.27	0.31	0.26	0.23	0.31	0.26	0.23	0.30	0.26	0.23	0.22
25			EFF = 66.2%			% DN = 100			% UP = 0			Lamp = (3) F40T12 SC (along, across, 45°) = 1.3, 1.6, 1.5						
		0	0.79	0.79	0.79	0.77	0.77	0.77	0.74	0.74	0.74	0.70	0.70	0.70	0.68	0.68	0.68	0.66
		1	0.74	0.72	0.69	0.72	0.70	0.68	0.67	0.66	0.64	0.65	0.64	0.62	0.62	0.61	0.61	0.59
		2	0.69	0.64	0.61	0.67	0.63	0.60	0.61	0.58	0.56	0.59	0.57	0.55	0.57	0.55	0.53	0.52
		3	0.63	0.58	0.53	0.62	0.57	0.53	0.55	0.51	0.48	0.53	0.50	0.48	0.52	0.49	0.47	0.46
		4	0.59	0.52	0.47	0.57	0.51	0.47	0.50	0.46	0.42	0.48	0.45	0.42	0.47	0.44	0.41	0.40
		5	0.54	0.47	0.42	0.53	0.46	0.41	0.45	0.41	0.37	0.44	0.40	0.37	0.43	0.39	0.37	0.35
		6	0.50	0.43	0.37	0.49	0.42	0.37	0.41	0.36	0.33	0.40	0.36	0.33	0.39	0.35	0.33	0.31
		7	0.47	0.39	0.33	0.46	0.38	0.33	0.37	0.33	0.30	0.36	0.32	0.29	0.35	0.32	0.29	0.28
2 × 4, 3-Lamp parabolic troffer with 4″ semi-spec. louvers, 18 cells		8	0.44	0.35	0.30	0.43	0.35	0.30	0.34	0.30	0.26	0.33	0.29	0.26	0.33	0.29	0.26	0.25
		9	0.41	0.33	0.27	0.40	0.32	0.27	0.31	0.27	0.24	0.31	0.27	0.24	0.30	0.26	0.24	0.23
		10	0.38	0.30	0.25	0.37	0.30	0.25	0.29	0.25	0.22	0.28	0.24	0.22	0.28	0.24	0.22	0.20

(*continued*)

TABLE 16.3.2 (*Continued*)

Typical Luminaire	Typical Intensity Distribution	ρcc →	80			70			50			30			10			0
		ρw →	70	50	30	70	50	30	50	30	10	50	30	10	50	30	10	0
		RCR ↓	EFF = 67.8%			% DN = 100%			% UP = 0%			Lamp = (3) FT40 SC (along, across, 45°) = 1.3, 1.6, 1.6						
26		0	0.81	0.81	0.81	0.79	0.79	0.79	0.75	0.75	0.75	0.72	0.72	0.72	0.69	0.69	0.69	0.68
		1	0.75	0.73	0.71	0.74	0.71	0.69	0.69	0.67	0.66	0.66	0.65	0.64	0.64	0.63	0.62	0.60
		2	0.70	0.65	0.62	0.68	0.64	0.61	0.62	0.59	0.57	0.60	0.57	0.55	0.58	0.56	0.54	0.53
		3	0.65	0.59	0.54	0.63	0.58	0.53	0.56	0.52	0.49	0.54	0.51	0.48	0.52	0.50	0.47	0.46
		4	0.60	0.53	0.48	0.58	0.52	0.47	0.50	0.46	0.43	0.49	0.45	0.42	0.47	0.44	0.42	0.40
		5	0.55	0.47	0.42	0.54	0.47	0.42	0.45	0.41	0.38	0.44	0.40	0.37	0.43	0.40	0.37	0.35
		6	0.51	0.43	0.38	0.50	0.42	0.37	0.41	0.37	0.33	0.40	0.36	0.33	0.39	0.36	0.33	0.31
		7	0.47	0.39	0.34	0.46	0.39	0.33	0.38	0.33	0.30	0.37	0.32	0.29	0.36	0.32	0.29	0.28
		8	0.44	0.36	0.30	0.43	0.35	0.30	0.34	0.30	0.27	0.34	0.29	0.26	0.33	0.29	0.26	0.25
2 × 2, 3-Lamp with 3″ semi-spec. louvers, 9 cells		9	0.41	0.33	0.28	0.40	0.32	0.27	0.32	0.27	0.24	0.31	0.27	0.24	0.30	0.26	0.24	0.23
		10	0.39	0.30	0.25	0.38	0.30	0.25	0.29	0.25	0.22	0.29	0.25	0.22	0.28	0.24	0.22	0.20
27			EFF = 50.8%			% DN = 100%			% UP = 0%			Lamp = (2) F31T8/U/6 SC (along, across, 45°) = 1.2, 1.5, 1.4						
		0	0.61	0.61	0.61	0.59	0.59	0.59	0.56	0.56	0.56	0.56	0.56	0.56	0.54	0.54	0.54	0.52
		1	0.57	0.55	0.53	0.55	0.54	0.52	0.52	0.50	0.49	0.52	0.50	0.49	0.50	0.49	0.48	0.47
		2	0.53	0.49	0.47	0.51	0.48	0.46	0.47	0.45	0.43	0.47	0.45	0.43	0.45	0.43	0.42	0.42
		3	0.49	0.44	0.41	0.48	0.43	0.40	0.42	0.39	0.37	0.42	0.39	0.37	0.41	0.38	0.37	0.38
		4	0.45	0.40	0.36	0.44	0.39	0.36	0.38	0.35	0.32	0.38	0.35	0.32	0.37	0.34	0.32	0.34
		5	0.42	0.36	0.32	0.41	0.35	0.32	0.34	0.31	0.29	0.34	0.31	0.29	0.33	0.31	0.28	0.30
		6	0.39	0.33	0.29	0.38	0.32	0.28	0.31	0.28	0.25	0.31	0.28	0.25	0.30	0.27	0.25	0.27
		7	0.36	0.30	0.26	0.35	0.29	0.25	0.29	0.25	0.23	0.29	0.25	0.23	0.28	0.25	0.22	0.24
2 × 2, 2-Lamp (U) parabolic troffer with 3″ semi-spec. louver, 16 cells		8	0.33	0.27	0.23	0.33	0.27	0.23	0.26	0.23	0.20	0.26	0.23	0.20	0.26	0.22	0.20	0.22
		9	0.31	0.25	0.21	0.31	0.25	0.21	0.24	0.21	0.18	0.24	0.21	0.18	0.24	0.20	0.18	0.20
		10	0.29	0.23	0.19	0.29	0.23	0.19	0.22	0.19	0.17	0.22	0.19	0.17	0.22	0.19	0.17	0.19
28			EFF = 67.2%			% DN = 100%			% UP = 0%			Lamp = (2) F32T8 SC (along, across, 45°) = 1.3, 1.6, 1.5						
		0	0.80	0.80	0.80	0.78	0.78	0.78	0.75	0.75	0.75	0.72	0.72	0.72	0.69	0.69	0.69	0.67
		1	0.75	0.73	0.70	0.73	0.71	0.69	0.68	0.67	0.65	0.66	0.64	0.63	0.63	0.62	0.61	0.60
		2	0.70	0.65	0.61	0.68	0.64	0.61	0.62	0.59	0.56	0.59	0.57	0.55	0.58	0.56	0.54	0.53
		3	0.64	0.58	0.54	0.63	0.57	0.53	0.55	0.52	0.49	0.54	0.51	0.48	0.52	0.49	0.47	0.46
		4	0.59	0.52	0.47	0.58	0.52	0.47	0.50	0.46	0.43	0.49	0.45	0.42	0.47	0.44	0.42	0.40
		5	0.55	0.47	0.42	0.54	0.47	0.42	0.45	0.41	0.37	0.44	0.40	0.37	0.43	0.39	0.37	0.35
		6	0.51	0.43	0.37	0.50	0.42	0.37	0.41	0.36	0.33	0.40	0.36	0.33	0.39	0.35	0.33	0.31
		7	0.47	0.39	0.34	0.46	0.38	0.33	0.37	0.33	0.29	0.36	0.32	0.29	0.36	0.32	0.29	0.28
1 × 4, 2-Lamp parabolic troffer with 3″ semi-spec. louver, 8 or 9 cells		8	0.44	0.36	0.30	0.43	0.35	0.30	0.34	0.30	0.26	0.33	0.29	0.26	0.33	0.29	0.26	0.25
		9	0.41	0.33	0.27	0.40	0.32	0.27	0.31	0.27	0.24	0.31	0.27	0.24	0.30	0.26	0.24	0.22
		10	0.38	0.30	0.25	0.38	0.30	0.25	0.29	0.25	0.22	0.28	0.24	0.21	0.28	0.24	0.21	0.20
29			EFF = 67.2%			% DN = 100			% UP = 0			Lamp = (3) F32T8 SC (along, across, 45°) = 1.3, 1.5, 1.5						
		0	0.80	0.80	0.80	0.78	0.78	0.78	0.75	0.75	0.75	0.72	0.72	0.72	0.69	0.69	0.69	0.67
		1	0.76	0.74	0.72	0.74	0.72	0.70	0.69	0.68	0.67	0.67	0.66	0.65	0.64	0.64	0.63	0.62
		2	0.71	0.67	0.64	0.70	0.66	0.63	0.64	0.62	0.59	0.62	0.60	0.58	0.60	0.58	0.57	0.56
		3	0.67	0.62	0.58	0.65	0.61	0.57	0.59	0.56	0.53	0.57	0.54	0.52	0.55	0.53	0.51	0.50
		4	0.62	0.56	0.52	0.61	0.56	0.51	0.54	0.50	0.48	0.53	0.50	0.47	0.51	0.49	0.47	0.45
		5	0.58	0.52	0.47	0.57	0.51	0.47	0.50	0.46	0.43	0.49	0.45	0.43	0.47	0.44	0.42	0.41
		6	0.55	0.47	0.43	0.53	0.47	0.42	0.46	0.42	0.39	0.45	0.41	0.39	0.44	0.41	0.38	0.37
		7	0.51	0.44	0.39	0.50	0.43	0.39	0.42	0.38	0.35	0.41	0.38	0.35	0.41	0.37	0.35	0.34
		8	0.48	0.40	0.36	0.47	0.40	0.35	0.39	0.35	0.32	0.38	0.35	0.32	0.38	0.34	0.32	0.31
2 × 4, 3-Lamp parabolic troffer, spec. louvers, 18 cells, RP-1		9	0.45	0.37	0.33	0.44	0.37	0.32	0.36	0.32	0.29	0.36	0.32	0.29	0.35	0.32	0.29	0.28
		10	0.42	0.35	0.30	0.42	0.34	0.30	0.34	0.30	0.27	0.33	0.29	0.27	0.33	0.29	0.27	0.26
30			EFF = 51.4%			% DN = 100			% UP = 0			Lamp = (3) F32T8 SC (along, across, 45°) = 1.3, 1.3, 1.5						
		0	0.68	0.61	0.55	0.67	0.60	0.55	0.58	0.53	0.50	0.56	0.52	0.49	0.54	0.51	0.48	0.46
		1	0.63	0.55	0.49	0.61	0.54	0.48	0.52	0.47	0.43	0.50	0.46	0.42	0.49	0.45	0.42	0.40
		2	0.58	0.49	0.43	0.57	0.48	0.42	0.47	0.42	0.38	0.45	0.41	0.37	0.44	0.40	0.37	0.35
		3	0.54	0.44	0.38	0.52	0.44	0.38	0.42	0.37	0.33	0.41	0.37	0.33	0.40	0.36	0.33	0.31
		4	0.50	0.40	0.34	0.49	0.40	0.34	0.39	0.34	0.30	0.38	0.33	0.29	0.31	0.33	0.29	0.28
		5	0.47	0.37	0.31	0.46	0.37	0.31	0.36	0.30	0.27	0.35	0.30	0.26	0.34	0.30	0.26	0.25
		6	0.44	0.34	0.28	0.43	0.34	0.28	0.33	0.28	0.24	0.32	0.27	0.24	0.31	0.27	0.24	0.23
		7	0.41	0.32	0.26	0.40	0.31	0.26	0.30	0.25	0.22	0.30	0.25	0.22	0.29	0.25	0.22	0.21
2 × 4, 3-Lamp parabolic troffer, 1.5 × 1.5 × 1.0″ silver louvers, RP-1		8	0.00	0.00	0.00	0.00	0.00	0.00	0.00	0.00	0.00	0.00	0.00	0.00	0.00	0.00	0.00	0.00
		9	0.00	0.00	0.00	0.00	0.00	0.00	0.00	0.00	0.00	0.00	0.00	0.00	0.00	0.00	0.00	0.00
		10	0.77	0.77	0.77	0.76	0.76	0.76	0.72	0.72	0.72	0.69	0.69	0.69	0.66	0.66	0.66	0.65

(continued)

TABLE 16.3.2 (*Continued*)

Typical Luminaire	Typical Intensity Distribution	ρcc →	80			70			50			30			10			0
		ρw →	70	50	30	70	50	30	50	30	10	50	30	10	50	30	10	0
		RCR ↓	EFF = 64.6%			% DN = 100%			% UP = 0%			Lamp = (3) F31T8/U/6 SC (along, across, 45°) = 1.3, 1.5, 1.3						
31		0	0.77	0.77	0.77	0.75	0.75	0.75	0.72	0.72	0.72	0.69	0.69	0.69	0.66	0.66	0.66	0.65
		1	0.73	0.71	0.69	0.71	0.69	0.68	0.67	0.65	0.64	0.64	0.63	0.62	0.62	0.61	0.60	0.59
		2	0.68	0.65	0.62	0 67	0.64	0.61	0.61	0.59	0.57	0.59	0.58	0.56	0.58	0.56	0.55	0.54
		3	0.64	0.59	0.55	0.63	0.58	0.55	0.56	0.54	0.51	0.55	0.52	0.50	0.53	0.51	0.50	0.48
		4	0.60	0.54	0.50	0.59	0.53	0.50	0.52	0.49	0.46	0.51	0.48	0.45	0.49	0.47	0.45	0.44
		5	0.56	0.50	0.45	0.55	0.49	0.45	0.48	0.44	0.41	0.47	0.43	0.41	0.46	0.43	0.41	0.39
		6	0.53	0.46	0.41	0.51	0.45	0.41	0.44	0.40	0.37	0.43	0.40	0.37	0.42	0.39	0.37	0.36
		7	0.49	0.42	0.38	0.48	0.42	0.37	0.41	0.37	0.34	0.40	0.36	0.34	0.39	0.36	0.34	0.33
		8	0.46	0.39	0.34	0.45	0.39	0.34	0.38	0.34	0.31	0.37	0.33	0.31	0.36	0.33	0.31	0.30
2 × 2, 3-Lamp troffer, spec. louvers, 12 cells, RP-1		9	0.43	0.36	0.32	0.43	0.36	0.31	0.35	0.31	0.28	0.34	0.31	0.28	0.34	0.31	0.28	0.27
		10	0.41	0.33	0.29	0.40	0.33	0.29	0.33	0.29	0.26	0.32	0.28	0.26	0.32	0.28	0.26	0.25
32			EFF = 75.6%			% DN = 100%			% UP = 0%			Lamp = (3) F32T8 SC (along, across, 45°) = 1.3, 1.3, 1.4						
		0	0.90	0.90	0.90	0.88	0.88	0.88	0.84	0.84	0.84	0.80	0.80	0.80	0.77	0.77	0.77	0.76
		1	0.83	0.79	0.76	0.81	0.78	0.75	0.75	0.72	0.70	0.72	0.70	0.68	0.69	0.67	0.66	0.65
		2	0.76	0.70	0.65	0.74	0.69	0.64	0.66	0.62	0.59	0.64	0.61	0.58	0.61	0.59	0.57	0.55
		3	0.70	0.62	0.57	0.68	0.61	0.56	0.59	0.54	0.51	0.57	0.53	0.50	0.55	0.52	0.49	0.47
		4	0.64	0.56	0.49	0.63	0.55	0.49	0.53	0.48	0.44	0.51	0.47	0.43	0.49	0.46	0.43	0.41
		5	0.59	0.50	0.44	0.58	0.49	0.43	0.48	0.42	0.38	0.46	0.41	0.38	0.45	0.41	0.37	0.36
		6	0.55	0.45	0.39	0.53	0.45	0.38	0.43	0.38	0.34	0.42	0.37	0.33	0.41	0.37	0.33	0.32
		7	0.51	0.41	0.35	0.50	0.41	0.35	0.39	0.34	0.30	0.38	0.33	0.30	0.37	0.33	0.30	0.28
		8	0.48	0.38	0.31	0.46	0.37	0.31	0.36	0.31	0.27	0.35	0.30	0.27	0.34	0.30	0.27	0.25
2 × 4, 3-Lamp troffer with A12 lens		9	0.44	0.35	0.29	0.43	0.34	0.28	0.33	0.28	0.24	0.33	0.28	0.24	0.32	0.27	0.24	0.23
		10	0.42	0.32	0.26	0.41	0.32	0.26	0.31	0.26	0.22	0.30	0.25	0.22	0.29	0.25	0.22	0.21
33			EFF = 72.4%			% DN = 100%			% UP = 0%			Lamp = (3) F32T8 SC (along, across, 45°) = 1.3, 1.3, 1.3						
		0	0.86	0.86	0.86	0.84	0.84	0.84	0.80	0.80	0.80	0.77	0.77	0.77	0.74	0.74	0.74	0.72
		1	0.80	0.77	0.75	0.78	0.76	0.73	0.73	0.71	0.69	0.70	0.68	0.67	0.67	0.66	0.65	0.63
		2	0.74	0.69	0.65	0.72	0.68	0.64	0.65	0.62	0.59	0.63	0.60	0.58	0.61	0.59	0.57	0.55
		3	0.69	0.62	0.57	0.67	0.61	0.56	0.59	0.55	0.52	0.57	0.54	0.51	0.55	0.52	0.50	0.48
		4	0.64	0.56	0.50	0.62	0.55	0.50	0.53	0.49	0.45	0.52	0.48	0.45	0.50	0.47	0.44	0.43
		5	0.59	0.51	0.45	0.58	0.50	0.45	0.48	0.44	0.40	0.47	0.43	0.40	0.46	0.42	0.39	0.38
		6	0.55	0.46	0.40	0.54	0.45	0.40	0.44	0.39	0.36	0.43	0.39	0.35	0.42	0.38	0.35	0.34
		7	0.51	0.42	0.36	0.50	0.42	0.36	0.41	0.36	0.32	0.40	0.35	0.32	0.39	0.35	0.32	0.30
		8	0.48	0.39	0.33	0.47	0.38	0.33	0.37	0.32	0.29	0.36	0.32	0.29	0.36	0.32	0.29	0.27
2 × 4, 3-Lamp troffer with A19 lens		9	0.45	0.36	0.30	0.44	0.35	0.30	0.35	0.30	0.26	0.34	0.29	0.26	0.33	0.29	0.26	0.25
		10	0.42	0.33	0.28	0.41	0.33	0.28	0.32	0.27	0.24	0.31	0.27	0.24	0.31	0.27	0.24	0.23
34			EFF = 68.4%			% DN = 100			% UP = 0			Lamp = (3) FT40 SC (along, across, 45°) = 1.2, 1.3, 1.3						
		0	0.81	0.81	0.81	0.80	0.80	0.80	0.76	0.76	0.76	0.73	0.73	0.73	0.70	0.70	0.70	0.68
		1	0.75	0.72	0.70	0.73	0.71	0.69	0.68	0.66	0.64	0.65	0.64	0.62	0.63	0.62	0.60	0.59
		2	0.69	0.64	0.60	0.68	0.63	0.59	0.61	0.58	0.55	0.59	0.56	0.54	0.56	0.54	0.52	0.51
		3	0.64	0.57	0.52	0.62	0.56	0.52	0.54	0.50	0.47	0.53	0.49	0.46	0.51	0.48	0.46	0.44
		4	0.59	0.52	0.46	0.58	0.51	0.46	0.49	0.45	0.41	0.47	0.44	0.41	0.46	0.43	0.40	0.39
		5	0.55	0.47	0.41	0.53	0.46	0.41	0.44	0.40	0.36	0.43	0.39	0.36	0.42	0.38	0.35	0.34
		6	0.51	0.42	0.37	0.49	0.42	0.36	0.40	0.36	0.32	0.39	0.35	0.32	0.38	0.35	0.32	0.30
		7	0.47	0.39	0.33	0.46	0.38	0.33	0.37	0.32	0.29	0.36	0.32	0.29	0.35	0.31	0.28	0.27
		8	0.44	0.35	0.30	0.43	0.35	0.30	0.34	0.29	0.26	0.33	0.29	0.26	0.32	0.29	0.26	0.24
2 × 2, 3-Lamp troffer with A12 lens		9	0.41	0.33	0.27	0.40	0.32	0.27	0.31	0.27	0.24	0.31	0.27	0.23	0.30	0.26	0.23	0.22
		10	0.39	0.30	0.25	0.38	0.30	0.25	0.29	0.25	0.22	0.29	0.24	0.21	0.28	0.24	0.21	0.20
35			EFF = 57.1%			% DN = 100			% UP = 0			Lamp = (2) F31T8/U/6 SC (along, across, 45°) = 1.2, 1.3, 1.4						
		0	0.68	0.68	0.68	0.66	0.66	0.66	0.63	0.63	0.63	0.61	0.61	0.61	0.58	0.58	0.58	0.57
		1	0.64	0.62	0.60	0.62	0.60	0.59	0.58	0.57	0.55	0.56	0.55	0.54	0.54	0.53	0.52	0.51
		2	0.59	0.55	0.52	0.58	0.54	0.51	0.52	0.50	0.48	0.51	0.49	0.47	0.49	0.47	0.46	0.45
		3	0.55	0.50	0.46	0.53	0.49	0.45	0.47	0.44	0.42	0.46	0.43	0.41	0.44	0.42	0.40	0.39
		4	0.51	0.45	0.40	0.49	0.44	0.40	0.43	0.39	0.36	0.41	0.38	0.36	0.40	0.38	0.36	0.34
		5	0.47	0.40	0.36	0.46	0.40	0.36	0.39	0.35	0.32	0.38	0.34	0.32	0.37	0.34	0.31	0.30
		6	0.43	0.37	0.32	0.42	0.36	0.32	0.35	0.31	0.28	0.34	0.31	0.28	0.33	0.30	0.28	0.27
		7	0.40	0.33	0.29	0.39	0.33	0.29	0.32	0.28	0.25	0.31	0.28	0.25	0.31	0.27	0.25	0.24
		8	0.38	0.31	0.26	0.37	0.30	0.26	0.29	0.26	0.23	0.29	0.25	0.23	0.28	0.25	0.23	0.22
2 × 2, 2-Lamp troffer with A12 lens		9	0.35	0.28	0.24	0.34	0.28	0.24	0.27	0.23	0.21	0.27	0.23	0.21	0.26	0.23	0.21	0.20
		10	0.33	0.26	0.22	0.32	0.26	0.22	0.25	0.21	0.19	0.25	0.21	0.19	0.24	0.21	0.19	0.18

(*continued*)

TABLE 16.3.2 (*Continued*)

Typical Luminaire / Typical Intensity Distribution	ρcc →	80			70			50			30			10			0
	ρw →	70	50	30	70	50	30	50	30	10	50	30	10	50	30	10	0
36	RCR ↓	EFF = 65.1%			% DN = 100%			% UP = 0%			Lamp = (2) F32T8 SC (along, across, 45°) = 1.3, 1.3, 1.3						
	0	0.77	0.77	0.77	0.76	0.76	0.76	0.72	0.72	0.72	0.69	0.69	0.69	0.66	0.66	0.66	0.65
	1	0.71	0.69	0.66	0.70	0.67	0.65	0.64	0.62	0.61	0.62	0.60	0.59	0.59	0.58	0.57	0.56
	2	0.66	0.61	0.57	0.64	0.59	0.56	0.57	0.54	0.51	0.55	0.52	0.50	0.53	0.51	0.49	0.48
	3	0.60	0.54	0.49	0.59	0.53	0.48	0.51	0.47	0.44	0.49	0.46	0.43	0.48	0.45	0.42	0.41
	4	0.55	0.48	0.43	0.54	0.47	0.42	0.46	0.41	0.38	0.44	0.40	0.37	0.43	0.40	0.37	0.36
	5	0.51	0.43	0.38	0.50	0.43	0.37	0.41	0.37	0.33	0.40	0.36	0.33	0.39	0.35	0.33	0.31
	6	0.47	0.39	0.34	0.46	0.39	0.33	0.38	0.33	0.29	0.36	0.32	0.29	0.35	0.32	0.29	0.28
	7	0.44	0.36	0.30	0.43	0.35	0.30	0.34	0.30	0.26	0.33	0.29	0.26	0.33	0.29	0.26	0.25
	8	0.41	0.33	0.27	0.40	0.32	0.27	0.31	0.27	0.24	0.31	0.27	0.23	0.30	0.26	0.23	0.22
1 × 4, 2-Lamp troffer with A12 lens	9	0.39	0.30	0.25	0.38	0.30	0.25	0.29	0.25	0.21	0.28	0.24	0.21	0.28	0.24	0.21	0.20
	10	0.36	0.28	0.23	0.35	0.28	0.23	0.27	0.23	0.19	0.26	0.22	0.19	0.26	0.22	0.19	0.18
37		EFF = 68.9%			% DN = 91.4%			% UP = 8.6%			Lamp = (2) F32T8 SC (along, across, 45°) = 1.3, 1.5, 1.5						
	0	0.81	0.81	0.81	0.78	0.78	0.78	0.73	0.73	0.73	0.69	0.69	0.69	0.65	0.65	0.65	0.63
	1	0.74	0.71	0.68	0.71	0.68	0.66	0.64	0.62	0.60	0.61	0.59	0.58	0.57	0.56	0.55	0.53
	2	0.68	0.62	0.58	0.65	0.61	0.56	0.57	0.54	0.51	0.54	0.51	0.49	0.51	0.49	0.47	0.45
	3	0.62	0.55	0.50	0.60	0.54	0.49	0.51	0.47	0.44	0.48	0.45	0.42	0.46	0.43	0.41	0.39
	4	0.57	0.50	0.44	0.55	0.48	0.43	0.46	0.41	0.38	0.44	0.40	0.37	0.41	0.38	0.36	0.34
	5	0.53	0.45	0.39	0.51	0.43	0.38	0.41	0.37	0.33	0.39	0.35	0.32	0.38	0.34	0.31	0.30
	6	0.49	0.40	0.35	0.47	0.39	0.34	0.38	0.33	0.29	0.36	0.32	0.29	0.34	0.31	0.28	0.26
	7	0.46	0.37	0.31	0.44	0.36	0.31	0.34	0.30	0.26	0.33	0.29	0.26	0.31	0.28	0.25	0.24
	8	0.42	0.34	0.28	0.41	0.33	0.28	0.32	0.27	0.23	0.30	0.26	0.23	0.29	0.25	0.23	0.21
1 × 4, 2-Lamp lensed wrap-around, surface mounted	9	0.40	0.31	0.26	0.38	0.30	0.25	0.29	0.24	0.21	0.20	0.24	0.21	0.27	0.23	0.20	0.19
	10	0.37	0.29	0.23	0.36	0.28	0.23	0.27	0.22	0.19	0.26	0.22	0.19	0.25	0.21	0.19	0.17
38		EFF = 54.2%			% DN = 99.2%			% UP = 0.5%			Lamp = (2) FT36* SC (along, across, 45°) = 1.3, 1.5, 1.5						
	0	0.65	0.65	0.65	0.63	0.63	0.63	0.60	0.60	0.60	0.58	0.58	0.58	0.55	0.55	0.55	0.54
	1	0.58	0.55	0.53	0.57	0.54	0.52	0.52	0.50	0.48	0.49	0.48	0.46	0.47	0.46	0.45	0.44
	2	0.53	0.48	0.44	0.51	0.47	0.43	0.45	0.41	0.39	0.43	0.40	0.38	0.41	0.39	0.37	0.36
	3	0.48	0.42	0.37	0.46	0.41	0.36	0.39	0.35	0.32	0.37	0.34	0.31	0.36	0.33	0.31	0.30
	4	0.43	0.37	0.32	0.42	0.36	0.31	0.34	0.30	0.27	0.33	0.29	0.27	0.32	0.29	0.26	0.25
	5	0.40	0.33	0.27	0.39	0.32	0.27	0.31	0.26	0.23	0.30	0.26	0.23	0.28	0.25	0.23	0.21
	6	0.37	0.29	0.24	0.36	0.29	0.24	0.28	0.23	0.20	0.27	0.23	0.20	0.26	0.22	0.20	0.19
	7	0.34	0.26	0.21	0.33	0.26	0.21	0.25	0.21	0.18	0.24	0.20	0.18	0.23	0.20	0.17	0.16
	8	0.32	0.24	0.19	0.31	0.24	0.19	0.23	0.19	0.16	0.22	0.18	0.16	0.21	0.18	0.15	0.14
	9	0.29	0.22	0.17	0.29	0.22	0.17	0.21	0.17	0.14	0.20	0.17	0.14	0.20	0.16	0.14	0.13
2 × 2, semi-recessed troffer	10	0.28	0.20	0.16	0.27	0.20	0.16	0.19	0.15	0.13	0.19	0.15	0.13	0.18	0.15	0.13	0.12
39		EFF = 84.1%			% DN = 0			% UP = 100			Lamp = (2) F32T8 SC (along, across, 45°) = N/A						
	0	0.80	0.80	0.80	0.68	0.68	0.68	0.47	0.47	0.47	0.27	0.27	0.27	0.09	0.09	0.09	0.00
	1	0.73	0.69	0.66	0.62	0.59	0.57	0.41	0.39	0.38	0.23	0.23	0.22	0.07	0.07	0.07	0.00
	2	0.66	0.61	0.56	0.56	0.52	0.48	0.36	0.33	0.31	0.21	0.19	0.18	0.07	0.06	0.06	0.00
	3	0.60	0.53	0.48	0.51	0.46	0.41	0.31	0.28	0.26	0.18	0.17	0.15	0.06	0.05	0.05	0.00
	4	0.55	0.47	0.41	0.47	0.40	0.35	0.28	0.25	0.22	0.16	0.14	0.13	0.05	0.05	0.04	0.00
	5	0.50	0.41	0.35	0.43	0.36	0.30	0.24	0.21	0.19	0.14	0.12	0.11	0.05	0.04	0.04	0.00
	6	0.46	0.37	0.31	0.39	0.32	0.27	0.22	0.19	0.16	0.13	0.11	0.10	0.04	0.04	0.03	0.00
	7	0.42	0.33	0.27	0.36	0.28	0.23	0.20	0.16	0.14	0.11	0.10	0.08	0.04	0.03	0.03	0.00
	8	0.39	0.30	0.24	0.33	0.26	0.21	0.18	0.14	0.12	0.10	0.09	0.07	0.03	0.03	0.02	0.00
9″ Wide, thin profile, wide spread indirect	9	0.36	0.27	0.21	0.31	0.23	0.18	0.16	0.13	0.11	0.09	0.08	0.06	0.03	0.02	0.02	0.00
	10	0.34	0.24	0.19	0.29	0.21	0.16	0.15	0.11	0.09	0.08	0.07	0.06	0.03	0.02	0.02	0.00
40		EFF = 88.3%			% DN = 0			% UP = 100			Lamp = (2) F32T8 SC (along, across, 45°) = N/A						
	0	0.84	0.84	0.84	0.72	0.72	0.72	0.49	0.49	0.49	0.28	0.28	0.28	0.09	0.09	0.09	0.00
	1	0.76	0.73	0.70	0.65	0.62	0.60	0.43	0.41	0.40	0.25	0.24	0.23	0.08	0.08	0.07	0.00
	2	0.70	0.64	0.59	0.59	0.54	0.50	0.37	0.35	0.33	0.22	0.20	0.19	0.07	0.07	0.06	0.00
	3	0.63	0.56	0.50	0.54	0.48	0.43	0.33	0.30	0.27	0.19	0.17	0.16	0.06	0.06	0.05	0.00
	4	0.58	0.49	0.43	0.49	0.42	0.37	0.29	0.26	0.23	0.17	0.15	0.14	0.05	0.05	0.05	0.00
	5	0.53	0.43	0.37	0.45	0.37	0.32	0.26	0.22	0.20	0.15	0.13	0.12	0.05	0.04	0.04	0.00
	6	0.48	0.39	0.32	0.41	0.33	0.28	0.23	0.19	0.17	0.13	0.11	0.10	0.04	0.04	0.03	0.00
	7	0.45	0.35	0.28	0.38	0.30	0.24	0.21	0.17	0.15	0.12	0.10	0.09	0.04	0.03	0.03	0.00
	8	0.41	0.31	0.25	0.35	0.27	0.22	0.19	0.15	0.13	0.11	0.09	0.08	0.03	0.03	0.03	0.00
	9	0.38	0.28	0.22	0.32	0.24	0.19	0.17	0.13	0.11	0.10	0.08	0.07	0.03	0.03	0.02	0.00
V-shaped, completely indirect	10	0.35	0.26	0.20	0.30	0.22	0.17	0.15	0.12	0.10	0.09	0.07	0.06	C.03	0.02	0.02	0.00

(*continued*)

TABLE 16.3.2 (*Continued*)

Typical Luminaire	Typical Intensity Distribution	ρcc →	80			70			50			30			10			0
		ρw →	70	50	30	70	50	30	50	30	10	50	30	10	50	30	10	0
		RCR ↓	EFF = 82.4%			% DN = 5.4%			% UP = 94.6%			Lamp = (2) F32T8 SC (along, across, 45°) = N/A						
41		0	0.80	0.80	0.80	0.69	0.69	0.69	0.48	0.48	0.48	0.30	0.30	0.30	0.12	0.12	0.12	0.04
		1	0.72	0.69	0.66	0.62	0.59	0.57	0.42	0.40	0.39	0.27	0.25	0.24	0.11	0.10	0.10	0.04
		2	0.66	0.60	0.55	0.56	0.52	0.48	0.37	0.34	0.32	0.22	0.21	0.20	0.09	0.09	0.09	0.03
		3	0.60	0.53	0.47	0.51	0.45	0.41	0.32	0.29	0.27	0.20	0.18	0.17	8.08	0.08	0.07	0.02
		4	0.55	0.46	0.40	0.47	0.40	0.35	0.28	0.25	0.23	0.17	0.16	0.14	3 07	0.07	0.06	0.02
		5	0.50	0.41	0.35	0.43	0.36	0.30	0.25	0.22	0.19	0.16	0.14	0.12	0.07	0.06	0.05	0.02
		6	0.46	0.37	0.30	0.39	0.32	0.27	0.23	0.19	0.17	0.14	0.12	0.11	0.06	0.05	0.05	0.02
		7	0.42	0.33	0.27	0.36	0.28	0.23	0.20	0.17	0.14	0.13	0.11	0.09	0.05	0.05	0.04	0.01
		8	0.39	0.29	0.24	0.33	0.26	0.21	0.18	0.15	0.13	0.11	0.09	0.08	0.05	0.04	0.04	0.01
Indirect with performated metal underside		9	0.36	0.27	0.21	0.31	0.23	0.18	0.17	0.13	0.11	0.10	0.08	0.07	0.04	0.04	0.03	0.01
		10	0.33	0.24	0.19	0.29	0.21	0.16	0.15	0.12	0.10	0.05	0.08	0.06	0.04	0.03	0.03	0.01
42			EFF = 83.2%			% DN = 21.6%			% UP = 78.4%			Lamp = (2) F32T8 SC (along, across, 45°) = N/A						
		0	0.83	0.83	0.83	0.74	0.74	0.74	0.56	0.56	0.56	0.39	0.39	0.39	0.24	0.24	0.24	0.17
		1	0.76	0.73	0.70	0.67	0.65	0.62	0.49	0.48	0.46	0.35	0.34	0.33	0.22	0.21	0.21	0.15
		2	0.70	0.64	0.59	0.62	0.57	0.53	0.44	0.41	0.39	0.31	0.30	0.28	0.20	0.19	0.18	0.13
		3	0.64	0.57	0.51	0.56	0.50	0.46	0.39	0.36	0.33	0.28	0.26	0.24	0.18	0.17	0.16	0.12
		4	0.58	0.50	0.44	0.52	0.45	0.40	0.35	0.31	0.28	0.25	0.23	0.21	0.16	0.15	0.14	0.10
		5	0.54	0.45	0.39	0.47	0.40	0.35	0.31	0.27	0.25	0.23	0.20	0.18	0.15	0.13	0.12	0.09
		6	0.49	0.40	0.34	0.44	0.36	0.31	0.28	0.24	0.21	0.20	0.18	0.16	0.13	0.12	0.11	0.08
		7	0.46	0.36	0.30	0.40	0.32	0.27	0.25	0.22	0.19	0.19	0.16	0.14	0.12	0.11	0.10	0.07
		8	0.42	0.33	0.27	0.37	0.29	0.24	0.23	0.19	0.17	0.17	0.15	0.13	0.11	0.10	0.09	0.07
Semi-indirect, 2-lamp, v-shape, parabolic baffles		9	0.39	0.30	0.24	0.35	0.27	0.22	0.21	0.17	0.15	0.16	0.13	0.11	0.10	0.09	0.08	0.06
		10	0.37	0.27	0.22	0.32	0.24	0.20	0.19	0.16	0.13	0.14	0.12	0.10	0.10	0.08	0.07	0.06
43			EFF = 85.2%			% DN = 28.7%			% UP = 71.3%			Lamp = (2) F32T8 SC (along, across, 45°) = N/A						
		0	3.87	0.87	0.87	0.78	0.78	0.78	0.61	0.61	0.61	0.45	0.45	0.45	0.31	0.31	0.31	0.24
		1	3.80	0.77	0.74	0.72	0.69	0.66	0.54	0.53	0.51	0.41	0.40	0.39	0.29	0.28	0.28	0.22
		2	0.73	0.68	0.63	0.66	0.61	0.57	0.48	0.46	0.44	0.37	0.35	0.34	0.26	0.25	0.24	0.20
		3	0.67	0.60	0.55	0.60	0.54	0.50	0.43	0.40	0.37	0.33	0.31	0.29	0.24	0.23	0.22	0.18
		4	0.62	0.54	0.48	0.55	0.49	0.43	0.39	3.35	0.32	0.30	0.28	0.26	0.22	0.20	0.19	0.16
		5	0.57	0.48	0.42	0.51	0.44	0.38	0.35	0.31	0.28	0.27	0.25	0.23	0.23	0.18	0.17	0.14
		6	0.53	0.43	0.37	0.47	0.39	0.34	0.32	0.28	0.25	0.25	0.22	0.20	0.18	0.17	0.15	0.13
		7	0.49	0.39	0.33	0.44	0.36	0.30	0.29	0.25	0.22	0.23	0.20	0.18	0.17	0.15	0.14	0.11
		8	0.45	0.35	0.29	0.41	0.32	0.27	0.26	0.22	0.20	0.21	0.18	0.16	0.15	0.14	0.12	0.10
Semi-indirect, 2-lamp, thin profile, parabolic baffles, 70% up		9	0.42	0.32	0.26	0.38	0.30	0.24	0.24	0.20	0.18	0.19	0.16	0.14	0.14	0.13	0.11	0.09
		10	0.39	0.30	0.24	0.35	0.27	0.22	0.22	0.18	0.16	0.18	0.15	0.13	0.13	0.12	0.10	0.09
44			EFF = 88.3%			% DN = 40.3			% UP = 59.7			Lamp = (2) F32T8 SC (along, across, 45°) = N/A						
		0	0.93	0.93	0.93	0.84	0.84	0.84	0.69	0.69	0.69	0.55	0.55	0.55	0.42	0.42	0.42	0.36
		1	0.85	0.82	0.79	0.77	0.75	0.72	0.61	0.60	0.58	0.49	0.48	0.47	0.38	0.37	0.37	0.31
		2	0.78	0.72	0.67	0.71	0.66	0.62	0.54	0.51	0.49	0.44	0.42	0.40	0.34	0.33	0.32	0.27
		3	0.71	0.64	0.58	0.65	0.58	0.53	0.48	0.45	0.42	0.39	0.37	0.34	0.31	0.29	0.28	0.24
		4	0.65	0.57	0.50	0.60	0.52	0.46	0.43	0.39	0.36	0.35	0.32	0.30	0.28	0.26	0.24	0.21
		5	0.60	0.50	0.44	0.55	0.46	0.40	0.39	0.34	0.31	0.32	0.28	0.26	0.25	0.23	0.21	0.18
		6	0.55	0.45	0.38	0.50	0.42	0.36	0.35	0.30	0.27	0.28	0.25	0.23	0.23	0.20	0.19	0.16
		7	0.51	0.41	0.34	0.47	0.38	0.32	0.32	0.27	0.24	0.26	0.23	0.20	0.21	0.18	0.16	0.14
		8	0.47	0.37	0.30	0.43	0.34	0.28	0.29	0.24	0.21	0.24	0.20	0.18	0.19	0.16	0.15	0.13
Direct/indirect, 2-lamp, thin profile, parabolic baffles, 60% up		9	0.44	0.34	0.27	0.40	0.31	0.25	0.26	0.22	0.19	0.22	0.18	0.16	0.17	0.15	0.13	0.11
		10	0.41	0.31	0.25	0.37	0.28	0.23	0.24	0.20	0.17	0.20	0.17	0.14	0.16	0.14	0.12	0.10

(*continued*)

TABLE 16.3.2 (*Continued*)

Typical Luminaire / Typical Intensity Distribution	ρcc →	80			70			50			30			10			0
	ρw →	70	50	30	70	50	30	50	30	10	50	30	10	50	30	10	0
	RCR ↓	EFF = 53.4%			% DN = 81.7%			% UP = 18.3%			Lamp = (2) F32T8 SC (along, across, 45°) = 1.2, 1.2, 1.3						
45	0	0.61	0.61	0.61	0.59	0.59	0.59	0.54	0.54	0.54	0.50	0.50	0.50	0.46	0.46	0.46	0.44
	1	0.57	0.55	0.53	0.55	0.53	0.51	0.49	0.48	0.47	0.45	0.44	0.44	0.42	0.41	0.41	0.39
	2	0.53	0.49	0.46	0.51	0.48	0.45	0.44	0.42	0.40	0.41	0.39	0.38	0.38	0.37	0.36	0.34
	3	0.49	0.44	0.41	0.47	0.43	0.39	0.40	0.37	0.35	0.37	0.35	0.33	0.35	0.33	0.32	0.30
	4	0.45	0.40	0.36	0.43	0.38	0.35	0.36	0.33	0.31	0.34	0.31	0.29	0.32	0.30	0.28	0.27
	5	0.42	0.36	0.32	0.40	0.35	0.31	0.33	0.30	0.27	0.31	0.28	0.26	0.29	0.27	0.25	0.24
	6	0.39	0.33	0.29	0.37	0.32	0.28	0.30	0.27	0.24	0.28	0.25	0.23	0.26	0.24	0.22	0.21
	7	0.36	0.30	0.26	0.35	0.29	0.25	0.27	0.24	0.22	0.26	0.23	0.21	0.24	0.22	0.20	0.19
	8	0.34	0.27	0.23	0.32	0.27	0.23	0.25	0.22	0.19	0.24	0.21	0.19	0.22	0.20	0.18	0.17
	9	0.32	0.25	0.21	0.30	0.24	0.21	0.23	0.20	0.18	0.22	0.19	0.17	0.21	0.18	0.17	0.16
6″ square x-section, mostly downlight	10	0.30	0.23	0.19	0.29	0.23	0.19	0.21	0.18	0.16	0.20	0.18	0.16	0.19	0.17	0.15	0.14
46		EFF = 73%			% DN = 90%			% UP = 100%			Lamp = M400/U (coated)* SC (along, across, 45°) = N/A						
	0	0.69	0.69	0.69	0.59	0.59	0.59	0.41	0.54	0.41	0.23	0.23	0.23	0.07	0.07	0.07	0.00
	1	0.63	0.60	0.58	0.54	0.52	0.43	0.35	0.34	0.33	0.20	0.20	0.19	0.06	0.06	0.06	0.00
	2	0.57	0.53	0.48	0.49	0.45	0.42	0.31	0.29	0.27	0.18	0.17	0.16	0.06	0.05	0.05	0.00
	3	0.52	0.46	0.41	0.45	0.39	0.36	0.27	0.25	0.23	0.16	0.14	0.13	0.05	0.05	0.01	0.00
	4	0.48	0.41	0.35	0.41	0.35	0.31	0.24	0.21	0.19	0.14	0.12	0.11	0.04	0.04	0.04	0.00
	5	0.44	0.36	0.31	0.37	0.31	0.26	0.21	0.18	0.16	0.12	0.11	0.10	0.04	0.04	0.03	0.00
	6	0.40	0.32	0.27	0.34	0.27	0.23	0.19	0.16	0.14	0.11	0.09	0.08	0.04	0.03	0.03	0.00
	7	0.37	0.29	0.23	0.31	0.25	0.20	0.17	0.14	0.12	0.10	0.08	0.07	0.03	0.03	0.02	0.00
	8	0.34	0.26	0.21	0.29	0.22	0.18	0.15	0.13	0.11	0.09	0.07	0.06	0.03	0.02	0.02	0.00
	9	0.31	0.23	0.18	0.27	0.20	0.16	0.14	0.11	0.09	0.08	0.07	0.05	0.03	0.02	0.02	0.00
Completely indirect HID	10	0.29	0.21	0.16	0.25	0.18	0.14	0.13	0.10	0.08	0.07	0.06	0.05	0.02	0.02	0.02	0.00
47		EFF = 71.6%			% DN = 0.3%			% UP = 99.7%			Lamp = (1) F32T8 SC (along, across, 45°) = N/A						
	0	0.42	0.35	0.30	0.40	0.34	0.30	0.32	0.28	0.26	0.30	0.27	0.25	0.29	0.26	0.24	0.23
	1	0.40	0.32	0.28	0.38	0.31	0.27	0.30	0.26	0.24	0.28	0.25	0.23	0.27	0.24	0.22	0.21
	2	0.37	0.30	0.26	0.36	0.29	0.25	0.28	0.24	0.22	0.26	0.23	0.21	0.25	0.22	0.20	0.19
	3	0.35	0.28	0.24	0.34	0.27	0.23	0.26	0.22	0.20	0.25	0.22	0.19	0.23	0.21	0.19	0.18
	4	0.00	0.00	0.00	0.00	0.00	0.00	0.00	0.00	0.00	0.00	0.00	0.00	0.00	0.00	0.00	0.00
	5	0.00	0.00	0.00	0.00	0.00	0.00	0.00	0.00	0.00	0.00	0.00	0.00	0.00	0.00	0.00	0.00
	6	0.67	0.67	0.67	0.64	0.64	0.64	0.58	0.58	0.58	0.53	0.53	0.53	0.48	0.48	0.48	0.46
	7	0.62	0.59	0.57	0.59	0.57	0.55	0.52	0.50	0.49	0.48	0.46	0.45	0.43	0.43	0.42	0.40
	8	0.57	0.53	0.50	0.55	0.51	0.48	0.47	0.45	0.43	0.43	0.41	0.40	0.40	0.38	0.37	0.35
Fluorescent cove with specular	9	0.53	0.48	0.44	0.51	0.46	0.43	0.43	0.40	0.37	0.40	0.37	0.35	0.36	0.35	0.33	0.31
reflector	10	0.50	0.44	0.39	0.47	0.42	0.38	0.39	0.36	0.33	0.36	0.34	0.32	0.34	0.32	0.30	0.28

Notes: EFF = efficiency; SC = spacing criteria.
For 6″ reflector with PAR-38, multiply by 0.98.
For A12 lens, 2 × 4's, ratio of F40 to T8, 0.96, 0.94, and 0.89 for 2, 3, and 4-lamp versions.
For A12 lens, 2 × 4's, ratio of 2 to 3-lamp effic. = 1.05; ratio of 4 to 3-lamp effic. = 0.98.
For parabolics, 2 lamp versions have 12 cells, 4 lamp versions have 32 cells.
For 3″ parabolic, 2 × 4's, ratio of 2 to 3-lamp effic. = 1.06; ratio of 4 to 3-lamp effic. = 0.96.
For 4″ parabolic, 2 × 4's, ratio of 2 to 3-lamp effic. = 0.98; ratio of 4 to 3-lamp effic. = 0.90.
CUs do not consider shadowing of the lighting fixture or cove.
* Different lamp types have been considered in determining the listed photometric report.

(Reprinted from the IESNA Lighting Handbook, 9th Edition, 2000, Courtesy of the Illuminating Engineering Society of North America.)

TABLE 16.3.4 Light-Loss Factors by Groups

Nonrecoverable
Luminaire ambient temperature factor
Heat extraction thermal factor
Voltage-to-luminaire factor
Ballast factor
Ballast-lamp photometric factor
Equipment operating factor
Lamp position (tilt) factor
Luminaire surface depreciation factor
Recoverable
Lamp lumen depreciation factor
Luminaire dirt depreciation factor
Room surface dirt depreciation factor
Lamp burnout factor

(Reprinted from the IESNA Lighting Handbook, 9th Edition, 2000, Courtesy of the Illuminating Engineering Society of North America.)

0.95. The ballast factor for highly loaded rapid-start lamps is 0.95 and for various low-wattage lamps is 0.90. A conservative estimate for a CBM-certified ballast would be 0.93.

5. *Ballast-lamp photometric factor.* In general, refer to the manufacturer's data or use a factor of 1.
6. *Equipment operating factor.* For HID lamp-ballast-luminaire combination only, refer to the manufacturer's data or use a factor of 1.
7. *Lamp-position (tilt) factor.* Part of the equipment operating factor. Refer to Figure 16.3.6 for typical average data and to manufacturers for specific lamp types.
8. *Luminaire surface depreciation factor.* Over time, the various surfaces of a light fixture will change (some plastic lenses yellow, for example). In the absence of data, use a value of 1.

Recoverable Factors

1. *Lamp lumen depreciation factor.* All lamps put out less light as they age. Specific information is available from each manufacturer, or you can use the figures in Table 16.3.7 for preliminary calculations.
2. *Luminaire dirt depreciation (LDD) factor.* The accumulation of dirt on luminaires results in a loss of light output and therefore a loss of light on the workplane. This loss is known as the *luminaire dirt depreciation LDD* factor and is determined as follows:
   a. The luminaire maintenance category is selected from the manufacturer's data or by using Table 16.3.8.
   b. The atmosphere (one of five degrees of dirt conditions) in which the luminaire operates is found as follows. Dirt in the atmosphere comes from two sources—that passed from adjacent air and that generated by work done in the vicinity. Dirt may be classified as adhesive, attracted, or inert, and it may come from intermittent or constant sources. Adhesive dirt clings to luminaire surfaces by its stickiness, whereas attracted dirt is held by electrostatic force. Inert dirt varies in accumulation from practically nothing on vertical surfaces to as much as a horizontal surface holds before the dirt is dislodged by gravity or air circulation. Examples of adhesive dirt are grease from cooking, particles

from machine operation borne by oil vapor, particles borne by water vapor as in a laundry, and fumes from metal-pouring operations or plating tanks. Examples of attracted dirt are hair, lint, fibers, and dry particles that are electrostatically charged from machine operations. Examples of inert dirt are nonsticky, uncharged particles such as dry flour, sawdust, and fine cinders. Tables 16.3.9 and 16.3.10 may be useful for evaluating the atmosphere. Table 16.3.10 is intended to evaluate the atmosphere-dirt category. Factors 1 to 5 should be assessed and inserted into the spaces in Table 16.3.9 because they are required to describe the conditions of the space. The "Area Adjacent to Task Area" column represents the area separated from but adjacent to the area in which the luminaire operates (which is the "Area Surrounding Task"). The "Filter Factor" column contains the percentages of dirt allowed to pass from the adjacent atmosphere to the surrounding atmosphere. The "From Adjacent" column indicates the net amount of such dirt that can pass through. This category might include, for example, an open window with a filter factor of 1.0 (no filtering at all) or an air-conditioning system with a filter factor of 0.1 (90 percent of the dirt is filtered out). The total of all the numbers in the "Subtotal" column is a number up to 60 and can be translated into the applicable atmosphere-dirt category listed at the bottom of the table.

c. From the appropriate luminaire maintenance category curve of Figure 16.3.11, the applicable dirt condition curve, and the proper elapsed time in months of the planned cleaning cycle, the *LDD* factor is found. For example, if the category is I, the atmosphere is dirty, and the cleaning occurs every 20 months, the *LDD* is approximately 0.80.

3. *Room surface dirt depreciation (RSDD) factor.* The accumulation of dirt on room surfaces reduces the amount of luminous flux reflected and interreflected to the workplane. To take this into account, Figure 16.3.12 has been developed to provide *RSDD* factors for use in calculating maintained average illuminance levels. These factors are determined as follows:
   a. From one of the five curves in Figure 16.3.12, find the expected dirt depreciation using Table 16.3.9 or 16.3.10 as a guide to atmospheric dirt conditions, together with an estimate of the time between cleanings. For example, if the atmosphere is dirty and room surfaces are cleaned every 24 months, the expected dirt depreciation is 30 percent.
   b. Knowing the expected dirt depreciation (step *a*), the type of luminaire distribution, and the room cavity ratio (*RCR*), determine the *RSDD* factor from Figure 16.3.12. For example, for a dirt depreciation of 30 percent, a direct luminaire, and a room cavity ratio of 4, the *RSDD* would be 0.92.
4. *Lamp burnout (LBO) factor.* If lamps are replaced as they burn out, use a factor of 0.95. If a group replacement maintenance program is employed, use a factor of 1.

16.3.5 Light Output Change Due to Voltage Change (*see page 16.20*)

16.3.6 Lumen Output for HID Lamps as a Function of Operating Position (*see page 16.20*)

16.3.7 Lamp Lumen Depreciation (*see page 16.20*)

16.3.5

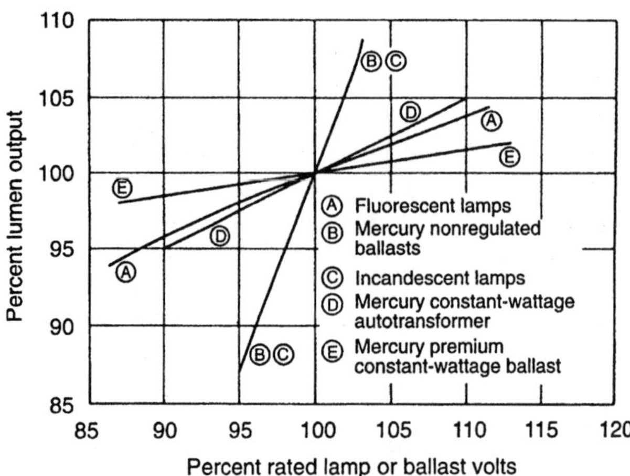

(Reprinted from the IESNA Lighting Handbook, 9th Edition, 2000, Courtesy of the Illuminating Engineering Society of North America.)

16.3.6

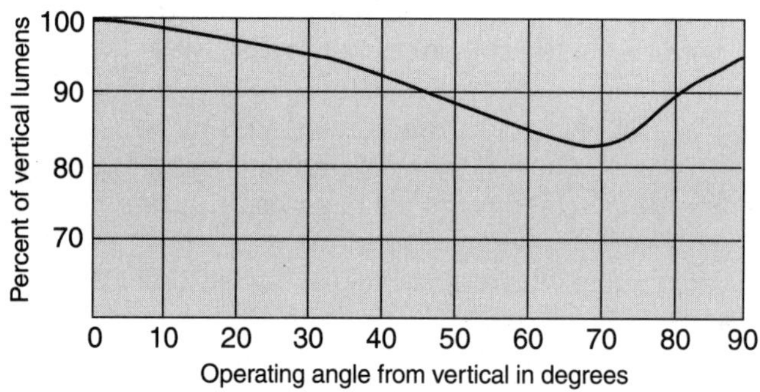

(Reprinted from the IESNA Lighting Handbook, 9th Edition, 2000, Courtesy of the Illuminating Engineering Society of North America.)

TABLE 16.3.7

Lamp Type	Group Replacement	Burnout Replacement
Fluorescent	0.90	0.85
Incandescent	0.94	0.88
Metal-halide	0.87	0.80
Mercury	0.82	0.74
Tungsten-halogen	0.94	0.88
High-pressure sodium	0.94	0.88

16.3.8 Procedure for Determining Luminaire Maintenance Categories

16.3.8

To assist in determining Luminaire Dirt Depreciation (LDD) factors, luminaires are separated into six maintenance categories (I through VI). To arrive at categories, luminaires are arbitrarily divided into sections, a Top Enclosure and a Bottom Enclosure, by drawing a horizontal line through the light center of the lamp or lamps. The characteristics listed for the enclosures are then selected as best describing the luminaire. Only one characteristic for the top enclosure and one for the bottom enclosure should be used in determining the category of a luminaire. Percentage of uplight is based on 100% for the luminaire.

The maintenance category is determined when there are characteristics in both enclosure columns. If a luminaire falls into more than one category, the lower numbered category is used.

Maintenance Category	Top Enclosure	Bottom Enclosure
I	1. None	1. None
II	1. None 2. Transparent with 15% or more uplight through apertures. 3. Translucent with 15% or more uplight through apertures. 4. Opaque with 15% or more uplight through apertures.	1. None 2. Louvers or baffles
III	1. Transparent with less than 15% upward light through apertures. 2. Translucent with less than 15% upward light through apertures. 3. Opaque with less than 15% upward light through apertures.	1. None 2. Louvers or baffles
IV	1. Transparent unapertured. 2. Translucent unapertured. 3. Opaque unapertured.	1. None 2. Louvers
V	1. Transparent unapertured. 2. Translucent unapertured. 3. Opaque unapertured.	1. Transparent unapertured 2. Translucent unapertured
VI	1. None 2. Transparent unapertured. 3. Translucent unapertured. 4. Opaque unapertured.	1. Transparent unapertured 2. Translucent unapertured 3. Opaque unapertured

(Reprinted from the IESNA Lighting Handbook, 9th Edition, 2000, Courtesy of the Illuminating Engineering Society of North America.)

16.3.9 Evaluation of Operating Atmosphere

TABLE 16.3.9

Type of Dirt*	Area Adjacent to Task Area			Filter Factor (percent of dirt passed)	Area Surrounding Task			Sub Total
	Intermittent Dirt	Constant Dirt	Total		From Adjacent	Intermittent Dirt	Constant Dirt	
Adhesive Dirt	+	=	×	=	+	+	=	
Attracted Dirt	+	=	×	=	+	+	=	
Inert Dirt	+	=	×	=	+	+	=	
Total of Dirt Factors:								
0–12 = Very Clean	13–24 = Clean			25–36 = Medium		37–49 = Dirty		49–60 = Very Dirty

*See step 2 under Luminaire Dirt Depreciation.
Factors for use in the table are 1: Cleanest conditions imaginable; 2: Clean, but not the cleanest; 3: Average; 4: Dirty, but not the dirtiest; 5: Dirtiest conditions imaginable.

(Reprinted from the IESNA Lighting Handbook, 9th Edition, 2000, Courtesy of the Illuminating Engineering Society of North America.)

16.3.10 Five Degrees of Dirt Conditions

TABLE 16.3.10

	Very Clean	Clean	Medium	Dirty	Very Dirty
Generated Dirt	None	Very little	Noticeable but not heavy	Accumulates rapidly	Constant accumulation
Ambient Dirt	None (or none enters area)	Some (almost none enters)	Some enters area	Large amount enters area	Almost none excluded
Removal or Filtration	Excellent	Better than average	Poorer than average	Only fans or blowers if any	None
Adhesion	None	Slight	Enough to be visible after some months	High—probably due to oil, humidity, or static	High
Examples	High grade offices, not near production; laboratories; clean rooms	Offices in older buildings or near production; light assembly; inspection	Mill offices; paper processing; light machining	Heat treating; high-speed printing; rubber processing	Similar to dirty but luminaires within immediate area of contamination

(Reprinted from the IESNA Lighting Handbook, 9th Edition, 2000, Courtesy of the Illuminating Engineering Society of North America.)

16.3.11 Luminaire Dirt Depreciation (*LDD*) Factors for Six Luminaire Categories (I through VI) and for the Five Degrees of Dirtiness as Determined from Figure 16.3.8 or Table 16.3.9 (*see page 16.23*)

16.3.12 Room Surface Dirt Depreciation (*RSDD*) Factors (*see page 16.23*)

16.3.11

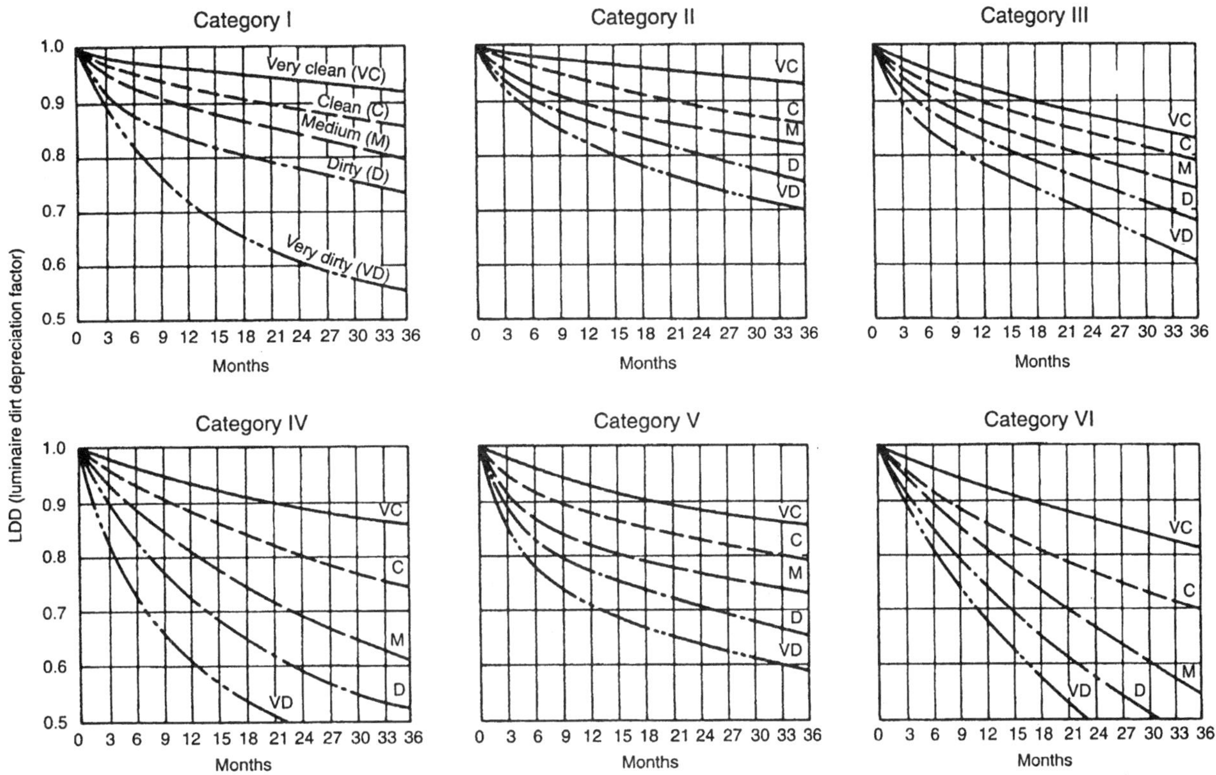

(Reprinted from the IESNA Lighting Handbook, 9th Edition, 2000, Courtesy of the Illuminating Engineering Society of North America.)

TABLE 16.3.12

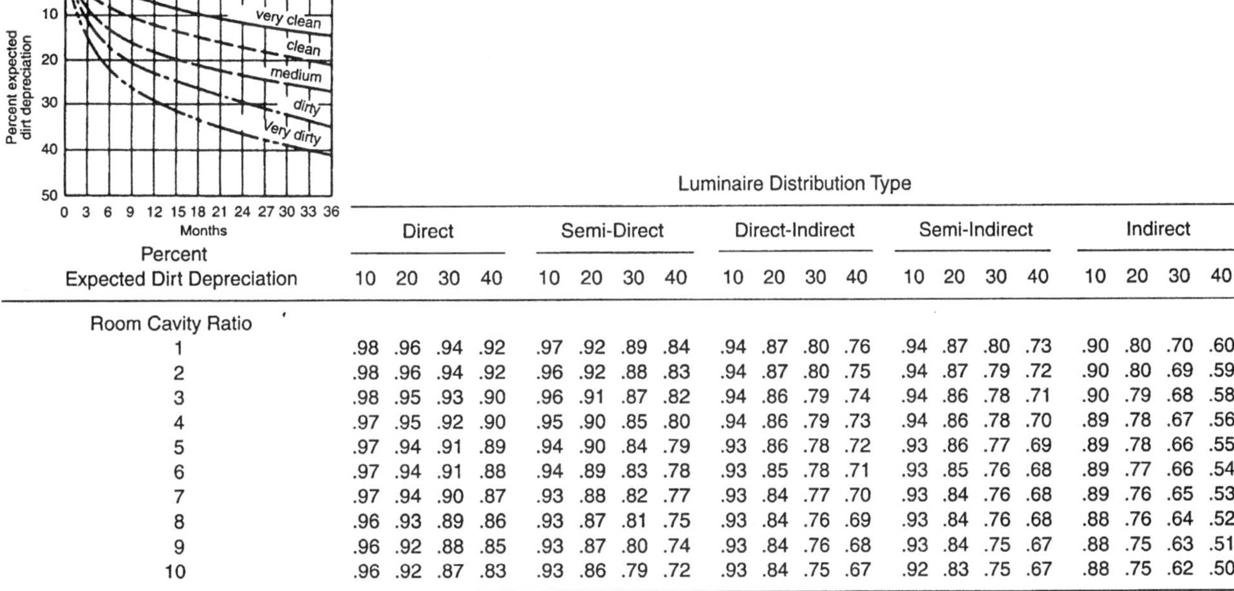

Percent Expected Dirt Depreciation	Direct 10	Direct 20	Direct 30	Direct 40	Semi-Direct 10	Semi-Direct 20	Semi-Direct 30	Semi-Direct 40	Direct-Indirect 10	Direct-Indirect 20	Direct-Indirect 30	Direct-Indirect 40	Semi-Indirect 10	Semi-Indirect 20	Semi-Indirect 30	Semi-Indirect 40	Indirect 10	Indirect 20	Indirect 30	Indirect 40
Room Cavity Ratio																				
1	.98	.96	.94	.92	.97	.92	.89	.84	.94	.87	.80	.76	.94	.87	.80	.73	.90	.80	.70	.60
2	.98	.96	.94	.92	.96	.92	.88	.83	.94	.87	.80	.75	.94	.87	.79	.72	.90	.80	.69	.59
3	.98	.95	.93	.90	.96	.91	.87	.82	.94	.86	.79	.74	.94	.86	.78	.71	.90	.79	.68	.58
4	.97	.95	.92	.90	.95	.90	.85	.80	.94	.86	.79	.73	.94	.86	.78	.70	.89	.78	.67	.56
5	.97	.94	.91	.89	.94	.90	.84	.79	.93	.86	.78	.72	.93	.86	.77	.69	.89	.78	.66	.55
6	.97	.94	.91	.88	.94	.89	.83	.78	.93	.85	.78	.71	.93	.85	.76	.68	.89	.77	.66	.54
7	.97	.94	.90	.87	.93	.88	.82	.77	.93	.84	.77	.70	.93	.84	.76	.68	.89	.76	.65	.53
8	.96	.93	.89	.86	.93	.87	.81	.75	.93	.84	.76	.69	.93	.84	.76	.68	.88	.76	.64	.52
9	.96	.92	.88	.85	.93	.87	.80	.74	.93	.84	.76	.68	.93	.84	.75	.67	.88	.75	.63	.51
10	.96	.92	.87	.83	.93	.86	.79	.72	.93	.84	.75	.67	.92	.83	.75	.67	.88	.75	.62	.50

Luminaire Distribution Type

(Reprinted from the IESNA Lighting Handbook, 9th Edition, 2000, Courtesy of the Illuminating Engineering Society of North America.)

16.3.13 Step-by-Step Calculations for the Number of Luminaires Required for a Particular Room

1. Compile the Following Information:

- Length and width of room.
- Height of floor cavity; the distance from the floor to the work surface (usually taken as 2.5 ft).
- Height of the ceiling cavity; the distance from the ceiling to the light fixture. If the fixture is recessed or ceiling- (surface-) mounted, the value is zero.
- Height of the room cavity; the distance from the work surface to the light fixture.
- Surface reflectances of the ceiling, the walls, and the floor. If the wall surface of the floor cavity is different from the room cavity wall surface (as with a wainscot, for example) obtain both figures. Surface reflectances are usually available from paint companies, ceiling tile manufacturers, and manufacturers of other finishes. If these are not readily available, use the values in Table 16.3.14.

2. Determine Cavity Ratios.

$$CR = 2.5 \times \frac{\text{Area of cavity wall}}{\text{Area of base of cavity}}$$

For rectangular spaces, the formula becomes

$$CR = 5h \times \frac{l + w}{l \times w}$$

where h = height of the cavity
l = length of the room
w = width of the room

Note that if the work surface is the floor, or if the luminaires are surface-mounted, the floor cavity ratio and ceiling cavity ratio, respectively, are zero. Also, since the three cavity ratios are related, after finding one, you can find the other two ratios by:

$$CCR = RCR\left(\frac{h_{cc}}{h_{rc}}\right)$$

$$FCR = RCR\left(\frac{h_{fc}}{h_{rc}}\right)$$

where CCR = ceiling cavity ratio
FCR = floor cavity ratio
RCR = room cavity ratio
h_{cc} = height of ceiling cavity
h_{fc} = height of floor cavity
h_{rc} = height of room cavity

You can find the cavity ratios by calculation or use the values given in Table 16.3.15. First, find the *RCR,* and then use the ratios to find the values of the *CCR* and *FCR*.

3. Determine the Effective Ceiling Cavity Reflectance and the Effective Floor Cavity Reflectance. These are values of the imaginary planes at the height of the luminaire and the work surface that will be used in finding the coefficient of utilization of a particular light fixture. If the luminaires are recessed or surface-mounted, the effective ceiling cavity reflectance is the same as the reflectance of the ceiling itself. Use Table 16.3.16 to find the effective reflectances, knowing the cavity ratios you determined in step 2.

4. Determine the Coefficient of Utilization of the Fixture under Consideration by Using the *CU* Tables from the Manufacturer's Literature or Table 16.3.2. Straight-line interpolation probably will be necessary. Most tables are set up for a floor reflectance of 20 percent. If the effective floor reflectance varies significantly from this, use the correction factors given in Table 16.3.17 and multiply by the CU for the fixture.

5. Determine the Recommended Illumination for the Space Being Designed. Refer to Table 16.2.1.

6. Determine the Lumen Output of the Lamps That Will Be Used in the Luminaire You Have Selected. Values for lumen output for some representative lamps are given in Table 16.3.18. More accurate data can be obtained from the fixture manufacturer or a lamp manufacturer. Determine the number of lamps that will be used in each luminaire.

7. With the Information Compiled in the Previous Steps and with the Light-Loss Factor (*LLF*), Use the Following Formula:

$$\text{Number of luminaires} = \frac{\text{Footcandles required} \times \text{area of room}}{N \times \text{lumens per lamp} \times CU \times LLF}$$

You also can determine the area per luminaire using the formula given at the beginning of this section.

16.3.14 Reflectance Values of Various Materials and Colors

TABLE 16.3.14

Material	Approximate Reflectance (in %)
Acoustical ceiling tile	75–85
Aluminum, brushed	55–58
Aluminum, polished	60–70
Clear glass	8–10
Granite	20–25
Marble	30–70
Stainless steel	55–65
Wood	
Light oak	25–35
Dark oak	10–15
Mahogany	6–12
Walnut	5–10
Color	
White	80–85
Light gray	45–70
Dark gray	20–25
Ivory white	70–80
Ivory	60–70
Pearl gray	70–75
Buff	40–70
Tan	30–50
Brown	20–40
Green	25–50
Azure blue	50–60
Sky blue	35–40
Pink	50–70
Cardinal red	20–25
Red	20–40

16.3.15 Room Cavity Ratios (*see page 16.27*)

TABLE 16.3.15

Room W	Room L	Cavity Depth 2.5	5.5	6.0	6.5	7.0	7.5	8.0	8.5	9.0	10.0	12.0	14.0	16.0	18.0
10	10	2.5	5.5	6.0	6.5	7.0	7.5	8.0	8.5	9.0					
	12	2.3	5.0	5.5	6.0	6.4	6.9	7.3	7.8	8.3					
	14	2.1	4.7	5.1	5.6	6.0	6.4	6.9	7.3	7.7	8.6				
	15	2.1	4.6	5.0	5.4	5.8	6.3	6.7	7.1	7.5	8.3				
	16	2.0	4.5	4.9	5.3	5.7	6.1	6.5	6.9	7.3	8.1				
12	12	2.1	4.6	5.0	5.4	5.8	6.3	6.7	7.1	7.5	8.3				
	14	1.9	4.3	4.6	5.0	5.4	5.8	6.2	6.6	7.0	7.7				
	16	1.8	4.0	4.4	4.7	5.1	5.5	5.8	6.2	6.6	7.3				
	18	1.7	3.8	4.2	4.5	4.9	5.2	5.6	5.9	6.3	6.9				
	20	1.7	3.7	4.0	4.3	4.7	5.0	5.3	5.7	6.0	6.7				
14	14	1.8	3.9	4.3	4.6	5.0	5.4	5.7	6.1	6.4	7.1	8.6			
	16	1.7	3.7	4.0	4.4	4.7	5.0	5.4	5.7	6.0	6.7	8.0			
	18	1.6	3.5	3.8	4.1	4.4	4.8	5.1	5.4	5.7	6.3	7.6			
	20	1.5	3.3	3.6	3.9	4.3	4.6	4.9	5.2	5.5	6.1	7.3			
	22	1.5	3.2	3.5	3.8	4.1	4.4	4.7	5.0	5.3	5.8	7.0			
16	16	1.6	3.4	3.8	4.1	4.4	4.7	5.0	5.3	5.6	6.3	7.5	8.8		
	18	1.5	3.2	3.5	3.8	4.1	4.4	4.7	5.0	5.3	5.9	7.1	8.3		
	20	1.4	3.1	3.4	3.7	3.9	4.2	4.5	4.8	5.1	5.6	6.8	7.9		
	22	1.3	3.0	3.2	3.5	3.8	4.0	4.3	4.6	4.9	5.4	6.5	7.6		
	24	1.3	2.9	3.1	3.4	3.6	3.9	4.2	4.4	4.7	5.2	6.3	7.3		
18	18	1.4	3.1	3.3	3.6	3.9	4.2	4.4	4.7	5.0	5.6	6.7	7.8	8.9	
	22	1.3	2.8	3.0	3.3	3.5	3.8	4.0	4.3	4.5	5.1	6.1	7.1	8.1	
	26	1.2	2.6	2.8	3.1	3.3	3.5	3.8	4.0	4.2	4.7	5.6	6.6	7.5	
	30	1.1	2.4	2.7	2.9	3.1	3.3	3.6	3.8	4.0	4.4	5.3	6.2	7.1	
	34	1.1	2.3	2.5	2.8	3.0	3.2	3.4	3.6	3.8	4.2	5.1	5.9	6.8	
20	20	1.3	2.8	3.0	3.3	3.5	3.8	4.0	4.3	4.5	5.0	6.0	7.0	8.0	9.0
	24	1.1	2.5	2.8	3.0	3.2	3.4	3.7	3.9	4.1	4.6	5.5	6.4	7.3	8.3
	28	1.1	2.4	2.6	2.8	3.0	3.2	3.4	3.6	3.9	4.3	5.1	6.0	6.9	7.7
	32	1.0	2.2	2.4	2.6	2.8	3.0	3.3	3.5	3.7	4.1	4.9	5.7	6.5	7.3
	36	1.0	2.1	2.3	2.5	2.7	2.9	3.1	3.3	3.5	3.9	4.7	5.4	6.2	7.0
24	24	1.0	2.3	2.5	2.7	2.9	3.1	3.3	3.5	3.8	4.2	5.0	5.8	6.7	7.5
	28	1.0	2.1	2.3	2.5	2.7	2.9	3.1	3.3	3.5	3.9	4.6	5.4	6.2	7.0
	32	0.9	2.0	2.2	2.4	2.6	2.7	2.9	3.1	3.3	3.6	4.4	5.1	5.8	6.6
	36	0.9	1.9	2.1	2.3	2.4	2.6	2.8	3.0	3.1	3.5	4.2	4.9	5.6	6.3
	40	0.8	1.8	2.0	2.2	2.3	2.5	2.7	2.8	3.0	3.3	4.0	4.7	5.3	6.0
28	34	0.8			2.1	2.3	2.4	2.6	2.8	2.9	3.3	3.9	4.6	5.2	5.9
	40	0.8			2.0	2.1	2.3	2.4	2.6	2.7	3.0	3.6	4.3	4.9	5.5
	46	0.7			1.9	2.0	2.2	2.3	2.4	2.6	2.9	3.4	4.0	4.6	5.2
	52	0.7			1.8	1.9	2.1	2.2	2.3	2.5	2.7	3.3	3.8	4.4	4.9
32	38	0.7					2.2	2.3	2.4	2.6	2.9	3.5	4.0	4.6	5.2
	44	0.7					2.0	2.2	2.3	2.4	2.7	3.2	3.8	4.3	4.9
	50	0.6					1.9	2.1	2.2	2.3	2.6	3.1	3.6	4.1	4.6
	56	0.6					1.8	2.0	2.1	2.2	2.5	2.9	3.4	3.9	4.4
38	46	0.6					1.8	1.9	2.0	2.2	2.4	2.9	3.4	3.8	4.3
	54	0.6					1.7	1.8	1.9	2.0	2.2	2.7	3.1	3.6	4.0
	62	0.5					1.6	1.7	1.8	1.9	2.1	2.5	3.0	3.4	3.8
	70	0.5							1.7	1.8	2.0	2.4	2.8	3.2	3.7
44	50	0.5							1.8	1.9	2.1	2.6	3.0	3.4	3.8
	60	0.5							1.7	1.8	2.0	2.4	2.8	3.2	3.5
	70	0.5							1.6	1.7	1.9	2.2	2.6	3.0	3.3
	80	0.4							1.5	1.6	1.8	2.1	2.5	2.8	3.2

16.3.16 Percent Effective Ceiling or Floor Cavity Reflectances for Various Reflectance Combinations

(*see pages 16.28 –16.29*)

TABLE 16.3.16 Percent Effective Ceiling or Floor Cavity Reflectances for Various Reflectance Combinations*

Percent Base† Reflectance: 90

Percent Wall Reflectance / Cavity Ratio	90	80	70	60	50	40	30	20	10	0
0.2	89	88	88	87	86	85	85	84	84	82
0.4	88	87	86	85	84	83	81	80	79	76
0.6	87	86	84	82	80	79	77	76	74	73
0.8	87	85	82	80	77	75	73	71	69	67
1.0	86	83	80	77	75	72	69	66	64	62
1.2	85	82	78	75	72	69	66	63	60	57
1.4	85	80	77	73	69	65	62	59	57	52
1.6	84	79	75	71	67	63	59	56	53	50
1.8	83	78	73	69	64	60	56	53	50	48
2.0	83	77	72	67	62	56	53	50	47	43
2.2	82	76	70	65	59	54	50	47	44	40
2.4	82	75	69	64	58	53	48	45	41	37
2.6	81	74	67	62	56	51	46	42	38	35
2.8	81	73	66	60	54	49	44	40	36	34
3.0	80	72	64	58	52	47	42	38	34	30
3.2	79	71	63	56	50	45	40	36	32	28
3.4	79	70	62	54	48	43	38	34	30	27
3.6	78	69	61	53	47	42	36	32	28	25
3.8	78	69	60	51	45	40	35	31	27	23
4.0	77	69	58	51	44	39	33	29	25	22
4.2	77	62	57	50	43	37	32	28	24	21
4.4	76	61	56	49	42	36	31	27	23	20
4.6	76	60	55	47	40	35	30	26	22	19
4.8	75	59	54	46	39	34	28	25	21	18
5.0	75	59	53	45	38	33	28	24	20	16
6.0	73	61	49	41	34	29	24	20	16	11
7.0	70	58	45	38	30	27	21	18	14	08
8.0	68	55	42	35	27	23	18	15	12	06
9.0	66	52	38	31	25	21	16	14	11	05
10.0	65	51	36	29	22	19	15	11	09	04

Percent Base† Reflectance: 80

Percent Wall Reflectance / Cavity Ratio	90	80	70	60	50	40	30	20	10	0
0.2	79	78	78	77	77	76	76	75	74	72
0.4	79	77	76	75	74	73	72	71	70	68
0.6	78	76	75	73	71	70	68	66	65	63
0.8	78	75	73	71	69	67	65	63	61	57
1.0	77	74	72	69	67	65	62	60	57	55
1.2	76	73	70	67	64	61	58	55	53	51
1.4	76	72	68	65	62	59	55	53	50	48
1.6	75	71	67	63	60	57	53	50	47	44
1.8	75	70	66	62	58	54	50	47	44	41
2.0	74	69	64	60	56	52	48	45	41	38
2.2	74	68	63	58	54	49	45	42	38	35
2.4	73	67	61	56	52	47	43	40	36	33
2.6	73	66	60	55	50	45	41	38	34	31
2.8	73	65	59	53	48	43	39	36	32	29
3.0	72	65	58	52	47	42	37	34	30	27
3.2	72	65	57	51	45	40	35	33	28	25
3.4	71	64	56	49	44	39	34	32	27	24
3.6	71	63	54	48	43	38	32	30	25	23
3.8	70	62	53	47	41	36	31	28	24	22
4.0	70	61	53	46	40	35	30	26	22	20
4.2	69	60	52	45	39	34	29	25	21	18
4.4	69	60	51	44	38	33	28	24	20	17
4.6	69	59	50	43	37	32	27	23	19	15
4.8	68	58	49	42	36	31	26	22	18	14
5.0	68	58	48	41	35	30	25	21	18	14
6.0	66	55	44	38	31	27	22	19	15	10
7.0	64	53	41	35	28	24	19	16	12	07
8.0	62	50	38	32	25	21	17	14	11	05
9.0	61	49	36	30	23	19	15	13	10	04
10.0	59	46	33	27	21	18	14	11	08	03

Percent Base† Reflectance: 70

Percent Wall Reflectance / Cavity Ratio	90	80	70	60	50	40	30	20	10	0
0.2	70	69	68	68	67	67	66	66	65	64
0.4	69	68	67	66	65	64	63	62	61	58
0.6	69	67	65	64	63	61	59	58	57	54
0.8	68	66	64	62	60	58	56	55	53	50
1.0	68	65	62	60	58	55	53	52	50	47
1.2	67	64	61	59	57	54	50	48	46	44
1.4	67	63	60	58	55	51	47	45	44	41
1.6	67	62	59	56	53	47	45	43	41	38
1.8	66	61	58	54	51	46	42	40	38	35
2.0	66	60	56	52	49	45	40	38	36	33
2.2	66	60	55	51	48	43	38	36	34	32
2.4	65	60	54	50	46	41	37	35	32	30
2.6	65	59	54	49	45	40	35	33	30	28
2.8	65	59	53	48	43	38	33	30	28	26
3.0	64	58	52	47	42	37	32	29	27	24
3.2	64	58	51	46	40	36	31	28	25	23
3.4	64	57	50	45	39	35	29	27	24	22
3.6	63	56	49	44	38	33	28	25	22	20
3.8	63	56	49	43	37	32	27	24	21	19
4.0	63	55	48	42	36	31	26	23	20	17
4.2	62	55	47	41	35	30	25	22	19	16
4.4	62	54	46	40	34	29	24	21	18	15
4.6	62	53	45	39	33	28	24	21	17	14
4.8	62	53	45	38	32	27	23	20	16	13
5.0	61	52	44	36	31	26	22	19	16	12
6.0	60	51	41	35	28	24	19	16	13	09
7.0	58	43	38	32	26	22	17	14	11	06
8.0	57	46	35	29	23	19	15	13	10	05
9.0	56	45	33	27	21	18	14	12	09	04
10.0	55	43	31	25	19	16	12	10	08	03

Percent Base† Reflectance: 60

Percent Wall Reflectance / Cavity Ratio	90	80	70	60	50	40	30	20	10	0
0.2	60	59	59	59	58	57	56	56	55	53
0.4	60	59	59	58	57	55	54	53	52	50
0.6	60	58	57	56	55	53	51	51	50	46
0.8	59	57	56	55	54	51	48	47	46	43
1.0	59	57	55	53	51	48	45	44	43	41
1.2	59	56	54	51	49	46	44	42	40	38
1.4	59	56	53	49	47	44	41	39	38	36
1.6	59	55	52	48	45	42	39	37	35	33
1.8	58	55	51	47	44	40	37	35	33	31
2.0	58	54	50	46	43	39	35	33	31	29
2.2	58	53	49	45	42	37	34	31	29	28
2.4	58	53	48	44	41	36	32	30	27	26
2.6	58	53	48	43	39	35	31	28	26	24
2.8	58	53	47	43	38	34	29	27	24	22
3.0	57	52	46	42	37	32	28	25	23	20
3.2	57	51	45	41	36	31	27	23	22	18
3.4	57	51	45	40	35	30	26	23	20	17
3.6	57	50	44	39	34	29	25	22	19	16
3.8	57	50	43	38	33	29	24	21	19	15
4.0	57	49	42	37	32	28	23	20	18	14
4.2	56	49	42	37	32	27	22	19	17	14
4.4	56	49	42	36	31	27	22	19	16	13
4.6	56	49	41	35	30	26	21	18	16	13
4.8	56	48	41	34	29	25	21	18	15	12
5.0	56	48	40	34	28	24	20	17	14	11
6.0	55	45	37	31	25	21	17	14	11	07
7.0	54	43	35	30	24	20	15	12	09	05
8.0	53	42	33	28	22	18	14	11	08	04
9.0	52	40	31	26	20	16	12	10	07	03
10.0	51	39	29	24	18	15	11	09	07	02

Percent Base† Reflectance: 50

Percent Wall Reflectance / Cavity Ratio	90	80	70	60	50	40	30	20	10	0
0.2	50	50	49	49	48	48	47	46	46	44
0.4	50	49	48	48	47	46	45	45	44	42
0.6	50	48	47	46	45	44	43	42	41	38
0.8	50	48	47	45	44	42	40	39	38	36
1.0	50	48	46	44	43	41	38	37	36	34
1.2	50	47	45	43	41	39	36	35	34	29
1.4	50	47	45	42	40	38	35	34	35	27
1.6	50	47	44	41	39	36	33	32	30	26
1.8	50	46	43	40	38	35	31	30	28	25
2.0	50	46	43	40	37	34	30	28	26	24
2.2	50	46	42	38	36	33	29	27	24	22
2.4	50	46	42	37	35	31	27	25	23	21
2.6	50	46	41	37	34	30	26	23	21	20
2.8	50	46	41	36	33	29	25	22	20	19
3.0	50	45	40	36	32	28	24	21	19	17
3.2	50	44	39	35	31	27	23	20	18	16
3.4	50	44	39	35	30	26	22	19	17	15
3.6	50	44	39	34	29	25	21	18	16	14
3.8	50	44	38	34	29	25	21	17	15	13
4.0	50	44	38	33	28	24	20	17	15	12
4.2	50	43	37	32	28	24	20	17	14	12
4.4	50	43	37	32	27	23	19	16	13	11
4.6	50	43	36	31	26	22	18	15	13	10
4.8	50	43	36	31	26	22	18	15	12	09
5.0	50	42	35	30	25	21	17	14	12	09
6.0	50	42	34	29	23	19	15	13	10	06
7.0	49	41	32	27	21	18	14	11	08	05
8.0	49	40	30	25	19	16	12	10	07	03
9.0	48	39	29	24	18	15	11	09	07	03
10.0	47	37	27	22	17	14	10	08	06	02

* Values in this table are based on a length to width ratio of 1.6.

†Ceiling, floor, or floor of cavity.

(continued)

TABLE 16.3.16

Percent Base† Reflectance: 40

Cavity Ratio \ Percent Wall Reflectance	90	80	70	60	50	40	30	20	10	0
0.2	40	40	39	39	39	38	38	37	36	36
0.4	41	40	39	39	38	37	36	35	34	34
0.6	41	40	39	38	37	36	34	33	32	31
0.8	41	40	38	37	36	35	33	32	31	29
1.0	42	40	38	37	35	33	32	31	29	27
1.2	42	40	38	36	34	32	30	29	27	25
1.4	42	39	37	35	33	31	29	27	25	23
1.6	42	39	37	35	32	30	27	25	23	22
1.8	42	39	36	34	31	29	26	24	22	21
2.0	42	39	36	34	31	28	25	23	21	19
2.2	42	39	36	33	30	27	24	22	19	18
2.4	43	39	35	33	29	27	24	21	18	17
2.6	43	39	35	32	29	26	23	20	17	15
2.8	43	39	35	32	28	25	22	19	16	14
3.0	43	39	35	31	27	24	21	18	16	13
3.2	43	39	35	31	27	23	20	71	15	13
3.4	43	39	34	30	26	23	20	17	14	12
3.6	44	39	34	30	26	22	19	16	14	11
3.8	44	38	33	29	25	22	18	16	13	10
4.0	44	38	33	29	25	21	18	15	12	10
4.2	44	38	33	29	24	21	17	15	12	10
4.4	44	38	33	28	24	20	17	14	11	09
4.6	44	38	32	28	23	19	16	14	11	08
4.8	44	38	32	27	22	19	16	13	10	08
5.0	45	38	31	27	22	19	15	13	10	07
6.0	44	37	30	25	20	17	13	11	08	05
7.0	44	36	29	24	19	16	12	10	07	04
8.0	44	35	28	23	18	15	11	09	06	03
9.0	44	35	26	21	16	13	10	08	05	02
10.0	43	34	25	20	15	12	08	07	05	02

Percent Base† Reflectance: 30

Cavity Ratio \ Percent Wall Reflectance	90	80	70	60	50	40	30	20	10	0
0.2	31	31	30	30	29	29	29	28	28	27
0.4	31	31	30	30	29	28	28	27	26	25
0.6	32	31	30	29	28	27	26	26	25	23
0.8	32	31	30	29	28	26	25	25	23	22
1.0	33	32	30	29	27	25	24	23	22	20
1.2	33	32	30	28	27	25	23	22	21	19
1.4	34	32	30	28	26	24	22	21	19	18
1.6	34	33	29	27	25	23	22	20	18	17
1.8	35	33	29	27	25	23	21	19	17	16
2.0	35	33	29	26	24	22	20	18	16	14
2.2	36	32	29	26	24	22	19	17	15	13
2.4	36	32	29	26	24	22	19	16	14	12
2.6	36	32	29	25	23	21	18	16	14	12
2.8	37	33	29	25	23	21	17	15	13	11
3.0	37	33	29	25	22	20	17	15	12	10
3.2	37	33	29	25	22	19	16	14	12	10
3.4	37	33	29	25	22	19	16	14	11	09
3.6	38	33	29	24	21	18	15	13	10	09
3.8	38	33	28	24	21	18	15	13	10	08
4.0	38	33	28	24	21	18	14	12	09	07
4.2	38	33	28	24	20	17	14	12	09	07
4.4	39	33	28	24	20	17	14	11	09	06
4.6	39	33	28	24	20	17	13	10	08	06
4.8	39	33	28	24	20	17	13	10	08	05
5.0	39	33	28	24	19	16	13	10	08	05
6.0	39	33	27	23	18	15	11	09	06	04
7.0	40	33	26	22	17	14	10	08	05	03
8.0	40	33	26	21	16	13	09	07	04	02
9.0	40	33	25	20	15	12	09	07	04	02
10.0	40	32	24	19	14	11	08	06	03	01

Percent Base† Reflectance: 20

Cavity Ratio \ Percent Wall Reflectance	90	80	70	60	50	40	30	20	10	0
0.2	21	20	20	20	20	20	19	19	19	17
0.4	22	21	20	20	20	19	19	18	18	16
0.6	23	21	21	20	19	19	18	18	17	15
0.8	24	22	21	20	19	19	18	17	16	14
1.0	25	23	22	20	19	18	17	16	15	13
1.2	25	23	22	20	19	17	17	16	14	12
1.4	26	24	22	20	18	17	16	15	13	12
1.6	26	24	22	20	18	17	16	15	13	11
1.8	27	25	23	20	18	17	15	14	12	10
2.0	28	25	23	20	18	16	15	13	11	09
2.2	28	25	23	20	18	16	14	12	10	09
2.4	29	26	23	20	18	16	14	12	10	08
2.6	29	26	23	20	18	16	14	11	09	08
2.8	30	27	23	20	18	15	13	11	09	07
3.0	30	27	23	20	17	15	13	11	09	07
3.2	31	27	23	20	17	15	12	11	09	06
3.4	31	27	23	20	17	15	12	10	08	06
3.6	32	27	23	20	17	15	12	10	08	05
3.8	32	28	23	20	17	15	12	10	07	05
4.0	33	28	23	20	17	14	11	09	07	05
4.2	33	28	23	20	17	14	11	09	07	04
4.4	34	28	24	20	17	14	11	09	07	04
4.6	34	29	24	20	17	14	11	09	07	04
4.8	35	29	24	20	17	13	10	08	06	04
5.0	35	29	24	20	16	13	10	08	06	04
6.0	36	30	24	20	16	13	10	08	05	02
7.0	36	30	24	20	15	12	03	07	04	02
8.0	37	30	23	19	15	12	08	06	03	01
9.0	37	29	23	19	14	11	08	06	03	01
10.0	37	29	22	18	13	10	07	05	03	01

Percent Base† Reflectance: 10

Cavity Ratio \ Percent Wall Reflectance	90	80	70	60	50	40	30	20	10	0
0.2	11	11	11	10	10	10	10	09	09	09
0.4	12	11	11	11	11	10	10	09	09	08
0.6	13	13	12	11	11	10	10	09	08	08
0.8	15	14	13	12	11	10	10	09	08	07
1.0	16	14	13	12	12	11	10	09	08	07
1.2	17	15	14	13	12	11	10	09	07	06
1.4	18	16	14	13	12	11	10	09	07	06
1.6	19	17	15	14	12	11	09	08	07	06
1.8	19	17	15	14	13	11	09	08	06	05
2.0	20	18	16	14	13	11	09	08	06	05
2.2	21	19	16	14	13	11	09	07	06	05
2.4	22	19	17	15	13	11	09	07	06	05
2.6	23	20	17	15	13	11	09	07	06	04
2.8	23	20	18	16	13	11	09	07	05	03
3.0	24	21	18	16	13	11	09	07	05	03
3.2	25	21	18	16	13	11	09	07	05	03
3.4	26	22	18	16	13	11	09	07	05	03
3.6	26	22	19	16	13	11	09	06	04	03
3.8	27	23	19	17	14	11	09	06	04	02
4.0	27	23	20	17	14	11	09	06	04	02
4.2	28	24	20	17	14	11	09	06	04	02
4.4	28	24	20	17	14	11	08	06	04	02
4.6	29	25	20	17	14	11	08	06	04	02
4.8	29	25	20	17	14	11	08	06	04	02
5.0	30	25	20	17	14	11	08	06	04	02
6.0	31	26	21	18	14	11	08	06	03	01
7.0	32	27	21	17	13	11	08	06	03	01
8.0	33	27	21	17	13	10	07	05	03	01
9.0	34	28	21	17	13	10	07	05	02	01
10.0	34	28	21	17	12	10	07	05	02	01

Percent Base† Reflectance: 0

Cavity Ratio \ Percent Wall Reflectance	90	80	70	60	50	40	30	20	10	0
0.2	02	02	02	01	01	01	01	00	00	0
0.4	04	03	03	02	02	02	01	01	00	0
0.6	05	05	04	03	03	02	02	01	01	0
0.8	07	06	05	04	04	03	02	02	01	0
1.0	08	07	06	05	04	03	02	02	01	0
1.2	10	08	07	06	05	04	03	02	01	0
1.4	11	09	08	07	06	04	03	02	01	0
1.6	12	10	09	07	06	05	03	02	01	0
1.8	13	11	09	08	07	05	04	03	01	0
2.0	14	12	10	09	07	05	04	03	01	0
2.2	15	13	11	09	07	06	04	03	01	0
2.4	16	13	11	09	08	06	04	03	01	0
2.6	17	14	12	10	08	06	05	03	02	0
2.8	17	15	13	10	08	07	05	03	02	0
3.0	18	16	13	11	09	07	05	03	02	0
3.2	19	16	14	11	09	07	05	03	02	0
3.4	20	17	14	12	09	07	05	03	02	0
3.6	20	17	15	12	10	08	05	04	02	0
3.8	21	18	15	12	10	08	05	04	02	0
4.0	22	18	15	13	10	08	05	04	02	0
4.2	22	19	16	13	10	08	06	04	02	0
4.4	23	19	16	13	10	08	06	04	02	0
4.6	23	20	17	13	11	08	06	04	02	0
4.8	24	20	17	14	11	08	06	04	02	0
5.0	25	21	17	14	11	08	06	04	02	0
6.0	27	23	18	15	12	09	06	04	02	0
7.0	28	24	19	15	12	09	06	04	02	0
8.0	30	25	20	15	12	09	06	04	02	0
9.0	31	25	20	15	12	09	06	04	02	0
10.0	31	25	20	15	12	09	06	04	02	0

* Values in this table are based on a length to width ratio of 1.6.
†Ceiling, floor, or floor of cavity.

(Reprinted from the IESNA Lighting Handbook, 9th Edition, 2000, Courtesy of the Illuminating Engineering Society of North America.)

16.3.17 Multiplying Factors for Effective Floor Cavity Reflectances Other than 20 percent (0.2)

TABLE 16.3.17

% Effective Ceiling Cavity Reflectance, ρ_{cc}	80				70				50			30			10		
% Wall Reflectance, ρ_w	70	50	30	10	70	50	30	10	50	30	10	50	30	10	50	30	10
For 30% Effective Floor Cavity Reflectance (20% = 1.00)																	
Room Cavity Ratio																	
1	1.092	1.082	1.075	1.068	1.077	1.070	1.064	1.059	1.049	1.044	1.040	1.028	1.026	1.023	1.012	1.010	1.008
2	1.079	1.066	1.055	1.047	1.068	1.057	1.048	1.039	1.041	1.033	1.027	1.026	1.021	1.017	1.013	1.010	1.006
3	1.070	1.054	1.042	1.033	1.061	1.048	1.037	1.028	1.034	1.027	1.020	1.024	1.017	1.012	1.014	1.009	1.005
4	1.062	1.045	1.033	1.024	1.055	1.040	1.029	1.021	1.030	1.022	1.015	1.022	1.015	1.010	1.014	1.009	1.004
5	1.056	1.038	1.026	1.018	1.050	1.034	1.024	1.015	1.027	1.018	1.012	1.020	1.013	1.008	1.014	1.009	1.004
6	1.052	1.033	1.021	1.014	1.047	1.030	1.020	1.012	1.024	1.015	1.009	1.019	1.012	1.006	1.014	1.008	1.003
7	1.047	1.029	1.018	1.011	1.043	1.026	1.017	1.009	1.022	1.013	1.007	1.018	1.010	1.005	1.014	1.008	1.003
8	1.044	1.026	1.015	1.009	1.040	1.024	1.015	1.007	1.020	1.012	1.006	1.017	1.009	1.004	1.013	1.007	1.003
9	1.040	1.024	1.014	1.007	1.037	1.022	1.014	1.006	1.019	1.011	1.005	1.016	1.009	1.004	1.013	1.007	1.002
10	1.037	1.022	1.012	1.006	1.034	1.020	1.012	1.005	1.017	1.010	1.004	1.015	1.009	1.003	1.013	1.007	1.002
For 10% Effective Floor Cavity Reflectance (20% = 1.00)																	
Room Cavity Ratio																	
1	.923	.929	.935	.940	.933	.939	.943	.948	.956	.960	.963	.973	.976	.979	.989	.991	.993
2	.931	.942	.950	.958	.940	.949	.957	.963	.962	.968	.974	.976	.980	.985	.988	.991	.995
3	.939	.951	.961	.969	.945	.957	.966	.973	.967	.975	.981	.978	.983	.988	.988	.992	.996
4	.944	.958	.969	.978	.950	.963	.973	.980	.972	.980	.986	.980	.986	.991	.987	.992	.996
5	.949	.964	.976	.983	.954	.968	.978	.985	.975	.983	.989	.981	.988	.993	.987	.992	.997
6	.953	.969	.980	.986	.958	.972	.982	.989	.977	.985	.992	.982	.989	.995	.987	.993	.997
7	.957	.973	.983	.991	.961	.975	.985	.991	.979	.987	.994	.983	.990	.996	.987	.993	.998
8	.960	.976	.986	.993	.963	.977	.987	.993	.981	.988	.995	.984	.991	.997	.987	.994	.998
9	.963	.978	.987	.994	.965	.979	.989	.994	.983	.990	.996	.985	.992	.998	.988	.994	.999
10	.965	.980	.989	.995	.967	.981	.990	.995	.984	.991	.997	.986	.993	.998	.988	.994	.999
For 0% Effective Floor Cavity Reflectance (20% = 1.00)																	
Room Cavity Ratio																	
1	.859	.870	.879	.886	.873	.884	.893	.901	.916	.923	.929	.948	.954	.960	.979	.983	.987
2	.871	.887	.903	.919	.886	.902	.916	.928	.926	.938	.949	.954	.963	.971	.978	.983	.991
3	.882	.904	.915	.942	.898	.918	.934	.947	.936	.950	.964	.958	.969	.979	.976	.984	.993
4	.893	.919	.941	.958	.908	.930	.948	.961	.945	.961	.974	.961	.974	.984	.975	.985	.994
5	.903	.931	.953	.969	.914	.939	.958	.970	.951	.967	.980	.964	.977	.988	.975	.985	.995
6	.911	.940	.961	.976	.920	.945	.965	.977	.955	.972	.985	.966	.979	.991	.975	.986	.996
7	.917	.947	.967	.981	.924	.950	.970	.982	.959	.975	.988	.968	.981	.993	.975	.987	.997
8	.922	.953	.971	.985	.929	.955	.975	.986	.963	.978	.991	.970	.983	.995	.976	.988	.998
9	.928	.958	.975	.988	.933	.959	.980	.989	.966	.980	.993	.971	.985	.996	.976	.988	.998
10	.933	.962	.979	.991	.937	.963	.983	.992	.969	.982	.995	.973	.987	.997	.977	.989	.999

(Reprinted from the IESNA Lighting Handbook, 9th Edition, 2000, Courtesy of the Illuminating Engineering Society of North America.)

16.3.18 Characteristics of Typical Lamps

TABLE 16.3.18

Standard Incandescent

Bulb Description	Watts	Length/ Size (in in.)	Lamp Life (in hours) (1)	Color Temp. °K (1)	Initial Lumens (1)	Lamp Lumen Depreciation (1)
A-19	60		1000	2790	860	0.93
A-19	75		750	2840	1180	0.92
A-19	100		750	2900	1740	0.91
A-19	100		2500		1490	0.93
A-21	100		750	2880	1690	0.90
A-21	150		750	2960	2880	0.89
A-23	150		2500		2350	0.89
PS-25	150		750	2900	2660	0.88
A-23	200		750	2980	4000	0.90
A-23	200		2500		3400	0.88
PS-25	300		750	3010	6360	0.88
PS-30	300		2500		5200	0.79
PS-35	500		1000	3050	10600	0.89

R, PAR, and ER Lamps

Bulb Description	Watts	Length/ Size (in in.)	Lamp Life (in hours) (1)	Color Temp. (1)	Initial Lumens (1,2)	Lamp Lumen Depreciation (1)
R-30 Spot/Flood	75		2000		850	
R-40 Spot/Flood	150		2000		1825	
R-40 Spot/Flood	300		2000		3600	
PAR-38 Spot/Flood	100		2000		1250	
PAR-38 Spot/Flood	150		2000		1730	
ER-30	50		2000		525	
ER-30	75		2000		850	
ER-30	90		5000		950	
ER-40	120		2000		1475	

Fluorescent

Bulb Description	Watts	Length/ Size (in in.)	Lamp Life (in hours) (1,3)	Color Temp. (1,4)	Initial Lumens (1,5)	Lamp Lumen Depreciation (1)
F40T12CW/RS	40	48	20000	4300	3150	0.84
F40T12WW/RS	40	48	20000	3100	3170	0.84
F40T12CWX/RS	40	48	20000	4100	2200	0.84
F40T12WWX/RS	40	48	20000	3000	2170	0.84
F40T12D/RS	40	48	20000	6500	2600	0.84
F40T12W/RS	40	48	20000	3600	3180	0.84
F96T12CW	75	96	12000	4300	6300	0.89
F96T12WW	75	96	12000	3100	6335	0.89
F96T12CWX	75	96	12000	4100	4465	0.89
F96T12WWX	75	96	12000	3000	4365	

(continued)

TABLE 16.3.18 (*Continued*)

Tungsten-Halogen (Quartz-Iodine) Bulb Description	Watts	Length/ Size (in in.)	Lamp Life (in hours) (1)	Color Temp. (1)	Initial Lumens (1)	Lamp Lumen Depreciation (1)
T-4	100		1000		1800	0.93
T-4	150		1500	3000	2900	0.93
T-4	250		2000	2950	5000	0.97
PAR-38	250		6000		3500	0.95
Mercury						
Bulb Description	**Watts**	**Length/ Size (in in.)**	**Lamp Life (in hours) (1)**	**Color Temp. (1)**	**Initial Lumens (1)**	**Lamp Lumen Depreciation (1)**
H45AY-40/50 DX	50		16000		1680	
H43AY-75/DX	75		24000		3000	
H38BP-100/DX	100		24000+		2865	
H38JA-100/WDX	100		24000+		4000	
H38MP-100/DX	100		24000		4275	
H39BN-175/DX	175		24000		5800	
H39KC-175/DX	175		24000+		8600	
H37KC-250/DX	250		24000+		12775	
Metal-Halide						
Bulb Description	**Watts**	**Length/ Size (in in.)**	**Lamp Life (in hours) (1)**	**Color Temp. (1)**	**Initial Lumens (1)**	**Lamp Lumen Depreciation (1)**
M57PF-175	175		7500	3600	14000	
M58PH-250	250		10000		20500	
M59PK-400	400		1500	3800	34000	
High-Pressure Sodium						
Bulb Description	**Watts**	**Length/ Size (in in.)**	**Lamp Life (in hours) (1)**	**Color Temp. (1)**	**Initial Lumens (1)**	**Lamp Lumen Depreciation (1)**
S68MT-50	50		24000		3800	
S54MC-100	100		24000		8800	
S55MD-150	150		24000		15000	

(1) Figures listed are approximate. Exact values vary with manufacturer.

(2) Initial lumens for R, PAR, and ER lamps is for total lumens.

(3) Lamp life for fluorescent depends on number of hours per start; figures given are for approximately 10 hours per start.

(4) Technically, "color temperature" applies only to incandescent sources, but it is often used to describe the degree of whiteness of other light sources.

(5) Lumens at 40% of rated life.

16.3.19 Guide to Lamp Selection

(*see pages 16.33–16.34*)

TABLE 16.3.19

Lamp Type and Efficacy (1)	Lamp Appearance Effect on Neutral Surfaces	Effect on "Atmosphere"	Colors Strengthened	Colors Grayed	Effect on Complexions	Remarks
Fluorescent						
Cool white (#4) (2)	White	Neutral to moderately cool	Orange, blue, yellow	Red	Pale pink	Blends with natural daylight—good color acceptance
Deluxe cool white (#2) (2)	White	Neutral to moderately cool	All nearly equal	None appreciably	Most natural	Best overall color rendition, simulates natural daylight
Warm white (#4) (3)	Yellowish white	Warm	Orange, yellow	Red, green, blue	Sallow	Blends with incandescent light—poor color acceptance
Deluxe warm white (#2) (3)	Yellowish white	Warm	Red, orange, yellow, green	Blue	Ruddy	Good color rendition; simulates incandescent light
Daylight (#3)	Bluish white	Very cool	Green, blue	Red, orange	Grayed	Usually replaceable with cool white
White (#4)	Pale yellowish white	Moderately warm	Orange, yellow	Red, green, blue	Pale	Usually replaceable with cool white or warm white

(1) Efficacy (lumens/watt): #1 = low; #2 = medium; #3 = medium high; #4 = high.
(2) Greater preference at higher levels.
(3) Greater preference at lower levels.

(continued)

TABLE 16.3.19 (*Continued*)

Lamp Type and Efficacy (1)	Lamp Appearance Effect on Neutral Surfaces	Effect on "Atmosphere"	Colors Strengthened	Colors Grayed	Effect on Complexions	Remarks
Incandescent, Tungsten-Halogen						
Filament (#1) (3)	Yellowish white	Warm	Red, orange, yellow	Blue	Ruddiest	Good color rendering
High-Intensity Discharge						
Clear mercury (#2)	Greenish blue-white	Very cool, greenish	Yellow, green, blue	Red, orange	Greenish	Very poor color rendering
White mercury (#2)	Greenish white	Moderately cool, greenish	Yellow, green, blue	Red, orange	Very pale	Moderate color rendering
Deluxe white mercury (#2)	Purplish white	Warm, purplish	Red, yellow, blue	Green	Ruddy	Color acceptance similar to cool white fluorescent
Metal-Halide (#4) (2)	Greenish white	Moderately cool greenish	Yellow, blue, green	Red	Grayed	Color acceptance similar to cool white
High-pressure sodium (#4)	Yellowish	Warm, yellowish	Yellow, green, orange	Red, blue	Yellowish	Color acceptance approaches warm white fluorescent

(1) Efficacy (lumens/watt): #1 = low; #2 = medium; #3 = medium high; #4 = high.
(2) Greater preference at higher levels.
(3) Greater preference at lower levels.

Source: GE Lighting Business Group.

16.3.20 Recommended Reflectances of Interior Surfaces

TABLE 16.3.20

	Recommended Reflectances in Percent					
	Ceilings	**Walls**	**Floors**	**Furniture**	**Other**	
Offices	80+	50-70	20-40	25-45	40-70	Partitions
Schools	70-90	40-60	30-50	35-50	up to 20	Chalkboards
Industrial	80-90	40-60	20+		25-45	Benchtops, machines, etc.
Residential	60-90	35-60 (1)	15-35 (1)		45-85	Large drapery areas

(1) Where specific visual tasks are more important than lighting for environment, minimum reflectances should be 40% for walls and 25% for floors.

16.3.21 Recommended Luminance Ratios

TABLE 16.3.21

		Recommended Ratios (1)				
		Between task and immediate darker surroundings		Between task and immediate lighter surroundings	Between task and general surroundings	
Use	**Task**	**Minimum**	**Desired**	**Maximum**	**Minimum**	**Desired**
Residential	1	1/5	1/3		0.1-10	0.2-5
Office	1		1/3			0.1-10
Classroom	1	1/3		5 (2)	1/3	
Merchandising	1	1/3	1/5			
Industrial	1		1/3	3	0.5-20	0.1-10

(1) These are recommended guidelines for most applications. Ratios higher or lower are acceptable if they do not exceed a significant portion of the visual field.

(2) Any significant surface normally viewed directly should be no greater than five times the luminance of the task.

16.3.22 Average Illuminance Calculation Sheet

16.3.22

GENERAL INFORMATION

Project identification: ______________________________
(Give name of area and/or building and room number)

Average maintained illuminance for design: __ lux or __ footcandles

Luminaire data:

Manufacturer: ______________

Catalog number: ______________

Lamp data:

Type and color: ________

Number per luminaire: ________

Total lumens per luminaire: ________

SELECTION OF COEFFICIENT OF UTILIZATION

Step 1: Fill in sketch at right

Step 2: Determine Cavity Ratios

Room Cavity Ratio, RCR = ________

Ceiling Cavity Ratio, CCR = ________

Floor Cavity Ratio, FCR = ________

Step 3: Obtain Effective Ceiling Cavity Reflectance (ρ_{CC}) ρ_{CC} = ____

Step 4: Obtain Effective Floor Cavity Reflectance (ρ_{FC}) ρ_{FC} = ____

Step 5: Obtain Coefficient of Utilization (CU) from Manufacturer's Data CU = ____

SELECTION OF LIGHT-LOSS FACTORS

Nonrecoverable		Recoverable	
Luminaire ambient temperature	____	Room surface dirt depreciation RSDD	____
Voltage to luminaire	____	Lamp lumen depreciation LLD	____
Ballast factor	____	Lamp burnouts factor LBO	____
Luminaire surface depreciation	____	Luminaire dirt depreciation LDD	____

Total light-loss factor, LLF (product of individual factors above) = ______

CALCULATIONS

(Average Maintained Illuminance)

$$\text{Number of Luminaires} = \frac{(\text{Illuminance}) \times (\text{Area})}{(\text{Lumens per Luminaire}) \times (\text{CU}) \times (\text{LLF})}$$

= ______________________ =

$$\text{Illuminance} = \frac{(\text{Number of Luminaires}) \times (\text{Lumens per Luminaire}) \times (\text{CU}) \times (\text{LLF})}{(\text{Area})}$$

= ______________________ =

Calculated by: ______________________ Date: ____________

(Reprinted from the IESNA Lighting Handbook, 9th Edition, 2000, Courtesy of the Illuminating Engineering Society of North America.)

Section

17

Hazardous (Classified) Locations

17.1.0 Introduction
17.1.1 Table Summary Classification of Hazardous Atmospheres (*NEC* Articles 500 through 516)
17.1.2 Classification of Hazardous Atmospheres
17.1.3 Prevention of External Ignition and Explosion
17.1.4 Equipment for Hazardous Areas
17.1.5 Wiring Methods and Materials
17.1.6 Maintenance Principles
17.1.7 Gases and Vapors: Hazardous Substances Used in Business and Industry
17.1.8 Dusts: Hazardous Substances Used in Business and Industry
17.1.9 *NEC* Table 500.8(B), Classification of Maximum Surface Temperature
17.1.10 *NEC* Table 500.8(C)(2), Class II Ignition Temperatures
17.1.11 *NEC* Article 505, Class I, Zone 0, 1, and 2 Locations
17.1.12 *NEC* Article 511, Commercial Garages, Repair and Storage
17.1.13 *NEC* Article 513, Aircraft Hangars
17.1.14 *NEC* Article 514, Motor Fuel Dispensing Facilities
17.1.15 *NEC* Article 515, Bulk Storage Plants
17.1.16 *NEC* Article 516, Spray Application, Dipping, and Coating Processes
17.1.17 Installation Diagram for Sealing
17.1.18 Diagram for Class 1, Zone 1 Power and Lighting Installation
17.1.19 Diagram for Class 1, Division 1 Lighting Installation
17.1.20 Diagram for Class 1, Division 1 Power Installation
17.1.21 Diagram for Class 1, Division 2 Power and Lighting Installation
17.1.22 Diagram for Class 2 Lighting Installation
17.1.23 Diagram for Class 2 Power Installation
17.1.24 Crouse-Hinds "Quick Selector": Electrical Equipment for Hazardous Locations
17.1.25 Worldwide Explosion Protection Methods, Codes, Categories, Classifications, and Testing Authorities

17.1.0 Introduction

This chapter contains selected key abstracts of *NEC* Articles 500 through 516 requirements to meet the needs of design personnel for a ready reference to equipment and installation ideas in hazardous locations.

Diagrams of recommended power and lighting installations have been included to assist those involved in the design of such systems for hazardous locations. Tables included are those of most frequent applicability and use. Many of the diagrams, illustrations, examples, equipment and component designations, etc. are those of Cooper Crouse-Hinds, by way of example as a leading industry manufacturer of hazardous equipment and components.

The *NEC,* which is widely used for classification purposes, divides atmospheric explosion hazards into three broad classes. However, it must be understood that considerable skill and judgment must be applied when deciding to what degree an area contains hazardous concentrations of vapors, combustible dusts, or easily ignitible fibers and flyings. Many factors such as temperature, barometric pressure, quantity of release, humidity, ventilation, distance from the vapor source, specific gravity, etc. must be considered. When information on all factors concerned is properly evaluated, a consistent classification for the selection and location of electrical equipment can be developed.

Many of the illustrations which follow are based on the 1999 *NEC*. These have been verified to conform to the newly released 2002 edition of the *NEC*. The only variation from the 2002 edition of the *NEC* may be that illustrations may not reflect the dimensions in their metric equivalent.

17.1.1 Table Summary Classification of Hazardous Atmospheres (*NEC* Articles 500 through 504) (*see pages 17.3–17.4*)

17.1.2 Classification of Hazardous Atmospheres

Definitions. The definitions that follow are extracted from the 2002 *National Electrical Code (NEC)*. These definitions are included to define and clarify the meaning and intent of such terms as *explosion-proof* and *dust-ignition-proof.*

Explosion-proof apparatus Apparatus enclosed in a case that is capable of withstanding an explosion of a specified gas or vapor surrounding the enclosure by sparks, flashes, or explosion of the gas or vapor within and that operates at such an external temperature that a surrounding flammable atmosphere will not be ignited thereby. For further information, see Explosion-Proof and Dust-Ignition-Proof Electrical Equipment for use in Hazardous (Classified) Locations, ANSI/UL 1203-1994.

Dust-ignition-proof As used in this Article, shall mean enclosed in a manner that will exclude dusts and, where installed and protected in accordance with this Code, will not permit arcs, sparks, or heat otherwise generated or liberated inside the enclosure to cause ignition of exterior accumulations or atmospheric suspensions of a specified dust on or in the vicinity of the enclosure. For further information on dust-ignition-proof enclosures, see Type 9 enclosure in Enclosures for Electrical Equipment, ANSI/NEMA 250-1991, and Explosion-Proof and Dust-Ignition-Proof Electrical Equipment for Hazardous (Classified) Locations, ANSI/UL 1203-1994.

For further information on classification of areas, see NFPA 497, Classification of Flammable Liquids, Gases or Vapors and of Hazardous (Classified) Locations for Electrical Installations in Chemical Process Areas, 1997, and NFPA 499, Classification of Combustible Dusts and of Hazardous (Classified) Locations for Electrical Installations in Chemical Process Areas, 1997.

The *National Electrical Code* and Underwriters Laboratories, Inc. The *NEC* is considered the definitive classification tool and contains explanatory data about flammable gases and combustible dusts as they may apply to storage areas, garages, or gasoline stations. Specific installation practices have been set up for heavier-than-air vapors. However, there are no such specific data for acetylene, whose vapor density is very near that of air. In the case of hydrogen, which has a low vapor density and is used indoors, the most hazardous concentrations are likely to be in the upper portion of the room.

Many states, municipalities, and public service companies use the *NEC* as a requirement for their inspectors.

Underwriters Laboratories, Inc. (UL), is a not-for-profit independent organization testing for public safety. Its function is to determine whether or not devices and equipment submitted to it are safe and can be used in the *NEC* category for which they were designed. To do this, it maintains extensive laboratory and testing facilities.

TABLE 17.1.1 Summary of Hazardous Atmospheres

Class	Division	Group	Typical Atmosphere/Ignition Temps.	Devices Covered	Temperature Measured	Limiting Value
I Gases, vapors	1 Normally hazardous	A	Acetylene (305C, 581F)	All electrical devices and wiring	Maximum external temperature in 40C ambient	Shall not exceed the ignition temp. of the specific gas or vapor (See 500-5 (e) of NEC)
		B	1,3-Butadiene[1] (420C, 788F) Ethylene Oxide[2] (429C, 804F) Fuel and Combustible Process Gas (containing more than 30 percent H_2 by volume) Hydrogen (520C, 968F) Manufactured Gas (see Fuel and Combustible Process Gas) Propylene Oxide[2] (449C, 840F)			
		C	Acetaldehyde (175C, 347F) Diethyl Ether (160C, 320F) Ethylene (450C, 842F) Unsymmetrical Dimethyl Hydrazine (UDMH) (249C, 480F)			
		D	Acetone (465C, 869F) Acrylonitrile (481C, 898F) Ammonia[3] (498C, 928F) Benzene (498C, 928F) Butane (288C, 550F) 1-Butanol (343C, 650F) 2-Butanol (405C, 761F) n-Butyl Acetate (421C, 790F) Cyclopropane (503C, 938F) Ethane (472C, 882F) Ethanol (363C, 685F) Ethyl Acetate (427C, 800F) Ethylene Dichloride (413C, 775F) Gasoline (280-471C, 536-880F) Heptane (204C, 399F) Hexane (225C, 437F) Isoamyl Alcohol (350C, 662F) Isoprene (220C, 428F) Methane (537C, 999F) Methanol (385C, 725F) Methyl Ethyl Ketone (404C, 759F) Methyl Isobutyl Ketone (440C, 840F) 2-Methyl-1-Propanol (416C, 780F) 2-Methyl-2-Propanol (478C, 892F) Naphtha (Petroleum)[4] (288C, 550F) Octane (206C, 403F) Pentane (243C, 470F) 1-Pentanol (300C, 572F) Propane (450C, 842F) 1-Propanol (413C, 775F) 2-Propanol (399C, 750F) Propylene (455C, 851F) Styrene (490C, 914F) Toluene (480C, 896F) Vinyl Acetate (402C, 756F) Vinyl Chloride (472C, 882F) Xylenes (464-529C, 867-984F)			

(continued)

TABLE 17.1.1 Summary of Hazardous Atmospheres (*Continued*)

Class	Division	Group	Typical Atmosphere/Ignition Temps.	Devices Covered	Temperature Measured	Limiting Value
I Gases, vapors	2 Not normally hazardous	A B C D	Same as Division 1 Same as Division 1 Same as Division 1 Same as Division 1	Lamps, resistors, coils, etc., other than arcing devices. (see Div. 1)	Max. internal or external temp. in 40°C ambient	Same as Division 1
II Combustible dusts	1 Normally hazardous	E	Atmospheres containing combustible metal dusts, or other combustible dusts of similarly hazardous characteristics	All electrical equipment	Max. external temp. in 40C ambient with a dust blanket	Shall not exceed the ignition temperature of the specific dust (See 500-5(f) of NEC)
		F	Atmospheres containing combustible carbonaceous dusts			Also shall not exceed 165°C for Group G dusts that may dehydrate or carbonize
		G	Atmospheres containing combustible dusts not included in Group E or F, including flour, grain, wood, plastic and chemicals			
	2 Not normally hazardous	F	Same as Division 1	All electrical equipment	Max. external temp. under conditions of use	Same as Division 1
		G	Same as Division 1			
III Easily ignitible fibers and flyings	1 & 2			Lighting fixtures	Max. external temp. under conditions of use	165C (329F)

[1]*Group D equipment may be used for this atmosphere if such equipment is isolated in accordance with Section 501-5(a) by sealing all conduit 1/2-inch size or larger.*

[2]*Group C equipment may be used for this atmosphere if such equipment is isolated in accordance with Section*
[3]*For classification of areas involving ammonia atmosphere, see Safety Code for Mechanical Refrigeration (ANSI/ASHRAE 15-1989) and Safety Requirements for the Storage and Handling of Anhydrous Ammonia (ANSI/CGA G2.1-1981).*

[4]*A saturated hydrocarbon mixture boiling in the range 20 – 135°C (68 – 275°F). Also known by the synonyms benzene, ligroin, petroleum ether, or naphtha.*

† *For a more complete list of flammable liquids, gases, and solids, see NFPA 497, NFPA499 and Appendix I and II.*

**For alternative area classification system utilizing zones and their corresponding groups (Article 505) please turn to page 17.19.*

UL's function does not include actual enforcement of the *NEC*. However, as indicated previously, inspection authorities use UL's listing criteria in carrying out their inspections of hazardous areas.

Class I Atmospheric Hazards. Class I atmospheric hazards are divided not only into the four groups shown in Table 17.1.1 but also into two divisions. Division 1 covers locations where flammable gases and vapors may exist under normal operating conditions, under frequent repair or maintenance operations, or where breakdown or faulty operation of process equipment also might cause simultaneous failure of electrical equipment.

Division 2 covers locations where flammable gases, vapors, or volatile liquids are handled either in a closed system or confined within suitable enclosures or where hazardous concentrations normally are prevented by positive mechanical ventilation. Areas adjacent to Division 1 locations, into which gases might flow occasionally, would also be Division 2.

An alternate "zone classification" system was adopted into the *NEC*. The new method of classification using zones is for locations classified in accordance with the International Electrotechnical Commission (IEC), defined under *NEC* Article 505.

Class II Atmospheric Hazards. Class II atmospheric hazards cover three groups of combustible dusts. The groupings are based on the type of material: metallic, carbonaceous, or organic. Whether an area is Division 1 or 2 depends on the quantity of dust present, except that for group E, there is only Division 1.

Class III Atmospheric Hazards. Class III atmospheric hazards cover locations where combustible fibers or flyings are present but not likely to be in suspension in air in quantities sufficient to produce ignitible mixtures. Division 1 is where they are manufactured, and Division 2 is where they are stored.

17.1.3 Prevention of External Ignition and Explosion

Sources of Ignition. A source of energy is all that is needed to touch off an explosion when flammable gases or combustible dusts are mixed in the proper proportion with air.

One prime source of energy is electricity. Equipment such as switches, circuit breakers, motor starters, pushbutton stations, or plugs and receptacles can produce arcs or sparks in normal operation when contacts are opened and closed. This could easily cause ignition.

Other hazards are devices that produce heat, such as lighting fixtures and motors. Here, surface temperatures may exceed the safe limits of many flammable atmospheres.

Finally, many parts of the electrical system can become potential sources of ignition in the event of insulation failure. This group would include wiring (particularly splices in the wiring), transformers, impedance coils, solenoids, and other low-temperature devices without make-or-break contacts.

Nonelectrical hazards such as sparking metal also can easily cause ignition. A hammer, file, or other tool that is dropped on masonry or on a ferrous surface is thus a hazard unless the tool is made of nonsparking material. For this reason, portable electrical equipment is usually made from aluminum or other material that will not produce sparks if the equipment is dropped.

Electrical safety, therefore, is of crucial importance. The electrical installation must prevent accidental ignition of flammable liquids, vapors, and dusts released to the atmosphere. In addition, since much of this equipment is used outdoors or in corrosive atmospheres, the material and finish much be such that maintenance costs and shutdowns are minimized.

Combustion Principles. Three basic conditions must be satisfied for a fire or explosion to occur:

1. A flammable liquid, vapor, or combustible dust must be present in sufficient quantity.
2. The flammable liquid, vapor, or combustible dust must be mixed with air or oxygen in the proportions required to produce an explosive mixture.
3. A source of energy must be applied to the explosive mixture.

In applying these principles, the quantity of the flammable liquid or vapor that may be liberated and its physical characteristics must be recognized.

Vapors from flammable liquids also have a natural tendency to disperse into the atmosphere and rapidly become diluted to concentrations below the lower explosion limit, particularly when there is natural or mechanical ventilation.

The possibility that the gas concentration may be above the upper explosion limit does not afford any degree of safety, since the concentration must first pass through the explosive range to reach the upper explosion limit.

Evaluation of Hazardous Areas. Each area that contains gases or dusts that are considered hazardous must be evaluated carefully to make certain the correct electrical equipment is selected. Many hazardous atmospheres are Class I, Group D, or Class I, Group G. However, certain areas may involve other groups, particularly Class I, Groups B and C. Conformity with the *NEC* requires the use of fittings and enclosures approved for the specific hazardous gas or dust involved.

Enclosures. In Class I, Division 1 and 2 locations, conventional relays, contactors, and switches that have arcing contacts must be enclosed in explosion-proof housings, except for those few cases where general-purpose enclosures are permitted by the *NEC.*

By definition, enclosures for these locations must prevent the ignition of an explosive gas or vapor that may surround it. In other words, an explosion inside the enclosure must be prevented from starting a larger explosion on the outside.

Adequate strength is one requirement for such an enclosure. For explosion-proof equipment, a safety factor of 4 is used; i.e., the enclosure must withstand a hydrostatic pressure test of four times the maximum pressure from an explosion within the enclosure.

In addition to being strong, the enclosure must be *flame tight.* This term does not imply that the enclosure is hermetically sealed but rather that the joints or flanges are held within narrow tolerances. These carefully machined joints cool the hot gases resulting from an internal explosion so that by the time they reach the outside hazardous atmosphere, they are not hot enough to cause ignition.

The strains and stresses caused by internal explosive pressures are illustrated in Figure 1. Dotted lines indicate the shape that a rectangular enclosure strives to attain under these conditions. Openings in an enclosure for these applications can be threaded-joint type (Figure 2) or flat-joint type (Figure 3).

Figure 1

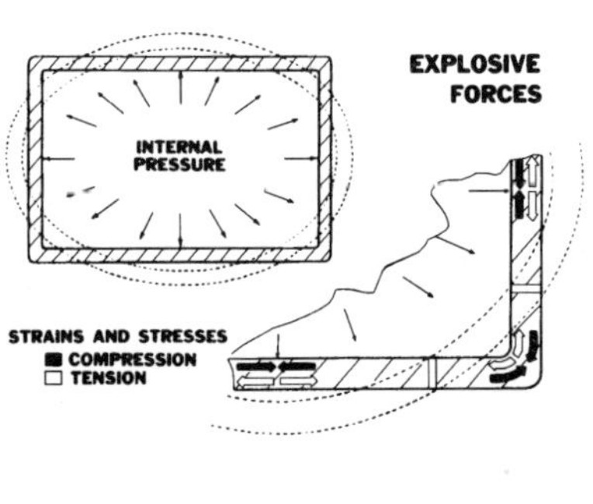

Figure 2

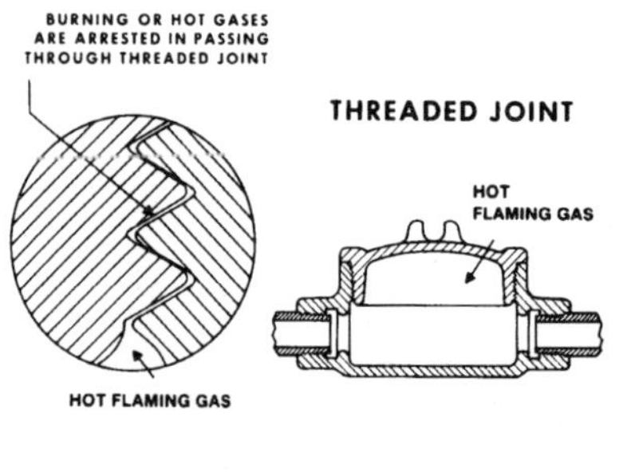

Figure 3

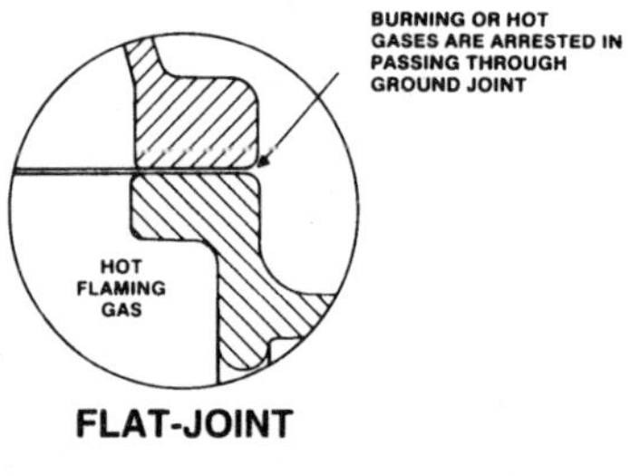

In Class II locations, the enclosure must keep the dust out of the interior and operate at a safe surface temperature. Since there will be no internal explosions, the enclosure may have thinner wall sections. The construction of these enclosures is known as dust-ignition-proof.

Purging/Pressurization Systems. Purging/pressurization systems permit the safe operation of electrical equipment under conditions of hazard for which approved equipment may not be available commercially. For instance, most switchgear units and many large-size motors do not come in designs listed for Class I, Groups A and B.

Whether cast-metal enclosures for hazardous locations or sheet-metal enclosures with pressurization should be used is mainly a question of economics if both types are available. As a typical example, if an installation had many electronic instruments that could be enclosed in a single sheet-metal enclosure, the installation would lend itself to the purging/pressurization system. However, if the instruments, due to their nature, had to be installed in separate enclosures, then a cast-metal hazardous location housing almost invariably would prove more economical.

Pressurized enclosures require

- A source of clean air or inert gas
- A compressor to maintain the required pressure on the system
- Pressure control valves to prevent the power from being applied before the enclosures have been purged and to deenergize the system should pressure fall below a safe value

In addition, door-interlock switches are required to prevent access to the equipment while the circuits are energized. It can readily be seen that all these accessories can add up to a considerable expenditure.

For a detailed description of purging/pressurizing systems, see NFPA 496-1998, "Purged and Pressurized Enclosures for Electrical Equipment."

Intrinsically Safe Equipment. The use of intrinsically safe equipment is limited primarily to process-control instrumentation because these electrical systems lend themselves to low energy requirements. ANSI/UL 913-1997 provides information on the design, testing, and evaluation of this equipment. Installation requirements are covered in *NEC* Article 504. The definition of *intrinsically safe equipment and wiring* is equipment and wiring that are incapable of releasing sufficient electrical or thermal energy under normal or abnormal conditions to cause ignition of a specific hazardous atmospheric mixture in its most easily ignited concentration.

Intrinsically safe energy levels are sufficient for most instruments. This operating energy is supplied from the safe area to the protected instrument. Output from the instrument is returned to a processor back in a nonhazardous location. Preventing increased energy levels such as faults or spikes from the hazardous area, an energy-bleeding interface is used in the circuitry. These devices safely bleed excess energy to an electrical ground.

Underwriters Laboratories, Inc., the Canadian Standards Association, and Factory Mutual list several devices in this category. The equipment and its associated wiring must be installed so that they are positively separated from the non-intrinsically safe circuits. Induced voltages could defeat the concept of intrinsically safe circuits.

17.1.4 Equipment for Hazardous Areas

Switchgear and Industrial Controls. A wide variety of explosion-proof or dust-ignition-proof electrical control equipment is available for Class I and II areas. Explosion-proof pushbutton stations, motor controls, and branch-circuit breakers are suitable for use in both these locations.

In exposed but unclassified areas, industrial controls frequently are installed in cast-metal enclosures selected for maximum protection against corrosion and the weather. Additional coatings and vapor-phase inhibitors enhance this protection.

Lighting Fixtures. Hazardous area lighting is concerned primarily with functional illumination without regard to the symmetry of the installation. The present practice is to classify many lighting areas as Division 2.

While incandescent lighting is still used, the more efficient high-intensity discharge and fluorescent-type fixtures is being specified for most new installations.

Local lighting is required in many areas. If these areas are Class I, Division 1, fixtures suitable for use in these locations must be used. In Class I, Division 2 areas, a fixture specifically designed and tested for this location is frequently used. It is also permitted to use a fixture suitable for Class I, Division 1.

Since lighting fixtures are heat-producing devices, operating temperatures are very important to consider when designing a hazardous location lighting system. Section 500.8(C)(1) of the *NEC* requires the temperature of the fixture to not exceed the ignition temperature of the specific gas or vapor to be encountered. The limits are based on a 40°C (104°F) ambient temperature while the device is operating continuously at full rated load, voltage, and frequency.

When lighting fixtures are used in Class I, Division 2 locations, the *NEC* permits the fixtures to operate up to the ignition temperature of the gas or vapor involved if they have been tested and found incapable of igniting the gas or vapor. If the fixture has not been tested, it must not operate over 80 percent of the ignition temperature of the gas or vapor involved. Information on operating temperatures may be found on the product or obtained from the manufacturer.

Standard fluorescent fixtures generally are used for control room lighting, whereas strategically located floodlights have found wide use in general area lighting for outdoor areas.

In Class II, Division 1, a dust-ignition-proof fixture must be used. The maximum surface temperature of the fixture must be in accordance with Section 500.8(C)(2) when covered with a layer of dust.

In locations where a flammable gas and a combustible dust are present simultaneously, heat-producing equipment such as lighting fixtures must operate safely in the presence of the gas and with a dust blanket. *This rating is quite different from being approved for Class I or II locations only.*

Motors and Generators. Since electric motors are needed to drive pumps, compressors, fans, blowers, and conveyors, their presence in hazardous atmospheres is frequently unavoidable. The selection of the proper type of motor is important, since this has a considerable effect on the initial cost. The types of hazardous atmospheres and the corrosive conditions are both major factors in this selection because they dictate the degree of protection needed to avoid excessive maintenance and expensive shutdowns.

Corrosive and environmental conditions vary between areas in plants; consequently, no single type of motor construction will suffice for all applications. The types available vary all the way from drip-proof to totally enclosed and fan-cooled motors. In Class I, Division 1 locations, only the explosion-proof, totally enclosed and pressurized with clean air, totally enclosed inert-gas-filled, and special submerged-type motors may be used.

It should not be assumed that the motors and controls designed for one group are suited for use in hazardous locations of a different group. Motors for use in Class I, Division 2 locations in which sliding contacts, switching mechanisms, or integral resistance devices are employed also must be explosion-proof or pressurized. Open-type motors such as squirrel-cage induction motors without any arcing devices may be used in Class I, Division 2 locations.

UL has issued a procedure for the repair of listed explosion-proof motors. The manufacturer of the motor should be consulted as to which repair shops have been authorized to make the necessary repairs. Unauthorized maintenance of an explosion-proof motor may result in voiding the manufacturer's liability.

Plugs and Receptacles. In the majority of explosion-proof devices, all the current-carrying parts are inside the enclosure. However, in plugs and receptacles, contact must be made outside the enclosure. The problem is to make such a device safe for use in explosive atmospheres.

Two different methods can be used:

1. *Interlocked, dead front.* Receptacle contacts are interlocked with a switch located in an explosion-proof enclosure. Receptacle contacts will not be live when the plug is inserted or withdrawn.
2. *Delayed action.* Plug and receptacle are so constructed that any electrical arcs that may occur at the contacts will be confined inside the explosion-proof chambers. This design also prevents rapid withdrawal of the plug from the receptacle, thereby giving any heated metal parts or particles time to cool before they come in contact with the surrounding explosive atmosphere.

Both designs are practical and widely used, although the interlocked dead front is prevalent. There is also a wide variety of plugs and receptacles that are suitable for Class II locations.

Portable Devices. The design of portable units for use in hazardous locations must permit ready replacement of approved types of flexible cord when the cord becomes damaged. Hence it is usual to have a separate compartment or connector for the cord connections outside the explosion-proof compartment.

In many plants, the use of portable equipment is restricted as much as possible. When it is used, explosion-proof construction is specified. The *NEC* requires that all portable equipment operated in hazardous locations be grounded by means of a separate grounding in cord approved for extrahard use.

17.1.5 Wiring Methods and Materials

Conduit. In Class I locations, all conduits must be rigid metal or steel IMC with at least five full tapered threads tightly engaged in the enclosure. All factory-drilled and tapped enclosures satisfy this requirement. When field drilling and tapping are performed, it may be required to drill and tap deeper than standard NPT to ensure engagement of five full threads.

A common method of wiring employs thick-walled conduit with a corrosion-resistant finish. In addition to the protective finish on the conduit, various types of paints or special finishes are used extensively to give extra protection from corrosive atmospheres.

Alternate changes in temperature and barometric pressure cause *breathing*—the entry and circulation of air throughout the conduit. Since joints in a conduit system and its components are seldom tight enough to prevent this breathing, moisture in the air condenses and collects at the base of vertical conduit runs and equipment enclosures. This could cause equipment shorts or grounds. To eliminate this condition, inspection fittings should be installed and equipped with explosion-proof drains to automatically drain off the water.

Seals for Conduit System. *NEC* Article 501.5 requires that sealing fittings filled with an approved compound be installed in conduits connecting explosion-proof enclosures. Seals are necessary to limit volume, to prevent an explosion from traveling throughout the conduit system, to block gases or vapors from moving from a hazardous to a nonhazardous area through connecting raceways or from enclosure to enclosure, and to stop pressure piling—the buildup of pressure inside conduit lines caused by precompression as the explosion travels through the conduit.

The standard-type seals are not intended to prevent the passage of liquids, gases, or vapors at pressures continuously above atmospheric. Temperature extremes and highly corrosive liquids and vapors may affect the ability of the seals to perform their intended function.

In hazardous locations, seals are needed in the following instances:

- Where the conduit enters an enclosure that houses arcing or high-temperature equipment. (A seal must be within 450 mm or 18 in of the enclosure it isolates.)
- Where the conduit enters enclosures that house terminals, splices, or taps if the conduit is 2 in or more in diameter.
- Where the conduit leaves a Division 1 area or passes from a Division 2 hazardous area to a nonhazardous location.

NEC Section 501-5(a)(1) permits unions, couplings, reducers, elbows, small GUAs, and OEs to be placed between the seal and the explosion-proof enclosure.

Mineral-Insulated Cable. Another type of wiring system suitable for Division 1 is mineral-insulated (MI) cable. Mineral-insulated wiring consists of copper conductors properly spaced and encased in tightly compressed magnesium oxide and clad in an overall copper sheath.

Below the melting temperature of the copper sheath, MI cable is impervious to fire. Because of limitations on end connections, its operating range is generally considered to be −40 to 80°C with standard terminals and up to 250°C with special terminals.

When installed properly, MI cable is suitable and approved for all Class I and Class II locations.

MI cable is available in 1 to 17 conductors, making it most suitable for wiring of control boards, control components, and instrumentation circuits where crowded conditions make conduit installations difficult and expensive.

MI cable is hygroscopic; therefore, moisture can be a problem when the ends are left exposed. Care must be taken to install and seal the end fittings on the job as soon as possible to prevent moisture accumulation. If moisture enters, the end must be cut off or dried out with a torch.

Metal-Clad Cable. Metal-clad cable (type MC) is permitted by the *NEC* for application in Class I, Division 2 locations. Use of this type of cable is not limited to any voltage class. The armor itself is available in various metals. When further protection from chemical attack is needed, a supplemental protective jacket may be used.

MC cable must be terminated properly because the armor may be used as a grounding conductor.

The *NEC* also permits, under certain restrictions, a particular kind of MC cable to be used in Class I, Division 1 locations. This is detailed in 501.4(A)(1)(c). Similarly, 501.4(A)(1)(d) permits a certain type of instrumentation tray cable (ITC).

Tray Cable. Power and control tray cable (type TC) is permitted in Class I, Division 2 locations. It is a factory assembly of two or more insulated conductors with or without the grounding conductor under the nonmetallic sheath.

Other Permitted Cables. In Class I, Division 1 locations, the *NEC* also recognizes the use of type PLTC, similar to TC except the conductors are limited to no. 22 through no. 16 AWG, as well as type MV, a single or multiconductor solid dielectric insulated cable rated 2001 V or higher. The *NEC* also permits type ITC cable, as covered by Article 727, "Instrumentation Tray Cable," which details its construction and use.

Cable Sealing. In Class I, Division 1 locations, the use of cable, except type MI and certain type MCs and ITCs, is limited to installation in conduit. Multiconductor cables that cannot transmit gases through the cores are sealed as single conductors. If a cable can transmit gases through its core, the outer jacket must be removed so that the sealing compound surrounds each individual insulated conductor and the jacket.

In Class I, Division 2 locations, cables must be sealed where they enter enclosures required to be explosion-proof. EYS seals with appropriate cable terminators are recommended for this application. If the cable core can transmit gases, the outer jacket must be removed so that the sealing compound surrounds each conductor to prevent the passage of gases. Type TMCX fittings are recommended where type MC or ITC cables are used.

Cables without a gas-tight continuous sheath must be sealed at the boundary of the Division 2 and nonhazardous location.

If attached to equipment that may cause pressure at a cable end, a sheathed cable that can transmit gases through its core must be sealed to prevent migration of gases into an unclassified area.

17.1.6 Maintenance Principles

Chapter 5 of the *NEC* requires equipment to be constructed and installed in such a way as to ensure safe performance under conditions of proper use and maintenance. It is important that the following points be checked carefully:

- *Electric circuits.* Electrical equipment should be serviced or disassembled only after deenergizing the electric supply circuits. This also applies when lighting fixtures or units are partially disassembled for relamping. All electrical enclosures should be tightly reassembled before the electric supply circuits are reenergized.
- *Assembly or disassembly of enclosures.* Hammers or prying tools must not be allowed to damage the flat-joint surfaces. Do not handle covers roughly or place them on surfaces that might damage or scratch the flat-joint surfaces. Protect all surfaces that form a part of the flame path from mechanical injury. In storing equipment, always make sure that covers are assembled to their mating bodies.
- *Cover attachment screws.* All cover screws and bolts intended to hold explosion-proof joints firmly together always must be tight while circuits are alive. Leaving screws or bolts loose may render the equipment unsafe. Care should be taken to use only bolts or screws provided by the equipment manufacturer because the substitution of other types of material may weaken the assembly and render it unsafe.

- *Cleaning and lubrication.* Particles of foreign material should not be allowed to accumulate on flat and threaded joints because these materials tend to prevent a close fit and may permit dangerous arcs, sparks, or flames to propagate through them.

When assembling, remove all old grease, dirt, paint, or other foreign material from the surfaces using a brush and kerosene or a similar solvent with a flash point higher than 38°C (100°F). A film of light oil or lubricant of a type recommended by the original equipment manufacturer should be applied to both body and cover joint.

Any lubricated joints exposed for long periods of time may attract small particles of dirt or other foreign material. To avoid this, body and cover joints should be reassembled immediately.

Threaded joints should be tightened sufficiently to prevent accidental loosening due to vibration, but they should not be forced. If the threads are kept clean and lubricated, safe operation can be assured with a minimum of maintenance.

Shaft and Bearing Surfaces. Because a rotating shaft must turn freely, the clearance between the shaft and bearing is established carefully within close tolerances by the equipment manufacturer. This clearance should be maintained to prevent flames or sparks from escaping to the outside hazardous atmosphere. Always follow the manufacturer's recommendations with respect to lubrication and other servicing.

Corrosive Locations. Threaded covers, flat joints, surfaces, rotating shafts, bearings, and operating shafts should be well lubricated. If corrosion products have accumulated on explosion-proof joints or surfaces and cannot be removed readily with solvents, the parts should be discarded and replaced. Never use an abrasive material or a file to remove the corrosion products from threaded or flat-joint surfaces. In extremely corrosive locations, equipment should be inspected periodically to guard against unusual deterioration and possible porosity, since this may weaken the enclosure structurally.

Portable Equipment. The extrahard-use flexible cord that should be used with this equipment must be examined frequently and replaced at the first indication of mechanical damage or deterioration. Terminal connections to the cord must be maintained properly. In general, where portable equipment is necessary, avoid rough handling and inspect the assembly frequently.

Overall Safety. Safety in hazardous locations may be endangered if additional openings or other alterations are made in assemblies specifically designed for use in these locations.

In painting the exterior of housings for hazardous locations, care should be taken not to obscure the nameplate, which may contain cautionary or other information of importance to maintenance personnel.

Plug-In Replacement Units. One technique that speeds and eases the work of the maintenance department is the use of plug-in-type electrical equipment that allows the substitution of a replacement unit while the original unit is being repaired outside the hazardous area.

17.1.7 Gases and Vapors: Hazardous Substances Used in Business and Industry (*see pages 17.13–17.15*)

TABLE 17.1.7

Class I* Group	Substance	Auto-* Ignition Temp. °F	°C	Flash** Point °F	°C	Explosive Limits** Per Cent by Volume Lower	Upper	Vapor** Density (Air Equals 1.0)
C	Acetaldehyde	347	175	-38	-39	4.0	60	1.5
D	Acetic Acid	867	464	103	39	4.0	19.9 @ 200°F	2.1
D	Acetic Anhydride	600	316	120	49	2.7	10.3	3.5
D	Acetone	869	465	-4	-20	2.5	13	2.0
D	Acetone Cyanohydrin	1270	688	165	74	2.2	12.0	2.9
D	Acetonitrile	975	524	42	6	3.0	16.0	1.4
A	Acetylene	581	305	gas	gas	2.5	100	0.9
B(C)	Acrolein (inhibited)[1]	455	235	-15	-26	2.8	31.0	1.9
D	Acrylic Acid	820	438	122	50	2.4	8.0	2.5
D	Acrylonitrile	898	481	32	0	3.0	17	1.8
D	Adiponitrile	—	—	200	93	—	—	—
C	Allyl Alcohol	713	378	70	21	2.5	18.0	2.0
D	Allyl Chloride	905	485	-25	-32	2.9	11.1	2.6
B(C)	Allyl Glycidyl Ether[1]	—	—	—	—	—	—	—
D	Ammonia[2]	928	498	gas	gas	15	28	0.6
D	n-Amyl Acetate	680	360	60	16	1.1	7.5	4.5
D	sec-Amyl Acetate	—	—	89	32	—	—	4.5
D	Aniline	1139	615	158	70	1.3	11	3.2
D	Benzene	928	498	12	-11	1.3	7.9	2.8
D	Benzyl Chloride	1085	585	153	67	1.1	—	4.4
B(D)	1,3-Butadiene[1]	788	420	gas	gas	2.0	12.0	1.9
D	Butane	550	288	-76	-60	1.6	8.4	2.0
D	1-Butanol	650	343	98	37	1.4	11.2	2.6
D	2-Butanol	761	405	75	24	1.7 @ 212°F	9.8 @ 212°F	2.6
D	n-Butyl Acetate	790	421	72	22	1.7	7.6	4.0
D	iso-Butyl Acetate	790	421	—	—	—	—	—
D	sec-Butyl Acetate	—	—	88	31	1.7	9.8	4.0
D	t-Butyl Acetate	—	—	—	—	—	—	—
D	n-Butyl Acrylate (inhibited)	559	293	118	48	1.5	9.9	4.4
C	n-Butyl Formal	—	—	—	—	—	—	—
B(C)	n-Butyl Glycidyl Ether[1]	—	—	—	—	—	—	—
C	Butyl Mercaptan	—	—	35	2	—	—	3.1
D	t-Butyl Toluene	—	—	—	—	—	—	—
D	Butylamine	594	312	10	-12	1.7	9.8	2.5
D	Butylene	725	385	gas	gas	1.6	10.0	1.9
C	n-Butyraldehyde	425	218	-8	-22	1.9	12.5	2.5
D	n-Butyric Acid	830	443	161	72	2.0	10.0	3.0
-[3]	Carbon Disulfide	194	90	-22	-30	1.3	50.0	2.6
C	Carbon Monoxide	1128	609	gas	gas	12.5	74.0	1.0
C	Chloroacetaldehyde	—	—	—	—	—	—	—
D	Chlorobenzene	1099	593	82	28	1.3	9.6	3.9
C	1-Chloro-1-Nitropropane	—	—	144	62	—	—	4.3
D	Chloroprene	—	—	-4	-20	4.0	20.0	3.0
D	Cresol	1038-1110	559-599	178-187	81-86	1.1-1.4	—	—
C	Crotonaldehyde	450	232	55	13	2.1	15.5	2.4
D	Cumene	795	424	96	36	0.9	6.5	4.1
D	Cyclohexane	473	245	-4	-20	1.3	8.0	2.9
D	Cyclohexanol	572	300	154	68	—	—	3.5
D	Cyclohexanone	473	245	111	44	1.1 @ 212°F	9.4	3.4
D	Cyclohexene	471	244	<20	<-7	—	—	2.8
D	Cyclopropane	938	503	gas	gas	2.4	10.4	1.5
D	p-Cymene	817	436	117	47	0.7 @ 212°F	5.6	4.6
C	n-Decaldehyde	—	—	—	—	—	—	—
D	n-Decanol	550	288	180	82	—	—	5.5
D	Decene	455	235	<131	<55	—	—	4.84
D	Diacetone Alcohol	1118	603	148	64	1.8	6.9	4.0
D	o-Dichlorobenzene	1198	647	151	66	2.2	9.2	5.1
D	1,1-Dichloroethane	820	438	22	-6	5.6	—	—
D	1,2-Dichloroethylene	860	460	36	2	5.6	12.8	3.4
C	1,1-Dichloro-1-Nitroethane	—	—	168	76	—	—	5.0
D	1,3-Dichloropropene	—	—	95	35	5.3	14.5	3.8
C	Dicyclopentadiene	937	503	90	32	—	—	—
D	Diethyl Benzene	743-842	395-450	133-135	56-57	—	—	4.6
C	Diethyl Ether	320	160	-49	-45	1.9	36.0	2.6
C	Diethylamine	594	312	-9	-23	1.8	10.1	2.5
C	Diethylaminoethanol	—	—	—	—	—	—	—
C	Diethylene Glycol Monobutyl Ether	442	228	172	78	0.85	24.6	5.6
C	Diethylene Glycol Monomethyl Ether	465	241	205	96	—	—	—
D	Di-isobutyl Ketone	745	396	120	49	0.8 @ 200°F	7.1 @ 200°F	4.9
D	Di-isobutylene	736	391	23	-5	0.8	4.8	3.9
C	Di-isopropylamine	600	316	30	-1	1.1	7.1	3.5
C	N-N-Dimethyl Aniline	700	371	145	63	—	—	4.2
D	Dimethyl Formamide	833	455	136	58	2.2 @ 212°F	15.2	2.5
D	Dimethyl Sulfate	370	188	182	83	—	—	4.4
C	Dimethylamine	752	400	gas	gas	2.8	14.4	1.6
C	1,4-Dioxane	356	180	54	12	2.0	22	3.0
D	Dipentene	458	237	113	45	0.7 @ 302°F	6.1 @ 302°F	4.7
C	Di-n-propylamine	570	299	63	17	—	—	3.5
C	Dipropylene Glycol Methyl Ether	—	—	186	86	—	—	5.11
D	Dodecene	491	255	—	—	—	—	—
C	Epichlorohydrin	772	411	88	31	3.8	21.0	3.2
D	Ethane	882	472	gas	gas	3.0	12.5	1.0

(continued)

TABLE 17.1.7 (*Continued*)

Class I* Group	Substance	Auto-* Ignition Temp. °F	Auto-* Ignition Temp. °C	Flash** Point °F	Flash** Point °C	Explosive Limits** Per Cent by Volume Lower	Explosive Limits** Per Cent by Volume Upper	Vapor** Density (Air Equals 1.0)
D	Ethanol	685	363	55	13	3.3	19	1.6
D	Ethyl Acetate	800	427	24	-4	2.0	11.5	3.0
D	Ethyl Acrylate (inhibited)	702	372	50	10	1.4	14	3.5
D	Ethyl sec-Amyl Ketone	—	—	—	—	—	—	—
D	Ethyl Benzene	810	432	70	21	0.8	6.7	3.7
D	Ethyl Butanol	—	—	—	—	—	—	—
D	Ethyl Butyl Ketone	—	—	115	46	—	—	4.0
D	Ethyl Chloride	966	519	-58	-50	3.8	15.4	2.2
D	Ethyl Formate	851	455	-4	-20	2.8	16.0	2.6
D	2-Ethyl Hexanol	448	231	164	73	0.88	9.7	4.5
D	2-Ethyl Hexyl Acrylate	485	252	180	82	—	—	—
C	Ethyl Mercaptan	572	300	<0	<-18	2.8	18.0	2.1
C	n-Ethyl Morpholine	—	—	—	—	—	—	—
C	2-Ethyl-3-Propyl Acrolein	—	—	155	68	—	—	4.4
D	Ethyl Silicate	—	—	125	52	—	—	7.2
D	Ethylamine	725	385	<0	<-18	3.5	14.0	1.6
C	Ethylene	842	450	gas	gas	2.7	36.0	1.0
D	Ethylene Chlorohydrin	797	425	140	60	4.9	15.9	2.8
D	Ethylene Dichloride	775	413	56	13	6.2	16	3.4
C	Ethylene Glycol Monobutyl Ether	460	238	143	62	1.1 @ 200°F	12.7 @ 275°F	4.1
C	Ethylene Glycol Monobutyl Ether Acetate	645	340	160	71	0.88 @ 200°F	8.54 @ 275°F	—
C	Ethylene Glycol Monoethyl Ether	455	235	110	43	1.7 @ 200°F	15.6 @ 200°F	3.0
C	Ethylene Glycol Monoethyl Ether Acetate	715	379	124	52	1.7	—	4.72
D	Ethylene Glycol Monomethyl ether	545	285	102	39	1.8 @ STP	14 @ STP	2.6
B(C)	Ethylene Oxide[1]	804	429	-20	-28	3.0	100	1.5
D	Ethylenediamine	725	385	104	40	2.5	12.0	2.1
C	Ethylenimine	608	320	12	-11	3.3	54.8	1.5
C	2-Ethylhexaldehyde	375	191	112	44	0.85 @ 200°F	7.2 @ 275°F	4.4
B	Formaldehyde (Gas)	795	429	gas	gas	7.0	73	1.0
D	Formic Acid (90%)	813	434	122	50	18	57	1.6
B	Fuel and Combustible Process Gas (containing more than 30 percent H_2 by volume)	—	—	—	—	—	—	—
D	Fuel Oils	410-765	210-407	100-336	38-169	0.7	5	—
C	Furfural	600	316	140	60	2.1	19.3	3.3
C	Furfuryl Alcohol	915	490	167	75	1.8	16.3	3.4
D	Gasoline	536-880	280-471	-36 to -50	-38 to -46	1.2-1.5	7.1-7.6	3-4
D	Heptane	399	204	25	-4	1.05	6.7	3.5
D	Heptene	500	260	<32	<0	—	—	3.39
D	Hexane	437	225	-7	-22	1.1	7.5	3.0
D	Hexanol	—	—	145	63	—	—	3.5
D	2-Hexanone	795	424	77	25	—	8	3.5
D	Hexenes	473	245	<20	<-7	—	—	3.0
D	sec-Hexyl Acetate	—	—	—	—	—	—	—
C	Hydrazine	74-518	23-270	100	38	2.9	9.8	1.1
B	Hydrogen	968	520	gas	gas	4.0	75	0.1
C	Hydrogen Cyanide	1000	538	0	-18	5.6	40.0	0.9
C	Hydrogen Selenide	—	—	—	—	—	—	—
C	Hydrogen Sulfide	500	260	gas	gas	4.0	44.0	1.2
D	Isoamyl Acetate	680	360	77	25	1.0 @ 212°F	7.5	4.5
D	Isoamyl Alcohol	662	350	109	43	1.2	9.0 @ 212°F	3.0
D	Isobutyl Acrylate	800	427	86	30	—	—	4.42
C	Isobutyraldehyde	385	196	-1	-18	1.6	10.6	2.5
C	Isodecaldehyde	—	—	185	85	—	—	5.4
C	Iso-octyl Alcohol	—	—	180	82	—	—	—
C	Iso-octyl Aldehyde	387	197	—	—	—	—	—
D	Isophorone	860	460	184	84	0.8	3.8	—
D	Isoprene	428	220	-65	-54	1.5	8.9	2.4
D	Isopropyl Acetate	860	460	35	2	1.8 @ 100°F	8	3.5
D	Isopropyl Ether	830	443	-18	-28	1.4	7.9	3.5
C	Isopropyl Glycidyl Ether	—	—	—	—	—	—	—
D	Isopropylamine	756	402	-35	-37	—	—	2.0
D	Kerosene	410	210	110-162	43-72	0.7	5	—
D	Liquefied Petroleum Gas	761-842	405-450	—	—	—	—	—
	Manufactured Gas (see Fuel and Combustible Process Gas)	—	—	—	—	—	—	—
D	Mesityl Oxide	652	344	87	31	1.4	7.2	3.4
D	Methane	999	537	gas	gas	5.0	15.0	0.6
D	Methanol	725	385	52	11	6.0	36	1.1
D	Methyl Acetate	850	454	14	-10	3.1	16	2.8
D	Methyl Acrylate	875	468	27	-3	2.8	25	3.0
D	Methyl Amyl Alcohol	—	—	106	41	1.0	5.5	—
D	Methyl n-Amyl Ketone	740	393	102	39	1.1 @ 151°F	7.9 @ 250°F	3.9
C	Methyl Ether	662	350	gas	gas	3.4	27.0	1.6
D	Methyl Ethyl Ketone	759	404	16	-9	1.7 @ 200°F	11.4 @ 200°F	2.5
D	2-Methyl-5-Ethyl Pyridine	—	—	155	68	1.1	6.6	4.2
C	Methyl Formal	460	238	—	—	—	—	—
D	Methyl Formate	840	449	-2	-19	4.5	23	2.1

(continued)

TABLE 17.1.7 (*Continued*)

Class I* Group	Substance	Auto-* Ignition Temp. °F	°C	Flash** Point °F	°C	Explosive Limits** Per Cent by Volume Lower	Upper	Vapor** Density (Air Equals 1.0)
D	Methyl Isobutyl Ketone	840	440	64	18	1.2 @ 200°F	8.0 @ 200°F	**3.5**
D	Methyl Isocyanate	994	534	19	-7	5.3	26	**1.97**
C	Methyl Mercaptan	—	—	—	—	3.9	21.8	**1.7**
D	Methyl Methacrylate	792	422	50	10	1.7	8.2	**3.6**
D	2-Methyl-1-Propanol	780	416	82	28	1.7 @ 123°F	10.6 @ 202°F	**2.6**
D	2-Methyl-2-Propanol	892	478	52	11	2.4	8.0	**2.6**
D	alpha-Methyl Styrene	1066	574	129	54	1.9	6.1	—
C	Methylacetylene	—	—	gas	gas	1.7	—	1.4
C	Methylacetylene-Propadiene (stabilized)	—	—	—	—	—	—	—
D	Methylamine	806	430	gas	gas	4.9	20.7	1.0
D	Methylcyclohexane	482	250	25	-4	1.2	6.7	3.4
D	Methylcyclohexanol	565	296	149	65	—	—	3.9
D	o-Methylcyclohexanone	—	—	118	48	—	—	3.9
D	Monoethanolamine	770	410	185	85	—	—	2.1
D	Monoisopropanolamine	705	374	171	77	—	—	2.6
C	Monomethyl Aniline	900	482	185	85	—	—	3.7
C	Monomethyl Hydrazine	382	194	17	-8	2.5	92	1.6
C	Morpholine	590	310	98	37	1.4	11.2	3.0
D	Naphtha (Coal Tar)	531	277	107	42	—	—	—
D	Naphtha (Petroleum)[4]	550	288	<0	<-18	1.1	5.9	2.5
D	Nitrobenzene	900	482	190	88	1.8 @ 200°F	—	4.3
C	Nitroethane	778	414	82	28	3.4	—	2.6
C	Nitromethane	785	418	95	35	7.3	—	2.1
C	1-Nitropropane	789	421	96	36	2.2	—	3.1
C	2-Nitropropane	802	428	75	24	2.6	11.0	3.1
D	Nonane	401	205	88	31	0.8	2.9	4.4
D	Nonene	—	—	78	26	—	—	4.35
D	Nonyl Alcohol	—	—	165	74	0.8 @ 212°F	6.1 @ 212°F	5.0
D	Octane	403	206	56	13	1.0	6.5	3.9
D	Octene	446	230	70	21	—	—	3.9
D	n-Octyl Alcohol	—	—	178	81	—	—	4.5
D	Pentane	470	243	<-40	<-40	1.5	7.8	2.5
D	1-Pentanol	572	300	91	33	1.2	10.0 @ 212°F	3.0
D	2-Pentanone	846	452	45	7	1.5	8.2	3.0
D	1-Pentene	527	275	0	-18	1.5	8.7	2.4
D	Phenylhydrazine	—	—	190	88	—	—	—
D	Propane	842	450	gas	gas	2.1	9.5	1.6
D	1-Propanol	775	413	74	23	2.2	13.7	2.1
D	2-Propanol	750	399	53	12	2.0	12.7 @ 200°F	2.1
D	Propiolactone	—	—	165	74	2.9	—	2.5
C	Propionaldehyde	405	207	-22	-30	2.6	17	2.0
D	Propionic Acid	870	466	126	52	2.9	12.1	2.5
D	Propionic Anhydride	545	285	145	63	1.3	9.5	4.5
D	n-Propyl Acetate	842	450	55	13	1.7 @ 100°F	8	3.5
C	n-Propyl Ether	419	215	70	21	1.3	7.0	3.53
B	Propyl Nitrate	347	175	68	20	2	100	—
D	Propylene	851	455	gas	gas	2.0	11.1	1.5
D	Propylene Dichloride	1035	557	60	16	3.4	14.5	3.9
B(C)	Propylene Oxide[1]	840	449	-35	-37	2.3	36	2.0
D	Pyridine	900	482	68	20	1.8	12.4	2.7
D	Styrene	914	490	88	31	0.9	6.8	3.6
C	Tetrahydrofuran	610	321	6	-14	2.0	11.8	2.5
D	Tetrahydronaphthalene	725	385	160	71	0.8 @ 212°F	5.0 @ 302°F	4.6
C	Tetramethyl Lead	—	—	100	38	—	—	6.5
D	Toluene	896	480	40	4	1.1	7.1	3.1
D	Tridecene	—	—	—	—	—	—	—
C	Triethylamine	480**	249**	16	-9	1.2	8.0	3.5
D	Triethylbenzene	—	—	181	83	—	—	5.6
D	Tripropylamine	—	—	105	41	—	—	4.9
D	Turpentine	488	253	95	35	0.8	—	—
D	Undecene	—	—	—	—	—	—	—
C	Unsymmetrical Dimethyl Hydrazine (UDMH)	480	249	5	-15	2	95	2.0
C	Valeraldehyde	432	222	54	12	—	—	3.0
D	Vinyl Acetate	756	402	18	-8	2.6	13.4	3.0
D	Vinyl Chloride	882	472	-108.4	-78	3.6	33.0	2.2
D	Vinyl Toluene	921	494	127	53	0.8	11.0	4.1
D	Vinylidene Chloride	1058	570	-19	-28	6.5	15.5	3.4
D	Xylenes	867-984	464-529	81-90	27-32	1.0-1.1	7.0	3.7

[1] If equipment is isolated by sealing all conduit 1/2 in. or larger, in accordance with Section 501.5(A) of NFPA 70, *National Electrical Code,* equipment for the group classification shown in parentheses is permitted.

[2] For classification of areas involving Ammonia, see *Safety Code for Mechanical Refrigeration,* ANSI/ASHRAE 15, and *Safety Requirements for the Storage and Handling of Anhydrous Ammonia,* ANSI/CGA G2.1.

[3] Certain chemicals may have characteristics that require safeguards beyond those required for any of the above groups. Carbon disulfide is one of these chemicals because of its low autoignition temperature and the small joint clearance to arrest its flame propagation.

[4] Petroleum Naphtha is a saturated hydrocarbon mixture whose boiling range is 20° to 135°C. It is also known as benzine, ligroin, petroleum ether, and naphtha.

*Data from NFPA 497M-1991, *Classification of Gases, Vapors, and Dusts for Electrical Equipment in Hazardous (Classified) Locations.*

**Data from NFPA 325M-1991, *Fire Hazard Properties of Flammable Liquids, Gases, and Volatile Solids.*

17.1.8 Dusts: Hazardous Substances Used in Business and Industry

TABLE 17.1.8

Class II, Group E	Minimum Cloud or Layer Ignition Temp.[1]		
Material[2]	°F		°C
Aluminum, atomized collector fines	1022	CI	550
Aluminum, A422 flake	608		320
Aluminum — cobalt alloy (60-40)	1058		570
Aluminum — copper alloy (50-50)	1526		830
Aluminum — lithium alloy (15% Li)	752		400
Aluminum — magnesium alloy (Dowmetal)	806	CI	430
Aluminum — nickel alloy (58-42)	1004		540
Aluminum — silicon alloy (12% Si)	1238	NL	670
Boron, commercial-amorphous (85% B)	752		400
Calcium Silicide	1004		540
Chromium, (97%) electrolytic, milled	752		400
Ferromanganese, medium carbon	554		290
Ferrosilicon (88%, 9% Fe)	1472		800
Ferrotitanium (19% Ti, 74.1% Fe, 0.06% C)	698	CI	370
Iron, 98%, H_2 reduced	554		290
Iron, 99%, Carbonyl	590		310
Magnesium, Grade B, milled	806		430
Manganese	464		240
Silicon, 96%, milled	1436	CI	780
Tantalum	572		300
Thorium, 1.2%, O_2	518	CI	270
Tin, 96%, atomized (2% Pb)	806		430
Titanium, 99%	626	CI	330
Titanium Hydride, (95% Ti, 3.8% H_2)	896	CI	480
Vanadium, 86.4%	914		490
Zirconium Hydride, (93.6% Zr, 2.1% H_2)	518		270
Class II, Group F			
CARBONACEOUS DUSTS			
Asphalt, (Blown Petroleum Resin)	950	CI	510
Charcoal	356		180
Coal, Kentucky Bituminous	356		180
Coal, Pittsburgh Experimental	338		170
Coal, Wyoming	—		—
Gilsonite	932		500
Lignite, California	356		180
Pitch, Coal Tar	1310	NL	710
Pitch, Petroleum	1166	NL	630
Shale, Oil	—		—
Class II, Group G			
AGRICULTURAL DUSTS			
Alfalfa Meal	392		200
Almond Shell	392		200
Apricot Pit	446		230
Cellulose	500		260
Cherry Pit	428		220
Cinnamon	446		230
Citrus Peel	518		270
Cocoa Bean Shell	698		370
Cocoa, natural, 19% fat	464		240
Coconut Shell	428		220
Corn	482		250
Corncob Grit	464		240
Corn Dextrine	698		370
Cornstarch, commercial	626		330
Cornstarch, modified	392		200
Cork	410		210
Cottonseed Meal	392		200
Cube Root, South Amer.	446		230
Flax Shive	446		230
Garlic, dehydrated	680	NL	360
Guar Seed	932	NL	500
Gum, Arabic	500		260
Gum, Karaya	464		240
Gum, Manila (copal)	680	CI	360
Gum, Tragacanth	500		260
Hemp Hurd	428		220
Lycopodium	590		310
Malt Barley	482		250
Milk, Skimmed	392		200
Pea Flour	500		260
Peach Pit Shell	410		210
Peanut Hull	410		210
Peat, Sphagnum	464		240
Pecan Nut Shell	410		210

Class II, Group G (cont'd)	Minimum Cloud or Layer Ignition Temp.		
	°F		°C
Pectin	392		200
Potato Starch, Dextrinated	824	NL	440
Pyrethrum	410		210
Rauwolfia Vomitoria Root	446		230
Rice	428		220
Rice Bran	914	NL	490
Rice Hull	428		220
Safflower Meal	410		210
Soy Flour	374		190
Soy Protein	500		260
Sucrose	662	CI	350
Sugar, Powdered	698	CI	370
AGRICULTURAL DUSTS			
Tung, Kernels, Oil-Free	464		240
Walnut Shell, Black	428		220
Wheat	428		220
Wheat Flour	680		360
Wheat Gluten, gum	968	NL	520
Wheat Starch	716	NL	380
Wheat Straw	428		220
Woodbark, Ground	482		250
Wood Flour	500		260
Yeast, Torula	500		260
CHEMICALS			
Acetoacetanilide	824	M	440
Acetoacet-p-phenetidide	1040	NL	560
Adipic Acid	1022	M	550
Anthranilic Acid	1076	M	580
Aryl-nitrosomethylamide	914	NL	490
Azelaic Acid	1130	M	610
2,2-Azo-bis-butyronitrile	662		350
Benzoic Acid	824	M	440
Benzotriazole	824	M	440
Bisphenol-A	1058	M	570
Chloroacetoacetanilide	1184	M	640
Diallyl Phthalate	896	M	480
Dicumyl Peroxide (suspended on $CaCO_3$), 40-60	356		180
Dicyclopentadiene Dioxide	788	NL	420
Dihydroacetic Acid	806	NL	430
Dimethyl Isophthalate	1076	M	580
Dimethyl Terephthalate	1058	M	570
3,5 - Dinitrobenzoic Acid	860	NL	460
Dinitrotoluamide	932	NL	500
Diphenyl	1166	M	630
Ditertiary Butyl Paracresol	878	NL	470
Ethyl Hydroxyethyl Cellulose	734	NL	390
Fumaric Acid	968	M	520
Hexamethylene Tetramine	770	S	410
Hydroxyethyl Cellulose	770	NL	410
Isotoic Anhydride	1292	NL	700
Methionine	680		360
Nitrosoamine	518	NL	270
Para-oxy-benzaldehyde	716	CI	380
Paraphenylene Diamine	1148	M	620
Paratertiary Butyl Benzoic Acid	1040	M	560
Pentaerythritol	752	M	400
Phenylbetanaphthylamine	1256	NL	680
Phthalic Anydride	1202	M	650
Phthalimide	1166	M	630
Salicylanilide	1130	M	610
Sorbic Acid	860		460
Stearic Acid, Aluminum Salt	572		300
Stearic Acid, Zinc Salt	950	M	510
Sulfur	428		220
Terephthalic Acid	1250	NL	680
DRUGS			
2-Acetylamino-5-nitrothiazole	842		450
2-Amino-5-nitrothiazole	860		460
Aspirin	1220	M	660
Gulasonic Acid, Diacetone	788	NL	420
Mannitol	860	M	460
Nitropyridone	806	M	430
1-Sorbose	698	M	370
Vitamin B1, mononitrate	680	NL	360
Vitamin C (Ascorbic Acid)	536		280

(continued)

TABLE 17.1.8 (*Continued*)

Class II, Group G (cont'd)	Minimum Cloud or Layer Ignition Temp.[1] °F		°C
DYES, PIGMENTS, INTERMEDIATES			
Beta-naphthalene-azo-Dimethylaniline	347		175
Green Base Harmon Dye	347		175
Red Dye Intermediate	347		175
Violet 200 Dye	347		175
PESTICIDES			
Benzethonium Chloride	716	CI	380
Bis(2-Hydroxy-5-chlorophenyl) methane	1058	NL	570
Crag No. 974	590	CI	310
Dieldrin (20%)	1022	NL	550
2, 6-Ditertiary-butyl-paracresol	788	NL	420
Dithane	356		180
Ferbam	302		150
Manganese Vancide	248		120
Sevin	284		140
∝, ∝ - Trithiobis (N,N-Dimethylthio-formamide)	446		230
THERMOPLASTIC RESINS AND MOLDING COMPOUNDS			
Acetal Resins			
Acetal, Linear (Polyformaldehyde)	824	NL	440
Acrylic Resins			
Acrylamide Polymer	464		240
Acrylonitrile Polymer	860		460
Acrylonitrile-Vinyl Pyridine Copolymer	464		240
Acrylonitrile-Vinyl Chloride-Vinylidene Chloride Copolymer (70-20-10)	410		210
Methyl Methacrylate Polymer	824	NL	440
Methyl Methacrylate-Ethyl Acrylate Copolymer	896	NL	480
Methyl Methacrylate-Ethyl Acrylate-Styrene Copolymer	824	NL	440
Methyl Methacrylate-Styrene-Butadiene-Acrylonitrile Copolymer	896	NL	480
Methacrylic Acid Polymer	554		290
Cellulosic Resins			
Cellulose Acetate	644		340
Cellulose Triacetate	806	NL	430
Cellulose Acetate Butyrate	698	NL	370
Cellulose Propionate	860	NL	460
Ethyl Cellulose	608	CI	320
Methyl Cellulose	644		340
Carboxymethyl Cellulose	554		290
Hydroxyethyl Cellulose	644		340
Chlorinated Polyether Resins			
Chlorinated Polyether Alcohol	860		460
Nylon (Polyamide) Resins			
Nylon Polymer (Polyhexa-methylene Adipamide)	806		430
Polycarbonate Resins			
Polycarbonate	1310	NL	710
Polyethylene Resins			
Polyethylene, High Pressure Process	716		380
Polyethylene, Low Pressure Process	788	NL	420
Polyethylene Wax	752	NL	400
Polymethylene Resins			
Carboxypolymethylene	968	NL	520

[1]Normally, the minimum ignition temperature of a layer of a specific dust is lower than the minimum ignition temperature of a cloud of that dust. Since this is not universally true, the lower of the two minimum ignition temperatures is listed. If no symbol appears between the two temperature columns, then the layer ignition temperature is shown. "CI" means the cloud ignition temperature is shown. "NL" means that no layer ignition temperature is available and the cloud ignition temperature is shown. "M" signifies that the dust layer melts before it ignites; the cloud ignition temperature is shown. "S" signifies that the dust layer sublimes before it ignites; the cloud ignition temperature is shown.

Class II, Group G (cont'd)	Minimum Cloud or Layer Ignition Temp. °F		°C
THERMOPLASTIC RESINS AND MOLDING COMPOUNDS			
Polypropylene Resins			
Polypropylene (No Antioxidant)	788	NL	420
Rayon Resins			
Rayon (Viscose) Flock	482		250
Styrene Resins			
Polystyrene Molding Cmpd.	1040	NL	560
Polystyrene Latex	932		500
Styrene-Acrylonitrile (70-30)	932	NL	500
Styrene-Butadiene Latex (>75% Styrene; Alum Coagulated)	824	NL	440
Vinyl Resins			
Polyvinyl Acetate	1022	NL	550
Polyvinyl Acetate/Alcohol	824		440
Polyvinyl Butyral	734	NL	390
Vinyl Chloride-Acrylonitrile Copolymer	878		470
Polyvinyl Chloride-Dioctyl Phthalate Mixture	608	NL	320
Vinyl Toluene-Acrylonitrile Butadiene Copolymer	936	NL	530
THERMOSETTING RESINS AND MOLDING COMPOUNDS			
Allyl Resins			
Allyl Alcohol Derivative (CR-39)	932	NL	500
Amino Resins			
Urea Formaldehyde Molding Compound	860	NL	460
Urea Formaldehyde-Phenol Formaldehyde Molding Compound (Wood Flour Filler)	464		240
Epoxy Resins			
Epoxy	1004	NL	540
Epoxy - Bisphenol A	950	NL	510
Phenol Furfural	590		310
Phenolic Resins			
Phenol Formaldehyde	1076	NL	580
Phenol Formaldehyde Molding Cmpd. (Wood Flour Filler)	932	NL	500
Phenol Formaldehyde, Polyalkylene-Polyamine Modified	554		290
Polyester Resins			
Polyethylene Terephthalate	932	NL	500
Styrene Modified Polyester-Glass Fiber Mixture	680		360
Polyurethane Resins			
Polyurethane Foam, No Fire Retardant	824		440
Polyurethane Foam, Fire Retardant	734		390
SPECIAL RESINS AND MOLDING COMPOUNDS			
Alkyl Ketone Dimer Sizing Compound	320		160
Cashew Oil, Phenolic, Hard	356		180
Chlorinated Phenol	1058	NL	570
Coumarone-Indene, Hard	968	NL	520
Ethylene Oxide Polymer	662	NL	350
Ethylene-Maleic Anhydride Copolymer	1004	NL	540
Lignin, Hydrolized, Wood-Type, Fines	842	NL	450
Petrin Acrylate Monomer	428	NL	220
Petroleum Resin (Blown Asphalt)	932		500
Rosin, DK	734	NL	390
Rubber, Crude, Hard	662	NL	350
Rubber, Synthetic, Hard (33% S)	608	NL	320
Shellac	752	NL	400
Sodium Resinate	428		220
Styrene — Maleic Anhydride Copolymer	878	CI	470

[2]Certain metal dusts may have characteristics that require safeguards beyond those required for atmospheres containing the dusts of aluminum, magnesium, and their commercial alloys. For example, zirconium, thorium, and uranium dusts have extremely low ignition temperatures (as low as 20°C) and minimum ignition energies lower than any material classified in any of the Class I or Class II groups.

17.1.9 *NEC* Table 500.8(B), Classification of Maximum Surface Temperature

TABLE 17.1.9

Maximum Temperature		Identification Number
°C	°F	
450	842	T1
300	572	T2
280	536	T2A
260	500	T2B
230	446	T2C
215	419	T2D
200	392	T3
180	356	T3A
165	329	T3B
160	320	T3C
135	275	T4
120	248	T4A
100	212	T5
85	185	T6

17.1.10 *NEC* Table 500.8(C)(2), Class II Ignition Temperatures

TABLE 17.1.10

Class II Group	Equipment that Is Not Subject to Overloading		Equipment (Such as Motors or Power Transformers) that May Be Overloaded			
			Normal Operation		Abnormal Operation	
	°C	°F	°C	°F	°C	°F
E	200	392	200	392	200	392
F	200	392	150	302	200	392
G	165	329	120	248	165	329

NEC 500.8(C)(2), Class II temperature: The temperature marking specified in (B) (Table 17.1.9 above) shall be less than the ignition temperature of the specific dust to be encountered. For organic dusts that may dehydrate or carbonize, the temperature marking shall not exceed the lower of either the ignition temperature or 165°C (329°F). See "Recommended Practice for the Classification of Combustible Dusts and of Hazardous (Classified) Locations for Electrical Installations in Chemical Process Areas," NFPA 499-1997, for minimum ignition temperatures of specific dusts.

The ignition temperature for which equipment was approved prior to this requirement shall be assumed to be as shown in the preceding table.

17.1.11 *NEC* Article 505, Class I, Zone 0, 1, and 2 Locations

This will serve as a brief introduction to the zone classification system. Refer to the complete text of *NEC* Article 505 for further details. This alternate zone classification is based on the IEC three-zone system; Table 17.1.11 briefly illustrates the relationship between zones and divisions. The definitions of Zones 1 and 2 in this article are significantly different from the IEC definitions, so it is not entirely clear how closely the two systems will be harmonized.

TABLE 17.1.11 Area Classification

Area Classification			
Class	**Zone**	**Description**	**Typical Example**
Class I Liquids, Vapors & Gases	Zone 0	Similar to Division 1, continuously hazardous	vapor space in a vented tank
	Zone 1	Similar to Division 1, frequently hazardous under normal conditions	container filling area in a refinery
	Zone 2	Similar to Division 2, hazardous under abnormal conditions	container storage area

Since the introduction of the zone classification system into the 1996 *NEC,* three new or revised national documents now contain guidelines for its application:

- ANSI/ISA-S12.24.01, "Recommended Practice for Classification of Locations for Electrical Installations Classified as Class I, Zone 0, Zone 1, or Zone 2"
- ANSI/NFPA 497
- ANSI/API RP505, "Recommended Practice"

ISA-S12.24.01 is based on IEC 60079-10, a general guide to the method of zone classifications, not specific to any industry and not giving specific guidance on the extent of any classified areas. NFPA 497 was expanded to include zone classifications as an alternative to divisions for areas and processes typical to the chemical industry. API RP505 is a companion to RP500 for divisions and contains recommended zone classifications related to the petroleum industry. The NFPA and API documents contain numerous diagrams with specific dimensions of classified areas, and it should be noted that for a given process or piece of equipment, the recommended extent of Zone 1 areas is the same as for Division 1 and similarly for Zone 2 and Division 2. In the end, the only essential difference between the two systems is that a few of the zone diagrams contain a small Zone 0.

It has been estimated that less than 1 percent of classified areas would qualify as Zone 0. It is also generally accepted that Division 1 constitutes 10 percent or less of classified areas in the United States. In countries using the IEC classification system, however, Zone 1 commonly represents up to 60 percent of classified areas. The disparity between these two ratios is another indication that Zone 0 is not the only difference between the division and the IEC zone system.

17.1.12 *NEC* Article 511, Commercial Garages, Repair and Storage

17.1.12 *A.* Classification of locations in commercial garages; *B.* Raceways embedded in a masonry wall or buried beneath the floor are not considered located in the Class I location only under conditions where there are not extensions into the Class I location; *C.* Recommended installation for commercial garages.

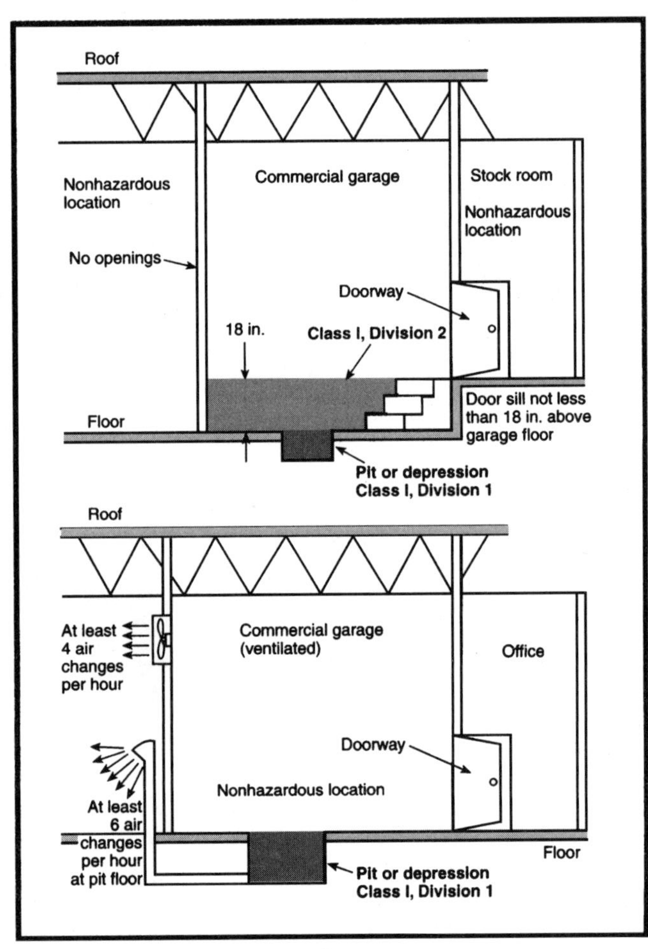

A

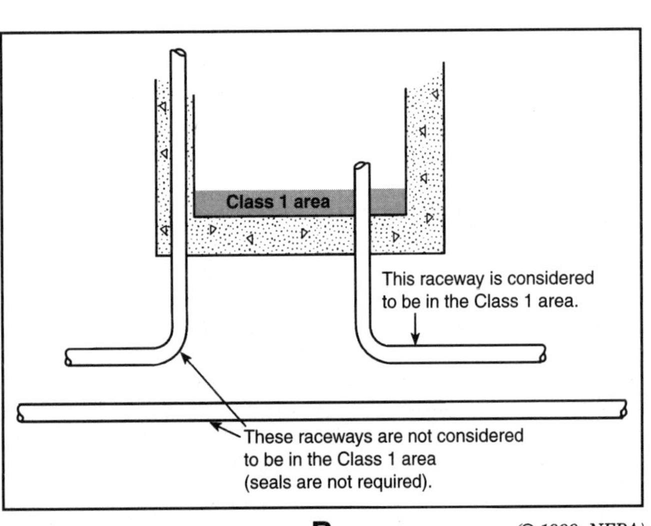

B

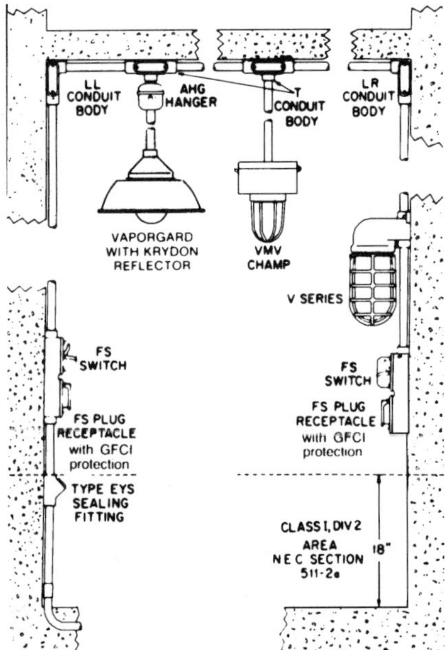

C

17.1.13 *NEC* Article 513, Aircraft Hangars

17.1.13 Aircraft hangars. *A.* Area classification. *B.* Recommended installation for aircraft hangars.

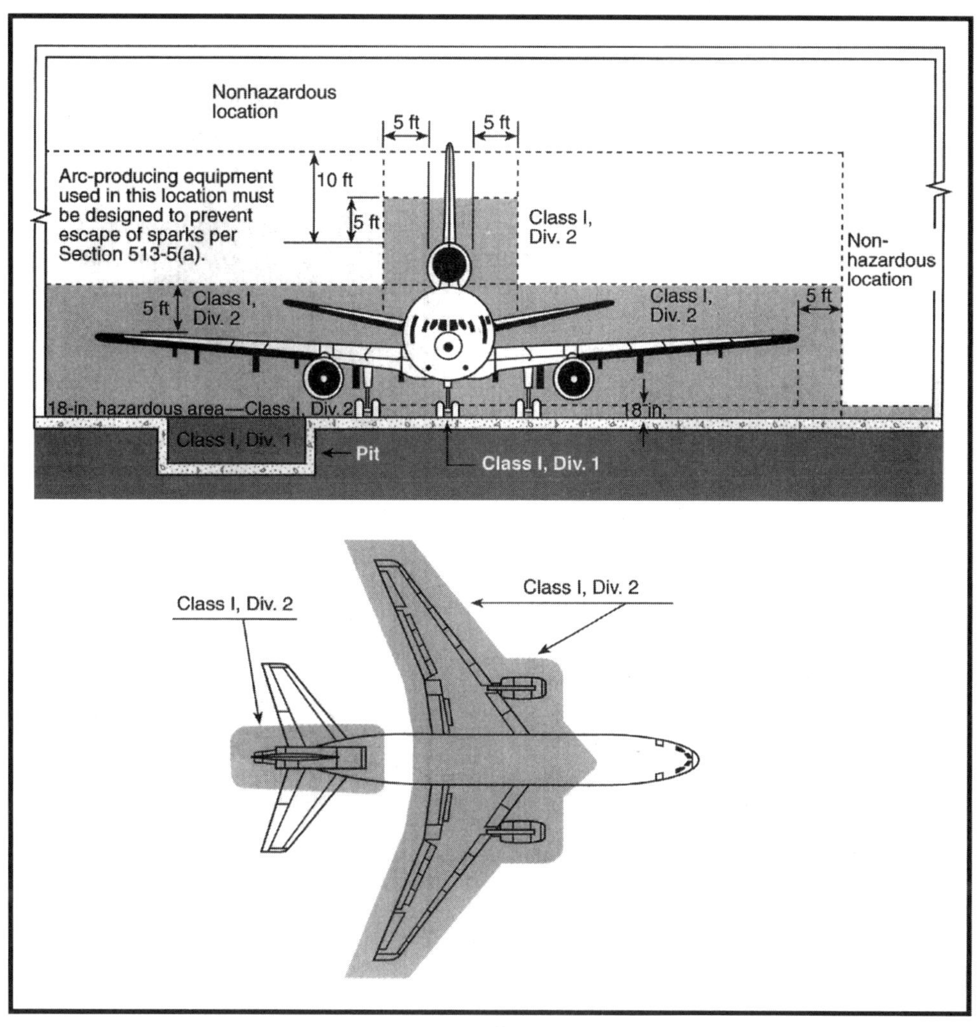

A

(© 1999, NFPA)

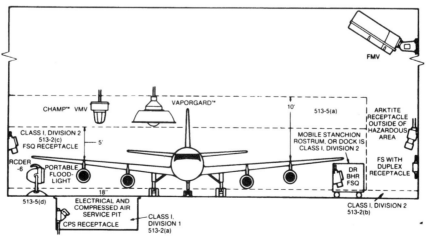

FMV
CHAMP™ VMV
VAPORGARD™
10′
513-5(a)
ARKTITE RECEPTACLE OUTSIDE OF HAZARDOUS AREA
CLASS I, DIVISION 2 513-2(c) FSQ RECEPTACLE
5′
MOBILE STANCHION ROSTRUM, OR DOCK IS CLASS I, DIVISION 2
RCDER -6
PORTABLE FLOOD-LIGHT
DR BHR FSQ
FS WITH DUPLEX RECEPTACLE
18″
513-5(d)
ELECTRICAL AND COMPRESSED AIR SERVICE PIT
CPS RECEPTACLE
CLASS I, DIVISION 1 513-2(a)
CLASS I, DIVISION 2 513-2(b)

B

(© 1999, NFPA)

17.1.14 *NEC* Article 514, Motor Fuel Dispensing Facilities. *A. NEC* Table 514.3(B)(1), Class 1 Locations—Motor Fuel Dispensing Facilities and Commercial Garages. *B. NEC* Table 514.3(B)(2), Electrical Equipment Classified Areas for Dispensing Devices (see pages 17.23–*17.27*)

TABLE 17.1.14 *A. NEC* Table 514.3(B)(1), Class 1 Locations—Motor Fuel Dispensing Facilities and Commercial Garages

Location	Class I, Group D Division	Extent of Classified Location
Underground Tank		
Fill opening	1	Any pit, box, or space below grade level, any part of which is within the Division 1 or Division 2, Zone 1 or Zone 2 classified location
	2	Up to 450 mm (18 in.) above grade level within a horizontal radius of 3.0 m (10 ft) from a loose fill connection and within a horizontal radius of 1.5 m (5 ft) from a tight fill connection
Vent — discharging upward	1	Within 900 mm (3 ft) of open end of vent, extending in all directions
	2	Space between 900 mm (3 ft) and 1.5 m (5 ft) of open end of vent, extending in all directions
Dispensing Device[1,4] (except overhead type)[2]		
Pits	1	Any pit, box, or space below grade level, any part of which is within the Division 1 or Division 2, Zone 1 or Zone 2 classified location
Dispenser		FPN: Space classification inside the dispenser enclosure is covered in ANSI/UL 87-1995, *Power Operated Dispensing Devices for Petroleum Products.*
	2	Within 450 mm (18 in.) horizontally in all directions extending to grade from the dispenser enclosure or that portion of the dispenser enclosure containing liquid-handling components FPN: Space classification inside the dispenser enclosure is covered in ANSI/UL 87-1995, *Power Operated Dispensing Devices for Petroleum Products.*
Outdoor	2	Up to 450 mm (18 in.) above grade level within 6.0 m (20 ft) horizontally of any edge of enclosure.
Indoor with mechanical ventilation	2	Up to 450 mm (18 in.) above grade or floor level within 6.0 m (20 ft) horizontally of any edge of enclosure
with gravity ventilation	2	Up to 450 mm (18 in.) above grade or floor level within 7.5 m (25 ft) horizontally of any edge of enclosure
Dispensing Device[4]		
Overhead type [2]	1	The space within the dispenser enclosure, and all electrical equipment integral with the dispensing hose or nozzle
	2	A space extending 450 mm (18 in.) horizontally in all directions beyond the enclosure and extending to grade
	2	Up to 450 mm (18 in.) above grade level within 6.0 m (20 ft) horizontally measured from a point vertically below the edge of any dispenser enclosure
Remote Pump — Outdoor	1	Any pit, box, or space below grade level if any part is within a horizontal distance of 3.0 m (10 ft) from any edge of pump
	2	Within 900 mm (3 ft) of any edge of pump, extending in all directions. Also up to 450 mm (18 in.) above grade level within 3.0 m (10 ft) horizontally from any edge of pump
Remote Pump — Indoor	1	Entire space within any pit
	2	Within 1.5 m (5 ft) of any edge of pump, extending in all directions. Also up to 900 mm (3 ft) above grade level within 7.5 m (25 ft) horizontally from any edge of pump
Lubrication or Service Room — with Dispensing	1	Any pit within any unventilated space
	2	Any pit with ventilation
	2	Space up to 450 mm (18 in.) above floor or grade level and 900 mm (3 ft) horizontally from a lubrication pit

(continued)

TABLE 17.1.14 *A. NEC* Table 514.3(B)(1), Class 1 Locations—Motor Fuel Dispensing Facilities and Commercial Garages (*Continued*)

Location	Class I, Group D Division	Extent of Classified Location
Dispenser for Class I liquids	2	Within 900 mm (3 ft) of any fill or dispensing point, extending in all directions
Lubrication or Service Room — Without Dispensing	2	Entire area within any pit used for lubrication or similar services where Class I liquids may be released
	2	Area up to 450 mm (18 in.) above any such pit and extending a distance of 900 mm (3 ft) horizontally from any edge of the pit
	2	Entire unventilated area within any pit, belowgrade area, or subfloor area
	2	Area up to 450 mm (18 in.) above any such unventilated pit, belowgrade work area, or subfloor work area and extending a distance of 900 mm (3 ft) horizontally from the edge of any such pit, belowgrade work area, or subfloor work area
	Unclassified	Any pit, belowgrade work area, or subfloor work area that is provided with exhaust ventilation at a rate of not less than 0.3 $m^3/min/m^2$ (1 cfm/ft^2) of floor area at all times that the building is occupied or when vehicles are parked in or over this area and where exhaust air is taken from a point within 300 mm (12 in.) of the floor of the pit, belowgrade work area, or subfloor work area
Special Enclosure Inside Building[3]	1	Entire enclosure
Sales, Storage, and Rest Rooms	Unclassified	If there is any opening to these rooms within the extent of a Division 1 location, the entire room shall be classified as Division 1
Vapor Processing Systems Pits	1	Any pit, box, or space below grade level, any part of which is within a Division 1 or Division 2 classified location or that houses any equipment used to transport or process vapors
Vapor Processing Equipment Located Within Protective Enclosures FPN: See 10.1.7 of NFPA 30A-2000, *Code for Motor Fuel Dispensing Facilities and Repair Garages.*	2	Within any protective enclosure housing vapor processing equipment
Vapor Processing Equipment Not Within Protective Enclosures (excluding piping and combustion devices)	2	The space within 450 mm (18 in.) in all directions of equipment containing flammable vapor or liquid extending to grade level. Up to 450 mm (18 in.) above grade level within 3.0 m (10 ft) horizontally of the vapor processing equipment
Equipment Enclosures	1	Any space within the enclosure where vapor or liquid is present under normal operating conditions
Vacuum-Assist Blowers	2	The space within 450 mm (18 in.) in all directions extending to grade level. Up to 450 mm (18 in.) above grade level within 3.0 m (10 ft) horizontally

[1]Refer to Figure 514.3 for an illustration of classified location around dispensing devices.
[2]Ceiling mounted hose reel.
[3]FPN: See 4.3.9 of NFPA 30A-2000, *Code for Motor Fuel Dispensing Facilities and Repair Garages.*
[4]FPN: Area classification inside the dispenser enclosure is covered in ANSI/UL 87-1995, *Power-Operated Dispensing Devices for Petroleum Products.* [NFPA 30A, Table 8-3]

[NFPA 30A, Table 12.6]

(*continued*)

TABLE 17.1.14 *B. NEC* Table 514.3(B)(2), Electrical Equipment Classified Areas for Dispensing Devices

	Extent of Classified Area	
Dispensing Device	Class I, Division 1	Class I, Division 2
Compressed Natural Gas	Entire space within the dispenser enclosure	1.5 m (5 ft) in all directions from dispenser enclosure
Liquefied Natural Gas	Entire space within the dispenser enclosure and 1.5 m (5 ft) in all directions from the dispenser enclosure	From 1.5 m to 3.0 m (5 ft to 10 ft) in all directions from the dispenser enclosure
Liquefied Petroleum Gas	Entire space within the dispenser enclosure; 450 mm (18 in.) from the exterior surface of the dispenser enclosure to an elevation of 1.2 m (4 ft) above the base of the dispenser; the entire pit or open space beneath the dispenser and within 6.0 m (20 ft) horizontally from any edge of the dispenser when the pit or trench is not mechanically ventilated.	Up to 450 mm (18 in.) aboveground and within 6.0 m (20 ft) horizontally from any edge of the dispenser enclosure, including pits or trenches within this area when provided with adequate mechanical ventilation

17.1.14 Motor fuel dispensing facilities. *A.* Extent of Class I location around overhead gasoline dispensing units, in accordance with *NFPA 30A.*

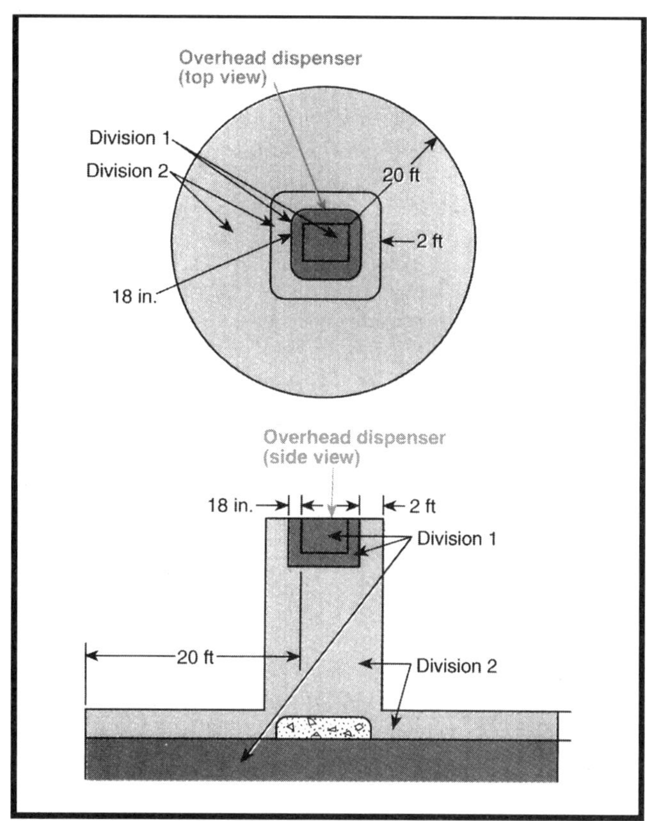

A

17.1.14 Motor fuel dispensing facilities (*Continued*). *B.* Classified locations adjacent to dispensers as detailed in *NEC* Table 514.3(B)(1)(*NFPA30A*, Figure 8.3). *C.* A gasoline dispensing installation indicating location for sealing fittings. Emergency controls are required for self-service stations. See *NFPA 30A*, Sections 514.5(A) and (B).

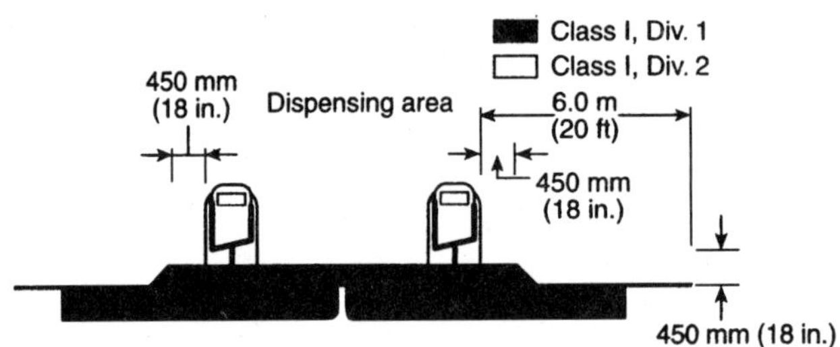

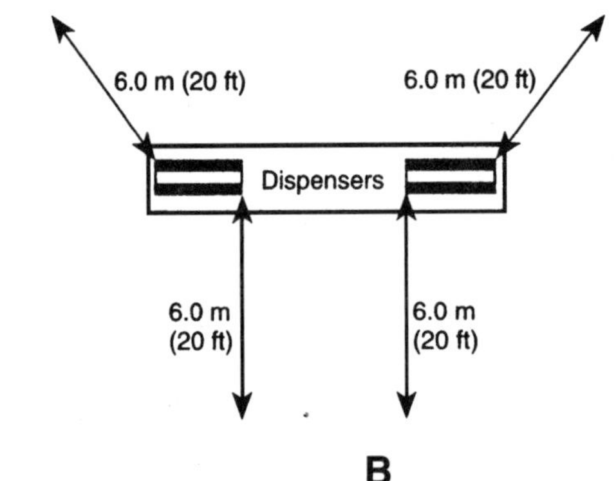

B

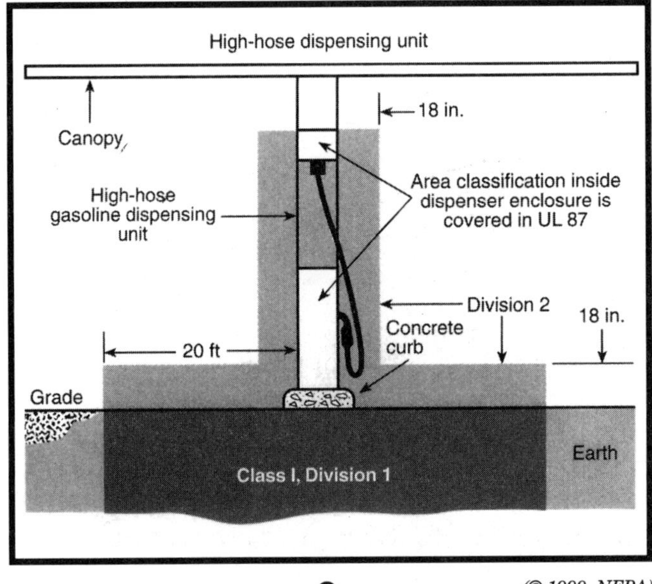

C

(© 1999, NFPA)

(continued)

17.1.14 Motor fuel dispensing facilities (*Continued*). *D.* Seals are required at points marked "S." Seals are not required at the sign and two of the lights because conduit runs do not pass through a hazardous location. *E.* Use of rigid nonmetallic conduit in accordance with *NFPA 30A* Exception No. 2 to Section 514.8.

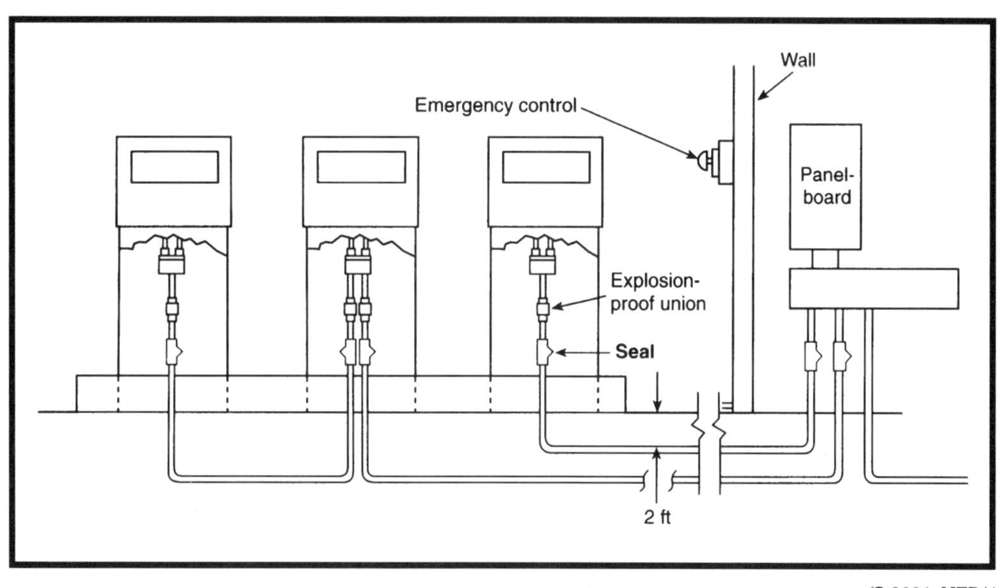

D

Below grade area classification

Sign

Light

Light

Dispensers

20 ft

Vents discharging upward with ends elevated at least 5 ft above grade

Extent of hazardous location

Tight fill connections

5 ft

Light

20 ft

Dispensers

Light

Panelboard in nonhazardous location (S = conduit seal)

E

17.1.15 *NEC* Article 515, Bulk Storage Plants

TABLE 17.1.15 Bulk Storage Plants, *NEC* Table 515.3, Electrical Area Classifications

Location	NEC Class I Division	Zone	Extent of Classified Area
Indoor equipment installed in accordance with 5.3 of NFPA 30 where flammable vapor–air mixtures can exist under normal operation	1	0	The entire area associated with such equipment where flammable gases or vapors are present continuously or for long periods of time
	1	1	Area within 1.5 m (5 ft) of any edge of such equipment, extending in all directions
	2	2	Area between 1.5 m and 2.5 m (5 ft and 8 ft) of any edge of such equipment, extending in all directions; also, space up to 900 mm (3 ft) above floor or grade level within 1.5 m to 7.5 m (5 ft to 25 ft) horizontally from any edge of such equipment[1]
Outdoor equipment of the type covered in 5.3 of NFPA 30 where flammable vapor–air mixtures may exist under normal operation	1	0	The entire area associated with such equipment where flammable gases or vapors are present continuously or for long periods of time
	1	1	Area within 900 mm (3 ft) of any edge of such equipment, extending in all directions
	2	2	Area between 900 mm (3 ft) and 2.5 m (8 ft) of any edge of such equipment, extending in all directions; also, space up to 900 mm (3 ft) above floor or grade level within 900 mm to 3.0 m (3 ft to 10 ft) horizontally from any edge of such equipment
Tank storage installations inside buildings	1	1	All equipment located below grade level
	2	2	Any equipment located at or above grade level
Tank – aboveground	1	0	Inside fixed roof tank
	1	1	Area inside dike where dike height is greater than the distance from the tank to the dike for more than 50 percent of the tank circumference
Shell, ends, or roof and dike area	2	2	Within 3.0 m (10 ft) from shell, ends, or roof of tank; also, area inside dike to level of top of tank
Vent	1	0	Area inside of vent piping or opening
	1	1	Within 1.5 m (5 ft) of open end of vent, extending in all directions
	2	2	Area between 1.5 m and 3.0 m (5 ft and 10 ft) from open end of vent, extending in all directions
Floating roof with fixed outer roof	1	0	Area between the floating and fixed roof sections and within the shell
Floating roof with no fixed outer roof	1	1	Area above the floating roof and within the shell
Underground tank fill opening	1	1	Any pit, or space below grade level, if any part is within a Division 1 or 2, or Zone 1 or 2, classified location
	2	2	Up to 450 mm (18 in.) above grade level within a horizontal radius of 3.0 m (10 ft) from a loose fill connection, and within a horizontal radius of 1.5 m (5 ft) from a tight fill connection
Vent – discharging upward	1	0	Area inside of vent piping or opening
	1	1	Within 900 mm (3 ft) of open end of vent, extending in all directions
	2	2	Area between 900 mm and 1.5 m (3 ft and 5 ft) of open end of vent, extending in all directions

(continued)

TABLE 17.1.15 Bulk Storage Plants, *NEC* Table 515.3, Electrical Area Classifications (*Continued*)

Location	NEC Class I Division	Zone	Extent of Classified Area
Drum and container filling – outdoors or indoors	1	0	Area inside the drum or container
	1	1	Within 900 mm (3 ft) of vent and fill openings, extending in all directions
	2	2	Area between 900 mm and 1.5 m (3 ft and 5 ft) from vent or fill opening, extending in all directions; also, up to 450 mm (18 in.) above floor or grade level within a horizontal radius of 3.0 m (10 ft) from vent or fill opening
Pumps, bleeders, withdrawal fittings,			
Indoors	2	2	Within 1.5 m (5 ft) of any edge of such devices, extending in all directions; also, up to 900 mm (3 ft) above floor or grade level within 7.5 m (25 ft) horizontally from any edge of such devices
Outdoors	2	2	Within 900 mm (3 ft) of any edge of such devices, extending in all directions. Also, up to 450 mm (18 in.) above grade level within 3.0 m (10 ft) horizontally from any edge of such devices
Pits and sumps			
Without mechanical ventilation	1	1	Entire area within a pit or sump if any part is within a Division 1 or 2, or Zone 1 or 2, classified location
With adequate mechanical ventilation	2	2	Entire area within a pit or sump if any part is within a Division 1 or 2, or Zone 1 or 2, classified location
Containing valves, fittings, or piping, and not within a Division 1 or 2, or Zone 1 or 2, classified location	2	2	Entire pit or sump
Drainage ditches, separators, impounding basins			
Outdoors	2	2	Area up to 450 mm (18 in.) above ditch, separator, or basin; also, area up to 450 mm (18 in.) above grade within 4.5 m (15 ft) horizontally from any edge
Indoors			Same classified area as pits
Tank vehicle and tank car[2] loading through open dome	1	0	Area inside of the tank
	1	1	Within 900 mm (3 ft) of edge of dome, extending in all directions
	2	2	Area between 900 mm and 4.5 m (3 ft and 15 ft) from edge of dome, extending in all directions
Loading through bottom connections with atmospheric venting	1	0	Area inside of the tank
	1	1	Within 900 mm (3 ft) of point of venting to atmosphere, extending in all directions
	2	2	Area between 900 mm and 4.5 m (3 ft and 15 ft) from point of venting to atmosphere, extending in all directions; also, up to 450 mm (18 in.) above grade within a horizontal radius of 3.0 m (10 ft) from point of loading connection
Office and rest rooms	Ordinary		If there is any opening to these rooms within the extent of an indoor classified location, the room shall be classified the same as if the wall, curb, or partition did not exist.

(*continued*)

TABLE 17.1.15 Bulk Storage Plants, *NEC* Table 515.3, Electrical Area Classifications (*Continued*)

Location	NEC Class I Division	Zone	Extent of Classified Area
Loading through closed dome with atmospheric venting	1	1	Within 900 mm (3 ft) of open end of vent, extending in all directions
	2	2	Area between 900 mm and 4.5 m (3 ft and 15 ft) from open end of vent, extending in all directions; also, within 900 mm (3 ft) of edge of dome, extending in all directions
Loading through closed dome with vapor control	2	2	Within 900 mm (3 ft) of point of connection of both fill and vapor lines extending in all directions
Bottom loading with vapor control or any bottom unloading	2	2	Within 900 mm (3 ft) of point of connections, extending in all directions; also up to 450 mm (18 in.) above grade within a horizontal radius of 3.0 m (10 ft) from point of connections
Storage and repair garage for tank vehicles	1	1	All pits or spaces below floor level
	2	2	Area up to 450 mm (18 in.) above floor or grade level for entire storage or repair garage
Garages for other than tank vehicles	Ordinary		If there is any opening to these rooms within the extent of an outdoor classified location, the entire room shall be classified the same as the area classification at the point of the opening.
Outdoor drum storage	Ordinary		
Inside rooms or storage lockers used for the storage of Class I liquids	2	2	Entire room
Indoor warehousing where there is no flammable liquid transfer	Ordinary		If there is any opening to these rooms within the extent of an indoor classified location, the room shall be classified the same as if the wall, curb, or partition did not exist.
Piers and wharves			See Figure 515.3.

[1]The release of Class I liquids may generate vapors to the extent that the entire building, and possibly an area surrounding it, should be considered a Class I, Division 2 or Zone 2 location.

[2]When classifying extent of area, consideration shall be given to fact that tank cars or tank vehicles may be spotted at varying points. Therefore, the extremities of the loading or unloading positions shall be used. [NFPA 30: Table 5-9.5.3]

17.1.15 Bulk storage plants. *A.* Marine terminal handling flammable liquids. *B.* Tank car/tank truck loading and unloading via closed system. Transfer through dome only.

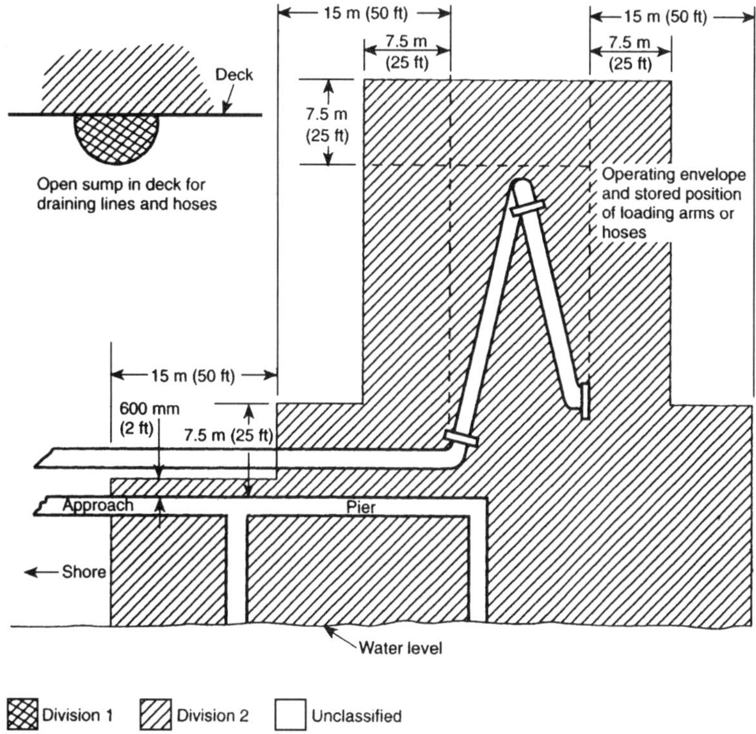

Notes:
(1) The "source of vapor" shall be the operating envelope and stored position of the outboard flange connection of the loading arm (or hose).
(2) The berth area adjacent to tanker and barge cargo tanks is to be Division 2 to the following extent:
a. 7.6 m (25 ft) horizontally in all directions on the pier side from that portion of the hull containing cargo tanks
b. From the water level to 7.6 m (25 ft) above the cargo tanks at their highest position
(3) Additional locations may have to be classified as required by the presence of other sources of flammable liquids on the berth, by Coast Guard, or other regulations.

A

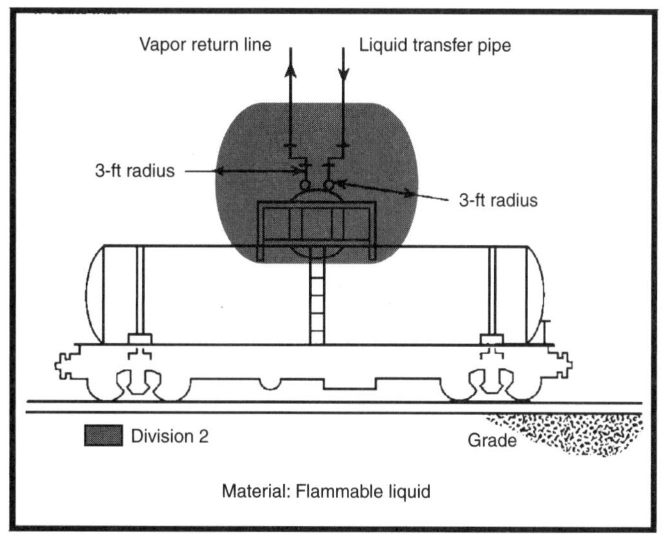

B

(continued)

17.1.15 Bulk storage plants (*Continued*). *C.* Tank car/tank truck loading and unloading via closed system. Bottom product transfer only. *D.* Tank car/tank truck loading and unloading via open system. Top or bottom product transfer.

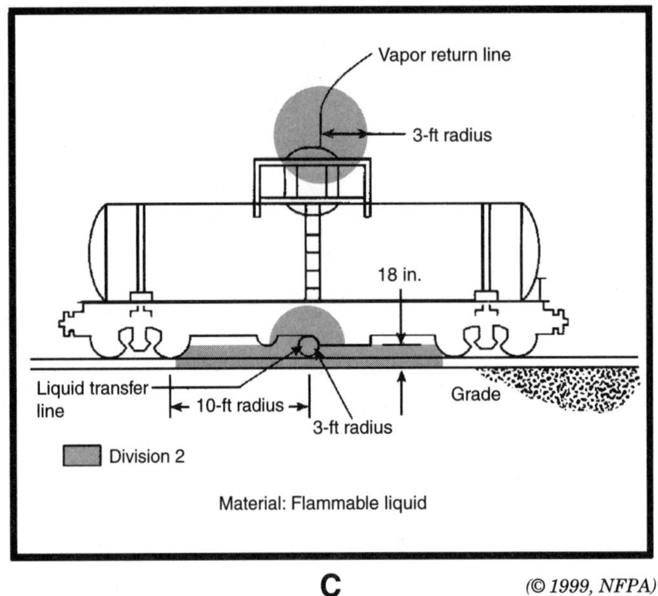

C (© 1999, NFPA)

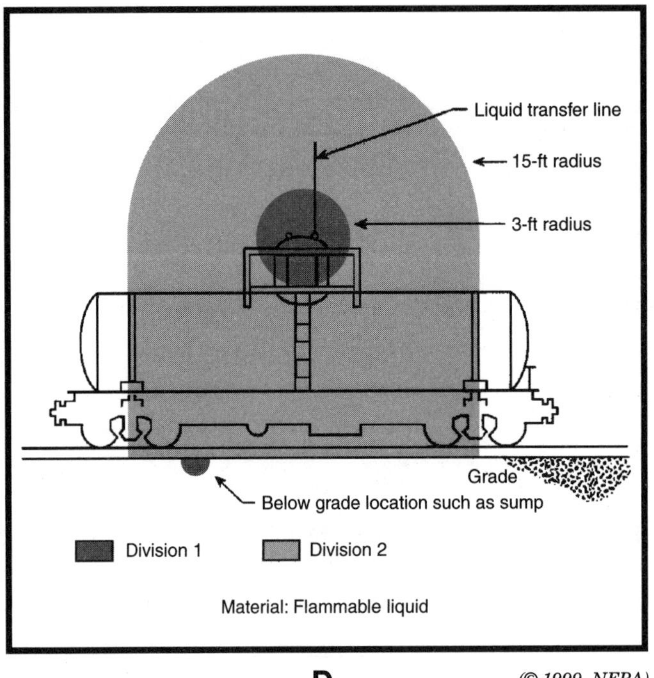

D (© 1999, NFPA)

(continued)

17.1.15 Bulk storage plants (*Continued*). *E.* Tank car/tank truck loading and unloading via closed system. Transfer through dome only. *F.* Drum filling station, outdoors or indoors, with adequate ventilation.

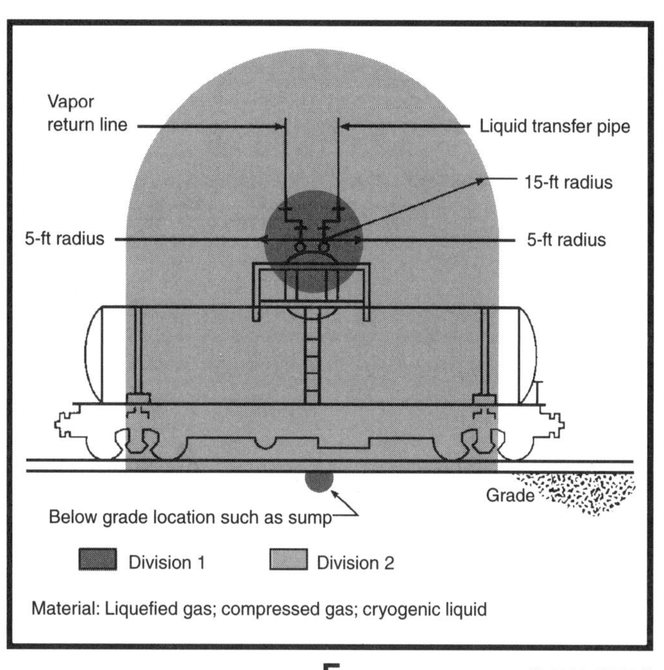

Vapor return line
Liquid transfer pipe
15-ft radius
5-ft radius
5-ft radius
Grade
Below grade location such as sump
Division 1
Division 2
Material: Liquefied gas; compressed gas; cryogenic liquid

E (© 1999, NFPA)

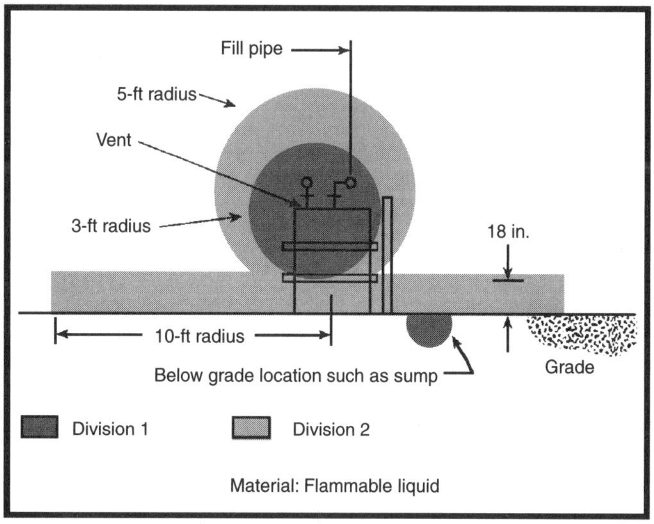

Fill pipe
5-ft radius
Vent
3-ft radius
18 in.
10-ft radius
Below grade location such as sump
Grade
Division 1
Division 2
Material: Flammable liquid

F (© 1999, NFPA)

(continued)

17.1.15 Bulk storage plants (*Continued*). *G.* Storage tanks for cryogenic liquids. (*From NFPA 59A-1996, Standard for the Production, Storage, and Handling of Liquefied Natural Gas (LNG).*) *H.* Storage tanks, outdoors at grade. (*From API Recommended Practice 500A.*)

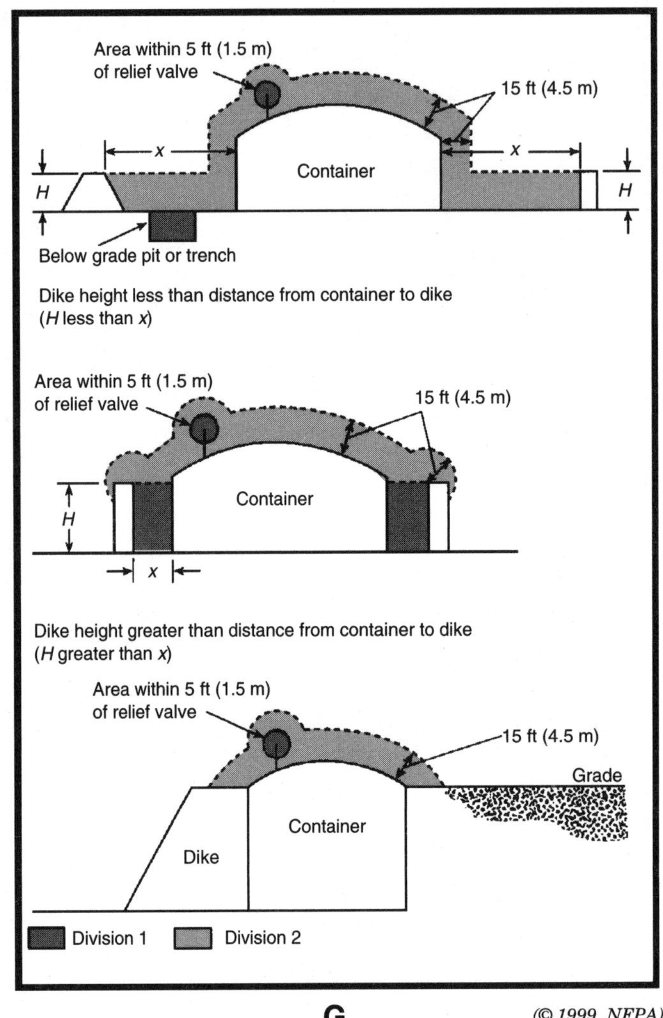

G

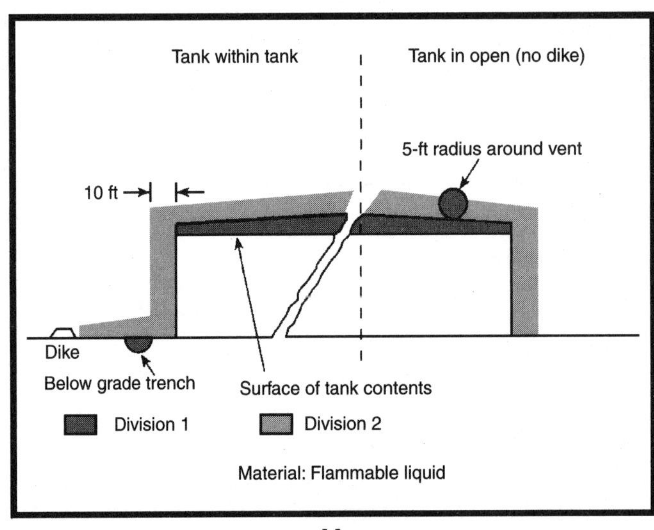

H

17.1.16 *NEC* Article 516, Spray Application, Dipping, and Coating Processes

17.1.16 Spray application, dipping, and coating processes. *A.* Electrical area classification for open spray areas. *B.* Class I or Class II, Division 2 locations adjacent to a closed top, open face, or open front spray booth or room.

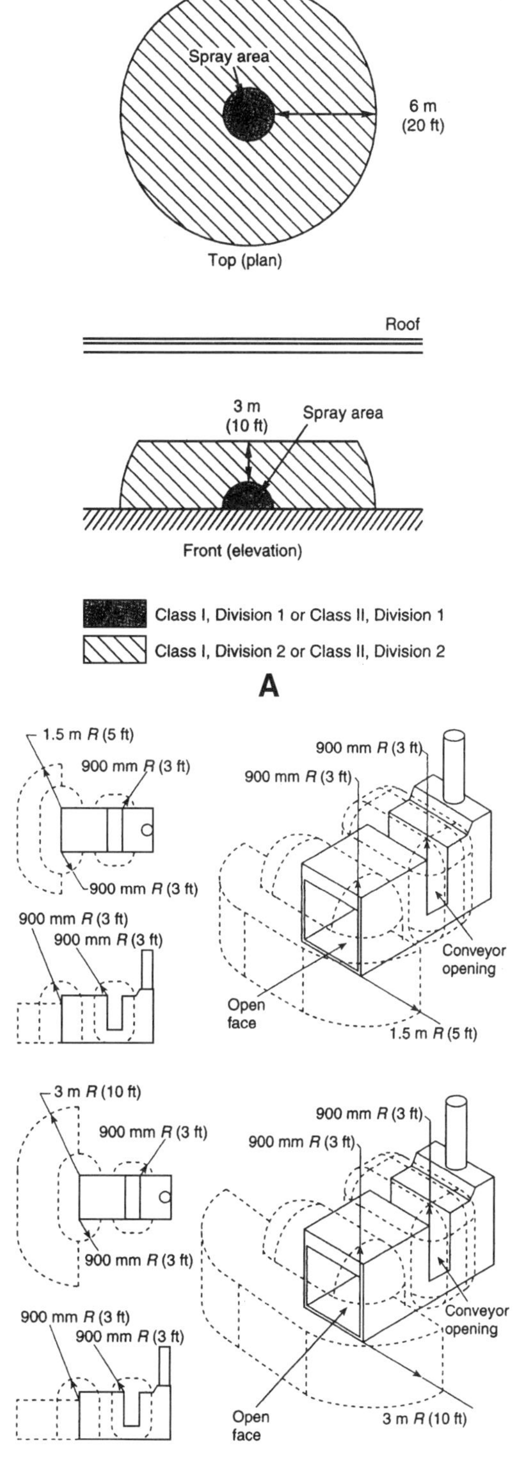

(continued)

17.1.16 Spray application, dipping, and coating processes (*Continued*). *C.* Class I or Class II, Division 2 locations adjacent to an enclosed spray booth or spray room. *D.* Electrical area classification for open processes without vapor containment or ventilation.

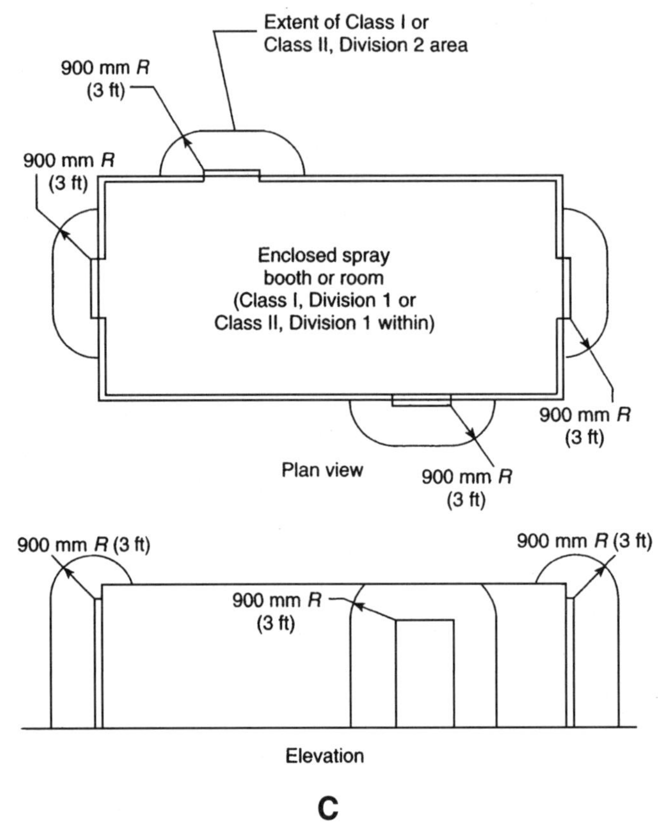

C

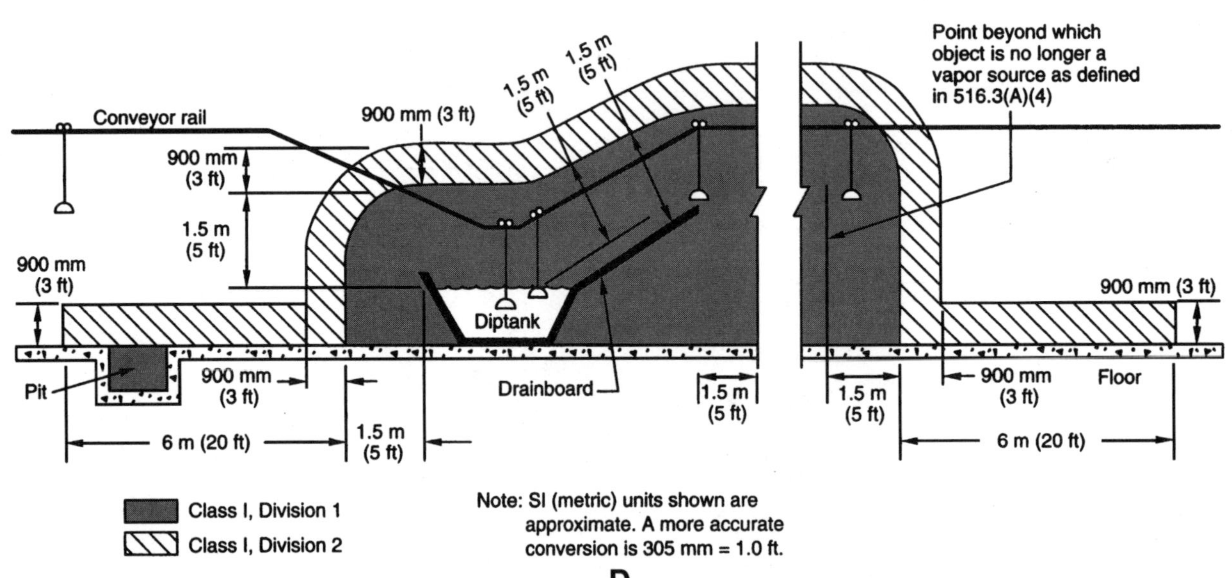

D

17.1.17 Installation Diagram for Sealing

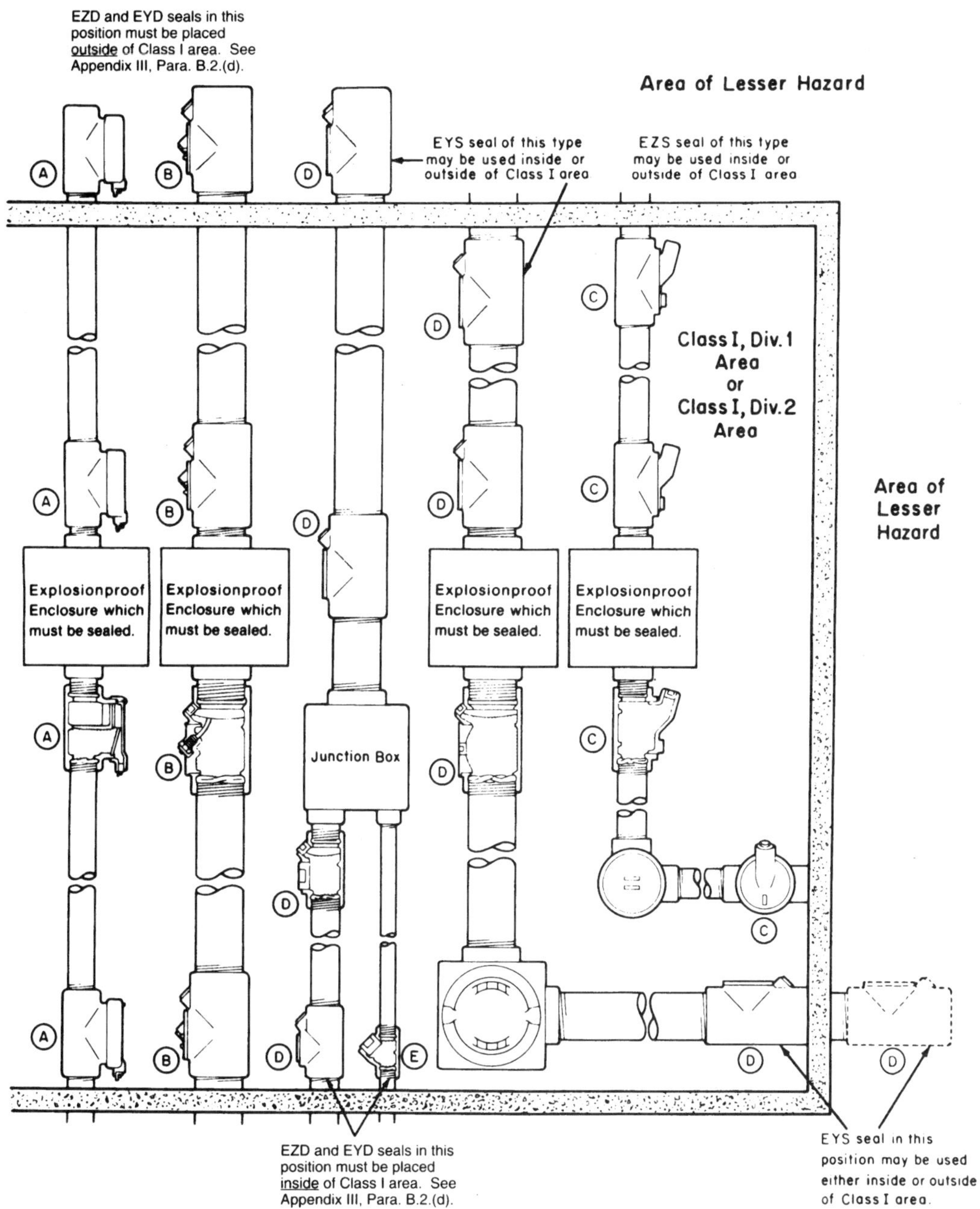

(A) EZD drain seals are available in 1/2 – 2 inch conduit sizes for vertical conduits only.

(B) EYD drain seals are available in 1/2 – 4 inch conduit sizes for vertical conduits only.

(C) EZS seals of this form are available in 1/2 – 3 inch conduit sizes for vertical or horizontal conduits.

(D) EYS seals of this form are available in 1/2 – 6 inch conduit sizes for vertical or horizontal conduits.

(E) EYS seals of this form are available in 1/2 – 1 inch conduit sizes for vertical conduits.

17.1.18 Diagram for Class I, Zone 1 Power and Lighting Installation

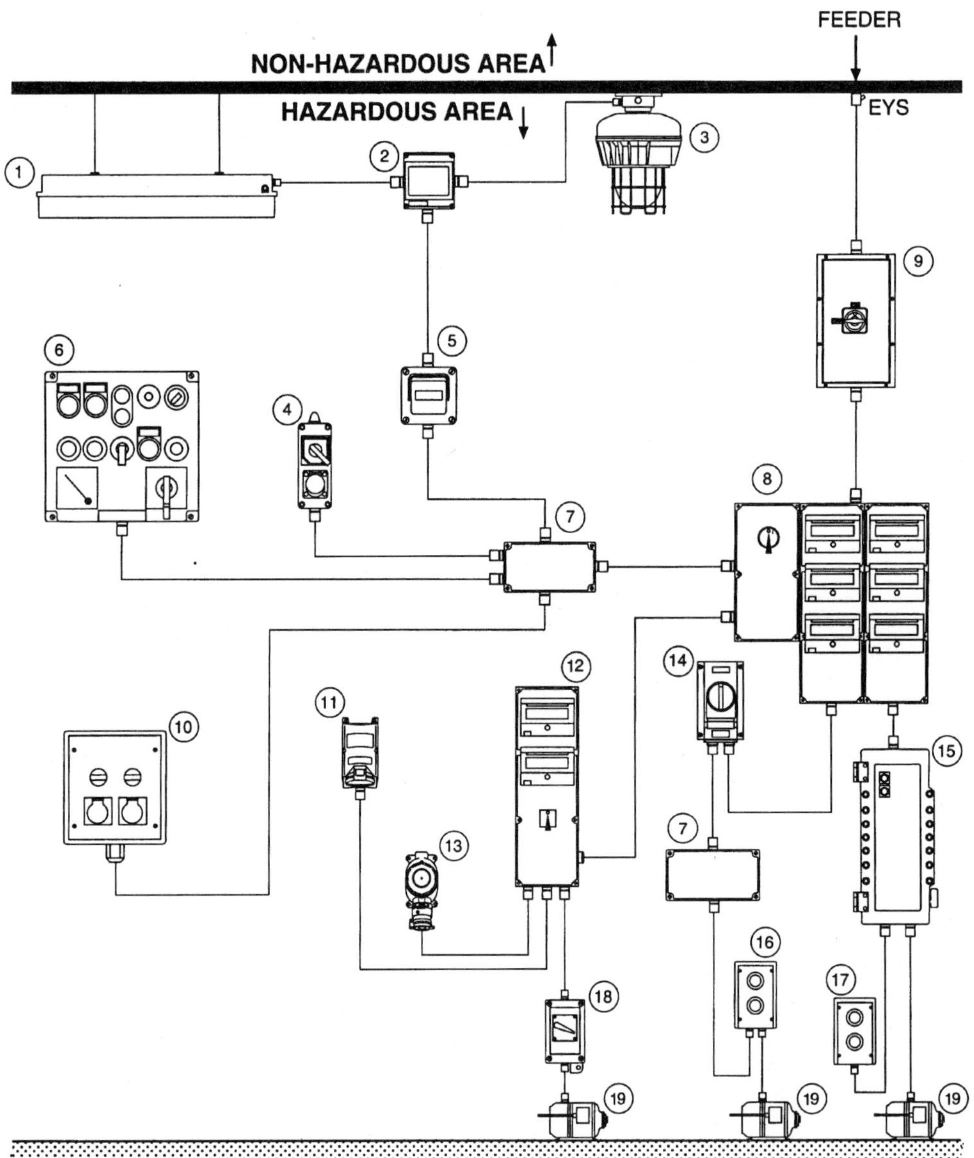

Key to Numerals*

1. eLLK or FVS Fluorescent Lighting Fixture with TMC Terminator™ cable gland

2. eAZK Lighting Junction Box with Zone 1 Myers® hubs

3. EVLP or EVMA HID Lighting Fixture with TMC Terminator cable gland

4. GHG Switched IEC 309 Receptacle with Zone 1 Myers hubs

5. GHG 273 Ex Push-Button Control Station with Zone 1 Myers hubs

6. GHG 44 Control Panel with Zone 1 Myers hubs

7. GHG 744 Terminal Box with Zone 1 Myers hubs

8. EXKO Molded Plastic Distribution Panel with Zone 1 Myers hubs

9. N2RS Control Switch with TMC Terminator cable gland

10. GHG 981 Socket Distribution Panel with Zone 1 Myers hubs

11. GHG Interlocked IEC 309 Receptacle with Zone 1 Myers hubs

12. EXKO Molded Plastic Distribution Panel with Zone 1 Myers hubs

13. FSQC Arktite Interlocked Receptacle with TMCX Terminator cable gland

14. GHG 635 Manual Motor Starter with Zone 1 Myers hubs

15. EBMS Magnetic Motor Starter with TMCX Terminator cable gland

16. N2SCU SpecOne™ Control Station with Zone 1 Myers hubs

17. GHG 43 SpecOne Control Station with Zone 1 Myers hubs

18. EDS Motor Starter with TMCX Terminator cable gland

19. Zone 1 Rated Motor with TMC or TMCX Terminator cable gland

* North American wiring practices allow the use of MC cable or rigid conduit for Zone 1 hazardous areas. When entering a non-metallic enclosure, use a Zone 1 Myers hub. If the installation requires MC cable, use TMC/TMCX Terminator connectors.

17.1.19 Diagram for Class I, Division 1 Lighting Installation

17.1.19

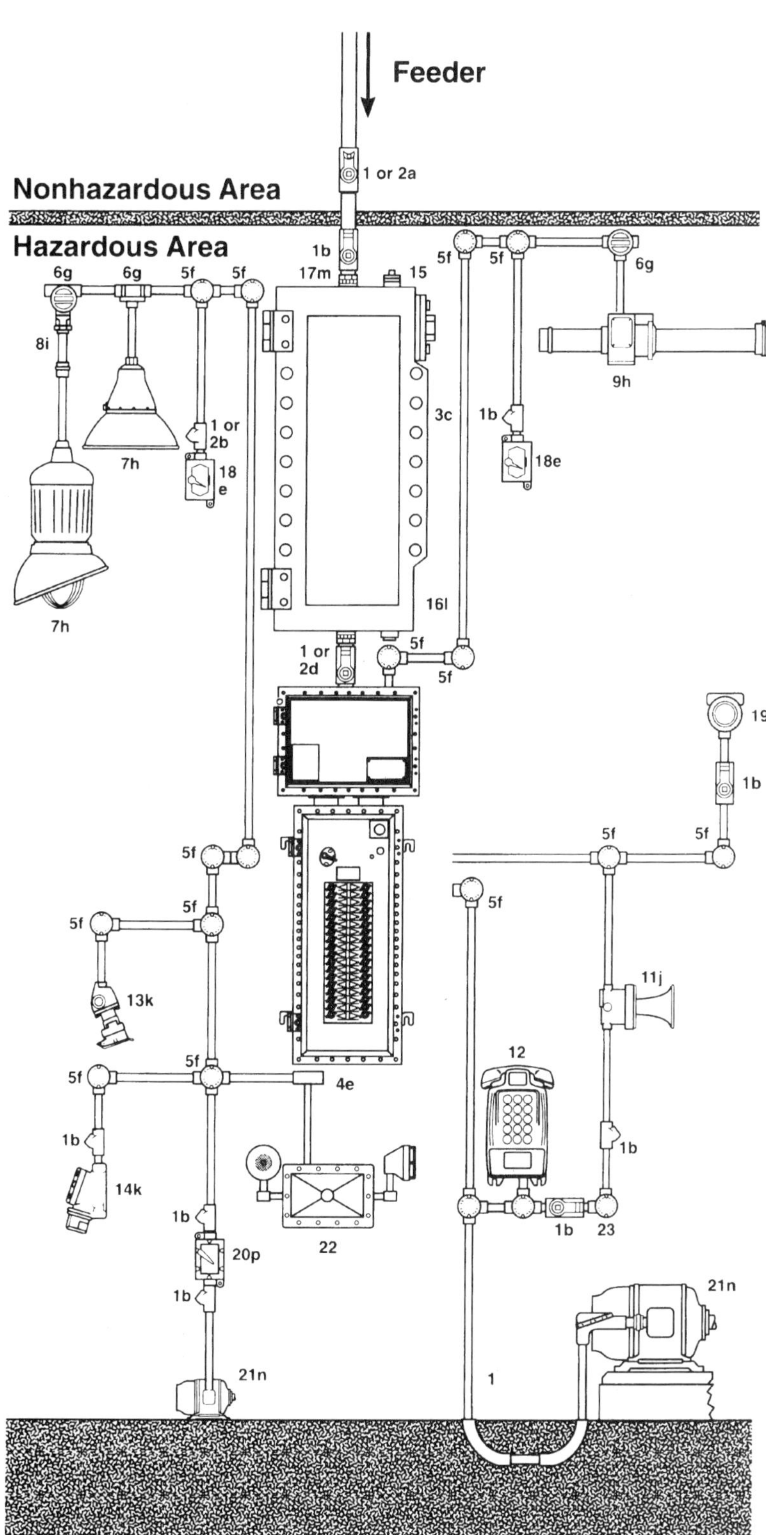

Key to Numerals

1 Sealing fitting. EYS for horizontal or vertical.
2 Sealing fitting. EZS for vertical or horizontal conduits.
3 Circuit breaker. Type EBM.
4 Panelboard. EXDC/EPL. Branch circuits are factory sealed. No seals required in mains or branches unless 2" or over in size.
5 Junction box. Series GUA, GUB, GUJ, have threaded covers. Series CPS has ground flat surface covers.
6 Fixture hanger. EFHC, GUAC, or EFH.
7 Lighting fixture. EV Series incandescent and EVM Series. H.I.D.
8 Flexible fixture support. ECHF.
9 Fluorescent fixture. EVFT.
11 Signal. ETH horns and sirens. ESR bells, Flex•Tone™ signals.
12 ETW explosionproof telephone.
13 Plug receptacle. CES delayed action.
14 Plug receptacle. FSQ. Interlocked with switch.
15 Breather. ECD.
16 Drain. ECD.
17 Union. UNY.
18 Switch. Series EFS.
19 Instrument enclosure. EIH.
20 Manual line starter. EMN.
21 Motors. Explosionproof.
22 Emergency lighting system. ELPS.
23 ETC Power Relay.

National Electrical Code References

a Sec. 501-5(a)(4). Seal required where conduit passes from hazardous to nonhazardous area.
b Sec. 501-5(a)(1). Seals required within 18 inches of all arcing devices.
c Sec. 384-16. Circuit breaker protection required ahead of panelboard.
d Sec. 501-5(a)(2). Seals required if conduit is 2 inches or larger.
e Sec. 501-6(a). All arcing devices must be explosionproof.
f Sec. 501-4(a). All boxes must be explosionproof and threaded for rigid or IMC conduit.
g Sec. 501-9(a)(4). All boxes and fittings for support of lighting fixtures must be approved for Class I locations.
h Sec. 501-9(a)(1). All lighting fixtures fixed or portable must be explosionproof.
i Sec. 501-9(a)(3). Pendant fixture stems must be threaded rigid or IMC conduit. Conduit stems if over 12 inches must have flexible connector, or must be braced.
j Sec. 501-14(a). All signal and alarm equipment irrespective of voltage must be approved for Class I, Division 1 locations.
k Sec. 501-12. Receptacles and plugs must be explosionproof and provide grounding connections for portable equipment.
l Sec. 501-5(f)(1). Breathers and drains needed in all humid locations.
m Sec. 501-4(a). All joints and fittings must be explosionproof.
n Sec. 501-8(a). Motor must be suitable for Class I.
p Art. 430. Motor overcurrent protection.

**Also applicable for Class I, Zone 1, see 505-15(b)*

17.1.20 Diagram for Class I, Division 1 Power Installation

17.1.20

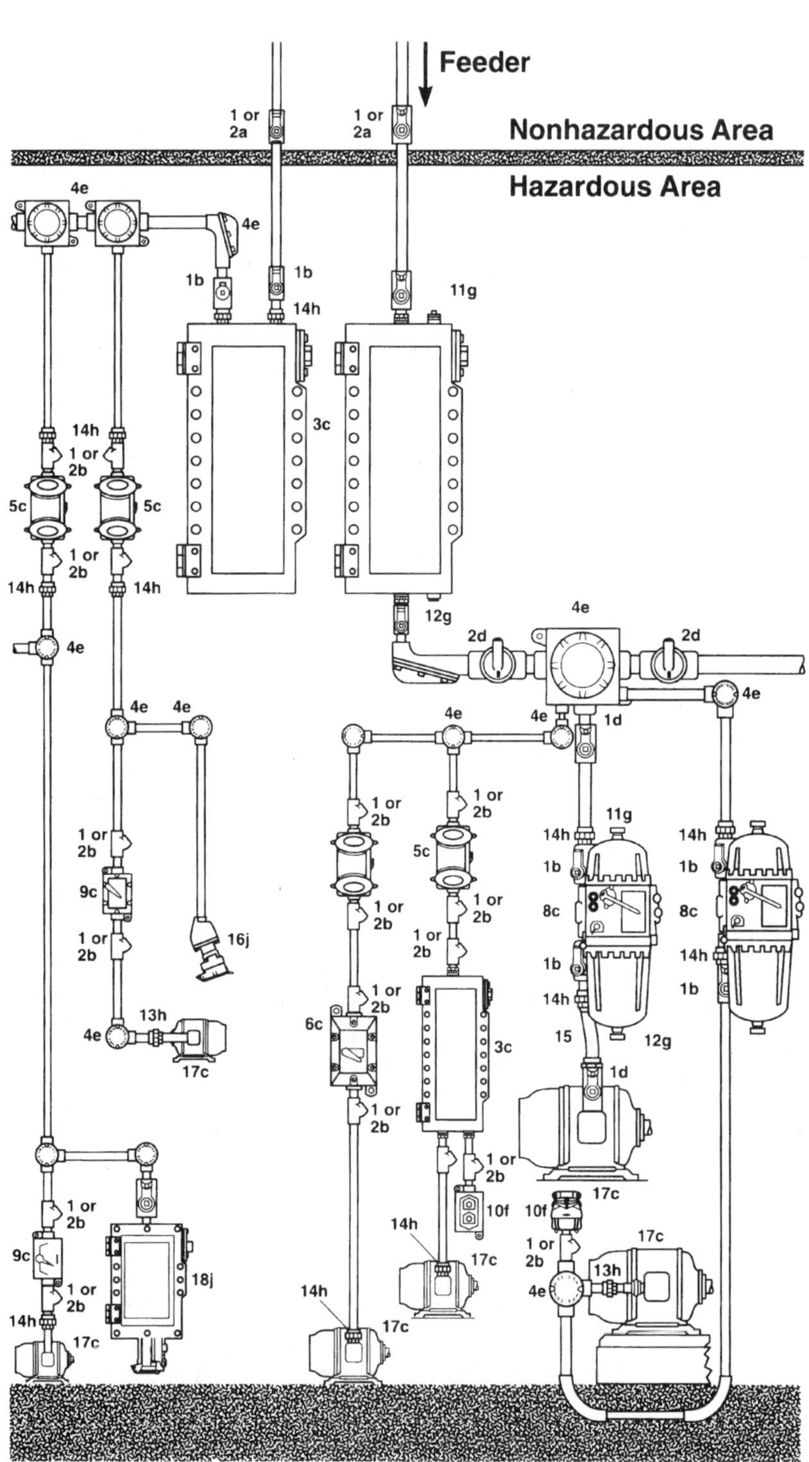

Key to Numerals

1 Sealing fitting. EYS for horizontal or vertical.
2 Sealing fitting. EZS for vertical or horizontal conduits.
3 Circuit breaker EBMB.
4 Junction box Series GUA, GUB, and GUJ have threaded covers. Series CPS and Type LBH have ground flat surface covers.
5 Circuit breaker FLB.
6 Manual line starter EMN.
7 Magnetic line starter EBMS.
8 Combination circuit breaker and line starter EPC.
9 Switch or motor starter. Series EFS, EDS, or EMN.
10 Pushbutton station. Series EFS or OAC.
11 Breather. ECD.
12 Drain. ECD.
13 Union. UNF.
14 Union. UNY.
15 Flexible coupling. EC.
16 Plug receptacle. CES. Factory sealed.
17 Motor for hazardous location.
18 Plug receptacle. EBBR Interlocked Arktite® receptacle with circuit breaker.

***National Electrical Code* References**

a Sec. 501-5(a)(4). Seals required where conduits pass from hazardous to nonhazardous area.
b Sec. 501-5(a)(1). Seals required within 18 inches of all arcing devices.
c Art. 430 should be studied for detailed requirements for conductors, motor feeders, motor feeder and motor branch circuit protection, motor overcurrent protection, motor controllers, and motor disconnecting means.
d Sec. 501-5(a)(2). Seals required if conduit is 2 inches or larger.
e Sec. 501-4(a). All boxes must be explosionproof and threaded for rigid or IMC conduit.
f Sec. 501-6(a). Pushbutton stations must be explosionproof.
g Sec. 501-5(f)(1). Breathers and drains needed in all humid locations.
h Sec. 501-4(a). All joints and fittings must be explosionproof.
i Sec. 501-4(a). Flexible connections must be explosionproof.
j Sec. 501-12. Receptacles and plugs must be explosionproof, and provide grounding connections for portable devices.

**Also applicable for Class I, Zone 1, see 505-15(b)*

17.1.21 Diagram for Class I, Division 2 Power and Lighting Installation

17.1.21

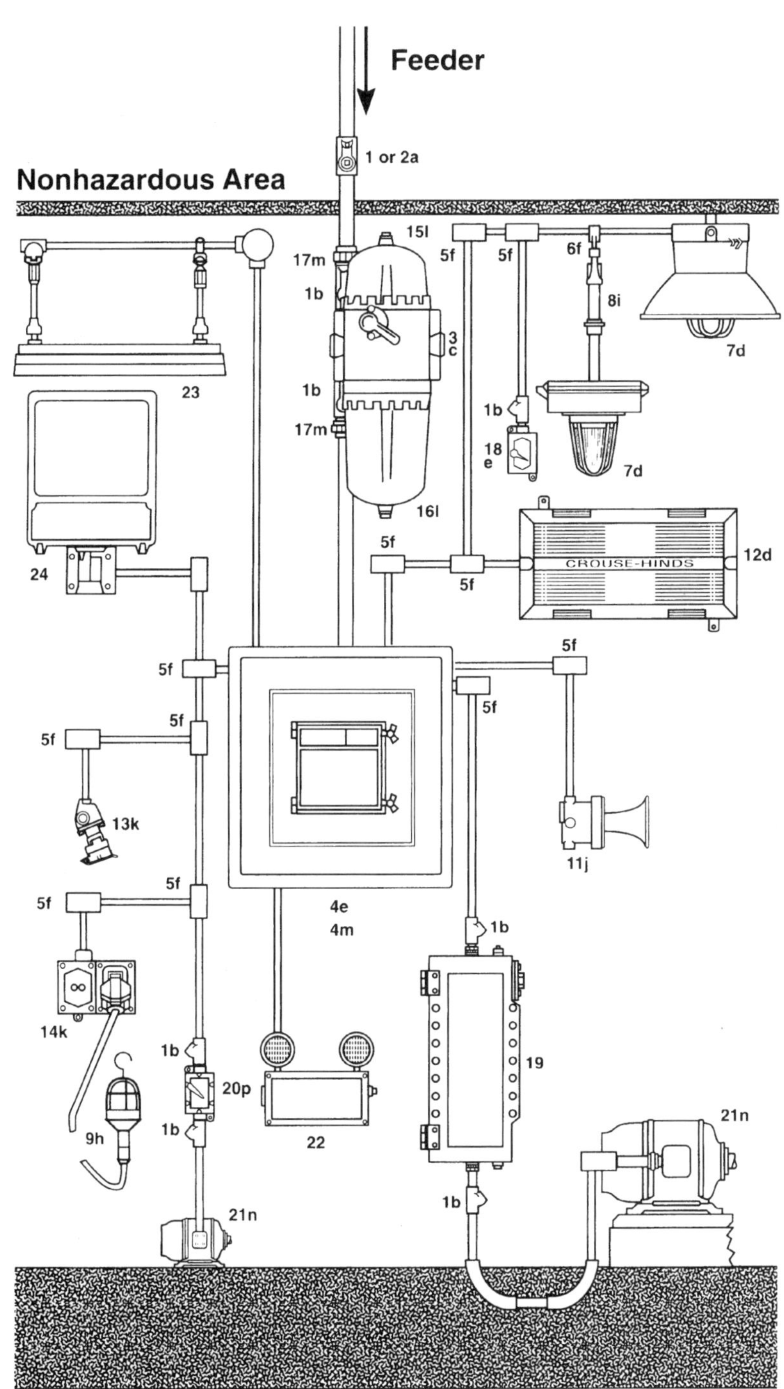

**Also applicable for Class I, Zone 2, see 505-15(c)*

Key to Numerals

1 Sealing fitting. EYS for horizontal or vertical.
2 Sealing fitting. EZS for vertical or horizontal conduits.
3 Circuit breaker. Type EPC.
4 Panelboard. D2PB, N2PB. Branch circuits are factory sealed.
5 Junction box or conduit fitting. NJB, Condulet®.
6 Fixture hanger. AHG, GS, UNJ.
7 Lighting fixture. VMV, DMV, and LMV (CHAMP®).
8 Flexible fixture support. ECHF.
9 Handlamp. EVH.
11 Signal. ETH horns and sirens. ESR bells, Flex•Tone™, and W2H.
12 Compact fluorescent lighting fixture. FVS.
13 Plug receptacle. CES delayed action.
14 Plug receptacle. ENR or CPS delayed action with GFS-1 ground fault circuit interrupter.
15 Breather. ECD.
16 Drain. ECD.
17 Union. UNY.
18 Switch. Series EFS.
19 Magnetic line starter. EBMS.
20 Manual line starter. EMN.
21 Motors. Suitable for Class I, Division 2 locations.
22 Emergency lighting system. N2LPS.
23 Fluorescent fixture. FVN.
24 Floodlight. FMV.

***National Electrical Code* References**

a Sec. 501-5(b)(2). Seal required where conduit passes from hazardous to nonhazardous area.
b Sec. 501-5(b)(1). Seals required within 18 inches of all arcing devices.
c Sec. 384-16. Circuit breaker protection required ahead of panelboard.
d Sec. 501-9(b)(2). All fixed lighting fixtures shall be enclosed and gasketed and not exceed ignition temperature of the gas.
e Sec. 501-6(b)(1). Most arcing devices must be explosionproof.
f Sec. 501-4(b). All boxes must be threaded for rigid or IMC conduit.
h Sec. 501-9(b)(1). All portable lighting fixtures must be explosionproof.
i Sec. 501-9(b)(3). Pendant fixture stems must be threaded rigid conduit or IMC. Rigid stems if over 12 inches must have flexible connector, or must be braced.
j Sec. 501-14(b). All signaling equipment must be approved for Class I location.
k Sec. 501-12. Receptacles and plugs must be explosionproof and provide grounding connections for portable equipment.
l Sec. 501-5(f)(1). Breathers and drains needed in all humid locations.
m Sec. 501-4(b). Not all joints and fittings are required to be explosionproof.
n Sec. 501-8(b). Motor shall be suitable for Division 2.
p Art. 430. Motor overcurrent protection.

17.1.22 Diagram for Class II Lighting Installation

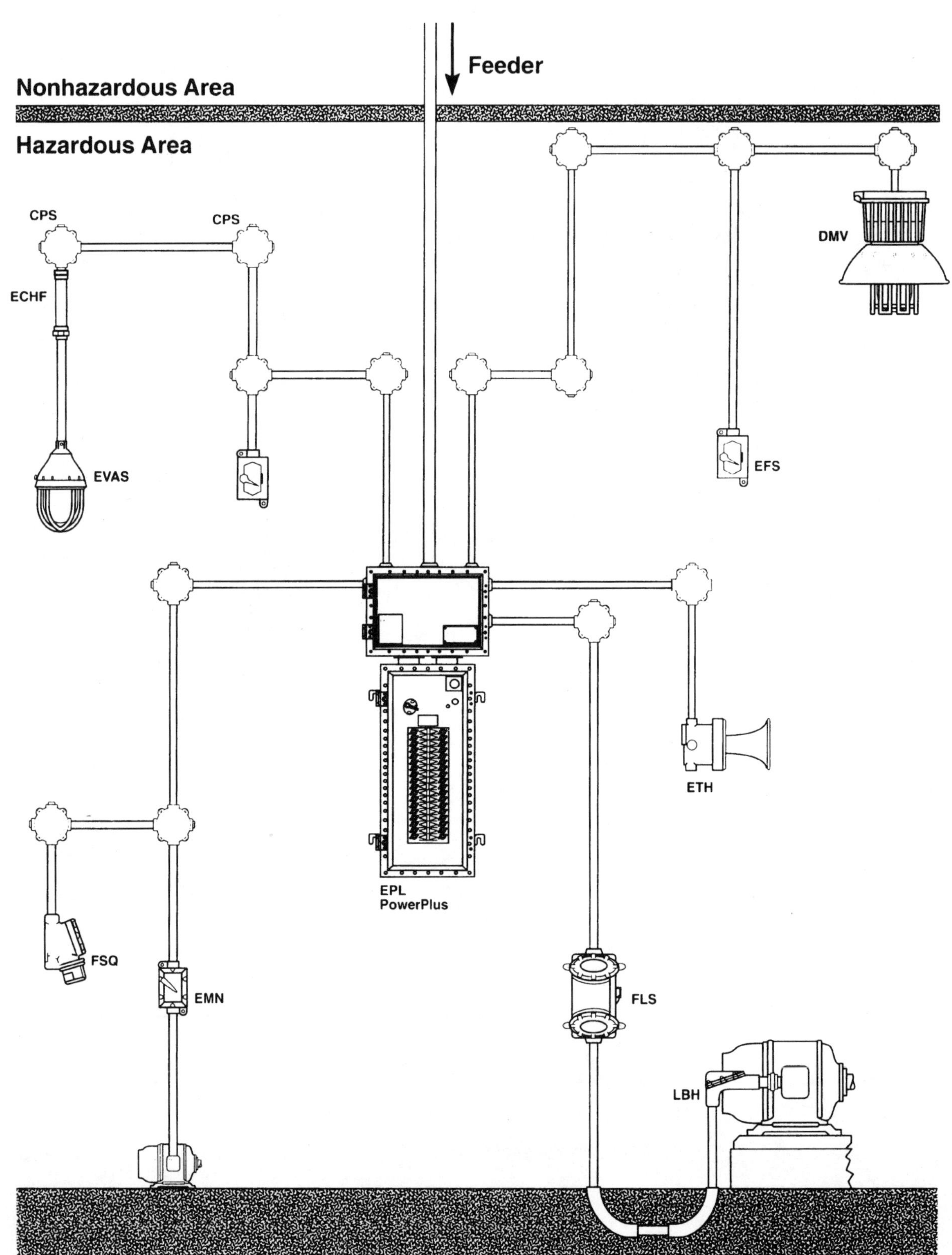

17.1.23 Diagram for Class II Power Installation

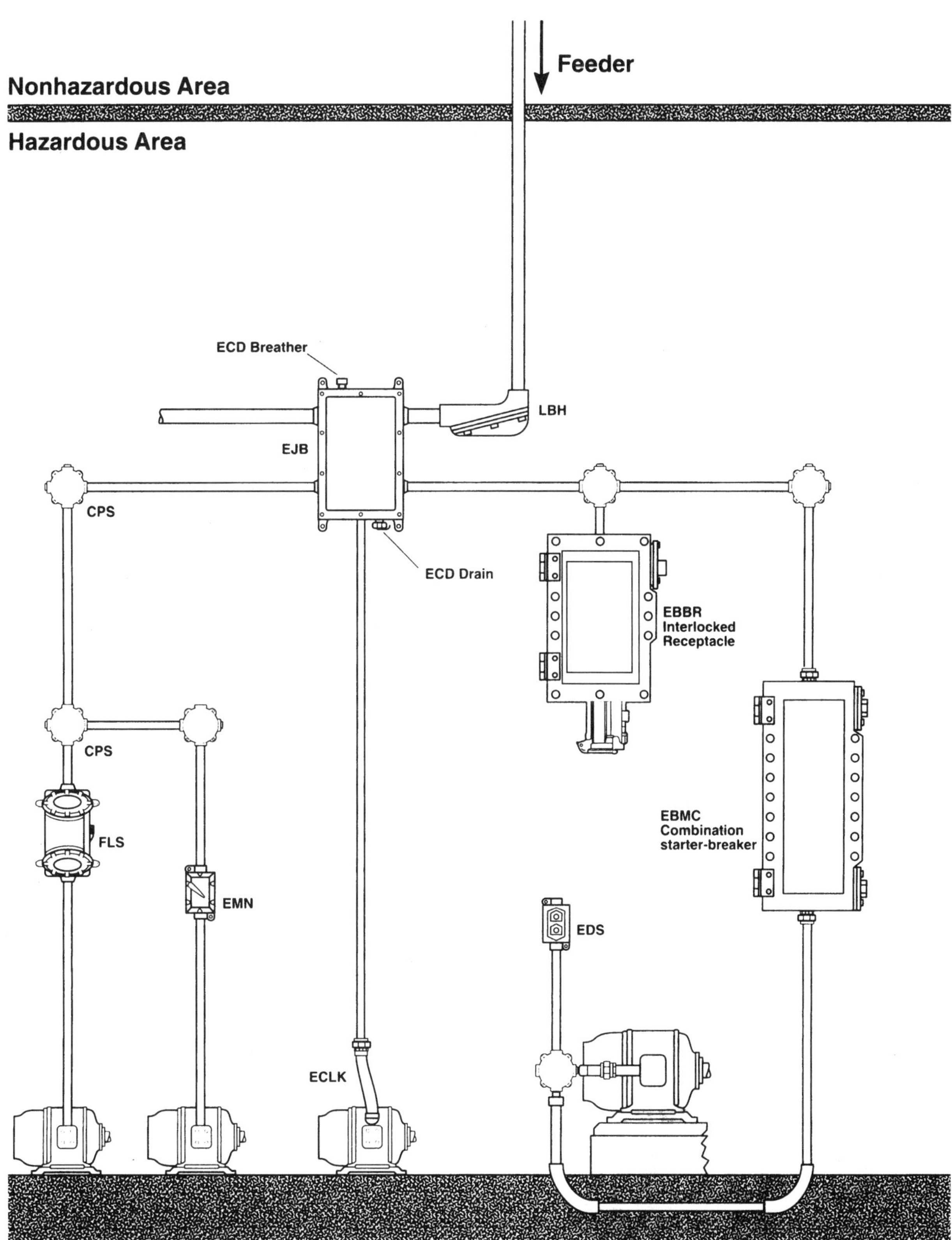

17.1.24 Crouse-Hinds "Quick-Selector": Electrical Equipment for Hazardous Locations

TABLE 17.1.24

Note: Not all types listed are suitable for use in all four groups for Class I hazardous locations. Consult catalog listing pages for applicable classes and groups.	**Class I Div. 1***	**Class I Div. 2†**	**Class II Div. 1** Groups E & F	**Class II Div. 1** Group G	**Class II Div. 2**	**Class III Div. 1**	**Class III Div. 2**
Conduit bodies and junction boxes	CPS EC EJB EJH EKC GUB LBH GUA, EAB EAJ UNF/UNY UNFL/UNYL	Form 7 Series, etc., W-Series, NJB and all products shown under Class I, Division 1	CPS EC EJB EJH EKC GUB LBH GUA, EAB, EAJ UNF/UNY UNFL/UNYL	Same if splices, taps, etc.	Form 7 Series, etc., W-Series, and all products shown under Class II, Division 1		
Switches	EDS, EFD, EFS, EHS, FLS, FSPC, GUSC, OAC, AF Series						
Panelboards	EPL ESPBH EWP EXD GUSC	D2PB D2D D2L ESPBH GUSC N2PB	EPL ESPBH EXD GUSC		All products shown under Class II, Division 1		
Lighting fixtures	ELPS EV EVF EVFDR EVH EVP EXL *HAZARD•GARD®* RCDE EVFT	FMV *CHAMP®* DMV *CHAMP®* FVN LMV *CHAMP®* NDA *Vaporgard™* VMV *CHAMP®* V Series N2LPS FVS N2MV	EVFT EVP DMV *CHAMP®* EV EVF EVFDR EXL FVN *HAZARD•GARD®* LMV *CHAMP®* VMV *CHAMP®* ELPS FVS				
Fixture hangers	COUP CPS EAHC EC EFHC EFHX GUA UNR	AHG EC UNE UNH UNJ UNJC	CPS EC EAHC EFHC EFHX GUA UNR				
Plugs and receptacles	APJ, BHP, BHR, CES, CESD, CPH, CPP, CPS, DR, ENP, ENR, FP, FSQ, NPJ, SRD, SP		Groups F, G only APJ, BHP, BHR, DR, ENP, ENR, FP, FSQ, NPJ, SRD, SP				
Interlocked plug receptacles	BHR, C2SR, EPC, FSQ, SRD EBBR, EPCB		Groups F, G only BHR, DBR, EPC, FSQ, SRD, EBBR				
Control stations and pilot lights	EDS, EFD, EFS, EMP, OAC, N2S (Class I, Division 2 only)						
Industrial control	EDS, EFD, EFS, EMN, EPC, EPCB, FSPC, OAC, EBM						
Circuit breakers	EDS, EFD, EPC, EPCB, FLB, GUSC		EDS, EFD, EPC, FLB, GUSC, EBM				
Telephones, instruments, and signals	EMH, ESR, ETW and D2TW Telephones, ETH, ETH *Flex•Tone™* Device, EV Strobe, GUB (W2H for Class I, Division 2)						

▲Denotes revision.

**Also suitable for Class I, Zone 1, see 505-15(b) and 505-20 (b)*

†Also suitable for Class I, Zone 2, see 505-15(c) and 505-20 (c)

17.1.25 Worldwide Explosion Protection Methods, Codes, Categories, Classifications, and Testing Authorities

TABLE 17.1.25

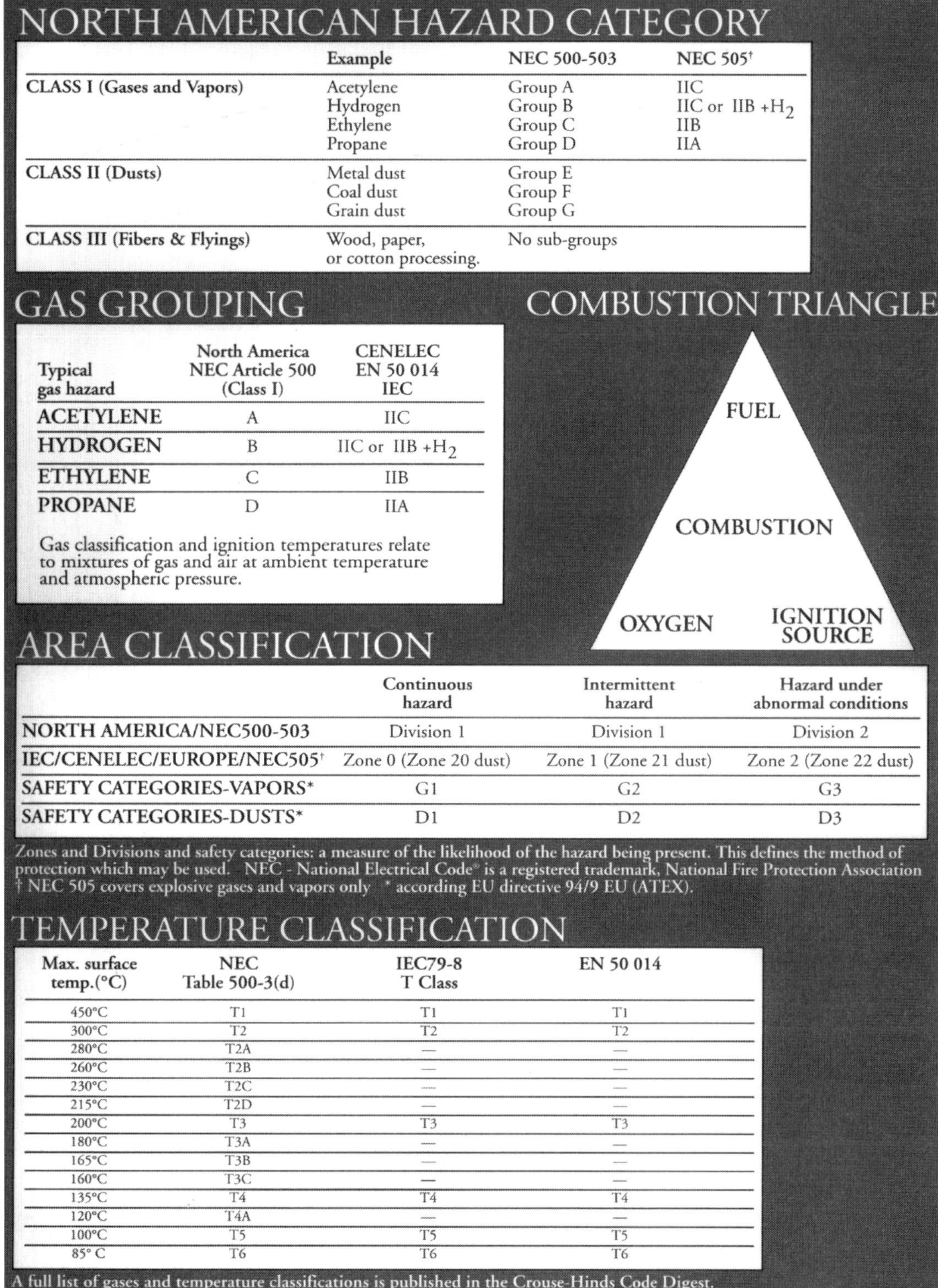

NORTH AMERICAN HAZARD CATEGORY

	Example	NEC 500-503	NEC 505[†]
CLASS I (Gases and Vapors)	Acetylene Hydrogen Ethylene Propane	Group A Group B Group C Group D	IIC IIC or IIB $+H_2$ IIB IIA
CLASS II (Dusts)	Metal dust Coal dust Grain dust	Group E Group F Group G	
CLASS III (Fibers & Flyings)	Wood, paper, or cotton processing.	No sub-groups	

GAS GROUPING

Typical gas hazard	North America NEC Article 500 (Class I)	CENELEC EN 50 014 IEC
ACETYLENE	A	IIC
HYDROGEN	B	IIC or IIB $+H_2$
ETHYLENE	C	IIB
PROPANE	D	IIA

Gas classification and ignition temperatures relate to mixtures of gas and air at ambient temperature and atmospheric pressure.

COMBUSTION TRIANGLE

AREA CLASSIFICATION

	Continuous hazard	Intermittent hazard	Hazard under abnormal conditions
NORTH AMERICA/NEC500-503	Division 1	Division 1	Division 2
IEC/CENELEC/EUROPE/NEC505[†]	Zone 0 (Zone 20 dust)	Zone 1 (Zone 21 dust)	Zone 2 (Zone 22 dust)
SAFETY CATEGORIES-VAPORS*	G1	G2	G3
SAFETY CATEGORIES-DUSTS*	D1	D2	D3

Zones and Divisions and safety categories: a measure of the likelihood of the hazard being present. This defines the method of protection which may be used. NEC - National Electrical Code® is a registered trademark, National Fire Protection Association † NEC 505 covers explosive gases and vapors only * according EU directive 94/9 EU (ATEX).

TEMPERATURE CLASSIFICATION

Max. surface temp.(°C)	NEC Table 500-3(d)	IEC79-8 T Class	EN 50 014
450°C	T1	T1	T1
300°C	T2	T2	T2
280°C	T2A	—	—
260°C	T2B	—	—
230°C	T2C	—	—
215°C	T2D	—	—
200°C	T3	T3	T3
180°C	T3A	—	—
165°C	T3B	—	—
160°C	T3C	—	—
135°C	T4	T4	T4
120°C	T4A	—	—
100°C	T5	T5	T5
85° C	T6	T6	T6

A full list of gases and temperature classifications is published in the Crouse-Hinds Code Digest.

(continued)

TABLE 17.1.25 (*Continued*)

METHODS OF EXPLOSION PROTECTION

Method of protection	Permitted in Division	Permitted in Zone	IEC Standard 79-	CENELEC Standard EN 50...	Code letter IEC Ex... CENELEC EEx...
GENERAL REQUIREMENTS	-	-	0	014	-
OIL IMMERSION	1 or 2	1 or 2	6	015	o
PRESSURIZATION	1 or 2	1 or 2	2	016	p
POWDER/SAND FILLING	2	1 or 2	5	017	q
FLAMEPROOF	-	1 or 2	1	018	d
INCREASED SAFETY	2	1 or 2	7	019	e
INTRINSIC SAFETY	1 or 2	0*,1 or 2	11	020 (apparatus) 039(systems)	ia or ib
FACTORY SEALING	1 or 2	1 or 2	-	-	-
ELECTRICAL APPARATUS with type of protection "n"^	2	2	15	021 (pending)	n
ENCAPSULATION	-	1 or 2	18	028	m
SPECIAL	-	1 or 2	None **	None	s

*ia: Zone 0,1,2; ib: Zone 1,2, not Zone 0 ** mentioned in IEC 79-0
^ includes non-sparking, restricted breathing, hermetically-sealed, non-incendive, etc.

NORTH AMERICAN AND EU TESTING AUTHORITIES

Country	Testing Authority	Country	Testing Authority
USA	Underwriters Laboratories	France	INERIS
USA	Factory Mutual	France	LCIE
USA	ETL	Italy	CESI
Canada	CSA	Netherlands	KEMA
Mexico	ANCE	Norway	NEMKO
Austria	TÜV-A	Spain	LOM
Austria	BVFA	Sweden	SP
Belgium	ISSEP	United Kingdom	BASEEFA/EECS
Denmark	DEMKO	United Kingdom	SCS
Finland	VTT	Germany	PTB
		Germany	BVS

(*continued*)

TABLE 17.1.25 (*Continued*)

INGRESS PROTECTION: (IP) CODES AND NEMA TYPES

FIRST NUMERAL Protection against solid bodies
0 - NO PROTECTION
1 - OBJECTS EQUAL TO OR GREATER THAN 50 mm
2 - OBJECTS EQUAL TO OR GREATER THAN 12.5 mm
3 - OBJECTS EQUAL TO OR GREATER THAN 2.5 mm
4 - OBJECTS EQUAL TO OR GREATER THAN 1.0 mm
5 - DUST-PROTECTED
6 - DUST-TIGHT

SECOND NUMERAL Protection against liquid
0 - NO PROTECTION
1 - VERTICALLY DRIPPING WATER
2 - 75 TO 105°-ANGLED DRIPPING WATER
3 - SPRAYING WATER
4 - SPLASHING WATER
5 - WATER JETS
6 - HEAVY SEAS, POWERFUL WATER JETS
7 - EFFECTS OF IMMERSION
8 - INDEFINITE IMMERSION

Example: IP56

Conversion of Enclosure Type numbers to IEC Classification Designations
(Cannot be used to convert IEC Classification Designations to NEMA Type numbers)

NEMA ENCLOSURE TYPE NUMBER	IEC ENCLOSURE CLASSIFICATION DESIGNATION
3	IP54
3R	IP54
3S	IP54
4 and 4X	IP56
5	IP52
6 and 6P	IP67
12 and 12K	IP52

NEMA and UL Types of Enclosures.

- **Type 3 Enclosure**
Type 3 enclosures are intended for outdoor use primarily to provide a degree of protection against windblown dust, rain, sleet and external ice formation.

- **Type 3R Enclosure**
Type 3R enclosures are intended for outdoor use primarily to provide a degree of protection against falling rain and external ice formation.

- **Type 3S Enclosure**
Enclosures are intended for outdoor use primarily to provide a degree of protection against rain, sleet, windblown dust, and to provide for operation of external mechanisms when ice laden.

- **Type 4 Enclosure**
Type 4 enclosures are intended for indoor or outdoor use primarily to provide a degree of protection against windblown dust and rain, splashing water, hose-directed water and external ice formation.

- **Type 4X Enclosure**
Type 4X enclosures are intended for indoor or outdoor use primarily to provide a degree of protection against corrosion, windblown dust and rain, splashing water, hose-directed water and external ice formation.

- **Type 7 Enclosure**
Type 7 enclosures are for use indoors in locations classified as Class I, Groups A, B, C or D, as defined in the National Electrical Code®.

- **Type 9 Enclosure**
Type 9 enclosures are for use in indoor locations classified as Class II, Groups E, F or G, as defined in the National Electrical Code®.

- **Type 12 Enclosure**
Type 12 enclosures are intended for indoor use primarily to provide a degree of protection against dust, falling dirt and dripping noncorrosive liquids.

Section

18

Telecommunications Structured Cabling Systems

18.1.0 Introduction
18.1.1 Important Codes and Standards
18.1.2 Comparison of ANSI/TIA/EIA, ISO/IEC, and CENELEC Cabling Standards
18.2.0 Major Elements of a Telecommunications Structured Cabling System.
18.2.1 Typical Ranges of Cable Diameter
18.2.2 Conduit Sizing-Number of Cables
18.2.3 Bend Radii Guidelines for Conduits
18.2.4 Guidelines for Adapting Designs to Conduits with Bends
18.2.5 Recommended Pull Box Configurations
18.2.6 Minimum Space Requirements in Pull Boxes Having One Conduit Each in Opposite Ends of the Box
18.2.7 Cable Tray Dimensions (Common Types)
18.2.8 Topology
18.2.9 Horizontal Cabling to Two Individual Work Areas
18.2.10 Cable Lengths
18.2.11 Twisted-Pair (Balanced) Cabling Categories
18.2.12 Optical Fiber Cable Performance
18.2.13 Twisted-Pair Work Area Cable
18.2.14 Eight-Position Jack Pin/Pair Assignments (TIA-568A)(Front View of Connector)
18.2.15 Optional Eight-Position Jack Pin/Pair Assignments (TIA-568B)(Front View of Connector)
18.2.16 Termination Hardware for Category-Rated Cabling Systems
18.2.17 Patch Cord Wire Color Codes
18.2.18 ANSI/TIA/EIA-568A Categories of Horizontal Copper Cables (Twisted-Pair Media)
18.2.19 Work Area Copper Cable Lengths to a Multi-User Telecommunications Outlet Assembly (MUTOA)
18.2.20 U.S. Twisted-Pair Cable Standards
18.2.21 Optical Fiber Sample Connector Types
18.2.22 Duplex SC Interface
18.2.23 Duplex SC Adapter with Simplex and Duplex Plugs
18.2.24 Duplex SC Patch Cord Crossover Orientation
18.2.25 Optical Fibers
18.2.26 Backbone System Components
18.2.27 Backbone Star Wiring Topology
18.2.28 Example of Combined Copper/Fiber Backbone Supporting Voice and Data Traffic
18.2.29 Backbone Distances
18.2.30 Determining 100 mm (4 in) Floor Sleeves

18.2.31 Determining Size of Floor Slots
18.2.32 Conduit Fill Requirements for Backbone Cable
18.2.33 TR Cross-Connect Field Color Codes
18.2.34 TR Temperature Ranges
18.2.35 TR Size Requirements
18.2.36 Allocating Termination Space in TRs
18.2.37 Typical Telecommunications Room (TR) Layout
18.2.38 TR Industry Standards
18.2.39 TR Regulatory and Safety Standards
18.2.40 Environmental Control Systems Standards for Equipment Rooms (ERs)
18.2.41 Underground Entrance Conduits for Entrance Facilities (EFs)
18.2.42 Typical Underground Installation to EF
18.2.43 Equipment Room (ER) Floor Space (Special-Use Buildings)
18.2.44 Entrance Facility (EF) Wall Space (Minimum Equipment and Termination Wall Space)
18.2.45 Entrance Facility (EF) Floor Space (Minimum Equipment and Termination Floor Space)
18.2.46 Separation of Telecommunications Pathways from 480-Volt or Less Power Lines
18.2.47 Cabling Standards Document Summary
18.3.0 Blown Optical Fiber Technology (BOFT) Overview
18.3.1 Diagram Showing Key Elements of BOFT System
18.3.2 BOFT Indoor Plenum 5-mm Multiduct
18.3.3 BOFT Outdoor 8-mm Multiduct
18.3.4 BOFT Installation Equipment

18.1.0 Introduction

Structured cabling is a term widely used to describe a generic voice, data, and video (telecommunications) cabling system design that supports a multiproduct, multivendor, and multimedia environment. It is an information technology (IT) infrastructure which provides direction for the cabling system design based on the end user's requirements, and it enables cabling installations where there is little or no knowledge of the active equipment to be installed.

The following provides an overview of the industry standards.

18.1.1 Important Codes and Standards

- American National Standards Institute (ANSI)
- Canadian Standards Association (CSA)
- Comité Européen de Normalisation Electrotechnique (CENELEC)
- Federal Communications Commission (FCC)
- Insulated Cable Engineers Association (ICEA)
- International Electrotechnical Commission (IEC)
- Institute of Electrical and Electronics Engineers, Inc. (IEEE)
- International Organization for Standardization (ISO)
- International Organization for Standardization/International Electrotechnical Commission Joint Technical Committee Number 1 (ISO/IEC JTC1)
- U.S. National Fire Protection Association (NFPA)
- National Research Council of Canada, Institute for Research in Construction

(NRC-IRC)

- Telecommunications Industry Association/Electronic Industries Alliance (TIA/EIA)

18.1.2 Comparison of ANSI/TIA/EIA, ISO/IEC, and CENELEC Cabling Standards

TABLE 18.1.2*

	ANSI/TIA/EIA-568-A, TSBs and addenda	ISO/IEC 11801:1995 and amendments	CENELEC EN 50173:1995 and amendments
100 ohm balanced cable	Supported	Supported	Supported
120 ohm balanced cable	Not supported	Supported	Supported
150 ohm STP cable	Supported[1]	Supported[1]	Supported[1]
50/125 μm multimode fiber	Not supported[2]	Supported	Supported
62.5/125 μm multimode fiber	Supported	Supported	Supported
Singlemode fiber	Supported	Supported	Supported
Component categories	Category 3, 4[3], 5[4]	Category 3, 4[3], 5[5]	Category 3, 5[5]
Link and channel specifications	Category 3, 4[3], 5[4], 5e	Class A, B, C, D[5]	Class A, B, C, D[5]
Backbone cable types	100 ohm 150 ohm STP[1] 62.5 μm fiber[2] singlemode fiber	100 or 120 or 150[1] ohm (100 ohm preferred) 50 or 62.5 μm fiber (62.5 μm preferred) singlemode fiber	100 or 120 ohm (100 ohm preferred) 50 or 62.5 μm fiber (62.5 μm preferred) singlemode fiber
Horizontal cable types	100 ohm 150 ohm STP[1] 62.5 μm fiber[2] (choice depends on application)	100 or 120 or 150[1] ohm (100 ohm preferred) 50 or 62.5 μm fiber (62.5 μm preferred) singlemode fiber	100 or 120 or 150[1] ohm (100 ohm preferred) 50 or 62.5 μm fiber (62.5 μm preferred) singlemode fiber

(continued)

TABLE 18.1.2 (*Continued*)*

	ANSI/TIA/EIA-568-A, TSBs and addenda	ISO/IEC 11801:1995 and amendments	CENELEC EN 50173:1995 and amendments
TO cable recommendations	1st TO: 100 ohm (Category 3 minimum) + 2nd TO: 100 ohm (Category 5[4] required) or 150 ohm STP[1] or 62.5 μm multimode[2]	1st TO: 100 or 120 ohm (Category 3 minimum) + 2nd TO: 100 or 120 ohm (Category 5[5] recommended) or 150 ohm STP[1] or 62.5 μm multimode	1st TO: 100 or 120 ohm (Category 5[5] recommended) + 2nd TO: 100 or 120 ohm (Category 5[5] recommended) or 150 ohm STP[1] or 62.5 μm multimode
Twisted-pair outlet configuration	4 pairs required Configured either T568A or T568B	2 or 4 pairs allowed (4 pairs recommended) Configured pairs to pins	2 or 4 pairs allowed (no preference) Configured pairs to pins
Attenuation of flexible (stranded) cordage	Up to 120% of horizontal cable allowed	Up to 150% of horizontal cable allowed	Up to 150% of horizontal cable allowed
Application mapping	None included[6]	Comprehensive guidance in Annex G	Guidance in Annex F

[1] STP-A cabling and components will not be recommended for new installations in ANSI/TIA/EIA-568-B.1 and will be deleted from the next editions of ISO/IEC 11801 and EN 50173. Requirements for 100 ohm ScTP are provided in TIA/EIA IS 729.

[2] Requirements for 50/125 μm fiber will be specified in ANSI/TIA/EIA-568-B.1.

[3] Category 4 requirements will not be provided in ANSI/TIA/EIA-568-B.1 or in the next edition of ISO/IEC 11801.

[4] Specifications for Category 5 cabling and components will be replaced by Category 5e in ANSI/TIA/EIA-568-B.1 and ANSI/TIA/EIA-568-B.2. Category 5 values will be provided for information only.

[5] ISO/IEC and CENELEC Category 5 and Class D requirements will be aligned with TIA Category 5e component and cabling specifications in the next editions of ISO/IEC 11801 and EN 50173.

[6] ANSI/TIA/EIA-568-B.1 will provide application mapping for optical fiber cabling.

**Here, and throughout chapter, indicates that this material is reprinted with permission from BICSI's Telecommunications Distribution Methods Manual, 9th Edition.*

18.2.0 Major Elements of a Telecommunications Structured Cabling System.

- Horizontal pathway systems
- Horizontal cabling systems
- Backbone distribution systems

- Backbone building pathways
- Backbone building cabling
- Work areas (WAs)
- Telecommunications Outlets (TOs)
- Telecommunications Rooms (TRs)
- Equipment Rooms (ERs)
- Telecommunications Entrance Facilities (EFs)

The data which follows provides key data and details for these major elements.

18.2.1 Typical Ranges of Cable Diameter

TABLE 18.2.1*

Horizontal Cable Type	Typical Range of Overall Diameter
4-pair 100 Ω UTP or ScTP (FTP)	3.6 mm to 6.3 mm (0.14 in to 0.25 in)
2-fiber optical cable	2.8 mm to 4.6 mm (0.11 in to 0.18 in)
4-pair 100 Ω STP	7.9 mm to 11 mm (0.31 in to 0.43 in)

NOTES: FTP = Foiled twisted-pair; ScTP = Screened twisted-pair; STP = Shielded twisted-pair; UTP = Unshielded twisted-pair

18.2.2 Conduit Sizing-Number of Cables

TABLE 18.2.2*

Inside Diameter mm	Trade Size	Cable Outside Diameter mm (in)									
		3.3 (0.13)	4.6 (0.18)	5.6 (0.22)	6.1 (0.24)	7.4 (0.29)	7.9 (0.31)	9.4 (0.37)	13.5 (0.53)	15.8 (0.62)	17.8 (0.70)
16	1/2	1	1	0	0	0	0	0	0	0	0
21	3/4	6	5	4	3	2	2	1	0	0	0
27	1	8	8	7	6	3	3	2	1	0	0
35	1-1/4	16	14	12	10	6	4	3	1	1	1
41	1-1/2	20	18	16	15	7	6	4	2	1	1
53	2	30	26	22	20	14	12	7	4	3	2
63	2-1/2	45	40	36	30	17	14	12	6	3	3
78	3	70	60	50	40	20	20	17	7	6	6
91	3-1/2							22	12	7	6
103	4							30	14	12	7

NOTE: These conduit sizes are typical in the United States and Canada, and may vary in other countries.

18.2.3 Bend Radii Guidelines for Conduits

TABLE 18.2.3*

If the Conduit Has an Internal Diameter of...	The Bend Radius Must Be at Least...
50 mm (2 in) or less	6 times the internal conduit diameter.
More than 50 mm (2 in)	10 times the internal conduit diameter.

NOTE: For additional information on conduit bend radius requirements and recommendations in the United States, see specifications in the *NEC* (Chapter 9) and ANSI/TIA/EIA-569-A, (Chapter 5, Table 5.2-1). In Canada, refer to CSA-C22.1 (Sections 12-900 through 12-2502) and CSA-T530. These specifications provide bend radius guidelines for standard trade-size conduits.

18.2.4 Guidelines for Adapting Designs to Conduits with Bends

TABLE 18.2.4*

If a Conduit Run Requires...	Then...
More than two 90 degree bends	Provide a pull box (PB) between sections with two bends or less.
A reverse bend (between 100 degree and 180 degree)	Insert a pull point or PB at each bend having an angle from 100 degree to 180 degree.
A third 90 degree bend (between pull points or PBs)	For this additional bend, derate the design capacity by 15 percent.

NOTE: Consider an offset as equivalent to a 90 degree bend.

18.2.5 Recommended Pull Box Configurations

18.2.5*

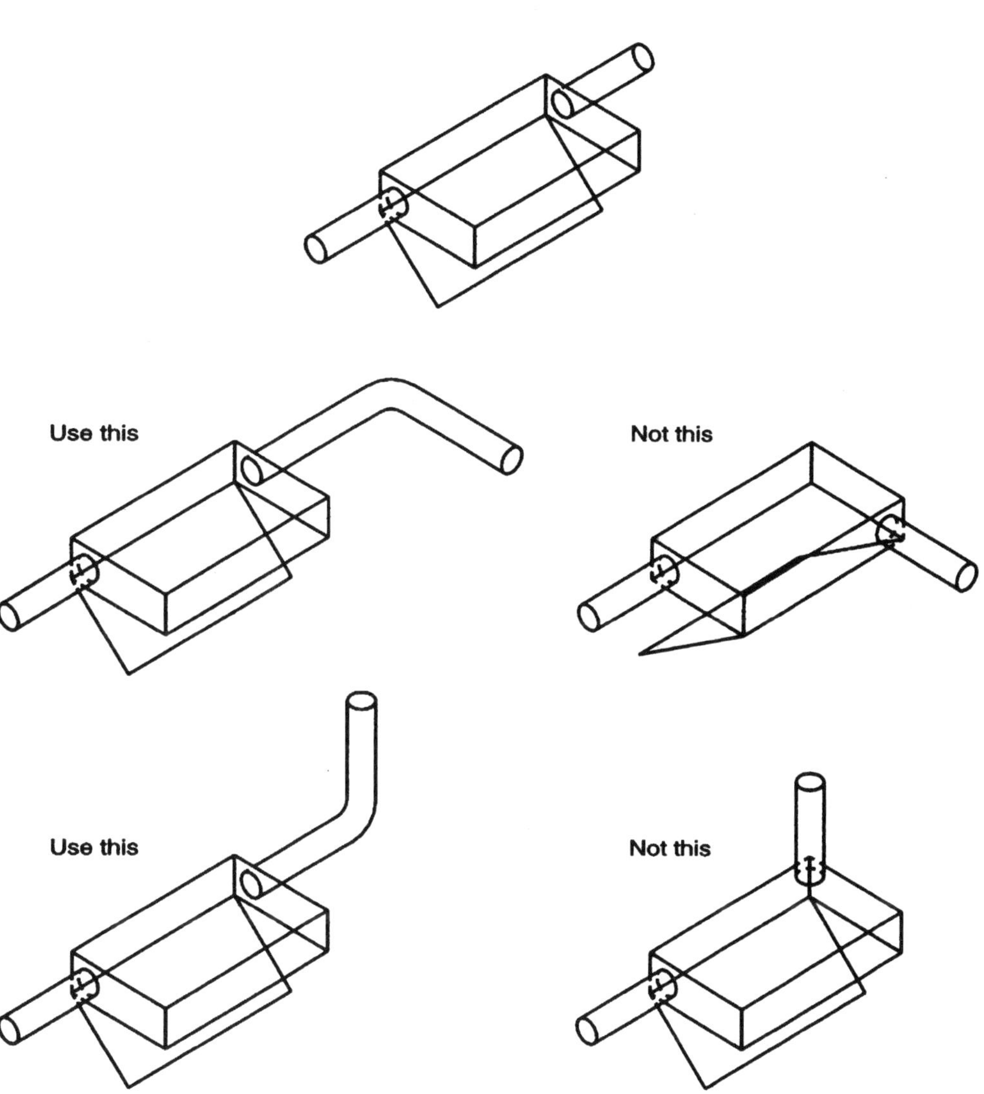

18.2.6 Minimum Space Requirements in Pull Boxes Having One Conduit Each in Opposite Ends of the Box

TABLE 18.2.6*

Maximum Trade Size of Conduit	Size of Box			For Each Additional Conduit Increase Width
	Width	Length	Depth	
21 mm (3/4)	100 mm (4 in)	300 mm (12 in)	75 mm (3 in)	50 mm (2 in)
27 mm (1)	100 mm (4 in)	400 mm (16 in)	75 mm (3 in)	50 mm (2 in)
35 mm (1-1/4)	150 mm (6 in)	500 mm (20 in)	75 mm (3 in)	75 mm (3 in)
41 mm (1-1/2)	200 mm (8 in)	675 mm (27 in)	100 mm (4 in)	100 mm (4 in)
53 mm (2)	200 mm (8 in)	900 mm (36 in)	100 mm (4 in)	125 mm (5 in)
63 mm (2-1/2)	250 mm (10 in)	1050 mm (42 in)	125 mm (5 in)	150 mm (6 in)
78 mm (3)	300 mm (12 in)	1200 mm (48 in)	125 mm (5 in)	150 mm (6 in)
91 mm (3-1/2)	300 mm (12 in)	1350 mm (54 in)	150 mm (6 in)	150 mm (6 in)
103 mm (4)	375 mm (15 in)	1520 mm (60 in)	200 mm (8 in)	200 mm (8 in)

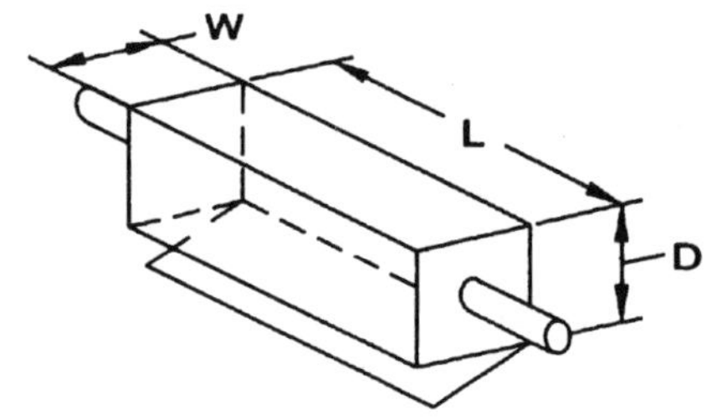

18.2.7 Cable Tray Dimensions (Common Types)

TABLE 18.2.7*

	Ladder	Ventilated Trough	Ventilated Channel	Solid-Bottom
Lengths	3.7 m (12 ft) 7.3 m (24 ft)	3.7 m (12 ft) 7.3 m (24 ft)	3.7 m (12 ft) 7.3 m (24 ft)	3.7 m (12 ft) 7.3 m (24 ft)
Widths (Inside)	150 mm (6 in) 300 mm (12 in) 450 mm (18 in) 600 mm (24 in) 750 mm (30 in) 900 mm (36 in)	150 mm (6 in) 300 mm (12 in) 450 mm (18 in) 600 mm (24 in) 750 mm (30 in) 900 mm (36 in)	75 mm (3 in) 100 mm (4 in) 150 mm (6 in) — — — — — —	150 mm (6 in) 300 mm (12 in) 450 mm (18 in) 600 mm (24 in) 750 mm (30 in) 900 mm (36 in)
	NOTE: The side rail outside depths (height) can be as much as 32 mm (1-1/4 in) more than the inside loading depth for ladder, ventilated trough, and solid bottom cable tray.			
Depths	75 mm (3 in) 100 mm (4 in) 125 mm (5 in) 150 mm (6 in)	75 mm (3 in) 100 mm (4 in) 125 mm (5 in) 150 mm (6 in)	32 mm (1-1/4 in) 45 mm (1-3/4 in) — — — —	75 mm (3 in) 100 mm (4 in) 125 mm (5 in) 150 mm (6 in)
Rung spacing	150 mm (6 in) 225 mm (9 in) 300 mm (12 in) 450 mm (18 in)	— — — — — — — —	— — — — — — — —	— — — — — — — —
Radii	300 mm (12 in) 600 mm (24 in) 900 mm (36 in)	300 mm (12 in) 600 mm (24 in) 900 mm (36 in)	300 mm (12 in) 600 mm (24 in) 900 mm (36 in)	300 mm (12 in) 600 mm (24 in) 900 mm (36 in)
Degrees of arc	30° 45° 60° 90°	30° 45° 60° 90°	30° 45° 60° 90°	30° 45° 60° 90°
Transverse element spacing	— —	100 mm (4 in)	— —	— —

18.2.8 Topology

ANSI/EIA/TIA-568A specifies a star topology—a hierarchical series of distribution levels. Each WA outlet must be cabled directly to a horizontal cross-connect {HC [floor distributor (FD)]} in the telecommunications room (TR) except when a consolidation point (CP) is required to open office cabling, or a transition point (TP) is required to connect undercarpet cable. Horizontal cabling should be terminated in a TR that is on the same floor as the area being served.

NOTES: Splices are not permitted for twisted-pair horizontal cabling.

Bridged taps (multiple appearances of the same cable pairs at several distribution points) are not permitted in horizontal cabling.

Cabling between TRs is considered part of the backbone cabling. Such connections between TRs may be used for configuring "virtual bus" and "virtual ring" cabling schemes using a star topology.

18.2.9 Horizontal Cabling to Two Individual Work Areas

18.2.9*

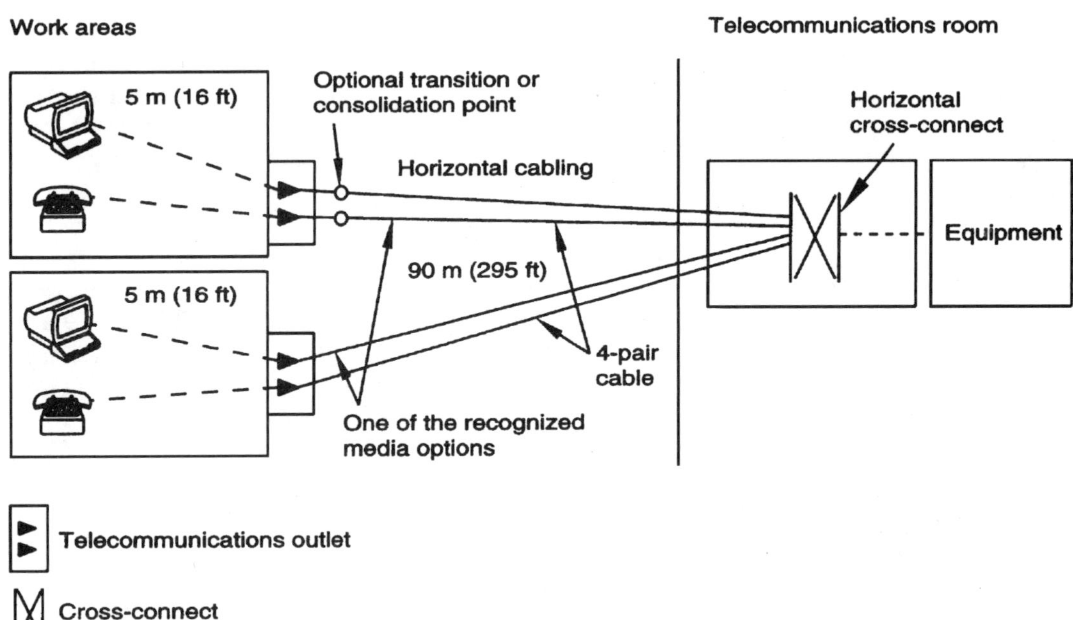

NOTES: Provided that the minimum requirements are met for horizontal cabling to each individual WA, additional cables and outlets may be provided to support other applications such as CATV.

Label all cables that are left unterminated in walls or other horizontal spaces according to the requirements of ANSI/TIA/EIA-606 (see Chapter 2: Codes, Standards, and Regulations). In Canada, refer to CSA-T528. Cables that extend to outlet boxes must be covered with an outlet faceplate and identified for telecommunications use only.

18.2.10 Cable Lengths

TABLE 18.2.10*

Horizontal Cables...	Must Be No More Than...
From the HC (FD) to the outlet/connector	90 m (295 ft) long.
Used for patch cords and cross-connect jumpers in the HC (FD)	5 m (16 ft) long. (See Note.)

NOTE: In establishing limits on horizontal cable lengths, a 10 m (33 ft) allowance was made for the combined length of patch cords and cables used to connect equipment in the WA and TR. All equipment cords should meet the same performance requirements as the patch cords. Equipment cords differ from patch cables and cross-connect jumpers in that they attach directly to active equipment; patch cords and cross-connect jumpers do not attach directly to active equipment.

18.2.11 Twisted-Pair (Balanced) Cabling Categories

TABLE 18.2.11*

Category	Definition
Category 3	This category consists of cables and connecting hardware specified up to 16 MHz. The performance of Category 3 cabling links corresponds to application Class C links as originally specified in ISO/IEC 11801 and CENELEC EN 50173.
Category 5	This category consists of cables and connectors specified up to 100 MHz. The performance of Category 5 cabling links corresponds to application Class D links as originally specified in ISO/IEC 11801 and CENELEC EN 50173.
Category 5e	This category consists of cables and connectors specified up to 100 MHz. Category 5e transmission performance of Category 5e cabling is specified in ANSI/TIA/EIA-568-A-5 and is intended to support applications that use more than one pair to transmit in each direction.
Category 6	This category consists of cables and connectors specified up to 250 MHz. The performance of Category 6 cabling links corresponds to application Class E links to be specified in ISO/IEC 11801 and CENELEC EN 50173.
Category 7	This category consists of shielded cables and connectors specified up to 600 MHz. The performance of Category 7 cabling links corresponds to application Class F links to be specified in ISO/IEC 11801 and CENELEC EN 50173.

NOTES:
Categories 1 and 2 are not recognized cables.
Category 4 is not recommended.
Categories 3 and 5e meet ANSI/TIA/EIA-568-B.1 and B.2.
Categories 6 and 7 specifications are under development in TIA and ISO/IEC.

18.2.12 Optical Fiber Cable Performance

TABLE 18.2.12*

Fiber Type	Fiber Performance
62.5/125 µm	Minimum bandwidth of 160 and 500 MHz • km at 850 and 1300 nm respectively.
50/125 µm	Minimum bandwidth of 500 and 500 MHz • km at 850 and 1300 nm respectively.

18.2.13 Twisted-Pair Work Area Cable

18.2.13*

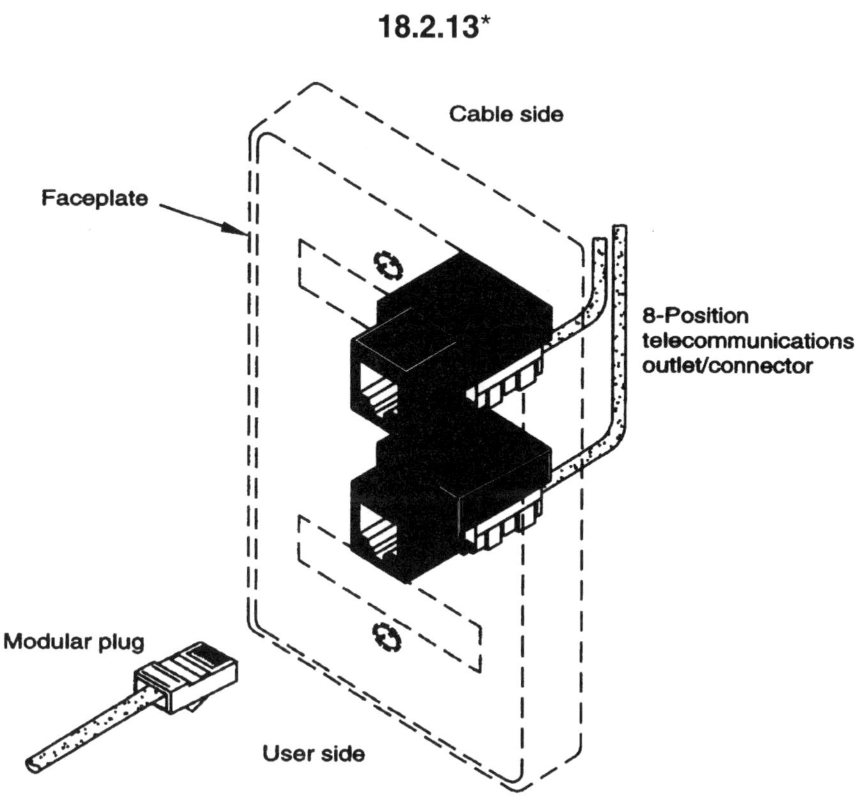

18.2.14 Eight-Position Jack Pin/Pair Assignments (TIA-568A)(Front View of Connector)

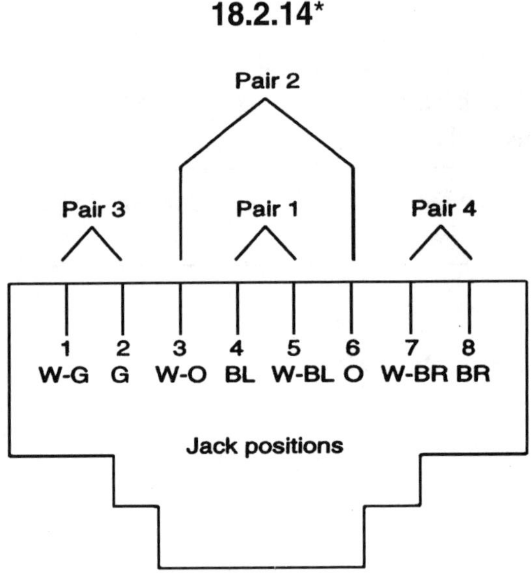

18.2.15 Optional Eight-Position Jack Pin/Pair Assignments (TIA-568B)(Front View of Connector)

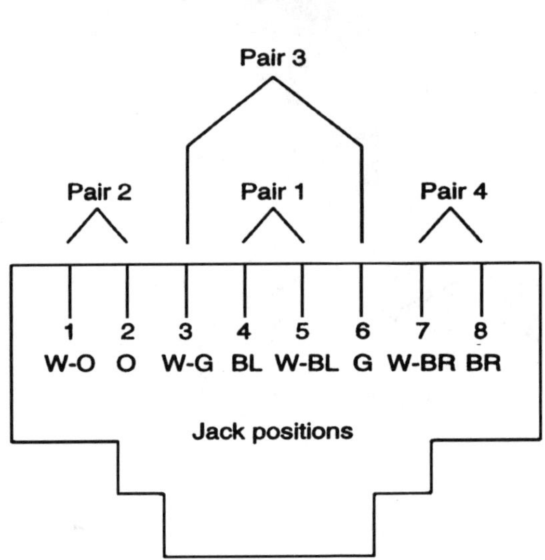

NOTE: The colors indicated are associated with horizontal twisted-pair cable. Color-coding for equipment cables, WA cords, patch cords, and jumpers may vary.

18.2.16 Termination Hardware for Category-Rated Cabling Systems

TABLE 18.2.16

Termination Hardware	Category 3	Category 4	Category 5
Screw terminals	(1)	–	–
25 pair connector	(2)	(2)	(2)
66-clip	Yes	Yes	(2)
110	Yes	Yes	Yes
Krone®	Yes	Yes	(2)
BIX®	Yes	Yes	(2)

Note (1): If the application specifically requires it.
Note (2): Some versions comply; check with the manufacturer.

18.2.17 Patch Cord Wire Color Codes

TABLE 18.2.17

Conductor Identification (1)	Wire Color
Pair 1 + Pair 1 -	White (2) Blue (3)
Pair 2 + Pair 2 -	White (2) Orange (3)
Pair 3 + Pair 3 -	White (2) Green (3)
Pair 4 + Pair 4 -	White (2) Brown (3)

Notes: (1) + = Tip, - = Ring
(2) Mostly white wire may have the associate color as a band or stripe.
(3) Mostly colored wire may have white as a band or stripe.

18.2.18 ANSI/TIA/EIA-568A Categories of Horizontal Copper Cables (Twisted-Pair Media)

TABLE 18.2.18*

Designation	Definition
Category 1, 2	These twisted-pair cables are not recognized in the ANSI/TIA/EIA-568-A standard. They are typically used for voice and low speed data (9600 b/s or less) transmission rates.
Category 3	This designation applies to twisted-pair cable and connection hardware currently specified in the ANSI/TIA/EIA-568-A standard. The characteristics of these cables are specified up to 16 MHz. They are typically used for voice and data transmission rates up to 10 Mb/s (e.g., IEEE 802.5 4 Mb/s twisted-pair annex and IEEE 802.3 10BASE-T).
Category 4	The characteristics of these twisted-pair cabling components are specified up to 20 MHz. They are intended to be used for voice and data transmission rates up to and including, 16 Mb/s (e.g., IEEE 802.5 16 Mb/s twisted-pair standard).
Category 5	The characteristics of these twisted-pair cabling components are specified up to 100 MHz. They are intended to be used for voice and data transmission rates up to and greater than, 16 Mb/s (e.g., IEEE 802.5 16 Mb/s twisted-pair standard and ANSI X3T9.5 100 Mb/s twisted-pair physical-media dependent [TP-PMD]).
Category 5e	The characteristics of Category 5e cabling components are specified up to 100 MHz, with additional transmission parameters necessary to support applications that make use of all four pairs in the cable for simultaneous bidirectional transmission (such as IEEE 802.3 1000BASE-T).
Category 6*	Continued development of high-speed applications drove the need for more bandwidth than Category 5e cabling systems. Category 6 channels have a power sum ACR that is greater than zero at 200 MHz, and parameters are specified up to 250 MHz.
Category 7**	Cabling consists of four individually shielded twisted-pairs having nominal impedance of 100 Ω. Category 7 cable requires a new fully-shielded connector design, which is still under development. Category 7 cabling has a bandwidth of 500 MHz (PSACR > 0) and the parameters are specified to 600 MHz.
STP-A	The characteristics of these 150 Ω STP cabling components are specified up to 300 MHz. These cables consist of two individually twisted-pairs of 22 AWG [0.64 mm (0.025 in)] conductors enclosed by a shield and an overall jacket.

* Proposed

** Under consideration in ISO/IEC 11801

18.2.19 Work Area Copper Cable Lengths to a Multi-User Telecommunications Outlet Assembly (MUTOA)

TABLE 18.2.19*

Length of Horizontal Cables H	Maximum Length of Work Area Cords W	Maximum Combined Length of Work Area Cords, Patch Cords, and Equipment Cables C
90 m (295 ft)	5 m (16 ft)	10 m (33 ft)
85 m (279 ft)	9 m (30 ft)	14 m (46 ft)
80 m (262 ft)	13 m (44 ft)	18 m (60 ft)
75 m (246 ft)	17 m (57 ft)	22 m (71 ft)
70 m (230 ft)	22 m (71 ft)	27 m (89 ft)

18.2.20 U.S. Twisted-Pair Cable Standards

TABLE 18.2.20*

Parameter	EIA	IBM	UL	NEMA	Telcordia	ICEA
Published specification	ANSI/TIA/ EIA-568-A	GA27-3773-1	200-131A	WC63	TA-NWT-000133	S80-576
Conductor sizes (AWG)	22, 24	22, 24, 26	22, 24	22, 24, 26	24	22, 24, 26
Impedance (ohms)	100	150	100	100, 150	100	Not specified
Cable sizes (Pairs)	4 to 25-Pair Subunits	2 to 6	25 or less	6 or less	Any	3600 or less
Shielding	UTP/STP-A	STP	STP/UTP	STP/UTP	UTP*	STP/UTP
Performance	Category: 1-5e	Type: 1-9	Category: 1-5e	Standard; low loss; low loss extended frequency	Category: 1-5e	Not specified
Equivalence to ANSI/TIA/ EIA-568-A	1 2 3 4 5 5e	(none) Type 3 (none) (none) (none)	1 2 3 4 5 5e	(none) (none) Standard low loss low loss, extended frequency	1 2 3 4 5	(none) (none) (none) (none) (none)

* The technical advisory does not preclude STP.

18.2.21 Optical Fiber Sample Connector Types

18.2.21*

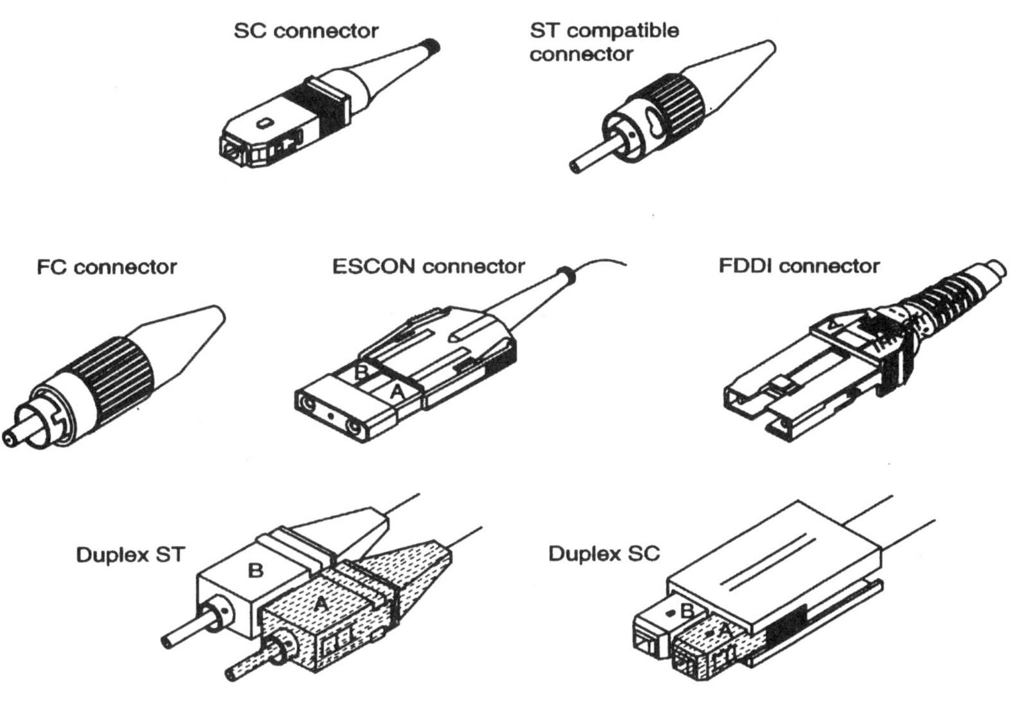

18.2.22 Duplex SC Interface

18.2.22*

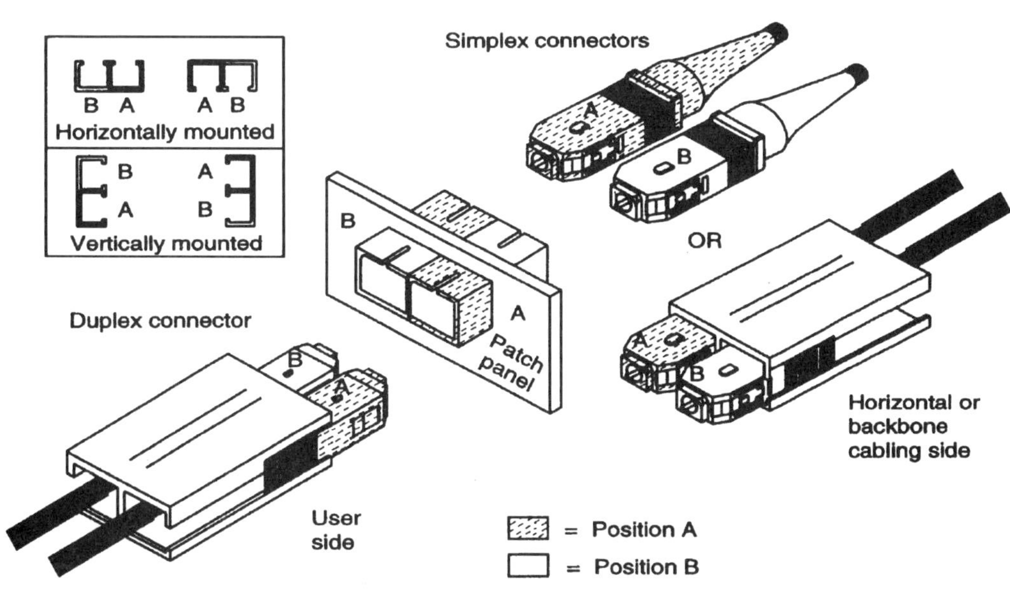

NOTE: Shading for clarification only.

18.2.23 Duplex SC Adapter with Simplex and Duplex Plugs

18.2.23*

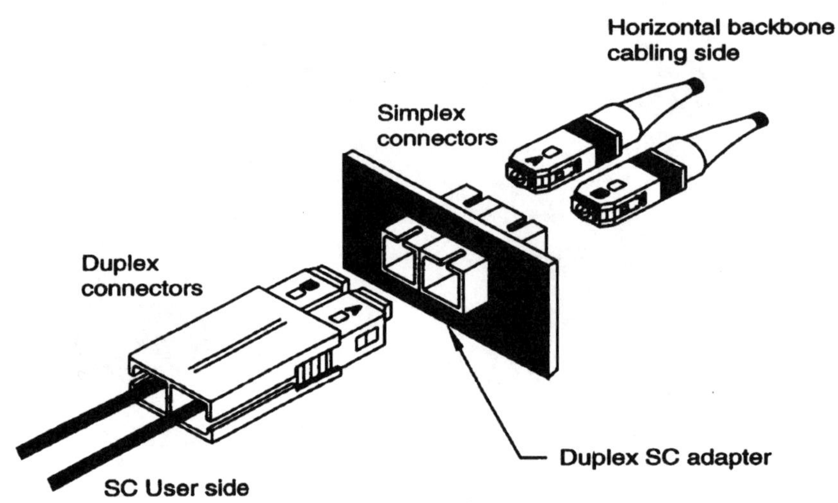

18.2.24 Duplex SC Patch Cord Crossover Orientation

18.2.24*

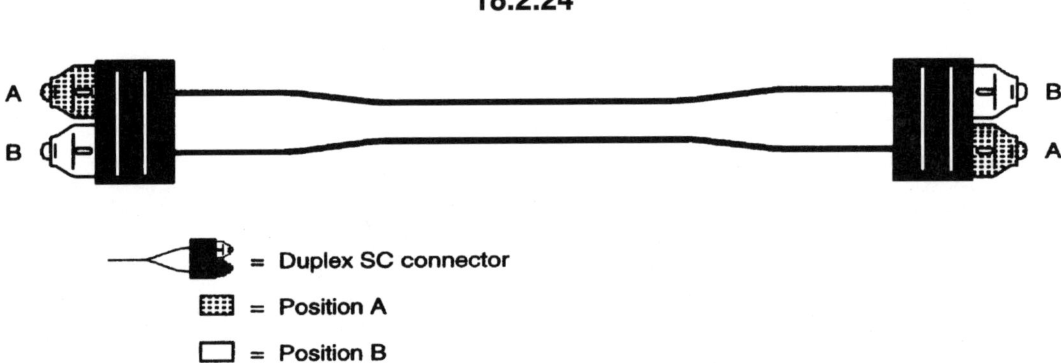

= Duplex SC connector

= Position A

= Position B

18.2.25 Optical Fibers

18.2.25

62.5/125 Multimode

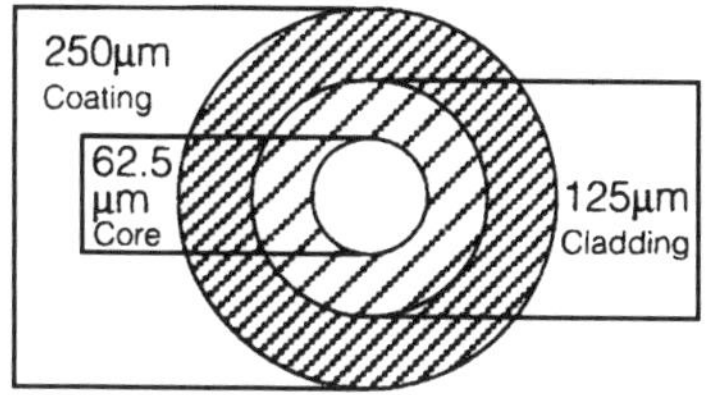

Singlemode

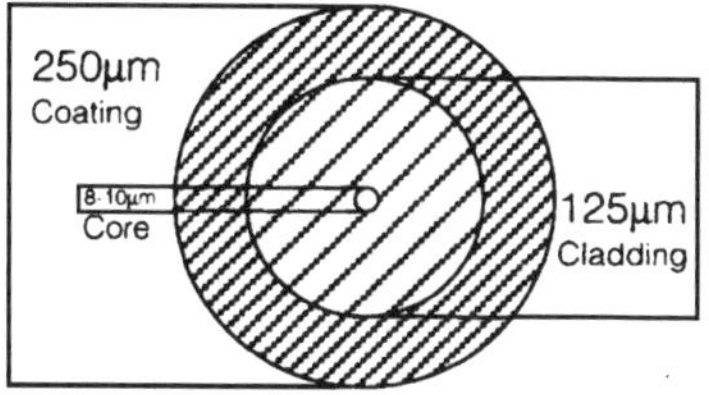

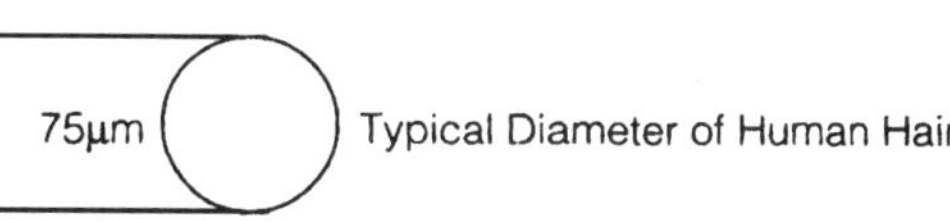

Notes: Coating protects glass from abrasion and assures high strength.

Cladding keeps the optical signal within the core.

18.2.26 Backbone System Components

TABLE 18.2.26*

Component	Description
Cable pathways	Shafts, conduits, raceways, and floor penetrations (e.g., sleeves or slots) that provide routing space for cables.
ERs	Areas where telecommunications systems are housed and connected to the telecommunications wiring system (see Chapter 8: Equipment Rooms).
TRs	Areas or locations that contain telecommunications equipment for connecting the horizontal cabling to the backbone cabling systems (see Chapter 7: Telecommunications Rooms).
Telecommunications service entrance facility	An area or location where outside plant cables enter a building (see Chapter 9: Telecommunications Entrance Facilities and Termination).

TABLE 18.2.26 (*Continued*)*

Component	Description
Transmission media	The actual cables, which may be: • Optical fiber. • Twisted-pair copper. • Coaxial copper. Connecting hardware, which may be: • Connecting blocks. • Patch panels. • Interconnections. • Cross-connections. NOTE: Backbone cabling can also be a combination of media.
Miscellaneous support facilities	Materials needed for the proper termination and installation of the backbone cables. These include: • Cable support hardware. • Firestopping (see Chapter 15: Firestopping). • Grounding hardware (see Chapter 17: Grounding, Bonding, and Electrical Protection). • Protection and security.

18.2.27 Backbone Star Wiring Topology

18.2.27*

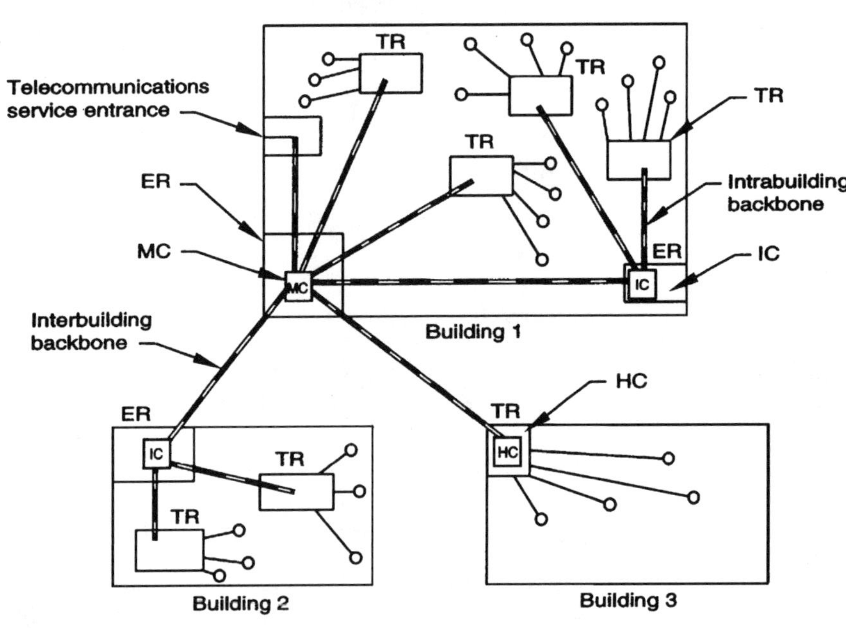

ER = Equipment room
HC = Horizontal cross-connect (floor distributor [FD])
IC = Intermediate cross-connect (building distributor [BD])
MC = Main cross-connect (campus distributor [CD])
TR = Telecommunications room

NOTE: Bridged taps are not permitted as part of the backbone wiring.

18.2.28 Example of Combined Copper/Fiber Backbone Supporting Voice and Data Traffic

18.2.28*

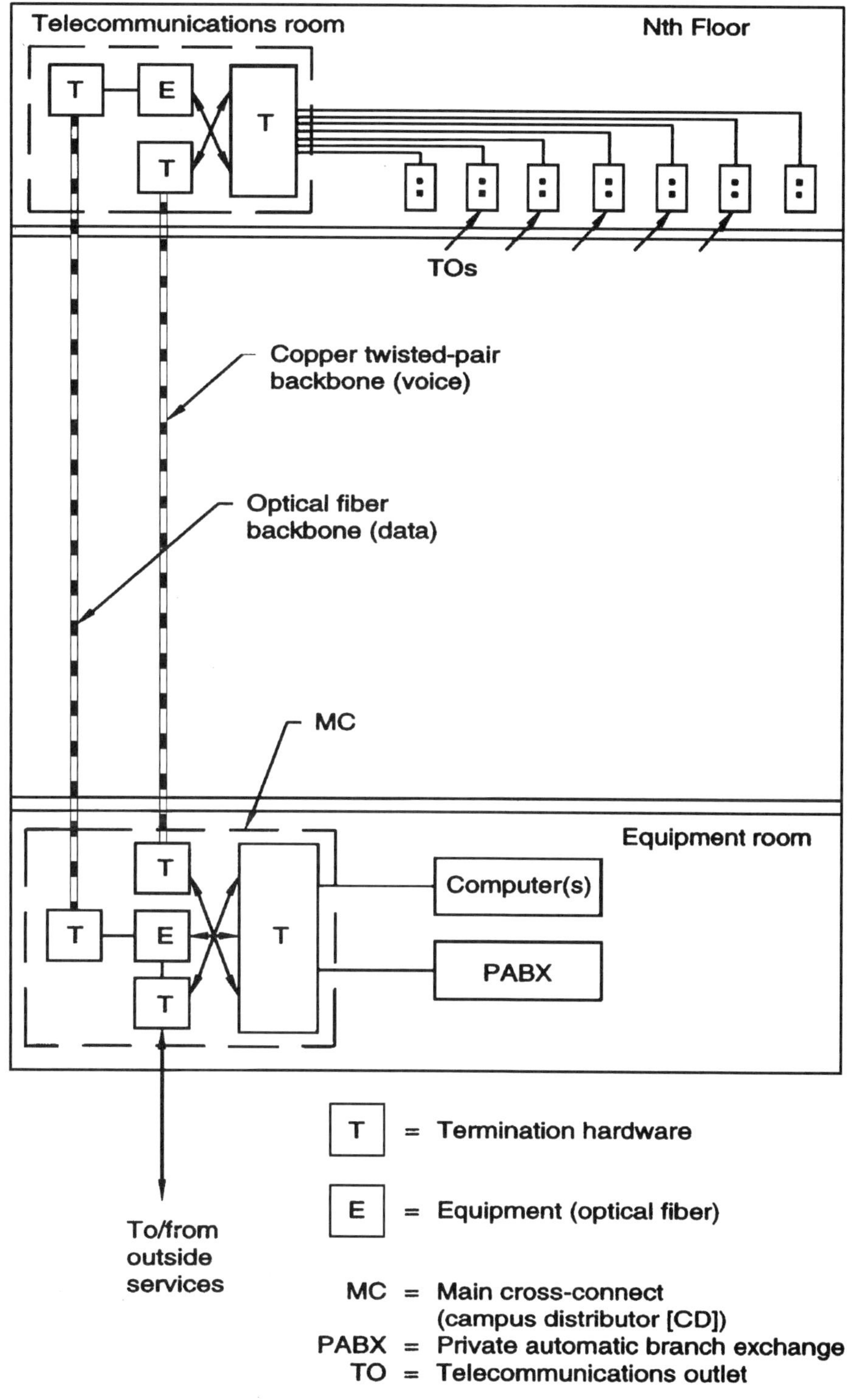

18.2.29 Backbone distances

18.2.29*

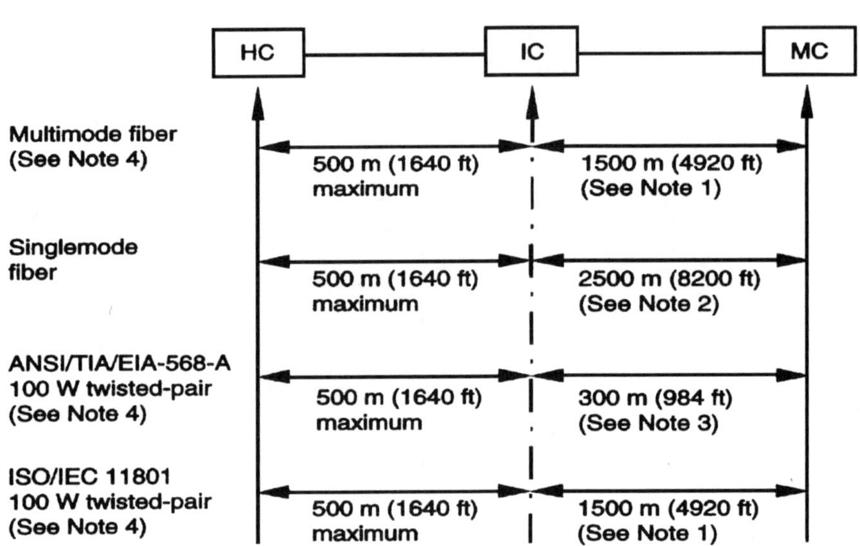

HC = Horizontal cross-connect (floor distributor [FD])
IC = Intermediate cross-connect (building distributor [BD])
MC = Main cross-connect (campus distributor [CD])

NOTES:

1. When the HC to IC distance is less than maximum, the IC to MC distance can be increased accordingly to a maximum of 2000 m (6560 ft).
2. When the HC to IC distance is less than maximum, the IC to MC distance can be increased accordingly to a maximum of 3000 m (9840 ft).
3. When the HC to IC distance is less than maximum, the IC to MC distance can be increased accordingly to a maximum of 800 m (2625 ft).
4. Actual backbone distances supported will depend on the performance of cabling installed and the applications being supported.

18.2.30 Determining 100 mm (4 in) Floor Sleeves

TABLE 18.2.30*

Total Usable Floor Area Served in Sleeves m^2 (ft^2)	Quantity of Sleeves
Up to 5000 (50,000)	3
> 5000 (50,000) to 10 000 (100,000)	4
> 10 000 (100,000) to 30 000 (300,000)	5-8
> 30 000 (300,000) to 50 000 (500,000)	9-12

18.2.31 Determining Size of Floor Slots

TABLE 18.2.31*

Total Usable Floor Area Served by Slot m^2 (ft^2)	Size of Slot mm (in)
≤ 25 000 (250,000)	150 (6) x 225 (9)
> 25 000 (250,000) to 50 000 (500,000)	150 (6) x 450 (18)
> 50 000 (500,000) to 100 000 (1,000,000)	225 (9) x 500 (20)
> 100 000 (1,000,000) to 140 000 (1,400,000)	300 (12) x 500 (20)
> 140 000 (1,400,000) to 200 000 (2,000,000)	375 (15) x 600 (24)

WARNING: In general, all structural changes and floor penetrations must be approved by a licensed engineer of the same state in which the work is performed.

18.2.32 Conduit Fill Requirements for Backbone Cable

TABLE 18.2.32*

Conduit			Area of Conduit								Minimum Radius of Bends			
					Maximum Occupancy Recommended									
					A		B		C		D		E	
Trade Size mm (in)	Internal Diameter*		Area = .79D^2 Total 100%		1 Cable 50% Fill		2 Cables 50% Fill		3 Cables or More 40% Fill		Layers of Steel within Sheath		Other Sheath	
	mm	in	mm^2	in^2	mm^2	in^2	mm^2	in^2	mm^2	in^2	mm	in	mm	in
21 (3/4)	20.9	0.82	345	0.53	183	0.28	107	0.16	138	0.21	210	8	130	5
27 (1)	26.6	1.05	559	0.87	296	0.46	173	0.27	224	0.35	270	11	160	6
35 (1-1/4)	35.1	1.38	973	1.51	516	0.80	302	0.47	389	0.60	350	14	210	8
41 (1-1/2)	40.9	1.61	1322	2.05	701	1.09	410	0.64	529	0.82	410	16	250	10
53 (2)	52.5	2.07	2177	3.39	1154	1.80	675	1.05	871	1.36	530	21	320	12
63 (2-1/2)	62.7	2.47	3106	4.82	1646	2.56	963	1.49	1242	1.93	630	25	630	25
78 (3)	77.9	3.07	4794	7.45	2541	3.95	1486	2.31	1918	2.98	780	31	780	31
91 (3-1/2)	90.1	3.55	6413	9.96	3399	5.28	1988	3.09	2565	3.98	900	36	900	36
103 (4)	102.3	4.03	8268	12.83	4382	6.80	2563	3.98	3307	5.13	1020	40	1020	40
129 (5)	128.2	5.05	12 984	20.15	6882	10.68	4025	6.25	5194	8.06	1280	50	1280	50
155 (6)	154.1	6.07	18 760	29.11	9943	15.43	5816	9.02	7504	11.64	1540	60	1540	60

* Internal diameters are taken from the manufacturing standard for electrical metallic tubing (EMT) and rigid metal conduit.

Apply these fill percentages to straight runs with nominal offsets equivalent to no more than two 90-degree bends.

Column D indicates a bend of 10 times (10x) the conduit diameter for cable sheaths consisting partly of steel tape.

Column E indicates a bend of six times (6x) the conduit diameter up to 53 mm (2 trade size), and 10 times (10x) the conduit diameter above 53 mm (2 trade size).

NOTE: For additional information, see Conduit Guidelines in this section.

18.2.33 TR Cross-Connect Field Color Codes

TABLE 18.2.33*

The Color...	Identifies...
Orange	Demarcation point (e.g., central office terminations).
Green	Network connections (e.g., network and auxiliary equipment).
Purple	Common equipment, private branch exchange (PBX), local area networks (LANs), multiplexers (e.g., switching and data equipment).
White	First-level backbone (e.g., MC [CD] to a HC [FD] or to an IC [BD]).
Gray	Second-level backbone (e.g., IC [BD] to a HC [FD]).
Blue	Horizontal cable (e.g., horizontal connections to telecommunications outlets).
Brown	Interbuilding backbone (campus cable terminations). NOTE: Brown takes precedence over white or gray for interbuilding runs.
Yellow	Miscellaneous (e.g., auxiliary, alarms, security).
Red	Reserved for future use (also, key telephone systems).

NOTE: These color codes are aligned with ANSI/TIA/EIA-606.

18.2.34 TR Temperature Ranges

TABLE 18.2.34*

For Telecommunications Rooms That...	The Temperature Range Should Be...
Do not contain active equipment	10 °C to 35 °C (50 °F to 95 °F). It is preferable that temperature be maintained to within ± 5 °C (± 9 °F) of the adjoining office space and that humidity be kept below 85% relative humidity.
House active equipment	18 °C to 24 °C (64 °F to 75 °F). The humidity range should be 30% to 55% relative humidity.

18.2.35 TR Size Requirements

TABLE 18.2.35*

If the Serving Area Is...	Then the Interior Dimensions of the Room Must Be at Least...
500 m^2 (5000 ft^2) or less	3.0 m x 2.4 m (10 ft x 8 ft). (See note below.)
Larger than 500 m^2 and less than or equal to 800 m^2 (>5000 ft^2 to 8000 ft^2)	3.0 m x 2.7 m (10 ft x 9 ft).
Larger than 800 m^2 and less than or equal to 1000 m^2 (>8000 ft^2 to 10,000 ft^2)	3.0 m x 3.4 m (10 ft x 11 ft).

NOTE: ANSI/TIA/EIA-569-A recommends a minimum TR size of 3.0 m x 2.1 m (10 ft x 7 ft). The size of 3 m x 2.4 m (10 ft x 8 ft) is specified here to allow a center rack configuration (see Figure 7.1).

18.2.36 Allocating Termination Space in TRs

TABLE 18.2.36*

For...	Allocate...
Twisted-pair cross-connections (see Notes)	2600 mm^2 (4 in^2) for each 4-pair circuit to be patched or cross-connected (allows for two 4-pair cable terminations and two 4-pair modular patch connections per circuit).
Optical fiber cross-connections	1300 mm^2 (2.0 in^2) for each fiber pair to be patched or cross-connected (allows for two cable/patch connections per channel). This space allocation is also appropriate for coaxial cable.

NOTES: For twisted-pair cross-connections using insulation displacement connector (IDC) connecting blocks and jumpers, cross-connect field density may be considerably greater.

When cabling requires surge protection, the recommended space allocation is two to four times larger than the space for regular cross-connections.

These space allocations do not include cable runs to and from the cross-connect fields. Up to 20 percent more space may be required for proper routing of cables, jumpers, and patch cords.

18.2.37 Typical Telecommunications Room (TR) Layout

18.2.37*

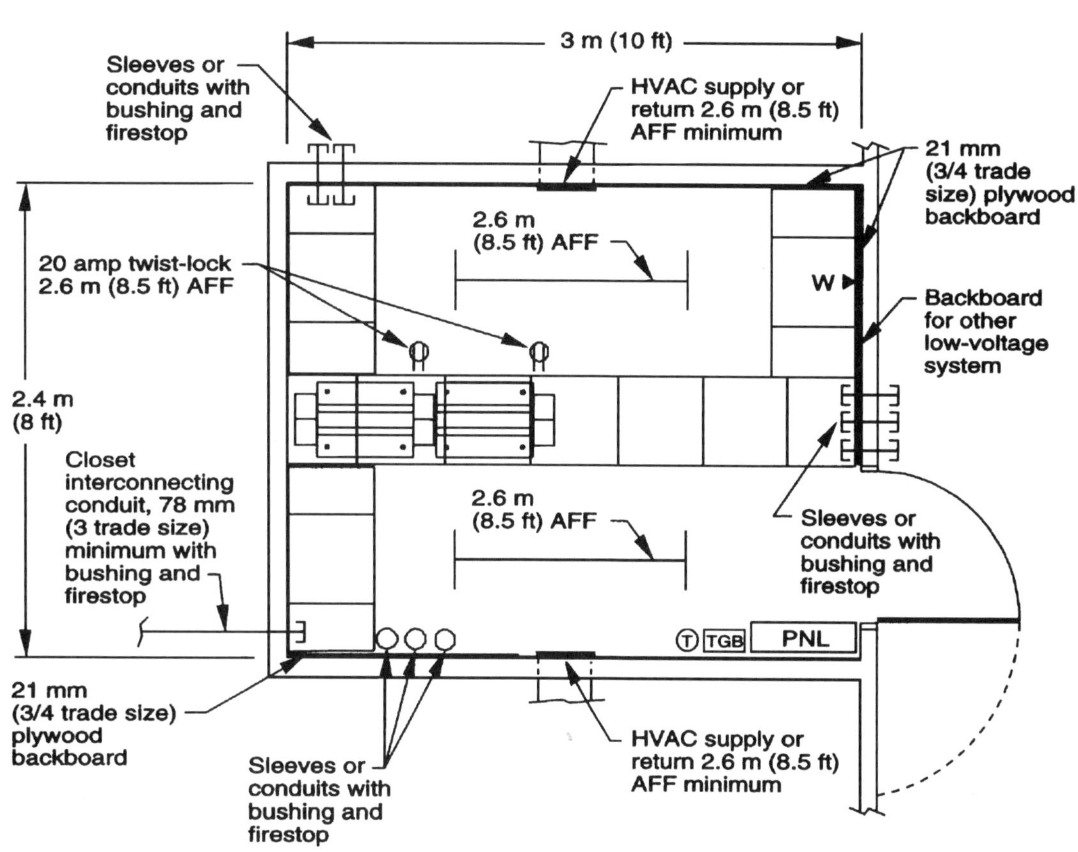

Ⓣ = Thermostat
AFF = Above finished floor
HVAC = Heating, ventilating, and air-conditioning
PNL = Panel
TGB = Telecommunications grounding busbar
W = Wall outlet

NOTES: If better cable reel access from the hallway is desirable, the three floor sleeves and the electrical panel (and TGB) could be interchanged.

Power outlets are shown for illustration only. Place convenience outlets at 1.8 m (6 ft) intervals around perimeter walls.

18.2.38 TR Industry Standards

TABLE 18.2.38*

Specification	Title
ANSI/TIA/EIA-568-A	*Commercial Building Telecommunications Cabling Standard.* (In Canada, see specification CSA T529-1996.)
ANSI/TIA/EIA-569-A	*Commercial Building Standard for Telecommunications Pathways and Spaces.* (In Canada, see specification CSA T530-1997.)
ANSI/TIA/EIA-570-A	*Residential Telecommunications Cabling Standard.*
ANSI/TIA/EIA-606	*Administration Standard for the Telecommunications Infrastructure of Commercial Buildings.* (In Canada, see specification CSA T528.)
ANSI/TIA/EIA-607	*Commercial Building Grounding and Bonding Requirements for Telecommunications.* (In Canada, see specification CSA T527.)
ISO/IEC 11801	*Generic Cabling for Customer Premises.*

The portions of the above-referenced specifications that relate directly to the content of this chapter include: Chapter 7 of ANSI/TIA/EIA-568-A; Chapter 7 of ANSI/TIA/EIA-569-A; Chapter 8 of ANSI/TIA/EIA-606; Chapter 7 of ANSI/TIA/EIA-607; and Chapter 5 of ISO/IEC 11801.

18.2.39 TR Regulatory and Safety Standards

TABLE 18.2.39*

Specification	Title
ANSI/NFPA 70	*The National Electrical Code®*, current edition.
CSA C22.1	*Canadian Electrical Code®, Part 1.*
FCC Part 68	*Code of Federal Regulations, Title 47, Telecommunications.*
UL 1459	*Underwriters Laboratories Standard for Safety—Telephone Equipment.*
UL 1863	*Underwriters Laboratories Standard for Safety—Communication Circuit Accessories.*

18.2.40 Environmental Control Systems Standards for Equipment Rooms (ERs)

TABLE 18.2.40*

Environmental Factor	Requirement
Temperature	18 °C to 24 °C (64 °F to 75 °F)
Relative humidity	30% to 55%
Heat dissipation	750 to 5,000 Btu per hour per cabinet

NOTES: Filtration systems may be required to minimize particle levels in the air.

Keep changes in temperature and humidity to a minimum.

HVAC sensors and controls must be located in the ER. Ideally, the sensors are placed 1.5 m (5 ft) above the finished floor.

18.2.41 Underground Entrance Conduits for Entrance Facilities (EFs)

TABLE 18.2.41*

Telephone Entrance Pairs...	Require...
1-99	One 53 mm (2 trade size) conduit plus 1 spare.
100-300	One 78 mm (3 trade size) conduit plus 1 spare.
301-1000	One 103 mm (4 trade size) conduit plus 1 spare.
1001-2000	Two 103 mm (4 trade size) conduits plus 1 spare.
2001-3000	Three 103 mm (4 trade size) conduits plus 1 spare.
3001-5000	Four 103 mm (4 trade size) conduits plus 1 spare.
5001-7000	Five 103 mm (4 trade size) conduits plus 1 spare.
7001-9000	Six 103 mm (4 trade size) conduits plus 1 spare.

18.2.42 Typical Underground Installation to EF

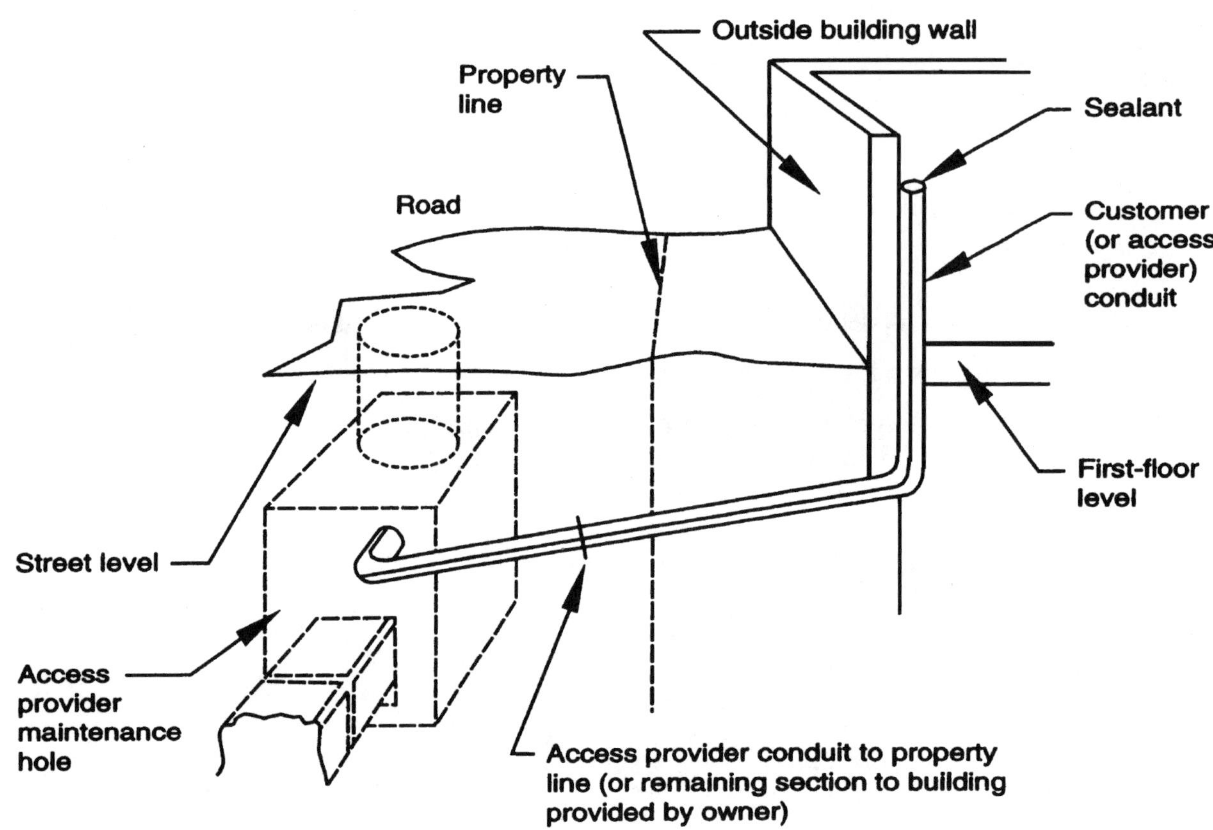

18.2.43 Equipment Room (ER) Floor Space (Special-Use Buildings)

TABLE 18.2.43

Work Areas	AREA (m^2)	(ft^2)
Up to 100	14	150
101 to 400	37	400
401 to 800	74	800
801 to 1,200	111	1,200

18.2.44 Entrance Facility (EF) Wall Space (Minimum Equipment and Termination Wall Space)

TABLE 18.2.44

GROSS FLOOR SPACE		WALL LENGTH	
m^2	ft^2	mm	in
500	5,000	990	39
1,000	10,000	990	39
2,000	20,000	1,060	42
4,000	40,000	1,725	68
5,000	50,000	2,295	90
6,000	60,000	2,400	96
8,000	80,000	3,015	120
10,000	100,000	3,630	144

18.2.45 Entrance Facility (EF) Floor Space (Minimum Equipment and Termination Floor Space)

TABLE 18.2.45

GROSS FLOOR SPACE		ROOM DIMENSIONS	
m^2	ft^2	mm	ft
7,000	70,000	3,660 x 1,930	12 x 6.3
10,000	100,000	3,660 x 1,930	12 x 6.3
20,000	200,000	3,660 x 2,750	12 x 9
40,000	400,000	3,660 x 3,970	12 x 13
50,000	500,000	3,660 x 4,775	12 x 15
60,000	600,000	3,660 x 5,588	12 x 18.3
80,000	800,000	3,660 x 6,810	12 x 22.3
100,000	1,000,000	3,660 x 8,440	12 x 27.7

18.2.46 Separation of Telecommunications Pathways from 480-Volt or Less Power Lines

TABLE 18.2.46

Condition	Minimum Separation Distance		
	< 2 kVA	2-5 kVA	> 5 kVA
Unshielded power lines or electrical equipment in proximity to open or nonmetal pathways.	127 mm (5 in)	305 mm (12 in)	610 mm (24 in)
Unshielded power lines or electrical equipment in proximity to a grounded metal conduit pathway	64 mm (2.5 in)	152 mm (6 in)	305 mm (12 in)
Power lines enclosed in a grounded metal conduit (or equivalent shielding) in proximity to a grounded metal conduit pathway.	- -	76 mm (3 in)	152 mm (6 in)

18.2.47 Cabling Standards Document Summary

TABLE 18.2.47*

Cabling Standards Document Summary

Several standards documents specify and/or recommend transmission parameters for the different cabling systems. Following is a summary of the most common documents:

ANSI/TIA/EIA-568-A, *Commercial Building Telecommunications Cabling Standard*

- Released October 1995.
- Covers Categories 3, 4, 5, and STP-A.
- Specifies:
  - Attenuation for cable and connecting hardware.
  - NEXT loss for cable and connecting hardware.

ANSI/TIA/EIA-568A-1, *Propagation Delay and Delay Skew Specifications for 100-Ohm 4-Pair Cable*

- Released September 1997.
- Covers Categories 3, 4, 5, and screened twisted-pair (ScTP).
- Specifies:
  - Propagation delay for cable.
  - Delay skew for cable.

ANSI/TIA/EIA-568A-3, *Hybrid Cables*

- Released September 1998.
- Covers hybrid and bundled cables.

ANSI/TIA/EIA-568-A-4, *Production Modular Cord NEXT Loss Test Method and Requirements for Unshielded Twisted-Pair Cabling*

- Released August 1999.
- Covers patch cords.

ANSI/TIA/EIA-568A-5, *Additional Transmission Performance Specifications for 4-Pair 100-Ohm Category 5e Cabling*

- Released in 1999.
- Covers Category 5e.
- Specifies:
  - NEXT loss for cable, connecting hardware, basic link, and channel.
  - PSNEXT loss for cable and cabling.
  - ELFEXT loss for cable and cabling.
  - FEXT loss for connecting hardware.
  - PSELFEXT loss for cable and cabling.
  - Return loss for cable, connecting hardware, basic link, and channel.
  - Propagation delay for basic link and channel.
  - Delay skew for basic link and channel.

18.3.0 Blown Optical Fiber Technology (BOFT) Overview

Reprinted with permission of General Cable Corporation (*www.generalcable.com*). BloLite is the trademark of BICC, PLC and is used under license.

Blown optical fiber technology is an exciting method of delivering a fiber solution that provides unmatched flexibility and significant cost savings when compared with conventional fiber cables. In a blown optical fiber system, the fiber route is "plumbed" with small tubes. The small tubes, known as *microduct,* come in 5- and 8-mm diameters and are approved for riser, plenum, or outside-plant applications. They are currently available as a single *microduct*, or with two, four, and seven *microducts* bundled (straight, not twisted) and covered with an outer sheath, called *multiducts.* They are lightweight and easy to handle. Splicing along the route is accomplished through simple push-pull connectors. These *microducts* are empty during installation, thereby eliminating the possibility of damaging the fibers during installation.

Fiber is then installed, or "blown," into the *microduct*. The fiber is fed into the *microduct* and rides on a current of compressed air. Carried by viscous drag, the fibers are lifted into the airstream and away from the wall of the microduct, thereby eliminating friction even around tight bends.

In a relatively short period, coated fibers can be blown for distances up to 1 km (3281 ft) in a single run of 8-mm-diameter *microduct*, up to 1000 ft vertical, or through any network architecture or topology turning up to 300 tight corners with 90 bends of 1-in radius for over 1000 ft, utilizing 5-mm-diameter *microduct.*

The practical benefits of BOFT systems translate directly into financial benefits for the end user. For most installations, a BOFT infrastructure is similar to or slightly higher than the cost for conventional fiber cabling. Savings can be realized during the initial installation because (1) it simplifies the cable installation by allowing the pulling of empty or unpopulated *microduct*, (2) fewer, if any, fiber splices may be required, and (3) you only pay upfront for those fibers that you need immediately. The additional expense of hybrid cables is eliminated.

True cost savings and the convenience of blown optical fiber are realized during the first fiber upgrade or moves, adds, and changes. An upgrade of an existing fiber backbone generally will incur workplace disruptions such as removing a ceiling grid, moving office furniture, and network downtime that requires the work to be done outside normal business hours. New fibers can be added to a BOFT system simply by accessing an existing unpopulated *microduct* and blowing in the fibers. There is no disruption to the workplace, and the process requires a minimal amount of time to complete. In the event that there are no empty *microducts*, the existing fiber can be blown out in minutes and replaced with the new fiber(s) immediately.

The flexibility of BOFT makes it particularly amenable to renovation and retrofit applications.

18.3.1 Diagram Showing Key Elements of BOFT System (*page 18.37*)

18.3.1 Diagram Showing Key Elements of BOFT System

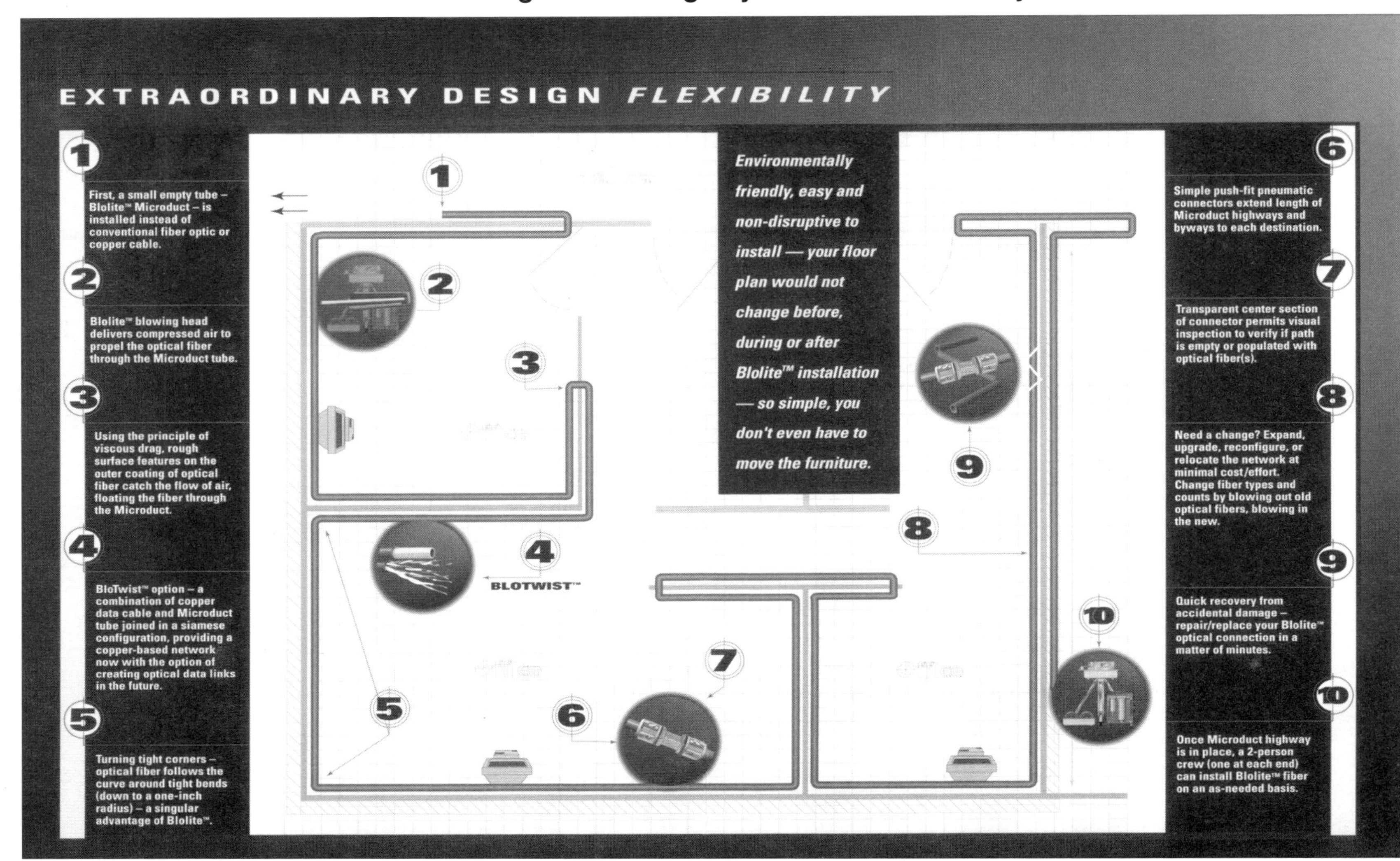

EXTRAORDINARY DESIGN FLEXIBILITY
1
First, a small empty tube – Blolite™ Microduct – is installed instead of conventional fiber optic or copper cable.
2
Blolite™ blowing head delivers compressed air to propel the optical fiber through the Microduct tube.
3
Using the principle of viscous drag, rough surface features on the outer coating of optical fiber catch the flow of air, floating the fiber through the Microduct.
4
BloTwist™ option – a combination of copper data cable and Microduct tube joined in a siamese configuration, providing a copper-based network now with the option of creating optical data links in the future.
5
Turning tight corners – optical fiber follows the curve around tight bends (down to a one-inch radius) – a singular advantage of Blolite™.
Environmentally friendly, easy and non-disruptive to install — your floor plan would not change before, during or after Blolite™ installation — so simple, you don't even have to move the furniture.
BLOTWIST™
6
Simple push-fit pneumatic connectors extend length of Microduct highways and byways to each destination.
7
Transparent center section of connector permits visual inspection to verify if path is empty or populated with optical fiber(s).
8
Need a change? Expand, upgrade, reconfigure, or relocate the network at minimal cost/effort. Change fiber types and counts by blowing out old optical fibers, blowing in the new.
9
Quick recovery from accidental damage – repair/replace your Blolite™ optical connection in a matter of minutes.
10
Once Microduct highway is in place, a 2-person crew (one at each end) can install Blolite™ fiber on an as-needed basis.

18.3.2 BOFT Indoor Plenum 5-mm Multiduct

18.3.2

BLOLITE™ INDOOR PLENUM 5mm MULTIDUCT

Plenum-rated Multiduct consist of a number of 5mm OD/3.5mm ID Microducts (*see 5mm Plenum Microduct data sheet*). The Microduct tubing is covered by an outer jacket. The number of Microduct constructions available are 2, 4 or 7. All plenum Microducts are orange in color and are printed with a unique number at regular intervals. The overall jacket is a plenum-rated PVC material colored orange. The outer jacket surface has product identification printing and sequential length marking at one-meter intervals.

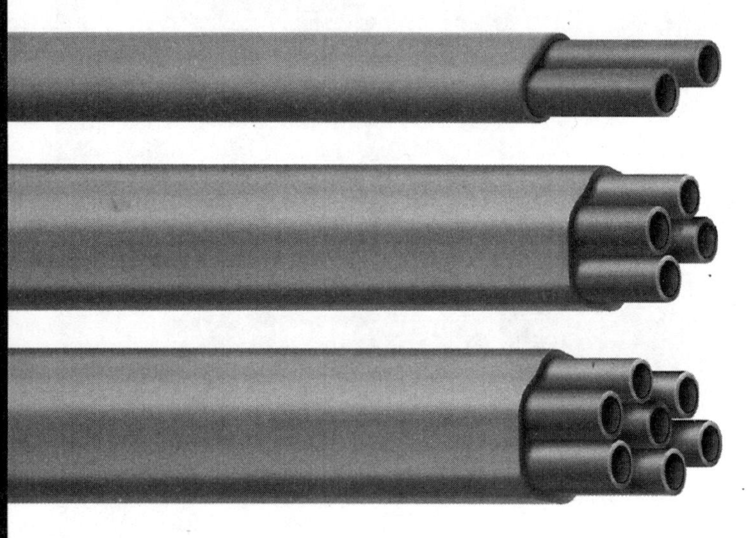

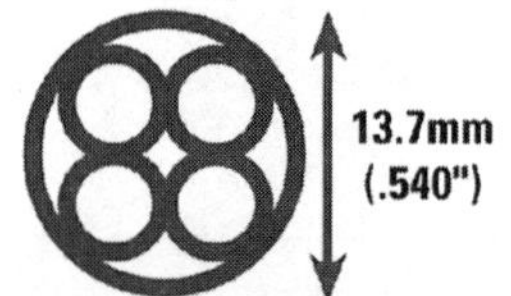

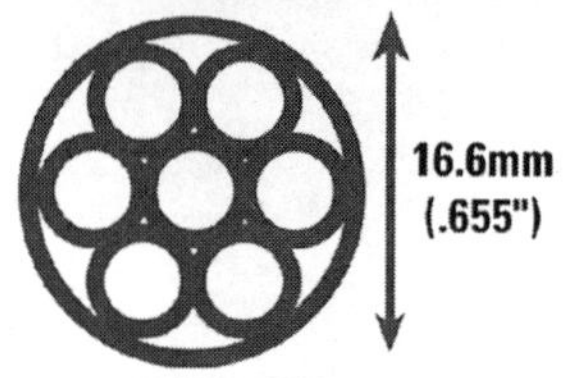

INSTALLATION TENSION	2 way 4 way 7 way	300N (67 lbs) 500N (112 lbs) 700N (157 lbs)
MINIMUM BEND RADIUS (UNLOADED)	2 way 4 way 7 way	60mm 140mm 180mm
MAX. INTERNAL PRESSURE	140 PSI	
COMPLIANCE	ETL TYPE OFNP c(ETL) TYPE OFNP	

TEMPERATURE RANGE	Storage Installation Operating	-30°C to + 80°C 0°C to + 30°C -20°C to + 70°C
MAX. CRUSH RESISTANCE	2 Way 4 Way 7 way	1200N/cm (685 lbs/in) 1200N/cm (685 lbs/in) 1200N/cm (685 lbs/in)

		kg/km	lbs/1000'
NOMINAL WEIGHT	2 way	58.2	39.1
	4 way	120.3	80.8
	7 way	185.3	124.5

ORDERING INFORMATION

PART NUMBER	DESCRIPTION
706500	2 way 5mm Multiduct, plenum
707270	4 way 5mm Multiduct, plenum
706350	7 way 5mm Multiduct, plenum

NOTES

1. Standard lengths are 500 and 1000 feet, supplied on non-returnable reels. Ends are sealed to prevent the penetration of moisture prior to shipping.

18.3.3 BOFT Outdoor 8-mm Multiduct

18.3.3

BLOLITE™ OUTDOOR 8mm MULTIDUCT

Outdoor-rated 8mm Multiduct consist of a number of 8mm OD/6mm ID Microducts *(see 8mm Non-Plenum Microduct data sheet)*. The Microduct tubing is covered by an aluminum moisture barrier and an outer jacket. The number of Microduct constructions available are 2, 4 or 7. All Microducts are blue in color and are printed with a unique number at regular intervals. The overall jacket is a black medium-density polyethylene. The outer jacket surface has product identification printing and sequential length marking at one-meter intervals.

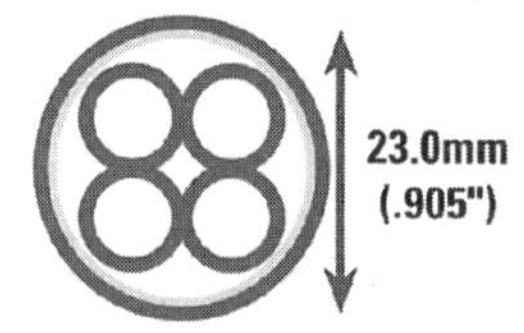

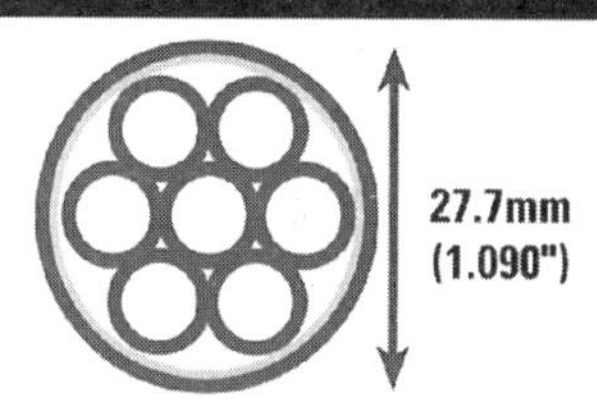

INSTALLATION TENSION		
	2 way	450N (101 lbs)
	4 way	750N (191 lbs)
	7 way	1100N (247 lbs)

MINIMUM BEND RADIUS (UNLOADED)		
	2 way	140mm
	4 way	280mm
	7 way	380mm

MAX. INTERNAL PRESSURE	140 PSI

TEMPERATURE RANGE		
	Storage	-30°C to + 80°C
	Installation	0°C to + 30°C
	Operating	-20°C to + 70°C

MAX. CRUSH RESISTANCE		
	2 Way	1800N/cm (1030 lbs/in)
	4 Way	1800N/cm (1030 lbs/in)
	7 way	1800N/cm (1030 lbs/in)

NOMINAL WEIGHT		kg/km	lbs/1000'
	2 way	150.4	101.1
	4 way	256.8	172.6
	7 way	383.7	257.9

ORDERING INFORMATION

PART NUMBER	DESCRIPTION
706240	2 way 8mm Multiduct, outside plant
706250	4 way 8mm Multiduct, outside plant
706260	7 way 8mm Multiduct, outside plant

NOTES

1. Standard lengths are 500 and 1000 feet, supplied on non-returnable reels. Ends are sealed to prevent the penetration of moisture prior to shipping.

18.3.4 BOFT Installation Equipment

18.3.4

BLOLITE™ INSTALLATION EQUIPMENT

The efficient installation of optical fibers into Blolite Microduct requires the use of specially designed equipment. The Fiber Installation Equipment kit provides two units housed in sturdy carrying cases as well as a lightweight tripod.

An Air Supply Conditioning Unit (ACU)–complete with filtration and air-drying units, in addition to the component parts of the payoff system, is housed in one case. The Installation Module–a blowing head–utilizing a mechanically driven system to introduce the fibers into the Microduct along with a fiber installation control device–is housed in another case.

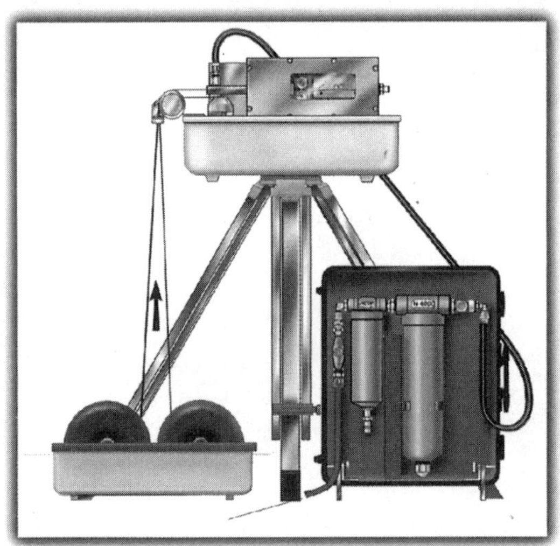

FEED FROM COMPRESSOR OR AIR CYLINDER

TYPICAL INSTALLATION TIMES

In order to establish repeatable maximum blowing distances a series of tests have been conducted. All tests are based on four fibers being installed.

DUCT Size (mm)/Length (m)	ROUTE	INSTALLATION TIME Typical Spec.
5mm / 100m	standard	4 minutes
5mm / 100m	challenging	7 minutes
5mm / 500m	standard	40 minutes
8mm / 500m	standard	40 minutes
8mm /1000m	standard	90 minutes

ORDERING INFORMATION

Blolite installation equipment is leased through a licensing agreement to certified installers. This lease includes permission to blow fiber under the original patent, the supply of the necessary equipment, training and certification, and technical support by BICCGeneral.

Section

19

Miscellaneous Special Applications

19.1.0 Fire Alarm Systems: Introduction
19.1.1 Fire Alarm Systems: Common Code Requirements
19.1.2 Fire Alarm System Classifications
19.1.3 Fire Alarm Fundamentals: Basic Elements (Typical Local Protective Signaling System)
19.1.4 Fire Alarm System Circuit Designations
19.1.5 Fire Alarm System: Class
19.1.6 Fire Alarm System: Style
19.1.7 Performance of Initiating-Device Circuits (IDCs)
19.1.8 Performance of Signaling-Line Circuits (SLCs)
19.1.9 Notification-Appliance Circuits (NACs)
19.1.10 Installation of Class A Circuits
19.1.11 Secondary Supply Capacity and Sources
19.1.12 Audible Notification Appliances to Meet the Requirements of ADA, NFPA 72 (1993), and BOCA
19.1.13 Visual Notification Appliances to Meet the Requirements of ADA, NFPA 72 (1993), and BOCA
19.1.14 ADA-Complying Mounting Height for Manual Pull Stations (High Forward-Reach Limit)
19.1.15 ADA-Complying Mounting Height for Manual Pull Stations (High and Low Side-Reach Limits)
19.1.16 Application Tips
19.2.1 Fire Pump Applications
19.2.2 Typical One-Line Diagram of Fire Pump System with Separate ATS
19.2.3 Typical One-Line Diagram of Fire Pump System with ATS Integrated with the Fire Pump Controller
19.3.1 Wiring of Packaged Rooftop AHUs with Remote VFDs
19.4.1 Wye-Delta Motor Starter Wiring
19.5.1 Elevator Recall Systems
19.5.2 Typical Elevator Recall/Emergency Shutdown Schematic
19.5.3 Typical Elevator Hoistway/Machine Room Device Installation Detail
19.6.0 Harmonic Effects and Mitigation

19.1.0 Fire Alarm Systems: Introduction

Fire alarm systems have become increasingly sophisticated and functionally more capable and reliable in recent years. They are designed to fulfill two general requirements: protection of property and assets and protection of life. As a result of state and local codes, the life-safety aspect of fire protection has become a major factor in the last two decades.

There are a number of reasons for the substantial increases in the life-safety form of fire protection during recent years, foremost of which are

1. The proliferation of high-rise construction and the concern for life safety within these buildings.
2. A growing awareness of the life-safety hazard in residential, institutional, and educational occupancies.
3. Increased hazards caused by new building materials and furnishings that create large amounts of toxic combustion products (i.e., plastics, synthetic fabrics, etc.).
4. Vast improvements in smoke detection and related technology made possible through quantum advances in electronic technology.
5. The passing of the Americans with Disabilities Act (ADA), signed into law on July 26, 1990, providing comprehensive civil rights protection for individuals with disabilities. With an effective date of January 26, 1992, these requirements included detailed accessability standards for both new construction and renovation toward the goal of equal usability of buildings for everyone, regardless of limitations of sight, hearing, and mobility. This had a significant impact on fire alarm system signaling devices, power requirements, and device locations.

19.1.1 Fire Alarm Systems: Common Code Requirements

The following codes apply to fire alarm systems:

NFPA 70, *National Electrical Code*

NFPA 72, *National Fire Alarm Code*

NFPA 90A, *Standard for the Installation of Air Conditioning and Ventilation Systems*

NFPA 101, *Life Safety Code*

BOCA, SBCCI, ICBO. The *National Basic Building Code* and *National Fire Prevention Code,* published by the Building Officials Code Administrators International (BOCA), the *Uniform Building* and *Uniform Fire Code* of the International Conference of Building Officials (ICBO), and the *Standard Building Code* and the *Standard Fire Prevention Code* of the Southern Building Code Congress International (SBCCI) all have reference to fire alarm requirements.

Many states and municipalities have adopted these model building codes in full or in part. You should consult with the local authority having jurisdiction (AHJ) to verify the requirements in your area.

19.1.2 Fire Alarm System Classifications

NFPA 72 classifies fire alarm systems as follows:

- *Household fire alarm system.* A system of devices that produces an alarm signal in the household for the purpose of notifying the occupants of the presence of fire so that they will evacuate the premises.
- *Protected-premises (local) fire alarm system.* A protected-premises system that sounds an alarm at the protected premises as the result of the manual operation of a fire alarm box or the operation of protection equipment or systems, such as

water flowing in a sprinkler system, the discharge of carbon dioxide, the detection of smoke, or the detection of heat.

- *Auxiliary fire alarm system.* A system connected to a municipal fire alarm system for transmitting an alarm of fire to the public fire service communications center. Fire alarms from an auxiliary fire alarm system are received at the public fire service communications center on the same equipment and by the same methods as alarms transmitted manually from municipal fire alarm boxes located on streets. There are three subtypes of this system: local energy, parallel telephone, and shunt type.
- *Remote supervising station fire alarm system.* A system installed in accordance with NFPA 72 to transmit alarm, supervisory, and trouble signals from one or more protected premises to a remote location at which appropriate action is taken.
- *Proprietary supervising station fire alarm system.* An installation of fire alarm systems that serves contiguous and noncontiguous properties, under one ownership, from a proprietary supervising station located at the protected property, at which trained, competent personnel are in constant attendance. This includes the proprietary supervising station, power supplies, signal-initiating devices, initiating-device circuits, signal-notification appliances, equipment for the automatic and permanent visual recording of signals, and equipment for initiating the operation of emergency building control services.
- *Central station fire alarm system.* A system or group of systems in which the operations of circuits and devices are transmitted automatically to, recorded in, maintained by, and supervised from a listed central station having competent and experienced servers and operators who, on receipt of a signal, take such action as required by NFPA 72. Such service is to be controlled and operated by a person, firm, or corporation whose business is the furnishing, maintaining, or monitoring of supervised fire alarm systems.
- *Municipal fire alarm system.* A system of alarm-initiating devices, receiving equipment, and connecting circuits (other than a public telephone network) used to transmit alarms from street locations to the public fire service communications center.

19.1.3 Fire Alarm Fundamentals: Basic Elements (Typical Local Protective Signaling System) (*see page 19.4*)

Regardless of type, application, complexity, or technology level, any fire alarm system is comprised of four basic elements:

1. Initiating devices
2. Control panel
3. Signaling devices
4. Power supply

These components must be electrically compatible and are interconnected by means of suitable wiring circuits to form a complete functional system, as illustrated in Figure 19.1.3.

The figure shows a conventional version of a protected premises (local) fire alarm system, which is the most widely used classification type in commercial and institutional buildings. The requirements for this type of system are detailed in Chapter 3 of NFPA 72. Some highlights of this chapter's requirements are worthy of note and are given in abridged form hereafter.

19.1.3

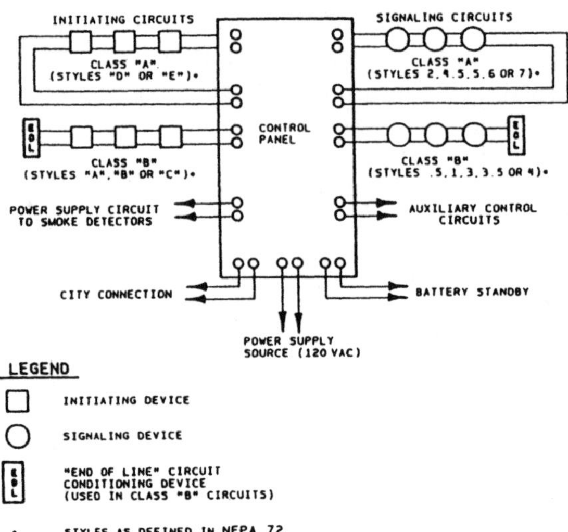

19.1.4 Fire Alarm System Circuit Designations

Initiating-device, notification-appliance, and signaling-line circuits shall be designated by class or style, or both, depending on the circuits' capability to operate during specified fault conditions.

19.1.5 Fire Alarm System: Class

Initiating-device, notification-appliance, and signaling-line circuits shall be permitted to be designated as either class A or class B depending on the capability of the circuit to transmit alarm and trouble signals during nonsimultaneous single-circuit-fault conditions as specified by the following:

1. Circuits capable of transmitting an alarm signal during a single open or a non-simultaneous single ground fault on a circuit conductor shall be designated as class A.
2. Circuits not capable of transmitting an alarm beyond the location of the fault conditions specified in 1 above shall be designated as class B.

Faults on both class A and class B circuits shall result in a trouble condition on the system in accordance with the requirements of NFPA 72, Article 1-5.8.

19.1.6 Fire Alarm System: Style

Initiating-device, notification-appliance, and signaling-line circuits shall be permitted to be designated by style also, depending on the capability of the circuit to transmit alarm and trouble signals during specified simultaneous multiple-circuit-fault conditions in addition to the single-circuit-fault conditions considered in the designation of the circuits by class.

19.1.7 Performance of Initiating-Device Circuits (IDCs) (*see page 19.5*)

TABLE 19.1.7

Class	B			B			B			A			A		
Style	A			B			C			D			Eα		
R = Required capability X = Indication required at protected premises and as required by Chapter 4 α = Style exceeds minimum requirements for Class A	Alarm	Trouble	Alarm receipt capability during abnormal condition	Alarm	Trouble	Alarm receipt capability during abnormal condition	Alarm	Trouble	Alarm receipt capability during abnormal condition	Alarm	Trouble	Alarm receipt capability during abnormal condition	Alarm	Trouble	Alarm receipt capability during abnormal condition
Abnormal Condition	1	2	3	4	5	6	7	8	9	10	11	12	13	14	15
A. Single open		X			X			X			X	X		X	X
B. Single ground		R			X	R		X	R		X	R		X	R
C. Wire-to-wire short	X			X				X		X				X	
D. Loss of carrier (if used)/channel interface								X						X	

The assignment of class designations or style designations, or both, to initiating circuits shall be based on their performance capabilities under abnormal (fault) conditions in accordance with Table 19.1.7.

An initiating-device circuit shall be permitted to be designated as either style A, B, C, D, or E depending on its ability to meet the alarm and trouble performance requirements shown during a single-open, single-ground, wire-to-wire short and loss of carrier fault condition.

19.1.8 Performance of Signaling-Line Circuits (SLCs) (*see page 19.6*)

The assignment of class designations or style designations, or both, to signaling-line circuits shall be based on their performance capabilities under abnormal (fault) conditions in accordance with Table 19.1.8.

A signaling-line circuit shall be permitted to be designated as either style 0.5, 1, 2, 3, 3.5, 4, 4.5, 5, 6, or 7 depending on its ability to meet the alarm and trouble performance requirements during a single-open, single-ground, wire-to-wire short, simultaneous wire-to-wire short and open, simultaneous wire-to-wire short and ground, simultaneous open and ground, and loss of carrier fault conditions.

19.1.9 Notification-Appliance Circuits (NACs) (*see page 19.6*)

The assignment of class designations or style designations, or both, to notification-appliance circuits shall be based on their performance capabilities under abnormal (fault) conditions in accordance with Table 19.1.9.

TABLE 19.1.8

Class	B			B			A			B			B			B			B			A			A			A		
Style	0.5			1			2α			3			3.5			4			4.5			5α			6α			7α		
M = May be capable of alarm with wire-to-wire short R = Required capability X = Indication required at protected premises and as required by Chapter 4 α = Style exceeds minimum requirements for Class A	Alarm	Trouble	Alarm receipt capability during abnormal condition	Alarm	Trouble	Alarm receipt capability during abnormal condition	Alarm	Trouble	Alarm receipt capability during abnormal condition	Alarm	Trouble	Alarm receipt capability during abnormal condition	Alarm	Trouble	Alarm receipt capability during abnormal condition	Alarm	Trouble	Alarm receipt capability during abnormal condition	Alarm	Trouble	Alarm receipt capability during abnormal condition	Alarm	Trouble	Alarm receipt capability during abnormal condition	Alarm	Trouble	Alarm receipt capability during abnormal condition	Alarm	Trouble	Alarm receipt capability during abnormal condition
Abnormal Condition	1	2	3	4	5	6	7	8	9	10	11	12	13	14	15	16	17	18	19	20	21	22	23	24	25	26	27	28	29	30
A. Single open		X			X			X	R		X			X			X			X	R		X	R		X	R		X	R
B. Single ground		X			X	R		X	R		X	R		X			X	R		X			X	R		X	R		X	R
C. Wire-to-wire short									M		X			X			X			X			X			X			X	R
D. Wire-to-wire short & open									M		X			X			X			X			X			X			X	
E. Wire-to-wire short & ground								X	M		X			X			X			X			X			X			X	
F. Open and ground								X	R		X			X			X			X			X			X	X		X	R
G. Loss of carrier (if used)/ channel interface														X			X			X			X			X			X	

TABLE 19.1.9

Class	B		B		B		A	
Style	W		X		Y		Z	
X = Indication required at protected premises	Trouble indication at protected premises	Alarm capability during abnormal conditions	Trouble indication at protected premises	Alarm capability during abnormal conditions	Trouble indication at protected premises	Alarm capability during abnormal conditions	Trouble indication at protected premises	Alarm capability during abnormal condition
Abnormal condition	1	2	3	4	5	6	7	8
Single open	X		X	X	X		X	X
Single ground	X		X		X	X	X	X
Wire-to-wire short	X		X		X		X	

A notification-appliance circuit shall be permitted to be designated as either style W, X, Y, or Z depending on its ability to meet the alarm and trouble performance requirements shown during a single-open, single-ground, and wire-to-wire short fault condition.

19.1.10 Installation of Class A Circuits

All styles of class A circuits using physical conductors (e.g., metallic, optical fiber) shall be installed such that the outgoing and return conductors, exiting from and returning to the control unit, respectively, are routed separately. The outgoing and return (redundant) circuit conductors shall not be run in the same cable assembly (i.e., multiconductor cable), enclosure or raceway.

Exception No. 1. For a distance not to exceed 10 ft (3 m) where the outgoing and return conductors enter or exit the initiating device, notification appliance, or control-unit enclosures; or

Exception No. 2. Where the vertically run conductors are contained in a 2-hour rated cable assembly or enclosed (installed) in a 2-hour rated enclosure other than a stairwell; or

Exception No. 3. Where permitted and where the vertically run conductors are enclosed (installed) in a 2-hour rated stairwell in a building fully sprinklered in accordance with NFPA 13, "Standard for the Installation of Sprinkler Systems."

Exception No. 4. Where looped conduit/raceway systems are provided, single conduit/raceway drops to individual devices or appliances shall be permitted.

Exception No. 5. Where looped conduit/raceway systems are provided, single conduit/raceway drops to multiple devices or appliances installed within a single room not exceeding 1000 ft^2 (92.9 m^2) in area shall be permitted.

19.1.11 Secondary Supply Capacity and Sources

From NFPA 72, Chapter 1, "Fundamentals," the secondary source for a protected premises system should have a secondary supply source capacity of 24 hours and, at the end of that period, shall be capable of operating all alarm notification appliances used for evacuation or to direct aid to the location of an emergency for 5 minutes. The secondary power supply for emergency voice/alarm communications service shall be capable of operating the system under maximum load for 24 hours and then shall be capable of operating the system during a fire or other emergency condition for a period of 2 hours. Fifteen minutes of evacuation alarm operation at maximum connected load shall be considered the equivalent of 2 hours of emergency operation.

19.1.12 Audible Notification Appliances to Meet the Requirements of ADA, NFPA 72 (1993), and BOCA (*see page 19.8*)

19.1.13 Visual Notification Appliances to Meet the Requirements of ADA, NFPA 72 (1993), and BOCA (*see page 19.9*)

TABLE 19.1.12

ADA	NFPA	BOCA	Design Criteria	Design Comments	Available Devices

• Intensity and frequency that can attract individuals who have partial hearing loss
• Periodic element to its signal such as:
  • Single stroke bell
  • Hi-Low
  • Fast whoop
• Avoid continuous or reverberating tones. Select a signal which has a sound characterized by three or four clear tones without a great deal of noise in between.
	• To insure that audible public mode signals are clearly heard, it shall be required that their sound level be at least 15 dBA above the average ambient sound level, or 5 dBA above the maximum sound level having a duration of at least 60 seconds, whichever is greater, measured at 5' above the floor in the occupiable area
	• Mechanical Equipment Rooms
	  • Design for a minimum of 85 dBA for all type occupancies
	• Sleeping Areas
	  • Design for a minimum of 70 dBA at any point in the sleeping area
	• Mounting location
	  • Wall mounted appliances -not less than 90" AFF -not less than 6" BFC
	  • Combination A/V Units -Bottoms 80"- 96" AFF
	• Effective July 1, 1996, the fire alarm signal used to notify building occupants shall be in accordance with ANSI S3.41 (NFPA 3-7.2)
	• Temporal Slow Whoop <u>or</u>
	• Temporal Code 3-3, 1 second bursts of signal with 2 seconds quiet before repeating the 3 bursts
		• Minimum of 15 dBA over average ambient
		• Every occupied space within the building
		• Minimum of 70 dBA in use groups R, I-1
		• Minimum of 90 dBA in Mechanical Rooms
		• Minimum of 60 dBA in all other use groups
		• Maximum of 130 dBA at minimum hearing distance from audible appliance
			• Ratings/listings - most devices are rated for dBA output at 10' from device;
			• Doubling the distance from the device - drop of 6 dBA
			  • A device with an output of 96 dBA at 10' will have 90 dBa at 20', 84 dBA at 40', 78 dBA at 80', etc.
			• Acustic tile ceiling causes approximately a 3 dBA drop in sound levels;
			• Rug on floor - causes approximately 3 dBA drop in sound levels;
			• An open door: 8-12 dBA drop;
			• Closed hollow core door: 12-20 dBA drop;
			• Closed solid core, rated door: 20-30 dBA drop;
			• 4" Partition: 40-45 dBA drop;
			• Multiple signals effect: add approximately 3 dBA at mid-point of signals;
			<u>Typical ambient sound levels</u>:
			• High School Office: 60 dBA
			• Corridor with back- ground music: 60 dBA
			• Classroom with students "Under Control": 62 dBA
			• Classroom with TV set turned on: 65 dBA
			• Classroom with students "out of control" end of day: 70 dBA
			• Corridor with students at end of day: 80 dBA
			• Normal Business Office: 55 to 60 dBA (air diff., computer on, 1 person talking on phone)
			• Hotel Room with A/C unit running in room and TV turned on: 65 dBA
				• It is good fire alarm system design engineering to provide audible devices that allow for adjusting their sound level output to accommodate the sound level environment they are installed in;
				• "OVER KILL" in dBA output can be a disaster for the END USER (installing horns, mini-horns in all spaces)
				• No more than one type of Fire Alarm Signaling Device may be used in an area (PA Labor & Industry). All audible alarm notification appliance devices in a facility shall be distinguishable from all other audible devices in the building (BOCA):
				• Horns or bells in the corridor and buzzers in the rooms may not meet this criteria;
				• Under most circumstances, the only practical way to achieve the required sound level to meet the ADA and applicable codes, is to install an audible notification appliance in every room and occupied space in the facility
				• Presently, the only practical approved audible device available, with a wide range of dBA adjustments to meet these requirements is the Fire Alarm Speaker.
				• Present technology allows tones to be generated on the speakers to meet the desired sound characteristics
					<u>Fire Alarm Horn</u>
					• Ratings from 88 dBA to 110 dBA
					• Settings of "loud to louder"
					• Normally mid to high frequency
					• Multi-tone settings in field available
					• Relatively low current draw
					• Low profile - standard mounting
					• Low to moderate price
					<u>Fire Alarm Bell</u>
					• Ratings of 87 dBA to 92 dBA
					• Output not adjustable
					• Low to mid range frequency
					• Low current draw
					• Approximate same cost as a horn
					• Surface mounting
					• Large in size than a horn
					<u>Fire Alarm Speakers</u>
					• Ratings from 75 dBA to 120 dBA
					• Wide range of adjustment
					• Frequency of low to high
					• Flush and surface mount
					• Slightly higher cost when supplied with variable taps
					<u>Speakers</u>
					• Speakers are available with outputs adjustable from 75 dBA to 120 dBA
					• A common tone can be generated at the main control and amplified and distributed to all speaker circuits
					• Emergency paging can normally be added as an option
					• Speakers can be re-taped if changes in ambient sound level occur in the area they are installed in
					• Design circuits to a maximum of 75% to 80% of rated capacity to allow for ambient sound level changes

TABLE 19.1.13

ADA	ADA (continued)

• Xenon strobe or equivalent
• Clear or nominal white lens color
• Minimum of 75 caldela or equivalent facilitation
• 1 to 3 Hz flash rate
• 80" AFF or 6" BFC
• No place in any room or space required to have a visual signal shall be mote than 50' from the visual signal
• In large open spaces, such as auditoriums exceeding 100' across, mount 6' AFF, spaced a maximum of 100' apart
• No place in corridors or hallways shall be more than 50' from a visual signal
• Install in restrooms, general use areas, meeting rooms, hallways, lobbies and other common use area
• ADA does not mandate emergency alarm systems
• In existing buildings, the update of the fire alarm system requires ADA compliance
• Common Use areas include:
  • Meeting and conference rooms
  • Employee break rooms
  • Classrooms
  • Cafeterias
  • Filing and photocopy rooms
  • Dressing rooms
  • Examination rooms
  • Treatment rooms
  • Similar space not used solely as employee work areas
	• Not required in individual offices and work stations
	• Visual units not required in:
	  • Mechanical, electrical, telephone rooms
	  • Janitor's closets
	  • Similar non-occupiable spaces
	  • Non-assigned work areas
	• Lamps tested at 1/3 Hz were judged ineffective. Requires a flash rate of from1 to 3 Hz
	• Mounting heights from 80" to 96" AFF are considered equivalent
	• Recommend 100' spacing in corridors and installed on alternate walls
	• Maximize lamp intensity to minimize number of fixtures
	• Lesser intensity may be sufficient as an equivalent facilitation
	• Equivalent facilitation permits alternate designs
	• Where a single lamp can provide the necessary intensity and coverage, multiple lamps should not be installed because of their potential effect on persons with photosensitivity
	• Health Care Facilities: modify to suit industry-accepted practices (NFPA 101).
	Mounting Heights
	• Forward Reach: 15"-48" AFF
	• Side Reach: 9"-54" AFF
	
	UL 1971: - 1/3 Hz rate; - Allows ceiling mount.; - 15 cd corridor units
	ADA: - 1 to 3 Hz rate; - No ceiling mounting; - Equivalent facilitation

NFPA	BOCA	Design Criteria	Design Comments
• NFPA accepts the requirements of UL 1971 to determine compliance for visual units; • It is important to determine if the system is designed to meet the ADA or UL 1971 Guidelines Mounting Heights • Minimum of 42" - Maximum of 54"	• Required in public and common areas of all buildings housing the hearing impaired. • In Use Groups I-1 and R-1, in required accessible sleeping rooms and suites. • Sleeping room visual units shall be activated by the in-room smoke detector and building fire alarm system Mounting Heights • Minimum of 42" - Maximum of 54"	• Synchronization of strobes when more than two strobes are installed in the same room • Keep tuned: ADA is considering changes Mounting Heights • PAL&I - Minimum of 36"-Maximum of 44"	• Check with the strobe manufacturer's data sheets to determine coverage and compliance with the ADA for corridor strobes. • Some manufacturer's 15 cd strobes may be spaced 100' apart in corridors; others require closer spacing.

19.1.14 ADA-Complying Mounting Height for Manual Pull Stations (High Forward-Reach Limit)

19.1.14

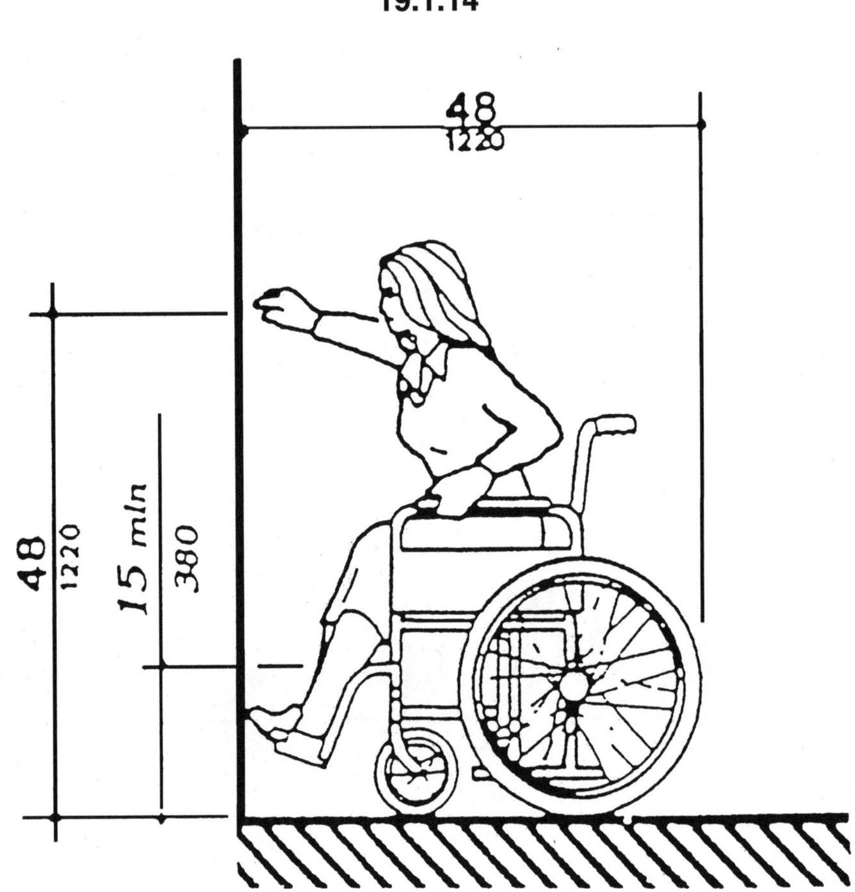

19.1.15 ADA-Complying Mounting Height for Manual Pull Stations (High and Low Side-Reach Limits) *(see page 19.11)*

19.1.16 Application Tips

A very general rule of thumb for spacing automatic fire detectors is to allow 900 ft^2 per head. This is good for very rough estimating in preliminary stages of design. There are many factors to consider for each specific application, e.g., architectural and structural features such as beams and coves, special-use spaces, ambient temperature, and other environmental considerations, etc. It is therefore prudent to refer to and become familiar with NFPA 72, Appendix B, "Application Guide for Automatic Fire Detector Spacing," coupled with your own experience.

In the design of any fire alarm system, it is necessary to determine what codes and other requirements are applicable to the project site and what editions of same have been adopted and are in effect at the time of design (sometimes states and/or municipalities do not adopt the latest edition of codes until several years later), and it is good practice to review the design with the AHJ periodically throughout the design process. This latter step also will be beneficial in resolving any conflicts

19.1.15 ADA-Complying Mounting Height for Manual Pull Stations (High and Low Side-Reach Limits)

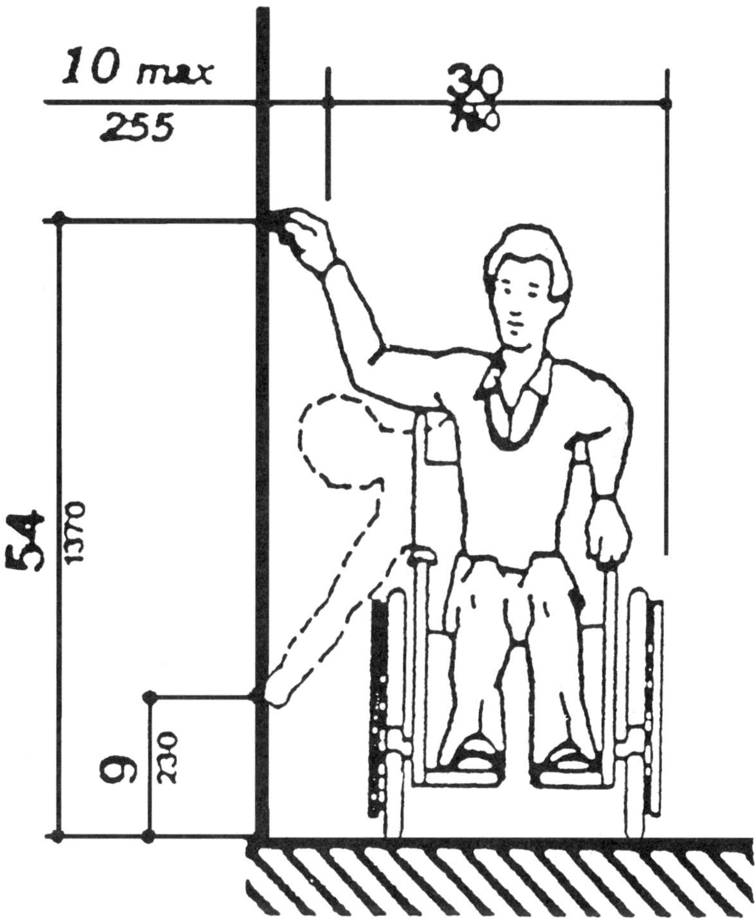

between codes and the ADAAG (these do occur) through equivalent facilitation, thus achieving compliance with all codes and regulations that apply.

It is also essential to coordinate with the architect, structural engineer, and other trade disciplines (e.g., sprinkler systems) to determine their effects on fire alarm system requirements.

Fire alarm system technology today has reached a profoundly high level, with multiplexed digital communication, 100 percent addressable systems, and even "smart" automatic fire detectors that can be programmed with profiles of their ambient environmental conditions, thus preventing nuisance alarms by being able to discriminate between "normal" and "abnormal" conditions for their specific environment. These capabilities provide the designer with a lot of flexibility to design safe and effective fire alarm systems.

19.2.1 Fire Pump Applications

The electrical requirements for electric-drive fire pumps are discussed in detail in Chapters 6 and 7 and Appendix A of NFPA 20. These requirements are supplemented by NFPA 70 (*NEC*), in particular, Articles 230, 430, and 700. The following guideline items are design highlights (based on CT and MA requirements). Please refer to any different or additional codes or requirements that may be applicable in your state. However, the following generally should be applicable.

1. All electric fire pumps shall be provided with emergency power in accordance with Article 700 of NFPA 70. State of Connecticut requirement (add to Chapter 7, C.L.S.).
2. State of Massachusetts (add to 780 CMR, Item 924.3): Electrical fire pumps in many occupancies require emergency power per NFPA 20 and *NEC,* Articles 695 and 700.
3. State of Massachusetts (add to 527 CMR, *NEC,* Article 700): Emergency system feeders and generation and distribution equipment, including fire pumps, shall have a 2-hour fire separation from all other spaces and equipment.
4. The fire pump feeder conductors shall be physically routed outside the building or enclosed in 2 in of concrete (1-hour equivalent fire resistance) except in the electrical switchgear or fire pump rooms (NFPA 20, 6-3.1.1).
5. All pump room wiring shall be in rigid, intermediate, or liquid-tight flexible metal conduit (NFPA 20, 6-3.1.2; MI cable is added to this in the 1993 version).
6. Maximum permissible voltage drop at the fire pump input terminals is 15 percent (NFPA 20, 6-3.1.4).
7. Protective devices (fuses or circuit breakers) ahead of the fire pump shall not open at the sum of the locked-rotor currents of the facility or the fire pump auxiliaries (NFPA 20, 6-3.4).
8. The pump room feeder minimum size shall be 125 percent of the sum of the fire pump(s), jockey pump, and pump auxiliary full-load currents (NFPA 20, 6-3.5).
9. Automatic load shed and sequencing of fire pumps is permitted (NFPA 20, 6-7).
10. Remote annunciation of the fire pump controller is permitted per NFPA 20, 7-4.6 and 7-4.7. *Note:* A good practice is to assume that this will happen and make provisions for it, i.e., fire alarm connections or wiring to the appropriate location.
11. When necessary, automatic transfer switch may be used. It must be listed for fire pump use. It may be a separate unit or integrated with the fire pump controller in a barriered compartment (NFPA 20, 7-8.2).
12. A jockey pump is not required to be on emergency power.
13. Step loading the fire pump onto an emergency generator can help control the generator size. A time-delay relay (0 to 60 seconds) to start or restart a fire pump when on generator power will help coordinate generator loading. The relay should be a part of the fire pump controller. (See Item 9 above.)
14. Reduced-voltage starters (i.e., autotransformer or wye-delta) for fire pumps are recommended.
15. Fire pumps, fire pump controllers, and fire pump listed automatic transfer switches generally are provided under division 15. Division 16 is responsible for powering, wiring, and connecting this equipment.

19.2.2 Typical One-Line Diagram of Fire Pump System with Separate ATS

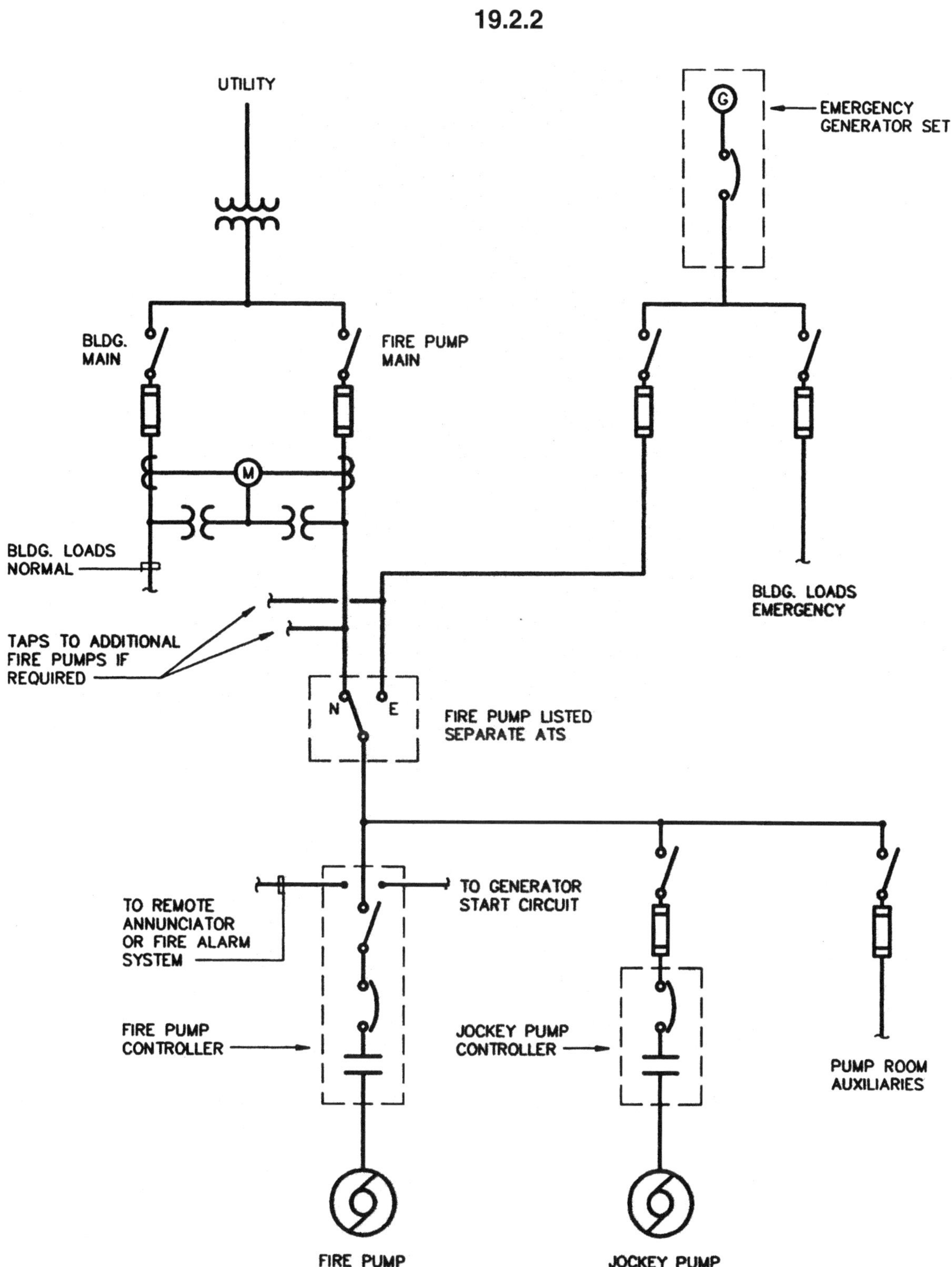

19.2.3 Typical One-Line Diagram of Fire Pump System with ATS Integrated with the Fire Pump Controller

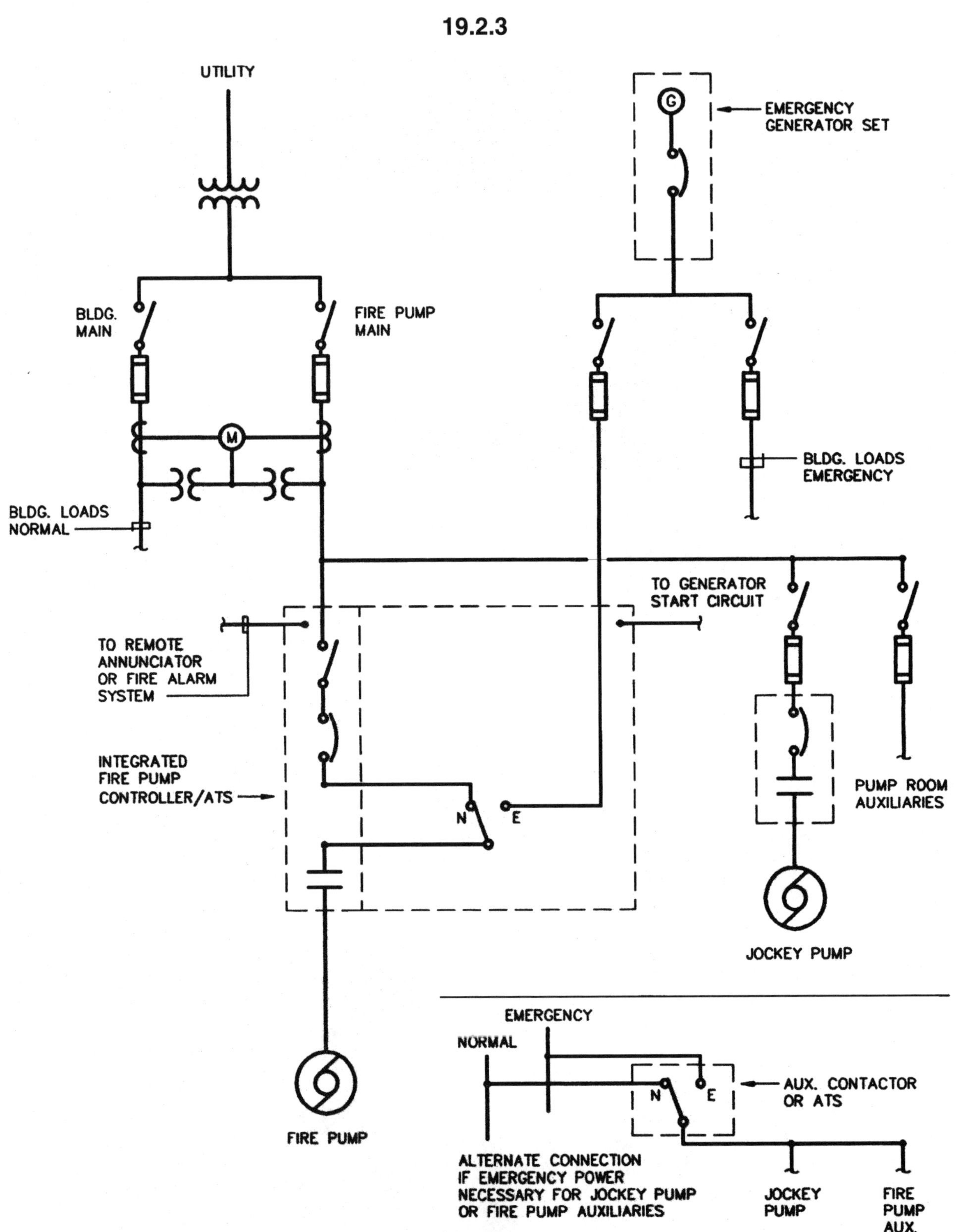

19.3.1 Wiring of Packaged Rooftop AHUs with Remote VFDs

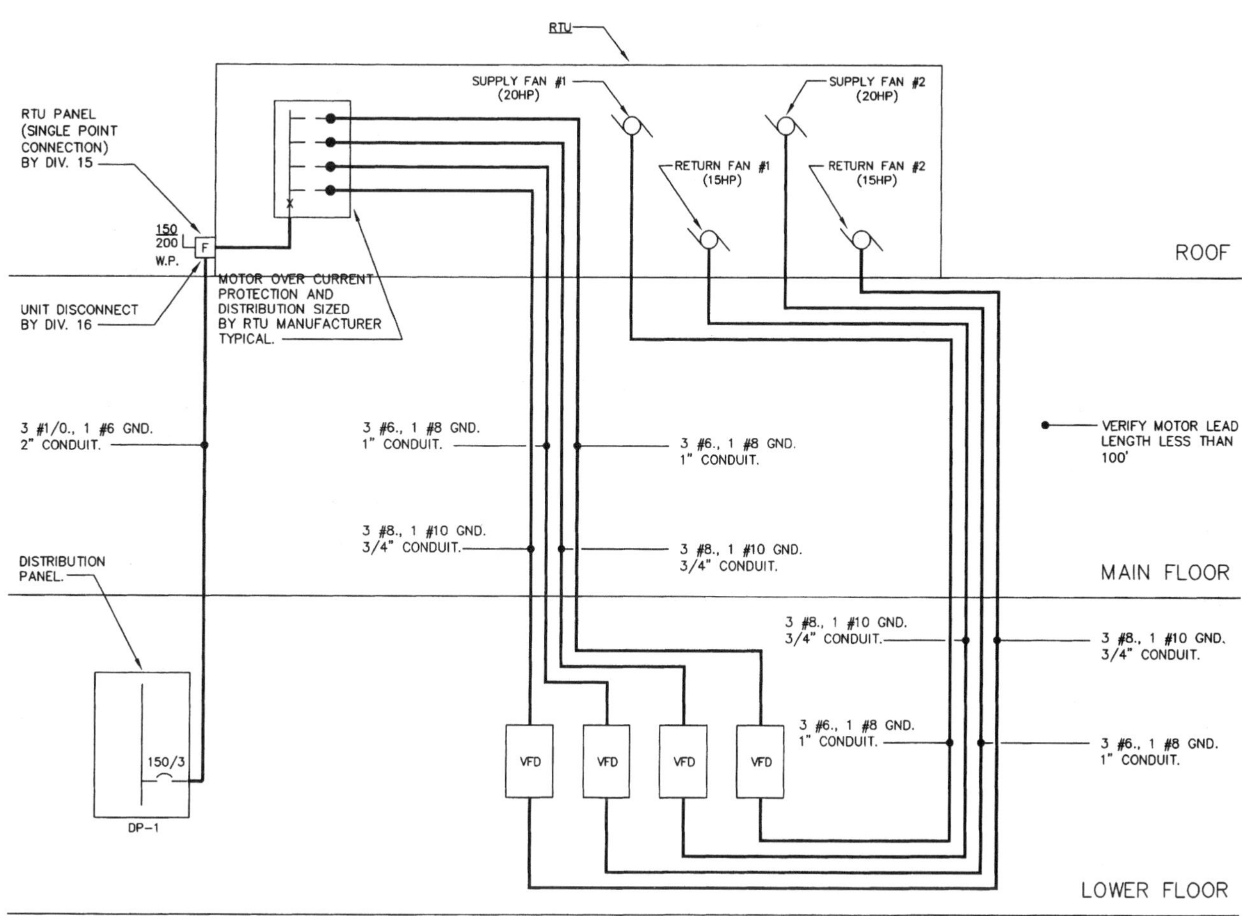

An emerging trend in HVAC design is the use of packaged rooftop air-handling units (AHUs) with remote mounted variable-frequency drives (VFDs). In this circumstance, multiple electrical connections and significant additional wiring are required, not the traditional single point of connection needed previously. It is therefore critically important to coordinate closely with the mechanical design professionals to ensure that complete and proper wiring is provided.

19.4.1 Wye-Delta Motor Starter Wiring

A common misapplication that is encountered is the improper sizing of the six motor leads between the still very popular wye-delta reduced-voltage motor starter and the motor. This is best demonstrated by an example.

Assume that you have a 500-ton electrical centrifugal chiller operating at 460 V, three-phase, 60 Hz, with a nameplate rating of 588 full-load amps (FLA). You normally would apply the correct factor of 125 percent required by *NEC* Article 440 to arrive at the required conductor ampacity: 588 × 1.25 = 735 ampacity for each of the three conductors. Since there will be six conductors between the load side of the starter and the compressor motor terminals, the 735 ampacity is divided by 2, and you would select six conductors, each having an ampacity of not less than 368 A. Referring to *NEC* Article 310, Table 310.16 for insulated copper conductors at 75°C would result in the selection of 500-kcmil conductors.

This wire size is incorrect when used between the wye-delta starter and motor terminals.

The problem is caused by a common failure to recognize that the motor may consist of a series of single-phase windings. To permit the transition from wye-start to delta-run configuration, the motor is wound without internal connections. Each end of the three internal motor windings is brought out to a terminal, as shown in Figure 19.4.1(A).

19.4.1A Wye-to-delta internal motor windings brought out to terminals.

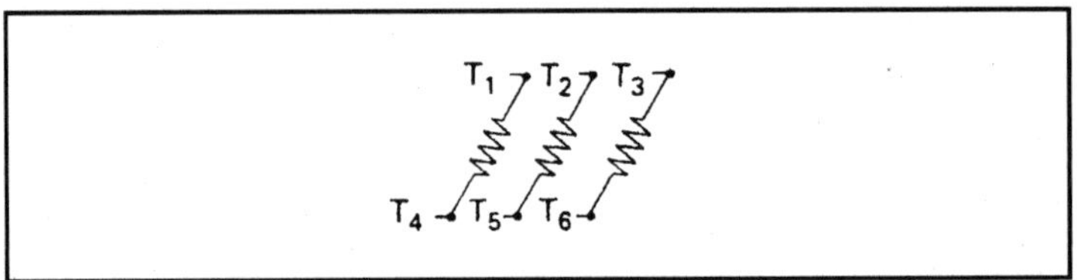

The motor windings are configured as required for either starting or running at the starter, as shown in Figure 19.4.1(B).

19.4.1B Wye-start, delta-run motor winding configuration.

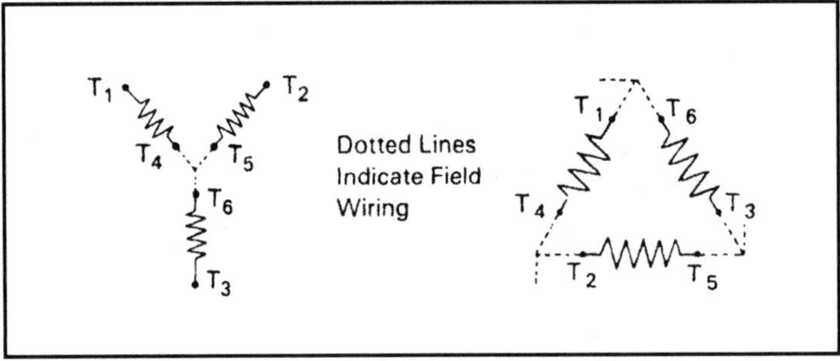

In the running-delta configuration, the field wiring from the load side of the starter to the compressor motor terminals consists of six conductors, electrically balancing the phases to each of the internal motor windings, as described in Figure 19.4.1(C).

Note, for example, that motor winding $T_1 - T_4$ is connected to the line voltage across phase $L_1 - L_2$.

It should be apparent that the windings within the motor are single-phase connected to the load side of the starter. Thus the interconnecting field wiring between the starter and motor must be sized as though the motor were single-phase. Electrical

19.4.1C Field wiring between starter and motor in wye-start, delta-run configuration.

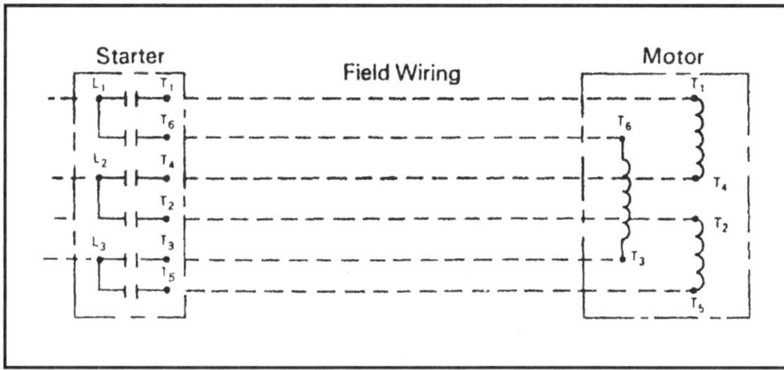

terminology simply describes this motor as being *phase connected,* and the current carried by the interconnecting conductors as *phase amps.*

To size the conductors between the motor starter and the motor correctly, therefore, it is necessary to calculate the ampacity with the 125 percent feeder sizing factor required by the *NEC* on a single-phase basis as follows:

$$\text{Ampacity per terminal conductor} = \frac{\text{3-phase FLA} \times 1.25}{1.73}$$

For the example given,

$$\text{Ampacity per terminal conductor} = \frac{588 \times 1.25}{1.73} = 424$$

Thus it is clear that the current in the conductors between the starter and the motor on a single-phase basis is 58 percent of the three-phase value, not 50 percent, as originally assumed, because the current in one phase of a three-phase system in the delta-connected winding is 1 divided by $\sqrt{3}$ due to the vector relationship.

In the original example, the conductors were sized for a minimum ampacity of 368 A. From the *NEC,* 500-kcmil copper conductors at 75°C have a maximum allowable ampacity of 380 A. The preceding calculation discloses that the conductors should be selected for not less than 424 A. Referring to the *NEC* again, 600-kcmil conductors have a maximum allowable ampacity of 420 A. In many cases, depending on the interpretation of the local electrical inspector, 600-kcmil would be acceptable (usually within 3 percent is acceptable). However, 500-kcmil wire would not be.

Almost needless to say, the conductors supplying the line side of the wye-delta starter are sized as conventional three-phase motor conductors.

19.5.1 Elevator Recall Systems

Elevator recall systems are discussed here rather than under fire alarm systems in Section 19.1 because they can be installed as a stand-alone system, even though they are generally a part of a fire alarm system. Also, several codes are applicable to the installation of these systems, specifically ANSI/ASME A17.1, *Safety Code for Elevators and Escalators,* NFPA 72, *National Fire Alarm Code,* NFPA 13, *Standard for Installation of Sprinklers,* and NFPA 101, *Life Safety Code*—to which the reader is referred for complete details.

Further, applying these codes properly in combination can be problematic (e.g., whether sprinklers are present or not), coupled with the requirements of the authority having jurisdiction (which generally are more stringent).

Briefly stated, ANSI/ASME A17.1 is written so as to ensure that an elevator car will not stop and open the door on a fire-involved floor by requiring elevators to be recalled nonstop to a designated safe floor when smoke detectors located in elevator lobbies, other than the designated level, are actuated. When the smoke detector at the designated level is activated, the cars return to an alternate level approved by the enforcing authority.

If the elevator is equipped with front and rear doors, it is necessary to have smoke detectors in both lobbies at the designated level.

Activation of a smoke detector in any elevator machine room, except a machine room at the designated level, shall cause all elevators having any equipment located in that machine room and any associated elevators of a group automatic operation to return nonstop to the designated level. When a smoke detector in an elevator machine room is activated that is at the designated level, with the other conditions being the same as above, the elevators shall return nonstop to the alternate level or the appointed level when approved by the authority having jurisdiction.

NFPA 72 requires that in facilities without a building fire alarm system, these smoke detectors shall be connected to a dedicated fire alarm system control unit that shall be designated as "elevator recall control and supervisory panel." Thus the stand-alone operation first noted above.

As noted, the foregoing is by no means complete but captures the intent and basic cause-and-effect relationship between an elevator recall system's smoke detectors and elevator operation under the various stated conditions.

Figure 19.5.2 shows a typical elevator recall/emergency shutdown schematic. Please note that the authority having jurisdiction required that the elevator recall smoke detectors in this application be independent of the building fire alarm system smoke detectors. Figure 19.5.3 shows a typical elevator hoistway/machine room device installation detail for the same project application shown in Figure 19.5.2. Note that the fire alarm system is fully addressable and that the elevator machine rooms are at the designated level for egress.

19.5.2 Typical Elevator Recall/Emergency Shutdown Schematic (*see page 19.19*)

19.5.3 Typical Elevator Hoistway/Machine Room Device Installation Detail (*see page 19.20*)

19.6.0 Harmonic Effects and Mitigation

Harmonics are the result of nonlinear loads so prevalent with late-twentieth-century technology. Personal computers, adjustable-speed drives, uninterruptable power supplies, to name a few, all have nonlinear load characteristics. What all nonlinear loads have in common is that they convert ac to dc and contain some kind of rectifier.

A sinusoidal system can supply nonsinusoidal current demands because any nonsinusoidal waveform can be generated by the proper combination of harmonics of the fundamental frequency. Each harmonic in the combination has a specific amplitude and phase relative to the fundamental. The particular harmonics drawn by a nonlinear load are a function of the rectifier circuit and are not affected by the type of load.

19.5.2 Typical elevator recall/emergency shutdown schematic. No scale.

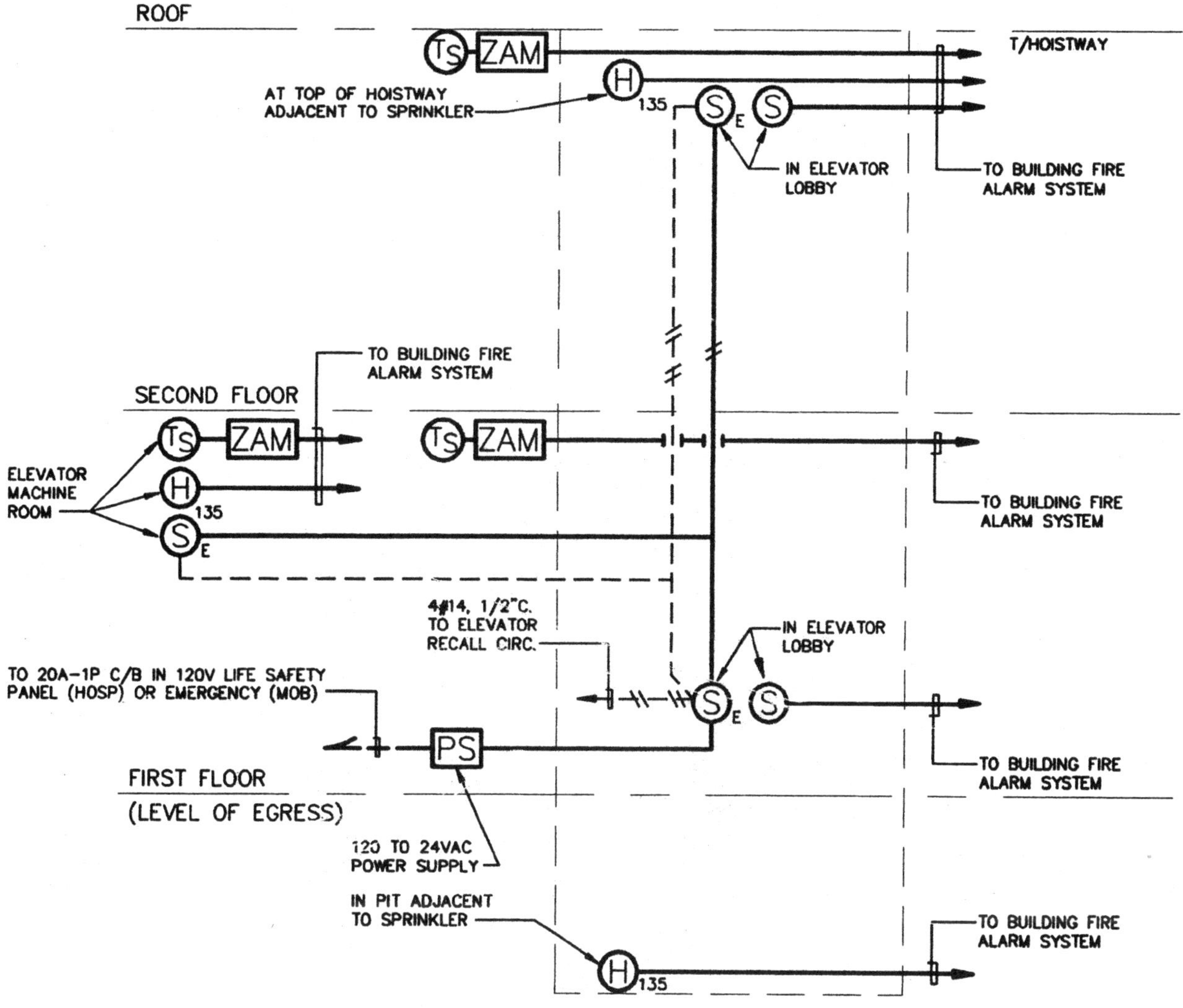

Harmonic Origins. Harmonics have two basic origins: current-wave distortion and voltage-wave distortion.

- Harmonics-producing equipment (voltage distortion)
  - Uninterruptable power supplies
  - Variable-frequency drives
  - Large battery chargers
  - Elevators
  - Synchronous clock systems
  - Radiology equipment
  - Large electronic dimming systems
  - Arc heating devices
- Harmonics-producing equipment (current distortion)
  - Personal computers

19.5.3 Typical elevator hoistway/machine room device installation detail. Scale: ¼″ = 1′-0″.

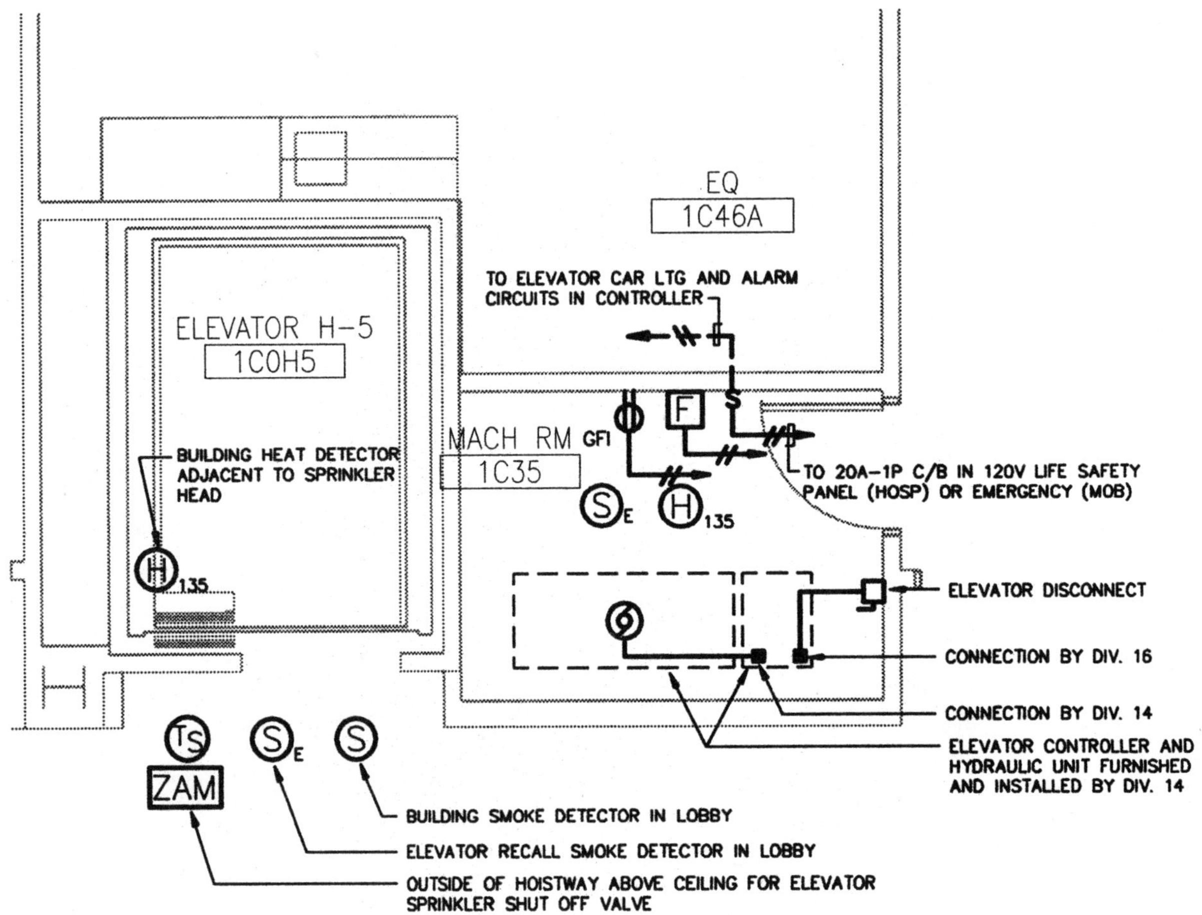

Desktop printers

Small battery chargers

Electric discharge lighting

Electronic/electro-magnetic ballasts

Small electronic dimming systems

It should be noted that voltage distortion is more difficult to deal with because it is systemwide.

Harmonic Characteristics

- Harmonics are integer multiples of the fundamental frequency.
- First order is the fundamental frequency (e.g., 60 Hz); the second order is $2 \times 60 = 120$ Hz; the third order is $3 \times 60 = 180$ Hz, etc.
- In three-phase systems, even harmonics cancel; odd harmonics are additive in the neutral and ground paths.
- Harmonics that are multiples of three are called *triplens* (i.e., third, ninth, fifteenth, etc.).
- Triplen harmonics, particularly the third, cause major problems in electrical distribution systems

Problems with Harmonics

- Harmonics do no work but contribute to the rms current the system must carry.
- Triplen harmonics are additive in the system neutral.
- These currents return to the transformer source over the neutral and are dissipated as heat in the transformer, cables, and load devices.

Symptoms of Harmonic Problems

- Overheated neutral conductors, panels, and transformers
- Premature failure of transformers, generators, and UPS systems
- Lost computer data
- Interference on communications lines
- Operation of protective devices without overload or short circuit
- Random component failure in electronic devices
- Operating problems with electronic devices not traceable to component problems
- Interaction between multiple VFDs throwing off setpoints
- Interaction between UPSs and their supplying generators
- System power factor reduction and related system capacity loss
- Problems with capacitor operation and life

Harmonic Mitigation. Currently, there are no devices that completely eliminate harmonics and thus their effects; however, they can be mitigated substantially to control their deleterious consequences. Essentially, current techniques consist of accommodating harmonics and include the following:

- Increasing neutral sizes, usually doubling feeder neutral sizes and installing a separate neutral with each single-phase branch circuit of a three-phase system, effectively a triple-neutral, rather than a single common neutral of the same size as the phase conductor
- *k*-Rated transformers
- Harmonic-rated distribution equipment such as switchgear and panelboards
- Passive filters such as phase shifters, phase cancelers, zig-zag transformers, and zero-sequence transformers
- Active filters, electronic, primarily protects upstream equipment/devices
- Proper grounding
- Isolation transformers (electrostatically shielded)
- Motor-generator sets
- Oversizing equipment

Most of the preceding involve "beefing up" to accommodate harmonics.

Active versus Passive Devices

Active devices:

Pros	Cons
Works well for mitigation of harmonics upstream of the device.	Expensive.
Protects the transformer.	High maintenance costs.
	Uses power.
	Works only upstream.

Passive devices:

Pros	Cons
No electronic circuitry.	Work only upstream to accommodate harmonics.
Very reliable.	Location is critical.
	Phase loads must be balanced.
	Can be overloaded.
	Dissipate heat.
	Require fused disconnect.

Ultimate/Ideal Solution. The ultimate ideal solution would

- Eliminate the production of harmonics at the source (not just accommodate them).
- Be passive and therefore cost-effective, reliable, and efficient.
- Be easily installed and not require protection.
- Handle any load on the distribution system (not require load balancing to be effective).
- Resist overloading (not become a harmonic sink for the rest of the distribution system).

Section

20

Metrification

20.0.0 Introduction to the 1975 Metric Conversion Act
20.1.0 What Will Change and What Will Remain the Same
20.2.0 How Metric Units Will Apply in the Construction Industry
20.3.0 Metrification of Pipe Sizes
20.4.0 Metrification of Standard Lumber Sizes
20.5.0 Metric Rebar Conversions
20.6.0 Metric Conversion of ASTM Diameter and Wall-Thickness Designations
20.7.0 Metric Conversion Scales (Temperature and Measurements)
20.8.0 Approximate Metric Conversions
20.9.0 Quick Imperial (Metric Equivalents)
20.10.0 Metric Conversion Factors

20.0.0 Introduction to the 1975 Metric Conversion Act

20.0.0

As the federal government moves to convert the inch-pound units to the metric system, in accordance with the 1975 Metric Conversion Act, various parts of the construction industry will begin the conversion to this more universal method of measurement.

Metric units are often referred to as *SI units*, an abbreviation taken from the French Le Système International d'Unités. Another abbreviation that will be seen with more frequency is ISO - the International Standards Organization charged with supervising the establishment of a universal standards system. For everyday transactions it may be sufficient to gain only the basics of the metric system.

Name of metric unit	Symbol	Approximate size (length/pound)
meter	m	39½ inches
kilometer	km	0.6 mile
centimeter	cm	width of a paper clip
millimeter	mm	thickness of a dime
hectare	ha	2½ acres
square meter	m2	1.2 square yards
gram	g	weight of a paper clip
kilogram	kg	2.2 pounds
metric ton	t	long ton (2240 pounds)
liter	L	one quart and two ounces
milliliter	mL	⅕ teaspoon
kilopascal	kPa	atmospheric pressure is about 100 kPa

The Celsius temperature scale is used. Instead of referring to its measurement as *degree centigrade*, the term *degree Celsius* is the correct designation. Using this term, familiar points are:

- Water freezes at 0 degrees
- Water boils at 100 degrees
- Normal body temperature is 37 degrees (98.6 F)
- Comfortable room temperature 20 to 35 (68 to 77 F)

20.1.0 What Will Change and What Will Remain the Same

20.1.0

Metric Module and Grid

What will change

- The basic building module, from 4 inches to 100 mm.
- The planning grid, from 2' × 2' to 600 × 600 mm.

What will stay the same

- A module and grid based on rounded, easy-to-use dimensions. The 100 mm module is the global standard.

Drawings

What will change

- Units, from feet and inches to millimeters for all building dimensions and to meters for site plans and civil engineering drawings. Unit designations are unnecessary: if there's no decimal point, it's millimeters; if there's a decimal point carried to one, two, or three places, it's meters. In accordance with ASTM E621, centimeters are not used in construction because (1) they are not consistent

(continued)

20.1.0 *(Continued)*

with the preferred use of multiples of 1000, (2) the order of magnitude between a millimeter and centimeter is only 10 and the use of both units would lead to confusion and require the use of unit designations, and (3) the millimeter is small enough to almost entirely eliminate decimal fractions from construction documents.

- Drawing scales, from inch-fractions-to-feet to true ratios. Preferred metric scales are:

  1:1 (full size)
  1:5 (close to 3" = 1'-0")
  1:10 (between 1"= 1'-0" and 1½" = 1'-0")
  1:20 (between ½" = 1'-0" and ¾" = 1'-0")
  1:50 (close to ¼" = 1'-0")
  1:100 (close to ⅛" = 1'-0")
  1:200 (close to 1⁄16" = 1'-0")
  1:500 (close to 1" = 40'-0")
  1:1000 (close to 1" = 80'-0")

As a means of comparison, inch-fraction scales may be converted to true ratios by multiplying a scale's divisor by 12; for example, for ¼" = 1'-0", multiply the 4 by 12 for a true ratio of 1:48.

- Drawing sizes, to ISO "A" series:

  A0 (1189 × 841 mm, 46.8 × 33.1 inches)
  A1 (841 × 594 mm, 33.1 × 23.4 inches)
  A2 (594 × 420 mm, 23.4 × 16.5 inches)
  A3 (420 × 297 mm, 16.5 × 11.7 inches)
  A4 (297× 210 mm, 11.7 × 8.3 inches)

  Of course, metric drawings can be made on any size paper.

  What will stay the same.

- Drawing contents

Never use dual units (both inch-pound and metric) on drawings. It increases dimensioning time, doubles the chance for errors, makes drawings more confusing, and only postpones the learning process. An exception is for construction documents meant to be viewed by the general public.

Specifications

What will change

- Units of measure, from feet and inches to millimeters for linear dimensions, from square feet to square meters for area, from cubic yards to cubic meters for volume (except use liters for fluid volumes), and from other inch-pound measures to metric measures as appropriate.

  What will stay the same

- Everything else in the specifications

Do not use dual units in specifications except when the use of an inch-pound measure serves to clarify an otherwise unfamiliar metric measure; then place the inch-pound unit in parentheses after the metric. For example, "7.5 kW (10 horsepower) motor." All unit conversions should be checked by a professional to ensure that rounding does not exceed allowable tolerances.

For more information, see the July–August 1994 issue of *Metric in Construction*.

Floor Loads

What will change

- Floor load designations, from "psf" to kilograms per square meter (kg/m^2) for everyday use and kilonewtons per square meter (kN/m^2) for structural calculations.

(continued)

20.1.0 *(Continued)*

What will stay the same

- Floor load requirements

Kilograms per square meter often are used to designate floor loads because many live and dead loads (furniture, filing cabinets, construction materials, etc.) are measured in kilograms. However, kilonewtons per square meter or their equivalent, kilopascals, are the proper measure and should be used in structural calculations.

Construction Products

What will change

- Modular products: brick, block, drywall, plywood, suspended ceiling systems, and raised floor systems. They will undergo "hard" conversion; that is, their dimensions will change to fit the 100 mm module.
- Products that are custom-fabricated or formed for each job (for example, cabinets, stairs, handrails, ductwork, commercial doors and windows, structural steel systems, and concrete work). Such products usually can be made in any size, inch-pound or metric, with equal ease; therefore, for metric jobs, they simply will be fabricated or formed in metric.

What will stay the same

- All other products, since they are cut-to-fit at the jobsite (for example, framing lumber, woodwork, siding, wiring, piping, and roofing) or are not dimensionally sensitive (for example, fasteners, hardware, electrical components, plumbing fixtures, and HVAC equipment). Such products will just be "soft" converted—that is, relabeled in metric units. A 2¾" × 4½" wall switch face plate will be relabeled 70 × 115 mm and a 30 gallon tank, 114 L. Manufacturers eventually may convert the physical dimensions of many of these products to new rational "hard" metric sizes but only when it becomes convenient for them to do so.

"2-By-4" Studs and Other "2-By" Framing (Both Wood and Metal)

What will change

- Spacing, from 16" to 400 mm, and 24" to 600 mm.

What will stay the same

- Everything else.

"2-bys" are produced in "soft" fractional inch dimensions so there is no need to convert them to new rounded "hard" metric dimensions. 2-by-4s may keep their traditional name or perhaps they'll eventually be renamed 50 by 100 (mm), or, more exactly, 38 × 89.

Drywall, Plywood, and Other Sheet Goods

What will change

- Widths, from 4'-0" to 1200 mm.
- Heights, from 8'-0" to 2400 mm, 10'-0" to 3000 mm.

What will stay the same

- Thicknesses, so fire, acoustic, and thermal ratings won't have to be recalculated.

Metric drywall and plywood are readily available but may require longer lead times for ordering and may cost more in small amounts until their use becomes more common.

(continued)

20.1.0 (*Continued*)

Batt Insulation

What will change

- Nominal width labels, from 16" to 16"/400 mm and 24" to 24"/600 mm.

What will stay the same

- Everything else.

Batts will not change in width; they'll just have a tighter "friction fit" when installed between metric-spaced framing members.

Doors

What will change

- Height, from 6'-8" to 2050 mm or 2100 mm and from 7'-0" to 2100 mm.
- Width, from 2'-6" to 750 mm, from 2'-8" to 800 mm, from 2'-10" to 850 mm, from 3'-0" to 900 mm or 950 mm, and from 3'-4" to 1000 mm.

What will stay the same

- Door thicknesses.
- Door materials and hardware.

For commercial work, doors and door frames can be ordered in any size since they normally are custom-fabricated.

Ceiling Systems

What will change

- Grids and lay-in ceiling tile, air diffusers and recessed lighting fixtures, from 2' × 2' to 600 × 600 mm and from 2' × 4' to 600 × 1200 mm.

What will stay the same

- Grid profiles, tile thicknesses, air diffuser capacities, florescent tubes, and means of suspension.

On federal building projects, metric recessed lighting fixtures may be specified if their total installed costs are estimated to be no more than for inch-pound fixtures.

Raised Floor Systems

What will change

- Grids and lay-in floor tile, from 2' × 2' to 600 × 600 mm.

What will stay the same

- Grid profiles, tile thicknesses, and means of support.

HVAC Controls

What will change

- Temperature units, from Fahrenheit to Celsius.

(*continued*)

20.1.0 (*Continued*)

What will stay the same

- All other parts of the controls.

Controls are now digital so temperature conversions can be made with no difficulty.

Brick

What will change

- Standard brick, to 90 × 57 × 190 mm.
- Mortar joints, from ⅜" and ½" to 10 mm.
- Brick module, from 2' × 2' to 600 × 600 mm.

What will stay the same

- Brick and mortar composition.

Of the 100 or so brick sizes currently made, 5 to 10 are within a millimeter of a metric brick so the brick industry will have no trouble supplying metric brick.

For more information, see the March-April 1995 issue of *Metric in Construction.*

Concrete Block

What will change

- Block sizes, to 190 × 190 × 390 mm.
- Mortar joints, from ½" to 10 mm.
- Block module, from 2' × 2' to 600 × 600 mm.

What will stay the same

- Block and mortar composition.

On federal building projects, metric block may be specified if its total installed cost is estimated to be no more than for inch-pound block. The Construction Metrication Council recommends that, wherever possible, block walls be designed and specified in a manner that permits the use of either inch-pound or metric block, allowing the final decision to be made by the contractor.

Sheet Metal

What will change

- Designation, from "gage" to millimeters.

What will stay the same

- Thickness, which will be soft-converted to tenths of a millimeter.

In specifications, use millimeters only or millimeters with the gage in parentheses.

Concrete

What will change

- Strength designations, from "psi" to megapascals, rounded to the nearest 5 megapascals per ACI 318M as follows:

  2500 psi to 20 MPa
  3000 psi to 25 MPa
  3500 psi to 25 MPa

(continued)

20.1.0 (*Continued*)

4000 psi to 30 MPa
4500 psi to 35 MPa
5000 psi to 35 MPa

Depending on exact usage, however, the above metric conversions may be more exact than those indicated.

What will stay the same

- Everything else.

For more information, see the November-December 1994 issue of *Metric in Construction*.

Rebar

What will change

- Rebar will not change in size but will be renamed per ASTM A615M-96a and ASTM A706M-96a as follows:

No. 3 to No. 10 | No. 9 to No. 29
No. 4 to No. 13 | No. 10 to No. 32
No. 5 to No. 16 | No. 11 to No. 36
No. 6 to No. 19 | No. 14 to No. 43
No. 7 to No. 22 | No. 18 to No. 57
No. 8 to No. 25

What will stay the same

- Everything else.

For more information, see the July-August 1996 issue of *Metric in Construction*.

Glass

What will change

- Cut sheet dimensions, from feet and inches to millimeters.

What will stay the same

- Sheet thickness; sheet glass can be rolled to any dimension and often is rolled in millimeters now.

See ASTM C1036.

Pipe and Fittings

What will change

- Nominal pipe and fitting designations, from inches to millimeters

What will stay the same

- Pipe and fitting cross sections and threads.

Pipes and fittings are produced in "soft" decimal inch dimensions but are identified in nominal inch sizes a matter of convenience. A 2-inch pipe has neither an inside nor an outside diameter of 2 inches, a 1-inch fitting has no exact 1-inch dimension, and a ½-inch sprinkler head contains no ½-inch dimension anywhere; consequently, there is no need to "hard" convert pipes and fittings to rounded metric dimensions. Instead, they will not change size but simply be relabeled in metric as follows:

⅛" = 6 mm | 1½" = 40 mm
³⁄₁₆" = 7 mm | 2" = 50 mm

(continued)

20.1.0 (*Continued*)

¼" = 8 mm	2½" = 65 mm
⅜" = 10 mm	3" = 75 mm
½" = 15 mm	3½" = 90 mm
⅝" = 18 mm	4" = 100 mm
¾" = 20 mm	4½" = 115 mm
1" = 25 mm	1" = 25 mm for all larger sizes
1¼" = 32 mm	

For more information, see the September-October 1993 issue of *Metric in Construction.*

Electrical Conduit

What will change

- Nominal conduit designations, from inches to millimeters.

What will stay the same

- Conduit cross sections.

Electrical conduit is similar to piping: it is produced in "soft" decimal inch dimensions but is identified in nominal inch sizes. Neither metallic nor nometallic conduit will change size; they will be relabeled in metric units as follows:

½" = 16 (mm)	2½" = 63 (mm)
¾" = 21 (mm)	3" = 78 (mm)
1" = 27 (mm)	3½" = 91 (mm)
1¼" = 35 (mm)	4" = 103 (mm)
1½" = 41 (mm)	5" = 129 (mm)
2" = 53 (mm)	6" = 155 (mm)

These new metric names were assigned by the National Electrical Manufacturers Association.

Electrical Wire

What will change

- Nothing at this time.

What will stay the same

- Existing American Wire Gage (AWG) sizes.

Structural Steel

What will change

- Section designations, from inches to millimeters and from pounds per foot to kilograms per meter, in accordance with ASTM A6M.
- Bolts—to metric diameters and threads per ASTM A325M and A490M.

What will stay the same

- Cross sections.

Like pipe and conduit, steel sections are produced in "soft" decimal inch dimensions (with actual depths varying by weight) but are named in rounded inch dimensions so there is no need to "hard" convert them to metric units. Rather, their names will be changed to metric designations, and rounded to the nearest 10 mm. Thus, a 10-inch section is relabeled as a 250-mm section and a 24-inch section is relabeled as a 610-mm section.

20.2.0 How Metric Units Will Apply in the Construction Industry

20.2.0

	Quantity	Unit	Symbol
Masonry	length	meter, millimeter	m, mm
	area	square meter	m^2
	mortar volume	cubic meter	m^3
Steel	length	meter, millimeter	m, mm
	mass	megagram (metric ton) kilogram	Mg (t) kg
	mass per unit length	kilogram per meter	kg/m
Carpentry	length	meter, millimeter	m, mm
Plastering	length	meter, millimeter	m, mm
	area	square meter	m^2
	water capacity	liter (cubic decimeter)	L (dm^3)
Glazing	length	meter, millimeter	m, mm
	area	square meter	m^2
Painting	length	meter, millimeter	m, mm
	area	square meter	m^2
	capacity	liter (cubic decimeter) milliliter (cubic centimeter)	L (dm^3) mL (cm^3)
Roofing	length	meter, millimeter	m, mm
	area	square meter	m^2
	slope	percent ratio of lengths	% mm/mm, m/m
Plumbing	length	meter, millimeter	m, mm
	mass	kilogram, gram	kg, g
	capacity	liter (cubic decimeter)	L (dm^3)
	pressure	kilopascal	kPa
Drainage	length	meter, millimeter	m, mm
	area	hectare (10 000 m2) square meter	ha m^2
	volume	cubic meter	m^3
	slope	percent ratio of lengths	% mm/mm, m/m
HVAC	length	meter, millimeter	m, mm
	volume (capacity)	cubic meter liter (cubic decimeter)	m^3 L (dm^3)
	air velocity	meter/second	m/s
	volume flow	cubic meter/second liter/second (cubic decimeter per second)	m^3/s L/s (dm^3/s)
	temperature	degree Celsius	°C
	force	newton, kilonewton	N, kN
	pressure	pascal, kilopascal	Pa, kPa
	energy	kilojoule, megajoule	kJ, MJ
	rate of heat flow	watt, kilowatt	W, kW
Electrical	length	millimeter, meter, kilometer	mm, m, km
	frequency	hertz	Hz
	power	watt, kilowatt	W, kW
	energy	magajoule kilowatt hour	MJ kWh
	electric current	ampere	A
	electric potential	volt, kilovolt	V, kV
	resistance	milliohm, ohm	mΩ, Ω

20.3.0 Metrification of Pipe Sizes

20.3.0

Pipe diameter sizes can be confusing because their designated size does not correspond to their actual size. For instance, a 2-inch steel pipe has an inside diameter of approximately 2⅛ inches and an outside diameter of about 2⅜ inches.

The *2 inch* designation is very similar to the 2" × 4" designation for wood studs, neither dimensions are "actual", but they are a convenient way to describe these items.

Pipe sizes are identified as *NPS* (*nominal pipe size*) and their conversion to metric would conform to ISO (International Standards Organization) criteria and are referred to as *DN* (*diameter nominal*). These designations would apply to all plumbing, mechanical, drainage, and miscellaneous pipe commonly used in civil works projects.

NPS size	DN size
⅛"	6 mm
3/16"	7 mm
¼"	8 mm
⅜"	10 mm
½"	15 mm
⅝"	18 mm
¾"	20 mm
1"	25 mm
1¼"	32 mm
1½"	40 mm
2"	50 mm
2½"	65 mm
3"	80 mm
3½"	90 mm
4"	100 mm
4½"	115 mm
5"	125 mm
6"	150 mm
8"	200 mm
10"	250 mm
12"	300 mm
14"	350 mm
16"	400 mm
18"	450 mm
20"	500 mm
24"	600 mm
28"	700 mm
30"	750 mm
32"	800 mm
36"	900 mm
40"	1000 mm
44"	1100 mm
48"	1200 mm
52"	1300 mm
56"	1400 mm
60"	1500 mm

For all pipe over 60 inches nominal, use 1 inch equals 25 mm.

20.4.0 Metrification of Standard Lumber Sizes

20.4.0

Metric units: ASTM Standard E 380 was used as the authoritative standard in developing the metric dimensions in this standard. Metric dimensions are calculated at 25.4 millimeters (mm) times the actual dimension in inches. The nearest mm is significant for dimensions greater than ⅛ inch, and the nearest 0.1 mm is significant for dimensions equal to or less than ⅛ inch.

The rounding rule for dimensions greater than 1/8 inch: If the digit in the tenths of mm position (the digit after the decimal point) is less than 5, drop all fractional mm digits; if it is greater than 5 or if it is 5 followed by at least one nonzero digit, round one mm higher; if 5 followed by only zeroes, retain the digit in the unit position (the digit before the decimal point) if it is even or increase it one mm if it is odd.

The rounding rule for dimensions equal to or less than 1/8 inch: If the digit in the hundredths of mm position (the second digit after the decimal point) is less than 5, drop all digits to the right of the tenth position; if greater than or it is 5 followed by at least one nonzero digit, round one-tenth mm higher; if 5 followed by only zeros, retain the digit in the tenths position if it is even or increase it one-tenth mm if it is odd.

In case of a dispute on size measurements, the conventional (inch) method of measurement shall take precedence.

20.5.0 Metric Rebar Conversions

20.5.0

A615 M-96a & A706M-96a Metric Bar Sizes	Nominal Diameter	A615-96a & A706-96a Inch-Pound Bar Sizes
#10	9.5 mm/0.375"	#3
#13	12.7 mm/0.500"	#4
#16	15.9 mm/0.625"	#5
#19	19.1 mm/0.750"	#6
#22	22.2 mm/0.875"	#7
#25	25.4 mm/1.000"	#8
#29	28.7 mm/1.128"	#9
#32	32.3 mm/1.270"	#10
#36	35.8 mm/1.410"	#11
#43	43.0 mm/1.693"	#14
#57	57.3 mm/2.257"	#18

20.6.0 Metric Conversion of ASTM Diameter and Wall-Thickness Designations

20.6.0

METRIC CONVERSION OF ASTM DIAMETER DESIGNATIONS

in	mm	in	mm	in	mm	in	mm
6	150	30	750	57	1425	96	2400
8	200	33	825	60	1500	102	2550
10	250	36	900	63	1575	108	2700
12	300	39	975	66	1650	114	2850
15	375	42	1050	69	1725	120	3000
18	450	45	1125	72	1800	132	3300
21	525	48	1200	78	1950	144	3600
24	600	51	1275	84	2100	156	3900
27	675	54	1350	90	2250	168	4200

METRIC CONVERSION OF ASTM WALL THICKNESS DESIGNATIONS

in	mm	in	mm	in	mm	in	mm
1	25	3-1/8	79	5	125	8	200
1-1/2	38	3-1/4	82	5-1/4	131	8-1/2	213
2	50	3-1/2	88	5-1/2	138	9	225
2-1/4	56	3-3/4	94	5-3/4	144	9-1/2	238
2-3/8	59	3-7/8	98	6	150	10	250
2-1/2	63	4	100	6-1/4	156	10-1/2	263
2-5/8	66	4-1/8	103	6-1/2	163	11	275
2-3/4	69	4-1/4	106	6-3/4	169	11-1/2	288
2-7/8	72	4-1/2	113	7	175	12	300
3	75	4-3/4	119	7-1/2	188	12-1/2	313

20.7.0 Metric Conversion Scales (Temperature and Measurements)

°F: 210, 200, 190, 180, 170, 160, 150, 140, 130, 120, 110, 100, 90, 80, 70, 60, 50, 40, 30, 20, 10, 0, −10, −20

°C: 100, 90, 80, 70, 60, 50, 40, 30, 20, 10, 0, −10, −20, −30

yd: 0, 1, 2, 3, 4, 5, 6, 7, 8, 9, 10, 11

ft: 0, 1, 2, 3, 4, 5, 6, 7, 8, 9, 10, 11, 12, 13, 14, 15, 16, 17, 18, 19, 20, 21, 22, 23, 24, 25, 26, 27, 28, 29, 30, 31, 32, 33

m: 0, 1, 2, 3, 4, 5, 6, 7, 8, 9, 10

in: 0, 1, 2, 3, 4, 5

cm: 0, 1, 2, 3, 4, 5, 6, 7, 8, 9, 10, 11, 12, 13, 14

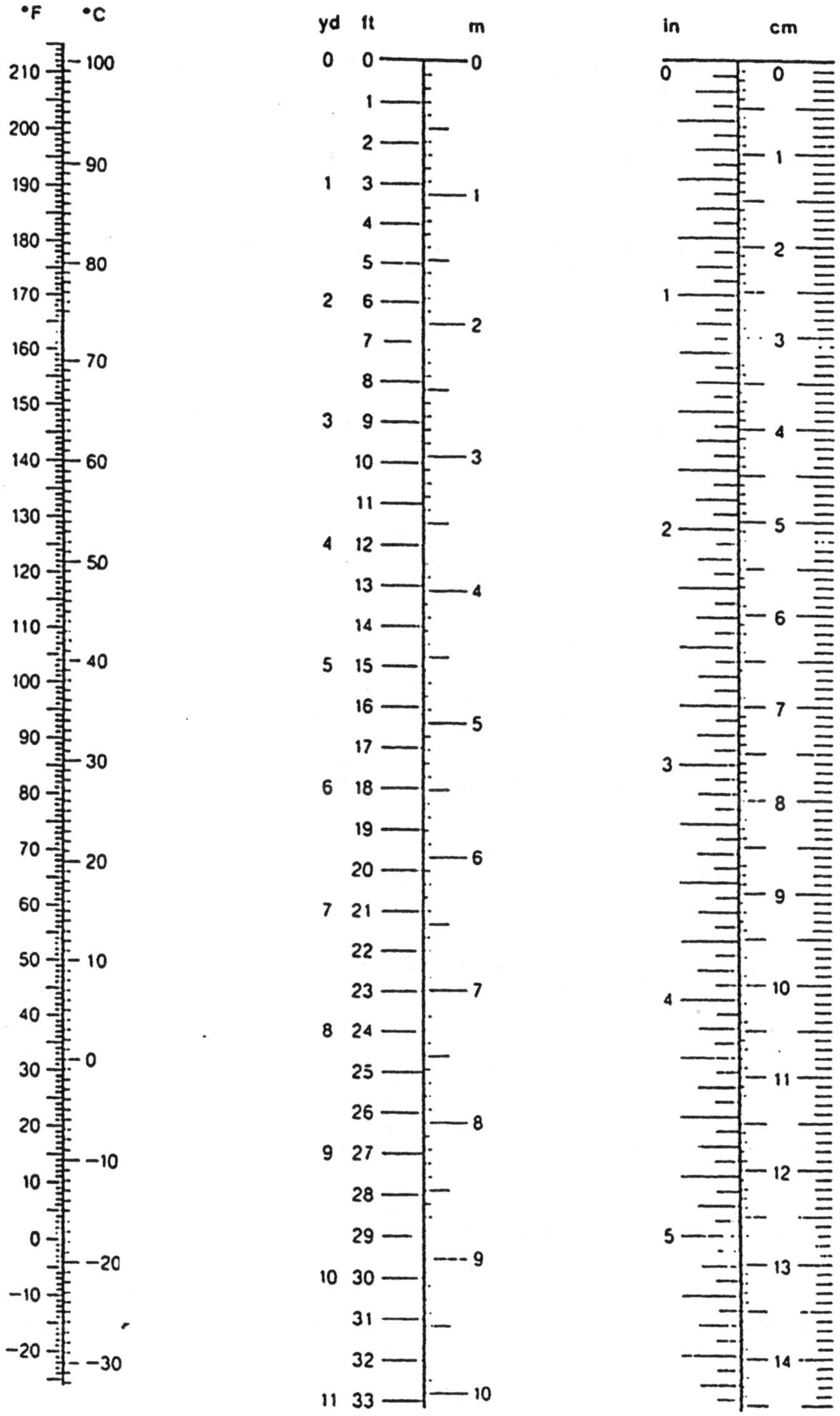

20.8.0 Approximate Metric Conversions

20.8.0

Symbol	*When You Know*	*Multiply by*	*To Find*	*Symbol*
		LENGTH		
mm	millimeters	0.04	inches	in
cm	centimeters	0.4	inches	in
m	meters	3.3	feet	ft
m	meters	1.1	yards	yd
km	kilometers	0.6	miles	mi
		AREA		
cm^2	square centimeters	0.16	square inches	in^2
m^2	square meters	1.2	square yards	yd^2
km^2	square kilometers	0.4	square miles	mi^2
ha	hectares (10,000 m^2)	2.5	acres	
		MASS (weight)		
g	grams	0.035	ounces	oz
kg	kilograms	2.2	pounds	lb
t	metric ton (1,000 kg)	1.1	short tons	
		VOLUME		
mL	milliliters	0.03	fluid ounces	fl oz
mL	milliliters	0.06	cubic inches	in^3
L	liters	2.1	pints	pt
L	liters	1.06	quarts	qt
L	liters	0.26	gallons	gal
m^3	cubic meters	35	cubic feet	ft^3
m^3	cubic meters	1.3	cubic yards	yd^3
		TEMPERATURE (exact)		
°C	degrees Celsius	multiply by 9/5, add 32	degrees Fahrenheit	°F

°C	-40	-20	0	20	37	60	80	100
°F	-40	0	32	80	98.6		160	212
			water freezes		body temperature			water boils

U.S. Department of Commerce Technology Administration, Office of Metric Programs, Washington, DC 20230

(continued)

20.8.0 *(Continued)*

Symbol	*When You Know*	*Multiply by*	*To Find*	*Symbol*
		LENGTH		
in	inches	2.5	centimeters	cm
ft	feet	30	centimeters	cm
yd	yards	0.9	meters	m
mi	miles	1.6	kilometers	km
		AREA		
in^2	square inches	6.5	square centimeters	cm^2
ft^2	square feet	0.09	square meters	m^2
yd^2	square yards	0.8	square meters	m^2
mi^2	square miles	2.6	square kilometers	km^2
	acres	0.4	hectares	ha
		MASS (weight)		
oz	ounces	28	grams	g
lb	pounds	0.45	kilograms	kg
	short tons (2000 lb)	0.9	metric ton	t
		VOLUME		
tsp	teaspoons	5	milliliters	mL
Tbsp	tablespoons	15	milliliters	mL
in^3	cubic inches	16	milliliters	mL
fl oz	fluid ounces	30	milliliters	mL
c	cups	0.24	liters	L
pt	pints	0.47	liters	L
qt	quarts	0.95	liters	L
gal	gallons	3.8	liters	L
ft^3	cubic feet	0.03	cubic meters	m^3
yd^3	cubic yards	0.76	cubic meters	m^3
		TEMPERATURE (exact)		
°F	degrees Fahrenheit	subtract 32, multiply by 5/9	degrees Celsius	°C

United States Department of Commerce, Technology Administration,
National Institute of Standards and Technology, Metric Program, Gaithersburg, MD 20899

20.9.0 Quick Imperial (Metric Equivalents)

20.9.0

Distance

Imperial	Metric
1 inch	= 2.540 centimetres
1 foot	= 0.3048 metre
1 yard	= 0.9144 metre
1 rod	= 5.029 metres
1 mile	= 1.609 kilometres

Metric	Imperial
1 centimetre	= 0.3937 inch
1 decimetre	= 0.3281 foot
1 metre	= 3.281 feet
	= 1.094 yard
1 decametre	= 10.94 yards
1 kilometre	= 0.6214 mile

Weight

1 ounce (troy)	= 31.103 grams
1 ounce (avoir)	= 28.350 grams
1 pound (troy)	= 373.242 grams
1 pound (avoir)	= 453.592 grams
1 ton (short)	= 0.907 tonne*

1 gram	= 0.032 ounce (troy)
1 gram	= 0.035 ounce (avoir)
1 kilogram	= 2.679 pounds (troy)
1 kilogram	= 2.205 pounds (avoir)
1 tonne	= 1.102 ton (short)

*1 tonne = 1000 kilograms

Capacity

Imperial	
1 pint	= 0.568 litre
1 gallon	= 4.546 litres
1 bushel	= 36.369 litres
1 litre	= 0.880 pint
1 litre	= 0.220 gallon
1 hectolitre	= 2.838 bushels

U.S.	
1 pint (U.S.)	= 0.473 litre
1 quart (U.S.)	= 0.946 litre
1 gallon (U.S.)	= 3.785 litres
1 barrel (U.S.)	= 158.98 litres

Area

1 square inch	= 6.452 square centimetres
1 square foot	= 0.093 square metre
1 square yard	= 0.836 square metre
1 acre	= 0.405 hectare*
1 square mile	= 259.0 hectares
1 square mile	= 2.590 square kilometres
1 square centimetre	= 0.155 square inch
1 square metre	= 10.76 square feet
1 square metre	= 1.196 square yard
1 hectare	= 2.471 acres
1 square kilometre	= 0.386 square mile

*1 hectare = 1 square hectometre

Volume

1 cubic inch	= 16.387 cubic centimetres
1 cubic foot	= 0.0283 cubic decimetres
1 cubic yard	= 0.765 cubic metre
1 cubic centimetre	= 0.061 cubic inch
1 cubic decimetre	= 35.314 cubic foot
1 cubic metre	= 1.308 cubic yard

20.10.0 Metric Conversion Factors

20.10.0

The following list provides the conversion relationship between U.S. customary units and SI (International System) units. The proper conversion procedure is to multiply the specified value on the left (primarily U.S. customary values) by the conversion factor exactly as given below and then round to the appropriate number of significant digits desired. For example, to convert 11.4 ft to meters: 11.4 × 0.3048 = 3.47472, which rounds to 3.47 meters. Do not round either value before performing the multiplication, as accuracy would be reduced. A complete guide to the SI system and its use can be found in ASTM E 380, Metric Practice.

To convert from	to	multiply by
Length		
inch (in.)	micron (μ)	25,400 E*
inch (in.)	centimeter (cm)	2.54 E
inch (in.)	meter (m)	0.0254 E
foot (ft)	meter (m)	0.3048 E
yard (yd)	meter (m)	0.9144
Area		
square foot (sq ft)	square meter (sq m)	0.09290304 E
square inch (sq in.)	square centimeter (sq cm)	6.452 E
square inch (sq in.)	square meter (sq m)	0.00064516 E
square yard (sq yd)	square meter (sq m)	0.8361274
Volume		
cubic inch (cu in.)	cubic centimeter (cu cm)	16.387064
cubic inch (cu in.)	cubic meter (cu m)	0.00001639
cubic foot (cu ft)	cubic meter (cu m)	0.02831685
cubic yard (cu yd)	cubic meter (cu m)	0.7645549
gallon (gal) Can. liquid	liter	4.546
gallon (gal) Can. liquid	cubic meter (cu m)	0.004546
gallon (gal) U.S. liquid**	liter	3.7854118
gallon (gal) U.S. liquid	cubic meter (cu m)	0.00378541
fluid ounce (fl oz)	milliliters (ml)	29.57353
fluid ounce (fl oz)	cubic meter (cu m)	0.00002957
Force		
kip (1000 lb)	kilogram (kg)	453.6
kip (1000 lb)	newton (N)	4,448.222
pound (lb) avoirdupois	kilogram (kg)	0.4535924
pound (lb)	newton (N)	4.448222
Pressure or stress		
kip per square inch (ksi)	megapascal (MPa)	6.894757
kip per square inch (ksi)	kilogram per square centimeter (kg/sq cm)	70.31
pound per square foot (psf)	kilogram per square meter (kg/sq m)	4.8824
pound per square foot (psf)	pascal (Pa)†	47.88
pound per square inch (psi)	kilogram per square centimeter (kg/sq cm)	0.07031
pound per square inch (psi)	pascal (Pa)†	6,894.757
pound per square inch (psi)	megapascal (MPa)	0.00689476
Mass (weight)		
pound (lb) avoirdupois	kilogram (kg)	0.4535924
ton, 2000 lb	kilogram (kg)	907.1848
grain	kilogram (kg)	0.0000648

To convert from	to	multiply by
Mass (weight) per length		
kip per linear foot (klf)	kilogram per meter (kg/m)	0.001488
pound per linear foot (plf)	kilogram per meter (kg/m)	1.488
Mass per volume (density)		
pound per cubic foot (pcf)	kilogram per cubic meter (kg/cu m)	16.01846
pound per cubic yard (lb/cu yd)	kilogram per cubic meter (kg/cu m)	0.5933
Temperature		
degree Fahrenheit (°F)	degree Celsius (°C)	$t_C = (t_F - 32)/1.8$
degree Fahrenheit (°F)	degree Kelvin (°K)	$t_K = (t_F + 459.7)/1.8$
degree Kelvin (°K)	degree Celsius (C°)	$t_C = t_K - 273.15$
Energy and heat		
British thermal unit (Btu)	joule (J)	1055.056
calorie (cal)	joule (J)	4.1868 E
Btu/°F · hr · ft²	W/m² · °K	5.678263
kilowatt-hour (kwh)	joule (J)	3,600,000. E
British thermal unit per pound (Btu/lb)	calories per gram (cal/g)	0.55556
British thermal unit per hour (Btu/hr)	watt (W)	0.2930711
Power		
horsepower (hp) (550 ft-lb/sec)	watt (W)	745.6999 E
Velocity		
mile per hour (mph)	kilometer per hour (km/hr)	1.60934
mile per hour (mph)	meter per second (m/s)	0.44704
Permeability		
darcy	centimeter per second (cm/sec)	0.000968
feet per day (ft/day)	centimeter per second (cm/sec)	0.000352

*E indicates that the factor given is exact.
**One U.S. gallon equals 0.8327 Canadian gallon.
†A pascal equals 1.000 newton per square meter.

Note:
One U.S. gallon of water weighs 8.34 pounds (U.S.) at 60°F.
One cubic foot of water weighs 62.4 pounds (U.S.).
One milliliter of water has a mass of 1 gram and has a volume of one cubic centimeter.
One U.S. bag of cement weighs 94 lb.

The prefixes and symbols listed below are commonly used to form names and symbols of the decimal multiples and submultiples of the SI units.

Multiplication Factor	Prefix	Symbol
$1{,}000{,}000{,}000 = 10^9$	giga	G
$1{,}000{,}000 = 10^6$	mega	M
$1{,}000 = 10^3$	kilo	k
$1 = 1$	—	—
$0.01 = 10^{-2}$	centi	c
$0.001 = 10^{-3}$	milli	m
$0.000001 = 10^{-6}$	micro	μ
$0.000000001 = 10^{-9}$	nano	n

Index

Basic spot network, 5.11

C values for conductors and busway, 7.6
Checklist
  project to-do (electrical), 1.2
  drawing design (electrical), 1.5
  site design (electrical), 1.8
  existing condition service and distribution, 1.10
  design coordination (electrical), 1.13
  fire alarm system, 1.16
Circuit breaker interrupting capacities, molded case, 9.11–9.17
Conductor conversion, based on using copper conductor, 7.10
Conduit weight comparisons (lb per 100 ft)
  empty, 1.47
  maximum cable fill, 1.47
Current limitation, protection through, 9.20
Current-limitation curves, Bussmann low-peak time-delay fuse KRP-C800SP, 9.23
Current-limiting fuse
  effect of, 9.21
  analysis of, 9.21

Distributed secondary network, 5.10

Electrical symbols, 1.20
Emergency power
  condensed general criteria, 14.2, 14.6–14.10
  emergency/standby lighting recommendations, typical, 14.11
  options and arrangements, 14.11
    elevator emergency power transfer system, 14.16–14.17
    multiengine automatic paralleling system, typical, 14.15–14.16
    multiple automatic double-throw transfer switches, 14.13, 14.14
    total transfer and critical-load transfer methods, 14.13–14.15
    two-utility-source system using one automatic transfer switch, 14.12
    two-utility-source system where any two circuit breakers can be closed, 14.12–14.13
    typical hospital installation, 14.18
  summary of codes for, in the United States, 14.2, 14.3–14.5
Engine and generator-set sizing, 14.17–14.26, 14.22–14.23*t*
  critical installation considerations, 14.24–14.25
  engine rating considerations, 14.18–14.19
  engine–generator set
    load factor, 14.19–14.20
    load management, 14.21
    standards, 14.21
  example, 14.22–14.24
  emergency standby generator installation, typical, **14.26**
Equipment sizes, 600-V class, typical, 1.45

Fault currents, assumptions for motor contributions to, 7.13
Fire alarm systems
  application tips, 19.10–19.11
  audible notification appliances, 19.8
  basic elements (typical local protective signaling system), 19.3–19.4
  circuit designations, 19.4
  class, 19.4
  class A circuits, installation of, 19.7
  classifications, 19.2–19.3
  common code requirements, 19.2
  elevator recall systems, 19.17–19.18
    elevator recall/emergency shutdown schematic, 19.19
    elevator hoistway/machine room device installation detail, 19.20
  fire pump
    applications, 19.11–19.12
    one-line with separate ATS, 19.13
    one-line with integrated ATS, 19.14
  harmonic effects and mitigation, 19.18–19.22
  initiating-device circuits (IDCs), performance of, 19.5
  manual pull stations, ADA-complying mounting height for
    high forward-reach limit, 19.10
    high and low side-reach limits, 19.11
  notification-appliance circuits (NACs), 19.5–19.7
  secondary supply capacity and sources, 19.7
  signaling-line circuits (SLCs), performance of, 19.5, 19.6
  style, 19.4
  visual notification appliances, 19.9
  wiring of packaged rooftop AHUs with remote VFDs, 19.15
  wye-delta motor starter wiring, 19.15–19.17
Formulas and terms, 1.70

Four-pole aluminum and copper bus duct by ampere rating, weight of, 1.47
Fuse class designations, 2.2

Generator weight, by kW, 1.46
Ground-fault protection, 13.5
  dual-source system, single-point grounding, 13.8–13.9
  sensing methods
    ground-return, 13.6
    residual, 13.7–13.8
    zero-sequence, 13.6–13.7
Grounded systems, 13.2–13.3
  grounding-electrode system (*NEC* Articles 250.50 and 250.52), 13.4
  grounding-electrode conductor for alternating-current systems (*NEC* Table 250.66), 13.5
  resistance-grounded systems, 13.3

Hazardous locations
  Class I, Zone 1 power and lighting installation, **17.38**
  Class I, Division 1 lighting installation, **17.39**
  Class I, Division 1 power installation, **17.40**
  Class I, Division 2 power and lighting installation, **17.41**
  Class II lighting installation, **17.42**
  Class II power installation, **17.43**
  classification of hazardous atmospheres, 17.2–17.5, 17.3–17.4*t*
  Crouse-Hinds "Quick Selector," electrical equipment for, 17.44
  equipment for, 17.8–17.9
  external ignition and explosion, prevention of, 17.5–17.7
  hazardous substances used in business and industry
    dusts, 17.16–17.17
    gases and vapors, 17.13–17.15
  installation for sealing, 17.37
  maintenance principles, 17.11–17.12
  wiring methods and materials, 17.9–17.11
  worldwide explosion protection methods, codes, categories, classifications, and testing authorities, 17.45–17.47

IEEE standard protective device numbers, 1.36
Illuminance categories, determination of, 16.5
Illumination
  units of, conversion factors for, 16.2
  ballast efficacy factor, U.S. and Canadian standards for, 16.2
  calculation of
    average illuminance calculation sheet, 16.36
    coefficients of utilization for typical luminaires, 16.7–16.17
    dirt conditions, five degrees of, 16.22
    guide to lamp selection, 16.33–16.34
    lamp lumen depreciation, 16.20
    light-loss factor (*LLF*), 16.6
    light-loss factors by groups, 16.6, 16.18–16.19
    light output change due to voltage change, 16.20
    light-source characteristics and, 16.4
    lumen output for HID lamps as a function of operating position, 16.20
    luminaire dirt depreciation (*LDD*) factors, 16.23
    luminaire maintenance categories, determining, procedure for, 16.21
    number of luminaires required for a particular room, 16.24–16.25
    operating atmosphere, evaluation of, 16.22
    recommended luminance ratios, 16.35
    reflectance
      multiplying factors for effective floor cavity reflectances, 16.30
      percent effective, for various reflectance combinations, 16.28–16.29
      recommended, of interior surfaces, 16.35
      of various materials and colors, 16.26
    room cavity ratios, 16.27
    room surface dirt depreciation (*RSDD*) factors, 16.23
    zonal cavity method, 16.5–16.6
  characteristics of typical lamps, 16.31–16.32
  effect on color, 16.3–16.4
  starting and restrike times, 16.3
IMC, RMC, and EMT, minimum support required for (with Exceptions), 4.53
Induction motor, characteristics, general effect of voltage variations on 11.8

Let-through current and clearing, maximum peak, 2.3
Let-through data
  pertinent to equipment withstand, 9.22
  charts, how to use, 9.22

Lightning protection, 13.9–13.11
average number of thunderstorm days per year, isokeraunic map showing, 13.12
cone of protection, 13.13
rolling-ball theory, 13.11, 13.13

Looped primary circuit arrangement, 5.9
LV busway, *R, X,* and *Z*, 7.8

Metrification
approximate conversions, 20.14–20.15
conversion of ASTM diameter and wall-thickness designations, 20.12
conversion scales (temperature and measurements), 20.13
how metric units will apply in the construction industry, 20.9
metric conversion factors, 20.17
of pipe sizes, 20.10
rebar conversions, 20.11
of standard lumber sizes, 20.11
quick imperial (metric equivalents), 20.16
what will change and what will remain the same, 20.2–20.8
Motor-circuit feeders
data sheets, 10.3
single-phase motor-circuit feeders
115-V, 10.6
200-V, 10.6
230-V, 10.7
sizing of, and their overcurrent protection, 10.1–10.3
three-phase motor-circuit feeders
480-V system (460-V motors) 10.4
208-V system (200-V motors), 10.5
Motor starter characteristics
reduced-voltage starter
characteristics, 10.8
selection table, 10.8
squirrel-cage motors, 10.7–10.8
Motors, standard, voltage ratings of, 11.7
Mounting heights for electrical devices, 1.31

NEC:
Annex B, Table B.310.1, Ampacities of Two or Three Insulated Conductors, Rated 0 through 2000 V, Within an Overall Covering (Multiconductor Cable), in Raceway in Free Air, 15.17
Annex B, Table B.310.3, Ampacities of Multiconductor Cables with Not More than Three Insulated Conductors, Rated 0 through 2000 V, in Free Air (for Type TC, MC, MI, UF, and USE Cables), 15.18
Annex B, Table B.310.5, Ampacities of Single-Insulated Conductors, Rated 0 through 2000 V, in Nonmagnetic Underground Electrical Ducts (One Conductor per Electrical Duct), 15.19
Annex B, Table B.310.6, Ampacities of Three Insulated Conductors, Rated 0 through 2000 V, Within an Overall Covering (Three-Conductor Cable) in Underground Electrical Ducts (One Cable per Duct), 15.20
Annex B, Table B.310.7, Ampacities of Three Single-Insulated Conductors, Rated 0 through 2000 V, in Underground Electrical Ducts (Three Conductors per Electrical Duct), 15.21
Annex B, Table B.310.8, Ampacities of Two or Three Insulated Conductors, Rated 0 through 2000 V, Cabled Within an Overall (Two- or Three-Conductor) Covering, Directly Buried in Earth, 15.22
Annex B, Table B.310.9, Ampacities of Three Triplexed Single-Insulated Conductors, Rated 0 through 2000 V, Directly Buried in Earth, 15.23
Annex B, Table B.310.10, Ampacities of Three Single-Insulated Conductors, Rated 0 through 2000 V, Directly Buried in Earth, 15.24
Annex B, Figure B.310.1, Interpolation Chart for Cables in a Duct Bank I_1 = Ampacity for Rho = 60, 50 LF; I_2 = Ampacity for Rho = 120, 100 LF (Load Factor); Desired Ampacity = $F \times I_1$, 15.25
Annex B, Figure B.310.2, Cable Installation Dimensions for Use with *NEC* Tables B.310.5 through B.310.10, 15.26
Annex B, Figure B.310.3, Ampacities of Single-Insulated Conductors, Rated 0 through 5000 V, in Underground Electrical Ducts (Three Conductors per Electrical Duct), Nine Single-Conductor Cables per Phase, 15.27
Annex B, Figure B.310.4, Ampacities of Single-Insulated Conductors, Rated 0 through 5000 V, in (r)MDNM-Nonmagnetic Underground Electrical Ducts (One Conductor per Electrical Duct), Four Single-Conductor Cables per Phase, 15.28
Annex B, Table B.310.5, Ampacities of Single-Insulated Conductors, Rated 0 through 5000 V, in Nonmagnetic Underground Electrical Ducts (One Conductor per Electrical Duct), Five Single-Conductor Cables per Phase, 15.29
Annex B, Table B.310.11, Adjustment Factors for More Than Three Current-Carrying Conductors in a Raceway or Cable with Load Diversity, 15.30
Annex C, Conduit and Tube Fill Tables for Conductors and Fixture Wires of the Same Size, 15.31

NEC (Cont.):
Article 505, Class I, Zone 0, 1, and 2 Locations, 17.19
Article 511, Commercial Garages, Repair and Storage, 17.20
Article 513, Aircraft Hangers, 17.21
Article 514, Gasoline Dispensing and Service Stations, 17.22–17.27
Article 515, Bulk Storage Plants, 17.28–17.34
Article 516, Spray Application, Dipping, and Coating Processes, 17.35–17.36
Chapter 9, Table 1, Percent of Cross Section of Conduit and Tubing for Conductor, 15.3
Chapter 9, Table 4, Dimensions and Percent Area of Conduit and Tubing (Areas of Conduit or Tubing for the Combinations of Wires Permitted in Table 1, Chapter 9), 15.4
Chapter 9, Table 5, Dimensions of Insulated Conductors and Fixture Wires, 15.9
Chapter 9, Table 5A, Compact Aluminum Building Wire Nominal Dimensions and Areas, 15.11
Chapter 9, Table 8, Conductor Properties, 15.12
Chapter 9, Table 9, Alternating-Current Resistance and Reactance for 600-V Cables, Three-Phase, 60-Hz, 75°C (167°F): Three Single Conductors in Conduit, 15.13
Chapter 9, Tables 11(A) and 11(B), Class 2 and Class 3, Alternating-Current and Direct-Current Power-Source Limitations, Respectively, 15.14
Chapter 9, Tables 12(A) and 12(B), PLFA Alternating-Current and Direct-Current Power-Source Limitations, Respectively, 15.16
Figure 310.60, Cable Installation Dimensions for Use with Tables 4.4.26 through 4.4.35 (*NEC* Tables 310-77 through 310-86), 4.31
Section 90.2, Scope of the *NEC*, 3.3
Section 110.3(A)(5), (6), and (8), Requirements for Equipment Selection, 3.3
Section 110.3(B), Requirements for Proper Installation of Listed and Labeled Equipment, 3.4
Section 110.9, Requirements for Proper Interrupting Rating of Overcurrent Protective Devices, 3.6
Section 110.10, Proper Protection of System Components from Short Circuits, 3.13
Section 110.22, Proper Marking and Identification of Disconnecting Means, 3.16
Section 210.20(A), Ratings of Overcurrent Devices on Branch Circuits Serving Continuous and Noncontinuous Loads, 3.16
Section 215.10, Requirements for Ground-Fault Protection of Equipment on Feeders, 3.16

Section 230.82, Equipment Allowed to Be Connected on the Line Side of the Service Disconnect, 3.17
Section 230.95, Ground-Fault Protection for Services, 3.17
Section 240.1, Scope of Article 240 on Overcurrent Protection, 3.18
Section 240.3, Protection of Conductors Other than Flexible Cords and Fixture Wires, 3.19
Section 240.4, Proper Protection of Fixture Wires and Flexible Cords, 3.20
Section 240.6, Standard Ampere Ratings, 3.21
Sections 240.8 and 380.17, Protective Devices Used in Parallel, 3.21
Section 240.9, Thermal Devices, 3.21
Section 240.10, Requirements for Supplementary Overcurrent Protection, 3.22
Section 240.11, Definition of Current-Limiting Overcurrent Protective Devices, 3.23
Section 240.12, System Coordination or Selectivity, 3.24
Section 240.13, Ground-Fault Protection of Equipment on Buildings or Remote Structures, 3.25
Section 240.21, Location Requirements for Overcurrent Devices and Tap Conductors, 3.25
Section 240.40, Disconnecting Means for Fuses, 3.27
Section 240.50, Plug Fuses, Fuseholders, and Adapters, 3.28
Section 240.51, Edison-Base Fuses, 3.28
Section 240.53, Type S Fuses, 3.28
Section 240.54, Type S Fuses, Adapters, and Fuseholders, 3.29
Section 240.60, Cartridge Fuses and Fuseholders, 3.29
Section 240.61, Classification of Fuses and Fuseholders, 3.29
Section 240.86, Series Ratings, 3.30
Sections 240.90 and 240.91, Supervised Industrial Installations, 3.30
Section 240.92(B), Transformer Secondary Conductors of Separately Derived Systems (Supervised Industrial Installations Only), 3.31
Section 240.92(B)(1), Short-Circuit and Ground-Fault Protection (Supervised Industrial Installations Only), 3.31
Section 240.92(B)(2), Overload Protection (Supervised Industrial Installations Only), 3.31
Section 240.92(C), Outside Feeder Taps (Supervised Industrial Installations Only), 3.31
Section 240.100, Feeder and Branch-Circuit Protection Over 600 V Nominal, 3.32
Section 240.100(C), Conductor Protection, 3.32
Section 250.2(D), Performance of Fault-Current Path, 3.32
Section 250.90, Bonding Requirements and Short-Circuit Current Rating, 3.32

NEC (*Cont.*):
Section 250.96(A), Bonding Other Enclosures and Short-Circuit Current Requirements, 3.32
Section 250.122, Sizing of Equipment Grounding Conductors, 3.33
Section 310.10, Temperature Limitation of Conductors, 3.34
Section 332.24 for Bends in Type MI Cable, illustration of, 4.48
Section 364.11, Protection at a Busway Reduction, 3.34
Section 384.16, Panelboard Overcurrent Protection, 3.34
Section 392.11(A)(3) for Multiconductor Cables, 2000 V or Less, with Not More Than Three Conductors per Cable (Ampacity to be Determined from Table B-310-3 in Appendix B), illustration of, 4.47
Section 392.11(B)(4) for Three Single Conductors Installed in a Triangular Configuration with Spacing Between Groups of Not Less Than 2.15 Times the Conductor Diameter (Ampacities to Be Determined from Table 310.20), illustration of, 4.47
Section 430.1, Scope of Motor Article, 3.35
Section 430.6, Ampacity of Conductors for Motor Branch Circuits and Feeders, 3.35
Section 430.8, Marking on Controllers, 3.35
Section 430.32, Motor Overload Protection, 3.36
Section 430.36, Fuses Used to Provide Overload and Single-Phasing Protection, 3.36
Section 430.52, Sizing of Various Overcurrent Devices for Motor Branch-Circuit Protection, 3.37
Section 430.53, Connecting Several Motors or Loads on One Branch Circuit, 3.38
Section 430.71, Motor Control-Circuit Protection, 3.38
Section 430.72(A), Motor Control-Circuit Overcurrent Protection, 3.39
Section 430.72(B), Motor Control-Circuit Conductor Protection, 3.39
Section 430.72(C), Motor Control-Circuit Transformer Protection, 3.41
Section 430.94, Motor Control-Center Protection, 3.42
Section 430.109(A)(6), Manual Motor Controller as a Motor Disconnect, 3.42
Section 440.5, Marking Requirements on HVAC Controllers, 3.42
Section 440.22, Application and Selection of the Branch-Circuit Protection for HVAC Equipment, 3.43
Section 450.3, Protection Requirements for Transformers, 3.43
Section 450.3(A), Protection Requirements for Transformers Over 600 V, 3.45
Section 450.3(B), Protection Requirements for Transformers 600 V or Less, 3.45
Section 450.6(A)(3), Tie-Circuit Protection, 3.45
Section 455.7, Overcurrent Protection Requirements for Phase Converters, 3.46
Section 460.8(B), Overcurrent Protection of Capacitors, 3.46
Section 501.6(B), Fuses for Class 1, Division 2 Locations, 3.46
Section 517.17, Requirements for Ground-Fault Protection and Coordination in Health Care Facilities, 3.47
Section 520.53(F)(2), Protection of Portable Switchboards on Stage, 3.47
Section 550.6(B), Overcurrent Protection Requirements for Mobile Homes and Parks, 3.48
Section 610.14(C), Conductor Sizes and Protection for Cranes and Hoists, 3.48
Section 620.62, Selective Coordination of Overcurrent Protective Devices for Elevators, 3.48
Section 670.3, Industrial Machinery, 3.49
Section 700.5, Emergency Systems: Their Capacity and Rating, 3.50
Section 700.16, Emergency Illumination, 3.50
Section 700.25, Emergency System Overcurrent Protection Requirements (FPN), 3.51
Section 705.16, Interconnected Electric Power Production Sources: Interrupting and Short-Circuit Current Rating, 3.51
Section 725.23, Overcurrent Protection for Class 1 Circuits, 3.52
Section 760.23, Requirements for Non-Power-Limited Fire Alarm Signaling Circuits, 3.52
Table C1, Maximum Number of Conductors and Fixture Wires in Electrical Metallic Tubing, 15.31–33
Table C1(A), Maximum Number of Compact Conductors in Electrical Metallic Tubing, 15.34
Table C2, Maximum Number of Conductors and Fixture Wires in Electrical Nonmetallic Tubing, 15.35–37
Table C2(A), Maximum Number of Compact Conductors in Electrical Nonmetallic Tubing, 15.38
Table C3, Maximum Number of Conductors and Fixture Wires in Flexible Metal Conduit, 15.39–41
Table C3(A), Maximum Number of Compact Conductors in Flexible Metal Conduit, 15.42
Table C4, Maximum Number of Conductors and Fixture Wires in Intermediate Metal Conduit, 15.43–45
Table C4(A), Maximum Number of Compact Conductors in Intermediate Metal Conduit, 15.46
Table C5, Maximum Number of Conductors and Fixture Wires in Liquidtight Flexible Nonmetallic Conduit (Type FNMC-B), 15.47–49
Table C5(A), Maximum Number of Compact Conductors in Liquidtight Flexible Nonmetallic Conduit (Type FNMC-B), 15.50

NEC (*Cont.*):
Table C6, Maximum Number of Conductors and Fixture Wires in Liquidtight Flexible Nonmetallic Conduit (Type FNMC-A), 15.51–53
Table C6(A), Maximum Number of Compact Conductors in Liquidtight Flexible Nonmetallic Conduit (Type FNMC-A), 15.54
Table C7, Maximum Number of Conductors and Fixture Wires in Liquidtight Flexible Metal Conduit, 15.55–57
Table C7(A), Maximum Number of Compact Conductors in Liquidtight Flexible Metal Conduit, 15.58
Table C8, Maximum Number of Conductors and Fixture Wires in Rigid Metal Conduit, 15.59–61
Table C8(A), Maximum Number of Compact Conductors in Rigid Metal Conduit, 15.62
Table C9, Maximum Number of Conductors and Fixture Wires in Rigid PVC Conduit, Schedule 80, 15.63–65
Table C9(A), Maximum Number of Compact Conductors in Rigid PVC Conduit, Schedule 80, 15.66
Table C10, Maximum Number of Conductors and Fixture Wires in Rigid PVC Conduit, Schedule 40 and HDPE Conduit, 15.67–69
Table C10(A), Maximum Number of Compact Conductors in Rigid PVC Conduit, Schedule 40 and HDPE Conduit, 15.70
Table C11, Maximum Number of Conductors and Fixture Wires in Type A Rigid PVC Conduit, 15.71–73
Table C11(A), Maximum Number of Compact Conductors in Type A Rigid PVC Conduit, 15.74
Table C12, Maximum Number of Conductors in Type EB PVC Conduit, 15.75–76
Table C12(A), Maximum Number of Compact Conductors in Type EB, PVC Conduit, 15.77
Table 250.122, Minimum Size Equipment Grounding Conductors for Grounding Raceway and Equipment, 13.2
Table 300.1(C), Metric designator and trade sizes, 4.4
Table 300.5, Minimum Cover Requirements, 0 to 600 V, Nominal, 4.5
Table 300.19(A), Spacings for Conductor Supports (Maximum Spacing Intervals in Vertical Raceways), 4.6
Table 300.50, Minimum Cover Requirements (Over 600 V), 4.8
Table 310.5, Minimum Size of Conductors, 4.9–4.14
Table 310.13, Conductor Application and Insulations, 4.10
Conductor Characteristics, 4.15
Table 310.15(B)(2)(a), Adjustment Factors for More Than Three Current-Carrying Conductors in a Raceway or Cable, 4.15–4.16
Table 310.16, Allowable Ampacities of Insulated Conductors Rated 0 through 2000 V, 60°C through 90°C (140°F through 194°F), Not More Than Three Current-Carrying Conductors in a Raceway, Cable, or Earth (Directly Buried), Based on Ambient Air Temperature of 30°C (86°F), 4.16–4.17
Table 310.17, Allowable Ampacities of Single-Insulated Conductors Rated 0 through 2000 V in Free Air, Based on Ambient Air Temperature of 30°C (86°F), 4.18–4.19
Table 310.18, Allowable Ampacities of Three Single-Insulated Conductors, Rated 0 through 2000 V, 150°C through 250°C (302°F through 482°F), in Raceway or Cable, Based on Ambient Air Temperature of 40°C (104°F), 4.20
Table 310.19, Allowable Ampacities of Single-Insulated Conductors, Rated 0 through 2000 V, 150°C through 250°C (302°F through 482°F), in Free Air, Based on Ambient Air Temperature of 40°C (104°F), 4.21
Table 310.20, Ampacities of Two or Three Single-Insulated Conductors, Rated 0 through 2000 V, Supported on a Messenger, Based on Ambient Air Temperature of 40°C (104°F), 4.22
Table 310.21, Ampacities of Bare or Covered Conductors, Based on 40°C (104°F) Ambient, 80°C (176°F) Total Conductor Temperature, 2 ft/s (610 mm/s) Wind Velocity, 4.23
Table 310.61, Conductor Application and Insulation, 4.23
Table 310.62, Thickness of Insulation for 601- to 2000-V Nonshielded Types RHH and RHW, in Mils, 4.24
Table 310.63, Thickness of Insulation and Jacket for Nonshielded Solid-Dielectric Insulated Conductors Rated 2001 to 8000 V, in Mils, 4.24
Table 310.64, Thickness of Insulation for Shielded Solid-Dielectric Insulated Conductors Rated 2001 to 35,000 V, in Mils, 4.25
Table 310.67, Ampacities of Insulated Single Copper Conductor Cables Triplexed in Air Based on Conductor Temperatures of 90°C (194°F) and 105°C (221°F) and Ambient Air Temperature of 40°C (104°F), 4.25
Table 310.68, Ampacities of Insulated Single Aluminum Conductor Cables Triplexed in Air Based on Conductor Temperatures of 90°C (194°F) and 105°C (221°F) and Ambient Air Temperature of 40°C (104°F), 4.26
Table 310.69, Ampacities of Insulated Single Copper Conductor Isolated in Air Based on Conductor Temperatures of 90°C (194°F) and 105°C (221°F) and Ambient Air Temperature of 40°C (104°F), 4.26
Table 310.70, Ampacities of Insulated Single Aluminum Conductor Isolated in Air Based on Conductor Temperatures of 90°C (194°F) and 105°C (221°F) and Ambient Air Temperature of 40°F (104°F), 4.27

NEC (*Cont.*):
Table 310.71, Ampacities of an Insulated Three-Conductor Copper Cable Isolated in Air Based on Conductor Temperatures of 90°C (194°F) and 105°C (221°F) and Ambient Air Temperature of 40°C (104°F), 4.27
Table 310.72, Ampacities of Insulated Three-Conductor Aluminum Cable Isolated in Air Based on Conductor Temperatures of 90°C (194°F) and 105°C (221°F) and Ambient Air Temperature of 40°C (104°F), 4.28
Table 310.73, Ampacities of an Insulated Triplexed or Three Single-Conductor Copper Cables in Isolated Conduit in Air Based on Conductor Temperatures of 90°C (194°F) and 105°C (221°F) and Ambient Air Temperature of 40°C (104°F), 4.28
Table 310.74, Ampacities of an Insulated Triplexed or Three Single-Conductor Aluminum Cables in Isolated Conduit in Air Based on Conductor Temperatures of 90°C (194°F) and 105°C (221°F) and Ambient Air Temperature of 40°C (104°F), 4.29
Table 310.75, Ampacities of an Insulated Three-Conductor Copper Cable in Isolated Conduit in Air Based on Conductor Temperatures of 90°C (194°F) and 105°C (221°F) and Ambient Air Temperature of 40°C (104°F), 4.29
Table 310.76, Ampacities of an Insulated Three-Conductor Aluminum Cable in Isolated Conduit in Air Based on Conductor Temperatures of 90°C (194°F) and 105°C (221°F) and Ambient Air Temperature of 40°C (104°F), 4.30
Table 310.77, Ampacities of Three Single-Insulated Copper Conductors in Underground Electrical Ducts (Three Conductors Per Electrical Duct) Based on Ambient Earth Temperature of 20°C (68°F), Electrical Duct Arrangement per Figure 4.4.25 (*NEC* Figure 310.60), 100 Percent Load Factor, Thermal Resistance (RHO) of 90, Conductor Temperatures of 90°C (194°F) and 105°C (221°F), 4.32
Table 310.78, Ampacities of Three Single-Insulated Aluminum Conductors in Underground Electrical Ducts (Three Conductors per Electrical Duct) Based on Ambient Earth Temperature of 20°C (68°F), Electrical Duct Arrangement per Figure 4.4.25 (*NEC* Figure 310.60), 100 Percent Load Factor, Thermal Resistance (RHO) of 90, Conductor Temperatures of 90°C (194°F) and 105°C (221°F), 4.33
Table 310.79, Ampacities of Three Insulated Copper Conductors Cabled Within an Overall Covering (Three-Conductor Cable) in Underground Electrical Ducts (One Cable per Electrical Duct) Based on Ambient Earth Temperature of 20°C (68°F), Electrical Duct Arrangement per Figure 4.4.25 (*NEC* Figure 310.60), 100 Percent Load Factor, Thermal Resistance (RHO) of 90, Conductor Temperatures of 90°C (194°F) and 105°C (221°F), 4.34–4.35
Table 310.80, Ampacities of Three Insulated Aluminum Conductors Cabled Within an Overall Covering (Three-Conductor Cable) in Underground Electrical Ducts (One Cable per Electrical Duct) Based on Ambient Earth Temperature of 20°C (68°F), Electrical Duct Arrangement per Figure 4.4.25 (*NEC* Figure 310.60), 100 Percent Load Factor, Thermal Resistance (RHO) of 90, Conductor Temperatures of 90°C (194°F) and 105°C (221°F), 4.36
Table 310.81, Ampacities of Single-Insulated Copper Conductors Directly Buried in Earth Based on Ambient Earth Temperature of 20°C (68°F), Arrangement per Figure 4.4.25 (*NEC* Figure 310.60), 100 Percent Load Factor, Thermal Resistance (RHO) of 90, Conductor Temperatures of 90°C (194°F) and 105°C (221°F), 4.37
Table 310.82, Ampacities of Single-Insulated Aluminum Conductors Directly Buried in Earth Based on Ambient Earth Temperature of 20°C (68°F), Arrangement per Figure 4.4.25 (*NEC* Figure 310.60), 100 Percent Load Factor, Thermal Resistance (RHO) of 90, Conductor Temperatures of 90°C (194°F) and 105°C (221°F), 4.38
Table 310.83, Ampacities of Three Insulated Copper Conductors Cabled Within an Overall Covering (Three-Conductor Cable) Directly Buried in Earth Based on Ambient Earth Temperature of 20°C (68°F), Arrangement per Figure 4.4.25 (*NEC* Figure 310.60), 100 Percent Load Factor, Thermal Resistance (RHO) of 90, Conductor Temperatures of 90°C (194°F) and 105°C (221°F), 4.39

Table 310.84, Ampacities of Three Insulated Aluminum Conductors Cabled Within an Overall Covering (Three-Conductor Cable) Directly Buried in Earth Based on Ambient Earth Temperature of 20°C (68°F), Arrangement per Figure 4.4.25 (*NEC* Figure 310.60), 100 Percent Load Factor, Thermal Resistance (RHO) of 90, Conductor Temperatures of 90°C (194°F) and 105°C (221°F), 4.40
Table 310.85, Ampacities of Three Triplexed Single-Insulated Copper Conductors Directly Buried in Earth Based on Ambient Earth Temperature of 20°C (68°F), Arrangement per Figure 4.4.25 (*NEC* Figure 310.60), 100 Percent Load Factor, Thermal Resistance (RHO) of 90, Conductor Temperatures of 90°C (194°F) and 105°C (221°F), 4.41

NEC (*Cont.*):
Table 310.86, Ampacities of Three Triplexed Single-Insulated Aluminum Conductors Directly Buried in Earth Based on Ambient Earth Temperature of 20°C (68°F), Arrangement per Figure 4.4.25 (*NEC* Figure 310.60), 100 Percent Load Factor, Thermal Resistance (RHO) of 90, Conductor Temperatures of 90°C (194°F) and 105°C (221°F), 4.42
Table 314.16(A), Metal Boxes, 4.59
Table 314.16(B), Volume Allowance Required per Conductor, 4.59
Table 318.7(B)(2), Metal Area Requirements for Cable Trays Used as Equipment Grounding Conductors, 4.43
Example of Multiconductor Cables in Cable Trays with Conduit Runs to Power Equipment Where Bonding Is Provided in Accordance with Section 318.7(B)(4), 4.44
Table 344.30(B)(2), Supports for Rigid Metal Conduit, 4.54
Table 346.10 and Exception, Radius of Conduit Bends for IMC, RMC, RNC, and EMT, 4.53
Table 348.22, Maximum Number of Insulated Conductors in Metric Designator 12 (Trade Size ⅜-in) Flexible Metal Conduit, 4.56
Table 352.30(B), Support of Rigid Nonmetallic Conduit, 4.54
Table 352.44(A), Expansion Characteristics of PVC Rigid Nonmetallic Conduit, Coefficient of Thermal Expansion = 3.38×10^{-5} in/in/°F, 4.55
Table 352.44(B), Expansion Characteristics of Reinforced Thermosetting Resin Conduit (RTRC), Coefficient of Thermal Expansion = 2.7×10^{-5} mm/mm/°C (1.5×10^{-5} in/in/°F), 4.56
Table 384.22, Channel Size and Inside Diameter Area, 4.58
Table 392.9, Allowable Cable Fill Area for Multiconductor Cables in Ladder, Ventilated-Trough, or Solid-Bottom Cable Trays for Cables Rated 2000 V or Less, 4.45
Table 392.9(E), Allowable Cable Fill Area for Multiconductor Cables in Ventilated-Channel Cable Trays for Cables Rated 2000 V or Less, 4.45
Table 392.9(E), Allowable Cable Fill Area for Multiconductor Cables in Solid Channel Cable Trays for Cables Rated 2000 V or Less, 4.46
Table 392.10, Allowable Cable Fill Area for Single-Conductor Cables in Ladder or Ventilated-Trough Cable Trays for Cables Rated 2000 V or Less, 4.46
Table 400.4, Flexible Cords and Cables, 4.60–4.65
Table 400.5(A), Allowable Ampacity for Flexible Cords and Cables [Based on Ambient Temperature of 30°C (86°F); See Section 400.13 and Table 400.4], 4.66
Table 400.5(B), Ampacity of Cable Types SC, SCE, SCT, PPE, G, G-GC, and W [Based on Ambient Temperature of 30°C (86°F); See Table 400.4], 4.67
Ampacities for Flexible Cords and Cables with More Than Three Current-Carrying Conductors, 4.67
Table 402.3, Fixture Wires, 4.68–4.70
Table 402.5, Allowable Ampacities for Fixture Wires, 4.71
Table 430.7(B), Locked-Rotor-Indicating Code Letters, 10.3
Table 450.3(A), Maximum Rating or Setting of Overcurrent Protection for Transformers over 600 V (as a Percentage of Transformer-Rated Current), 9.18
Table 450.3(B), Maximum Rating or Setting of Overcurrent Protection for Transformers 600 V and Less (as a Percentage of Transformer-Rated Current), 9.19
U.L. 1008 Minimum Withstand Test Requirement (for Automatic Transfer Switches), 9.19
Table 500.8(B), Classification of Maximum Surface Temperature, 17.18
Table 500.8(C)(2), Class II Ignition Temperatures, 17.18
NEMA configuration chart
general-purpose nonlocking plugs and receptacles, 1.34
specific-purpose locking plugs and receptacles, 1.35
NEMA standard enclosures
comparison of specific applications of, for,
indoor nonhazardous locations, 1.42
outdoor nonhazardous locations, 1.42
indoor hazardous locations, 1.42
knockout dimensions for, 1.43

Power cable
MI cable versus conventional construction in hazardous (classified) locations, 4.51
600-V MI, size and ampacities, 4.49–4.50
300-V MI twisted-pair and shielded twisted-pair, sizes, 4.51
Power loads
all-weather comfort standard, recommended heat-loss values, 6.4
appliance/general-purpose receptacle loads, typical, 6.2
apartment, typical, 6.3
connected electrical load for air conditioning only, typical, 6.3
central air conditioning, 6.4
commercial kitchens, typical, 6.6
comparison of maximum demand, 6.6
connected load and maximum demand by tenant classification, 6.7
electric hot water–heating system, requirements for, 6.5

Power loads (*Cont.*):
fire pumps in commercial buildings (light hazard), requirements for, 6.5
high-rise building water pressure–boosting systems, requirements for, 6.5
major elements of the electrical system serving HVAC systems, factors used to establish, 6.8
service entrance peak demand
Veterans Administration, 6.8
Hospital Corporation of America, 6.9
sizing distribution-system components, factors used in, 6.7
Prescriptive unit lighting power allowance (ULPA), gross lighted area of total building, 6.2
Primary- and secondary-selective circuit arrangement, double-ended substation with selective primary, 5.8
Protective devices, selective coordination of
recommended procedure for, 8.1–8.3
one-line diagram for, **8.4**
time-current curves, **8.5–8.10**
shortcut ratio method selectivity guide, 8.11

Radial circuit arrangements
in commercial buildings, 5.2
common primary feeder to secondary unit substations, 5.3
individual primary feeder to secondary unit substations, 5.4
primary, 5.5

Secondary-selective circuit arrangement
double-ended substation with single tie, 5.6
individual substations with interconnecting ties, 5.7
Seismic requirements, 1.48
Short-circuit calculations
Three-phase, point-to-point method, basic calculation procedure and formulas, 7.2, **7.3**
shortcut method 1: adding *Z*s, 7.7
shortcut method 2: chart approximate method, 7.9
charts for calculating short-circuit currents, 7.10–7.12

system A to faults X_1 and X_2, calculations for, 7.4
system B to faults X_1 and X_2, calculations for, 7.5
Short-circuit capacity, secondary, of typical power transformers, 7.14
600-V conductors:
2 or 3 single conductors, average characteristics of, 7.7
3 conductor cables (and interlocked armored cable), average characteristics of, 7.8
Short-circuit current withstand chart
with paper, rubber, or varnished-cloth insulation, 9.6
with thermoplastic insulation, 9.7
with cross-linked polyethylene and ethylene-propylene-rubber insulation, 9.8
copper cable
with paper, rubber, or varnished-cloth insulation, 9.3
with thermoplastic insulation, 9.4
cross-linked polyethylene and ethylene-propylene-rubber insulation, 9.5
Short-circuit ratings
equipment grounding conductor, comparison of, 9.9
of busway, NEMA standard, 9.10
U.L. no. 508 motor controller, 9.10
Short-circuit test currents, HVAC equipment, 9.20
Support bushings and cleats, installed, examples of, 4.7
Surface raceways (based on Wiremold), conductor fill table for, 4.57

Telecommunications structured cabling systems
backbone system
cable, conduit fill requirements for, 18.26
combined copper/fiber, example of, 18.23
components, 18.21
distances, 18.24
floor sleeves, determining, 18.25
floor slots, determining size of, 18.25
Star wiring, topology, 18.22
blown optical fiber technology (BOFT), 18.36

key elements of, 18.37
BOFT indoor plenum 5-mm multiduct, 18.38
BOFT outdoor 8-mm multiduct, 18.39
BOFT installation equipment, 18.40
cable diameter, typical ranges of, 18.5
cable lengths, 18.11
cable tray dimensions, 18.9
codes and standards, 18.2–18.3
comparison of ANSI/TIA/EIA, ISO/IEC, and CENELEC cabling standards, 18.3–18.4
conduit
bend radii guidelines, 18.6
with bends, guidelines for adapting designs to, 18.6

Telecommunications structured cabling systems, conduit (*Cont.*):
sizing, number of cables, 18.5
entrance facility (EF)
floor space (minimum equipment and termination floor space), 18.33
typical underground installation to, 18.32
underground entrance conduits for, 18.31
wall space (minimum equipment and termination wall space), 18.33
equipment room (ER)
environmental control systems standards for, 18.31
floor space (special-use buildings), 18.32
horizontal cabling to two individual work areas, 18.10

horizontal copper cables (twisted-pair media), ANSI/TIA/EIA-568A categories of, 18.16
major elements of, 18.4–18.5
pull box
recommended configurations, 18.7
having one conduit each in opposite ends of the box, minimum space requirements in, 18.8
optical fiber cable performance, 18.13
optical fiber sample connector
duplex
interface, 18.19
adapter with simplex and duplex plugs, 18.20
patch cord crossover orientation, 18.20
types, 18.19
optical fibers, 18.21
power lines, separation of telecommunications pathways from, 18.34
telecommunications room (TR)
allocating termination space in, 18.28
cross-connect field color codes, 18.27
industry standards, 18.30
regulatory and safety standards, 18.30
size requirements, 18.28
temperature ranges, 18.27
typical layout, 18.29
termination hardware, 18.15
topology, 18.10
twisted-pair cable
categories, 18.12
standards, U.S., 18.18
work area, 18.13
work area copper cable lengths to a multi-user telecommunications outlet assembly (MUTOA), 18.17
Transformer weight, by kVA, 1.46
Transformers
approximate loss and impedance data, 12.3
auto zigzag
buck-boost
connection diagrams for, in autotransformer arrangement for single-phase system, 12.9
connection diagrams for, in autotransformer arrangement for three-phase system, 12.10
low-voltage single-phase, wiring diagrams for, 12.8
three-phase connection summary, 12.8
building ambient sound levels, typical, 12.11
electrical connection diagrams, 12.6
grounding, for deriving a neutral schematic and wiring diagram, 12.7
ratings, 12.7
full-load current, three-phase, self-cooled ratings, 12.2
insulation system temperature ratings, 12.11
k-rated, 12.11–12.12
overcurrent protection, maximum rating or setting of
for transformers over 600 V (as a percentage of transformer rated current), 12.5
for transformers 600 V and less (as a percentage of transformer rated current), 12.5
primary (480-V, three-phase, delta) and secondary (208-Y/120-V, three-phase, four-wire) overcurrent protection, conductors and grounding, 12.4
sound levels, maximum average, 12.10
three-phase, liquid-filled transformers, typical impedances, 12.3
typical weights, by kVA, 12.2

Ungrounded systems, 13.3
Uninterruptible power supply (UPS) systems, 14.26–14.40
application of UPS, 14.31
"cold" standby redundant UPS system, 14.27, 14.29
distribution systems, 14.37–
dual redundant UPS system, 14.32, 14.34
400-Hz power-system configuration, 14.39–14.40
hot tied-bus UPS system, 14.35–14.36
superredundant parallel system, 14.36–14.37
isolated redundant UPS system, 14.30–14.31, 14.34–14.35
nonredundant UPS system configuration, 14.27, 14.28
parallel redundant UPS system, 14.30, 14.32, 14.33

Uninterruptible power supply (UPS) systems (*Cont.*):
parallel tandem UPS system, 14.35
parallel-capacity UPS system, 14.32, 14.33
power-system configuration for 60-Hz distribution, 14.31–14.39
single-module UPS system, 14.31, 14.32
60-Hz power-system configuration, 14.37
uninterruptible power with dual utility sources and static transfer switches, 14.37, 14.38

Vertical raceways, intervals in, 4.6
Voltage classes, system, 11.2
ANSI C84.1-1989., voltage profile of the limits of range A, 11.7
application of, 11.4
principal transformer connections, 11.5
regulated power-distribution system, 120-V base, standard voltage profile for, 11.6
standard nominal, in the United States, 11.2
standard nominal, and voltage ranges, 11.3–11.4
tolerance limits, 11.5–11.6
voltage systems outside the United States, 11.4–11.5
Voltage dips
calculation of (momentary voltage variations), 11.38–11.39
recurrent, flicker of incandescent lamps caused by, 11.37
Voltage drop
application tips, 11.36
for Al conductor
direct current, 11.10
in magnetic conduit
70 percent PF, 11.11
80 percent PF, 11.12
90 percent PF, 11.13
95 percent PF, 11.14
100 percent PF, 11.15
in nonmagnetic conduit
70 percent PF, 11.16
80 percent PF, 11.17
90 percent PF, 11.18
95 percent PF, 11.19
100 percent PF, 11.20
calculations, 11.7
for Cu conductor
direct current, 11.22
in magnetic conduit
70 percent PF, 11.23
80 percent PF, 11.24

Voltage drop, for Cu conductor, in magnetic conduit (*Cont.*):
90 percent PF, 11.25
95 percent PF, 11.26
100 percent PF, 11.27
in nonmagnetic conduit
70 percent PF, 11.28
80 percent PF, 11.29
90 percent PF, 11.30
95 percent PF, 11.31
100 percent PF, 11.32
curves
for typical plug-in-type Cu busway at balanced rated load, 11.34
for typical Cu feeder busways at balanced rated load mounted flat horizontally, 11.35
versus power factor for typical light-duty trolley busway carrying rated load, 11.35
for three-phase transformers, 11.36
for typical interleaved construction of copper busway at rated load, 11.33
and feeder sizing, calculation of, 4.52
values
for three-phase busways with copper bus bars, at rated current with balanced entire load at end, 11.33
for three-phase busways with aluminum bus bars, line-to-line, at rated current with balanced entire load at end, 11.34
Voltage variations
effect of, on incandescent lamps, 11.37
general effect of, on induction motor characteristics, 11.38

Working spaces, 2.4–2.12
access to
NEC Section 110.26(C), basic rule, first paragraph, 2.8
NEC Section 110.26(C), basic rule, second paragraph, 2.8
conditions 1, 2, and 3, 2.5
and dedicated electrical space, 2.10
over and under a panelboard, 2.11
elevation of unguarded live parts above, 2.12
exception 1, 2.6, 2.9
exception 2, 2.10
exception 3, 2.6
in front of a panelboard, 2.11
minimum depth at electrical equipment, 2.12
required 30-in-wide front working space, 2.7
required full-90 degree opening of equipment doors, 2.7
unacceptable arrangement of a large switchboard, 2.9

ABOUT THE AUTHOR

Bob Hickey, a licensed professional engineer, is President and CEO of vanZelm, Heywood & Shadford, Inc., a leading northeast United States mechanical and electrical consulting engineering firm. His almost 40 years experience spans the electric utility, contracting, and consulting engineering areas of the industry spectrum. He has taught electrical engineering technology as an adjunct faculty member in Connecticut's community/technical college system. Mr. Hickey is the author of McGraw-Hill's *Electrical Engineer's Portable Handbook*.